CHICAGO PUBLIC LIBRARY
W9-AWG-748

DATE
DISCARD

# Numerical Methods in Engineering Practice

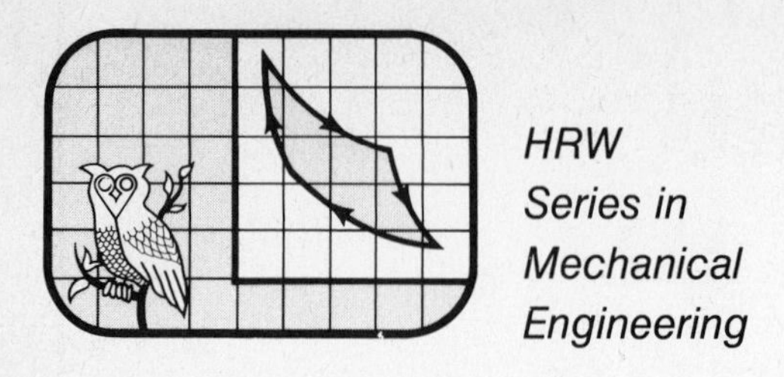

*HRW*
*Series in*
*Mechanical*
*Engineering*

**L. S. Fletcher, Series Editor**

A. W. Al-Khafaji and J. R. Tooley  NUMERICAL METHODS IN ENGINEERING PRACTICE
F. L. Stasa  APPLIED FINITE ELEMENT ANALYSIS FOR ENGINEERS
B. J. Torby  ADVANCED DYNAMICS FOR ENGINEERS

# Numerical Methods in Engineering Practice

**Amir Wadi Al-Khafaji**

*Associate Professor and Chairman of Civil Engineering*
*University of Evansville*

**John R. Tooley**

*Professor and Dean of Engineering*
*University of Evansville*

**HOLT, RINEHART AND WINSTON, INC.**
*New York Chicago San Francisco Philadelphia*
*Montreal Toronto London Sydney Tokyo*
*Mexico City Rio de Janeiro Madrid*

To My Brother Faiz,
Whose Ideals Are Like the Stars: I Could Never Reach Them,
But I Chart My Course by Them.

Amir Wadi Al-Khafaji

For My Sons,
Craig, Matthew, James, John W., for Their Constant Inspiration and Appreciation.

John R. Tooley



Address correspondence to:
383 Madison Avenue, New York, NY 10017

**Library of Congress Cataloging-in-Publication Data**

Al-Khafaji, Amir Wadi.
Numerical methods in engineering practice.

Bibliography: p.
Includes index.
1. Engineering mathematics. I. Tooley, John R.
II. Title.
TA330.A45 1986 620'.0042 85-24751
ISBN 0-03-001757-2

Printed in the United States of America
Published simultaneously in Canada

6 7 8 9 038 9 8 7 6 5 4 3 2 1

CBS College Publishing
Holt, Rinehart and Winston
The Dryden Press
Saunders College Publishing

# Contents

**Preface** *xiii*

**Chapter 1 INTRODUCTION** *1*

**1.1 Mathematical Models and Their Solutions** *2*
**1.2 The Need for Numerical Solutions** *2*
**1.3 Errors** *4*
**1.4 Taylor Series** *5*

**Chapter 2 MATRICES AND DETERMINANTS** *6*

**2.1 Introduction to Matrices** *7*
**2.2 Special Matrices** *8*
**2.3 Matrix Equality** *15*
**2.4 Matrix Addition and Subtraction** *16*
**2.5 Matrix Multiplication** *17*
2.5.1 Commutativity of Matrix Multiplication *21*
2.5.2 Associative Law of Matrix Multiplication *22*
2.5.3 Distributive Law of Matrix Multiplication *23*
**2.6 Manipulation of Partitioned Matrices** *25*
2.6.1 Transpose of a Partitioned Matrix *25*
2.6.2 Addition and Subtraction of Partitioned Matrices *27*
2.6.3 Multiplication of Partitioned Matrices *28*
**2.7 Rules for Combined Matrix Operations** *30*

**2.8 Application of Matrices to the Rotation of a Coordinate System** *31*
**2.9 Determinants and Their Evaluation** *33*
2.9.1 Properties of Determinants *34*
2.9.2 Laplace Method of Cofactors *35*
2.9.3 Upper-Triangle Elimination Method *37*
2.9.4 Method of Pivotal Condensation *40*
**2.10 Area and Volume Calculation Using Determinants** *44*
Recommended Reading *49*
Problems *49*

**Chapter 3 MATHEMATICAL MODELING OF TYPICAL ENGINEERING SYSTEMS** *53*

**3.1 Introduction** *54*
**3.2 Electrical Engineering Systems** *55*
3.2.1 Boundary-Value Problems *56*
3.2.2 Initial-Value Problems *59*
3.2.3 Eigenproblems *60*
**3.3 Mechanical Engineering Systems** *61*
3.3.1 Boundary-Value Problems *62*
3.3.2 Initial-Value Problems *64*
3.3.3 Eigenproblems *67*
**3.4 Civil Engineering Systems** *68*
3.4.1 Boundary-Value Problems *69*
3.4.2 Eigenproblems *71*
**3.5 Engineering System Response** *72*
**3.6 Models Involving Partial Differential Equations** *74*
3.6.1 Boundary-Value Problems *74*
3.6.2 Initial-Value Problems *77*
**3.7 Comparison of Engineering Models** *78*
Recommended Reading *79*
Problems *80*

**Chapter 4 SIMULTANEOUS LINEAR ALGEBRAIC EQUATIONS** *84*

**4.1 Introduction** *85*
**4.2 Cramer's Rule** *86*
**4.3 Gauss's Elimination Method** *90*
**4.4 Gauss–Jordan Elimination Method** *94*
**4.5 Crout's Method** *98*
**4.6 Square Root Method** *103*
**4.7 Reducing Matrix Method** *106*

**4.8 Solution of Tridiagonal Systems** *110*
**4.9 Iterative Methods** *113*
4.9.1 Jacobi's Method *114*
4.9.2 Gauss–Seidel Method *118*
**4.10 Ill-Conditioned Sets and Scaling** *121*
**4.11 Sets with More Unknowns Than Equations** *124*
**4.12 Linear Equations Involving Fewer Unknowns Than Equations** *126*
**4.13 Sets Involving Complex Coefficients** *129*
**4.14 Comparison of Method Efficiencies** *131*
Recommended Reading *132*
Problems *132*

**Chapter 5 MATRIX INVERSION** *137*

**5.1 Introduction** *138*
**5.2 Cramer's Rule** *139*
**5.3 Elimination Method** *143*
**5.4 Reducing Matrix Method** *145*
**5.5 Partitioning Method** *148*
**5.6 Matrices Involving Complex Coefficients** *151*
**5.7 Special Matrices** *153*
5.7.1 Triangular Matrices *153*
5.7.2 Symmetrical Matrices *155*
Recommended Reading *157*
Problems *157*

**Chapter 6 NONLINEAR ALGEBRAIC EQUATIONS** *160*

**6.1 Introduction** *161*
**6.2 Graphical Method** *161*
**6.3 Interval-Halving Method** *162*
**6.4 False-Position Method** *164*
**6.5 Newton–Raphson First Method** *167*
**6.6 Newton–Raphson Second Method** *174*
**6.7 Modified Newton–Raphson Methods** *177*
**6.8 Lin–Bairstow Method for Roots of Polynomials** *182*
**6.9 Newton–Raphson Method for Systems of Equations** *190*
**6.10 Practical Considerations** *198*
6.10.1 Errors *198*
6.10.2 Root Multiplicity *199*
Recommended Reading *200*
Problems *201*

**Chapter 7 EIGENPROBLEMS** *203*

**7.1 Introduction** *204*
**7.2 Characteristic Equation Determination** *206*
7.2.1 Direct Determinant Expansion *207*
7.2.2 Indirect Determinant Expansion *208*
**7.3 Eigenvalues and Eigenvectors** *212*
7.3.1 Column Buckling *215*
7.3.2 Mechanical Vibrations *218*
7.3.3 Electrical Circuits *223*
7.3.4 Eigenproblems and Dynamic Response *227*
**7.4 Vector Iteration Techniques** *234*
7.4.1 Largest Eigenvalue *234*
7.4.2 Smallest Eigenvalue *241*
7.4.3 Intermediate Eigenvalues *244*
**7.5 Polynomial Iteration Method** *247*
**7.6 Transformation Methods** *250*
7.6.1 Jacobi Method *251*
7.6.2 Householder Method *257*
**7.7 Functions of a Matrix** *261*
7.7.1 Caley–Hamilton Method *261*
7.7.2 Spectral Matrix Decomposition Method *265*
**7.8 Static Condensation** *266*
Recommended Reading *268*
Problems *268*

**Chapter 8 INTERPOLATION** *271*

**8.1 Introduction** *272*
**8.2 Interpolating Polynomials for Even Intervals** *272*
8.2.1 Forward Interpolation *273*
8.2.2 Backward Interpolation *274*
8.2.3 Central Interpolation *275*
**8.3 Difference Operators and Difference Tables** *281*
8.3.1 Forward Differences *281*
8.3.2 Backward Differences *284*
8.3.3 Central Differences *286*
8.3.4 Relationships between Difference Operators *287*
**8.4 Differences and Interpolating Polynomials** *290*
**8.5 Interpolating Polynomials for Uneven Intervals** *294*
**8.6 Interpolation Errors** *297*
**8.7 Inverse Interpolation** *299*
**8.8 Cubic Splines** *299*
Recommended Reading *307*
Problems *307*

**Chapter 9 CURVE FITTING** *309*

**9.1 Introduction** *310*
**9.2 Introduction to the Method of Least Squares** *310*
**9.3 What Type of Function to Fit** *312*
**9.4 Linear Regression** *312*
**9.5 Linearization** *315*
**9.6 Nonlinear Regression** *318*
**9.7 Multiple Regression** *320*
**9.8 Orthogonal Polynomials for Equal Intervals** *322*
**9.9 Goodness of Functional Approximations** *326*
9.9.1 Coefficients of Multiple Determination *327*
9.9.2 Standard Error of the Estimate *329*
Recommended Reading *333*
Problems *333*

**Chapter 10 NUMERICAL DIFFERENTIATION** *335*

**10.1 Introduction** *336*
**10.2 Review of Taylor Series** *336*
**10.3 Numerical Differentiation of Functions** *342*
10.3.1 Interpolating Polynomial Methods *343*
10.3.2 Taylor Series Method *345*
10.3.3 Undetermined Coefficients Method *349*
10.3.4 Errors *354*
**10.4 Numerical Differentiation of Data** *357*
**10.5 Mixed Derivatives** *357*
**10.6 Special Derivative Approximations** *360*
**10.7 Stencil Representation of Derivatives** *361*
**10.8 Stencil Representation of Mixed Derivatives** *369*
Recommended Reading *370*
Problems *370*

**Chapter 11 NUMERICAL INTEGRATION** *372*

**11.1 Introduction** *373*
**11.2 Trapezoidal Rule of Integration** *374*
**11.3 Simpson's ⅓ Rule of Integration Method** *380*
**11.4 Development of Special Integration Formulas** *382*
**11.5 Integration of Unevenly Spaced Base Points** *385*
**11.6 Stencil Representation of Integration Formulas** *388*
**11.7 Romberg's Integration** *390*
**11.8 Gauss Quadrature Formulas** *393*

11.9 **Double Integration** *406*
Recommended Reading *408*
Problems *409*

**Chapter 12 NUMERICAL SOLUTION OF ORDINARY DIFFERENTIAL EQUATIONS** *411*

12.1 **Introduction** *412*
12.2 **Taylor Series Method** *413*
12.3 **Least-Squares Method** *415*
12.4 **Galerkin Method** *417*
12.5 **Euler and Modified Euler Methods** *419*
12.6 **Runge–Kutta Methods** *424*
12.7 **Predictor–Corrector Methods** *428*
12.7.1 Adams Methods *429*
12.7.2 Milne Method *433*
12.7.3 Adams–Moulton Method *436*
12.7.4 Step Size and Errors *438*
12.8 **Stiff Equations** *440*
12.9 **High-Order Equations and Systems of First Order** *440*
12.10 **Initial-Value Problems Involving High-Order Equations** *444*
12.10.1 Finite Difference Methods *444*
12.10.2 Trapezoidal Rule Method *449*
12.11 **Initial-Value Problems Involving First-Order Systems** *465*
12.12 **Initial-Value Problems Involving Second-Order Systems** *467*
12.12.1 Decoupling Procedure *467*
12.12.2 Direct Integration Procedure *476*
12.13 **Boundary-Value Problems** *479*
Recommended Reading *493*
Problems *494*

**Chapter 13 NUMERICAL SOLUTION OF PARTIAL DIFFERENTIAL EQUATIONS** *500*

13.1 **Introduction** *501*
13.2 **Finite Difference Methods and Grid Patterns** *501*
13.3 **Transformation from Cartesian to Polar Coordinates** *503*
13.4 **Transformation from Cartesian to Skewed Coordinates** *507*
13.5 **Transformation from Cartesian to Triangular Coordinates** *511*
13.6 **Finite Difference Solution of Elliptic Equations** *513*
13.6.1 Regions Involving Rectangular Elements *513*
13.6.2 Circular Regions *524*
13.6.3 Regions Involving Skewed Elements *528*

13.6.4 Regions Involving Triangular Elements *532*
13.6.5 Special Considerations *535*
13.6.6 Irregular Boundaries *538*
**13.7 Parabolic Partial Differential Equations** *548*
13.7.1 Explicit Finite Difference Scheme *548*
13.7.2 Implicit Method *553*
13.7.3 Eigenproblem of the Explicit Scheme *558*
13.7.4 Eigenproblem of the Implicit Scheme *564*
13.7.5 Stability and Convergence *565*
13.7.6 Derivative Boundary Conditions *567*
13.7.7 Nonlinear Problems *568*
**13.8 Hyperbolic Partial Differential Equations** *576*
**13.9 Biharmonic Equation** *577*
**13.10 Introduction to the Finite Element Method** *584*
13.10.1 Introduction *584*
13.10.2 Discretization of a Region *584*
13.10.3 Interpolation and Shape Functions *586*
13.10.4 One-Dimensional Formulation *593*
Recommended Reading *601*
Problems *601*

**Chapter 14 ANALYTIC OPTIMIZATION** *606*

**14.1 Introduction** *607*
**14.2 Characteristics of Optimization Problems** *607*
**14.3 Unconstrained Optimization—Method of Calculus** *610*
14.3.1 Functions of One Variable *611*
14.3.2 Functions of Many Variables *615*
**14.4 Constrained Optimization—Lagrange Multipliers** *618*
**14.5 Optimization of Linear Models** *622*
14.5.1 Standard Equality Form *623*
14.5.2 Mathematical Search for an Optimal Solution *624*
14.5.3 Graphical Interpretation of an Optimal Solution *627*
**14.6 Simplex Method** *631*
Recommended Reading *637*
Problems *637*

**Index** *639*

# Preface

Numerical methods textbooks intended for undergraduates in engineering often fail to relate theory to application. *Numerical Methods in Engineering Practice* bridges this gap by presenting numerous practical examples drawn from various engineering disciplines. This text presents the principles of applied mathematics and extends them in a natural and systematic manner to numerical methods. It motivates the student and illustrates the methods with solutions to numerous practical problems drawn from civil, electrical, and mechanical engineering. It is organized in a manner roughly analogous to the familiar texts on classical engineering mathematics. The text develops the methods and provides descriptions of them in terms that can easily be implemented on computers of all sizes, including programmable calculators, and in any convenient high-order source language. A companion text containing many useful stand-alone programs incorporating the numerical methods of this text with two diskettes for an IBM PC is available.

The attitude of engineers toward mathematics has always been different from that of the mathematician. A mathematician may be interested in finding out whether a solution to a differential equation exists and the properties of such a solution. In contrast, an engineer simply assumes that the existence of a physical system is proof enough of the existence of a solution and focuses instead on finding it.

With the availability of computers of all sizes at affordable costs for engineers has also come a shift in the routine analysis and design methods used in engineering. While classical techniques based on closed-form mathematical modeling continue to be useful to gain introductory intuitive insight into the expected performance of engineering systems, these techniques are being largely replaced in the analysis and design of actual systems by numerical techniques applied to more general models—models not restricted to those which have closed-form solution and, therefore, models which can more accurately represent realistically complex systems; e.g., nonlinear, time-dependent state variable modeling techniques and finite element analysis techniques. The computer revolution, in short, has enabled engineers to successfully solve problems that were well beyond their reach just a few years ago.

This shift to much greater emphasis on numerical techniques in engineering makes it important to introduce these techniques and the related topics in applied mathematics early in the engineer's undergraduate education. Furthermore, they need to be introduced early enough to fully reveal the freedom they provide in solving engineering problems as compared to the classical techniques. In the authors' opinion, these techniques and topics should be first taught as soon as the student has completed a basic calculus sequence including differential equations and is being introduced to the various areas of engineering science of his discipline. Because we have aimed this text at students at this early point in their undergraduate education, we have taken special care to explicitly derive the useful end results from the elementary mathematical foundations known to these students. In particular, we have avoided the all too familiar practice of "leaving to the reader" the important intermediate steps of the development of the methods. We have been similarly explicit in the numerous examples.

This text should be suitable for a one-term course for engineering juniors and a one-term course in more advanced topics for seniors or first-year graduate students. At the University of Evansville, it serves as the text for a one-semester course for undergraduate engineers and a required course for beginning graduate students. If students have completed a sophomore course in differential equations and have a working knowledge of computer programming, our experience shows that the topics of Chapters 1–7 together with selected topics in Chapters 8–13 can be adequately covered in one course. A second graduate course can then emphasize the topics of Chapters 7–14. In addition, the text should be of interest to the practicing engineer.

The text is organized so as to provide a natural and sequential development of the topics, where each chapter serves as an appropriate foundation for the next chapter. Chapters 2–5 develop elementary concepts of matrices, systems of linear algebraic equations, and linear modeling. Chapters 6–9 extend these linear concepts to nonlinear equations, eigenvalue problems, interpolation, and regression as important applications. Chapters 10 and 11 employ these concepts to develop techniques for the central tasks of numerical differentiation and integration. Chapters 12 and 13 make application of these techniques to develop methods of solving systems of ordinary differential and partial differential equations. Chapter 14 is a nonexhaustive and less than comprehensive treatment of the optimization problem. It only includes selected classical analytical techniques for linear and nonlinear systems. The text develops an appropriate background for the study of finite element methods, which are briefly introduced in Chapter 13.

The authors are most grateful for their students' assistance in debugging the manuscript. We are forever indebted to the endless patience and support of Wilma, Pam, and Susan for typing the manuscript and to Mimi, whose courage was an inspiration.

A.W.A.
J.R.T.
*Evansville, Indiana*

# chapter 1

## Introduction

Earth as seen from space. *Courtesy of N.A.S.A.*

"From the dawn of civilization, man was concerned with putting scientific knowledge to practical uses."

# 1

## 1.1 MATHEMATICAL MODELS AND THEIR SOLUTIONS

Mathematical models in engineering and the applied sciences are normally classified as boundary-value, initial-value, or eigenvalue models. However, since engineering systems may have physical boundaries and time-dependent constraints, it is possible to encounter a model that is both a boundary-value and an initial-value model. In such cases, the classification is made in terms of model response (solution). The response is generally composed of two parts: steady state and transient. Steady-state models include stresses in structures, temperature distribution in plates, and static and dynamic displacement of structures. Transient models, which are also termed "unsteady," include heat propagation, electromagnetic energy field, and pore-water pressure in soil. It is evident that although transient solutions are time dependent, steady-state solutions may or may not be time dependent. For example, if a structure is subjected to a gust of wind, then we would expect the structure to respond to that force and come to rest soon after the wind has died down. This is called transient response. On the other hand, if that same structure is subjected to external forces produced by an eccentricity in rotating machinery within it, then we would expect the structure to vibrate (displace) as long as the force is in effect. The solution in this case is called the steady state.

The types of mathematical models encountered in engineering practice may be as simple as solving a system of linear algebraic equations or as difficult as solving partial differential equations in three spatial coordinates, in addition to the time coordinate. In elementary calculus and differential equations, one learns how to solve a variety of problems in a closed form (so-called exact solutions). Unfortunately, from a practical standpoint, the methods of calculus alone are not adequate to solve all the complex problems that an engineer might face. In fact, the reader may have already faced problems in calculus in which an integral cannot be evaluated in a closed form. For example, try to solve the following problem:

$$I = \int_a^b e^{x^2}\,dx$$

The purpose of this presentation is not to list difficult problems, but to highlight the need for solving practical problems in a simple and efficient manner. A number of such problems are illustrated in chapter 3.

## 1.2 THE NEED FOR NUMERICAL SOLUTIONS

While the closed-form solution of mathematical models is by far the most efficient method of tackling problems encountered in engineering, these are limited to physical systems with simple constraints. Furthermore, simplifying assumptions are often introduced so that a closed-form solution can be found. It is therefore possible to come up with a solution that may not be an accurate representation of the true behavior being studied. For example, the settlement of structures with time is modeled by the following partial differential equation:

$$\frac{\partial u}{\partial t} = c_v\left(\frac{\partial^2 u}{\partial x^2} + \frac{\partial^2 u}{\partial y^2} + \frac{\partial^2 u}{\partial z^2}\right)$$

where

$u$ = water pressure within a soil profile,

$t$ = time,

$x$, $y$, and $z$ = depth, width, and length coordinates, respectively,

$c_v$ = a physical constant which depends on soil type.

The closed-form solution to this problem is generally obtained by neglecting the changes in $u$ relative to width and length, and then assuming a constant ($c_v$) to give

$$\frac{\partial u}{\partial t} = c_v \frac{\partial^2 u}{\partial x^2}$$

While the first of these two simplifying assumptions may be appropriate in that a structure is expected to settle vertically, the second of these assumptions is not so realistic. This is because different soil types may be encountered in a given profile. One then may argue that the closed-form solution is still applicable when dealing with profiles involving one soil type. This is true but does show a rather significant limitation of the closed-form solution. Furthermore, the closed-form solution is obtained in terms of boundary conditions and initial values. These relate to the drainage conditions at the top and bottom of the layer and the initial water pressure caused by the structure itself. In essence, different closed-form solutions are obtained depending on the conditions assumed. For example, assuming free drainage at both ends of the soil layer and a constant initial pressure, the solution is given in closed form as follows:

$$u = \sum_{n=0}^{\infty}\left[\frac{2u_0}{\pi(m+\frac{1}{2})}\sin\left(\frac{\pi(m+\frac{1}{2})x}{H}\right)\right]\exp\left(-\frac{\pi^2(m+\frac{1}{2})^2 c_v t}{H^2}\right)$$

where $H$ is the thickness of the soil layer and $u_0$ is a constant. The question is, "If the initial conditions are not constant, can we use this solution?" The answer is obviously no. It is interesting that even for the simplest of boundary conditions the so-called exact solution is given as an infinite sum! The implication is that for all practical purposes, the solution is approximate since one can only include a few terms in calculating $u$ at the different depths and for different times. It is evident that realistic solutions should be more general, involve fewer simplifying assumptions, and be more versatile in handling different boundary and initial conditions. This is the essence of numerical solutions.

## 1.3 ERRORS

The numerical solution of engineering problems involves repeated operations. In this section we shall examine the question of errors in terms of their sources and how they may affect the solution of engineering problems. In general, errors may be related to mathematical formulation and/or simplification, measured physical parameters, and computational errors. Our presentation herein is limited to computational errors. Errors can arise from the following operations and limitations of computation.

### Significant Figures

The degree of precision of a number is commonly indicated by the number of digits known accurately or which, put simply, are significant. For example, when we say that someone's height is 6.0 ft, do we mean that it is approximately six feet? The answer is obviously yes. It could be 6.001 ft or even 6.01 ft. However, we just do not think it significant to include the additional decimal points. This concept is equally valid when dealing with engineering computations involving a large amount of number crunching.

### Roundoff

Suppose that the height of the individual is needed in centimeters. Then we can easily evaluate it by using the appropriate conversion factors 1 in. = 2.54 cm and 1 ft = 12 in. to give $2.54 \times 12 \times 6.0 = 182.88$ cm. However, for obvious reasons we may round off the height to 183 cm. Clearly, this situation occurs quite frequently when solving engineering problems and represents a rather significant source of errors. This is especially true when dealing with linear algebraic equations whose coefficients are of approximately the same order of magnitude.

### Truncation

The numerical solution of engineering problems are often based on a Taylor series which includes an infinite number of terms. For practical reasons, including cost of computation, only a few terms are used in the development of the various solution schemes. That is, we make a conscious decision to live with less accuracy in order to achieve an economical solution. In other words, we truncate the series in such a way that the solution is reasonably accurate.

### Computational Devices

Computers perform various tasks flawlessly at incredible speeds. However, they are limited in their capacity to retain significant figures as well as the amount of numbers they can store. Consequently, one must make a serious attempt to accommodate these limitations. The all too familiar single-precision and double-precision options available on mainframes are not common on hand-held calculators. Hence, if we wish to solve a problem involving numbers with large significant figures we should expect better accuracy using double precision than single precision. However, even double precision is normally limited to a finite number of digits. The implication is that besides human errors, the possibility always exists for computational device errors brought about mainly by the inability of the device to retain a large number of significant numbers of digits. For example, $\sqrt{2}$ was evaluated by using a pocket calculator, a programmable pocket computer, and a microcomputer. The computed values are listed in that order as follows:

$$\sqrt{2} = 1.4142136$$

$$\sqrt{2} = 1.414213562$$

$$\sqrt{2} = 1.4142135623731$$

It is evident that the pocket calculator rounded gives a value higher than the actual value, and the pocket computer gives a value less than the actual value. Obviously, we are in no position to pass a judgment on the value obtained form the microcomputer because we have nothing more accurate to compare it with. However, we do know that it is only an approximation of the real value. It is clear that even computational devices introduce errors to the solution of problems.

Overall, the practical trade-offs that must be made by the engineer are those related to the sources of error in his calculations and the costs of controlling them. As we examine the topics in this text, we shall look more closely at these error sources and identify and quantify them as appropriate to the topic. In this way, the reader will have a realistic basis for making the best trade-off for error control for his particular application.

## 1.4 TAYLOR SERIES

The basis for many of the techniques presented in this text can be traced back to the Taylor series. This series states simply that "any function can be expressed as a linear combination of an infinite number of rational functions." Hence, given an arbitrary function $f(x)$, the Taylor series for the function is as follows:

$$f(x) = f(x_0) + (x - x_0)f^{(1)}(x_0) + \frac{(x - x_0)^2}{2!}f^{(2)}(x_0) + \cdots + \frac{(x - x_0)^n}{n!}f^{(n)}(x_0)$$

or more simply

$$f(x) = f(x_0) + \sum_{k=1}^{\infty} \frac{(x - x_0)^k}{k!} f^{(k)}(x_0)$$

where $x_0$ is a point of expansion and $f^{(k)}(x_0)$ is the $k$th derivative of the function evaluated at $x_0$. The implication of the Taylor series is rather significant in that it permits the transformation of complicated functions into equivalent polynomials. This fact makes it possible to find solutions to some of the most complex problems involving mathematical models. The basis for this series is discussed more thoroughly in chapter 10.

# chapter 2

# Matrices and Determinants

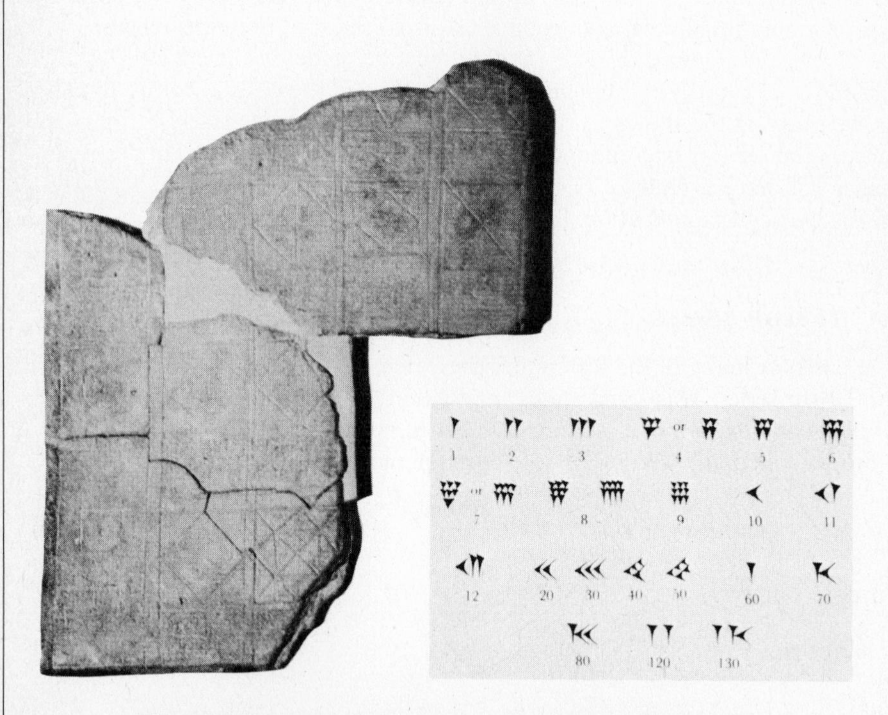


The Akkadian number system and a clay tablet (1800 B.C.) relating to geometrical problems. *Courtesy of the Iraqi government.* It reads:

"A square, the side is 1. I have made a border at each side, and I have drawn a second square and inside the square I have drawn a circle. This surface, what is it?"

# 2

## 2.1 INTRODUCTION TO MATRICES

Matrices and their algebra are vital to modern engineering mathematics. They represent an enormously effective means by which data and other mathematical quantities of interest can be manipulated in an elegant and efficient manner. The practical use of finite difference, finite elements, and many other numerical techniques are greatly facilitated by the use of matrices. While a complete treatment of this subject is beyond the scope of this text, most of the important basic concepts of matrices and matrix algebra are covered.

*A matrix is a rectangular array of elements arranged in rows and columns.* The elements may be numbers, variables, functions, integrals, derivatives, or even other matrices. The size of a matrix is specified by giving its number of rows ($m$) and number of columns ($n$). This is read as "$m$ by $n$" and written as "$m \times n$". The magnitude of a given element may be designated by a letter such as $a$ and its position by a pair of subscripts. Consequently, an element positioned in the $i$th row and $j$th column of a matrix $[A]$ is written as $a_{ij}$.

Matrices arise in the analysis of engineering problems whenever the system or phenomenon under study can be described with a set of *m linear algebraic equations in n variables.* While it is important to emphasize that we are restricting our attention to algebraic equations, it is equally important to emphasize that such equations may arise in the *numerical solution* of ordinary and partial differential equations pertaining to temperature distribution, deflections, fluid flow, voltage and current distribution, and many other situations.

The following set [equation (2.1)] of $m$ linear algebraic equations serves to illustrate the notational convention used in this text relative to matrix representation of a linear system of equations.

$$\begin{aligned}
a_{11}x_1 + a_{12}x_2 + \cdots + a_{1n}x_n &= b_1 \\
a_{21}x_1 + a_{22}x_2 + \cdots + a_{2n}x_n &= b_2 \\
\vdots \qquad \vdots \qquad\quad & \vdots \quad \vdots \\
a_{m1}x_1 + a_{m2}x_2 + \cdots + a_{mn}x_n &= b_m
\end{aligned} \tag{2.1}$$

The set of $n$ variables $x_1, x_2, \ldots, x_n$ and a set of $m$ dependent variables $b_1, b_2, \ldots, b_m$ are related through the $m \times n$ coefficients $a_{11}, a_{12}, \ldots, a_{mn}$. Using matrix notation we write equation (2.1) more compactly as follows:

$$\begin{bmatrix} a_{11} & a_{12} & \cdots & a_{1n} \\ a_{21} & a_{22} & \cdots & a_{2n} \\ \vdots & \vdots & & \vdots \\ a_{m1} & a_{m2} & \cdots & a_{mn} \end{bmatrix} \begin{Bmatrix} x_1 \\ x_2 \\ \vdots \\ x_n \end{Bmatrix} = \begin{Bmatrix} b_1 \\ b_2 \\ \vdots \\ b_m \end{Bmatrix} \tag{2.2}$$

The $m \times n$ array of $a$'s is referred to as the coefficient matrix $[A]$ or $[a_{ij}]$ and the independent variables as the $\{x\}$ or $\{x_j\}$ vector. Furthermore, the dependent quantities on the right-hand side are grouped in a separate array called the $\{B\}$ or $\{b_i\}$ vector. Clearly, it is our intention that equation (2.2) be defined to mean the same as equation (2.1). The details of this definition and the related algebraic rules for manipulating matrices and vectors are examined in section 2.3. Suffice it to say that using the definition of matrix multiplication one can express a set of linear algebraic equations in subscripted compact matrix form (2.3) or in unsubscripted compact matrix form (2.4):

$$[a_{ij}]\{x_j\} = \{b_i\}, \qquad i = 1, \ldots, m, \quad j = 1, \ldots, n \tag{2.3}$$

$$[A]_{m \times n}\{x\}_{n \times 1} = \{B\}_{m \times 1} \tag{2.4}$$

## 2.2 SPECIAL MATRICES

Engineering problems and their solutions may involve various types of special matrices. The unique properties of these matrices are important and are referred to throughout the text. For clarity and convenience, before examining the algebra of matrices, we provide the following definitions and descriptions.

### Vectors and Scalars

Matrices with only one row and $n$ columns $(1 \times n)$ or one column and $m$ rows $(m \times 1)$ are called vectors. The symbol $\{\ \}$ is used to indicate a row or column vector. Matrices with only one element $(1 \times 1)$ are called scalars. For example, the $3 \times 1$ matrix $\{A\}$

$$\{A\} = \begin{Bmatrix} 2 \\ 10 \\ -3 \end{Bmatrix}_{3 \times 1} = \{a_{ij}\}, \qquad i = 1, 2, 3, \quad j = 1$$

is a *column vector*. Similarly, the $1 \times 3$ matrix $\{B\}$ is a *row vector*:

$$\{B\} = \{7 \quad 30 \quad 1\}_{1 \times 3} = \{b_{ij}\}, \qquad i = 1, \quad j = 1, 2, 3$$

The $1 \times 1$ matrix $[C]$

$$[C] = [-2]_{1 \times 1} = [c_{ij}], \qquad i = 1, \quad j = 1$$

is a *scalar* and may be written without the brackets. It is evident that one may use a single subscript to define a vector. In fact this is exactly how the $\{x_i\}$ and $\{b_j\}$ vectors were defined in equation (2.3).

### The Null Matrix

This is a matrix whose elements are all equal to zero. Similarly, a vector in which all elements are zero is called a "null vector." For example,

$$[A]_{m\times n} = \begin{bmatrix} 0 & 0 & \cdots & 0 \\ 0 & 0 & \cdots & 0 \\ \vdots & \vdots & & \vdots \\ 0 & 0 & \cdots & 0 \end{bmatrix} = [a_{ij}]$$

The $[A]$ matrix is said to be a null matrix if and only if $a_{ij} = 0$ for $i = 1,\ldots,m$ and $j = 1,\ldots,n$. Note that a null matrix is not equal to the scalar quantity zero unless it is a $1 \times 1$ matrix.

### The Square Matrix

A matrix having an equal number of columns and rows is called a square matrix. The size of such a matrix is defined by the number of rows or columns. Therefore matrix $[A]$ is a square matrix of size $n$ while matrix $[B]$ is of size 2:

$$[A]_n = \begin{bmatrix} a_{11} & a_{12} & \cdots & a_{1n} \\ a_{21} & a_{22} & \cdots & a_{2n} \\ \vdots & \vdots & & \vdots \\ a_{n1} & a_{n2} & \cdots & a_{nn} \end{bmatrix} = [a_{ij}], \qquad i = 1,\ldots,n, \quad j = 1,\ldots,n$$

$$[B]_2 = \begin{bmatrix} b_{11} & b_{12} \\ b_{21} & b_{22} \end{bmatrix} = [b_{ij}], \qquad i = 1,2, \qquad j = 1,2$$

It is evident that a square matrix may be a null matrix, but a null matrix may not always be a square matrix. Square matrices play an important role in engineering and have properties that are unique. For example, the determinant and inverse can be defined only if the matrix is a square matrix.

### The Identity Matrix

The identity matrix is a square matrix whose off-diagonal elements are equal to zero and diagonal elements are equal to one. Matrix $[A]$ is an identity matrix of size $n$:

$$[A]_n = \begin{bmatrix} 1 & 0 & \cdots & 0 \\ 0 & 1 & \cdots & 0 \\ \vdots & \vdots & & \vdots \\ 0 & 0 & \cdots & 1 \end{bmatrix} = [a_{ij}], \qquad \begin{cases} a_{ij} = 1 & (i = j) \\ a_{ij} = a_{ji} = 0 & (i \neq j) \end{cases}$$

Since identity matrices are unique, they are designated by the letter $I$ and sometimes written as

$$[I]_n = \begin{bmatrix} 1 & & 0 \\ & 1 & \\ & & \ddots & \\ 0 & & & 1 \end{bmatrix}$$

Consequently matrices $[A]$, $[B]$, and $[C]$ represent identity matrices of size 3, 2, and 1, respectively:

$$[A] = \begin{bmatrix} 1 & 0 & 0 \\ 0 & 1 & 0 \\ 0 & 0 & 1 \end{bmatrix} = \begin{bmatrix} 1 & & 0 \\ & 1 & \\ 0 & & 1 \end{bmatrix} = [I]_3$$

$$[B] = \begin{bmatrix} 1 & 0 \\ 0 & 1 \end{bmatrix} = [I]_2$$

$$[C] = [1] = 1 = [I]_1$$

Note that matrix $[C]$ is a scalar equal to 1.0. Identity matrices have the property that when they multiply matrix equations the result is the same as multiplying by the scalar 1.0.

### The Diagonal Matrix

This is a square matrix whose off-diagonal elements are equal to zero and whose diagonal elements can take on any value. It is obvious that the identity matrix is a special case of a diagonal matrix. In general we say $[A]$ is a diagonal matrix if $a_{ij} = 0$ for $i \neq j$:

$$[A]_n = \begin{bmatrix} a_{11} & 0 & \cdots & 0 \\ 0 & a_{22} & \cdots & 0 \\ \vdots & \vdots & & \vdots \\ 0 & 0 & \cdots & a_{nn} \end{bmatrix} = [a_{ij}], \qquad i = 1, \ldots, n, \quad j = 1, \ldots, n$$

and $a_{ij}$ are specified arbitrarily for $i = j$. Diagonal matrices are generally expressed as follows:

$$[A]_n = \begin{bmatrix} a_{11} & & 0 \\ & a_{22} & \\ & & \ddots & \\ 0 & & & a_{nn} \end{bmatrix}$$

These matrices are extremely important and occur frequently when one is solving systems of ordinary differential equations. Furthermore, the term "diagonalizing" is used to mean that a nondiagonal square matrix is reduced to one which is diagonal.

### The Symmetrical Matrix

This is a matrix in which the corresponding off-diagonal elements are equal. That is, $a_{ij} = a_{ji}$ for $i = 1, \ldots, n$ and $j = 1, \ldots, n$. Consider the following matrix:

$$[A]_n = \begin{bmatrix} a_{11} & a_{12} & \cdots & a_{1n} \\ a_{21} & a_{22} & \cdots & a_{2n} \\ \vdots & \vdots & & \vdots \\ a_{n1} & a_{n2} & \cdots & a_{nn} \end{bmatrix}$$

Here matrix $[A]$ is symmetrical if and only if $a_{12} = a_{21}$, $a_{13} = a_{31}, \ldots, a_{n-1,n} = a_{n,n-1}$. That is, if one is to replace the corresponding columns of $[A]$ with its rows, then the same matrix $[A]$ would result. Therefore one need not write all elements of a symmetrical matrix, instead only the diagonal elements and either the elements appearing above or below the diagonal need be written. That is,

$$[A]_n = \begin{bmatrix} a_{11} & a_{12} & \cdots & a_{1n} \\ & a_{22} & \cdots & a_{2n} \\ & & \ddots & \vdots \\ & & & a_{nn} \end{bmatrix}$$

Symmetrical matrices play an important role in the solution of ordinary as well as partial differential equations pertaining to transient and steady-state system response.

### The Transpose Matrix

For any matrix $[A]$ there is a matrix $[B]$ called the transpose of $[A]$. The $[B]$ matrix is obtained by simply interchanging the corresponding columns of $[A]$ with its rows. For example, given

$$[A]_{m \times n} = \begin{bmatrix} a_{11} & a_{12} & \cdots & a_{1n} \\ a_{21} & a_{22} & \cdots & a_{2n} \\ \vdots & \vdots & & \vdots \\ a_{m1} & a_{m2} & \cdots & a_{mn} \end{bmatrix} = [a_{ij}], \qquad i = 1, \ldots, m, \quad j = 1, \ldots, n$$

then the transpose of $[A]$ is given by $[B]$ as follows:

$$[B]_{n \times m} = \begin{bmatrix} a_{11} & a_{21} & \cdots & a_{m1} \\ a_{12} & a_{22} & \cdots & a_{m2} \\ \vdots & \vdots & & \vdots \\ a_{1n} & a_{2n} & \cdots & a_{mn} \end{bmatrix} = [a_{ji}], \qquad i = 1, \ldots, m, \quad j = 1, \ldots, n$$

For convenience, one may write

$$[B]_{n \times m} \triangleq \text{transpose } [A]_{m \times n} \triangleq [A]^{t}_{n \times m}$$

where $\triangleq$ is read as "corresponds to." In general, given

$$[A]_{m \times n} = [a_{ij}]$$

then

$$[A]^{t}_{n \times m} = [a_{ji}], \qquad i = 1, \ldots, m, \quad j = 1, \ldots, n$$

It is evident that *the transpose of a symmetric matrix is equal to the matrix itself.* Note that the transpose of a row vector is the corresponding column vector.

Consider the following examples:

$$\{A\} = \{1 \quad 7 \quad 6\}, \qquad \{A\}^{t} = \begin{Bmatrix} 1 \\ 7 \\ 6 \end{Bmatrix}$$

$$[B] = \begin{bmatrix} 2 & -3 & 1 \\ 7 & 4 & -6 \end{bmatrix}, \qquad [B]^{t} = \begin{bmatrix} 2 & 7 \\ -3 & 4 \\ 1 & -6 \end{bmatrix}$$

$$[C] = \begin{bmatrix} 1 & -1 & 4 \\ -1 & 2 & 5 \\ 4 & 5 & 3 \end{bmatrix}, \qquad [C]^{t} = \begin{bmatrix} 1 & -1 & 4 \\ -1 & 2 & 5 \\ 4 & 5 & 3 \end{bmatrix} = [C]$$

Rules pertaining to the transpose of matrix equations, sums, products, and inverses are discussed in section 2.7.

### The Inverse Matrix

The matrix $[B]$ is called the inverse matrix of $[A]$ and denoted as $[B] = [A]^{-1}$. The product of $[A]$ and $[B]$ yields the identity matrix $[I]$. Only square matrices can have an inverse. Thus, given

$$[A][B] = [B][A] = [I]$$

then

$$[B] = [A]^{-1} \qquad \text{and} \qquad [A] = [B]^{-1}$$

Note that, just as the scalar 0 has no reciprocal or inverse, there are many different square matrices of size greater than $1 \times 1$ for which there is no inverse matrix.

### The Singular Matrix

This is a square matrix that has no inverse. Therefore a matrix for which the inverse can be defined is called a nonsingular matrix.

### The Orthogonal Matrix

This is a square matrix whose inverse can be determined by transposing it! That is, given $[A]$, then

$$[A]^{-1} = [A]^{t}$$

if and only if $[A]$ is orthogonal. Such matrices do occur in some engineering problems. The matrix used to obtain rotation of coordinates about the origin of a Cartesian system is one example of an orthogonal matrix.

### The Augmented Matrix

This is a matrix that can be formed from two other matrices. Consider the matrices $[A]$ and $[B]$:

$$[A] = \begin{bmatrix} a_{11} & a_{12} \\ a_{21} & a_{22} \end{bmatrix}, \qquad [B] = \begin{bmatrix} 1 & 0 \\ 0 & 1 \end{bmatrix}$$

Then a third matrix $[C]$ is the augmented matrix if

$$[C] = \left[\begin{array}{cc|cc} a_{11} & a_{12} & 1 & 0 \\ a_{21} & a_{22} & 0 & 1 \end{array}\right]$$

Note that $[C]$ can only be defined if both $[A]$ and $[B]$ have exactly the same number of rows. Augmented matrices are useful in solving sets of linear algebraic equations. The dashed line contained in $[C]$ is used to separate the elements of $[A]$ and $[B]$.

### The Submatrix

This is a matrix that belongs to a larger matrix. Furthermore, it contains a rectangular array of elements. Consequently, given the matrix $[A]$,

$$[A] = \begin{bmatrix} a_{11} & a_{12} & a_{13} \\ a_{21} & a_{22} & a_{23} \\ a_{31} & a_{32} & a_{33} \end{bmatrix}$$

then each element $a_{11}, \ldots, a_{33}$ can be seen as a submatrix of $[A]$. In addition, a whole column or a whole row may be taken as a submatrix of $[A]$. In general, one may subdivide a matrix into a variety of submatrices. Therefore matrix $[A]$ can be expressed in terms of submatrices as follows:

$$[A] = \begin{bmatrix} [A_{11}] & [A_{12}] \\ [A_{21}] & [A_{22}] \end{bmatrix}$$

where

$$[A_{11}] = \begin{bmatrix} a_{11} & a_{12} \\ a_{21} & a_{22} \end{bmatrix}, \qquad [A_{12}] = \begin{Bmatrix} a_{13} \\ a_{23} \end{Bmatrix},$$

$$[A_{21}] = \{a_{31} \quad a_{32}\}, \qquad [A_{22}] = [a_{33}]$$

Submatrices are used for convenience in expressing a matrix, especially when the matrix in question contains zero elements that can be grouped into a submatrix. They also play a role in the solution of matrix equations in which the unknown is not a vector but a matrix. Such equations arise in the solution of linear algebraic equations with complex coefficients. In fact, the inverse of a large matrix is often made easier by using submatrices.

### The Upper-Triangular Matrix

This is a square matrix in which all elements below the diagonal are equal to zero and the diagonal as well as the above-diagonal elements can take on any value. In general, given

$$[A]_{n \times n} = \begin{bmatrix} a_{11} & a_{12} & \cdots & a_{1n} \\ 0 & a_{22} & \cdots & a_{2n} \\ \vdots & \vdots & & \vdots \\ 0 & 0 & \cdots & a_{nn} \end{bmatrix} = [a_{ij}]$$

then we say $[A]$ is an upper-triangular matrix, since

$$a_{ij} = 0 \qquad \text{for} \quad i > j$$

and $a_{ij}$ can take on any value for $i \leq j$. The solution of linear algebraic equations using the Gauss elimination method results in an upper-triangular matrix.

### The Lower-Triangular Matrix

This is a square matrix in which all elements above the diagonal are zero and the rest of the elements can take on any value. That is,

$$[A]_{n \times n} = \begin{bmatrix} a_{11} & 0 & \cdots & 0 \\ a_{21} & a_{22} & \cdots & 0 \\ \vdots & \vdots & & \vdots \\ a_{n1} & a_{n2} & \cdots & a_{nn} \end{bmatrix} = [a_{ij}], \qquad i = 1, \ldots, n, \quad j = 1, \ldots, n$$

where $a_{ij} = 0$ for $i > j$ and $a_{ij}$ can take on any value for $i \leq j$. It is evident that the transpose of a lower-triangular matrix is an upper-triangular matrix and vice versa.

### The Upper-Unit-Triangular Matrix

This is an upper-triangular matrix whose diagonal elements are equal to one. Consider the matrix $[A]$ below:

$$[A] = \begin{bmatrix} 1 & a_{12} & a_{13} \\ 0 & 1 & a_{23} \\ 0 & 0 & 1 \end{bmatrix} = [a_{ij}], \qquad \begin{cases} i = 1, 2, 3 \\ j = 1, 2, 3 \\ a_{ii} = 1 \\ a_{ij} = 0 \quad (i > j) \end{cases}$$

Obviously this is an example where the $[A]$ matrix is an upper-triangular matrix of size $3 \times 3$ in which six out of the nine elements are specified and only three are not. That is, we can choose $a_{12}$, $a_{13}$, and $a_{23}$ (but not the rest of the entries) in any manner without changing the fact that $[A]$ is an upper-unit-triangular matrix. Similarly, the following matrix $[B]$ is a $2 \times 2$ upper-unit-triangular matrix. Note that, for a 2 × 2, only the element $b_{12}$ may be specified in an arbitrary way if [B] is to remain an upper-unit-triangular matrix.

$$[B] = \begin{bmatrix} 1 & b_{12} \\ 0 & 1 \end{bmatrix}$$

This type of matrix is used in solving linear algebraic equations with Crout's method. Furthermore, the transpose of an upper-unit-triangular matrix gives the lower-unit-triangular matrix.

## 2.3 MATRIX EQUALITY

A matrix [A] is said to be equal to a matrix [B] if and only if the corresponding elements in [A] and [B] are equal. In addition, matrix equality is defined for matrices or vectors only of equal size. Consider the following:

$$[A]_{m\times n} = \begin{bmatrix} a_{11} & a_{12} & \cdots & a_{1n} \\ a_{21} & a_{22} & \cdots & a_{2n} \\ \vdots & \vdots & & \vdots \\ a_{m1} & a_{m2} & \cdots & a_{mn} \end{bmatrix}$$

$$[B]_{m\times n} = \begin{bmatrix} b_{11} & b_{12} & \cdots & b_{1n} \\ b_{21} & b_{22} & \cdots & b_{2n} \\ \vdots & \vdots & & \vdots \\ b_{m1} & b_{m2} & \cdots & b_{mn} \end{bmatrix}$$

We say [A] = [B] if and only if

$$a_{ij} = b_{ij}$$

for $i = 1, \ldots, m$ and $j = 1, \ldots, n$. Otherwise we say $[A] \neq [B]$. This definition of matrix equality is extremely important despite its deceptive simplicity.

**E** EXAMPLE 2.1

Given the following matrices [A] and [B], determine the values of $x$ and $y$ such that [A] is equal to [B].

$$[A] = \begin{bmatrix} 2x & 1 \\ 3 & 10y^2 \end{bmatrix}, \quad [B] = \begin{bmatrix} 5 & 1 \\ 3 & 10 \end{bmatrix}$$

### Solution

Note that if $a_{12} \neq b_{12}$ and $a_{21} \neq b_{21}$ then there are no values for $x$ and $y$ that can make [A] = [B]. Since this is not the case in this example, let us proceed by equating $a_{11}$ to $b_{11}$ and $a_{22}$ to $b_{22}$ to give

$$2x = 5 \Rightarrow x = 2.5$$

$$10y^2 = 10 \Rightarrow y = \pm 1$$

Substituting $x$ and $y$ into matrix $[A]$ would indeed make $[A] = [B]$.

## 2.4 MATRIX ADDITION AND SUBTRACTION

The addition and/or subtraction of two or more matrices is accomplished by simply adding and/or subtracting corresponding elements. These operations are defined if and only if all matrices in question have equal size. Thus, given

$$[A]_{m \times n} = \begin{bmatrix} a_{11} & a_{12} & \cdots & a_{1n} \\ a_{21} & a_{22} & \cdots & a_{2n} \\ \vdots & \vdots & & \vdots \\ a_{m1} & a_{m2} & \cdots & a_{mn} \end{bmatrix} = [a_{ij}]$$

$$[B]_{m \times n} = \begin{bmatrix} b_{11} & b_{12} & \cdots & b_{1n} \\ b_{21} & b_{22} & \cdots & b_{2n} \\ \vdots & \vdots & & \vdots \\ b_{m1} & b_{m2} & \cdots & b_{mn} \end{bmatrix} = [b_{ij}]$$

Then

$$[A] + [B] = [C]_{m \times n} = [a_{ij} + b_{ij}] \tag{2.5a}$$

and

$$[A] - [B] = [D]_{m \times n} = [a_{ij} - b_{ij}] \tag{2.5b}$$

for $i = 1, \ldots, m$ and $j = 1, \ldots, n$. In general one may write

$$[A] \pm [B] = [a_{ij} \pm b_{ij}]$$

to indicate addition or subtraction of two matrices $[A]$ and $[B]$.

**E** EXAMPLE 2.2

Given the matrices $[A]$ and $[B]$ determine $[A] + [B]^t$ and $[A]^t - [B]$.

$$[A] = \begin{bmatrix} 2 & 5 \\ 0 & 7 \\ -1 & 4 \end{bmatrix}, \quad [B] = \begin{bmatrix} 1 & 1 & 0 \\ 2 & 0 & 0 \end{bmatrix}$$

Solution

Let $[C] = [A] + [B]^t$ and $[D] = [A]^t - [B]$; then

$$[C] = \begin{bmatrix} 2 & 5 \\ 0 & 7 \\ -1 & 4 \end{bmatrix} + \begin{bmatrix} 1 & 1 & 0 \\ 2 & 0 & 0 \end{bmatrix}^t$$

$$= \begin{bmatrix} 2 & 5 \\ 0 & 7 \\ -1 & 4 \end{bmatrix} + \begin{bmatrix} 1 & 2 \\ 1 & 0 \\ 0 & 0 \end{bmatrix}$$

$$= \begin{bmatrix} 2+1 & 5+2 \\ 0+1 & 7+0 \\ -1+0 & 4+0 \end{bmatrix}$$

$$= \begin{bmatrix} 3 & 7 \\ 1 & 7 \\ -1 & 4 \end{bmatrix}$$

Also,

$$[D] = \begin{bmatrix} 2 & 0 & -1 \\ 5 & 7 & 4 \end{bmatrix} - \begin{bmatrix} 1 & 1 & 0 \\ 2 & 0 & 0 \end{bmatrix}$$

$$= \begin{bmatrix} 2-1 & 0-1 & -1-0 \\ 5-2 & 7-0 & 4-0 \end{bmatrix}$$

$$= \begin{bmatrix} 1 & -1 & -1 \\ 3 & 7 & 4 \end{bmatrix}$$

## 2.5 MATRIX MULTIPLICATION

To explain the general rules of matrix multiplication, consider the product of the matrices $[A]$ and $[B]$, which is equal to a third matrix $[C]$:

$$[C] = [A][B]$$

Matrix $[C]$ can only be defined if the number of rows of matrix $[B]$ is equal to the number of columns of matrix $[A]$. That is,

$$[C]_{m \times L} = [A]_{m \times n}[B]_{n \times L}$$

or

$$[c_{ik}] = [a_{ij}][b_{jk}], \qquad i = 1, \ldots, m, \quad k = 1, \ldots, L$$

Obviously $[A]$ and $[B]$ do not have to be of equal size for their product to be defined. The elements of matrix $[C]$ are given as

$$c_{ik} = \sum_{j=1}^{n} a_{ij} b_{jk}, \qquad \begin{cases} i = 1, \ldots, m \\ k = 1, \ldots, L \end{cases} \tag{2.6}$$

Here $m$ is the number of rows of matrices $[A]$ and $[C]$; $n$ is the number of columns of $[A]$ and the number of rows of $[B]$; $L$ is the number of columns of $[B]$ and $[C]$.

To illustrate the application of equation (2.6) for finding the product of two matrices, consider the following example:

$$[A] = \begin{bmatrix} a_{11} & a_{12} & a_{13} \\ a_{21} & a_{22} & a_{23} \end{bmatrix}, \qquad [B] = \begin{bmatrix} b_{11} & b_{12} \\ b_{21} & b_{22} \\ b_{31} & b_{32} \end{bmatrix}$$

Let us define

$$[C] = [A]_{2\times 3}[B]_{3\times 2}$$

Clearly, the number of rows in $[B]$ is equal to the number of columns in $[A]$; therefore matrix $[C]$ can be determined. Matrix $[C]$ is a $2 \times 2$ matrix. Why?

$$\begin{bmatrix} a_{11} & a_{12} & a_{13} \\ a_{21} & a_{22} & a_{23} \end{bmatrix} \begin{bmatrix} b_{11} & b_{12} \\ b_{21} & b_{22} \\ b_{31} & b_{32} \end{bmatrix} = \begin{bmatrix} c_{11} & c_{12} \\ c_{21} & c_{22} \end{bmatrix}$$

where

$$c_{11} = \sum_{j=1}^{3} a_{1j}b_{j1}$$
$$= a_{11}b_{11} + a_{12}b_{21} + a_{13}b_{31}$$

$$c_{12} = \sum_{j=1}^{3} a_{1j}b_{j2}$$
$$= a_{11}b_{12} + a_{12}b_{22} + a_{13}b_{32}$$

$$c_{21} = \sum_{j=1}^{3} a_{2j}b_{j2}$$
$$= a_{21}b_{11} + a_{22}b_{21} + a_{23}b_{31}$$

and

$$c_{22} = \sum_{j=1}^{3} a_{2j}b_{j2}$$
$$= a_{21}b_{12} + a_{22}b_{22} + a_{23}b_{32}$$

The $[C]$ matrix is therefore given as follows:

$$[C] = \left[\begin{array}{c|c} a_{11}b_{11} + a_{12}b_{21} + a_{13}b_{31} & a_{11}b_{12} + a_{12}b_{22} + a_{13}b_{32} \\ \hline a_{21}b_{11} + a_{22}b_{21} + a_{23}b_{31} & a_{21}b_{12} + a_{22}b_{22} + a_{23}b_{32} \end{array}\right]$$

Careful consideration of this example confirms that the process involved in multiplying two matrices causes us to use an *entire row* of $[A]$ and an *entire column* of $[B]$ to obtain a *single element* of $[C]$. That is, "$c_{11}$" is determined by combining the products of the elements in the first column of $[B]$ with the elements of the first row in $[A]$; "$c_{12}$" is determined by combining the products of the elements in the second column of $[B]$ with the elements of the first row in $[A]$; . . . ; etc. That is,

$$c_{11} = \{a_{11} \quad a_{12} \quad a_{13}\} \begin{Bmatrix} b_{11} \\ b_{21} \\ b_{31} \end{Bmatrix}$$

$$c_{12} = \{a_{11} \quad a_{12} \quad a_{13}\} \begin{Bmatrix} b_{12} \\ b_{22} \\ b_{32} \end{Bmatrix}$$

$$= \{a_{21} \quad a_{22} \quad a_{23}\} \begin{Bmatrix} b_{11} \\ b_{21} \\ b_{31} \end{Bmatrix}$$

and

$$c_{22} = \{a_{21} \quad a_{22} \quad a_{23}\} \begin{Bmatrix} b_{12} \\ b_{22} \\ b_{32} \end{Bmatrix}$$

E

EXAMPLE 2.3

Given the following row and column matrices (vectors), determine their products:

(a) $\{B\}\{A\}$ and (b) $\{A\}\{B\}$

$$\{A\} = \begin{Bmatrix} 1 \\ 2 \\ 3 \end{Bmatrix}, \qquad \{B\} = \{3 \quad 2 \quad 1\}$$

Solution

(a) Obviously, the number of rows in $\{A\}$ is equal to one, which is also equal to the number of columns in $\{B\}$. Consequently, the product is possible and given by

$$[C]_{1\times 1} = \{3 \quad 2 \quad 1\} \begin{Bmatrix} 1 \\ 2 \\ 3 \end{Bmatrix}$$

$$= [(3)(1) + (2)(2) + (1)(3)]$$

$$= 10$$

(b) In this case the number of rows in $\{B\}$ is equal to three, which is once again equal to the number of columns in $\{A\}$. Hence the product is again defined and given by

$$[D]_{3\times 3} = \begin{Bmatrix} 1 \\ 2 \\ 3 \end{Bmatrix} \{3 \quad 2 \quad 1\}$$

$$= \begin{bmatrix} (1)(3) & (1)(2) & (1)(1) \\ (2)(3) & (2)(2) & (2)(1) \\ (3)(3) & (3)(2) & (3)(1) \end{bmatrix}$$

$$= \begin{bmatrix} 3 & 2 & 1 \\ 6 & 4 & 2 \\ 9 & 6 & 3 \end{bmatrix}$$

Note that the resulting product in (a) is a $1 \times 1$ matrix whereas a $3 \times 3$ matrix is obtained for (b).

EXAMPLE 2.4

Evaluate the products $[A][B]$ and $[B][A]$ for the following matrices, if possible:

$$[A] = \begin{bmatrix} 2 & 1 \\ 3 & -4 \end{bmatrix}, \qquad [B] = \begin{bmatrix} a & b & 0 \\ c & d & 0 \end{bmatrix}$$

Solution

$$[C]_{2\times 3} = \begin{bmatrix} 2 & 1 \\ 3 & -4 \end{bmatrix} \begin{bmatrix} a & b & 0 \\ c & d & 0 \end{bmatrix}$$

$$= \begin{bmatrix} 2a + c & 2b + d & 0 \\ 3a - 4c & 3b + 4d & 0 \end{bmatrix}$$

However, if we try to determine $[B][A]$, this product *cannot be defined* since the number of rows in $[A]$ is not equal to the number of columns in $[B]$.

It is evident that certain rules for multiplying two or more matrices must be established so that potential ambiguity is eliminated.

### 2.5.1 Commutativity of Matrix Multiplication

We have already seen in the examples of matrix multiplication that the product $[A][B]$ may be different from $[B][A]$; in fact it may even be undefined. This is a very important difference between the rules of matrix multiplication and the multiplication of scalars. In general, matrix multiplication is *noncommutative*. That is,

$$[A][B] \neq [B][A] \tag{2.7}$$

There are, however, many important examples of matrices which can commute. These include the product of a matrix with the identity matrix, with its inverse, or with a scalar. Furthermore, commutativity is not even possible for nonsquare matrices with the possible exception of vectors.

EXAMPLE 2.5

Show that the matrices $[A]$ and $[B]$ commute.

$$[A] = \begin{bmatrix} \cos\theta & \sin\theta \\ -\sin\theta & \cos\theta \end{bmatrix}, \quad [B] = \begin{bmatrix} \cos\theta & -\sin\theta \\ \sin\theta & \cos\theta \end{bmatrix}$$

Solution

$$[A][B] = \begin{bmatrix} \cos\theta & \sin\theta \\ -\sin\theta & \cos\theta \end{bmatrix}\begin{bmatrix} \cos\theta & -\sin\theta \\ \sin\theta & \cos\theta \end{bmatrix}$$

$$= \begin{bmatrix} \cos^2\theta + \sin^2\theta & 0 \\ 0 & \sin^2\theta + \cos^2\theta \end{bmatrix}$$

$$= \begin{bmatrix} 1 & 0 \\ 0 & 1 \end{bmatrix}$$

$$[B][A] = \begin{bmatrix} \cos\theta & -\sin\theta \\ \sin\theta & \cos\theta \end{bmatrix}\begin{bmatrix} \cos\theta & \sin\theta \\ -\sin\theta & \cos\theta \end{bmatrix}$$

$$= \begin{bmatrix} \cos^2\theta + \sin^2\theta & 0 \\ 0 & \sin^2\theta + \cos^2\theta \end{bmatrix}$$

$$= \begin{bmatrix} 1 & 0 \\ 0 & 1 \end{bmatrix}$$

In general, for commutativity to hold requires very strong constraints on the elements of the matrices. To see this more clearly consider the product of matrices $[A]$ and $[B]$ as follows:

$$[A][B] = \begin{bmatrix} a_{11} & a_{12} \\ a_{21} & a_{22} \end{bmatrix} \begin{bmatrix} b_{11} & b_{12} \\ b_{21} & b_{22} \end{bmatrix}$$

$$= \begin{bmatrix} a_{11}b_{11} + a_{12}b_{21} & a_{11}b_{12} + a_{12}b_{22} \\ a_{21}b_{11} + a_{22}b_{21} & a_{21}b_{12} + a_{22}b_{22} \end{bmatrix}$$

Now, let us evaluate $[B][A]$:

$$[B][A] = \begin{bmatrix} b_{11} & b_{12} \\ b_{21} & b_{22} \end{bmatrix} \begin{bmatrix} a_{11} & a_{12} \\ a_{21} & a_{22} \end{bmatrix}$$

$$= \begin{bmatrix} b_{11}a_{11} + b_{12}a_{21} & b_{11}a_{12} + b_{12}a_{22} \\ b_{21}a_{11} + b_{22}a_{21} & b_{21}a_{12} + b_{22}a_{22} \end{bmatrix}$$

It is clear that the resulting matrices are not going to be equal except in very special cases.

### 2.5.2 Associative Law of Matrix Multiplication

So far we have considered the product of two matrices only. Oftentimes the solution of engineering problems involves products of three or more matrices. The associative law of matrix multiplication provides the means by which three or more matrices can be multiplied. This law states that given the product

$$[D] = [A][B][C]$$

then

$$[D] = [[A][B]][C] \tag{2.8a}$$

or

$$[D] = [A][[B][C]] \tag{2.8b}$$

The implication here is that if the product $[A][B][C]$ is defined, then matrix $[D]$ can be evaluated by first multiplying $[A]$ and $[B]$; the resulting matrix is postmultiplied by matrix $[C]$ or by premultiplying the resulting matrix of $[B][C]$ by matrix $[A]$. Note, of course, that while equation (2.8) seems limited to products of three matrices, it is applicable regardless of the number of matrices involved. Note also that the order of matrices $[A]$, $[B]$, and $[C]$ must be maintained. That is, one may not interchange $[B]$ and $[C]$ and/or $[A]$, . . . , since this would require commutativity.

E

EXAMPLE 2.6

Given the following matrices, determine their product $[D]$ using equation (2.8):

$$[A]=\begin{bmatrix}1 & 2\\ -1 & 4\end{bmatrix},\qquad [B]=\begin{bmatrix}2 & 1\\ 3 & 3\end{bmatrix},\qquad [C]=\begin{bmatrix}0 & 1\\ 0 & 1\end{bmatrix}$$

where

$$[D]=[A][B][C]$$

Solution

Using equation (2.8a), we evaluate matrix $[D]$ by first multiplying $[A]$ and $[B]$. Thus

$$[D]=\begin{bmatrix}(1)(2)+(2)(3) & (1)(1)+(2)(3)\\ (-1)(2)+(4)(3) & (-1)(1)+(4)(3)\end{bmatrix}\begin{bmatrix}0 & 1\\ 0 & 1\end{bmatrix}$$

$$=\begin{bmatrix}8 & 7\\ 10 & 11\end{bmatrix}\begin{bmatrix}0 & 1\\ 0 & 1\end{bmatrix}$$

$$=\begin{bmatrix}0 & 15\\ 0 & 21\end{bmatrix}$$

This result can also be obtained by first multiplying $[B]$ and $[C]$ as follows:

$$[D]=\begin{bmatrix}1 & 2\\ -1 & 4\end{bmatrix}\begin{bmatrix}(2)(0)+(1)(0) & (2)(1)+(1)(1)\\ (3)(0)+(3)(0) & (3)(1)+(3)(1)\end{bmatrix}$$

$$=\begin{bmatrix}1 & 2\\ -1 & 4\end{bmatrix}\begin{bmatrix}0 & 3\\ 0 & 6\end{bmatrix}$$

$$=\begin{bmatrix}0 & 15\\ 0 & 21\end{bmatrix}$$

### 2.5.3 Distributive Law of Matrix Multiplication

This law states that if the product of the matrices appearing in equation (2.9a) is defined then equation (2.9b) holds:

$$[D]=[A][[B]+[C]] \tag{2.9a}$$

$$[D]=[A][B]+[A][C] \tag{2.9b}$$

In many engineering applications, the amount of work required is substantially reduced by taking advantage of equation (2.9b). On the other hand, one may wish to factor $[A]$ out of equation (2.9b), in which case equation (2.9a) would result.

E

EXAMPLE 2.7

Evaluate matrix $[D]$ using the distributive law as given by equation (2.9) for

$$[A] = \begin{bmatrix} 1 & 2 \\ 0 & 1 \end{bmatrix}, \qquad [B] = \begin{bmatrix} 1 & 1 \\ 1 & 1 \end{bmatrix}$$

$$[C] = \begin{bmatrix} -1 & 0 \\ 1 & 1 \end{bmatrix}, \qquad [D] = [A][[B] + [C]]$$

Solution

Let's evaluate the expression by summing matrix $[B]$ and $[C]$ first then premultiplying the resulting matrix by $[A]$:

$$[D] = \begin{bmatrix} 1 & 2 \\ 0 & 1 \end{bmatrix}\left[\begin{bmatrix} 1 & 1 \\ 1 & 1 \end{bmatrix} + \begin{bmatrix} -1 & 0 \\ 1 & 1 \end{bmatrix}\right]$$

$$= \begin{bmatrix} 1 & 2 \\ 0 & 1 \end{bmatrix}\begin{bmatrix} 0 & 1 \\ 2 & 2 \end{bmatrix}$$

$$= \begin{bmatrix} 4 & 5 \\ 2 & 2 \end{bmatrix}$$

Now, using equation (2.9b), the products of $[A][B]$ and $[A][C]$, then adding the resulting matrices, we should get the same result:

$$[D] = \begin{bmatrix} 1 & 2 \\ 0 & 1 \end{bmatrix}\begin{bmatrix} 1 & 1 \\ 1 & 1 \end{bmatrix} + \begin{bmatrix} 1 & 2 \\ 0 & 1 \end{bmatrix}\begin{bmatrix} -1 & 0 \\ 1 & 1 \end{bmatrix}$$

$$= \begin{bmatrix} 3 & 3 \\ 1 & 1 \end{bmatrix} + \begin{bmatrix} 1 & 2 \\ 1 & 1 \end{bmatrix}$$

$$= \begin{bmatrix} 4 & 5 \\ 2 & 2 \end{bmatrix}$$

The student should be aware that in this particular example, it is advantageous to add $[B]$ and $[C]$ first; however, there are instances in which it may be easier to use equation (2.9b). In fact, it may even be necessary to do so, especially when dealing with partitioned matrices.

## 2.6 MANIPULATION OF PARTITIONED MATRICES

It is clear that the amount of work required to perform the various matrix operations just outlined is directly related to the size of the matrices involved. Furthermore, memory storage for even a mainframe computer may become a limitation in solving problems involving large matrices. For this reason, it may be desirable to partition matrices into submatrices.

It is important to understand that one may perform operations on partitioned matrices according to the rules of matrix operations where the submatrices are treated as scalars, provided all the consequent matrix operations are defined. We emphasize that identical results can be obtained whether one works with submatrices or the original matrix, provided commutativity is observed.

### 2.6.1 Transpose of a Partitioned Matrix

Given a matrix $[A]$, one may partition it in a variety of ways. In fact, each element appearing may be looked at as a submatrix of size $1 \times 1$. The transpose of $[A]$ is equal to the transpose of the submatrices' transposes. That is, if matrix $[A]$ is defined as

$$[A]_{m \times n} = \begin{bmatrix} a_{11} & a_{12} & \cdots & a_{1n} \\ a_{21} & a_{22} & \cdots & a_{2n} \\ \vdots & \vdots & & \vdots \\ a_{m1} & a_{m2} & \cdots & a_{mn} \end{bmatrix}$$

or in a partitioned form

$$[A]_{m \times n} = \begin{bmatrix} [A_{11}] & [A_{12}] & \cdots & [A_{1L}] \\ [A_{21}] & [A_{22}] & \cdots & [A_{2L}] \\ \vdots & \vdots & & \vdots \\ [A_{K1}] & [A_{K2}] & \cdots & [A_{KL}] \end{bmatrix}$$

where $L$ is the number of submatrices appearing in a row and $K$ is the number of submatrices appearing in a column of $[A]$, then

$$[A]^{t}_{m \times n} = \begin{bmatrix} [A_{11}]^{t} & [A_{12}]^{t} & \cdots & [A_{1L}]^{t} \\ [A_{21}]^{t} & [A_{22}]^{t} & \cdots & [A_{2L}]^{t} \\ \vdots & \vdots & & \vdots \\ [A_{K1}]^{t} & [A_{K2}]^{t} & \cdots & [A_{KL}]^{t} \end{bmatrix}^{t} \tag{2.10a}$$

or, equivalently, using the definition of the transpose

$$[A]^{t} = \begin{bmatrix} [A_{11}]^{t} & [A_{21}]^{t} & \cdots & [A_{K1}]^{t} \\ [A_{12}]^{t} & [A_{22}]^{t} & \cdots & [A_{K2}]^{t} \\ \vdots & \vdots & & \vdots \\ [A_{1L}]^{t} & [A_{2L}]^{t} & \cdots & [A_{KL}]^{t} \end{bmatrix} \tag{2.10b}$$

E

EXAMPLE 2.8

Given the partitioned matrices $[A]$ and $[B]$, determine their transposes using equation (2.10).

$$[A] = \left[\begin{array}{c|cc} 1 & 0 & 0 \\ \hline 2 & 1 & -1 \\ 1 & 3 & 1 \end{array}\right] = \begin{bmatrix} [A_{11}] & \{A_{12}\} \\ \{A_{21}\} & [A_{22}] \end{bmatrix}.$$

$$[B] = \left[\begin{array}{cc} 1 & -2 \\ \hline -1 & 3 \\ 2 & 1 \end{array}\right] = \begin{bmatrix} \{B_{11}\} \\ [B_{21}] \end{bmatrix}$$

Solution

The dashed lines are used to define the various submatrices of $[A]$ and $[B]$. Thus

$$[A]^t = \begin{bmatrix} [1]^t & \{0 \quad 0\}^t \\ \begin{Bmatrix} 2 \\ 1 \end{Bmatrix}^t & \begin{bmatrix} 1 & -1 \\ 3 & 1 \end{bmatrix}^t \end{bmatrix}^t$$

$$= \begin{bmatrix} [1]^t & \begin{Bmatrix} 2 \\ 1 \end{Bmatrix}^t \\ \{0 \quad 0\}^t & \begin{bmatrix} 1 & -1 \\ 3 & 1 \end{bmatrix}^t \end{bmatrix}$$

$$= \begin{bmatrix} [1] & \{2 \quad 1\} \\ \begin{Bmatrix} 0 \\ 0 \end{Bmatrix} & \begin{bmatrix} 1 & 3 \\ -1 & 1 \end{bmatrix} \end{bmatrix}$$

Clearly, the result is equivalent to the transpose of the unpartitioned $[A]$ matrix.

Similarly,

$$[B]^t = \begin{bmatrix} \{1 \quad -2\}^t \\ \begin{bmatrix} -1 & 3 \\ 2 & 1 \end{bmatrix}^t \end{bmatrix}^t$$

$$= \begin{bmatrix} \{1 \quad -2\}^t & \begin{bmatrix} -1 & 3 \\ 2 & 1 \end{bmatrix}^t \end{bmatrix}$$

$$= \begin{bmatrix} \begin{Bmatrix} 1 \\ -2 \end{Bmatrix} & \begin{bmatrix} -1 & 2 \\ 3 & 1 \end{bmatrix} \end{bmatrix}$$

### 2.6.2 Addition and Subtraction of Partitioned Matrices

Two or more partitioned matrices can be added, provided they are of equal size. This is precisely the condition placed on unpartitioned matrices. Consequently, given matrices $[A]$ and $[B]$,

$$[A] = \begin{bmatrix} [A_{11}] & [A_{12}] & \cdots & [A_{1L}] \\ [A_{21}] & [A_{22}] & \cdots & [A_{2L}] \\ \vdots & \vdots & & \vdots \\ [A_{K1}] & [A_{K2}] & \cdots & [A_{KL}] \end{bmatrix}$$

$$[B] = \begin{bmatrix} [B_{11}] & [B_{12}] & \cdots & [B_{1L}] \\ [B_{21}] & [B_{22}] & \cdots & [B_{2L}] \\ \vdots & \vdots & & \vdots \\ [B_{K1}] & [B_{K2}] & \cdots & [B_{KL}] \end{bmatrix}$$

then their sum is given by a third matrix $[C]$,

$$[C]_{KL} = [A]_{KL} + [B]_{KL}$$

$$[C] = \begin{bmatrix} [A_{11}] + [B_{11}] & [A_{12}] + [B_{12}] & \cdots & [A_{1L}] + [B_{1L}] \\ [A_{21}] + [B_{21}] & [A_{22}] + [B_{22}] & \cdots & [A_{2L}] + [B_{2L}] \\ \vdots & \vdots & & \vdots \\ [A_{K1}] + [B_{K1}] & [A_{K2}] + [B_{K2}] & \cdots & [A_{KL}] + [B_{KL}] \end{bmatrix} \tag{2.11}$$

Furthermore, if one is interested in determining the resulting matrix when $[B]$ is subtracted from $[A]$, then the plus signs should be replaced by minus signs in (2.11). Note that addition of submatrices must be defined for equation (2.11) to hold.

**E**

EXAMPLE 2.9

Determine the sum of matrices $[A]$ and $[B]$, where

$$[A] = \left[\begin{array}{c|cc} 1 & -1 & 3 \\ \hline 5 & 6 & 2 \\ 9 & 0 & 4 \end{array}\right], \qquad [B] = \left[\begin{array}{c|cc} 7 & 2 & 1 \\ \hline -3 & 7 & 5 \\ 4 & 1 & 6 \end{array}\right]$$

Solution

Let $[C] = [A] + [B]$. Then

$$[C] = \begin{bmatrix} [1] & \{-1 \quad 3\} \\ \begin{Bmatrix} 5 \\ 9 \end{Bmatrix} & \begin{bmatrix} 6 & 2 \\ 0 & 4 \end{bmatrix} \end{bmatrix} + \begin{bmatrix} [7] & \{2 \quad 1\} \\ \begin{Bmatrix} -3 \\ 4 \end{Bmatrix} & \begin{bmatrix} 7 & 5 \\ 1 & 6 \end{bmatrix} \end{bmatrix}$$

Note that had we partitioned $[A]$ differently from $[B]$, then $[C]$ could not be determined.

$$[C] = \begin{bmatrix} [1] + [7] & \{-1 \quad 3\} + \{2 \quad 1\} \\ \begin{Bmatrix} 5 \\ 9 \end{Bmatrix} + \begin{Bmatrix} -3 \\ 4 \end{Bmatrix} & \begin{bmatrix} 6 & 2 \\ 0 & 4 \end{bmatrix} + \begin{bmatrix} 7 & 5 \\ 1 & 6 \end{bmatrix} \end{bmatrix}$$

$$= \begin{bmatrix} [8] & \{1 \quad 4\} \\ \begin{Bmatrix} 2 \\ 13 \end{Bmatrix} & \begin{bmatrix} 13 & 7 \\ 1 & 10 \end{bmatrix} \end{bmatrix}$$

Needless to say, this result would have been obtained had we operated on the unpartitioned $[A]$ and $[B]$ matrices.

### 2.6.3 Multiplication of Partitioned Matrices

It is evident that all of the rules of matrix operations pertaining to unpartitioned matrices can be applied to partitioned matrices. Recall that when multiplying two matrices $[A]$ and $[B]$, we stipulated that the number of rows in $[B]$ must equal the number of columns in $[A]$. This condition must now be satisfied for the products of all submatrices. Consequently, one must be careful in partitioning matrices so as not to violate this important principle. Consider the following:

$$[A] = \begin{bmatrix} [A_{11}] & [A_{12}] \\ [A_{21}] & [A_{22}] \end{bmatrix}$$

$$[B] = \begin{bmatrix} [B_{11}] & [B_{12}] \\ [B_{21}] & [B_{22}] \end{bmatrix}$$

The product of $[A][B]$ is expressed as

$$[D] = \begin{bmatrix} [A_{11}] & [A_{12}] \\ [A_{21}] & [A_{22}] \end{bmatrix} \begin{bmatrix} [B_{11}] & [B_{12}] \\ [B_{21}] & [B_{22}] \end{bmatrix}$$

$$= \begin{bmatrix} [A_{11}][B_{11}] + [A_{12}][B_{21}] & [A_{11}][B_{12}] + [A_{12}][B_{22}] \\ [A_{21}][B_{11}] + [A_{22}][B_{21}] & [A_{21}][B_{12}] + [A_{22}][B_{22}] \end{bmatrix} \tag{2.12}$$

Note here that while the submatrices are treated as if they were scalar quantities, the $[A]$ submatrices are premultiplying the $[B]$ submatrices. This is necessary in order not to violate the commutativity requirement. Furthermore, the number of rows in $[B_{11}]$ must equal the number of columns in $[A_{11}]$ and $[A_{21}]$. This is also true for $[B_{12}]$. Also, the number of rows in $[B_{11}]$ and $[B_{22}]$ must equal the number of both$[A_{12}]$ and $[A_{22}]$. At this stage one may be tempted to think that multiplication of partitioned matrices is a hopeless task. However, in the important special case in which $[A]$ and $[B]$ are square matrices we can guarantee that the requirements for multiplication of partitioned matrices are satisfied if we make sure that the matrices are partitioned in the same manner and their diagonal submatrices are also square. We make use of this concept when inverting a partitioned matrix in chapter 5.

**E** EXAMPLE 2.10

Given $[A]$ and $[B]$ determine their product $[C] = [A][B]$.

$$[A] = \left[\begin{array}{cc|c} 1 & 7 & 0 \\ -2 & 3 & 0 \\ \hline 6 & 5 & 9 \end{array}\right], \qquad [B] = \left[\begin{array}{cc|c} 2 & 4 & 1 \\ 5 & 6 & 2 \\ \hline 1 & -1 & 0 \end{array}\right]$$

Solution

$$[C] = \begin{bmatrix} \begin{bmatrix} 1 & 7 \\ -2 & 3 \end{bmatrix} & \begin{Bmatrix} 0 \\ 0 \end{Bmatrix} \\ \{6 \quad 5\} & [9] \end{bmatrix} \begin{bmatrix} \begin{bmatrix} 2 & 4 \\ 5 & 6 \end{bmatrix} & \begin{Bmatrix} 1 \\ 2 \end{Bmatrix} \\ \{1 \quad -1\} & [0] \end{bmatrix}$$

$$[C_{11}] = \begin{bmatrix} 1 & 7 \\ -2 & 3 \end{bmatrix} \begin{bmatrix} 2 & 4 \\ 5 & 6 \end{bmatrix} + \begin{Bmatrix} 0 \\ 0 \end{Bmatrix} \{1 \quad -1\} = \begin{bmatrix} 37 & 46 \\ 11 & 10 \end{bmatrix}$$

$$[C_{21}] = \{6 \quad 5\} \begin{bmatrix} 2 & 4 \\ 5 & 6 \end{bmatrix} + \{9\}\{1 \quad -1\} = \{46 \quad 45\}$$

$$[C_{12}] = \begin{bmatrix} 1 & 7 \\ -2 & 3 \end{bmatrix} \begin{Bmatrix} 1 \\ 2 \end{Bmatrix} + \begin{Bmatrix} 0 \\ 0 \end{Bmatrix} \{0\} = \begin{Bmatrix} 15 \\ 4 \end{Bmatrix}$$

$$[C_{22}] = \{6 \quad 5\} \begin{Bmatrix} 1 \\ 2 \end{Bmatrix} + \{9\}\{0\} = [16]$$

and

$$[C] = \begin{bmatrix} \begin{bmatrix} 37 & 46 \\ 11 & 10 \end{bmatrix} & \begin{Bmatrix} 15 \\ 4 \end{Bmatrix} \\ \{46 \quad 45\} & [16] \end{bmatrix}$$

Obviously, matrix operations on partitioned matrices are more involved than on unpartitioned matrices. However, the advantage of the partitioned matrix is that we did not have to deal with any matrix larger than $2 \times 2$ to calculate a product with a $3 \times 3$. This advantage may be even more significant when our matrices become larger. In fact, the product of two $n \times n$ matrices can be evaluated by evaluating products of submatrices of size not exceeding $\frac{1}{2}n \times \frac{1}{2}n$. This is extremely important when dealing with computers of limited storage capabilities.

## 2.7 RULES FOR COMBINED MATRIX OPERATIONS

The manipulation of matrix equations may involve addition, subtraction, transposition, multiplication, and/or inversion. Thus far, we have considered these operations separately. In this section the question of how these operations relate to one another is addressed. Given the matrices $[A]$, $[B]$, and $[C]$ and the scalar $k$ and assuming that the various operations are defined, the rules for combined matrix operations are summarized below.

| SUMMARY OF RULES FOR COMBINED MATRIX OPERATIONS | |
|---|---|
| ADDITION AND SUBTRACTION | EQUATION NUMBER |
| $[A]+[B]=[B]+[A]$ | **(2.13a)** |
| $[A]\pm[0]=[A]$ | **(2.13b)** |
| $[A]+[B]+[C]=[A]+[[B]+[C]]$ <br> $=[[A]+[B]]+[C]$ | **(2.13c)** |
| $k[[A]\pm[B]]=k[A]\pm k[B]$ | **(2.13d)** |
| $[A]+[A]=[A][[I]+[I]]=[A][I](2)=2[A]$ | **(2.13e)** |
| $[[A]\pm[B]]^t=[A]^t\pm[B]^t$ | **(2.13f)** |
| MULTIPLICATION AND SUMS | EQUATION NUMBER |
| $[A][I]=[I][A]=[A]$ | **(2.14a)** |
| $k[A][B]=[kA][B]=[A][kB]$ | **(2.14b)** |
| $k[A][[B]+[C]]=k[A][B]+k[A][C]$ | **(2.14c)** |
| $k[[A]+[B]][C]=k[A][C]+k[B][C]$ | **(2.14d)** |
| $[[A]+[B]]^2=[A]^2+2[A][B]+[B]^2$ | **(2.14e)** |
| $[[A]-[B]]^2=[A]^2-2[A][B]+[B]^2$ | **(2.14f)** |
| $[[A]-[B]][[A]+[B]]$ <br> $=[A]^2+[A][B]-[B][A]-[B]^2$ | **(2.14g)** |
| $k[A][B][C]=[kA][B][C]=[A][kB][C]$ <br> $=[A][B][kC]$ | **(2.14h)** |
| $[[A][B]]^t=[B]^t[A]^t$ | **(2.14i)** |
| $[[A][B][C]]^t=[C]^t[B]^t[A]^t$ | **(2.14j)** |

| INVERSION | EQUATION NUMBER |
|---|---|
| $[A][A]^{-1} = [A]^{-1}[A] = [I]$ | **(2.15a)** |
| $[[A][B]]^{-1} = [B]^{-1}[A]^{-1}$ | **(2.15b)** |
| $[kA]^{-1} = (1/k)[A]^{-1}$ | **(2.15c)** |
| $[[A]^{-1}]^t = [[A]^t]^{-1}$ | **(2.15d)** |
| $[[A]^n]^{-1} = [[A]^{-1}]^n$ | **(2.15e)** |

## 2.8 APPLICATION OF MATRICES TO THE ROTATION OF A COORDINATE SYSTEM

Consider the transformation resulting from rotating a two-dimensional Cartesian coordinate system by two successive angles, $\theta$ and $\beta$, as shown in figure 2.1.

Given a point $P$, we may describe its location as $(x_1, y_1)$ or $(x_2, y_2)$ or $(x_3, y_3)$ depending on which of the three coordinate systems is used. Suppose that we know its coordinates in the first system as $(x_1, y_1)$ and we now want to determine the coordinates $(x_2, y_2)$ if the system is rotated by an angle $\theta$. From figure 2.1 we know that

$$x_2 = d_1 + d_2 \tag{2.16}$$

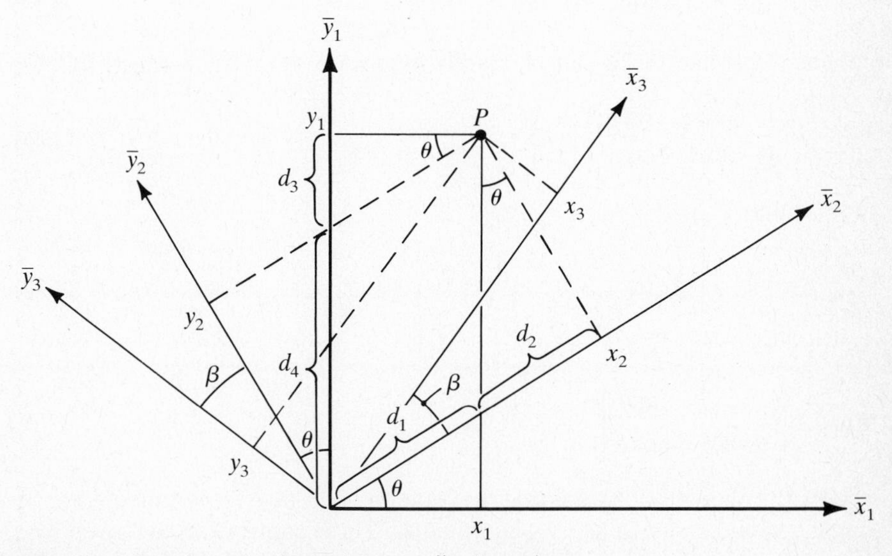


**Figure 2.1** Rotation of a Cartesian coordinate system.

where

$$d_1 = x_1/\cos\theta \tag{2.17}$$

and

$$d_2 = y_2 \tan\theta \tag{2.18}$$

Substituting equations (2.17) and (2.18) into (2.16) gives

$$x_2 = x_1/\cos\theta + y_2 \tan\theta \tag{2.19}$$

In addition,

$$y_1 = d_3 + d_4 \tag{2.20}$$

where

$$d_3 = x_1 \tan\theta \tag{2.21}$$

and

$$d_4 = y_2/\cos\theta \tag{2.22}$$

Substituting equations (2.21) and (2.22) into equation (2.20) gives

$$y_1 = x_1 \tan\theta + y_2/\cos\theta \tag{2.23}$$

Solving for $y_2$ in terms of $x_1$ and $y_1$, we get

$$y_2 = y_1 \cos\theta - x_1 \sin\theta \tag{2.24}$$

Now, substituting equation (2.24) into equation (2.23) and simplifying gives

$$x_2 = x_1/\cos\theta + (y_1 \cos\theta - x_1 \sin\theta)\tan\theta$$

or

$$x_2 = x_1 \cos\theta + y_1 \sin\theta \tag{2.25}$$

Equations (2.24) and (2.25) can be readily expressed in matrix form as follows:

$$\begin{Bmatrix} x_2 \\ y_2 \end{Bmatrix} = \begin{bmatrix} \cos\theta & \sin\theta \\ -\sin\theta & \cos\theta \end{bmatrix} \begin{Bmatrix} x_1 \\ y_1 \end{Bmatrix} \tag{2.26}$$

or more simply

$$\begin{Bmatrix} x_2 \\ y_2 \end{Bmatrix} = [A] \begin{Bmatrix} x_1 \\ y_1 \end{Bmatrix} \tag{2.27}$$

where

$$[A] = \begin{bmatrix} \cos\theta & \sin\theta \\ -\sin\theta & \cos\theta \end{bmatrix}$$

At this stage it should be obvious that the rotation of the $(\bar{x}_2, \bar{y}_2)$ coordinate axis by an angle $\beta$ would result in $(\bar{x}_3, \bar{y}_3)$ coordinates. These coordinates are given by a transformation similar to that obtained in equation (2.22). That is,

$$\begin{Bmatrix} x_3 \\ y_3 \end{Bmatrix} = \begin{bmatrix} \cos\beta & \sin\beta \\ -\sin\beta & \cos\beta \end{bmatrix} \begin{Bmatrix} x_2 \\ y_2 \end{Bmatrix} \tag{2.28}$$

or simply

$$\begin{Bmatrix} x_3 \\ y_3 \end{Bmatrix} = [B] \begin{Bmatrix} x_2 \\ y_2 \end{Bmatrix} \tag{2.29}$$

The transformation from $(x_1, y_1)$ to $(x_3, y_3)$ directly can be determined rather easily by substituting equation (2.27) into equation (2.29). Thus

$$\begin{Bmatrix} x_3 \\ y_3 \end{Bmatrix} = [B][A] \begin{Bmatrix} x_1 \\ y_1 \end{Bmatrix} \tag{2.30}$$

In general, one may carry out as many transformations as required by following the procedures just outlined.

Equation (2.30) is very interesting. This is because the transformation from $(x_1, y_1)$ to $(x_3, y_3)$ can also be accomplished by realizing that a single transformation may be used to cover the angle of rotation $\theta + \beta$. Hence

$$\begin{Bmatrix} x_3 \\ y_3 \end{Bmatrix} = \begin{bmatrix} \cos(\theta+\beta) & \sin(\theta+\beta) \\ -\sin(\theta+\beta) & \cos(\theta+\beta) \end{bmatrix} \begin{Bmatrix} x_1 \\ y_1 \end{Bmatrix} \tag{2.31}$$

The implication is that equation (2.31) must be equal to equation (2.30). Specifically,

$$\begin{bmatrix} \cos\beta & \sin\beta \\ -\sin\beta & \cos\beta \end{bmatrix} \begin{bmatrix} \cos\theta & \sin\theta \\ -\sin\theta & \cos\theta \end{bmatrix} = \begin{bmatrix} \cos(\theta+\beta) & \sin(\theta+\beta) \\ -\sin(\theta+\beta) & \cos(\theta+\beta) \end{bmatrix}$$

One can now prove the following useful and familiar trigonometric identities by using the principle of matrix equality. In particular,

$$\cos(\theta+\beta) = \cos\beta\cos\theta - \sin\beta\sin\theta$$

and

$$\sin(\theta+\beta) = \cos\beta\sin\theta + \sin\beta\cos\theta$$

## 2.9 DETERMINANTS AND THEIR EVALUATION

The determinant is a property of a square matrix. It plays a significant role in the solution of many engineering problems. These include the solution of sets of linear algebraic equations; determination of the stability of structural and mechanical systems; and solutions to the dynamic response of structures, circuits, and mechanical systems.

Every square matrix $[A]$ has a determinant (this is true even if the elements are not scalar constants). We denote the determinant of $[A]$ as $\det[A]$ or with vertical bars as

$$\det[A] = |a_{ij}|$$

For example, the determinant of a 3 × 3 matrix may be written

$$\det[A] = \begin{vmatrix} a_{11} & a_{12} & a_{13} \\ a_{21} & a_{22} & a_{23} \\ a_{31} & a_{32} & a_{33} \end{vmatrix}$$

The determinant of a square matrix is a scalar quantity or function irrespective of the number of elements in the matrix.

### 2.9.1 Properties of Determinants

The following are special properties which will be helpful in reducing the amount of work involved in evaluating determinants.

1. The determinant is zero if any two rows or two columns are identical or multiples of one another. For example,

$$\det[A] = \begin{vmatrix} 1 & 5 & 3 \\ -1 & 7 & 3 \\ 2 & 10 & 6 \end{vmatrix} = 0$$

since the third row is the same as the first row multiplied by 2.

2. A determinant is changed in sign if two rows or two columns are interchanged. For example,

$$\text{if} \quad \det[A] = \begin{vmatrix} 2 & 2 & 1 \\ 1 & 2 & 5 \\ -3 & 3 & 7 \end{vmatrix}, \qquad \text{then} \quad \det[A] = -1 \begin{vmatrix} 2 & 2 & 1 \\ -3 & 3 & 7 \\ 1 & 2 & 5 \end{vmatrix}$$

3. The determinant of a matrix is equal to the determinant of its transpose. For example,

$$\det[A] = \begin{vmatrix} 5 & -2 \\ 1 & 7 \end{vmatrix} = \begin{vmatrix} 5 & -2 \\ 1 & 7 \end{vmatrix}^{\text{t}}$$

that is,

$$\det[A] = \det[A]^{\text{t}}$$

Similarly, given that $\det[A] = |[A_1][A_2] \cdots [A_n]|$,

then

$$\det[A] = \det[A_1]^{\text{t}} \det[A_2]^{\text{t}} \cdots \det[A_n]^{\text{t}}$$

4. A term may be factored from all elements in a row or column; the product of the determinant of the resulting matrix and the factor is equal to the original determinant. Hence

$$\det[A] = \begin{vmatrix} 7 & 14 & 21 \\ 3 & 2 & 5 \\ 1 & 4 & 10 \end{vmatrix} = 7 \begin{vmatrix} (1 & 2 & 3) \\ 3 & 2 & 5 \\ 1 & 4 & 10 \end{vmatrix} = (7)(2) \begin{vmatrix} 1 & \frown{1} & 3 \\ 3 & 1 & 5 \\ 1 & \underset{\smile}{2} & 10 \end{vmatrix}$$

5. The value of a determinant is not changed if a multiple of a row or column is added to or subtracted from any parallel row or column:

$$\det[A] = \begin{vmatrix} 2 & 3 \\ 5 & 7 \end{vmatrix} = \begin{vmatrix} 2 & \begin{Bmatrix} 3 \\ 7 \end{Bmatrix} - 2\begin{Bmatrix} 2 \\ 5 \end{Bmatrix} \\ 5 & \end{vmatrix} = \begin{vmatrix} 2 & 3 - 2 \times 2 \\ 5 & 7 - 2 \times 5 \end{vmatrix} = \begin{vmatrix} 2 & -1 \\ 5 & -3 \end{vmatrix}$$

6. The determinant of a scalar matrix (1 × 1) is equal to the element itself. For example,

$$\det[A] = |-2| = -2$$

### 2.9.2 Laplace Method of Cofactors

The basis of this method for evaluating the determinant of a matrix is that corresponding to every element $a_{ij}$ there can be written a square matrix of order $n - 1$ consisting of the original matrix with the $i$th row and $j$th column removed. The determinant of this new matrix of order $n - 1$ is called the *minor of* $a_{ij}$ and designated as $M_{ij}$. Consider the following 3 × 3 determinant:

$$\det[A] = \begin{vmatrix} a_{11} & a_{12} & a_{13} \\ a_{21} & a_{22} & a_{23} \\ a_{31} & a_{32} & a_{33} \end{vmatrix}$$

The minor corresponding to the $a_{23}$ element is given as

$$M_{23} = \begin{vmatrix} a_{11} & a_{12} \\ a_{31} & a_{32} \end{vmatrix}$$

In addition, corresponding to each $a_{ij}$ element, there is also a *cofactor* $C_{ij}$, where

$$C_{ij} = (-1)^{i+j} M_{ij}$$

Thus the cofactor of $a_{23}$ is given as

$$C_{23} = (-1)^{2+3} \begin{vmatrix} a_{11} & a_{12} \\ a_{31} & a_{32} \end{vmatrix}$$

$$C_{23} = -\begin{vmatrix} a_{11} & a_{12} \\ a_{31} & a_{32} \end{vmatrix}$$

In terms of cofactors, the determinant is then given as follows:

$$\det[A] = \sum_{j=1}^{n} a_{ij} C_{ij} \tag{2.32}$$

This equation applies for any selected value if $i$ is between 1 and $n$ (summing about any row). Alternatively, one may evaluate a determinant by summing along any column. Thus

$$\det[A] = \sum_{i=1}^{n} a_{ij} C_{ij} \tag{2.33}$$

for any selected value of $j$ between 1 and $n$.

E

EXAMPLE 2.11

Find the determinant of any $2 \times 2$ matrix.

Solution

$$\det[A] = \begin{vmatrix} a_{11} & a_{12} \\ a_{21} & a_{22} \end{vmatrix}$$

First, determine the cofactors for a selected column or row. We will expand the determinant about the first column ($j = 1$). Hence we must compute the two following minors:

$$M_{11} = |a_{22}| = a_{22}, \qquad M_{21} = |a_{12}| = a_{12}$$

Using these we can compute the corresponding cofactors:

$$C_{11} = (-1)^{1+1}M_{11} = a_{22}, \qquad C_{21} = (-1)^{2+1}M_{21} = -a_{12}$$

Now we can calculate the determinant using equation (2.33) (recall that $j$ is equal to one):

$$\det[A] = \sum_{i=1}^{n} a_{ij}C_{ij} = \sum_{i=1}^{2} a_{i1}C_{i1} = a_{11}C_{11} + a_{21}C_{21}$$

Hence

$$\det[A] = a_{11}a_{22} - a_{21}a_{12} \tag{2.34}$$

This is an important result. Whenever a $2 \times 2$ determinant is evaluated, one need not go through all of the steps; one need only remember equation (2.34). The number of $2 \times 2$ determinants needed for calculating an $n \times n$ determinant is $\frac{1}{2}n!$.

E

EXAMPLE 2.12

Evaluate the determinant of the following $3 \times 3$ matrix.

$$[A] = \begin{bmatrix} 1 & 2 & 3 \\ 4 & 5 & 6 \\ 7 & 8 & 9 \end{bmatrix}$$

**Solution**

Again we will sum along the first column, which implies that $j = 1$. Hence the minors are

$$M_{11} = \begin{vmatrix} 5 & 6 \\ 8 & 9 \end{vmatrix} = 5 \times 9 - 8 \times 6 = -3$$

$$M_{21} = \begin{vmatrix} 2 & 3 \\ 8 & 9 \end{vmatrix} = 2 \times 9 - 8 \times 3 = -6$$

$$M_{31} = \begin{vmatrix} 2 & 3 \\ 5 & 6 \end{vmatrix} = 2 \times 6 - 3 \times 5 = -3$$

The corresponding cofactors are

$$C_{11} = (-1)^{1+1}(-3) = -3$$
$$C_{21} = (-1)^{2+1}(-6) = 6$$
$$C_{31} = (-1)^{3+1}(-3) = -3$$

Therefore, for $j = 1$,

$$\begin{aligned} \det[A] &= \sum_{i=1}^{n} a_{ij}C_{ij} = \sum_{i=1}^{3} a_{i1}C_{i1} \\ &= a_{11}C_{11} + a_{21}C_{21} + a_{31}C_{31} \\ &= 1(-3) + 4(6) + 7(-3) \\ &= 0 \end{aligned}$$

### 2.9.3 Upper-Triangle Elimination Method

The method of cofactors is highly inefficient for calculating determinants of large size. For example, a $15 \times 15$ determinant would require us to evaluate $\frac{1}{2}15!$ or $6.5383 \times 10^{11}$ determinants of size $2 \times 2$. Consequently a practical method is needed for large determinants. The upper-triangle elimination method takes advantage of the fact that the determinant of a triangular matrix is simply equal to the product of its diagonal elements. This fact can be readily illustrated by considering the following matrix:

$$[A] = \begin{bmatrix} a_{11} & a_{12} & \cdots & a_{1n} \\ a_{21} & a_{22} & \cdots & a_{2n} \\ \vdots & \vdots & & \vdots \\ a_{n1} & a_{n2} & \cdots & a_{nn} \end{bmatrix}$$

The determinant of $[A]$ is calculated by the method of cofactors and expanding about the first column:

$$\det[A] = \sum_{i=1}^{n} a_{i1} C_{i1}, \qquad j = 1$$

$$= a_{11}C_{11} + a_{21}C_{21} + \cdots + a_{n1}C_{n1}$$

But, if $[A]$ is an upper-triangle matrix then $a_{21} = a_{31} = \cdots = a_{n1} = 0$. Thus the determinant is reduced to

$$\det[A] = a_{11}C_{11}$$

Now, since the cofactor $C_{11}$ is a determinant of an $(n-1) \times (n-1)$ matrix which is also an upper-triangle matrix, the determinant of $[A]$ is given simply by the following:

$$\det[A] = (a_{11})(a_{22}) \cdots (a_{nn})$$

It is evident that $[A]$ may or may not be an upper-triangular matrix. Therefore it may be operated on so that it is an upper-triangular matrix of the form

$$\det[A] = K \begin{vmatrix} u_{11} & u_{12} & \cdots & u_{1n} \\ 0 & u_{22} & \cdots & u_{2n} \\ \vdots & \vdots & & \vdots \\ 0 & 0 & \cdots & u_{nn} \end{vmatrix}$$

in which case

$$\det[A] = \sum_{i=1}^{n} a_{ij} C_{ij} = K \prod_{i=1}^{n} u_{ii} \tag{2.35}$$

where $K$ is a constant resulting from the operations performed on the original $[A]$ matrix. It is clear that equation (2.35) has the advantage over the method of cofactors in that no cofactors $C_{ij}$ need be calculated. *Furthermore, the determinant of a lower-triangle matrix is also given by (2.35) since the transpose of an upper-triangle matrix is equal to a lower-triangle matrix.*

**E** EXAMPLE 2.13

Evaluate the determinant of matrix $[A]$ using the upper-triangle method:

$$[A] = \begin{bmatrix} 1 & 4 & 2 & -1 \\ 3 & -5 & 6 & 1 \\ 1 & 0 & 4 & 5 \\ -6 & 1 & 9 & 7 \end{bmatrix}$$

### Solution

Using the properties of determinants (not matrices), addressed earlier, multiples of row 1 ($R_1$) may be added to rows 2, 3, and 4 without altering the value of the determinant of matrix $[A]$. Hence

$$\det[A] = \begin{vmatrix} 1 & 4 & 2 & -1 \\ 3 & -5 & 6 & 1 \\ 1 & 0 & 4 & 5 \\ -6 & 1 & 9 & 7 \end{vmatrix} \begin{matrix} \\ -3R_1 + R_2 \\ -R_1 + R_2 \\ 6R_1 + R_2 \end{matrix}$$

The implication here is that adding $-3$ times row 1 to row 2, $-1$ times row 1 to row 3, and 6 times row 1 to row 4 reduces the first column to that of an upper triangle. Therefore

$$\det[A] = \begin{vmatrix} 1 & 4 & 2 & -1 \\ 0 & -17 & 0 & 4 \\ 0 & -4 & 2 & 6 \\ 0 & 25 & 21 & 1 \end{vmatrix}$$

Since $a_{22} = 1$, we can factor $-17$ from Row 2 and then proceed to reduce the second column to that of an upper triangle. That is,

$$\det[A] = -17\begin{vmatrix} 1 & 4 & 2 & -1 \\ 0 & 1 & 0 & -\frac{4}{17} \\ 0 & -4 & 2 & 6 \\ 0 & 25 & 21 & 1 \end{vmatrix} \begin{matrix} \\ \\ 4R_2 + R_3 \\ -25R_2 + R_4 \end{matrix}$$

$$\det[A] = -17\begin{vmatrix} 1 & 4 & 2 & -1 \\ 0 & 1 & 0 & -\frac{4}{17} \\ 0 & 0 & 2 & \frac{86}{17} \\ 0 & 0 & 21 & \frac{117}{17} \end{vmatrix}$$

Now factoring 2 from row 3 and continuing with the procedure gives

$$\det[A] = (2)(-17)\begin{vmatrix} 1 & 4 & 2 & -1 \\ 0 & 1 & 0 & -\frac{4}{17} \\ 0 & 0 & 1 & \frac{86}{17} \\ 0 & 0 & 21 & \frac{117}{17} \end{vmatrix} \begin{matrix} \\ \\ \\ -21R_3 + R_4 \end{matrix}$$

$$\det[A] = -34\begin{vmatrix} 1 & 4 & 2 & -1 \\ 0 & 1 & 0 & -\frac{4}{17} \\ 0 & 0 & 1 & \frac{86}{17} \\ 0 & 0 & 0 & -\frac{786}{17} \end{vmatrix}$$

Applying equation (2.35) yields the determinant of $[A]$. Thus

$$\det[A] = (-34)(1)(1)(1)(-\tfrac{786}{17}) = 1572$$

**SUMMARY OF UPPER-TRIANGLE METHOD FOR EVALUATING A DETERMINANT**

| STEP | OPERATION | SYMBOL |
|---|---|---|
| 1 | Check $a_{11}$. If it is equal to zero, then switch row 1 or column 1 with another row or column so that the new $a_{11} \neq 0$. Make sure that $-1$ is factored each time two rows or columns are switched. | |
| 2 | Factor $a_{11}$ from row 1, then add appropriate multiples of row 1 to rows 2–$n$ to obtain the first column of an upper-triangle form. | $-a_{ij}R_1 + R_i$, $i = 2, \ldots, n$, $j = 1$, $k = a_{11}$ |
| 3 | Repeat steps 1 and 2 for rows 2–$n$. | $-a_{ij}R_j + R_i$, $i = 4, \ldots, n$, $j = 3, \ldots, n-1$, |
| 4 | Calculate the determinant: $\det[A] = K(u_{11})(u_{22}) \cdots (u_{nn})$ | $\det[A] = K \sum_{i=1}^{n} u_{ii}$ |

### 2.9.4 Method of Pivotal Condensation

This method is a variation of the upper-triangle procedure, except that the triangular form is achieved implicitly by evaluating $2 \times 2$ determinants. We illustrate this method by first considering the following $3 \times 3$ matrix:

$$[A] = \begin{bmatrix} a_{11} & a_{12} & a_{13} \\ a_{21} & a_{22} & a_{23} \\ a_{31} & a_{32} & a_{33} \end{bmatrix}$$

The determinant $\det[A]$ is evaluated by reducing the first column to upper-triangle form. Therefore, factoring $a_{11}$ from the first row gives

$$\det[A] = a_{11} \begin{vmatrix} 1 & a_{12}/a_{11} & a_{13}/a_{11} \\ a_{21} & a_{22} & a_{23} \\ a_{31} & a_{32} & a_{33} \end{vmatrix}$$

Now, by multiplying row 1 first by $-a_{21}$ and then $-a_{31}$ and combining the results with rows 2 and 3 we obtain

$$\det[A] = a_{11}\begin{vmatrix} 1 & a_{12}/a_{11} & a_{13}/a_{11} \\ 0 & a_{22} - (a_{12}/a_{11})a_{21} & a_{23} - (a_{13}/a_{11})a_{21} \\ 0 & a_{32} - (a_{12}/a_{11})a_{31} & a_{33} - (a_{13}/a_{11})a_{31} \end{vmatrix}$$

But, by Laplace's method, only one $2 \times 2$ minor is needed; hence

$$\det[A] = a_{11}\begin{vmatrix} a_{22} - (a_{12}/a_{11})a_{21} & a_{23} - (a_{13}/a_{11})a_{21} \\ a_{32} - (a_{12}/a_{11})a_{31} & a_{33} - (a_{13}/a_{11})a_{31} \end{vmatrix}$$

Now if one is to factor out $1/a_{11}$ from both rows, then

$$\det[A] = \frac{a_{11}}{a_{11}a_{11}}\begin{vmatrix} a_{11}a_{22} - a_{12}a_{21} & a_{23}a_{11} - a_{13}a_{21} \\ a_{11}a_{32} - a_{12}a_{31} & a_{33}a_{11} - a_{13}a_{31} \end{vmatrix}$$

or

$$\det[A] = \frac{1}{a_{11}}\begin{vmatrix} \begin{vmatrix} a_{11} & a_{12} \\ a_{21} & a_{22} \end{vmatrix} & \begin{vmatrix} a_{11} & a_{13} \\ a_{21} & a_{23} \end{vmatrix} \\ \begin{vmatrix} a_{11} & a_{12} \\ a_{31} & a_{32} \end{vmatrix} & \begin{vmatrix} a_{11} & a_{13} \\ a_{31} & a_{33} \end{vmatrix} \end{vmatrix}$$

Careful consideration of the above expression clearly shows that we can reduce the problem of determining a $3 \times 3$ determinant to that of determining five $2 \times 2$ determinants. In general, for a determinant of size $n > 2$,

$$\det[A] = (a_{11})^{2-n}\begin{vmatrix} \begin{vmatrix} a_{11} & a_{12} \\ a_{21} & a_{22} \end{vmatrix} & \begin{vmatrix} a_{11} & a_{13} \\ a_{21} & a_{23} \end{vmatrix} & \cdots & \begin{vmatrix} a_{11} & a_{1n} \\ a_{21} & a_{2n} \end{vmatrix} \\ \vdots & \vdots & & \vdots \\ \begin{vmatrix} a_{11} & a_{12} \\ a_{n1} & a_{n2} \end{vmatrix} & \begin{vmatrix} a_{11} & a_{13} \\ a_{n1} & a_{n3} \end{vmatrix} & \cdots & \begin{vmatrix} a_{11} & a_{1n} \\ a_{n1} & a_{nn} \end{vmatrix} \end{vmatrix} \tag{2.36}$$

Equation (2.36) states that an $n \times n$ determinant can be reduced to an $(n-1) \times (n-1)$ determinant whose $(n-1)^2$ elements are the determinants of $2 \times 2$ matrices. This process can be repeated until a single $2 \times 2$ determinant is obtained.

E

EXAMPLE 2.14

Evaluate the determinant of matrix $[A]$.

$$[A] = \begin{bmatrix} 1 & 2 & 3 \\ 4 & 5 & 6 \\ 7 & 8 & 9 \end{bmatrix}$$

### Solution

Using equation (2.36) we have

$$\det[A] = 1^{2-3}\begin{vmatrix} \begin{vmatrix}1 & 2\\4 & 5\end{vmatrix} & \begin{vmatrix}1 & 3\\4 & 6\end{vmatrix} \\ \begin{vmatrix}1 & 2\\7 & 8\end{vmatrix} & \begin{vmatrix}1 & 3\\7 & 9\end{vmatrix}\end{vmatrix} = \begin{vmatrix}-3 & -6\\-6 & -12\end{vmatrix}$$

$$\det[A] = 1[(-3)(-12) - (-6)(-6)] = 0$$

E

EXAMPLE 2.15

Evaluate the following determinant using the method of pivotal condensation:

$$[A] = \begin{bmatrix} 4 & -1 & 2 & -2 & 3\\ 3 & 4 & -7 & 2 & 5\\ 1 & -9 & 1 & 4 & 7\\ 9 & 10 & 1 & 8 & 3\\ 5 & 4 & 3 & 2 & 1\end{bmatrix}$$

### Solution

$$\det[A] = 4^{2-5}\begin{vmatrix} \begin{vmatrix}4 & -1\\3 & 4\end{vmatrix} & \begin{vmatrix}4 & 2\\3 & -7\end{vmatrix} & \begin{vmatrix}4 & -2\\3 & 2\end{vmatrix} & \begin{vmatrix}4 & 3\\3 & 5\end{vmatrix} \\ \begin{vmatrix}4 & -1\\1 & -9\end{vmatrix} & \begin{vmatrix}4 & 2\\1 & 1\end{vmatrix} & \begin{vmatrix}4 & -2\\1 & 4\end{vmatrix} & \begin{vmatrix}4 & 3\\1 & 7\end{vmatrix} \\ \begin{vmatrix}4 & -1\\9 & 10\end{vmatrix} & \begin{vmatrix}4 & 2\\9 & 1\end{vmatrix} & \begin{vmatrix}4 & -2\\9 & 8\end{vmatrix} & \begin{vmatrix}4 & 3\\9 & 3\end{vmatrix} \\ \begin{vmatrix}4 & -1\\5 & 4\end{vmatrix} & \begin{vmatrix}4 & 2\\5 & 3\end{vmatrix} & \begin{vmatrix}4 & -2\\5 & 2\end{vmatrix} & \begin{vmatrix}4 & 3\\5 & 1\end{vmatrix}\end{vmatrix}$$

Simplifying, by evaluating the $4^3 = 64$ and the $2 \times 2$ determinants, yields

$$\det[A] = \frac{1}{64}\begin{vmatrix} 19 & -34 & 14 & 11\\ -35 & 2 & 18 & 25\\ 49 & -14 & 50 & -15\\ 21 & 2 & 18 & -11\end{vmatrix}$$

The size of the determinant is once again reduced by one. Thus

$$\det[A] = \frac{1}{64}\,19^{2-4}\begin{vmatrix} \begin{vmatrix} 19 & -34 \\ -35 & 2 \end{vmatrix} & \begin{vmatrix} 19 & 14 \\ -35 & 18 \end{vmatrix} & \begin{vmatrix} 19 & 11 \\ -35 & 25 \end{vmatrix} \\ \begin{vmatrix} 19 & -34 \\ 49 & -14 \end{vmatrix} & \begin{vmatrix} 19 & 14 \\ 49 & 50 \end{vmatrix} & \begin{vmatrix} 19 & 11 \\ 49 & -15 \end{vmatrix} \\ \begin{vmatrix} 19 & -34 \\ 21 & 2 \end{vmatrix} & \begin{vmatrix} 19 & 14 \\ 21 & 18 \end{vmatrix} & \begin{vmatrix} 19 & 11 \\ 21 & -11 \end{vmatrix} \end{vmatrix}$$

Further simplification by evaluating the $19^2 = 361$ and the new $2 \times 2$ determinants gives

$$\det[A] = \frac{1}{64}\frac{1}{361}\begin{vmatrix} -1152 & 832 & 860 \\ 1400 & 264 & -824 \\ 752 & 48 & -440 \end{vmatrix}$$

Finally,

$$\det[A] = \frac{1}{64}\frac{1}{361}(-1152)^{2-3}\begin{vmatrix} \begin{vmatrix} -1152 & 832 \\ 1400 & 264 \end{vmatrix} & \begin{vmatrix} -1152 & 860 \\ 1400 & -824 \end{vmatrix} \\ \begin{vmatrix} -1152 & 832 \\ 752 & 48 \end{vmatrix} & \begin{vmatrix} -1152 & 860 \\ 752 & -440 \end{vmatrix} \end{vmatrix}$$

or

$$\det[A] = \frac{1}{64}\frac{1}{361}\frac{-1}{1152}\begin{vmatrix} -1468928 & -254752 \\ -680960 & -139840 \end{vmatrix}$$

The determinant is now reduced to that of a single $2 \times 2$ which can be finally evaluated to give

$$\det[A] = \frac{1}{64}\frac{1}{361}\frac{-1}{1152}(3.1938 \times 10^{10}) = -1200$$

Note: this result required the evaluation of 30 $2 \times 2$ determinants.

A comparison of this method of evaluating determinants with the Laplace method of cofactors readily shows its versatility. Specifically, if we use the number of $2 \times 2$ determinants needed for evaluating an $n \times n$ determinant as a yardstick, then

$$S_{\text{p}} = \sum_{i=1}^{n-1} (n-i)^2 = \frac{1}{6}[2n^3 - 3n^2 + n]$$

where

$S_{\text{p}}$ = number of $2 \times 2$ determinant evaluations for the pivotal condensation,

$n$ = size of the determinant to be evaluated.

This compares favorably with the number of 2 × 2 determinant evaluations using Laplace's method of cofactors ($S_c$), where

$$S_c = \frac{n!}{2}$$

For example if $n = 3$, there are $S_p = 5$ and $S_c = 3$ determinants that must be evaluated. However, if $n = 15$, there are $S_p = 1015$ and $S_c = 6.5383 \times 10^{11}$ determinants of 2 × 2. Clearly, the method of pivotal condensation is far superior to the Laplace method of cofactors for even moderate values of $n$. Furthermore it is better suited to computer programming.

**SUMMARY OF PIVOTAL CONDENSATION**

| STEP | OPERATION | SYMBOL |
|---|---|---|
| 1 | Check $a_{11}$. If it is equal to zero, then switch the first row or column with another row or column so that the new $a_{11} \neq 0$. Make sure that $-1$ is factored each time two rows or columns are switched. | |
| 2 | Reduce $\lvert a_{ij}\rvert$ from an $n \times n$ to an $(n-1) \times (n-1)$ determinant $\lvert a'_{ij}\rvert$ by evaluating $(n-1)^2$ 2 × 2 determinants and factoring $(a_{11})^{n-2}$. | $a''_{ij} = \begin{vmatrix} a_{11} & \cdots & a_{1,j+1} \\ \vdots & & \vdots \\ a_{i+1,1} & \cdots & a_{i+1,j+1} \end{vmatrix}$, $i = 1, 2, \ldots, n-1$, $j = 1, 2, \ldots, n-1$ |
| 3 | Repeat steps 1 and 2 to reduce $\lvert a'_{ij}\rvert$ from $(n-1) \times (n-1)$ to the $(n-2) \times (n-2)$ determinant $\lvert a''_{ij}\rvert$ by evaluating $(n-2)^2$ determinants of 2 × 2. | $a''_{ij} = \begin{vmatrix} a'_{11} & \cdots & a'_{1,j+1} \\ \vdots & & \vdots \\ a'_{i+1,1} & \cdots & a'_{i+1,j+1} \end{vmatrix}$ $i = 1, 2, \ldots, n-2$, $j = 1, 2, \ldots, n-2$ |
| 4 | Continue until a single 2 × 2 determinant is evaluated. | |

## 2.10 AREA AND VOLUME CALCULATION USING DETERMINANTS

The determinant of a matrix can be interpreted graphically for 2 × 2 and 3 × 3 matrices. It is interesting that the determinant of a 2 × 2 matrix is actually equal to twice the area of a triangle formed by its two row vectors! In addition, the deter-

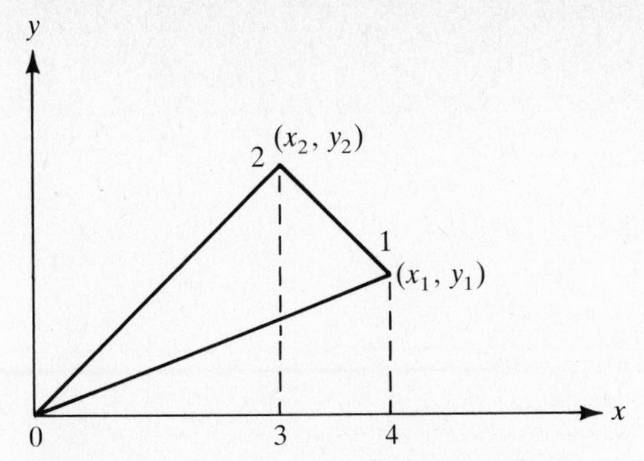


**Figure 2.2** Triangle formed by two row vectors.

minant of a 3 × 3 matrix is equal to six times the volume of a tetrahedron generated by its three row vectors. This graphical interpretation should help further students' intuitive understanding of the need for evaluating determinants. These relationships play a significant role in the finite elements method.

Let us begin by considering the following 2 × 2 determinant of matrix $[A]$:

$$\det[A] = \begin{vmatrix} x_1 & y_1 \\ x_2 & y_2 \end{vmatrix} = x_1 y_2 - x_2 y_1 \tag{2.37}$$

The row vectors of the determinant are represented graphically in figure 2.2.

The area of the triangle enclosed by nodes, 0, 1, and 2 is given as

$$A = A_{023} + A_{1234} - A_{014}$$

where $A_{023}$ is the area enclosed by the triangle with nodes 0, 2, and 3; $A_{1234}$ is the area of the trapezoid with nodes 1, 2, 3, and 4; and, finally, $A_{014}$ is the area of the triangle with nodes 0, 1, and 4. Consequently

$$A = \tfrac{1}{2}x_2 y_2 + \tfrac{1}{2}(x_1 - x_2)(y_1 + y_2) - \tfrac{1}{2}x_1 y_1$$

and

$$A = \tfrac{1}{2}(x_1 y_1 - x_2 y_1)$$

or

$$A = \frac{1}{2}\begin{vmatrix} x_1 & y_1 \\ x_2 & y_2 \end{vmatrix} \tag{2.38}$$

A comparison of (2.38) and (2.37) clearly shows that

$$\det[A] = 2A$$

thus

$$A = \tfrac{1}{2}\det[A] \tag{2.39}$$

So far, we have assumed that the triangle has a node at the origin of the coordinate system. In many problems this may not be the case. Consider the triangle of figure 2.3.

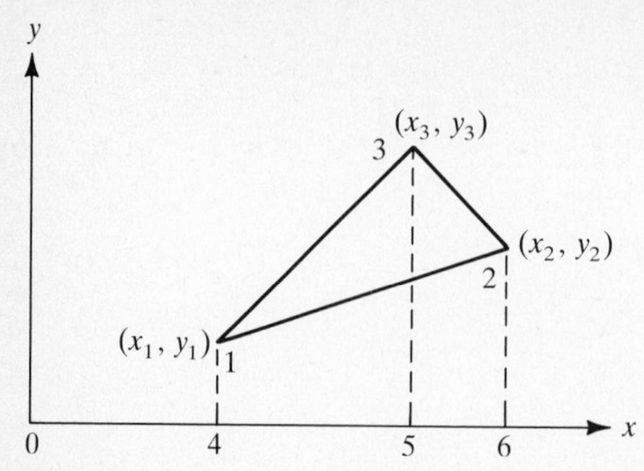


**Figure 2.3** Triangle formed by three row vectors.

The area of the triangle is given by the areas of the trapezoids enclosed by nodes 1, 2, 3, 4, 5, and 6. Hence

$$A = A_{1354} + A_{2356} - A_{1264}$$

and

$$A = \tfrac{1}{2}(x_3 - x_1)(y_1 + y_3) + \tfrac{1}{2}(x_2 - x_3)(y_2 + y_3) - \tfrac{1}{2}(x_2 - x_1)(y_1 + y_2)$$

Simplifying, we have

$$A = \tfrac{1}{2}[x_1(y_2 - y_3) - x_2(y_1 - y_3) + x_3(y_1 - y_2)]$$

or

$$A = \frac{1}{2}\begin{vmatrix} x_1 & y_1 & 1 \\ x_2 & y_2 & 1 \\ x_3 & y_3 & 1 \end{vmatrix} \tag{2.40}$$

Obviously, the area of a triangle whose nodes do not include the origin is given by a 3 × 3 determinant. In general, one may subdivide a region into many triangles and assign node numbers so that this procedure applies to areas of regions. Thus

$$A_r = \sum_{i=1}^{n} A_i$$

where $A_r$ is the area of the region in question, $A_i$ is the area of the $i$th triangle, and $n$ is the number of nodes. Therefore

$$A_r = A_1 + A_2 + \cdots + A_n$$

and

$$A_r = \frac{1}{2}\left\{\begin{vmatrix} x_1 & y_1 \\ x_2 & y_2 \end{vmatrix} + \begin{vmatrix} x_2 & y_2 \\ x_3 & y_3 \end{vmatrix} + \cdots + \begin{vmatrix} x_n & y_n \\ x_1 & y_1 \end{vmatrix}\right\} \tag{2.41}$$

Equation (2.41) is to be used if the region includes the origin of the coordinate system; otherwise use the following:

$$A_r = \frac{1}{2}\left\{\begin{vmatrix} x_1 & y_1 & 1 \\ x_2 & y_2 & 1 \\ x_3 & y_3 & 1 \end{vmatrix} + \begin{vmatrix} x_1 & y_1 & 1 \\ x_3 & y_3 & 1 \\ x_4 & y_4 & 1 \end{vmatrix} + \cdots + \begin{vmatrix} x_1 & y_1 & 1 \\ x_{n-1} & y_{n-1} & 1 \\ x_n & y_n & 1 \end{vmatrix}\right\} \tag{2.42}$$

It is evident that a proper scheme of node numbering must be followed for equations (2.41) and (2.42) to work; *it turns out that numbering nodes in a counterclockwise direction results in positive areas,* while a clockwise numbering yields negative areas.

The volume of a tetrahedron formed by the row vectors of a 3 × 3 determinant is given here without proof for reference:

$$V = \frac{1}{6}\begin{vmatrix} x_1 & y_1 & z_1 \\ x_2 & y_2 & z_2 \\ x_3 & y_3 & z_3 \end{vmatrix} \tag{2.43}$$

and

$$V = \frac{1}{6}\begin{vmatrix} x_1 & y_1 & z_1 & 1 \\ x_2 & y_2 & z_2 & 1 \\ x_3 & y_3 & z_3 & 1 \\ x_4 & y_4 & z_4 & 1 \end{vmatrix} \tag{2.44}$$

Note that equation (2.43) is to be used only if one of the nodes is the origin of the coordinate system; otherwise equation (2.44) should be used.

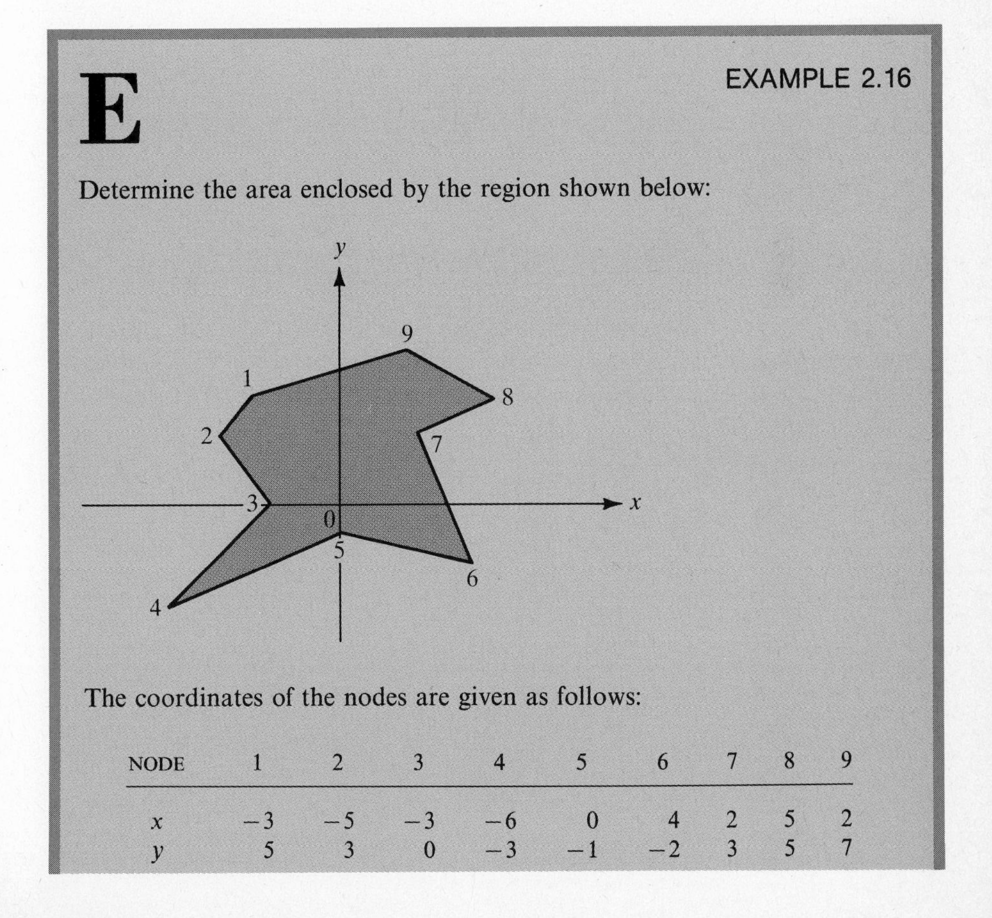

E

EXAMPLE 2.16

Determine the area enclosed by the region shown below:



The coordinates of the nodes are given as follows:

| NODE | 1 | 2 | 3 | 4 | 5 | 6 | 7 | 8 | 9 |
|---|---|---|---|---|---|---|---|---|---|
| $x$ | −3 | −5 | −3 | −6 | 0 | 4 | 2 | 5 | 2 |
| $y$ | 5 | 3 | 0 | −3 | −1 | −2 | 3 | 5 | 7 |

## Solution

Since the region in question includes the origin, we may use equation (2.41). Thus

$$2A = \begin{vmatrix} -3 & 5 \\ -5 & 3 \end{vmatrix} + \begin{vmatrix} -5 & 3 \\ -3 & 0 \end{vmatrix} + \begin{vmatrix} -3 & 0 \\ -6 & -3 \end{vmatrix} + \begin{vmatrix} -6 & -3 \\ 0 & -1 \end{vmatrix}$$

$$+ \begin{vmatrix} 0 & -1 \\ 4 & -2 \end{vmatrix} + \begin{vmatrix} 4 & -2 \\ 2 & 3 \end{vmatrix} + \begin{vmatrix} 2 & 3 \\ 5 & 5 \end{vmatrix} + \begin{vmatrix} 5 & 5 \\ 2 & 7 \end{vmatrix} + \begin{vmatrix} 2 & 7 \\ -3 & 5 \end{vmatrix}$$

$$2A = 111$$

$$A = 55.5$$

Note that while determinants are useful in calculating irregular areas, they can also be used to evaluate integrals. This concept is consistent with the definition of an integral and is illustrated in the following example.

# E

EXAMPLE 2.17

Approximate the following integral using equation (2.42):

$$I = \int_1^2 x^2\,dx = \frac{7}{3}$$

## Solution

While this integral is easy to evaluate, there are many instances of integrands where this is not the case. The function to be integrated, $f(x) = x^2$, is shown graphically as follows:

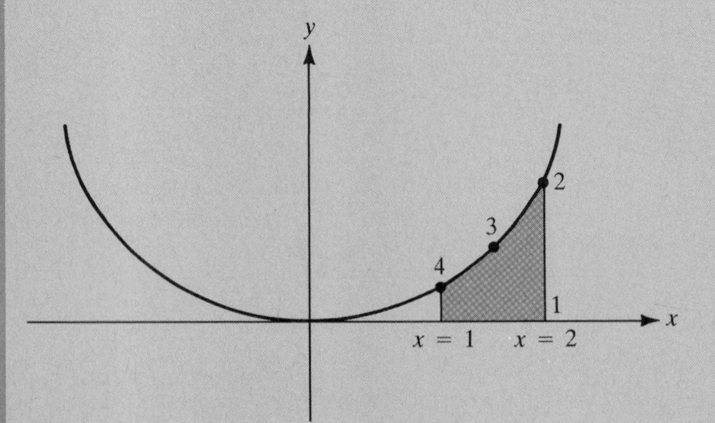

Since the method of determinants for area evaluation involves approximation of the actual curve with straight lines, then it would seem desirable to assume many nodes. Unfortunately, this would require evaluating many determinants. This is often a characteristic of numerical methods; that is, the more accuracy we desire the more work we will have to do. Therefore, assuming five nodes and evaluating the coordinates gives

| NODE | $x$ | $y$ |
|---|---|---|
| 1 | 2 | 0 |
| 2 | 2 | 4 |
| 3 | 1.5 | 2.25 |
| 4 | 1 | 1 |
| 5 | 1 | 0 |

The total area is then given as

$$2I = \begin{vmatrix} 2 & 0 & 1 \\ 2 & 4 & 1 \\ 1.5 & 2.25 & 1 \end{vmatrix} + \begin{vmatrix} 2 & 0 & 1 \\ 1.5 & 2.25 & 1 \\ 1 & 1 & 1 \end{vmatrix} + \begin{vmatrix} 2 & 0 & 1 \\ 1 & 1 & 1 \\ 1 & 0 & 1 \end{vmatrix}$$

$$2I = 4.75$$

$$I = 2.375$$

This value compares well with the exact value of $I = 2.333$, considering that we have only used five nodes.

## Recommended Reading

*Elementary Linear Algebra*, Paul C. Shields, Worth Publishing, Inc., New York, 1968.

*Finite Mathematics*, John G. Kemeny et al., Prentice-Hall, Inc., Englewood Cliffs, N.J., 1972.

*The Finite Element Method*, O. C. Zienkiewicz, McGraw-Hill Book Company, New York, 1977.

## PROBLEMS

**2.1** Express the following equations in matrix form:

**(a)** $2x_1 + x_2 - x_3 = 1$

**(b)** $x_1 + 3x_2 = 4$
$-x_1 + 3x_2 = K$

**(c)** $7x_1 - x_2 = y_1 + 2y_2 + y_3$
$-x_1 + x_2 = 4y_1 - y_2 + 3y_3$

**(d)** $x + \dot{x} + \ddot{x} + y + \dot{y} + \ddot{y} = F$
$-2x + 2\dot{x} + 3\ddot{x} - 7y + 6\dot{y} - 8\ddot{y} = 2F$

**2.2** Carry out the matrix additions, subtraction, and multiplication of the following:

**(a)** $\begin{bmatrix} K & 0 \\ 0 & L \end{bmatrix} + \begin{bmatrix} a_{11} & a_{12} \\ a_{21} & a_{22} \end{bmatrix}$ **(b)** $\begin{bmatrix} K & 0 \\ 0 & L \end{bmatrix} \begin{bmatrix} a_{11} & a_{12} \\ a_{21} & a_{22} \end{bmatrix}$

**(c)** $\begin{bmatrix} 1 & 0 & 0 \\ 2 & 1 & 0 \\ 0 & 0 & 1 \end{bmatrix} \begin{bmatrix} a_{11} & a_{12} & a_{13} \\ a_{21} & a_{22} & a_{23} \\ a_{31} & a_{32} & a_{33} \end{bmatrix}$ **(d)** $\begin{bmatrix} 0 & 0 & 1 \\ 0 & 1 & 0 \\ 1 & 0 & 0 \end{bmatrix} \begin{bmatrix} a_{11} & a_{12} & a_{13} \\ a_{21} & a_{22} & a_{23} \\ a_{31} & a_{32} & a_{33} \end{bmatrix}$

**(e)** $\begin{bmatrix} 2 & 1 & 2 \\ -3 & 5 & 1 \\ 0 & 4 & 1 \end{bmatrix} \begin{Bmatrix} 2 \\ 1 \\ -1 \end{Bmatrix} - \begin{Bmatrix} 3 \\ -2 \\ 3 \end{Bmatrix}$

**(f)** $\{1 \quad 1 \quad 1\} \begin{bmatrix} 1 & 2 & 3 \\ 1 & 2 & 3 \\ 1 & 2 & 3 \end{bmatrix} \begin{Bmatrix} 1 \\ 2 \\ 3 \end{Bmatrix} - \{1 \quad -1 \quad 1\} \begin{Bmatrix} 2 \\ 6 \\ 1 \end{Bmatrix}$

**(g)** $\begin{bmatrix} 2 & -1 & 4 & 3 \\ 2 & 0 & 0 & 1 \end{bmatrix} \begin{bmatrix} 6 & 1 & 3 \\ -1 & 0 & 10 \\ 2 & 6 & 1 \\ 4 & 1 & -7 \end{bmatrix}$

**2.3** Evaluate the following:

**(a)** $\begin{bmatrix} a_{11} & a_{12} & a_{13} \\ a_{21} & a_{22} & a_{23} \end{bmatrix}^t$ **(b)** $\left\{ \begin{bmatrix} 4 & 2 & -6 \\ 7 & 3 & 1 \\ 2 & 6 & 1 \end{bmatrix} \begin{Bmatrix} 1 \\ 2 \\ 1 \end{Bmatrix} \right\}^t$

**2.4** Evaluate the following partitioned matrices using the procedure outlined in sections 2.6.1 and 2.6.3:

**(a)** $\left[\begin{array}{cc|c} 2 & 4 & 1 \\ 4 & 2 & 0 \\ \hline 0 & 1 & 6 \end{array}\right]^t$ **(b)** $\left[\begin{array}{cc|c} 1 & 3 & -5 \\ 2 & 2 & 7 \\ \hline 3 & 9 & 3 \end{array}\right] \left[\begin{array}{cc|c} 1 & 3 & -5 \\ 2 & 2 & 7 \\ \hline -3 & 9 & 3 \end{array}\right]^t$

**2.5** Solve the following system of linear algebraic equations:

$$\frac{1}{64} \begin{bmatrix} 1302 & -818 & 62 \\ -818 & 518 & -42 \\ 62 & -42 & 6 \end{bmatrix} \begin{bmatrix} 1 & 2 & 4 \\ 2 & 3 & 7 \\ 1 & -1 & 9 \end{bmatrix} \begin{bmatrix} 1 & 2 & 1 \\ 2 & 3 & -1 \\ 4 & 7 & 9 \end{bmatrix} \begin{Bmatrix} x_1 \\ x_2 \\ x_3 \end{Bmatrix} = \begin{Bmatrix} 1 \\ 4 \\ 2 \end{Bmatrix}$$

(*Hint:* Use the associative law of matrix multiplication.)

**2.6** Evaluate the following determinants using the method of cofactors:

**(a)** $\begin{vmatrix} 1 & 2 & 3 \\ 0 & 4 & 5 \\ 0 & 0 & 6 \end{vmatrix}$ **(b)** $\begin{vmatrix} 1 & 2 & 3 \\ -4 & 5 & 6 \\ 7 & 8 & 9 \end{vmatrix}$

**(c)** $\begin{vmatrix} 2-\lambda & 1 & 7 \\ 3 & 2-\lambda & 6 \\ 1 & -4 & \lambda \end{vmatrix}$ **(d)** $\left| \begin{bmatrix} 4 & 9 & 10 \\ 12 & 6 & 8 \\ 5 & -7 & 1 \end{bmatrix} - \begin{bmatrix} 2 & 0 & 0 \\ 0 & 2 & 0 \\ 0 & 0 & 2 \end{bmatrix} (\lambda) \right|$

**2.7** Evaluate the determinants of problem 2.6 using the method of pivotal condensation.

**2.8** Evaluate the determinant of problem 2.6b using the upper-triangle method.

**2.9** At the University of Evansville, engineering sophomores are normally required to carry the following load during their fall term:

| DESIGNATION | COURSE | CREDIT HOURS |
|---|---|---|
| $C_1$ | Physics 211 | 5 |
| $C_2$ | Statics 212 | 3 |
| $C_3$ | Calculus 323 | 3 |
| $C_4$ | General Education | 4 |

Students receive 4 points for each A, 3 points for each B, 2 points for each C, 1 point for each D, and 0 for each F. The sum of these numbers divided by the total number of credit hours gives the average. Calculate the grade-point average for the following students.

**(a)** $C_1, C_2, C_3, C_4 = A, B, C, D$

**(b)** $C_1, C_2, C_3, C_4 = F, A, D, C$

**(c)** $C_1, C_2, C_3, C_4 = B, B, A, D$

**(d)** $C_1, C_2, C_3, C_4 = D, D, A, A$

What is their combined average? Use matrix algebra in solving this problem.

**2.10** The unit requirements for building three different houses are given by the following matrix.

| | STEEL | PAINT | WOOD | GLASS | LABOR |
|---|---|---|---|---|---|
| Colonial | 5 | 6 | 30 | 20 | 10 |
| Ranch | 6 | 10 | 20 | 17 | 18 |
| Contemporary | 8 | 4 | 25 | 12 | 15 |

**(a)** Compute the unit requirements for building 20 colonial, 30 ranch, and 8 contemporary houses.

**(b)** If the cost of steel is \$18/unit, paint \$2/unit, wood \$10/unit, glass \$7/unit, and labor \$6/unit calculate the cost of each house, then calculate the total cost to the contractor in (a). (*Note:* These figures are arbitrary.)

**2.11** Use the rotation of coordinate matrices to show that

**(a)** $\sin 2\theta = 2\cos\theta\sin\theta$

**(b)** $\cos 2\theta = \cos^2\theta - \sin^2\theta$

**(c)** $\sin 3\theta = \sin\theta(3\cos^2\theta - \sin^2\theta)$

**(d)** $\cos 3\theta = \cos\theta(\cos^2\theta - 3\sin^2\theta)$

**2.12** Determine the area enclosed by the following coordinates:

| $x$ | 0 | 1 | 2 | −1 | −2 |
|---|---|---|---|---|---|
| $y$ | 0 | 1 | 6 | 3 | −4 |

**2.13** Determine the area enclosed by the following coordinates:

| $x$ | 1 | 3 | 5 | 2 | $-1$ |
|---|---|---|---|---|---|
| $y$ | 2 | 3 | $-2$ | $-6$ | 7 |

**2.14** Approximate the following integral using $\Delta\chi = 0.5$:

$$I = \int_0^2 (x^2 + 2)\,dx$$

Compare your answer with the exact value.

**2.15** Rework problem 2.14 using $\Delta x = 0.2$.

# chapter 3

# Mathematical Modeling of Typical Engineering Systems

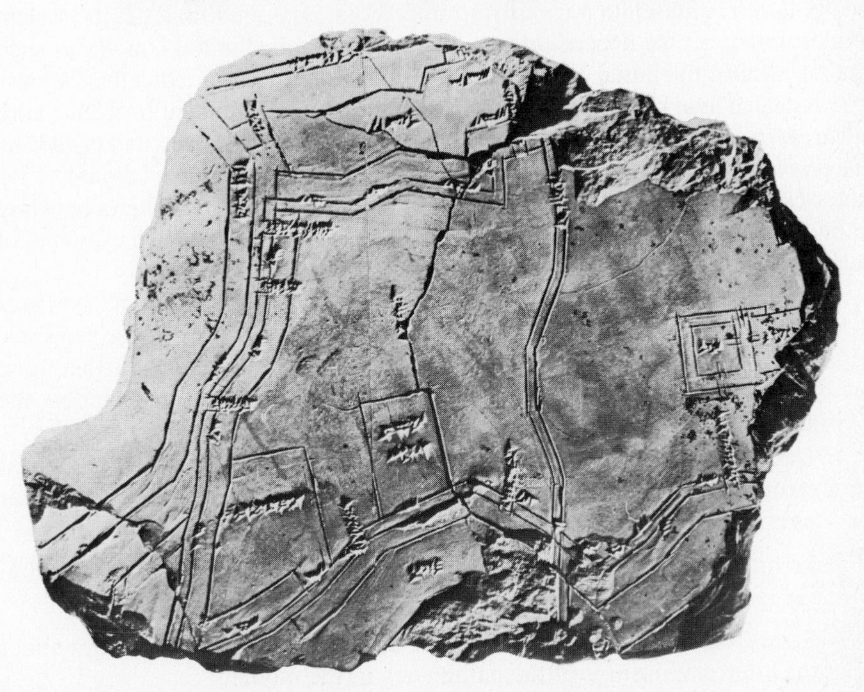

A clay tablet detailing Nippur, the holy city of the Sumerians. It is the most ancient urban project known to history. *Courtesy of the Iraqi government.*

"Mathematical modeling is a necessary first step in solving practical problems."

# 3

## 3.1 INTRODUCTION

This chapter covers basic concepts pertaining to mathematical modeling of some common systems occurring in electrical, mechanical, and civil engineering. In order to fully appreciate the applications covered in this chapter, the student is expected to be familiar with the principles of static and dynamic equilibrium ordinary differential equations, and basic linear circuits. Students who lack adequate preparation in these subjects may view this material as optional and proceed directly to the next chapter.

A mathematical model of a physical system is a mathematical description of an underlying physical process which relates external stimuli and intrinsic properties of the system to its response. We can predict the behavior of the system by analyzing the model. That is, for example, given a static force (external stimulus) acting on a beam of finite length and rigidity (intrinsic properties), we can predict its deflection (response) for various restraints (boundary conditions). If the system stimulus (force) is time dependent in nature, then so-called initial conditions such as initial velocity and initial displacement will also be needed to predict the response of the system. These kinds of time-dependent and time-independent problems arise in all areas of engineering and are referred to as *initial-value problems* and *boundary-value problems*, respectively. Other important problems that do not fall into either of these categories are called *eigenproblems* in this text. Predicting column buckling and finding the natural frequencies for some electrical circuits are two examples of eigenproblems.

As engineers, we are interested in the modeling, analysis, and design of systems in order to build systems which are both effective and economical. These processes are based on a mixture of science, experience, and common sense. Furthermore, none of them is exact. Errors are introduced by a variety of factors including the following:

1. Data derived from experimental measurements will contain measurement error—whether the data are obtained from suppliers, equipment, or other users:
2. Errors will be introduced through the choice of parameters interrelated in the model.
3. Errors will be introduced through the use of simplifying assumptions about the interrelationships of the parameters in the model.
4. An additional source of systematic error is inherent in the choice of analysis techniques and, in particular, the choice of numerical techniques of classical closed-form analysis.

Clearly, no method for predicting a system's behavior will be error free. This is, of course, precisely why safety factors are introduced in real engineering design problems. It is important for the engineer to realize that the most effective and economical designs will be obtained when the errors from all four of these sources are minimized. It is not necessarily true that lower overall error will result if choices are made to greatly reduce or even eliminate error in one of the categories but at a cost of allowing error in another of the categories to become excessive. In particular, while analysis techniques using numerical techniques as opposed to classical closed forms may be intrinsically less accurate analysis techniques, we shall see that the simplicity of numerical techniques permits us much greater freedom in choosing more realistically complicated mathematical descriptions of the interrelationships of system parameters; this alternative of choosing greater analysis error in exchange for less error due to simplifying modeling assumptions yields lower overall error in our design and generally permits more effective and more economical designs as a result.

## 3.2 ELECTRICAL ENGINEERING SYSTEMS

The formulation of mathematical models for some electrical engineering circuits depends on our knowledge of physical laws pertaining to voltage and current in linear circuits. In general, a linear circuit may involve resistors, capacitors, and inductors. The basic relationships relating voltage and current in these components are shown below.

*Resistors*

$$V_R = RI$$

$V_R$ = voltage across the resistor

$R$ = resistance (constant)

$I$ = current through the resistor

*Capacitors*

$$V_C = \frac{1}{C}\int I\,dt$$

$V_C$ = voltage across the capacitor

$C$ = capacitance (constant)

$I$ = current through the capacitor

*Inductors*

$$V_L = L\frac{dI}{dt}$$

$V_L$ = voltage across the inductor

$L$ = inductance (constant)

$I$ = current through the inductor

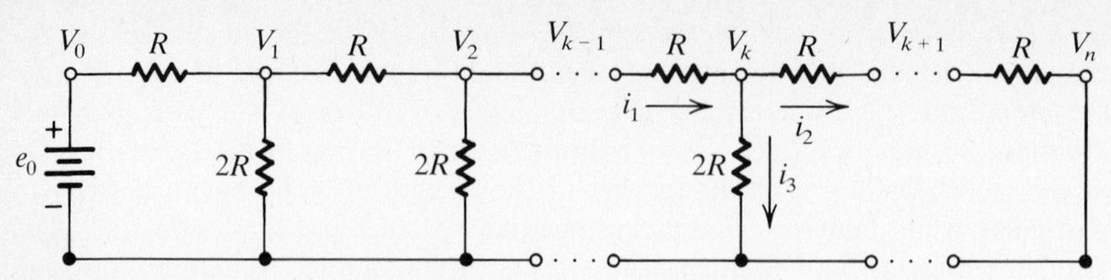


**Figure 3.1** Electrical resistor ladder circuit.

The physical properties of a circuit are specified by $R$, $C$, and $L$. Therefore the response of a circuit is expected to be a function of these physical parameters as well as the external stimuli.

### 3.2.1 Boundary-Value Problems

These are problems whose solutions are obtained by specifying input and/or output values at their geometrical boundaries. Consider the resistor ladder circuit shown in figure 3.1.

Suppose we want to determine the voltages $V_1, V_2, \ldots, V_n$ at the nodes in terms of the boundary voltages $e_0$ and $V_n$. Applying Kirchhoff's current law at the $k$th node, we obtain the following difference equation:

$$i_1 = i_2 + i_3$$

$$\frac{V_{k-1} - V_k}{R} = \frac{V_k - V_{k+1}}{R} + \frac{V_k}{2R} \tag{3.1}$$

Multiplying by $2R$ and collecting terms we get

$$2V_{k+1} - 5V_k + 2V_{k-1} = 0$$

or

$$2V_k - 5V_{k-1} + 2V_{k-2} = 0, \qquad k = 0, 2, \ldots, n \tag{3.2}$$

Equation (3.2) is called a linear homogeneous* difference equation. It represents the mathematical model for the circuit under consideration. The question we need to address at this stage is, "How suitable is this model?" Obviously it is an appropriate model if the resistances are known exactly and if we know for sure how they might change as a result of external factors, such as temperature and/or deformation.

The solution of a homogeneous difference equation is of the form $V_k = X^k$, where $V_k$ is the voltage at the $k$th node and $X$ is one of the roots of a polynomial derived from (3.2). Furthermore, if $V_k^1$ and $V_k^2$ are solutions to (3.2), then so is $a_1 V_k^1 + a_2 V_k^2$, where $a_1$ and $a_2$ are arbitrary constants. These constants are evaluated by

* A formal definition of homogeneity is that a function $F(x, y, \ldots)$ is homogeneous of degree $k$ in $x, y, \ldots$ if, and only if, $F(\lambda^x, \lambda^y, \ldots) = \lambda^k F(x, y), \ldots)$. When the function $F(x, y, \ldots)$ is linear, then $F$ is homogeneous whenever all terms not involving the variables are zero, as is the case in (3.2).

using the boundary voltages at nodes 0 and $n$. Therefore, since

$$V_k = X^k$$

substituting into (3.2) gives

$$2X^k - 5X^{k-1} + 2X^{k-2} = 0$$

or

$$X^k(2 - 5X^{-1} + 2X^{-2}) = 0$$

Clearly, $X^k = 0$ is a trivial solution and may be ignored. Alternatively, a nontrivial solution is obtained from

$$2 - 5X^{-1} + 2X^{-2} = 0$$

or more simply

$$2X^2 - 5X + 2 = 0 \tag{3.3}$$

Equation (3.3) has two roots,

$$X_1 = 4 \qquad \text{and} \qquad X_2 = 1$$

Therefore the solutions to (3.2) are given as

$$V'_k = X_1^k = 4^k$$

$$V''_k = X_2^k = 1^k = 1$$

The homogeneous solution is now determined as the linear combination of these solutions as follows:

$$V_k = a_1 V'_k + a_2 V''_k$$

or (3.4)

$$V_k = a_1 4^k + a_2$$

The arbitrary constants are fixed by using the following boundary conditions:

1. $k = 0$ implies $V_0 = e_0$
2. $k = n$ implies $V_n = 0$

Substituting into (3.4) yields two linear algebraic equations in the two unknown constants $a_1$ and $a_2$. Thus

$$e_0 = a_1 + a_2, \qquad 0 = a_1 4^n + a_2$$

which imply

$$a_1 = e_0(1 - 4^n)^{-1}, \qquad a_2 = -e_0 4^n(1 - 4^n)^{-1}$$

The solution to the model is then given for any node $k$ by substituting $a_1$ and $a_2$ into equation (3.4) to give

$$V_k = e_0(1 - 4^n)^{-1}(4^k - 4^n), \qquad k = 0, 1, \ldots, n \tag{3.5}$$

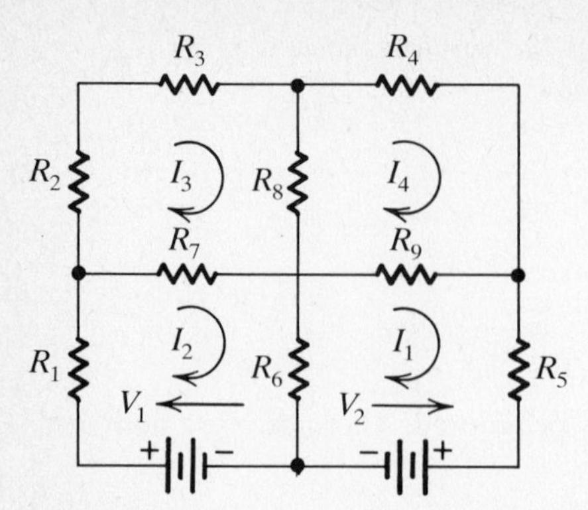


**Figure 3.2** A direct current electrical network.

Note that the solution requires, among other things, that we know how to solve linear as well as nonlinear algebraic equations. In addition, the solution given by (3.5) applies only if we know for sure that $V_n = 0$. This condition may not be satisfied for a small number of loops!

A more realistic boundary-value problem in electrical engineering may be of the form presented in figure 3.2.

This circuit is analyzed by using Kirchhoff's law, which states that "the net voltage drop in any closed loop must equal zero." Consequently, summing the voltages in loops 1–4 yields

Loop 1:

$$-V_2 - R_6(I_1 - I_2) - R_9(I_1 - I_4) - R_5 I_1 = 0$$

Loop 2:

$$V_1 - R_1 I_2 - R_7(I_2 - I_3) - R_6(I_2 - I_1) = 0$$

Loop 3:

$$-R_7(I_3 - I_2) - R_2 I_3 - R_3 I_3 - R_8(I_3 - I_4) = 0$$

Loop 4:

$$-R_9(I_4 - I_1) - R_8(I_4 - I_3) - R_4 I_4 = 0$$

Collecting terms and expressing the results in a matrix form gives

$$\begin{bmatrix} R_5 + R_6 + R_7 & -R_6 & 0 & -R_9 \\ -R_6 & R_1 + R_6 + R_7 & -R_7 & 0 \\ 0 & -R_7 & R_2 + R_3 + R_7 + R_8 & -R_8 \\ -R_9 & 0 & -R_8 & R_4 + R_8 + R_9 \end{bmatrix} \begin{Bmatrix} I_1 \\ I_2 \\ I_3 \\ I_4 \end{Bmatrix} = \begin{Bmatrix} -V_2 \\ V_1 \\ 0 \\ 0 \end{Bmatrix} \qquad \textbf{(3.6a)}$$

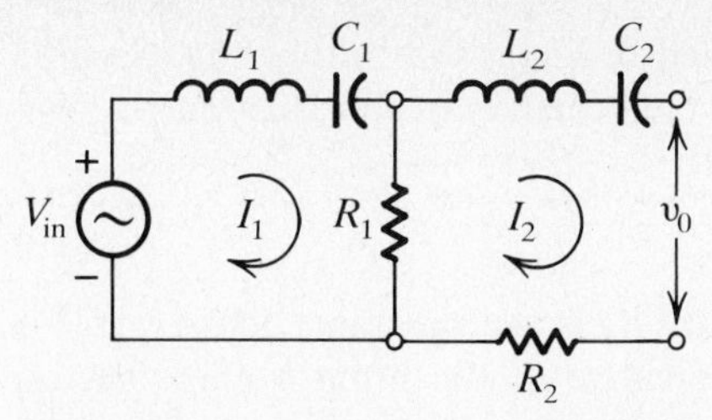


**Figure 3.3** A schematic representation of a filter circuit.

In compact matrix form

$$[R]\{I\} = \{V\} \tag{3.6b}$$

The $[R]$ matrix is symmetrical and is called the *resistance matrix*, $\{I\}$ is the unknown current vector, and $\{V\}$ is the known voltage vector representing the boundary conditions.

### 3.2.2 Initial-Value Problems

The design of filters for the purpose of reducing interference involves circuits that may include resistors, capacitors, and inductors. An example of one such network is shown in figure 3.3.

The mathematical model representing this circuit can be developed using the physical laws defined earlier for resistors, capacitors, inductors, and Kirchhoff's law for voltage. Thus

Loop 1:

$$V_{in} - L_1 \frac{dI_1}{dt} - \frac{1}{C_1}\int_{-\infty}^{t} I_1 dt - R_1(I_1 - I_2) = 0 \tag{3.7a}$$

Loop 2:

$$V_0 - R_1(I_2 - I_1) - L_2 \frac{dI_2}{dt} - \frac{1}{C_2}\int_{-\infty}^{t} I_2\, dt - R_2 I_2 = 0 \tag{3.7b}$$

Differentiating equation (3.7) with respect to time and letting $\dot{I} = dI/dt, \ddot{I} = d^2I/dt^2$, and $\dot{V} = dv/dt$ gives

$$L_1\ddot{I}_1 + I_1/C_1 + R_1\dot{I}_1 - R_1\dot{I}_2 = \dot{V}_{in}$$

$$-R_1\dot{I}_1 + R_1\dot{I}_2 + L_2\ddot{I}_2 + I_2/C_2 + R_2\dot{I}_2 = V_0$$

In matrix form

$$\begin{bmatrix} L_1 & 0 \\ 0 & L_2 \end{bmatrix}\begin{Bmatrix} \ddot{I}_1 \\ \ddot{I}_2 \end{Bmatrix} + \begin{bmatrix} R_1 & -R_1 \\ -R_1 & R_1 + R_2 \end{bmatrix}\begin{Bmatrix} \dot{I}_1 \\ \dot{I}_2 \end{Bmatrix} + \begin{bmatrix} \dfrac{1}{C_1} & 0 \\ 0 & \dfrac{1}{C_1} \end{bmatrix}\begin{Bmatrix} I_1 \\ I_2 \end{Bmatrix} = \begin{Bmatrix} \dot{V}_{in} \\ V_0 \end{Bmatrix} \tag{3.8}$$

Equation (3.8) represents a set of two coupled ordinary differential equations of the second order and can be expressed more conveniently in compact matrix form as follows:

$$[L]\{\ddot{I}\} + [R]\{\dot{I}\} + [C]\{I\} = \{\dot{V}\} \tag{3.9}$$

Note here that $[L]$ is the inductance matrix, $\{R\}$ is the resistance matrix, $[C]$ is the capacitance matrix, and $\{\dot{V}\}$ is the derivative of the input voltage vector. Furthermore, the output voltage can be easily evaluated from $V_0 = R_2 I_2$, where $I_2$ is the current in the second loop, which can be determined by solving (3.8). The solution involves four arbitrary constants whose values are specified by the initial currents and their derivatives in loop 1 and loop 2. The derivatives of $I$ are calculated from the voltages across the inductors $L_1$ and $L_2$ at the zero.

### 3.2.3 Eigenproblems

The general solution of the mathematical model given by equation (3.8) or (3.9) is the sum of the homogeneous solution $\{I\}_h$ (complementary) and the particular solution $\{I\}_p$. Thus

$$\{I\} = \{I\}_h + \{I\}_p \tag{3.10}$$

To illustrate this, consider the circuit modeled by (3.8) and assume constant voltage input. Consequently, the $\{\dot{V}\}$ vector is the null vector. This is because the derivative of a constant is zero. Substituting $\{\dot{V}\} = \{0\}$ into (3.9) gives

$$[L]\{\ddot{I}\} + [R]\{\dot{I}\} + [C]\{I\} = \{0\} \tag{3.11}$$

It is evident that the total solution of (3.11) is in fact equal to the homogeneous solution. Therefore we may proceed to solve for the current vector $\{I\}$ by assuming a solution of the form

$$\begin{Bmatrix} I_1 \\ I_2 \end{Bmatrix} = \begin{Bmatrix} A_1 \\ A_2 \end{Bmatrix} e^{st}$$

or

$$\{I\} = \{A\}e^{st} \tag{3.12a}$$

where $\{A\}$ is a vector of arbitrary constants whose values are dependent on initial conditions, $s$ is a parameter whose value(s) are dependent on the circuit's intrinsic properties, and $t$ is time. Therefore, by differentiating (3.12a),

$$\{\dot{I}\} = s\{A\}e^{st} \tag{3.12b}$$

and by differentiating (3.12b)

$$\{\ddot{I}\} = s^2\{A\}e^{st} \tag{3.12c}$$

Substituting (3.12) into (3.9) yields

$$s^2[L]\{A\}e^{st} + s[R]\{A\}e^{st} + [C]\{A\}e^{st} = \{0\}$$

Dividing through by $e^{st}$ gives

$$s^2[L]\{A\} + s[R]\{A\} + [C]\{A\} = \{0\}$$

or more simply

$$[s^2[L] + s[R] + [C]]\{A\} = \{0\} \qquad \textbf{(3.13a)}$$

Alternatively, letting $[Q] = [s^2[L] + s[R] + [C]]$ yields

$$[Q]\{A\} = \{0\} \qquad \textbf{(3.13b)}$$

Note here that $[L]$, $[R]$, $[C]$, and $[Q]$ are $2 \times 2$ matrices. Therefore an obvious solution to the set of linear equations given by (3.13) is the following trivial solution:

$$\{A\} = \begin{Bmatrix} A_1 \\ A_2 \end{Bmatrix} = \begin{Bmatrix} 0 \\ 0 \end{Bmatrix}$$

The implication here is that $\{I\} = \{0\}$. However, for constant input voltage, the currents $I_1$ and $I_2$ are *not necessarily zero.* Fortunately, the equality in (3.13) can also be met by choosing so that the resulting equations are dependent and (3.13) can be satisfied with a nonzero $\{A\}$. This is the essence of an eigenproblem. That is, determining a nontrivial solution to (3.13). This will be the subject of chapter 8.

## 3.3 MECHANICAL ENGINEERING SYSTEMS

A mechanical system may involve a spring, dashpot, and/or mass. The physics used in relating external effects (forces) to intrinsic physical properties (stiffness, damping, and mass) and response (displacement, velocity, and acceleration) is outlined as follows.

*Spring*

$$F_s = kx$$

$F_s$ = force in a spring

$k$ = stiffness (property)

$x$ = displacement

*Dashpot*

$$F_d = c\frac{dx}{dt}$$

$F_d$ = force in a dashpot

$c$ = damping (property)

$\frac{dx}{dt}$ = velocity

*Mass*

$$F_I = m\frac{d^2x}{dt^2}$$

$F_I$ = force in a mass (inertia)

$m$ = mass (property)

$\frac{d^2x}{dt^2}$ = acceleration

The intrinsic physical properties of a mechanical system are specified by $k$, $c$, and $m$. Note that these are related to displacement, velocity, and acceleration, respectively.

### 3.3.1 Boundary-Value Problems

These are problems that can be solved after specifying boundary conditions. An example of a typical mechanical system is shown in figure 3.4.

Suppose that we want to know the displacements $x_1$, $x_2$, and $x_3$ expressed in terms of the "boundary" forces $P_1$, $P_2$, and $P_3$ and the system's intrinsic properties $k_1$, $k_2$, $k_3$, $m_1$, $m_2$, and $m_3$. Assume that the external forces do not vary with time. The mathematical model for the system is easily developed by using Newton's second law of equilibrium. This law states that for a rigid body to be in

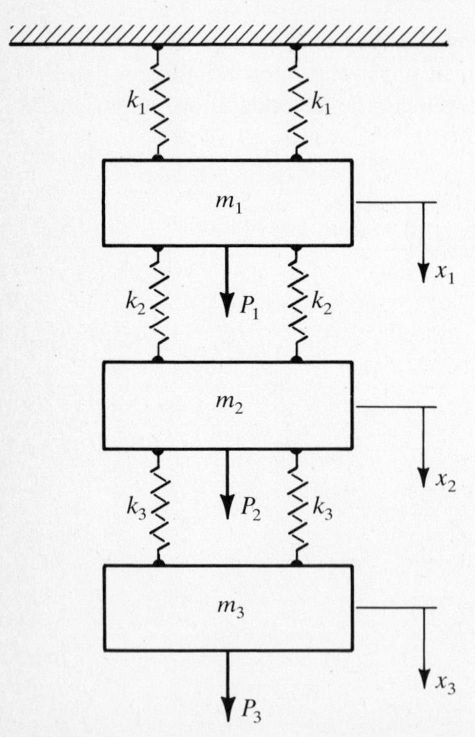


**Figure 3.4** An example of a mechanical system.

equilibrium the sum of forces acting on it must be zero. The application of this law is made easier by drawing the following free-body diagram:

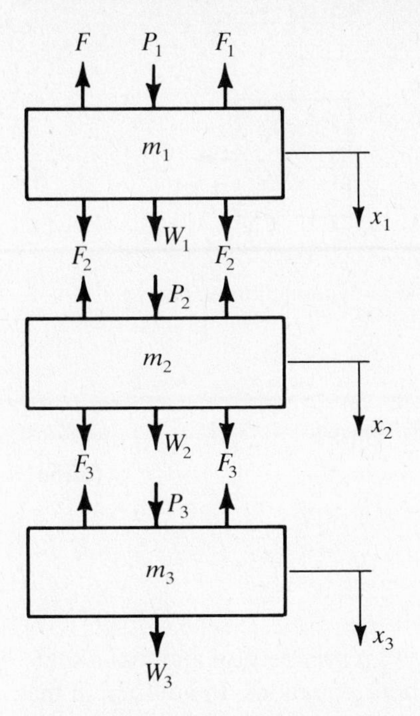


Applying Newton's law to the three masses yields

Mass 1:

$$+\uparrow\sum F_x = 0 = 2F_1 - P_1 - 2F_2 - W_1 \tag{3.14a}$$

Mass 2:

$$+\uparrow\sum F_x = 0 = 2F_2 - P_2 - 2F_3 - W_3 \tag{3.14b}$$

Mass 3:

$$+\uparrow\sum F_x = 0 = 2F_3 - P_3 - W_3 \tag{3.14c}$$

The forces appearing in (3.14) are determined by using the physics outlined earlier, i.e.,

$$F_1 = k_1 x_1, \qquad F_2 = k_2(x_2 - x_1), \qquad F_3 = k_3(x_3 - x_2)$$

$$W_1 = m_1 g, \qquad W_2 = m_2 g, \qquad W_3 = m_3 g$$

where $g$ is the acceleration due to gravity. Substituting these expressions into (3.14) and rearranging gives

$$\begin{aligned} 2(k_1 + k_2)x_1 - 2k_2 x_2 &= P_1 + m_1 g \\ -2k_1 x_1 + 2(k_2 + k_3)x_2 - 2k_3 x_3 &= P_2 + m_2 g \\ -2k_3 x_2 + 2k_3 x_3 &= P_3 + m_3 g \end{aligned}$$

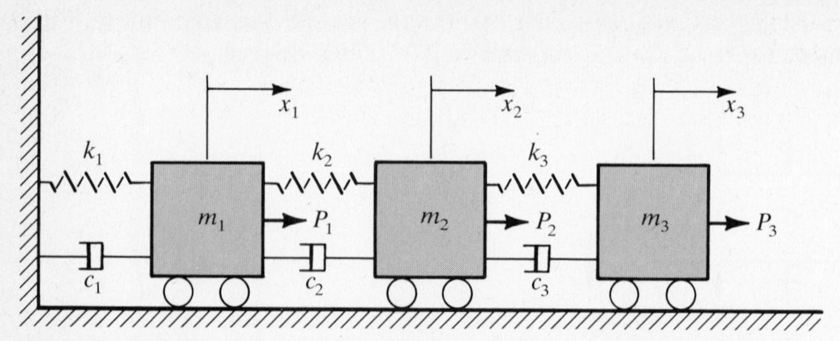


**Figure 3.5** Mass–dashpot–spring system subjected to dynamic forces and/or initial conditions.

In matrix form,

$$2\begin{bmatrix} k_1 + k_2 & -k_2 & 0 \\ -k_2 & k_2 + k_3 & -k_3 \\ 0 & -k_3 & k_3 \end{bmatrix} \begin{Bmatrix} x_1 \\ x_2 \\ x_3 \end{Bmatrix} = \begin{Bmatrix} P_1 \\ P_2 \\ P_3 \end{Bmatrix} + \begin{Bmatrix} m_1 g \\ m_2 g \\ m_3 g \end{Bmatrix} \tag{3.15a}$$

or in compact matrix form,

$$[k]\{x\} = \{P\} + \{W\} \tag{3.15b}$$

It is interesting to note that the stiffness matrix $[k]$ is symmetrical and that a solution is obtained by solving a set of linear algebraic equations. In addition, if the coordinates $x_1$, $x_2$, and $x_3$ are measured from positions after the system reaches its equilibrium position under its own weight, then $\{W\} = \{0\}$. Consequently, given the boundary forces $\{P\}$, the displacement vector $\{x\}$ is determined.

### 3.3.2 Initial-Value Problems

The determination of a mechanical system's dynamic response involves the solution of one or more ordinary differential equations. This type of problem arises when dynamic rather than static forces are applied to a system. Consider the system shown in figure 3.5.

This system has three degrees of freedom. That is, each of the three masses can be expected to move in the horizontal direction. Thus vertical motion is not allowed. The rollers are assumed to be frictionless. Consider the free-body diagram for the system shown in figure 3.6.

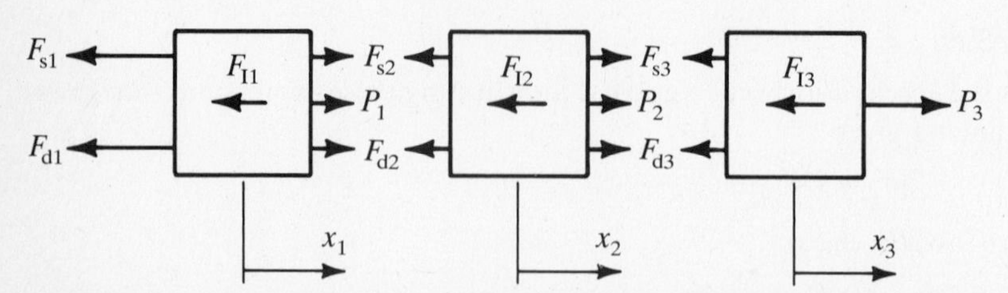


**Figure 3.6** Free-body diagram of a three-degrees-of-freedom *m-c-k* system.

It is interesting to note that while the forces shown in figure 3.6 are dynamic in nature they can be treated as static forces. This is an important observation in that it allows us to apply the static equilibrium equations to solve this problem. Therefore, summing the forces actively on each of the three masses, we get

Mass 1:

$$F_{s1} + F_{d1} + F_{I1} - P_1 - F_{s2} - F_{d2} = 0 \tag{3.16a}$$

Mass 2:

$$F_{s2} + F_{d2} + F_{I2} - P_2 - F_{s3} - F_{d3} = 0 \tag{3.16b}$$

Mass 3:

$$F_{s3} + F_{d3} + F_{I3} - P_3 = 0 \tag{3.16c}$$

The forces due to mass inertia ($F_I$), dashpot damping ($F_d$). and spring stiffness ($F_s$) are given below:

$$F_{s1} = k_1 x_1, \qquad F_{s2} = k_2(x_2 - x_1), \qquad F_{s3} = k_3(x_3 - x_2)$$

$$F_{d1} = c_1 \dot{x}_1, \qquad F_{d2} = c_2(\dot{x}_2 - \dot{x}_1), \qquad F_{d3} = c_3(\dot{x}_3 - \dot{x}_2)$$

$$F_{I1} = m_1 \ddot{x}_1, \qquad F_{I2} = m_2 \ddot{x}_2, \qquad F_{I3} = m_3 \ddot{x}_3$$

Substituting these forces into the equations of equilibrium, namely (3.16), permits us to express these equations in terms of system parameters. That is,

$$k_1 x_1 + c_1 \dot{x}_1 + m_1 \ddot{x}_1 - P_1 - k_2(x_2 - x_1) - c_2(\dot{x}_2 - \dot{x}_1) = 0$$

$$k_2(x_2 - x_1) + c_2(\dot{x}_2 - \dot{x}_1) + m_2 \ddot{x}_2 - P_2 - k_3(x_3 - x_2) - c_3(\dot{x}_3 - \dot{x}_2) = 0$$

$$k_3(x_3 - x_2) + c_3(\dot{x}_3 - \dot{x}_2) + m_3 \ddot{x}_3 - P_3 = 0$$

Rearranging and collecting terms, we obtain

$$(k_1 + k_2)x_1 - k_2 x_2 + (c_1 + c_2)\dot{x}_1 - c_2 \dot{x}_2 + m_1 \ddot{x}_1 = P_1$$

$$-k_2 x_1 + (k_2 + k_3)x_2 - k_3 x_3 - c_2 \dot{x}_1 + (c_2 + c_3)\dot{x}_2 - c_3 \dot{x}_3 + m_2 \ddot{x}_2 = P_2$$

$$-k_3 x_2 + k_3 x_3 - c_3 \dot{x}_2 + c_3 \dot{x}_3 + m_3 \ddot{x}_3 = P_3$$

A careful look at these equations clearly shows how difficult it is to make sense out of the messy algebraic form. Now let us see how these equations can be made more meaningful by using matrices.

$$\begin{bmatrix} k_1 + k_2 & -k_2 & 0 \\ -k_2 & k_2 + k_3 & -k_3 \\ 0 & -k_3 & k_3 \end{bmatrix} \begin{Bmatrix} x_1 \\ x_2 \\ x_3 \end{Bmatrix} + \begin{bmatrix} c_1 + c_2 & -c_2 & 0 \\ -c_2 & c_2 + c_3 & -c_3 \\ 0 & -c_3 & c_3 \end{bmatrix} \begin{Bmatrix} \dot{x}_1 \\ \dot{x}_2 \\ \dot{x}_3 \end{Bmatrix}$$

$$+ \begin{bmatrix} m_1 & 0 & 0 \\ 0 & m_2 & 0 \\ 0 & 0 & m_3 \end{bmatrix} \begin{Bmatrix} \ddot{x}_1 \\ \ddot{x}_2 \\ \ddot{x}_3 \end{Bmatrix} = \begin{Bmatrix} P_1 \\ P_2 \\ P_3 \end{Bmatrix} \tag{3.17a}$$

In compact matrix form,

$$[k]\{x\} + [c]\{\dot{x}\} + [M]\{\ddot{x}\} = \{P\} \tag{3.17b}$$

Obviously, the $[k]$, $[c]$, and $[M]$ matrices are symmetrical. In practice the $[k]$ matrix is referred to as the stiffness matrix, $[c]$ is the damping matrix, $[M]$ is the mass matrix, and $\{P\}$ is the force vector. In addition, $\{x\}$ is the displacement vector, $\{\dot{x}\}$ is the velocity vector, and $\{\ddot{x}\}$ is the acceleration vector. It should be evident to the student that equation (3.17) is of the same form as that developed for the electrical network initial-value problem. Therefore modeling this system involves solving an eigenproblem.

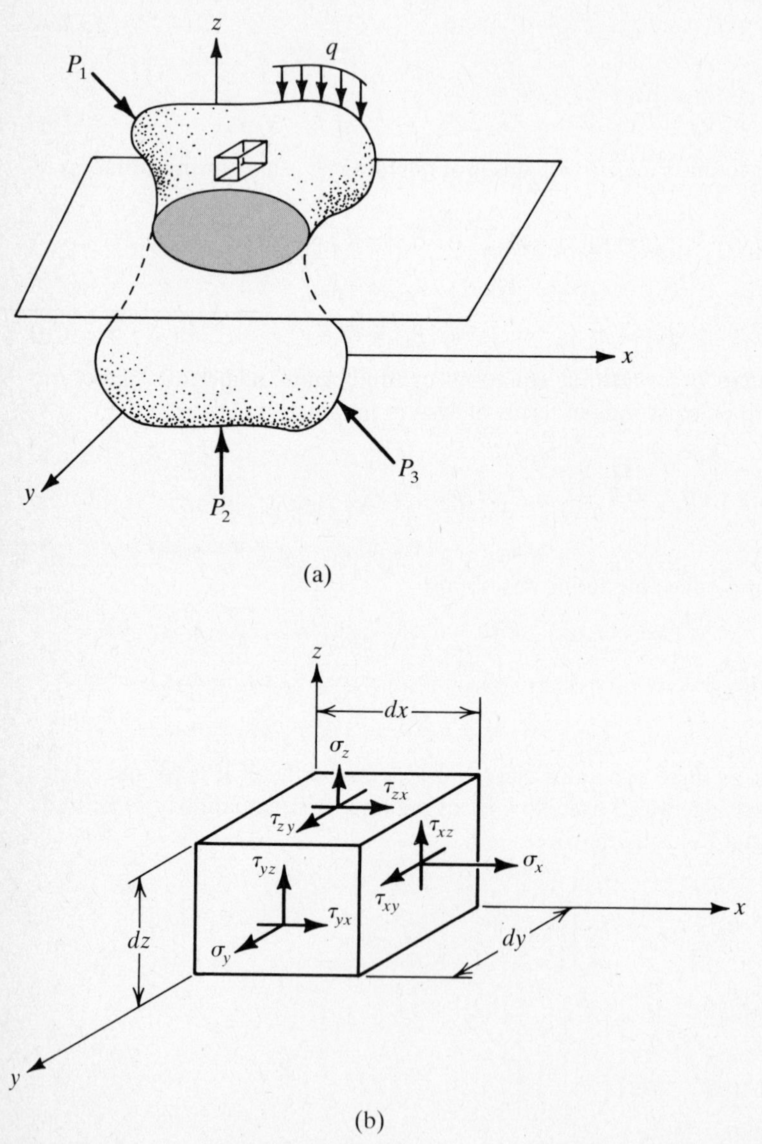


**Figure 3.7** (a) A three-dimensional rigid body subjected to an arbitrary external force. (b) An element within the body.

### 3.3.3 Eigenproblems

Oftentimes mechanical engineers as well as civil engineers are interested in determining the forces acting on a given surface cutting through a rigid body subjected to a set of external forces. Consider the rigid body and the element of finite dimension taken out of a given section as illustrated in figure 3.7.

We shall assume that the body is of a homogeneous material and that its physical properties are consequently independent of direction (isotropic). This simplifying assumption is significant in that it points out the difficulty involved in trying to model nonhomogeneous anisotropic solid bodies. Therefore the application of the model to be developed here must be limited to those materials whose properties can meet this requirement.

Figure 3.8 shows a segment of the element given in figure 3.7b. It is evident that as $r \to 0$ the tetrahedron will become of infinitesimal order with sides $dx$, $dy$, and $dz$. In this context we need to determine the components of the resultant stress $R_n$ acting on the plane whose normal is given by the $N$ axis. This is accomplished by considering the equilibrium of all forces in the $x$, $y$, and $z$ directions. That is, writing

$$\sum F_x = 0 = \tfrac{1}{2}\sigma_x\, dy\, dz + \tfrac{1}{2}\tau_{yx}\, dx\, dz + \tfrac{1}{2}\tau_{zx}\, dx\, dy - R_{nx}\, dA \qquad \textbf{(3.18a)}$$

$$\sum F_y = 0 = \tfrac{1}{2}\tau_{xy}\, dy\, dz + \tfrac{1}{2}\sigma_y\, dx\, dz + \tfrac{1}{2}\tau_{zy}\, dx\, dy - R_{ny}\, dA \qquad \textbf{(3.18b)}$$

$$\sum F_z = 0 = \tfrac{1}{2}\tau_{xz}\, dy\, dz + \tfrac{1}{2}\tau_{yz}\, dx\, dz + \tfrac{1}{2}\sigma_z\, dx\, dy - R_{nz}\, dA \qquad \textbf{(3.18c)}$$

where $dA$ is the area of the inclined face from figure 3.8, we have

$$\tfrac{1}{2}dy\, dz = dA\cos(N, x) = l\, dA$$

$$\tfrac{1}{2}dx\, dz = dA\cos(N, y) = m\, dA$$

$$\tfrac{1}{2}dx\, dy = dA\cos(N, z) = n\, dA$$

where $\cos(N, x)$ is the cosine of the angle between the normal ($N$) axis and the $x$ axis,

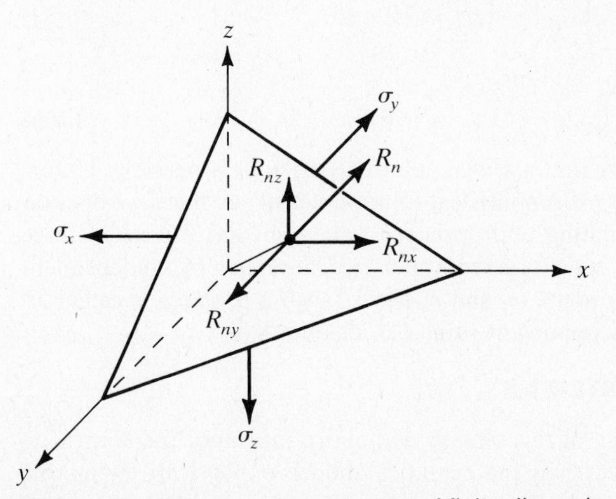


**Figure 3.8** Tetrahedron of an element of finite dimensions.

etc. Consequently, substituting into (3.18) yields

$$R_{nx} = \sigma_x l + \tau_{yx} m + \tau_{zx} n \tag{3.19a}$$

$$R_{ny} = \tau_{xy} l + \sigma_y m + \tau_{zy} n \tag{3.19b}$$

$$R_{nz} = \tau_{xz} l + \tau_{yz} m + \sigma_z n \tag{3.19c}$$

Equation (3.19) is a mathematical model which allows us to compute the resultant stress acting on any plane cutting through a rigid body.

Of particular interest to mechanical engineers is a plane on which there are no in-plane ("shearing") stresses. This plane is normally referred to as the "principal" plane and the normal stress acting on it is called the principal stress. Principal stresses play an important role in failure theories. For example, the Von Mises, Tresca, and Coulomb theories are expressed in terms of the principal stresses acting on a rigid body. Therefore, assuming that the inclined plane of figure 3.8 is a principal plane, we determine the components of the principal stress $\sigma$ acting on it as follows:

$$R_{nx} = \sigma l$$

$$R_{ny} = \sigma m$$

$$R_{nz} = \sigma n$$

Note that the direction of $\sigma$ is the normal axis $N$; substituting into (3.19) yields

$$\sigma l = \sigma_x l + \tau_{yx} m + \tau_{zx} n$$

$$\sigma m = \tau_{xy} l + \sigma_y m + \tau_{zy} n$$

$$\sigma n = \tau_{xz} l + \tau_{yz} m + \sigma_z n$$

These equations are now expressed in matrix form as follows.

$$\begin{bmatrix} \sigma_x - \sigma & \tau_{yx} & \tau_{zx} \\ \tau_{xy} & \sigma_y - \sigma & \tau_{zy} \\ \tau_{xz} & \tau_{yz} & \sigma_z - \sigma \end{bmatrix} \begin{Bmatrix} l \\ m \\ n \end{Bmatrix} = \begin{Bmatrix} 0 \\ 0 \\ 0 \end{Bmatrix} \tag{3.20a}$$

or in compact matrix form,

$$[T]\{x\} = \{0\} \tag{3.20b}$$

Equation (3.20) clearly shows that a trivial solution ($\{x\} = 0$) is possible. Unfortunately, this is not what we seek. A nontrivial solution is also possible if we operate on the $[T]$ matrix by calculating appropriate $\sigma$ values so that the set of three independent equations given in (3.20) is reduced to not more than two independent equations in the three unknowns $l$, $m$, and $n$. Again, such a problem is called an eigenproblem. (More on this important subject in chapter 8.)

## 3.4 CIVIL ENGINEERING SYSTEMS

As stated earlier, the purpose of this chapter is only to introduce the student to the processes involved in developing mathematical models of physical systems and to illustrate the importance of numerical solutions. Therefore it is by no means a complete treatment of this important and broad subject. With this caveat in mind, let us now consider some examples of mathematical models for certain civil engineering systems.

### 3.4.1 Boundary-Value Problems

Often civil engineers are faced with problems involving mathematical models in the form of ordinary differential equations. These equations describe a particular system subjected to specific physical constraints called boundary conditions. For example, the differential equation giving the vertical displacement of a beam of given length and physical properties when subjected to an arbitrary external load is of the form

$$EI\frac{d^4y}{dx^4} = -q(x) \tag{3.21}$$

where

$E$ = Young's modulus of elasticity,

$I$ = moment of inertia of the beam cross section,

$q(x)$ = load distribution along the beam.

Note that $E$ and $I$ are both intrinsic material properties that are assumed to be constants. The boundary conditions pertaining to beam deflection are summarized in table 3.1.

The ordinary differential equation given by (3.21) holds regardless of the boundary conditions. However, its solution depends on the type of constraints at the beam's ends (i.e., the boundary conditions).

Consider the cantilever beam shown in figure 3.9, which is subject to a uniformly distributed load. The deflection $y$ can be determined at any point $x$ $(0 \le x \le L)$ in terms of the applied external effects $q_0$ and the beam's intrinsic

Table 3.1 BOUNDARY CONDITIONS PERTAINING TO BEAM DEFLECTION

| TYPE | SYMBOL | CONDITIONS | EXPLANATION |
|---|---|---|---|
| Fixed end | | $y = 0$ | Deflection is zero |
| | | $\frac{dy}{dx} = 0$ | Slope is zero |
| Hinged end | | $y = 0$ | Deflection is zero |
| | | $EI\frac{d^2y}{dx^2} = 0$ | Moment is zero |
| Free end | | $EI\frac{d^2y}{dx^2} = 0$ | Moment is zero |
| | | $EI\frac{d^3y}{dx^3} = 0$ | Shear is zero |

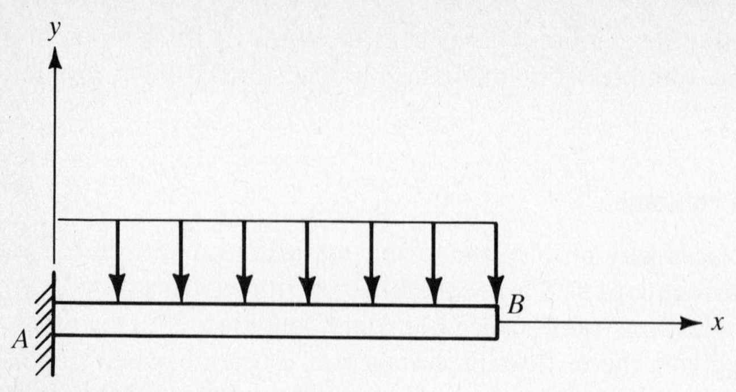


**Figure 3.9** A cantilever beam.

physical properties $EI$ by integrating (3.21) four times. That is,

$$EI\frac{d^3y}{dx^3} = -q_0x + C_1 \tag{3.22a}$$

$$EI\frac{d^2y}{dx^2} = -q_0\frac{x^2}{2} + C_1x + C_2 \tag{3.22b}$$

$$EI\frac{dy}{dx} = -q_0\frac{x^3}{6} = C_1\frac{x^2}{2} + C_2x + C_3 \tag{3.22c}$$

$$EIy = -q_0\frac{x^4}{24} + C_1\frac{x^3}{6} + C_2\frac{x^2}{2} + C_3x + C_4 \tag{3.22d}$$

The boundary conditions given in table 3.1 at the two ends are

At $x = 0$:

(1) $y = 0$

(2) $\dfrac{dy}{dx} = 0$

At $x = L$:

(3) $\dfrac{d^2y}{dx^2} = 0$

(4) $\dfrac{d^3y}{dx^3} = 0$

Substituting the four conditions into the four equations given in (3.22) permits the determination of the four constants of integration. Hence

(1) $0 = C_4$

(2) $0 = C_3$

(3) $0 = -q_0L^2/2 + C_1L + C_2$

(4) $0 = -q_0L + C_1$

These equations are easily solved for the unknowns to give

$$C_1 = q_0 L$$

$$C_2 = -q_0 L^2/2$$

$$C_3 = 0$$

$$C_4 = 0$$

The deflection is then given by

$$EIy = -q_0 x^4/24 + q_0 L x^3/6 - q_0 L^2 x^2/4 \quad \mathbf{(3.23)}$$

An important question that can be raised at this point is, "What if $q$ is not a continuous function?" We can then obtain the deflection by using numerical integration methods. This topic is covered in chapter 13.

### 3.4.2 Eigenproblems

Civil engineers are sometimes faced with the problem of how to determine the magnitude of an axial force that a column can safely carry before it buckles. The mathematical model used in solving this problem for a column of constant cross section is given by

$$\frac{d^4x}{dy^4} + K^2 \frac{d^2x}{dy^2} = 0 \quad \mathbf{(3.24)}$$

where

$K^2 = P/EI$,

$P$ = axial force,

$E$ = Young's modulus,

$I$ = moment of inertia.

Consider the special case of a column fixed at both ends as shown in figure 3.10.

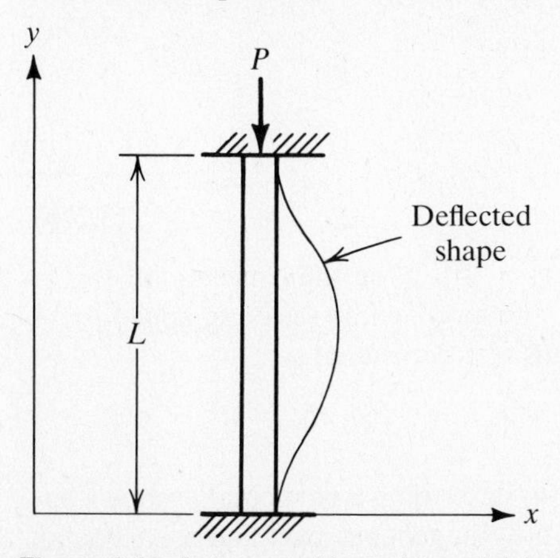


**Figure 3.10** Column subjected to axial force.

The critical load $P$ can be determined by integrating (3.10) in closed form to give

$$x = A_1 \sin ky + A_2 \cos ky + A_3 y/L + A_4 \qquad \textbf{(3.25)}$$

where $A_1$, $A_2$, $A_3$, and $A_4$ are constants of integration that depend on the type of boundary contraints at the column's ends. Therefore, using table 3.1 we have

At $x = 0$:

(1) $x = 0$

(2) $\dfrac{dx}{dy} = 0$

At $x = L$:

(3) $x = 0$

(4) $\dfrac{dx}{dy} = 0$

It is evident that we need to determine the first derivative of (3.25). Thus

$$\frac{dx}{dy} = A_1 k \cos ky - A_2 k \sin ky + \frac{A_3}{L} \qquad \textbf{(3.26)}$$

Substituting the first and third boundary conditions into (3.25) and the second and fourth boundary conditions into (3.26) yields

(1) $0 = A_2 + A_4$

(2) $0 = A_1 \sin kL + A_2 \cos kL + A_3 + A_4$

(3) $0 = A_1 k + A_3/L$

(4) $0 = A_1 k \cos kL - A_2 k \sin kL + A_3/L$

Letting $z = kL$ and expressing these homogeneous equations in a matrix form yields

$$\begin{bmatrix} 0 & 1 & 0 & 1 \\ \sin z & \cos z & 1 & 1 \\ z & 0 & 1 & 0 \\ z \cos z & -z \sin z & 1 & 0 \end{bmatrix} \begin{Bmatrix} A_1 \\ A_2 \\ -A_3 \\ A_4 \end{Bmatrix} = \begin{Bmatrix} 0 \\ 0 \\ -0 \\ 0 \end{Bmatrix} \qquad \textbf{(3.27a)}$$

or in compact matrix form

$$[S]\{A\} = \{0\} \qquad \textbf{(3.27b)}$$

Once again equation (3.27) is satisfied if $\{A\} = \{0\}$. Note that once again this is an eigenproblem. That is, the $[S]$ matrix can be modified by selecting appropriate $z$ values so that the resulting set of equations is a dependent set.

## 3.5 ENGINEERING SYSTEM RESPONSE

The solution of ordinary differential equations describing a particular system's behavior is generally expressed as the sum of a homogeneous part and a particular part. That is,

$$X(t) = X_h(t) + X_p(t) \tag{3.28}$$

where

$X(t)$ = total solution as a function of time,

$X_h(t)$ = solution of the homogeneous differential equation as a function of time,

$X_p(t)$ = a particular solution of the nonhomogeneous differential equation as a function of time.

Furthermore, for a system modeled by a set of ordinary differential equations, the solution is given in matrix form as follows:

$$\{X(t)\} = \{X(t)\}_h + \{X(t)\}_p \tag{3.29}$$

Equation (3.28) may be used to represent the solution of the systems described in sections 3.2.3 and 3.3.3. The homogeneous solution depends on the initial conditions and the particular part is related to the external effects. Both of these parts depend on the system's intrinsic properties. In addition, an initial-value problem may involve a boundary-value problem as well. For example, consider the one-story building shown in figure 3.11.

The free-body diagram for this structure is

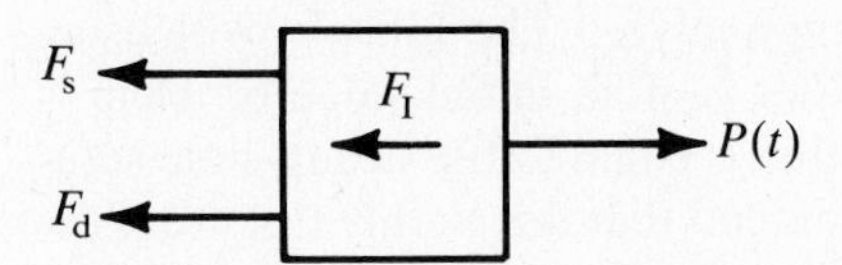


where $F_s$ is the force representing the stiffness of the columns, $F_d$ is the force representing the energy dissipation mechanism of the structure, and $F_I$ is the inertial force resulting from the mass resistance to acceleration. The stiffness is determined from a purely static analysis of the frame and its boundary conditions at A and B. The inertial force is related to the weight of the frame. Note that the columns and/or walls are assumed to be weightless. Finally, the damping force cannot be directly

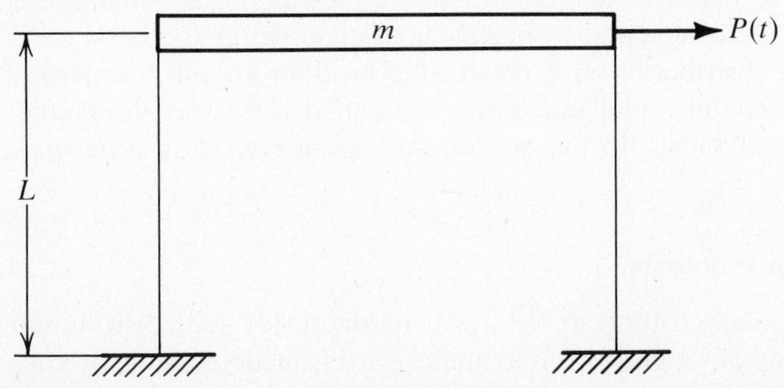


**Figure 3.11** A one-story building subjected to an external dynamic force.

measured, yet we know it exists. It is evident that if $P(t) = 0$ and only initial conditions, on say displacement, are introduced, then the frame is expected to vibrate but ultimately come to rest at its neutral position. The frame stops vibrating after a period of time as a result of the presence of damping forces resisting its motion. Consequently, the frame can be expected to continue vibrating indefinitely if no damping force is present!

The general solution can therefore be expressed more appropriately in terms of the system's damping characteristics as given by

$$X(t) = X_{ss}(t) + X_{tr}(t) \tag{3.30}$$

where

$X(t)$ = total solution as defined earlier,

$X_{ss}(t)$ = steady-state response as a function of time,

$X_{tr}(t)$ = transient response as a function of time.

Note that in general $X_h(t) \neq X_{ss}(t)$ and $X_p(t) \neq X_{tr}(t)$. Also, the transient response is simply equal to that part of the solution that vanishes as time approaches infinity. Thus

$$\lim_{t \to \infty} X_{tr}(t) = 0$$

This is why sometimes the terms "steady-state" problems and "transient" problems are used to refer to boundary-value and initial-value problems. The implication is that the dominant part of a solution—whether it is the steady-state or the transient part—determines the type of problem being analyzed. The difficulty with using this kind of terminology is rather obvious. That is, if an initial-value problem is to be analyzed, we may have a total solution that is equal to the steady-state solution (if damping is zero). However, one may contend that this is only true from the purely mathematical point of view since damping for physical systems is generally not equal to zero. With that in mind, it is a matter of preference as to which terminology is used. In this text, we prefer to use the type of physical constraints and/or initial values to classify a model rather than to use its response. The solution is, on the other hand, classified according to damping characteristics.

## 3.6 MODELS INVOLVING PARTIAL DIFFERENTIAL EQUATIONS

Partial differential equations are often used to model many important engineering systems. The distribution of temperature in a plate, the dissipation of ground water pressure, and voltage distribution in a conducting medium are all examples of systems involving partial differential equations. These models may involve boundary values and/or initial values. In this section only the temperature problem is formulated.

### 3.6.1 Boundary-Value Problems

The development of a mathematical model involving the steady-state distribution of temperatures throughout a body of finite dimensions is made possible by considering an infinitesimal element within the body. Consider figure 3.12.

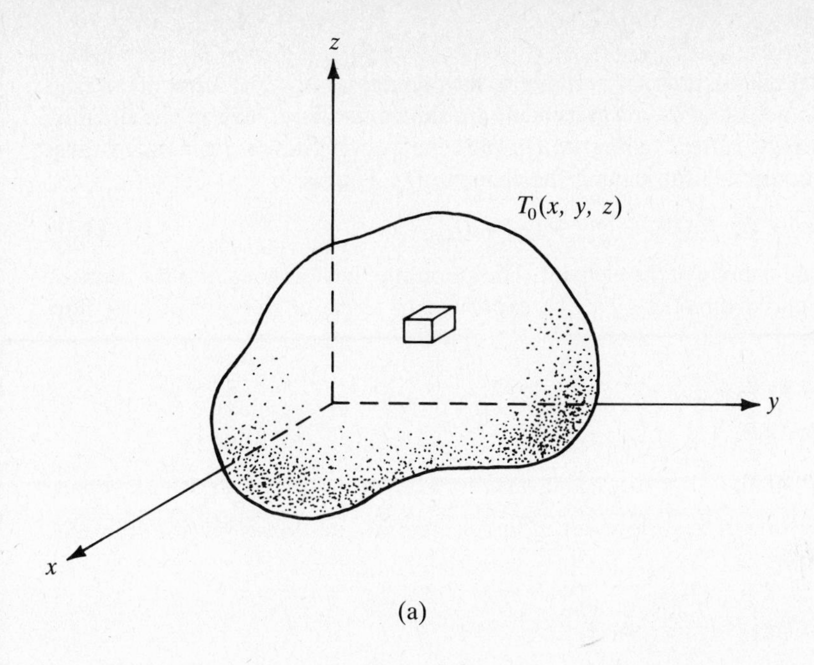


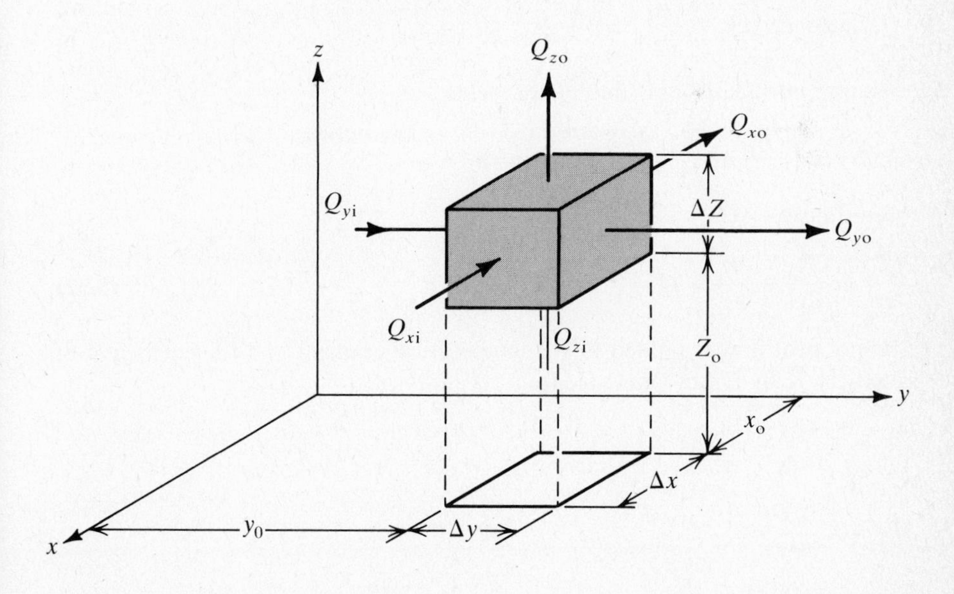


**Figure 3.12** Three-dimensional analysis of steady-state temperature distribution.

The body is subjected to boundary temperatures $T_0(x, y, z)$ as shown in figure 3.12a and, in general, internal volumetric heat generation $Q$. The amounts of heat flowing into and out of a typical element are shown in figure 3.12b. Consequently, the total amount of heat going in, $Q_i$, plus the volumetric heat generated must equal the amount of heat leaving the element, $Q_o$. That is,

$$Q_{xi} + Q_{yi} + Q_{zi} = QV_0 + Q_{xo} + Q_{yo} + Q_{zo} \tag{3.31}$$

where $V_0$ is the volume of the element. The amount of heat flowing into the element in the $x$, $y$, and $z$ directions can be expressed in terms of the rate of heat flow $R$ as follows:

$$Q_{xi} = \Delta z\,\Delta y\,R_x$$

$$Q_{yi} = \Delta z\,\Delta x\,R_y$$

$$Q_{zi} = \Delta x\,\Delta y\,R_z$$

Similarly, the rate of heat flow out of the element in question is given in the $x$, $y$, and $z$ directions as

$$Q_{xo} = \Delta z\,\Delta y\left(R_x + \frac{\partial R_x}{\partial x}\Delta x\right)$$

$$Q_{yo} = \Delta z\,\Delta x\left(R_y + \frac{\partial R_y}{\partial y}\Delta y\right)$$

$$Q_{zo} = \Delta x\,\Delta y\left(R_z + \frac{\partial R_z}{\partial z}\Delta z\right)$$

Substituting into (3.31) and simplifying yields

$$\Delta x\,\Delta y\,\Delta z\left(\frac{\partial R_x}{\partial x} + \frac{\partial R_y}{\partial y} + \frac{\partial R_z}{\partial z} + Q\right) = 0$$

or more simply

$$\frac{\partial R_x}{\partial x} + \frac{\partial R_y}{\partial y} + \frac{\partial R_z}{\partial z} + Q = 0 \tag{3.32}$$

The rate of heat flow is related to the temperature gradient and the coefficient of conductivity $K$ by Darcy's law. Hence

$$R_x = K_x\frac{\partial T}{\partial x}$$

$$R_y = K_y\frac{\partial T}{\partial y}$$

$$R_z = K_z\frac{\partial T}{\partial z}$$

where $K_x$, $K_y$, and $K_z$ are the coefficients of conductivity in the $x$, $y$, and $z$ directions, respectively, and $T$ is the temperature. Therefore, substituting into (3.32) gives

$$\frac{\partial}{\partial x}\left(K_x \frac{\partial T}{\partial x}\right) + \frac{\partial}{\partial y}\left(K_y \frac{\partial T}{\partial y}\right) + \frac{\partial}{\partial z}\left(K_z \frac{\partial T}{\partial z}\right) + Q = 0 \tag{3.33}$$

This equation may be simplified, if we assume that the body is homogeneous, to give

$$K_x \frac{\partial^2 T}{\partial x^2} + K_y \frac{\partial^2 T}{\partial y^2} + K_z \frac{\partial^2 T}{\partial z^2} + Q = 0 \tag{3.34}$$

Furthermore, if the conducting body is isotropic then $K_x = K_y = K_z$. This allows us to simplify (3.34) even further:

$$\frac{\partial^2 T}{\partial x^2} + \frac{\partial^2 T}{\partial y^2} + \frac{\partial^2 T}{\partial z^2} + \frac{1}{K} Q = 0 \tag{3.35}$$

For the case in which $Q = 0$, equation (3.35) reduces to the well-known Laplace equation. Note that while we are able to model accurately, simplifying assumptions are often introduced to help us find a closed-form solution. These assumptions need not be made when numerical methods are used. In fact, closed-form solutions are sometimes beyond reach when irregular boundary conditions are specified despite the simplifying assumptions.

### 3.6.2 Initial-Value Problems

The determination of the time-dependent distribution of temperature throughout a conductive medium is modeled by considering a term relating to the rate at which heat is being stored in or removed from a given element. For homogeneous anisotropic materials, equation (3.35) is modified to give

$$K_x \frac{\partial^2 T}{\partial x^2} + K_y \frac{\partial^2 T}{\partial y^2} + K_z \frac{\partial^2 T}{\partial z^2} + Q = C\rho \frac{\partial T}{\partial t} \tag{3.36}$$

where $C$ is the heat capacity and $\rho$ is the density of the material. Consequently, for homogeneous and isotropic material, further simplification of (3.36) is attained by letting $K_x = K_y = K_z = K$. Thus

$$\frac{\partial^2 T}{\partial x^2} + \frac{\partial^2 T}{\partial y^2} + \frac{\partial^2 T}{\partial z^2} + \frac{Q}{K} = \frac{C\rho}{K} \frac{\partial^2 T}{\partial t} \tag{3.37}$$

Obviously, if no heat is being generated within the element under consideration then $Q = 0$. Hence the model is simplified further to give

$$\frac{\partial^2 T}{\partial x^2} + \frac{\partial^2 T}{\partial y^2} + \frac{\partial^2 T}{\partial z^2} = \frac{C\rho}{K} \frac{\partial T}{\partial t} \tag{3.38}$$

When the medium being analyzed can be sufficiently modeled as a two-dimensional problem

$$\frac{\partial^2 T}{\partial x^2} + \frac{\partial^2 T}{\partial y^2} = \frac{C\rho}{K} \frac{\partial T}{\partial t} \tag{3.39}$$

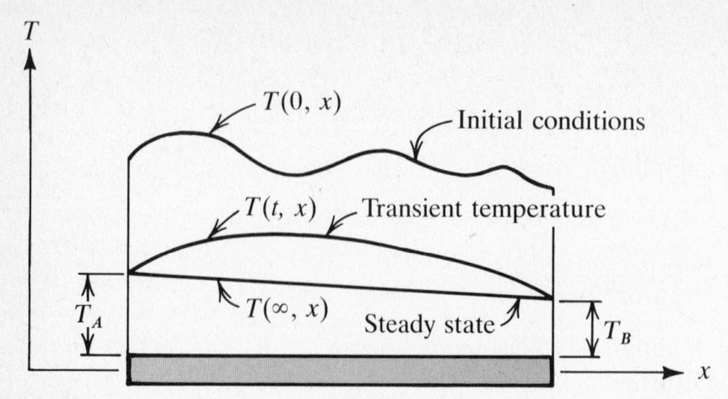


**Figure 3.13** Temperature distribution in a rod.

Furthermore, for one-dimensional problems the model becomes

$$\frac{\partial^2 T}{\partial x^2} = \frac{C\rho}{K}\frac{\partial T}{\partial t} \tag{3.40}$$

In this text, the solution of one- and two-dimensional models is considered. To fully appreciate the problem at hand, and to gain insight into the constraints we have dealt with, a schematic representation of the one-dimensional model is shown in figure 3.13.

The problem we are trying to analyze involves a rod of finite length $L$ whose end-point temperatures (boundary conditions) are specified by $T_A$ and $T_B$; the initial temperatures along its length (initial conditions) are given by discrete temperature values or by a continuous function $T(0, x)$. The question is how does $T(0, x)$ change with time so that at $t = \infty$ it becomes $T(\infty, x)$? It is evident that the solution depends on both the initial and boundary conditions. The steady-state solution is obtained by solving

$$\frac{\partial^2 T}{\partial x^2} = 0$$

and specifying $T_A$ and $T_B$, whereas the transient solution is obtained by specifying $T(0, x)$ and solving (3.40). The total solution is then given by

$$T(t, x) = T_{ss} + T_{tr}$$

Obviously as $t \to \infty$ the transient part goes to zero and the solution reduces to the steady-state part.

## 3.7 COMPARISON OF ENGINEERING MODELS

The purpose of this section is to reemphasize that while actual physical engineering systems can be very different, their behavior can be predicted by using the same mathematical procedures. Earlier we showed that a boundary-value problem may involve a set of linear algebraic equations; whereas, an initial-value problem may, in general, involve a set of ordinary differential equations. In addition, an eigenvalue problem involves solving a set of homogeneous linear algebraic equations whose coefficient matrix is not a scalar matrix. Models have been developed for electrical, mechanical, and civil engineering systems. However, the final form in each case is of the same general form regardless of the particulars of the system being analyzed. The conclusion to be drawn is that methods which are developed for solving a class of problems are equally applicable to all disciplines. This is an important observation in that it allows us to concentrate in this text on methods rather than disciplines. In fact, this is precisely the approach we pursue throughout this textbook. Furthermore, reference is made to particular applications so that the student can appreciate the physics and therefore have a better understanding of the problem at hand.

A complete engineering education should include both methods and their applications. However, we believe it will minimize confusion for the student if the study of methods for solving models can be separated from the development of the models for specific applications. That is why we want the student to know how to solve a set of linear algebraic equations, find eigenvalues, solve ordinary differential equations, and solve partial differential equations. On the other hand, we do not want the student to get lost in the specifics of a given method and confuse them with the particulars of a given application.

Consider the eigenproblems developed for the electrical, mechanical, and civil engineering systems as given by equations (3.13), (3.20), and (3.27), respectively. These can be considered in a general form as

$$[B(s)]\{X\} = \{0\} \tag{3.41}$$

where $[B(s)]$ is a matrix whose elements are functions of the variable $s$ and $\{X\}$ is a vector other than the null vector that satisfies the equality in equation (3.41). Therefore the general solution of a problem of the form given by (3.41) should be applicable to all three systems. This analogy can be extended to all of the other systems modeled thus far, including those involving partial differential equations. In fact, equation (3.35) may be used to model steady-state fluid flow through porous media by simply exchanging the total head $h$ for the temperature $T$. Similarly, by exchanging the voltage $V$ with $T$ we can model the steady-state voltage $m$ in a conductive media with the same equation.

## Recommended Reading

*Elementary Differential Equations*, Earl D. Rainville and Phillip E. Bedient, Macmillan Publishing Company, New York, 1969.

*Vector Mechanics for Engineers—Dynamics*, Ferdinand P. Beer and E. Russell Johnston, Jr., McGraw-Hill Book Company, New York, 1977.

*Signals and Linear Systems*, Robert A. Gabel and Richard A. Roberts, John Wiley and Sons, Inc., New York, 1980.

*Mechanics of Materials*, Nelson R. Bauld, Jr., Brooks/Cole Publishing Company, Monterey, California, 1982.

*Advanced Dynamics for Engineers*, Bruce J. Torby, Holt, Rienhart and Winston, New York, 1984.

## PROBLEMS

**3.1** Formulate a system of linear algebraic equations whose solutions give the currents in the three loops of the following circuit in terms of input and output voltages. Express your results in matrix form.

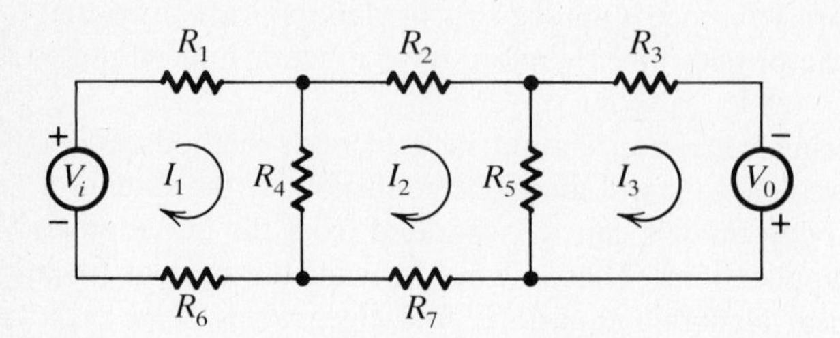


**3.2** Formulate a set of two second-order differential equations for determining the currents in the loops of the following electrical circuit. Express your results in matrix form.

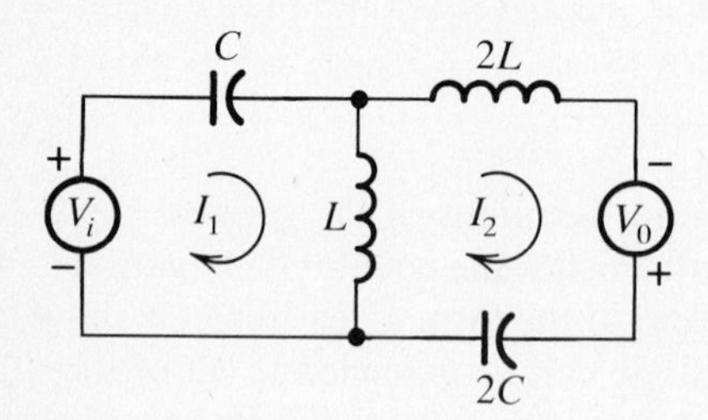


**3.3** Develop the transient response equations for the following circuit. Express your answer in matrix form.

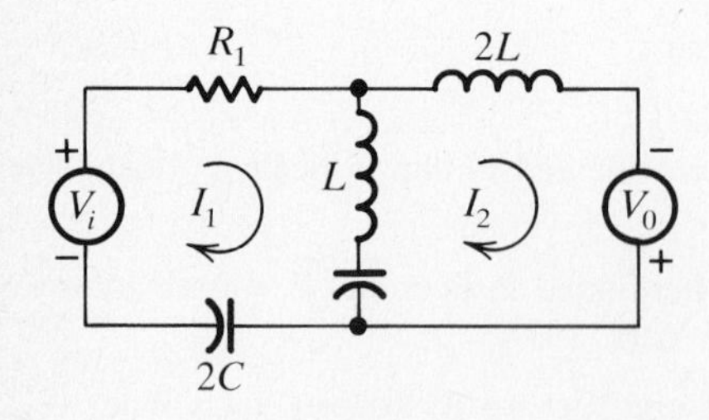


**3.4** Formulate a set of linear algebraic equations whose solution gives the displacement of the masses of the following mechanical system due to the application of the load $P$.

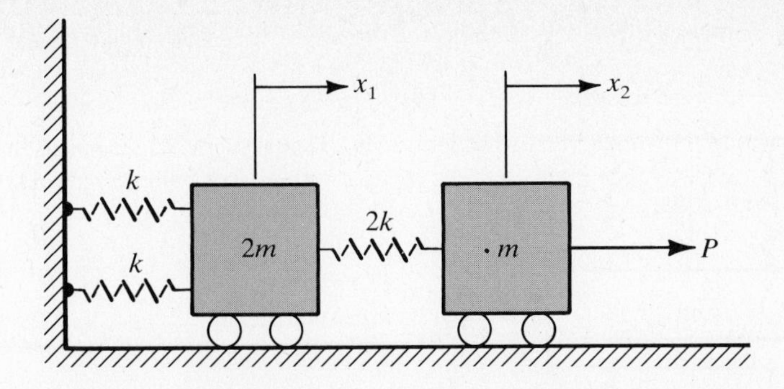


Note that $P$ is applied statically and that the rollers are frictionless.

**3.5** Formulate the equations needed to determine the steady-state displacement of the following system. Assume that the load $P$ is applied statically.

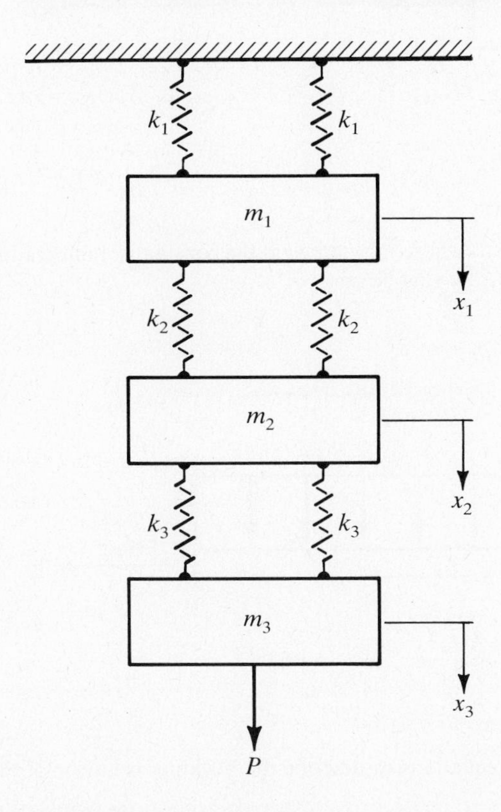


Give your results in matrix form.

**3.6** If the load $P$ in problem 3.4 is dynamic in nature, determine the set of differential equations whose solution gives the displacements $x_1$ and $x_2$. Express your results in matrix form.

**3.7** Determine the dynamic equilibrium equations of the transient response for the following system.

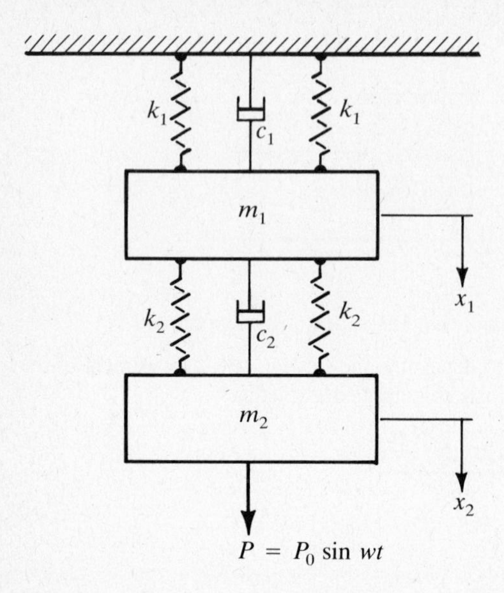


Express your results in matrix form.

**3.8** Set up the linear algebraic equations whose solution gives the constants of integration of the deflection equation for the following beam:

At $x = 0$:

(1) $\dfrac{d^2y}{dx^2} = 0$

(2) $y = 0$

At $x = L$:

(3) $\dfrac{dy}{dx} = 0$

(4) $y = 0$

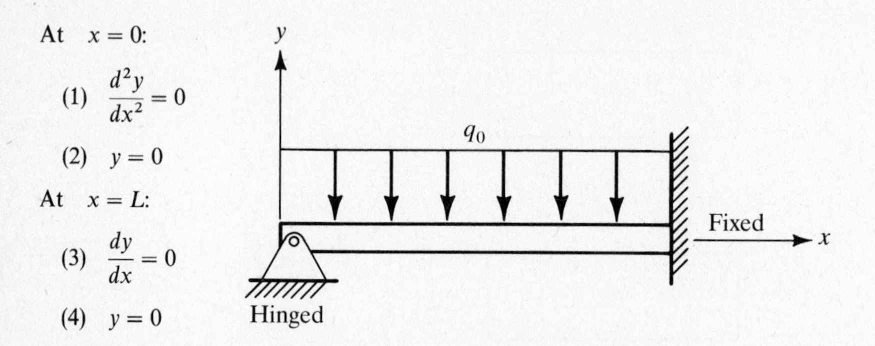


Express your results in matrix form.

**3.9** Develop a set of linear algebraic equations to describe the buckling behavior of the following column.

At $y = 0$:

(1) $x_2 = 0$

(2) $\dfrac{d^2x}{dy^2} = 0$

At $y = L$:

(3) $x = 0$

(4) $\dfrac{dx}{dy} = 0$

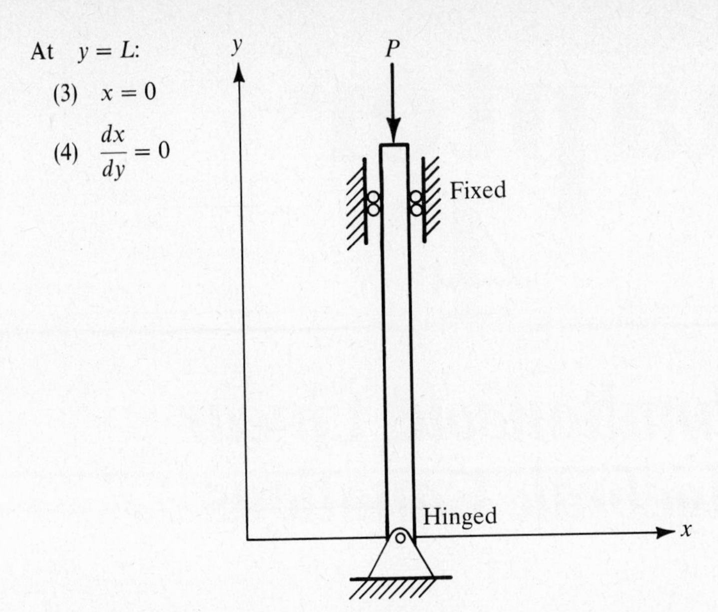


**3.10** Formulate the equations needed to determine the stability of the following column.

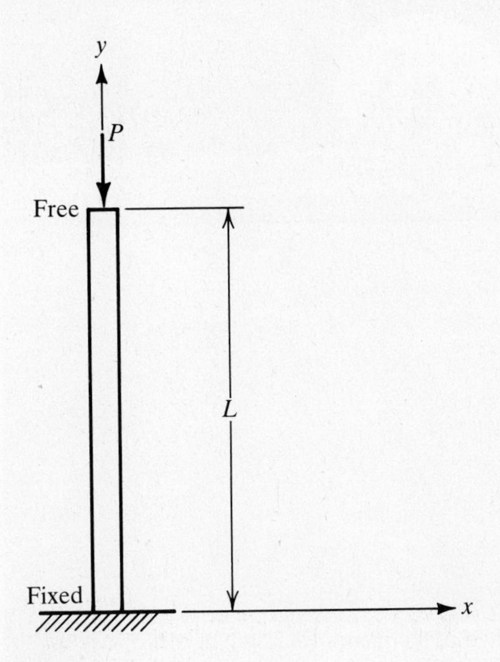


(*Note:* The boundary conditions at $y = L$ are $d^2x/dy^2$ and $d^3x/dy^3 + K^2\,dx/dy = 0$.)

# chapter 4

## Simultaneous Linear Algebraic Equations

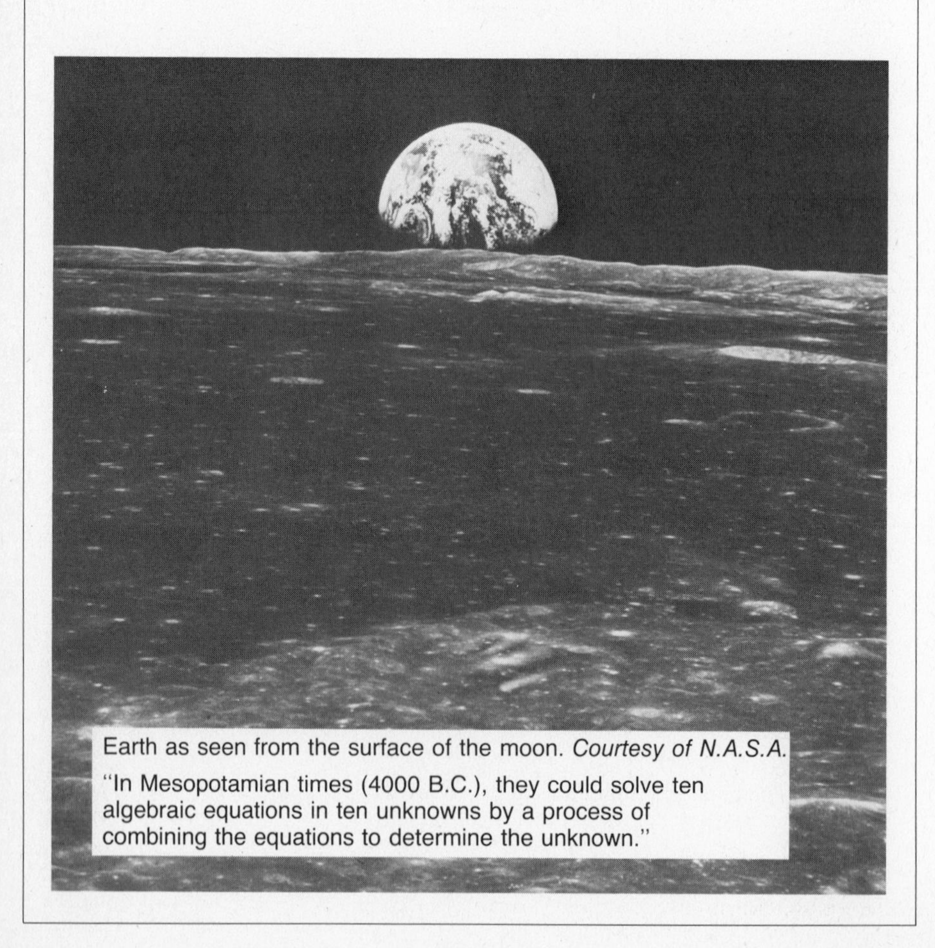

Earth as seen from the surface of the moon. *Courtesy of N.A.S.A.*

"In Mesopotamian times (4000 B.C.), they could solve ten algebraic equations in ten unknowns by a process of combining the equations to determine the unknown."

# 4

## 4.1 INTRODUCTION

The solution of a system of simultaneous linear algebraic equations is probably one of the most important topics in modern engineering computations. This is not an exaggeration if you consider that recent technological advances were made possible by our ability to solve larger and larger systems of equations. The finite element method; the finite difference method; the analysis of structural, mechanical, and electrical systems all involve the solution of systems of linear algebraic equations.

In the past, numerical methods which were used to solve systems of linear equations were not attractive for engineering use owing to the tremendous amount of calculation involved. However, computers have changed all that and altered our approach to engineering problem solving.

Most engineering students have, in one course or another, been introduced to one or more methods of solving a system of linear equations. The principal objectives of this chapter are to introduce various solution methods and to outline the advantages and limitations of each.

As a matter of definition, a "linear equation" is one in which a variable only appears to the first power in every term of a given equation. Thus a system of $m$ linear equations in $n$ unknowns $x_j$, $j = 1, 2, \ldots, n$, can be represented as follows:

$$\sum_{j=1}^{n} a_{ij} x_j = b_i, \qquad i = 1, 2, \ldots, m$$

In the expanded form, the equations are given as follows:

$$\begin{aligned} a_{11}x_1 + a_{12}x_2 + \cdots + a_{1n}x_n &= b_1 \\ a_{21}x_1 + a_{22}x_2 + \cdots + a_{2n}x_n &= b_2 \\ \vdots \qquad \vdots \qquad\qquad \vdots \quad &\quad \vdots \\ a_{m1}x_1 + a_{m2}x_2 + \cdots + a_{mn}x_n &= b_m \end{aligned} \tag{4.1}$$

Using the methods of matrix algebra, these equations can be yet put in a different form:

$$[A]_{m \times n}\{X\}_n = \{B\}_m \tag{4.2a}$$

or more explicitly

$$[a_{ij}]\{x_j\} = \{b_i\}, \qquad \begin{cases} i = 1, 2, \ldots, m \\ j = 1, 2, \ldots, n \end{cases} \tag{4.2b}$$

Whether in the form of equation (4.1) or (4.2), the problem we seek to solve is to find the values of the $n$ unknowns $x_j$, given values for $a_{ij}$ and $b_i$. In engineering, one may be faced with a variety of systems of equations. They are

1. A set of equations in which the number of unknowns is equal to the number of equations ($n = m$).
2. A set of equations in which the number of unknowns is less than the number of equations ($n < m$).
3. A set of equations in which the number of unknowns is greater than the number of equations ($n > m$).

Most engineering problems fall into the first category ($m = n$). That is, the problem is deterministic and has a unique solution, provided the determinant of the matrix $[A]$ is not zero. This is a necessary and sufficient condition for ensuring the existence of a unique solution. This class of problems may yet be subdivided into *homogeneous*—if the right-hand side of equations (4.2) is null ($\{b\} = \{0\}$), or *nonhomogeneous*—if the right-hand side is not zero ($\{b\} \neq \{0\}$). The eigenvalue problem involves the solution of homogeneous equations. Numerical solution of differential equations involves nonhomogeneous equations.

The category described in 2 may occur in problems involving experimental data. Then, for example, the least-squares procedure may be employed in getting a best solution because no unique solution otherwise exists.

The last category, given in 3 refers to problems that have an infinite number of solutions. The eigenvalue problem and many optimization problems fall into this category.

## 4.2 CRAMER'S RULE

This method is one of the least efficient for solving a large number of nonhomogeneous linear equations. It is, however, very useful for explaining some problems inherent in the solution of linear algebraic equations. The basis for this method may be explained by considering the following example of two nonhomogeneous equations:

$$\begin{bmatrix} a_{11} & a_{12} \\ a_{21} & a_{22} \end{bmatrix} \begin{Bmatrix} x_1 \\ x_2 \end{Bmatrix} = \begin{Bmatrix} b_1 \\ b_2 \end{Bmatrix} \tag{4.3}$$

Premultiplying equation (4.3) by the transpose matrix of the cofactors of the coefficient matrix gives

$$\begin{bmatrix} a_{22} & -a_{12} \\ -a_{21} & a_{11} \end{bmatrix} \begin{bmatrix} a_{11} & a_{12} \\ a_{21} & a_{22} \end{bmatrix} \begin{Bmatrix} x_1 \\ x_2 \end{Bmatrix} = \begin{bmatrix} a_{22} & -a_{12} \\ -a_{21} & a_{11} \end{bmatrix} \begin{Bmatrix} b_1 \\ b_2 \end{Bmatrix}$$

which simplifies to

$$\begin{bmatrix} a_{11}a_{22} - a_{12}a_{21} & 0 \\ 0 & a_{11}a_{22} - a_{12}a_{21} \end{bmatrix} \begin{Bmatrix} x_1 \\ x_2 \end{Bmatrix} = \begin{Bmatrix} a_{22}b_1 - a_{12}b_2 \\ a_{11}b_2 - a_{21}b_1 \end{Bmatrix}$$

Solving for the unknowns gives

$$x_1 = \frac{a_{22}b_1 - a_{12}b_2}{a_{11}a_{22} - a_{12}a_{21}} = \frac{\begin{vmatrix} b_1 & a_{12} \\ b_2 & a_{22} \end{vmatrix}}{\begin{vmatrix} a_{11} & a_{12} \\ a_{21} & a_{22} \end{vmatrix}}$$

$$x_2 = \frac{a_{11}b_2 - a_{21}b_1}{a_{11}a_{22} - a_{12}a_{21}} = \frac{\begin{vmatrix} a_{11} & b_1 \\ a_{21} & b_2 \end{vmatrix}}{\begin{vmatrix} a_{11} & a_{12} \\ a_{21} & a_{22} \end{vmatrix}}$$

This procedure can be employed for solving a set of equations of $n \times n$. Therefore, in general, given

$$[a_{ij}]\{x_j\} = \{b_i\}, \qquad \begin{cases} i = 1, \ldots, n \\ j = 1, \ldots, n \end{cases}$$

the solution is obtained by evaluating a set of determinants using one of the procedures outlined earlier. Thus

$$x_j = \frac{\det[A_j]}{\det[A]} = \frac{|A_j|}{|A|}, \qquad j = 1, \ldots, n \tag{4.4}$$

where $|A|$ is the determinant of the coefficient matrix $[a_{ij}]$ and $|A_j|$ is the determinant of the coefficient matrix with its $j$th column replaced by the $\{b_i\}$ vector. Note that while most textbooks on numerical methods use Laplace's method for expanding the determinants when using Cramer's rule this need not be the case. Both the pivotal condensation and elimination methods are more efficient for evaluating the determinants. The inefficiency of Cramer's rule is related in part to the method used. Clearly, a solution exists if, and only if, the determinant of the coefficient matrix $[A]$ is not zero. If $\det[A] = 0$, then the $[A]$ matrix is called singular and no solution can be defined to the set. In fact, the solution of a system of linear algebraic equations may be in jeopardy whenever the value of $|A|$ is close to zero. Such a system is called "ill-conditioned."

E

EXAMPLE 4.1

Use Cramer's rule to solve the following system of equations:

$$2x_1 - 3x_2 = 5$$

$$x_1 + x_2 = 5$$

Solution

Forming the various matrices

$$[A] = \begin{bmatrix} 2 & -3 \\ 1 & 1 \end{bmatrix}, \qquad \{B\} = \begin{Bmatrix} 5 \\ 5 \end{Bmatrix}, \qquad [A_1] = \begin{bmatrix} 5 & -3 \\ 5 & 1 \end{bmatrix},$$

$$[A_2] = \begin{bmatrix} 2 & 5 \\ 1 & 5 \end{bmatrix}$$

then using equation (4.4) gives

$$x_1 = \frac{|A_1|}{|A|} = \frac{\begin{vmatrix} 5 & -3 \\ 5 & 1 \end{vmatrix}}{\begin{vmatrix} 2 & -3 \\ 1 & 1 \end{vmatrix}} = \frac{5 \times 1 - (5)(-3)}{2 \times 1 - (1)(-3)} = \frac{20}{5} = 4$$

$$x_2 = \frac{|A_2|}{|A|} = \frac{\begin{vmatrix} 2 & 5 \\ 1 & 5 \end{vmatrix}}{5} = \frac{2 \times 5 - 1(5)}{5} = 1$$

E

EXAMPLE 4.2

Use Cramer's rule to solve the following ill-conditioned systems:

$$\text{(a)} \quad \begin{aligned} 2x_1 + 2.1x_2 &= 5 \\ x_1 + \quad x_2 &= 5 \end{aligned} \qquad \text{(b)} \quad \begin{aligned} 2x_1 + 1.9x_2 &= 5 \\ x_1 + \quad x_2 &= 5 \end{aligned}$$

Solution

$$\text{(a)} \quad x_1 = \frac{\begin{vmatrix} 5 & 2.1 \\ 5 & 1 \end{vmatrix}}{\begin{vmatrix} 2 & 2.1 \\ 1 & 1 \end{vmatrix}} = \frac{5 - 10.5}{2 - 2.1} = 55$$

$$x_2 = \frac{\begin{vmatrix} 2 & 5 \\ 1 & 5 \end{vmatrix}}{-0.1} = \frac{5}{-0.1} = -50$$

(b) $$x_1 = \frac{\begin{vmatrix} 5 & 1.9 \\ 5 & 1 \end{vmatrix}}{\begin{vmatrix} 2 & 1.9 \\ 1 & 1 \end{vmatrix}} = \frac{5 - 9.5}{2 - 1.9} = -45$$

$$x_2 = \frac{\begin{vmatrix} 2 & 5 \\ 1 & 5 \end{vmatrix}}{0.1} = \frac{10 - 5}{0.1} = 50$$

Note that the value of $x_1$ has changed from 55 to $-45$ and that of $x_2$ from $-50$ to 50 as a result of the small change in $a_{12}$ from 2.1 to 1.9. It is important to note this sensitivity, especially when dealing with empirical and/or experimental values.

Example 4.2 clearly illustrates that the coefficient matrix $[A]$ must be determined accurately and that its determinant must be significantly larger than zero, if a reliable solution is to be found. Keep in mind that whenever $|A|$ is close to zero two or more of the equations are nearly equal. (Why?)

Methods involving successive elimination of the unknowns in a given set constitute the most direct and efficient technique of solving simultaneous linear algebraic equations. There are many variations of this technique. We restrict our discussion in this text to the more popular ones.

**SUMMARY OF CRAMER'S RULE**

| STEP | OPERATION | SYMBOL |
|---|---|---|
| 1 | Form the coefficient matrix $[a_{ij}]$ and the right-hand vector $\{b_i\}$. | $\begin{bmatrix} a_{11} & \cdots & a_{1n} \\ \vdots & & \vdots \\ a_{n1} & \cdots & a_{nn} \end{bmatrix}$ and $\begin{Bmatrix} b_1 \\ \vdots \\ b_n \end{Bmatrix}$ |
| 2 | Evaluate the determinant of the coefficient matrix $[a_{ij}]$ using the upper-triangle method or pivotal condensation. If $\det[A] = 0$, then the set has no solution; otherwise proceed to the next step. | $\det[A] = \begin{vmatrix} a_{11} & \cdots & a_{1n} \\ \vdots & & \vdots \\ a_{n1} & \cdots & a_{nn} \end{vmatrix}$ |

| | | |
|---|---|---|
| 3 | Evaluate the determinant of the matrix $[A_j]$ by replacing the $j$th matrix with the column vector $\{b_i\}$. | $\det[A_j] = \begin{vmatrix} a_{11} & \cdots & b_1 & \cdots & a_{1n} \\ a_{21} & \cdots & b_2 & \cdots & a_{2n} \\ \vdots & & \vdots & & \vdots \\ a_{n1} & \cdots & b_n & \cdots & a_{nn} \end{vmatrix}$ |
| 4 | Repeat step 3 for $j = 1,\ldots,n$. | |
| 5 | Solve for the unknowns $x_j$ by dividing $\det[A_j]$ by $\det[A]$. | $x_j = \dfrac{\det[A_j]}{\det[A]}$, $j = 1,\ldots,n$ |

## 4.3 GAUSS'S ELIMINATION METHOD

This is one of the most popular and efficient methods of solving an $n \times n$ system of equations; with the exception of the factoring method, no method of solving linear equations requires fewer operations than Gauss's procedure.

This method is relatively simple and straightforward. It consists of a series of operations to transform the original set with $m = n$ [equation (4.1)] to a new system of $n$ simultaneous equations in $n$ unknowns having a triangular form from which each unknown is determined by back-substitution. The process begins with the arrangement of the system of equations given by equation (4.1), in such a manner that $a_{11} = 1$, to give

$$\begin{aligned} a_{11}x_1 + a_{12}x_2 + \cdots + a_{1n}x_n &= b_1 \\ a_{21}x_1 + a_{22}x_2 + \cdots + a_{2n}x_n &= b_2 \\ \vdots \qquad \vdots \qquad\qquad \vdots \quad &\quad \vdots \\ a_{n1}x_1 + a_{n2}x_2 + \cdots + a_{nn}x_n &= b_n \end{aligned} \tag{4.5}$$

dividing the first equation in (4.5) by $(a_{11})$ gives

$$\begin{aligned} x_1 + a'_{12}x_2 + \cdots + a'_{1n}x_n &= b'_1 \\ a_{21}x_1 + a_{22}x_2 + \cdots + a_{2n}x_n &= b_2 \\ \vdots \qquad \vdots \qquad\qquad \vdots \quad &\quad \vdots \\ a_{n1}x_1 + a_{n2}x_2 + \cdots + a_{nn}x_n &= b_n \end{aligned} \tag{4.6}$$

Multiplying the first equation in (4.6) by $-a_{i1}$ for $i = 2,\ldots,n$ then adding to the $i$th equation eliminates $x_1$ from all but the first equation to give

$$\begin{aligned} x_1 + a'_{12}x_2 + \cdots + a'_{1n}x_n &= b'_1 \\ a'_{22}x_2 + \cdots + a'_{2n}x_n &= b'_2 \\ \vdots \qquad\qquad \vdots \quad &\quad \vdots \\ a'_{n2}x_2 + \cdots + a'_{nn}x_n &= b'_n \end{aligned} \tag{4.7}$$

The second equation in (4.7) is now divided by $a_{22}$ to give

$$\begin{aligned} x_1 + a'_{12}x_2 + \cdots + a'_{1n}x_n &= b'_1 \\ x_2 + \cdots + a''_{2n}x_n &= b''_2 \\ \vdots \qquad \qquad \vdots \quad &\quad \vdots \\ a'_{n2}x_2 + \cdots + a'_{nn}x_n &= b'_n \end{aligned} \tag{4.8}$$

Multiplying the second equation in (4.8) by $-a'_{i2}$ for $i = 3, \ldots, n$ then adding to the $i$th equation eliminates $x_2$ from all but the first and second equations. This process is continued until one equation in one unknown remains. Note that at each stage the remaining equations may require rearranging to avoid a zero divisor in the $a_{ii}$ position. Once the process is completed, the system of equations given in (4.5) should have the following triangular form:

$$\begin{aligned} x_1 + u_{12}x_2 + \cdots + u_{1n}x_n &= c_1 \\ x_2 + \cdots + u_{2n}x_n &= c_2 \\ &\ \vdots \\ x_n &= c_n \end{aligned} \tag{4.9}$$

The unknowns are then determined by back-substitution. That is,

$$\begin{aligned} x_n &= c_n \\ x_{n-1} &= c_{n-1} - u_{n-1,n}x_n \\ &\ \vdots \\ x_1 &= c_1 - u_{12}x_2 - \cdots - u_{1n}x_n \end{aligned} \tag{4.10}$$

So far as the mechanics of the method is concerned, one can carry out the solution process without the unknowns $x_1, \ldots, x_n$ or the equality signs. This is easily accomplished by using matrix algebra. From equation (4.5), the following augmented matrix is formed:

$$\left[\begin{array}{cccc|c} a_{11} & a_{12} & \cdots & a_{1n} & b_1 \\ a_{21} & a_{22} & \cdots & a_{2n} & b_2 \\ \vdots & \vdots & & \vdots & \vdots \\ a_{n1} & a_{n2} & \cdots & a_{nn} & b_n \end{array}\right] \tag{4.11}$$

The operations used in Gauss's method can now be applied to the augmented matrix. The advantages are fairly obvious in that there is no need for writing the $x$'s or equal signs. In addition, matrix algebra is more elegant and neater to use than the algebraic approach. Consequently equation (4.9) is now written directly as follows:

$$\left[\begin{array}{cccc|c} 1 & u_{12} & \cdots & u_{1n} & c_1 \\ & 1 & \cdots & u_{2n} & c_2 \\ & & & \vdots & \vdots \\ 0 & & & 1 & c_n \end{array}\right] \tag{4.12}$$

from which the unknowns are determined as before by using equation (4.10).

E

EXAMPLE 4.3

Use the Gauss elimination procedure to solve for the unknowns of the following set:

$$2x_2 + 4x_3 = 6$$
$$4x_1 + x_2 - 3x_3 = 1$$
$$3x_1 - 8x_2 + 2x_3 = 2$$

Solution

Using the augmented matrix approach and switching the second and first equations (why?) gives

$$\left[\begin{array}{ccc|c} 4 & 1 & -3 & 1 \\ 0 & 2 & 4 & 6 \\ 3 & -8 & 2 & 2 \end{array}\right]$$

We begin by dividing the first row by 4; thus

$$\left[\begin{array}{ccc|c} 1 & 0.25 & -0.75 & 0.25 \\ 0 & 2 & 4 & 6 \\ 3 & -8 & 2 & 2 \end{array}\right] \begin{array}{l} R_1/4 \\ \\ \\ \end{array}$$

Multiplying the first row by $-3$ and adding it to the third row eliminates $x_1$ from all equations except the first. Hence

$$\left[\begin{array}{ccc|c} 1 & 0.25 & -0.75 & 0.25 \\ 0 & 2 & 4 & 6 \\ 0 & -8.75 & 4.25 & 1.25 \end{array}\right] \begin{array}{l} \\ \\ -3R_1 + R_3 \end{array}$$

The second row is now divided by 2 to give

$$\left[\begin{array}{ccc|c} 1 & 0.25 & -0.75 & 0.25 \\ 0 & 1 & 2 & 3 \\ 0 & -8.75 & 4.25 & 1.25 \end{array}\right] \begin{array}{l} \\ R_2/2 \\ \\ \end{array}$$

Finally, multiplying the second row by 8.75 and adding to the third row eliminates $x_2$ from the third equation. That is,

$$\left[\begin{array}{ccc|c} 1 & 0.25 & -0.75 & 0.25 \\ 0 & 1 & 2 & 3 \\ 0 & 0 & 21.75 & 27.5 \end{array}\right] \begin{array}{l} \\ \\ 8.75R_2 + R_3 \end{array}$$

Obviously, the original set of equations has now been transformed to an upper-triangular form. Hence, dividing row 3 by 21.75 and expressing the set in algebraic form yields

$$x_1 + 0.25x_2 - 0.75x_3 = 0.25$$
$$x_2 + \quad 2x_3 = 3$$
$$x_3 = 1.264$$

These equations are now solved by back-substitution to give

$$x_3 = 1.264$$
$$x_2 = 3 - 2(1.264) = 0.472$$
$$x_1 = 0.25 - 0.25(0.472) + 0.75(1.264) = 1.080$$

Note that operations performed on the original augmented matrix may be combined so that the number of steps are reduced. To do that, we shall adopt the symbol $R_i$ for the $i$th row. Consequently, dividing row number one by 2 is written as $\frac{1}{2}R_1$. Similarly, adding $-3$ times row number two to the third row is written as $-3R_2 + R_3$.

**SUMMARY OF GAUSS'S METHOD**

| STEP | OPERATION | SYMBOL |
|---|---|---|
| 1 | Form the augmented matrix of the $[a_{ij}]$ matrix and $\{b_i\}$ vector. | $[a_{ij} \vdots b_i]$, $i = 1,\ldots,n$, $j = 1,\ldots,n$ |
| 2 | Check $a_{11}$; if it is equal to zero then interchange rows so that $a_{11} \neq 0$. | |
| 3 | Divide row one by $a_{11}$ to get new coefficients $a'_{ij}$ where $a_{11} = 1.0$. | $a'_{ij} = \dfrac{a_{ij}}{a_{11}}$ |
| 4 | Multiply row one by $-a_{i1}$ and add to the $i$th row for $i = 2,\ldots,n$. | $-a_{i1}R_1 + R_i$, $i = 2,\ldots,n$ |
| 5 | Repeat steps 2, 3, and 4 for the second through $(n-1)$th rows. | |
| 6 | Solve for $x_n$ from the $n$th equation. | $x_n = c_n$ |
| 7 | Solve for $x_{n-1},\ldots,x_1$. | $x_j = c_j - \sum_{r=j+1}^{n} u_{jr}x_r$ |

## 4.4 GAUSS–JORDAN ELIMINATION METHOD

This method is basically an extension of the Gauss method. The difference is that there is no need for back-substitution using this method. This is accomplished quite simply by eliminating the coefficients above the diagonal in equation (4.12). Consequently, given an augmented matrix of the form

$$\left[\begin{array}{cccc|c} a_{11} & a_{12} & \cdots & a_{1n} & b_1 \\ a_{21} & a_{22} & \cdots & a_{2n} & b_2 \\ \vdots & \vdots & & \vdots & \vdots \\ a_{n1} & a_{n2} & & a_{nn} & b_n \end{array}\right]$$

using the Gauss–Jordan method, we transform it into the following:

$$\left[\begin{array}{cccc|c} 1 & & & 0 & b_1^* \\ & 1 & & & b_2^* \\ & & \ddots & & \vdots \\ 0 & & & 1 & b_n^* \end{array}\right] \tag{4.13}$$

in which case the solution is readily obtained as

$$\begin{aligned} x_1 &= b_1^* \\ x_2 &= b_2^* \\ \vdots \quad & \quad \vdots \\ x_n &= b_n^* \end{aligned}$$

The student should be aware that this method requires more computational effort than the Gauss method. However, it is cleaner and provides a direct method for solving the inverse problem discussed in the next chapter. Note also that rows may need to be switched to avoid dividing by zero after each iteration.

**E** EXAMPLE 4.4

Solve the following set using the Gauss–Jordan method:

$$\begin{aligned} 2x_1 - 4x_2 + 6x_3 &= 5 \\ x_1 + 3x_2 - 7x_3 &= 2 \\ 7x_1 + 5x_2 + 9x_3 &= 4 \end{aligned}$$

Solution

Forming the augmented matrix, we proceed by reducing the first column to that of an identity matrix. Thus

$$\left[\begin{array}{ccc|c} 2 & -4 & 6 & 5 \\ 1 & 3 & -7 & 2 \\ 7 & 5 & 9 & 4 \end{array}\right] \begin{array}{l} \tfrac{1}{2}R_1 \\ \\ \\ \end{array}$$

$$\Rightarrow \left[\begin{array}{ccc|c} 1 & -2 & 3 & 2.5 \\ 1 & 3 & -7 & 2 \\ 7 & 5 & 9 & 4 \end{array}\right] \begin{array}{l} \\ -R_1 + R_2 \\ -7R_1 + R_2 \end{array}$$

$$\left[\begin{array}{ccc|c} 1 & -2 & 3 & 2.5 \\ 0 & 5 & -10 & -0.5 \\ 0 & 19 & -12 & -13.5 \end{array}\right] \begin{array}{l} \\ \tfrac{1}{5}R_2 \\ \\ \end{array}$$

$$\Rightarrow \left[\begin{array}{ccc|c} 1 & -2 & 3 & 2.5 \\ 0 & 1 & -2 & -0.1 \\ 0 & 19 & -12 & -13.5 \end{array}\right] \begin{array}{l} 2R_2 + R_1 \\ \\ -19R_2 + R_3 \end{array}$$

$$\left[\begin{array}{ccc|c} 1 & 0 & -1 & 2.3 \\ 0 & 1 & -2 & -0.1 \\ 0 & 0 & 26 & -11.6 \end{array}\right] \begin{array}{l} \\ \\ \tfrac{1}{26}R_3 \end{array}$$

$$\Rightarrow \left[\begin{array}{ccc|c} 1 & 0 & -1 & 2.30 \\ 0 & 1 & -2 & -0.10 \\ 0 & 0 & 1 & -0.44 \end{array}\right] \begin{array}{l} R_3 + R_1 \\ 2R_3 + R_2 \\ \\ \end{array}$$

$$\left[\begin{array}{ccc|c} 1 & 0 & 0 & 1.85 \\ 0 & 1 & 0 & -0.99 \\ 0 & 0 & 1 & -0.45 \end{array}\right]$$

Note that this form is exactly the same as that given by equation (4.13). The solution to the set of equations is given as

$$x_1 = 1.85$$

$$x_2 = -0.99$$

$$x_3 = -0.45$$

Note that roundoff errors may be introduced depending on the number of significant figures used in manipulating the matrices. This may become significant if a large set of equations is to be solved.

**E**

EXAMPLE 4.5

Use the Gauss–Jordan method to solve the following set of equations:

$$
\begin{aligned}
x_1 - 4x_2 + 3x_3 &= A_1\\
3x_1 + 2x_2 - x_3 &= A_2\\
-5x_1 + x_2 + 3x_3 &= A_3
\end{aligned}
$$

### Solution

It is evident that this set of equations is harder to solve than the set solved in example 4.4. We begin by expressing these equations in matrix form as follows:

$$
\begin{bmatrix} 1 & -4 & 3 \\ 3 & 2 & -1 \\ -5 & 1 & 3 \end{bmatrix} \begin{Bmatrix} x_1 \\ x_2 \\ x_3 \end{Bmatrix} = \begin{bmatrix} 1 & 0 & 0 \\ 0 & 1 & 0 \\ 0 & 0 & 1 \end{bmatrix} \begin{Bmatrix} A_1 \\ A_2 \\ A_3 \end{Bmatrix}
$$

Now we form the augmented matrix to give

$$
\left[\begin{array}{ccc|ccc} 1 & -4 & 3 & 1 & 0 & 0 \\ 3 & 2 & -1 & 0 & 1 & 0 \\ -5 & 1 & 3 & 0 & 0 & 1 \end{array}\right]
$$

The matrix of coefficients can be reduced to the identity matrix as required by the Gauss–Jordan method. Therefore adding $-3R_1$ to $R_2$ and $5R_1$ to $R_3$ gives

$$
\left[\begin{array}{ccc|ccc} 1 & -4 & 3 & 1 & 0 & 0 \\ 0 & 14 & 8 & -3 & 1 & 0 \\ 0 & -19 & 18 & 5 & 0 & 1 \end{array}\right]
$$

Similarly, columns 2 and 3 can be reduced to the equivalent identity matrix form to give

$$
\left[\begin{array}{ccc|ccc} 1 & 0 & 0 & \frac{7}{62} & \frac{15}{62} & -\frac{2}{62} \\ 0 & 1 & 0 & -\frac{4}{62} & \frac{18}{62} & \frac{10}{62} \\ 0 & 0 & 1 & \frac{13}{62} & \frac{19}{62} & \frac{14}{62} \end{array}\right]
$$

from which the solution is determined as follows:

$$
\begin{Bmatrix} x_1 \\ x_2 \\ x_3 \end{Bmatrix} = \frac{1}{62} \begin{bmatrix} 7 & 15 & -2 \\ -4 & 18 & 10 \\ 13 & 19 & 14 \end{bmatrix} \begin{Bmatrix} A_1 \\ A_2 \\ A_3 \end{Bmatrix}
$$

or more simply

$$x_1 = \frac{1}{62}[7A_1 + 15A_2 - 2A_3]$$

$$x_2 = \frac{1}{62}[-4A_1 + 18A_2 + 10A_3]$$

$$x_3 = \frac{1}{62}[13A_1 + 19A_2 + 14A_3]$$

Note here that for any given values of $A_1$, $A_2$, and $A_3$ the solution is readily obtained. This concept is extremely important in engineering in that one may solve a set once and for all in terms of arbitrary constants. In order to appreciate the Gauss–Jordan method the student is encouraged to solve this example problem using Gauss's method; this will demonstrate the difficulty associated with back-substitution.

**SUMMARY OF THE GAUSS–JORDAN METHOD**

| STEP | OPERATION | SYMBOL |
|---|---|---|
| 1 | Form the augmented matrix of the $[a_{ij}]$ matrix and $\{b_i\}$ vector. | $[a_{ij} \vdots b_i]$ |
| 2 | Reduce the augmented matrix to unit upper-triangular form by using the Gauss procedure. | $\left[\begin{array}{cccc|c} 1 & u_{12} & \cdots & u_{1n} & c_1 \\ 0 & 1 & \cdots & u_{2n} & c_2 \\ \vdots & \vdots & & \vdots & \vdots \\ 0 & 0 & \cdots & 1 & c_n \end{array}\right]$ |
| 3 | Use the $n$th row to reduce the $n$th column to an equivalent identity matrix column. | |
| 4 | Repeat step 3 for rows $n-1$ through 2. | $\left[\begin{array}{cccc|c} 1 & 0 & \cdots & 0 & b_1^* \\ 0 & 1 & \cdots & & b_2^* \\ \vdots & \vdots & & \vdots & \vdots \\ 0 & 0 & \cdots & 1 & b_n^* \end{array}\right]$ |
| 5 | Solve for the unknowns $x_1, \ldots, x_n$. | $x_1 = b_1^*$<br>$x_2 = b_2^*$<br>$\vdots \quad \vdots$<br>$x_n = b_n^*$ |

## 4.5 CROUT'S METHOD

This method is a variation of the Gauss elimination technique except that the augmented matrix is decomposed into the product of a unit upper-triangular matrix and a lower-triangular matrix. Of course, this is possible only if the matrix containing the $a_{ij}$ coefficients is nonsingular. Consider the following augmented matrix for a $3 \times 3$ set:

$$\left[\begin{array}{ccc|c} a_{11} & a_{12} & a_{13} & b_1 \\ a_{21} & a_{22} & a_{23} & b_2 \\ a_{31} & a_{32} & a_{33} & b_3 \end{array}\right]$$

The Crout procedure assumes that the augmented matrix can be written as follows:

$$\left[\begin{array}{ccc|c} a_{11} & a_{12} & a_{13} & b_1 \\ a_{21} & a_{22} & a_{23} & b_2 \\ a_{31} & a_{32} & a_{33} & b_3 \end{array}\right] = \begin{bmatrix} L_{11} & 0 & 0 \\ L_{21} & L_{22} & 0 \\ L_{31} & L_{32} & L_{33} \end{bmatrix} \left[\begin{array}{ccc|c} 1 & T_{12} & T_{13} & c_1 \\ 0 & 1 & T_{23} & c_2 \\ 0 & 0 & 1 & c_3 \end{array}\right] \tag{4.14a}$$

or more simply

$$[a_{ij} \vdots b_i] = [L_{ij}][T_{ij} \vdots c_i] \tag{4.14b}$$

Equation (4.14) involves 12 known quantities $a_{11}, \ldots, a_{33}, b_1, b_2$, and $b_3$, and 12 unknown quantities $L_{11}, \ldots, L_{33}, T_{12}, T_{13}, T_{23}, c_1, c_2$, and $c_3$. Consequently one may solve for the unknowns. In general, for a set of $n \times n$ there are $n^2 + n$ known quantities in $n^2 + n$ unknown quantities. At first, one may question the merit in solving $n^2 + n$ equations when simply solving $n$ equations should be sufficient. However, the equations to be solved by using the Crout technique are extremely simple. In fact, we shall see that the solution can be obtained by inspection. Therefore let us proceed by multiplying the $[L]$ and $[T \vdots c]$ matrices then equate with the $[a_{ij} \vdots b_i]$ matrix to give

$$a_{11} = L_{11}$$
$$a_{21} = L_{21}$$
$$a_{31} = L_{31}$$

$$a_{12} = L_{11}T_{12}$$
$$a_{22} = L_{21}T_{12} + L_{22}$$
$$a_{32} = L_{31}T_{12} + L_{32}$$

$$a_{13} = L_{11}T_{13}$$
$$a_{23} = L_{21}T_{13} + L_{22}T_{23}$$
$$a_{33} = L_{31}T_{13} + L_{32}T_{23} + L_{33}$$

$$b_1 = L_{11}c_1$$
$$b_2 = L_{21}c_1 + L_{22}c_2$$
$$b_3 = L_{31}c_1 + L_{32}c_2 + L_{33}c_3$$

Solving these equations for the unknowns gives

$$L_{11} = a_{11}$$
$$L_{21} = a_{21}$$
$$L_{31} = a_{31}$$

$$T_{12} = a_{12}/L_{11}$$
$$L_{22} = a_{22} - L_{21}T_{12}$$
$$L_{32} = a_{32} - L_{31}T_{12}$$

$$T_{13} = a_{13}/L_{11}$$
$$T_{23} = (a_{23} - L_{21}T_{13})/L_{22}$$
$$L_{33} = a_{33} - L_{31}T_{13} - L_{32}T_{23}$$

$$c_1 = b_1/L_{11}$$
$$c_2 = (b_2 - L_{21}c_1)/L_{22}$$
$$c_3 = (b_3 - L_{31}c_1 - L_{32}c_2)/L_{33}$$

The unknowns $x_1$, $x_2$, and $x_3$ are then obtained by back-substitution. That is,

$$x_3 = c_3$$
$$x_2 = c_2 - T_{23}x_3$$
$$x_1 = c_1 - T_{13}x_3 - T_{12}x_2$$

In general, for an $n \times n$ set, the solution for the lower-triangle and the augmented unit upper-triangle matrices can be obtained directly from the following:

$$L_{ij} = a_{ij} - \sum_{k=1}^{j-1} L_{ik}T_{kj}, \qquad i \geq j, \quad i = 1, \ldots, n$$

$$T_{ij} = \frac{1}{L_{ii}}\left(a_{ij} - \sum_{k=1}^{i-1} L_{ik}T_{kj}\right), \qquad i < j, \quad j = 2, \ldots, n$$

$$c_i = \frac{1}{L_{ii}}\left(b_i - \sum_{k=1}^{i-1} L_{ik}c_k\right), \qquad i = 2, 3, \ldots, n$$

$$T_{ij} = a_{ij}/a_{11}, \qquad i = 1$$
$$L_{ij} = a_{i1}, \qquad j = 1$$

and

$$c_1 = b_1/L_{11}$$

The unknown $x_n$ is determined easily as

$$x_n = c_n$$

The remaining unknowns are then determined by back-substitution to give

$$x_j = c_j - \sum_{r=j+1}^{n} T_{jr}x_r, \qquad j = 1,\ldots,n-1$$

This method can be greatly simplified by noting that there is no need to store the zeros in $[L]$ or $[T \vdots c]$, and the ones on the diagonal of $[T \vdots c]$. Consequently, equation (4.14a) can be expressed in the following implicit form:

$$\begin{bmatrix} a_{11} & a_{12} & a_{13} & \vdots & b_1 \\ a_{21} & a_{22} & a_{23} & \vdots & b_2 \\ a_{31} & a_{32} & a_{33} & \vdots & b_3 \end{bmatrix} \Rightarrow \begin{bmatrix} L_{11} & T_{12} & T_{13} & \vdots & c_1 \\ L_{21} & L_{22} & T_{23} & \vdots & c_2 \\ L_{31} & L_{32} & L_{33} & \vdots & c_3 \end{bmatrix} \tag{4.15a}$$

$$[a_{ij} \vdots b_i] \Rightarrow [L/T \vdots c] \tag{4.15b}$$

Equation (4.15) permits the determination of the $[L]$ and $[T \vdots c]$ matrices directly without the need for formulas. This is especially useful for hand calculations. The solution should proceed by alternately determining a column and then a row until all coefficients are calculated.

*STEP 1* Determine $L_{11}$, $L_{21}$, and $L_{31}$. These are equal to $a_{11}$, $a_{21}$, and $a_{31}$, respectively.

$$[L/T \vdots c] = \begin{bmatrix} a_{11} & & & \vdots & \\ a_{21} & & & \vdots & \\ a_{31} & & & & \end{bmatrix}$$

*STEP 2* Determine $T_{12}$, $T_{13}$, and $c_1$. These are equal to $a_{12}$, $a_{13}$, and $b_1$ divided by $L_{11}$.

$$[L/T \vdots c] = \begin{bmatrix} L_{11} & a_{12}/a_{11} & a_{13}/a_{11} & \vdots & b_1/a_{11} \\ L_{21} & & & \vdots & \\ L_{31} & & & & \end{bmatrix}$$

*STEP 3* Determine $L_{22}$ and $L_{32}$, which are equal to their corresponding coefficients minus the product of the leading elements in the row and column of $|L/T \vdots c|$, respectively:

$$[L/T \vdots c] = \begin{bmatrix} L_{11} & T_{12} & T_{13} & \vdots & c_1 \\ L_{21} & a_{22} - L_{21}T_{12} & & \vdots & \\ L_{31} & a_{32} - L_{31}T_{12} & & & \end{bmatrix}$$

*STEP 4* Determine $T_{23}$ and $c_2$; these are equal to their corresponding coefficients minus the product of the leading elements in the row and column of $T_{23}$ and $c_2$, respectively, divided by the diagonal $L_{22}$.

$$[L/T \vdots c] = \begin{bmatrix} L_{11} & T_{12} & T_{13} & \vdots & c_1 \\ L_{21} & L_{22} & a_{23} - L_{21}T_{13} & \vdots & b_2 - L_{21}c_1 \\ L_{31} & L_{32} & & & \end{bmatrix}$$

*STEP 5.* Determine $L_{33}$, which is equal to $a_{33}$ minus the sum of the product of corresponding elements appearing in row and column of $L_{33}$.

$$[L/T \vdots c] = \begin{bmatrix} L_{11} & T_{12} & T_{13} & c_1 \\ L_{21} & L_{22} & T_{23} & c_2 \\ L_{31} & L_{32} & a_{33} - L_{31}T_{13} - L_{32}T_{23} & \end{bmatrix}$$

*STEP 6* Compute $c_3$, which is equal to $b_3$ minus the sum of the product of elements appearing in row and column of $c_3$ divided by $L_{33}$.

$$[L/T \vdots c] = \begin{bmatrix} L_{11} & T_{12} & T_{13} & c_1 \\ L_{21} & L_{22} & T_{23} & c_2 \\ L_{31} & L_{32} & L_{33} & (b_2 - L_{31}c_1 - L_{32}c_2)/L_{33} \end{bmatrix}$$

*STEP 7* Solve for the unknowns from the $[T \vdots c]$ matrix by back-substitution:

$$\begin{bmatrix} 1 & T_{12} & T_{13} \\ 0 & 1 & T_{23} \\ 0 & 0 & 1 \end{bmatrix} \begin{Bmatrix} x_1 \\ x_2 \\ x_3 \end{Bmatrix} = \begin{Bmatrix} c_1 \\ c_2 \\ c_3 \end{Bmatrix}$$

This procedure can be easily applied to solve an $n \times n$ set by continuing the process of determining a column and then a row until all coefficients are calculated.

## EXAMPLE 4.6

Solve the following set of equations using Crout's method:

$$2x_1 + x_2 + x_3 = 7$$

$$x_1 + 2x_2 + x_3 = 8$$

$$x_1 + x_2 + 2x_3 = 9$$

### Solution

Expressing these equations in matrix form gives

$$\begin{bmatrix} 2 & 1 & 1 \\ 1 & 2 & 1 \\ 1 & 1 & 2 \end{bmatrix} \begin{Bmatrix} x_1 \\ x_2 \\ x_3 \end{Bmatrix} = \begin{Bmatrix} 7 \\ 8 \\ 9 \end{Bmatrix}$$

or in augmented form

$$\begin{bmatrix} 2 & 1 & 1 & \vdots & 7 \\ 1 & 2 & 1 & \vdots & 8 \\ 1 & 1 & 2 & \vdots & 9 \end{bmatrix} \Rightarrow \begin{bmatrix} L_{11} & T_{12} & T_{13} & \vdots & c_1 \\ L_{21} & L_{22} & T_{23} & \vdots & c_2 \\ L_{31} & L_{32} & L_{33} & \vdots & c_3 \end{bmatrix}$$

Since first column equals first column,

$$L_{11} = 2, \qquad L_{21} = 1, \qquad L_{31} = 1$$

The unknown in the first row is the first row divided by $a_{11} = 2$:

$$T_{12} = \tfrac{1}{2}, \qquad T_{13} = \tfrac{1}{2}, \qquad c_1 = \tfrac{7}{2}$$

The unknowns in the second column can now be determined:

$$L_{22} = 2 - T_{12}L_{21} = \tfrac{3}{2}$$

$$L_{32} = 1 - T_{12}L_{31} = \tfrac{1}{2}$$

The second row is now determined:

$$T_{23} = \frac{1 - T_{13}L_{21}}{L_{22}} = \frac{1}{3}$$

$$c_2 = \frac{8 - c_1L_{21}}{L_{22}} = 3$$

The last element of the third column is readily determined as follows:

$$L_{33} = 2 - T_{13}L_{31} - T_{23}L_{32} = \tfrac{4}{3}$$

and the last element

$$c_3 = \frac{9 - c_1L_{31} - c_2L_{32}}{L_{33}} = 3$$

The augmented matrix of $[a_{ij} \vdots b_i]$ coefficients is now decoupled:

$$\begin{bmatrix} 2 & 1 & 1 & \vdots & 7 \\ 1 & 2 & 1 & \vdots & 8 \\ 1 & 1 & 2 & \vdots & 9 \end{bmatrix} = \begin{bmatrix} 2 & 0 & 0 \\ 1 & \frac{3}{2} & 0 \\ 1 & \frac{1}{2} & \frac{4}{3} \end{bmatrix} \begin{bmatrix} 1 & \frac{1}{2} & \frac{1}{2} & \vdots & \frac{7}{2} \\ 0 & 1 & \frac{1}{3} & \vdots & 3 \\ 0 & 0 & 1 & \vdots & 3 \end{bmatrix}$$

hence

$$\begin{bmatrix} 1 & \frac{1}{2} & \frac{1}{2} \\ 0 & 1 & \frac{1}{3} \\ 0 & 0 & 1 \end{bmatrix} \begin{Bmatrix} x_1 \\ x_2 \\ x_3 \end{Bmatrix} = \begin{Bmatrix} \frac{7}{2} \\ 3 \\ 3 \end{Bmatrix}$$

The solution can now be determined through back-substitutions:

$$x_3 = 3, \qquad x_2 = 2, \qquad x_1 = 1$$

This problem can be directly solved by using the step-by-step approach.

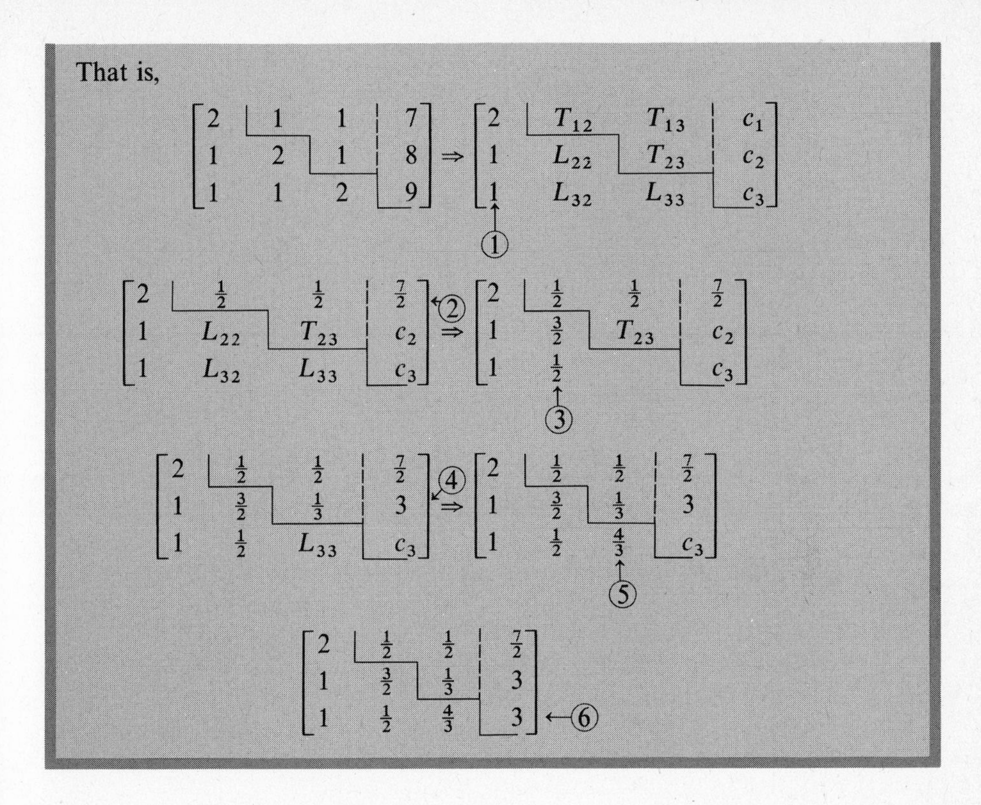


In some problems an $L_{ii}$ coefficient may become zero during the application of the solution procedure. For such problems, the reader is advised to simply switch the equation in which $L_{ii} = 0$ appears with one of the equations below it for which the new $L_{ii} \neq 0$.

Using Crout's method we can easily determine one additional piece of information, the determinant of the coefficient matrix (det[$A$]). This is possible since

$$\det[A] = \det[L]\det[T]$$

Since the determinant of $[T]$ is unity (because $T_{ii} = 1$),

$$\det[A] = \det[L] = (L_{11})(L_{22})(L_{33}) \cdots (L_{nn})$$

One of the advantages of knowing the determinant is that it sheds some light on whether or not the set is ill-conditioned det[$A$] $\approx 0$. We obviously cannot say that for the Gauss or Gauss–Jordan methods.

## 4.6 SQUARE ROOT METHOD

This method is of the same form as the Crout method except that it is limited to equations involving symmetrical coefficient matrices. Such equations often occur in engineering problems such as numerical solution of ordinary and partial differential equations. Consequently, the square root method is employed in order to save memory space and to reduce input data. Consider the following 3 × 3 set:

$$a_{11}x_1 + a_{12}x_2 + a_{13}x_3 = b_1$$
$$a_{12}x_1 + a_{22}x_2 + a_{23}x_3 = b_2$$
$$a_{13}x_1 + a_{23}x_2 + a_{33}x_3 = b_3$$

It is evident that the coefficient matrix is symmetrical. Therefore the square root method is employed to give

$$\left[\begin{array}{ccc|c} a_{11} & a_{12} & a_{13} & b_1 \\ & a_{22} & a_{23} & b_2 \\ \text{symm} & & a_{33} & b_3 \end{array}\right] = \begin{bmatrix} u_{11} & 0 & 0 \\ u_{12} & u_{22} & 0 \\ u_{13} & u_{23} & u_{33} \end{bmatrix} \left[\begin{array}{ccc|c} u_{11} & u_{12} & u_{13} & c_1 \\ 0 & u_{22} & u_{23} & c_2 \\ 0 & 0 & u_{33} & c_3 \end{array}\right] \tag{4.16a}$$

or more simply

$$[a_{ij} \vdots b_i] = [u_{ij}]^t [u_{ij} \vdots c_i] \tag{4.16b}$$

Invoking the principle of matrix equality after multiplying the two matrices appearing on the right-hand side of equation (4.16) yields

$$a_{11} = u_{11}^2 \quad \Rightarrow \quad u_{11} = \sqrt{a_{11}}$$
$$a_{12} = u_{11}u_{12} \quad \Rightarrow \quad u_{12} = a_{12}/u_{11}$$
$$a_{13} = u_{11}u_{13} \quad \Rightarrow \quad u_{13} = a_{13}/u_{11}$$

$$a_{22} = u_{11}^2 + u_{22}^2 \quad \Rightarrow \quad u_{22} = \sqrt{a_{22} - u_{12}^2}$$
$$a_{23} = u_{12}u_{13} + u_{22}u_{23} \quad \Rightarrow \quad u_{23} = (a_{23} - u_{12}u_{13})/u_{22}$$
$$a_{33} = u_{13}^2 + u_{23}^2 + u_{33}^2 \quad \Rightarrow \quad u_{33} = \sqrt{a_{33} - u_{13}^2 - u_{23}^2}$$

$$b_1 = c_1u_{11} \quad \Rightarrow \quad c_1 = b_1/u_{11}$$
$$b_2 = c_1u_{12} + c_2u_{22} \quad \Rightarrow \quad c_2 = (b_2 - c_1u_{12})/u_{22}$$
$$b_3 = c_1u_{13} + c_2u_{23} + c_3u_{33} \quad \Rightarrow \quad c_3 = (b_3 - c_1u_{13} - c_2u_{23})/u_{33}$$

The solution for the unknown is then determined by back-substitution using

$$\begin{bmatrix} u_{11} & u_{12} & u_{13} \\ 0 & u_{22} & u_{23} \\ 0 & 0 & u_{33} \end{bmatrix} \begin{Bmatrix} x_1 \\ x_2 \\ x_3 \end{Bmatrix} = \begin{Bmatrix} c_1 \\ c_2 \\ c_3 \end{Bmatrix}$$

It is evident that the method derives its name from the fact that the diagonal elements involve a square root. For a set of $n$ equations in $n$ unknowns ($n \times n$) the solution is given as follows:

$$u_{ii} = \sqrt{a_{ii} - \sum_{k=1}^{i-1} u_{ki}^2}, \qquad i = 2, \ldots, n$$

$$u_{ij} = \left(a_{ij} - \sum_{k=1}^{i-1} u_{ki}u_{kj}\right) \Big/ u_{ii}, \qquad j = 2, \ldots, n$$

$$c_i = \left(b_i - \sum_{k=1}^{i-1} u_{ki}c_k\right)\Big/u_{ii}, \qquad i = 2,\ldots,n$$

$$u_{11} = a_{11}$$

$$c_1 = b_1/u_{11}$$

$$u_{1j} = b_1/u_{11}, \qquad j = 2,\ldots,n$$

and the unknowns are determined as

$$x_n = c_n/u_{nn}$$

$$x_i = \left(c_i - \sum_{k=i+1}^{n} u_{ik}x_k\right)\Big/u_{ii}, \qquad i = 1,\ldots,n-1$$

This method is by far the most efficient for solving symmetrical sets of equations.

EXAMPLE 4.7

Solve the following set of equations using the square root method:

$$x_1 + 2x_2 = 5$$

$$2x_1 + x_2 = 4$$

### Solution

Express the set in augmented matrix form:

$$\left[\begin{array}{cc|c} 1 & 2 & 5 \\ 2 & 1 & 4 \end{array}\right] = \begin{bmatrix} u_{11} & 0 \\ u_{12} & u_{22} \end{bmatrix} \left[\begin{array}{cc|c} u_{11} & u_{12} & c_1 \\ 0 & u_{22} & c_2 \end{array}\right]$$

Equating corresponding elements after multiplying these yields

$$u_{11} = \sqrt{1} = 1$$

$$u_{12} = 2/u_{11} = 2$$

$$c_1 = 5/u_{11} = 5$$

$$u_{22} = \sqrt{1 - u_{12}^2} = \sqrt{-3} = j\sqrt{3}$$

$$c_2 = (4 - c_1u_{12})/u_{22} = -6/j\sqrt{3}$$

The solution is now given by back-substitution as

$$\begin{bmatrix} 1 & 2 \\ 0 & i\sqrt{3} \end{bmatrix} \begin{Bmatrix} x_1 \\ x_2 \end{Bmatrix} = \begin{Bmatrix} 5 \\ -6/j\sqrt{3} \end{Bmatrix}$$

from which

$$x_2 = -6/j^2(\sqrt{3})^2 = 2, \qquad x_1 = 5 - (2)(+2) = 1$$

## 4.7 REDUCING MATRIX METHOD

This method is a factoring method in which a given matrix is decomposed into a product of an upper- and a lower-triangular matrix. The method involves the formulation of reducing matrices to transform a given matrix into upper-triangular form. In fact, this method of solution is quite similar to the Crout method discussed earlier. However, it is the most efficient factoring procedure for solving a set of linear algebraic equations. In fact, this method requires exactly the same number of steps as the Gauss method. Furthermore, it provides a systematic and direct approach for transforming augmented matrices into the required form so that back-substitution can be employed. The basis for this method may be best explained by considering the following $3 \times 3$ set:

$$a_{11}x_1 + a_{12}x_2 + a_{13}x_3 = b_1$$

$$a_{21}x_1 + a_{22}x_2 + a_{23}x_3 = b_2$$

$$a_{31}x_1 + a_{32}x_2 + a_{33}x_3 = b_3$$

or in augmented matrix form

$$[A \vdots b] = \left[\begin{array}{ccc|c} a_{11} & a_{12} & a_{13} & b_1 \\ a_{21} & a_{22} & a_{23} & b_2 \\ a_{31} & a_{32} & a_{33} & b_3 \end{array}\right] \tag{4.17}$$

Recall that our objective in using the Gauss method was to reduce the augmented matrix to an upper-triangular matrix. This can be accomplished by reducing the first column, then the second, and finally the third column into their proper form so that the final augmented matrix is an upper triangle. Let us do just that, but by using reducing matrices. That is, to eliminate $a_{21}$ and $a_{31}$ from column 1 we premultiply the matrix in equation (4.17) by the following reducing matrix:

$$[R_1] = \begin{bmatrix} 1 & 0 & 0 \\ -a_{21}/a_{11} & 1 & 0 \\ -a_{31}/a_{11} & 0 & 1 \end{bmatrix}$$

Note here that the $[R_1]$ matrix is an identity matrix of the same size as the coefficient matrix except that its first column contains reducing coefficients that can be easily determined from the original matrix $[A \vdots b]$. Thus

$$[R_1][A \vdots b] = [A_1]$$

$$\begin{bmatrix} 1 & 0 & 0 \\ -a_{21}/a_{11} & 1 & 0 \\ -a_{31}/a_{11} & 0 & 1 \end{bmatrix} \begin{bmatrix} a_{11} & a_{12} & a_{13} & b_1 \\ a_{21} & a_{22} & a_{23} & b_2 \\ a_{31} & a_{32} & a_{33} & b_3 \end{bmatrix} = \begin{bmatrix} a_{11} & a_{12} & a_{13} & b_1 \\ 0 & a'_{22} & a'_{23} & b'_2 \\ 0 & a'_{32} & a'_{33} & b'_3 \end{bmatrix}$$

Note here that the first row in $[A_1]$ remains unchanged. Consequently, we now proceed by multiplying $[A_1]$ by the second reducing matrix $[R_2]$ to eliminate $a_{32}$.

Hence

$$[R_2][A_1] = [A_2]$$

$$\begin{bmatrix} 1 & 0 & 0 \\ 0 & 1 & 0 \\ 0 & -a'_{32}/a'_{22} & 1 \end{bmatrix} \left[\begin{array}{ccc|c} a_{11} & a_{12} & a_{13} & b_1 \\ 0 & a'_{22} & a'_{23} & b'_2 \\ 0 & a'_{32} & a'_{33} & b'_3 \end{array}\right] = \left[\begin{array}{ccc|c} a_{11} & a_{12} & a_{13} & b_1 \\ 0 & a'_{22} & a'_{23} & b'_2 \\ 0 & 0 & a''_{33} & b''_3 \end{array}\right]$$

Note once again that only the third row has changed as a result of premultiplying $[A_1]$ by the second reducing matrix. This is an extremely important observation in that we need not compute rows 1 and 2 again. It is apparent that for a $3 \times 3$ set two reducing matrices are required and for an $n \times n$ a total of $n - 1$ reducing matrices are needed. Keep in mind that only partial matrix multiplication is required. The solution for the unknowns is then determined by back-substitution. Note that the original matrix has now been decomposed into

$$\left[\begin{array}{ccc|c} a_{11} & a_{12} & a_{13} & b_1 \\ a_{21} & a_{22} & a_{23} & b_2 \\ a_{31} & a_{32} & a_{33} & b_3 \end{array}\right] = \begin{bmatrix} 1 & 0 & 0 \\ a_{21}/a_{11} & 1 & 0 \\ a_{31}/a_{11} & a'_{32}/a'_{22} & 1 \end{bmatrix} \left[\begin{array}{ccc|c} a_{11} & a_{12} & a_{13} & b_1 \\ 0 & a'_{22} & a'_{23} & b'_2 \\ 0 & 0 & a''_{33} & b''_3 \end{array}\right]$$

where the elements of the lower-triangle matrix are the negative of the corresponding reducing coefficients in matrices $[R_1]$ and $[R_2]$.

## E EXAMPLE 4.8

Solve the following $3 \times 3$ set using the reducing procedure:

$$\begin{aligned} x_1 + x_2 + x_3 &= 6 \\ -2x_1 - x_2 + 2x_3 &= 2 \\ 3x_1 - 2x_2 + x_3 &= 2 \end{aligned}$$

### Solution

$$\left[\begin{array}{ccc|c} 1 & 1 & 1 & 6 \\ -2 & -1 & 2 & 2 \\ 3 & -2 & 1 & 2 \end{array}\right] = \left[\begin{array}{ccc|c} a_{11} & a_{12} & a_{13} & b_1 \\ a_{21} & a_{22} & a_{23} & b_2 \\ a_{31} & a_{32} & a_{33} & b_3 \end{array}\right]$$

To reduce this matrix to the product of upper- and lower-triangular matrices, we proceed by reducing the element of the first column to zeros except the element on the diagonal. This is accomplished by forming a reducing matrix. The elements of the first column, namely $r_{i1}$, are given below:

$$r_{11} = 1.0$$

$$r_{21} = -\frac{a_{21}}{a_{11}} = -\frac{-2}{1} = 2$$

$$r_{31} = -\frac{a_{31}}{a_{11}} = -\frac{3}{1} = -3$$

Thus the first reducing matrix is formed as

$$[R_1] = \begin{bmatrix} 1 & 0 & 0 \\ 2 & 1 & 0 \\ -3 & 0 & 1 \end{bmatrix}$$

Premultiplying $[A \vdots b]$ by $[R_1]$ gives

$$[A_1] = \begin{bmatrix} 1 & 0 & 0 \\ 2 & 1 & 0 \\ -3 & 0 & 1 \end{bmatrix} \left[\begin{array}{ccc|c} 1 & 1 & 1 & 6 \\ -2 & -1 & 2 & 2 \\ 3 & -2 & 1 & 2 \end{array}\right]$$

$$= \left[\begin{array}{ccc|c} 1 & 1 & 1 & 6 \\ 0 & 1 & 4 & 14 \\ 0 & -5 & -2 & -16 \end{array}\right]$$

The second column of the matrix $[A_1]$ is now reduced by forming a second reducing matrix as follows:

$$r_{22} = 1$$

$$r_{32} = -\frac{a_{32}}{a_{22}} = -\frac{-5}{1} = 5$$

Therefore the reducing matrix takes the form

$$[R_2] = \begin{bmatrix} 1 & 0 & 0 \\ 0 & 1 & 0 \\ 0 & 5 & 1 \end{bmatrix}$$

Premultiplying $[A_1]$ by $[R_2]$ gives

$$[A_2] = \begin{bmatrix} 1 & 0 & 0 \\ 0 & 1 & 0 \\ 0 & 5 & 1 \end{bmatrix} \left[\begin{array}{ccc|c} 1 & 1 & 1 & 6 \\ 0 & 1 & 4 & 14 \\ 0 & -5 & -2 & 16 \end{array}\right] = \left[\begin{array}{ccc|c} 1 & 1 & 1 & 6 \\ 0 & 1 & 4 & 14 \\ 0 & 0 & 18 & 54 \end{array}\right]$$

Clearly, we have succeeded in reducing the original augmented matrix to a more manageable form that permits the solution of the original set of equations. Hence the solution is obtained by back-substitution as

$$x_1 = 1, \qquad x_2 = 2, \qquad x_3 = 3$$

Note that the corresponding lower-triangular matrix is given as

$$\begin{bmatrix} 1 & 0 & 0 \\ -2 & 1 & 0 \\ 3 & -5 & 1 \end{bmatrix}$$

Obviously, this matrix has the columns of the reducing matrices except that the elements below the diagonal have switched sign. Consequently,

$$\left[\begin{array}{rrr|r} 1 & 1 & 1 & 6 \\ -2 & -1 & 2 & 2 \\ 3 & -2 & 1 & 2 \end{array}\right] = \begin{bmatrix} 1 & 0 & 0 \\ -2 & 1 & 0 \\ 3 & -5 & 1 \end{bmatrix} \left[\begin{array}{rrr|r} 1 & 1 & 1 & 6 \\ 0 & 1 & 4 & 14 \\ 0 & 0 & 18 & 54 \end{array}\right]$$

E

EXAMPLE 4.9

Solve the following set of equations using the reducing procedures:

$$2x_1 - x_2 + 3x_3 - 2x_4 = 1$$

$$x_1 + 2x_2 - x_3 + x_4 = 6$$

$$3x_1 + 3x_2 + 4x_3 - 5x_4 = 1$$

$$5x_1 - 2x_2 + 7x_3 - 6x_4 = -2$$

Solution

Interchanging the first and second and expressing the set of equations in matrix form gives

$$\left[\begin{array}{rrrr|r} 1 & 2 & -1 & 1 & 6 \\ 2 & -1 & 3 & -2 & 1 \\ 3 & 3 & 4 & -5 & 1 \\ 5 & -2 & 7 & -6 & -2 \end{array}\right]$$

Premultiplying the above by the first reducing matrix gives

$$\begin{bmatrix} 1 & 0 & 0 & 0 \\ -2 & 1 & 0 & 0 \\ -3 & 0 & 1 & 0 \\ -5 & 0 & 0 & 1 \end{bmatrix} \left[\begin{array}{rrrr|r} 1 & 2 & -1 & 1 & 6 \\ 2 & -1 & 3 & -2 & 1 \\ 3 & 3 & 4 & -5 & 1 \\ 5 & -2 & 7 & -6 & -2 \end{array}\right]$$

$$= \left[\begin{array}{rrrr|r} 1 & 2 & -1 & 1 & 6 \\ 0 & -5 & 5 & -4 & -11 \\ 0 & -3 & 7 & -8 & -17 \\ 0 & -12 & 12 & -11 & -32 \end{array}\right]$$

Premultiplying the new matrix by the second reducing matrix gives

$$\begin{bmatrix} 1 & 0 & 0 & 0 \\ 0 & 1 & 0 & 0 \\ 0 & -\frac{3}{5} & 1 & 0 \\ 0 & -\frac{12}{5} & 0 & 1 \end{bmatrix} \left[\begin{array}{cccc|c} 1 & 2 & -1 & 1 & 6 \\ 0 & -5 & 5 & -4 & -11 \\ 0 & -3 & 7 & -8 & -17 \\ 0 & -12 & 12 & -11 & -32 \end{array}\right]$$

$$= \left[\begin{array}{cccc|c} 1 & 2 & -1 & 1 & 6 \\ 0 & -5 & 5 & -4 & -11 \\ 0 & 0 & 4 & -\frac{28}{5} & -\frac{52}{5} \\ 0 & 0 & 0 & -\frac{7}{5} & -\frac{28}{5} \end{array}\right]$$

This may be expressed in the original form as follows:

$$\begin{aligned} x_1 + 2x_2 - x_3 + x_4 &= 6 \\ -5x_2 + 5x_3 - 4x_4 &= -11 \\ 4x_3 - \tfrac{28}{5}x_4 &= -\tfrac{52}{5} \\ -\tfrac{7}{5}x_4 &= \tfrac{28}{5} \end{aligned}$$

Solving by back-substitutions gives

$$x_4 = 4$$

$$x_3 = \frac{-52 + 28(4)}{20} = 3$$

$$x_2 = \frac{-11 + (4)(4) - (5)(3)}{-5} = 2$$

$$x_1 = 6 - (2)(2) + (3)(1) - 4 = 1$$

## 4.8 SOLUTION OF TRIDIAGONAL SYSTEMS

The application of numerical methods to the solution of certain engineering problems may in some cases result in a set of tridiagonal linear algebraic equations. Heat conduction and fluid flow problems are a few of the many applications which generate such a system. These equations are characterized by the unique form of having nonzero elements on the diagonal and a maximum of one element to the left and to the right of the diagonal. The rest of the elements are zeros.

Obviously, one may use any one of the methods outlined in the previous sections for solving a tridiagonal set. However, such an approach will not utilize the relative simplicity of the set. Consequently, the Crout procedure is used to derive general equations for solving any tridiagonal set. This is the most efficient method for handling such a set. Consider the following:

$$\begin{bmatrix} d_1 & t_1 & 0 & 0 \\ l_2 & d_2 & t_2 & 0 \\ 0 & l_3 & d_3 & t_3 \\ 0 & 0 & l_4 & d_4 \end{bmatrix} \begin{Bmatrix} x_1 \\ x_2 \\ x_3 \\ x_4 \end{Bmatrix} = \begin{Bmatrix} b_1 \\ b_2 \\ b_3 \\ b_4 \end{Bmatrix}$$

Note that because of the unique form of a tridiagonal set the coefficient matrix is defined differently from the conventional form. That is, the $d$'s are the diagonal elements, the $l$'s are the coefficients below the diagonal, and the $t$'s are the coefficients above the diagonal. Applying the Crout procedure, we have

$$\left[\begin{array}{cccc|c} d_1 & t_1 & 0 & 0 & b_1 \\ l_1 & d_2 & t_2 & 0 & b_2 \\ 0 & l_3 & d_3 & t_3 & b_3 \\ 0 & 0 & l_4 & d_4 & b_4 \end{array}\right] = \begin{bmatrix} D_1 & 0 & 0 & 0 \\ L_2 & D_2 & 0 & 0 \\ 0 & L_3 & D_3 & 0 \\ 0 & 0 & L_4 & D_4 \end{bmatrix} \left[\begin{array}{cccc|c} 1 & T_1 & 0 & 0 & B_1 \\ 0 & 1 & T_2 & 0 & B_2 \\ 0 & 0 & 1 & T_3 & B_3 \\ 0 & 0 & 0 & 1 & B_4 \end{array}\right]$$

Multiplying the right-hand side then equating corresponding elements with the matrix on the left-hand side yields

$$\begin{aligned} d_1 &= D_1 & \qquad D_1 &= d_1 \\ l_2 &= L_2 & L_2 &= l_2 \\[6pt] t_1 &= T_1 D_1 & T_1 &= t_1/D_1 \\ d_2 &= T_1 L_2 + D_2 & D_2 &= d_2 - T_1 L_2 \\ l_3 &= L_3 & L_3 &= l_3 \\[6pt] t_2 &= D_2 T_2 & T_2 &= t_2/D_2 \\ d_3 &= L_3 T_2 + D_3 & D_3 &= d_3 - L_3 T_2 \\ l_4 &= L_4 & L_4 &= l_4 \\[6pt] t_3 &= T_3 D_3 & T_3 &= t_3/D_3 \\ d_4 &= T_3 L_4 + D_4 & D_4 &= d_4 - T_3 L_4 \\[6pt] b_1 &= B_1 D_1 & B_1 &= b_1/D_1 \\ b_2 &= B_1 L_2 + B_2 D_2 & B_2 &= (b_2 - B_1 L_2)/D_2 \\ b_3 &= B_2 L_3 + B_3 D_3 & B_3 &= (b_3 - B_2 L_3)/D_3 \\ b_4 &= B_3 L_4 + B_4 D_4 & B_4 &= (b_4 - B_3 L_4)/D_4 \end{aligned}$$

It is evident that the coefficients appearing below the diagonal in the lower-triangular matrix are equal to their corresponding coefficients in the augmented matrix. Consequently for an $n \times n$ system of equations, the decomposed matrices are given by

$$D_1 = d_1$$

$$T_i = t_i/D_i, \qquad i = 1, 2, \ldots, n-1$$

$$D_i = d_i - T_{i-1}L_i, \qquad i = 2, \ldots, n$$

$$B_1 = b_1/D_1$$

$$B_i = (b_i - B_{i-1}L_i)/D_i, \qquad i = 2, \ldots, n$$

The solution procedure is started by determining $D_1$, then $T_1$, then $D_2$, then $T_2$, etc. That is, we solve for the first column, then the first row, then the second column, etc. The unknowns $x_n, \ldots, x_1$ can then be determined by using back-substitution to give

$$x_n = B_1$$

$$x_i = B_i - T_i x_{i+1}, \qquad i = n-1, n-2, \ldots, 1$$

In order to show the efficiency of this method, the following example is included.

EXAMPLE 4.10

Solve the following set of tridiagonal algebraic equations using Crout's method:

$$\begin{aligned} x_1 + 4x_2 \qquad\qquad &= 10 \\ 2x_1 + 10x_2 - 4x_3 \qquad &= 7 \\ x_2 + 8x_3 - x_4 &= 6 \\ x_3 - 6x_4 &= 4 \end{aligned}$$

### Solution

Forming the augmented matrix and observing that the set is a tridiagonal one, we proceed by determining the coefficients of the upper- and lower-triangular matrices. Thus

$$\left[\begin{array}{cccc|c} 1 & 4 & 0 & 0 & 10 \\ 2 & 10 & -4 & 0 & 7 \\ 0 & 1 & 8 & -1 & 6 \\ 0 & 0 & 1 & -6 & 4 \end{array}\right] = \left[\begin{array}{cccc|c} D_1 & T_1 & 0 & 0 & B_1 \\ L_2 & D_2 & T_2 & 0 & B_2 \\ 0 & L_3 & D_3 & T_3 & B_3 \\ 0 & 0 & L_4 & D_4 & B_4 \end{array}\right]$$

Note that $L_2 = 2$, $L_3 = 1$, $L_4 = 1$.

For $i = 1$:

$$D_1 = d_1 = 1$$

$$T_1 = t_1/D_1 = \tfrac{4}{1} = 4$$

$$B_1 = b_1/D_1 = \tfrac{10}{1} = 10$$

For $i = 2$:

$$D_2 = d_2 - T_1 L_2 = 10 - 4(2) = 2$$

$$T_2 = t_2/D_2 = -\tfrac{4}{2} = 2$$

$$B_2 = (b_2 - B_1 L_2)/D_2 = [7 - 10(2)]/2 = -6.5$$

For $i = 3$:

$$D_3 = d_3 - L_3 T_2 = 8 - 1(-2) = 10$$

$$T_3 = t_3/D_3 = -\tfrac{1}{10} = -0.1$$

$$B_3 = (b_3 - B_2 L_3)/D_3 = [6 - (-6.5)(1)]/10 = 1.25$$

For $i = 4$:

$$D_4 = d_4 - L_4 T_3 = -6 - 1(-0.1) = -5.9$$

$$B_4 = (b_4 - B_3 L_4)/D_4 = [4 - 1(1.25)]/(-5.9) = -0.466$$

The set of equations is now reduced to the following upper-triangle form:

$$\begin{bmatrix} 1 & 4 & 0 & 0 \\ 0 & 1 & -2 & 0 \\ 0 & 0 & 1 & -0.1 \\ 0 & 0 & 0 & 1 \end{bmatrix} \begin{Bmatrix} x_1 \\ x_2 \\ x_3 \\ x_4 \end{Bmatrix} = \begin{Bmatrix} 10 \\ -6.5 \\ 1.25 \\ -0.466 \end{Bmatrix}$$

The solution is readily obtained by using back-substitution. That is,

$$x_4 = -0.466$$

$$x_3 = 1.25 - 0.1(-0.466) = 1.203$$

$$x_2 = -6.5 + 2(1.203) = -4.093$$

$$x_1 = 10 - 4(-4.093) = 26.373$$

## 4.9 ITERATIVE METHODS

These are methods by which an approximation to the solution of a system of linear algebraic equations may be obtained. Unlike the direct methods the iterative procedure may not always yield a solution, even if the determinant of the coefficient matrix is not zero. Consequently, for these techniques to work, certain additional conditions must be considered:

1. The set of equations must possess a strong diagonal. This is a necessary but not a sufficient condition for a solution to be found.
2. A sufficient condition for a solution to be found is that the absolute value of the diagonal coefficient in any equation must be greater than the sum of the absolute values of all other coefficients appearing in that equation.

These conditions can often be satisfied for a variety of practical problems encountered in engineering. The iterative techniques are generally used when dealing with sparse matrices (a lot of zeros). As a result less computer memory is required for solving a set of algebraic equations than would otherwise be required using the direct techniques discussed earlier.

### 4.9.1 Jacobi's Method

One of the easiest iteration methods is the Jacobi method. The following example illustrates the mechanics of this technique:

$$a_{11}x_1 + a_{12}x_2 + a_{13}x_3 = b_1 \tag{4.18a}$$

$$a_{21}x_1 + a_{22}x_2 + a_{23}x_3 = b_2 \tag{4.18b}$$

$$a_{31}x_1 + a_{32}x_2 + a_{33}x_3 = b_3 \tag{4.18c}$$

The solution process begins by solving for $x_1$ from equation (4.18a), $x_2$ is then determined from equation (4.18b), and finally $x_3$ is determined from equation (4.18c). In addition, a dummy index $k$ is introduced to number the iterations performed in approximating the solution. In particular, the equations used to get the next iteration from the last one are

$$x_1^{(k+1)} = \frac{b_1}{a_{11}} - \frac{a_{12}}{a_{11}} x_2^{(k)} - \frac{a_{13}}{a_{11}} x_3^{(k)} \tag{4.19a}$$

$$x_2^{(k+1)} = \frac{b_2}{a_{22}} \frac{a_{21}}{a_{22}} x_1^{(k)} - \frac{a_{23}}{a_{22}} x_3^{(k)} \tag{4.19b}$$

$$x_3^{(k+1)} = \frac{b_3}{a_{33}} - \frac{a_{31}}{a_{33}} x_1^{(k)} - \frac{a_{32}}{a_{33}} x_2^{(k)} \tag{4.19c}$$

Consequently, if we assume an initial vector $\{x\}_k = \{x_1^{(k)}, x_2^{(k)}, x_3^{(k)}\}$ and substitute into (4.19), a new vector $\{x\}_{k+1}$ is calculated. This process is continued until sufficient accuracy is achieved. That is, until

$$\{x\}_{k+1} \cong \{x\}_k, \qquad k = 0, \ldots, N$$

where $N$ is the total number of iterations performed.

In general, for an $n \times n$ set, equation (4.19) can be expressed in the following form:

$$x_i^{(k+1)} = \frac{1}{a_{ii}} \left( b_i - \sum_{\substack{j=1 \\ j \neq i}}^{n} a_{ij} x_j^{(k)} \right) \tag{4.20}$$

or in matrix form,

$$
\begin{Bmatrix} x_1 \\ x_2 \\ \vdots \\ x_n \end{Bmatrix}_{k+1} = \begin{Bmatrix} B_1 \\ B_2 \\ \vdots \\ B_n \end{Bmatrix} - \begin{vmatrix} 0 & A_{12} & \cdots & A_{1n} \\ A_{21} & 0 & \cdots & A_{2n} \\ \vdots & \vdots & & \vdots \\ A_{n1} & A_{n2} & \cdots & 0 \end{vmatrix} \begin{Bmatrix} x_1 \\ x_2 \\ \vdots \\ x_n \end{Bmatrix} \tag{4.21}
$$

where

$$
\begin{aligned}
B_i &= b_i/a_{ii}, \qquad i = 1,\ldots,n \\
A_{ij} &= a_{ij}/a_{ii}, \qquad i = 1,\ldots,n, \quad j = 1,\ldots,n \\
k &= 0,\ldots,N
\end{aligned}
$$

and $N$ is the total number of iterations.

The reader is to be reminded that equations (4.20) and (4.21) are guaranteed to converge if condition 2 is met. Otherwise, a solution may or may not be possible. This fact can be easily proven by considering equation (4.10). Let $\{x\}^t = \{x_1, x_2, \ldots, x_n\}$ be the exact solution to an $n \times n$ set of linear algebraic equations. Then

$$
x_i = \frac{1}{a_{ii}} \left( b_i - \sum_{\substack{j=1 \\ j \neq i}}^{n} a_{ij} x_j \right) \tag{4.22}
$$

The error of the $j$th component of the current solution $\{x\}_k = \{x_1^k, x_2^k, \ldots, x_n^k\}$ is denoted by

$$
E_j^{(k)} = x_j - x_j^{(k)}, \qquad j = 1, 2, \ldots, n \tag{4.23}
$$

Consequently, for the $(k + 1)$th iteration, the error is given as

$$
E_j^{(k+1)} = x_j - x_j^{(k+1)}, \qquad j = 1, 2, \ldots, n \tag{4.24}
$$

Substituting equations (4.20) and (4.22) into (4.24) yields

$$
E_j^{(k+1)} = -\frac{1}{a_{ii}} \left( \sum_{\substack{j=1 \\ j \neq i}}^{n} a_{ij}(x_j - x_j^{(k)}) \right)
$$

This equation can now be simplified by substituting equation (4.23) into the right-hand side. Hence

$$
E_j^{(k+1)} = -\frac{1}{a_{ii}} \left( \sum_{\substack{j=1 \\ j \neq i}}^{n} a_{ij} E_j^{(k)} \right)
$$

Since, in general, errors may be positive or negative, we should only be concerned with the absolute values of error magnitudes. Thus

$$
|E_j^{(k+1)}| = \frac{1}{|a_{ii}|} \left( \sum_{\substack{j=1 \\ j \neq i}}^{n} |a_{ij}|\,|E_j^{(k)}| \right) \tag{4.25}
$$

Denoting the largest error of $|E_j^{(k)}|$ by $|E^{(k)}|$ and substituting into equation (4.25) gives

$$E_j^{(k+1)} = \frac{|E^{(k)}|}{|a_{ii}|} \sum_{\substack{j=1 \\ j \neq i}}^{n} |a_{ij}|$$

Consequently, for the errors to converge,

$$\sum_{\substack{j=1 \\ j \neq i}}^{n} |a_{ij}| \Big/ |a_{ii}| < 1$$

or, more simply,

$$|a_{ii}| > \sum_{\substack{j=1 \\ j \neq i}}^{n} |a_{ij}| \tag{4.26}$$

This is precisely the sufficient condition noted earlier for the Jacobi method. A set of equations not meeting this requirement may converge because the maximum error in the $\{x\}$ vector was used rather than the actual errors in the derivation of equation (4.26).

E

EXAMPLE 4.11

Solve the following set of linear algebraic equations using the Jacobi method:

$$5x_1 - 2x_2 + x_3 = 4$$

$$x_1 + 4x_2 - 2x_3 = 3$$

$$x_1 + 2x_2 + 4x_3 = 17$$

Solution

We begin our solution by first checking the diagonal coefficients and determine whether the iterative procedure is suitable. Thus

$$|5| > |-2| + |1| \Rightarrow 5 > 3 \quad \text{O.K.}$$

$$|4| > |1| + |-2| \Rightarrow 4 > 3 \quad \text{O.K.}$$

$$|4| > |1| + |2| \Rightarrow 4 > 3 \quad \text{O.K.}$$

Obviously, the iterative approach will converge. Therefore expressing the set in the form of equation (4.19) yields

$$\begin{Bmatrix} x_1 \\ x_2 \\ x_3 \end{Bmatrix}_{k+1} = \begin{Bmatrix} 0.8 \\ 0.75 \\ 4.25 \end{Bmatrix} - \begin{bmatrix} 0 & -0.4 & 0.2 \\ 0.25 & 0 & -0.5 \\ 0.25 & 0.50 & 0 \end{bmatrix} \begin{Bmatrix} x_1 \\ x_2 \\ x_3 \end{Bmatrix}_k \tag{4.27}$$

Table 4.1 SUCCESSIVE APPROXIMATION OF SOLUTION (JACOBI'S METHOD)

| | ITERATION | | | | | | | | | |
|---|---|---|---|---|---|---|---|---|---|---|
| VARIABLE | 1 | 2 | 3 | 4 | 5 | 6 | 7 | 8 | 9 | 10 |
| $x_1$ | 0.8 | 0.25 | 1.14 | 1.24 | 1.02 | 0.92 | 0.98 | 1.02 | 1.01 | 0.99 |
| $x_2$ | 0.75 | 2.68 | 2.53 | 1.89 | 1.79 | 1.99 | 2.07 | 2.02 | 1.98 | 1.99 |
| $x_3$ | 4.25 | 3.68 | 2.85 | 2.70 | 2.99 | 3.10 | 3.02 | 2.97 | 2.98 | 3.01 |

Assuming an initial iteration $\{x\}_0 = \{0\}$ and substituting into equation (4.27) yields our first approximation to the solution. That is,

$$\begin{Bmatrix} x_1 \\ x_2 \\ x_3 \end{Bmatrix}_1 = \begin{Bmatrix} 0.8 \\ 0.75 \\ 4.25 \end{Bmatrix} - \begin{bmatrix} 0 & -0.4 & 0.2 \\ 0.25 & 0 & -0.5 \\ 0.25 & 0.5 & 0 \end{bmatrix} \begin{Bmatrix} 0 \\ 0 \\ 0 \end{Bmatrix}_0 = \begin{Bmatrix} 0.8 \\ 0.75 \\ 4.25 \end{Bmatrix}_1$$

Substituting $\{x\}_1$ into equation (4.27) gives the second approximation to the $\{x\}$ vector:

$$\begin{Bmatrix} x_1 \\ x_2 \\ x_3 \end{Bmatrix}_2 = \begin{Bmatrix} 0.8 \\ 0.75 \\ 4.25 \end{Bmatrix} - \begin{bmatrix} 0 & -0.4 & 0.2 \\ 0.25 & 0 & -0.5 \\ 0.25 & 0.50 & 0 \end{bmatrix} \begin{Bmatrix} 0.8 \\ 0.75 \\ 4.25 \end{Bmatrix}_1 = \begin{Bmatrix} 0.25 \\ 2.675 \\ 3.675 \end{Bmatrix}_2$$

This process is continued until successive values of each vector are sufficiently close in magnitude. For the example at hand, the first ten iterations are shown in table 4.1.

**SUMMARY OF JACOBI'S METHOD**

| STEP | OPERATION | SYMBOL |
|---|---|---|
| 1 | Check the diagonal coefficients in each of the equations making sure that it is the largest. If not, switch equations. | $\lvert a_{ii}\rvert > \lvert a_{ij}\rvert$, $i = 1,\ldots,n$, $j = 1,\ldots,n$ |
| 2 | An iterative solution is possible if the absolute value of the diagonal element in each equation is greater than the sum of the absolute values of the rest of the coefficients in that equation. This step is optional. | $\lvert a_{ii}\rvert > \sum_{\substack{j=1 \\ j\neq i}}^{n} \lvert a_{ij}\rvert$, $i = 1,2,\ldots,n$ |

| | | |
|---|---|---|
| 3 | Compute $B_i$ coefficients. | $B_i = b_i/a_{ii}$, $i = 1, \ldots, n$ |
| 4 | Compute $A_{ij}$ coefficients. | $A_{ij} = a_{ij}/a_{ii}$, $i = 1, \ldots, n$ $j = 1, \ldots, n$ |
| 5 | Solve for the first approximation of unknowns $x_1, \ldots, x_n$ using an initial set of values $\{x\}_0$. | $x_i^{(k+1)} = B_i - \sum_{\substack{j=1 \\ j \neq i}}^{n} A_{ij} x_j^{(k)}$, $i = 1, 2, \ldots, n$ $k = 0, 1, \ldots, N$ |
| 6 | Repeat step 5 for as many iterations as necessary. | |

### 4.9.2 Gauss–Seidel Method

This method converges more rapidly than the Jacobi method. However, it is constrained by the same conditions outlined in the previous section. The Gauss–Seidel solution procedure presupposes that the new solution values are a better approximation to the exact solution than the initial assumed values. This is true for most problems. Consequently, for the $3 \times 3$ set given by equation (4.18) the solutions for the new values become

$$x_1^{(k+1)} = \frac{b_1}{a_{11}} - \frac{a_{12}}{a_{11}} x_2^{(k)} - \frac{a_{13}}{a_{11}} x_3^{(k)} \tag{4.28a}$$

$$x_2^{(k+1)} = \frac{b_2}{a_{22}} - \frac{a_{21}}{a_{11}} x_1^{(k+1)} - \frac{a_{23}}{a_{22}} x_3^{(k)} \tag{4.28b}$$

$$x_3^{(k+1)} = \frac{b_3}{a_{33}} - \frac{a_{31}}{a_{33}} x_1^{(k+1)} - \frac{a_{32}}{a_{33}} x_2^{(k+1)} \tag{4.28c}$$

Comparison of equation (4.28) with (4.19) clearly shows that one need not assume an initial $x_1$ value. Instead $x_1$ is computed by assuming $x_2$ and $x_3$. Then the new $x_1$ value and the assumed $x_3$ value are used to calculate a new $x_2$ value. This process is continued by substituting the new $x_1$ and $x_2$ values to determine a new $x_3$ value.

In general for an $n \times n$ set, the solution is given by

$$x_i^{(k+1)} = \frac{b_i}{a_{ii}} - \sum_{j=1}^{i-1} \frac{a_{ij}}{a_{ii}} x_j^{(k+1)} - \sum_{j=i+1}^{n} \frac{a_{ij}}{a_{ii}} x_j^{(k)},$$

$$i = 1, \ldots, n, \quad j = 1, \ldots, m, \quad k = 1, \ldots, N \tag{4.29}$$

Equation (4.29) converges as long as the Jacobi method converges. That is, if

$$|a_{ii}| > |a_{ij}|, \qquad i = 1, \ldots, n, \quad j = 1, \ldots, n$$

For many engineering problems, this condition can be satisfied. The authors recommend that the Gauss–Seidel technique rather than the Jacobi method be used whenever an iterative solution is sought, because it converges faster.

E

EXAMPLE 4.12

Rework example 4.11 using the Gauss–Seidel method.

### Solution

Since the Jacobi method converged for this particular example, the Gauss–Seidel technique should also converge. Consequently, expressing the set in the form of equation (4.28) yields

$$x_1^{(k+1)} = 0.8 + 0.4x_2^{(k)} - 0.2x_3^{(k)} \quad \textbf{(4.30a)}$$

$$x_2^{(k+1)} = 0.75 - 0.25x_1^{(k+1)} + 0.5x_3^{(k)} \quad \textbf{(4.30b)}$$

$$x_3^{(k+1)} = 4.25 - 0.25x_1^{(k+1)} - 0.5x_2^{(k+1)} \quad \textbf{(4.30c)}$$

Now, assuming initial values of $x_2 = 0$ and $x_3 = 0$ and substituting into equation (4.30) gives for $k = 0$

$$x_1^{(1)} = 0.8 + 0 - 0 = 0.8$$

$$x_2^{(1)} = 0.75 - 0.25(0.8) + 0 = 0.55$$

$$x_3^{(1)} = 4.25 - 0.25(0.8) - 0.5(0.55) = 3.775$$

This process is continued by substituting the new $x_2^{(1)}$ and $x_3^{(1)}$ into (4.30), permitting the determination of yet another set of $x$ values. Computations for the first seven iterations are summarized in table 4.2.

Table 4.2 SUCCESSIVE APPROXIMATION OF SOLUTION (GAUSS–SEIDEL METHOD)

| | ITERATION | | | | | | |
|---|---|---|---|---|---|---|---|
| VARIABLE | 1 | 2 | 3 | 4 | 5 | 6 | 7 |
| $x_1$ | 0.8 | 0.265 | 1.249 | 0.956 | 1.002 | 1.001 | 0.999 |
| $x_2$ | 0.55 | 2.571 | 1.887 | 2.008 | 2.003 | 1.999 | 2.000 |
| $x_3$ | 3.775 | 2.898 | 2.994 | 3.007 | 2.998 | 3.000 | 3.000 |

Note that the results obtained in five iterations are more accurate than those obtained in ten iterations using the Jacobi method. A plot of the results given by tables 4.1 and 4.2 is shown in figure 4.1.

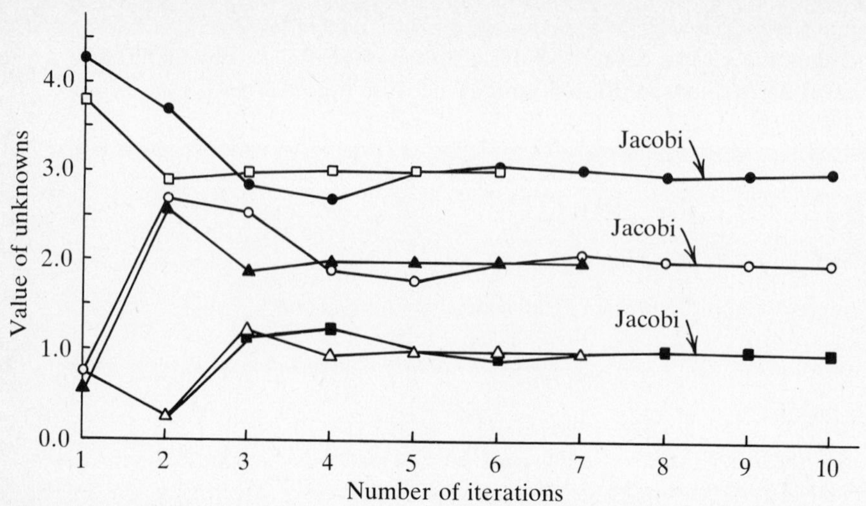


**Figure 4.1** Solution convergence rate for examples 4.11 and 4.12 using Jacobi and Gauss–Seidel techniques.

It is interesting to note that while the Gauss–Seidel method has resulted in marked improvement in the value of $x_3$ and to a lesser degree $x_2$ for the first few iterations, it did not appreciably improve the value of $x_1$. This is because $x_1$ was initially determined in terms of the assumed values of $x_2$ and $x_3$, while $x_3$ was calculated using the improved values of $x_1$ and $x_2$ and not the initial assumed values. The value of the new $x_2$ value was determined in terms of the assumed $x_3$ value and the improved $x_1$ value. Consequently the results are consistent with our intuitive expectations.

**SUMMARY OF THE GAUSS–SEIDEL METHOD**

| STEP | OPERATION | SYMBOL |
|---|---|---|
| 1 | Rearrange the set of $n \times n$ equations so that the diagonal coefficient is the largest in any equation. | $\lvert a_{ii}\rvert > \lvert a_{ij}\rvert$, $i = 1, \ldots, n$ $j = 1, \ldots, n$ |
| 2 | An iterative solution is possible if the absolute value of the diagonal coefficient is greater than the sum of the absolute values of the coefficients appearing in that equation. | $\lvert a_{ii}\rvert > \sum_{\substack{j=1 \\ j \neq 1}}^{n} \lvert a_{ij}\rvert$, $i = 1, 2, \ldots, n$ |

| | | |
|---|---|---|
| 3 | Assume an initial set of values for the unknowns $x_2, \ldots, x_n$. | $\{x\}_0 = \begin{Bmatrix} x_2^{(0)} \\ x_3^{(0)} \\ \vdots \\ x_n^{(0)} \end{Bmatrix}$ |
| 4 | Solve for $x_1^{(1)}$. | $x_1^{(1)} = \dfrac{b_1}{a_{11}} - \sum_{j=i+1}^{n} \dfrac{a_{ij}}{a_{11}} x_j^{(0)}$ |
| 5 | Solve for $x_2^{(1)}, \ldots, x_n^{(1)}$. | $x_i^{(1)} = \dfrac{b_i}{a_{ii}} - \sum_{j=1}^{i-1} \dfrac{a_{ij}}{a_{ii}} x_j^{(1)} - \sum_{j=i+1}^{n} \dfrac{a_{ij}}{a_{ii}} x_j^{(0)}$ |
| 6 | Repeat steps 4 and 5 for $k + 1$ iterations until changes in the $\{x\}$ vector become acceptably small. That is, $\{x\}_{k+1} \cong \{x\}_k$. | |

## 4.10 ILL-CONDITIONED SETS AND SCALING

Often, the solution of linear algebraic equations represents a single step in the analysis of real physical systems. Normally, a solution either exists (if $\det[A] = 0$) or does not exist (if $\det[A] \neq 0$). An ill-conditioned set is that set whose coefficient matrix has a determinant close to zero. That is,

$$\det[A] \approx 0$$

This situation may arise in the course of performing the computations in which roundoff of the significant digits becomes of paramount importance in calculating a determinant whose value is close to zero. Consequently, depending on whether the computations are being carried out by hand, by calculator, or by mainframe computer, roundoff errors accumulated during the intermediate steps could have significant influence on the final results. In other words, different solutions may be obtained depending on the number of significant digits retained. As a result no unique solution can be found. The stability of the solution is often that of the computation rather than the stability of the equations being solved. Unfortunately, there are situations in which the set of equations describing a particular physical system is not properly formulated in the first place. Keep in mind that generally the selection process of a set of variables is not unique when formulating a mathematical model. Hence it is conceivable that a different set of equations may be formulated for which a unique solution can be determined. To help illustrate the concepts involved, consider the following set of equations:

$$\begin{aligned} x + y &= 2 \\ 0.9x + y &= 1.9 \end{aligned} \qquad \textbf{(4.31)}$$

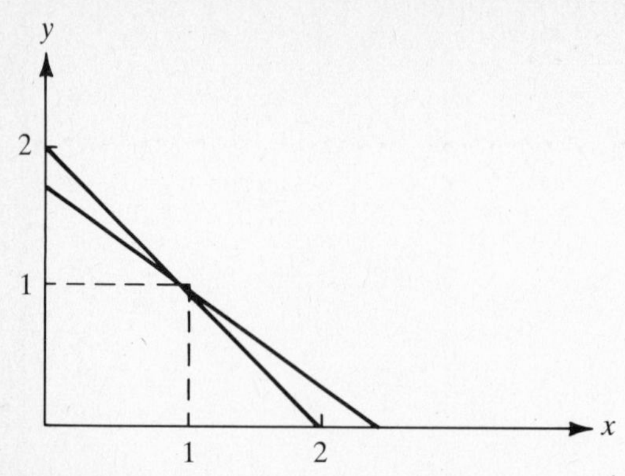


**Figure 4.2** Graphical representation of an ill-conditioned set of two linear algebraic equations.

which has the solution $x = 1$ and $y = 1$. These equations are shown graphically in figure 4.2. It is evident that the lines are nearly parallel. The implication is that a very small change in one or more of the coefficients could result in a significantly different point of intersection.

Suppose that these equations describe experimental data in which equipment or human error could be significant. That is, there is some uncertainty associated with the values of the coefficients. Theoretically, the lines representing equation (4.31) as given by figure 4.2 have zero width; however, these lines in reality now have widths whose magnitudes depend on the error involved in each of the coefficients. This concept is shown graphically in figure 4.3.

Obviously, we no longer have a unique solution. Instead, we have a region enclosed by the vertices 1, 2, 3, and 4 that defines an infinite set of solutions. This is why we can no longer speak of a unique solution with certainty. On the other hand, for a set of equations that are not nearly parallel, the effect of error associated with the coefficient is significantly less. This situation is shown in figure 4.4.

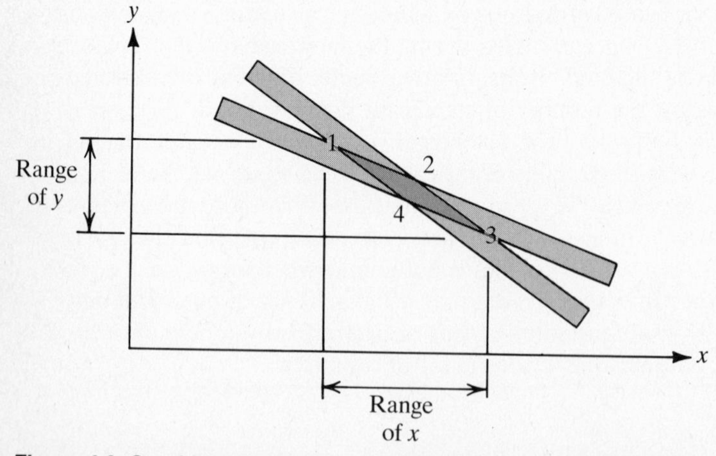


**Figure 4.3** Graphical illustration of coefficient error effects on the solution of a nearly parallel $2 \times 2$ set.

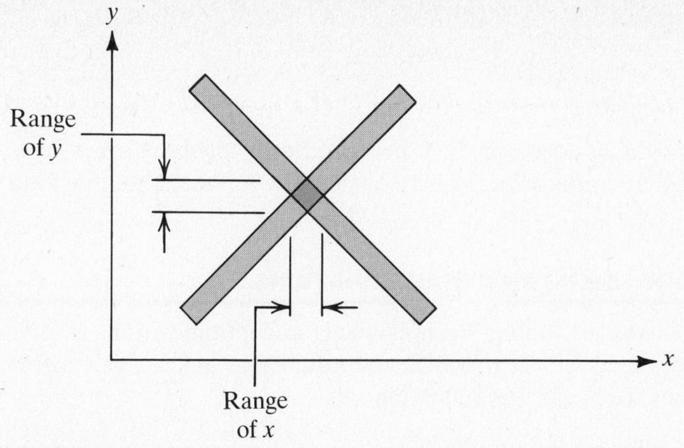


**Figure 4.4** Graphical illustration of coefficient error effects on the solution of an orthogonal 2 × 2 set.

Note that the region of feasible solutions (area enclosed by 1, 2, 3, and 4) is much smaller than that of an ill-conditioned set. One way of detecting an ill-conditioned set is to purposely change some of the coefficients and observe the corresponding changes in the solution. Alternatively, the condition of convergence discussed earlier in connection with the iterative procedure may be used as a guide. One way of combatting the problems associated with an ill-conditioned set is to retain more significant digits using double-precision arithmetic. Alternatively, one may transform the original set to a more stable one by reformulating the equations. This is possible if the physical problem being analyzed permits such a transformation.

For certain types of problems some coefficients may be significantly greater than others. This situation arises when one is formulating a model involving relationships between quantities having different units. Such problems can be easily handled by simply scaling the coefficients. Consider the following set:

$$x_1 - x_2 + 1000x_3 = 1$$

$$2x_1 + 7x_2 + 7500x_3 = 4$$

$$3x_1 + x_2 - 9600x_3 = 3$$

Evidently, the coefficients of $x_3$ are significantly larger than those of $x_1$ and $x_2$. This may cause problems, especially if the number of significant digits retained is small. Fortunately, this set of equations can be transformed into a more appropriate form by using a new variable as follows:

$$x_4 = 1000x_3$$

Substituting yields

$$x_1 - x_2 + x_4 = 1$$

$$2x_1 + 7x_2 + 7.5x_4 = 4$$

$$3x_1 + x_2 - 9.6x_4 = 3$$

This set can now be solved for the unknowns $x_1$, $x_2$, and $x_4$. The value of $x_3$ is then determined as

$$x_3 = \tfrac{1}{1000}x_4$$

While it is not possible for us to cover all aspects of linear algebraic equations, the reader is reminded that common sense may sometimes be more valuable than knowledge.

## 4.11 SETS WITH MORE UNKNOWNS THAN EQUATIONS

Some engineering problems, including eigenproblems and optimization, involve systems of linear algebraic equations in which the number of unknowns exceeds the number of equations. Consider the following set:

$$\begin{aligned}
a_{11}x_1 + a_{12}x_2 + \cdots + a_{1n}x_n &= b_1 \\
a_{21}x_1 + a_{22}x_2 + \cdots + a_{2n}x_n &= b_2 \\
\vdots \qquad \vdots \qquad\quad \vdots \quad &\quad \vdots \\
a_{m1}x_1 + a_{m2}x_2 + \cdots + a_{mn}x_n &= b_m
\end{aligned} \tag{4.32}$$

At first glance, one may erroneously assume that no solution exists. However, what if we rewrite the set in the following form

$$\begin{aligned}
a_{11}x_1 + a_{12}x_2 + \cdots + a_{1m}x_m &= b_1 - a_{1m+1}x_{m+1} - \cdots - a_{1n}x_n \\
a_{21}x_1 + a_{22}x_2 + \cdots + a_{2m}x_m &= b_2 - a_{2m+1}x_{m+1} - \cdots - a_{2n}x_n \\
\vdots \qquad \vdots \qquad\quad \vdots \quad &\quad \vdots \qquad \vdots \qquad\qquad \vdots \\
a_{m1}x_1 + a_{m2}x_2 + \cdots + a_{mm}x_m &= b_m - a_{mm+1}x_{m+1} - \cdots - a_{mn}x_n
\end{aligned} \tag{4.33}$$

Note that while we have arbitrarily transferred $n - m$ variables to the right-hand side of equation (4.32), care must be exercised in choosing the proper set of variables so that the determinant of the $a_{ij}$ coefficients appearing on the left-hand side of equation (4.33) is not zero. Therefore, using matrix notation, we have

$$\begin{bmatrix} a_{11} & a_{12} & \cdots & a_{1m} \\ a_{21} & a_{22} & \cdots & a_{2m} \\ \vdots & \vdots & & \vdots \\ a_{m1} & a_{m2} & \cdots & a_{mm} \end{bmatrix} \begin{Bmatrix} x_1 \\ x_2 \\ \vdots \\ x_m \end{Bmatrix} = \begin{bmatrix} b_1 & -a_{1m+1} & \cdots & -a_{1n} \\ b_2 & -a_{2m+1} & \cdots & -a_{2n} \\ \vdots & \vdots & & \vdots \\ b_m & -a_{mm+1} & \cdots & -a_{mn} \end{bmatrix} \begin{Bmatrix} 1 \\ x_{m+1} \\ \vdots \\ x_n \end{Bmatrix} \tag{4.34}$$

Equation (4.34) can now be solved for the unknowns $x_1, x_2, \ldots, x_n$ in terms of $x_{m+1}, \ldots, x_n$. The implication is that we have no unique solution. Instead, we actually have an infinite number of solutions. This is because one may assume any set of arbitrary values for $x_{m+1}, \ldots, x_n$ (independent variables) and then solve for $x_1, \ldots, x_m$ (dependent variables). The Gauss–Jordan method is well suited for solving this type of problem. Consequently, by forming the augmented matrix

$$\left[\begin{array}{cccc|cccc} a_{11} & a_{12} & \cdots & a_{1m} & b_1 & -a_{1m+1} & \cdots & -a_{1n} \\ a_{21} & a_{22} & \cdots & a_{2m} & b_2 & -a_{2m+1} & \cdots & -a_{2n} \\ \vdots & \vdots & & \vdots & \vdots & \vdots & & \vdots \\ a_{m1} & a_{m2} & \cdots & a_{mm} & b_m & -a_{mm+1} & \cdots & -a_{mn} \end{array}\right] \tag{4.35}$$

we seek to reduce equation (4.35) to the following:

$$\left[\begin{array}{cccc|cccc} 1 & 0 & \cdots & 0 & c_1 & c_{1,m+1} & \cdots & c_{1n} \\ 0 & 1 & \cdots & 0 & c_2 & c_{2,m+1} & \cdots & c_{2n} \\ \vdots & \vdots & & \vdots & \vdots & \vdots & & \vdots \\ 0 & 0 & \cdots & 1 & c_m & c_{m,m+1} & \cdots & c_{mn} \end{array}\right] \tag{4.36}$$

It is evident that the solution is now given by

$$\begin{Bmatrix} x_1 \\ x_2 \\ \vdots \\ x_m \end{Bmatrix} = \begin{bmatrix} c_1 & c_{1m+1} & \cdots & c_{1n} \\ c_2 & c_{2m+1} & \cdots & c_{2n} \\ \vdots & \vdots & & \vdots \\ c_m & c_{mm+1} & \cdots & c_{mn} \end{bmatrix} \begin{Bmatrix} 1 \\ x_{m+1} \\ \vdots \\ x_n \end{Bmatrix} \tag{4.37}$$

To illustrate the mechanics of the procedure just outlined, the following example is included.

EXAMPLE 4.13

Solve the following set of linear algebraic equations for $x_1$ and $x_2$ in terms of $x_3$:

$$x_1 + 2x_2 - 4x_3 = 6$$

$$2x_1 - 4x_2 + 3x_3 = 2$$

### Solution

Obviously this set of equations has an infinite number of solutions. Hence, rewriting in the form

$$\begin{bmatrix} 1 & 2 \\ 2 & -4 \end{bmatrix} \begin{Bmatrix} x_1 \\ x_2 \end{Bmatrix} = \begin{bmatrix} 6 & 4 \\ 2 & -3 \end{bmatrix} \begin{Bmatrix} 1 \\ x_3 \end{Bmatrix}$$

we can now proceed to use the Gauss–Jordan method. Thus

$$\left[\begin{array}{cc|cc} 1 & 2 & 6 & 4 \\ 2 & -4 & 2 & -3 \end{array}\right] \quad -2R_1 + R_2$$

$$\Rightarrow \left[\begin{array}{cc|cc} 1 & 2 & 6 & 4 \\ 0 & -8 & -10 & -11 \end{array}\right] \quad -1/8R_2$$

$$\left[\begin{array}{cc|cc} 1 & 2 & 6 & 4 \\ 0 & 1 & \frac{5}{4} & \frac{11}{8} \end{array}\right] \quad -2R_2 + R_1 \quad \Rightarrow \quad \left[\begin{array}{cc|cc} 1 & 0 & \frac{7}{2} & \frac{5}{4} \\ 0 & 1 & \frac{5}{4} & \frac{11}{8} \end{array}\right]$$

The solution is readily obtained as

$$\begin{bmatrix} 1 & 0 \\ 0 & 1 \end{bmatrix} \begin{Bmatrix} x_1 \\ x_2 \end{Bmatrix} = \begin{bmatrix} \frac{7}{2} & \frac{5}{4} \\ \frac{5}{4} & \frac{11}{8} \end{bmatrix} \begin{Bmatrix} 1 \\ x_3 \end{Bmatrix}$$

or more simply

$$x_1 = \tfrac{7}{2} + \tfrac{5}{4}x_3, \qquad x_2 = \tfrac{5}{4} + \tfrac{11}{8}x_3$$

Clearly, the set has no unique solution. The student is to be reminded that $x_2$ or $x_1$, rather than $x_3$, could have been selected as the independent variable.

## 4.12 LINEAR EQUATIONS INVOLVING FEWER UNKNOWNS THAN EQUATIONS

This type of problem generally arises when dealing with experimental data. It is also common in optimization-related problems. Consider the following set of equations:

$$a_{11}x_1 + a_{12}x_2 = b_1 \tag{4.38a}$$

$$a_{21}x_1 + a_{22}x_2 = b_2 \tag{4.38b}$$

$$a_{31}x_1 + a_{32}x_2 = b_3 \tag{4.38c}$$

Obviously, it is impossible to find a solution that can satisfy all of the equations unless two of the three equations are dependent. That is, if only two out of the three equations are unique, then a solution is possible. Otherwise, our best hope is to find a solution that minimizes the error. Since no solution is expected to satisfy all equations, one may assume an infinite number of such solutions. The question that we need to answer is, how can one find the so-called best solution out of the set of infinite possibilities? This question is addressed specifically and rather thoroughly in chapter 9. However, let us attempt to find an answer to the question using the mathematical tools we have learned thus far. Expressing equations (4.38) in matrix form gives

$$\begin{bmatrix} a_{11} & a_{12} \\ a_{21} & a_{22} \\ a_{31} & a_{32} \end{bmatrix} \begin{Bmatrix} x_1 \\ x_2 \end{Bmatrix} = \begin{bmatrix} b_1 \\ b_2 \\ b_3 \end{bmatrix} \tag{4.39a}$$

or in compact matrix form

$$[A]_{3\times 2}\{x\}_{2\times 1} = \{b\}_{3\times 1} \tag{4.39b}$$

Since the desired "best possible" solution is a unique one, we must somehow come up with two equations in the two unknowns $x_1$ and $x_2$. It turns out that this is indeed possible, if both sides of equation (4.39) are premultiplied by the transpose of the $[A]$ matrix. That is,

$$\begin{bmatrix} a_{11} & a_{21} & a_{31} \\ a_{12} & a_{22} & a_{32} \end{bmatrix} \begin{bmatrix} a_{11} & a_{12} \\ a_{21} & a_{22} \\ a_{31} & a_{32} \end{bmatrix} \begin{Bmatrix} x_1 \\ x_2 \end{Bmatrix} = \begin{bmatrix} a_{11} & a_{21} & a_{31} \\ a_{12} & a_{22} & a_{32} \end{bmatrix} \begin{Bmatrix} b_1 \\ b_2 \\ b_3 \end{Bmatrix}$$

Simplifying, we have

$$\begin{bmatrix} a_{11}^2 + a_{21}^2 + a_{31}^2 & a_{11}a_{12} + a_{21}a_{22} + a_{31}a_{32} \\ a_{12}a_{11} + a_{22}a_{21} + a_{32}a_{31} & a_{12}^2 + a_{22}^2 + a_{32}^2 \end{bmatrix} \begin{Bmatrix} x_1 \\ x_2 \end{Bmatrix} = \begin{Bmatrix} a_{11}b_1 + a_{21}b_2 + a_{31}b_3 \\ a_{12}b_1 + a_{22}b_2 + a_{32}b_3 \end{Bmatrix} \tag{4.40}$$

Equation (4.40) can now be solved uniquely for the unknown $\{x\}$ vector. In general, for $n$ equations in two unknowns, equation (4.40) can be given as

$$\begin{bmatrix} \sum_{i=1}^{n} a_{i1}^2 & \sum_{i=1}^{n} a_{i1}a_{i2} \\ \sum_{i=1}^{n} a_{i1}a_{i2} & \sum_{i=1}^{n} a_{i2}^2 \end{bmatrix} \begin{Bmatrix} x_1 \\ x_2 \end{Bmatrix} = \begin{Bmatrix} \sum_{i=1}^{n} a_{i1}b_i \\ \sum_{i=1}^{n} a_{i2}b_i \end{Bmatrix} \tag{4.41}$$

Similar expressions can be derived for $n$ equations in $m$ unknowns ($n \times m$). The reader is reminded that equation (4.41) was derived without any rational interpretation as to why this method works. However, the method does provide some insight into the various possible uses of matrices in solving engineering problems.

E

EXAMPLE 4.14

Determine the equation of the line that relates $y$ to $x$ for the following experimentally derived data.

| $x$ | 0 | 1 | 2 | 5 |
|---|---|---|---|---|
| $y$ | 0 | 2 | 3 | 7 |

Solution

The equation that relates $y$ to $x$ linearly is of the form

$$y = a + bx \tag{4.42}$$

Substituting the various corresponding data values gives

$$0 = a + b(0)$$

$$2 = a + b(1)$$

$$3 = a + b(2)$$

$$7 = a + b(5)$$

In matrix form

$$\begin{bmatrix} 1 & 0 \\ 1 & 1 \\ 1 & 2 \\ 1 & 5 \end{bmatrix} \begin{Bmatrix} a \\ b \end{Bmatrix} = \begin{Bmatrix} 0 \\ 2 \\ 3 \\ 7 \end{Bmatrix}$$

Premultiplying by the transpose of the coefficient matrix on the left-hand side gives

$$\begin{bmatrix} 1 & 1 & 1 & 1 \\ 0 & 1 & 2 & 5 \end{bmatrix} \begin{bmatrix} 1 & 0 \\ 1 & 1 \\ 1 & 2 \\ 1 & 5 \end{bmatrix} \begin{Bmatrix} a \\ b \end{Bmatrix} = \begin{bmatrix} 1 & 1 & 1 & 1 \\ 0 & 1 & 2 & 5 \end{bmatrix} \begin{Bmatrix} 0 \\ 2 \\ 3 \\ 7 \end{Bmatrix}$$

which simplifies to

$$\begin{bmatrix} 4 & 8 \\ 8 & 30 \end{bmatrix} \begin{Bmatrix} a \\ b \end{Bmatrix} = \begin{Bmatrix} 12 \\ 43 \end{Bmatrix}$$

Therefore solving for the unknowns yields

$$a = -\tfrac{49}{35}, \qquad b = \tfrac{76}{35}$$

Substituting into equation (4.42) yields the desired linear relationship between $y$ and $x$. That is,

$$y = -\tfrac{49}{35} + \tfrac{76}{35}x \tag{4.43}$$

A plot of this equation along with the data set used in deriving it is shown in figure 4.5.

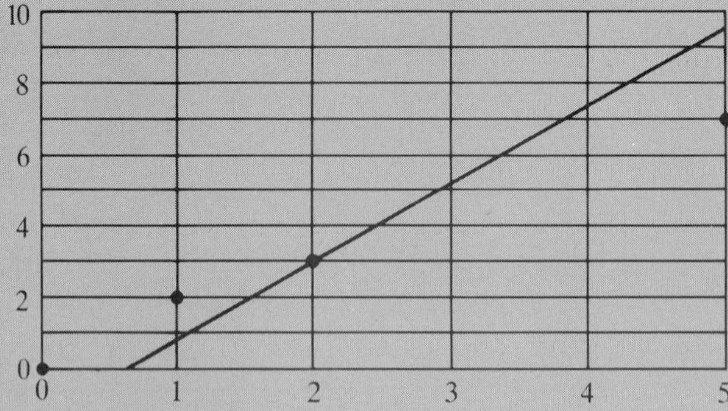


**Figure 4.5** Graphical representation of the best solution for example 4.14 using a linear relationship.

Obviously, the solution represents the best possible linear relationship between $y$ and $x$. However, one may correctly argue that a second-order polynomial would have resulted in a better approximation of the true relationship between the variables $x$ and $y$. On the other hand, if we are certain that the relationship is linear, then there is no question that equation (4.43) is a proper representation of the true behavior. Furthermore, the deviation of the data set from the derived linear function may be caused by equipment limitations and/or human errors in obtaining the laboratory data.

## 4.13 SETS INVOLVING COMPLEX COEFFICIENTS

So far, our discussion has been limited to linear algebraic equations with real coefficients. Some engineering applications may involve linear algebraic equations with complex coefficients. Such equations can be solved most efficiently by following the real algorithm and replacing all real operations by complex ones. This can be easily accomplished by using a programming language which permits the declaration of complex variables. In the absence of such a facility, this approach becomes cumbersome. Fortunately, an alternative procedure can be formulated when real numbers can be used to solve for the generally complex unknown variables. Consider the following complex $n \times n$ set:

$$[C_{ij}]\{Z_j\} = \{V_i\}, \qquad i = 1,\dots,n, \quad j = 1,\dots,n \tag{4.44}$$

where

$$[C_{ij}] = [a_{ij}] + [b_{ij}]J \tag{4.45a}$$

$$\{Z_j\} = \{x_j\} + \{y_j\}J \tag{4.45b}$$

$$\{V_i\} = \{r_i\} + \{S_i\}J \tag{4.45c}$$

$$J = \sqrt{-1}$$

Substituting equation (4.45) into equation (4.44) and carrying out the multiplication yields

$$[a_{ij}]\{x_j\} + [b_{ij}]\{x_j\}J + [a_{ij}]\{y_j\}J - [b_{ij}]\{y_j\} = \{r_i\} + \{S_i\}J$$

Equating real parts and imaginary parts gives

$$[a_{ij}]\{x_j\} - [b_{ij}]\{y_j\} = \{r_i\}$$

$$[b_{ij}]\{x_j\} + [a_{ij}]\{y_j\} = \{S_i\}$$

These two matrix equations can be combined into a single matrix equation by using submatrices. That is,

$$\begin{bmatrix} [a_{ij}] & -[b_{ij}] \\ [b_{ij}] & [a_{ij}] \end{bmatrix} \begin{Bmatrix} \{x_j\} \\ \{y_j\} \end{Bmatrix} = \begin{Bmatrix} \{r_i\} \\ \{S_i\} \end{Bmatrix} \tag{4.46}$$

Clearly, equation (4.46) is a $2n \times 2n$ set of linear algebraic equations involving real coefficients. This is exactly the price we have to pay for using real numbers instead of complex ones. Consequently, this approach is suitable in that no additional programming effort is needed to handle the complex arithmetic. However, if they occur frequently and the programming language in use does accommodate complex variables, then equation (4.46) is not recommended.

EXAMPLE 4.15

Solve the following set of equations, which describes the currents in the electrical circuit shown below:

$$(1.5 + 0.754J)I_1 - (0.5 + 0.754J)I_2 = 100$$

$$-(0.5 + 0.754J)I_1 + (2.5 + 0.223J)I_2 = 0$$

where $I_1$ and $I_2$ are the phasor currents in the two loops.

### Solution

Obviously, the student is not required to know how to formulate the set of equations. However, we thought it might be interesting to include the physical model we are trying to solve. Note that a total of four equations in four unknowns must be solved if equation (4.46) is used. Hence

$$[a_{ij}] = \begin{bmatrix} 1.5 & -0.5 \\ -0.5 & 2.5 \end{bmatrix}$$

$$[b_{ij}] = \begin{bmatrix} 0.754 & -0.754 \\ -0.754 & 0.223 \end{bmatrix}$$

$$\{r_i\} = \begin{Bmatrix} 100 \\ 0 \end{Bmatrix}$$

$$\{S_i\} = \begin{Bmatrix} 0 \\ 0 \end{Bmatrix}$$

The current vectors may be expressed as follows:

$$I_1 = x_1 + y_1 J$$

$$I_2 = x_2 + y_2 J$$

Therefore

$$\{x\} = \begin{Bmatrix} x_1 \\ x_2 \end{Bmatrix}$$

$$\{y\} = \begin{Bmatrix} y_1 \\ y_2 \end{Bmatrix}$$

Substituting into equation (4.46) gives the following set of linear algebraic equations with real coefficients:

$$\begin{bmatrix} \begin{bmatrix} 1.5 & -0.5 \\ -0.5 & 2.5 \end{bmatrix} & -\begin{bmatrix} 0.754 & -0.754 \\ -0.754 & 0.223 \end{bmatrix} \\ \begin{bmatrix} 0.754 & -0.754 \\ -0.754 & 0.223 \end{bmatrix} & \begin{bmatrix} 1.5 & -0.5 \\ -0.5 & 2.5 \end{bmatrix} \end{bmatrix} \begin{Bmatrix} \begin{Bmatrix} x_1 \\ x_2 \end{Bmatrix} \\ \begin{Bmatrix} y_1 \\ y_2 \end{Bmatrix} \end{Bmatrix} = \begin{Bmatrix} \begin{Bmatrix} 100 \\ 0 \end{Bmatrix} \\ \begin{Bmatrix} 0 \\ 0 \end{Bmatrix} \end{Bmatrix}$$

Obviously, the brackets of the submatrices can be dropped. These equations can now be solved for the unknowns by using any of the methods outlined earlier in connection with sets involving real coefficients to give

$$\begin{Bmatrix} x_1 \\ x_2 \\ y_1 \\ y_2 \end{Bmatrix} = \begin{Bmatrix} 58.05 \\ 17.60 \\ -16.09 \\ 12.72 \end{Bmatrix}$$

The currents are then given as

$$I_1 = 58.05 - 16.09J, \qquad I_2 = 17.60 + 12.72J$$

## 4.14 COMPARISON OF METHOD EFFICIENCIES

The computational effort required for solving a system of $n \times n$ linear algebraic equations varies substantially with the method being employed. While certain methods are better suited for a special class of problems, the fact remains that some methods are more efficient than others. Consider, for example, the square root method. Is it really the most efficient even if the coefficient matrix contains a significant number of zeros? The answer is obviously no. It is apparent that efficiency is relative in magnitude and depends on such factors as size of set, coefficient matrix, content, and in some cases ingenuity.

One criterion for comparing the efficiencies of the solution methods pertaining to linear algebraic equations is to determine the number of required arithmetic operations. This number can then be used to study the relative effect of the system's size on the amount of effort needed for a given set of equations. Table 4.3 summarizes the total number of multiplications and divisions as a function of system size $n$.

Obviously, the total number of operations required for the iterative methods described earlier depends on the number of nonzero coefficients involved in a given

Table 4.3 TOTAL NUMBER OF OPERATIONS REQUIRED FOR VARIOUS METHODS

| METHOD | NUMBER OF OPERATIONS |
|---|---|
| Cramer's rule | $(n+1)(n-1)(n!)+n$ |
| Gauss | $\frac{1}{3}(n^3+3n^2-n)$ |
| Gauss–Jordan | $\frac{1}{3}(n^3+3n^2-n)$ |
| Crout's | $\frac{1}{6}(4n^3+9n^2-7n)$ |
| Reducing matrix | $\frac{1}{3}(n^3+3n^2-n)$ |
| Complex coefficients | $2.3(n^3+3n^2-n)$ |
| Iterative | Not applicable |

set. Consequently, no general rule can be established. Furthermore, the given total number of operations required by Cramer's rule is based on the assumption that the determinants are expanded by using the method of cofactors.

## Recommended Reading

*Numerical Methods for Scientists and Engineers*, R. W. Hamming, McGraw-Hill Book Company, New York, 1962.

*Matrix Iterative Analysis*, R. S. Varga, Prentice-Hall, Inc., Englewood Cliffs, N.J., 1962.

*Rounding Errors in Algebraic Processes*, J. H. Wilkinson, Prentice-Hall, Inc., Englewood Cliffs, N.J., 1963.

*Iterative Methods for the Solution of Equations*, J. F. Traub, Prentice-Hall, Inc., Englewood Cliffs, N.J., 1964.

*Introduction to Matrix Computations*, B. W. Steward, Academic Press, New York, 1973.

## PROBLEMS

**4.1** Solve the following linear algebraic equation using Cramer's rule:

$$12y_1 - 4y_2 + y_3 = 1$$
$$y_1 + 4y_2 - 2y_3 = 2$$
$$-3y_1 + y_2 + 6y_3 = 7$$

**(a)** Expand determinants using the method of cofactors.
**(b)** Expand determinants using the method of pivotal condensation.
**(c)** Expand determinants using the upper-triangle method. Switch rows and/or columns to simplify the computations.

**4.2** Solve the set of equations given in problem 4.1 using the
**(a)** Gauss method,
**(b)** Gauss–Jordan method,
**(c)** Crout's method.

**4.3** Use the square root method to solve the following set of equations:

$$4x_1 + x_2 + 5x_3 = 5$$
$$x_1 - 4x_2 + 7x_3 = 2$$
$$5x_1 + 7x_2 - x_3 = 6$$

**4.4** Use the Gauss–Jordan method to solve for $x_1$, $x_2$, and $x_3$ in terms of $A_1$, $A_2$, and $A_3$:

$$x_1 - 7x_2 + 3x_3 = A_1$$
$$2x_1 + 6x_2 - x_3 = A_2$$
$$3x_1 + x_2 + 4x_3 = A_3$$

**4.5** Decompose the following matrix into upper- and lower-triangular matrices:

$$[A] = \begin{bmatrix} 4 & 1 & 3 & 1 \\ 6 & 6 & -1 & 7 \\ 1 & -8 & 8 & 3 \\ 2 & 2 & 6 & 9 \end{bmatrix}$$

**4.6** Solve problem 4.3 using the reducing matrix method.

**4.7** Determine $x_1$ and $x_2$ in terms of $y_1$ and $y_2$ for the following set:

$$2x_1 - x_2 = y_1 + y_2$$
$$x_1 + 2x_2 = 2y_1 - y_2$$

**(a)** using the Gauss method,
**(b)** using the Gauss–Jordan method.

**4.8** Solve the following set of equations using the tridiagonal method:

$$\begin{bmatrix} 1 & 2 & 0 \\ 3 & 4 & -1 \\ 0 & 1 & 5 \end{bmatrix} \begin{Bmatrix} x_1 \\ x_2 \\ x_3 \end{Bmatrix} = \begin{Bmatrix} 1 \\ 1 \\ 2 \end{Bmatrix}$$

**4.9** Approximate the solution to the following set of equations using the iterative procedure, beginning with the initial vector $\{x_1 \quad x_2\} = \{0 \quad 0\}$ and using three iterations:

$$4x_1 - x_2 = 3$$
$$x_1 + 4x_2 = 2$$

**(a)** using the Jacobi method,
**(b)** using the Gauss–Seidel method.

**4.10** Can the following sets of equations be solved by using the iterative methods? Try an initial vector of $\{x\} = \{0\}$ and ten iterations. Use the Jacobi method.

**(a)** $x_1 + x_2 = 1$, $2x_1 + x_2 = 1$ **(b)** $2x_1 - x_2 = 3$, $3x_1 + x_2 = 1$

**4.11** Consider the following set of equations:

$$x_1 + x_2 = 2$$
$$ax_1 + x_2 = 1$$

**(a)** Solve for $x_1$ and $x_2$ if $a = 1.1$.
**(b)** Solve for $x_1$ and $x_2$ if $a = 0.9$.
**(c)** Plot the regions of feasible solutions.

**4.12** Solve the following set of equations for $x_1$ and $x_2$ in terms of $x_3$:

$$x_1 + x_2 - x_3 = 3$$

$$2x_1 - 3x_2 + 2x_3 = 1$$

**(a)** using the Gauss method,
**(b)** using the Gauss–Jordan method.

**4.13** Use the procedure outlined in section 4.12 to determine the best solution for the following set of equations:

$$x_1 + 2x_2 = 1$$

$$2x_1 + 3x_2 = 2$$

$$7x_1 - 4x_2 = 5$$

**4.14** Determine the solution to the following complex set of linear algebraic equations:

$$(2 + J)Z_1 + (1 + 2J)Z_2 = 1$$

$$(1 - 2J)Z_1 + (2 + 3J)Z_2 = 3$$

**4.15** The reactions of the highway piers shown below are related to the applied load $P$ by the set of linear algebraic equations given below.

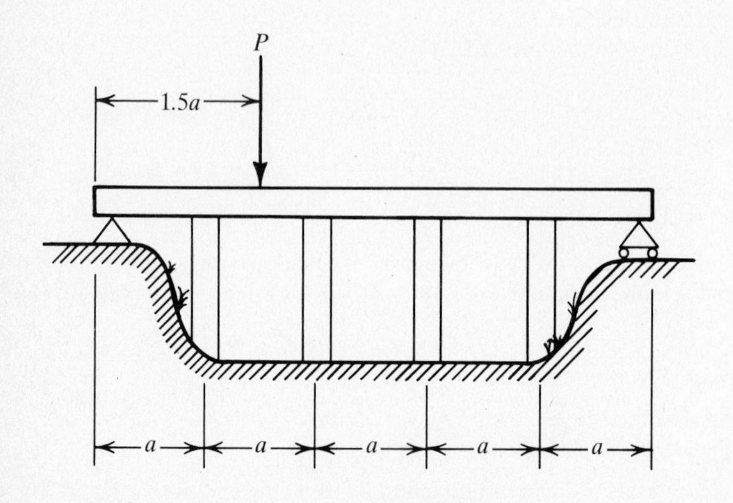


$$\begin{bmatrix} 1.067 & 1.500 & 1.333 & 0.767 \\ 1.500 & 2.400 & 2.267 & 1.333 \\ 1.333 & 2.267 & 2.400 & 1.500 \\ 0.767 & 1.333 & 1.500 & 1.067 \end{bmatrix} \begin{Bmatrix} R_1 \\ R_2 \\ R_3 \\ R_4 \end{Bmatrix} = \begin{Bmatrix} 1.370 \\ 2.063 \\ 1.875 \\ 1.088 \end{Bmatrix} P$$

Solve for the reactions $R_1, \ldots, R_4$ using
**(a)** the square root method,
**(b)** the Gauss–Jordan method.

**4.16** The deflection of the following beam is given in terms of the constants of integration $C_1$ and $C_2$ as follows:

$$y = \frac{q}{EI}\frac{x^4}{24} + C_1\frac{x^3}{6} + C_2\frac{x^2}{2}$$

where

$$\begin{bmatrix} L & 1 \\ L & 3 \end{bmatrix}\begin{Bmatrix} C_1 \\ C_2 \end{Bmatrix} = -\begin{Bmatrix} \frac{1}{2} \\ \frac{1}{4} \end{Bmatrix}\left(\frac{qL^2}{EI}\right)$$

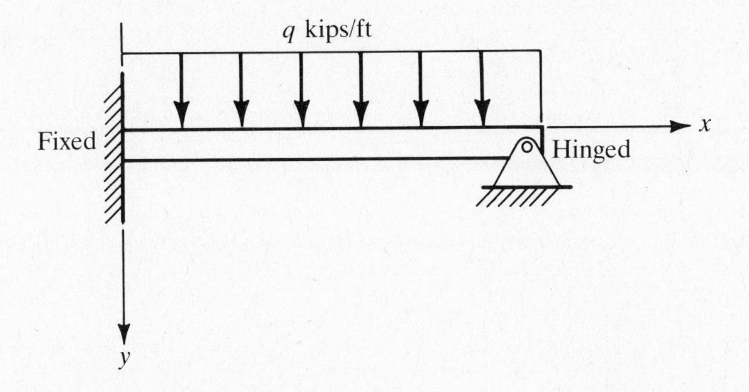


Determine the deflection equation. Note that $L$ denotes length, $q$ the distributed load per unit length, $E$ Young's modulus (material property), and $I$ the moment of inertia (property of beam cross section).

**4.17** The displacement of the following assembly due to the load $P$ and the weights corresponding to masses $m_1$ and $m_2$ is given by

$$(2k_1 + k_2)x_1 - k_2x_2 = m_1g$$
$$-k_2x_1 + k_2x_2 = m_2g + P$$

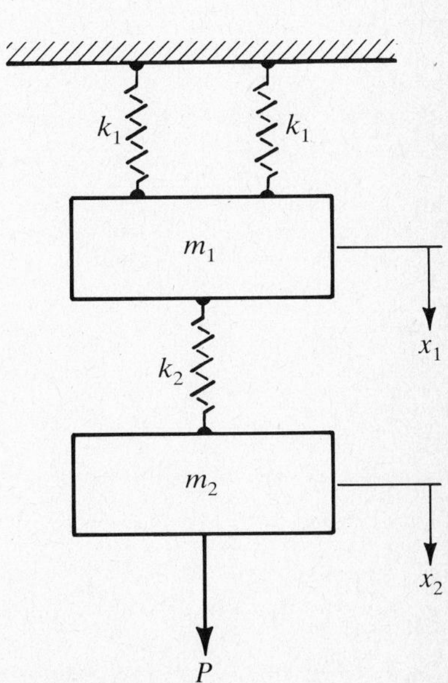

Note that $g$ is the gravitational acceleration. Use Cramer's rule to solve for $x_1$ and $x_2$.

**4.18** Determine the currents for the following electrical circuit:

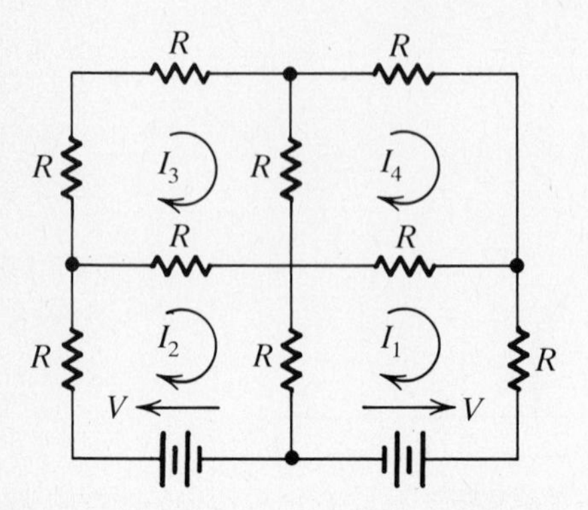


where the currents $I_1, \ldots, I_4$ are related to the resistance $R$ and the voltage $V$ as follows:

$$\begin{bmatrix} 3 & -1 & 0 & -1 \\ -1 & 3 & -1 & 0 \\ 0 & -1 & 4 & -1 \\ -1 & 0 & -1 & 3 \end{bmatrix} \begin{Bmatrix} I_1 \\ I_2 \\ I_3 \\ I_4 \end{Bmatrix} = \begin{Bmatrix} -1 \\ 1 \\ 0 \\ 0 \end{Bmatrix} \frac{V}{R}$$

**(a)** use the square root method,
**(b)** use the Gauss–Jordan method,
**(c)** use the Crout method.

# chapter 5

# Matrix Inversion

The space shuttle on a launching pad. *Courtesy of N.A.S.A.*

"Matrix inversion plays an important role in solving many engineering problems."

# 5

## 5.1 INTRODUCTION

This chapter is a continuation of the preceding chapter in that it deals with topics pertaining to the solution of linear algebraic equations. Many applications including numerical differentiation and numerical integration involve systems of linear algebraic equations whose right-hand sides are not necessarily constants, but variables. Furthermore, the solution of linear algebraic equations with complex coefficients using real analysis involves a set of two matrix equations whose unknowns are vectors rather than variables. The inverse matrix provides the tool needed for addressing these problems.

Recall that the inverse of a matrix $[A]$ is a property of square nonsingular matrices (determinant is not zero) and is denoted $[A]^{-1}$. In addition,

$$[A][A]^{-1} = [A]^{-1}[A] = [I] \tag{5.1}$$

The implication is that commutativity holds. The concept of an inverse matrix is similar to the inverse of a scalar. That is, the inverse of the scalar quantity $a$ is $a^{-1}$ and $aa^{-1} = 1$. Hence one may look at the identity matrix as a unity matrix. Consider the following set of linear algebraic equations:

$$\begin{aligned} a_{11}x_1 + a_{12}x_2 + a_{13}x_3 &= b_{11}y_1 + b_{12}y_2 \\ a_{21}x_1 + a_{22}x_2 + a_{23}x_3 &= b_{21}y_1 + b_{22}y_2 \\ a_{31}x_1 + a_{32}x_2 + a_{33}x_3 &= b_{31}y_1 + b_{32}y_2 \end{aligned} \tag{5.2}$$

This set can be easily transformed into a simpler form by letting

$$\begin{aligned} z_1 &= b_{11}y_1 + b_{12}y_2 \\ z_2 &= b_{21}y_1 + b_{22}y_2 \\ z_3 &= b_{31}y_1 + b_{32}y_2 \end{aligned}$$

Then substituting into equation (5.2) yields the following equivalent set:

$$\begin{bmatrix} a_{11} & a_{12} & a_{13} \\ a_{21} & a_{22} & a_{23} \\ a_{31} & a_{32} & a_{33} \end{bmatrix} \begin{Bmatrix} x_1 \\ x_2 \\ x_3 \end{Bmatrix} = \begin{Bmatrix} z_1 \\ z_2 \\ z_3 \end{Bmatrix} \tag{5.3a}$$

or in compact matrix form

$$[A]\{X\} = \{Z\} \tag{5.3b}$$

If the $[A]$ matrix is not singular, then its inverse $[A]^{-1}$ can be determined. Consequently, premultiplying equation (5.3) by $[A]^{-1}$ yields

$$[A]^{-1}[A]\{X\} = [A]^{-1}\{Z\} \tag{5.4}$$

The implication is rather significant in that we can now express the solution of a set of equations as a linear combination of the $\{Z\}$ vector. Consequently, if $[A]$ remains unchanged in a given application, the solution is directly given for any $\{Z\}$ vector by equation (5.4). The obvious question is, how do we determine the inverse?

## 5.2 CRAMER'S RULE

This method of inverting a matrix is also known as the method of cofactors. It is one of the most elegant methods available. Unfortunately, it suffers from the same limitations of inefficiency as Cramer's rule for solving linear algebraic equations. The advantage of the method lies in its simplicity, especially when dealing with matrices whose coefficients are variables and/or functions rather than scalars. The basis for this method can be easily illustrated by using Cramer's rule to solve the set of equations given by (5.3). Hence

$$x_j = |A_j|/|A|, \qquad j = 1, 2, 3$$

where

$|A|$ = determinant of the $[A]$ matrix

$|A_j|$ = determinant of the $[A]$ matrix whose $j$th column is replaced with the $\{Z\}$ vector.

If we are to evaluate $|A_j|$ using the method of cofactors by expanding about the column containing the $\{Z\}$ vector, then

$$x_j = \frac{1}{|A|} \sum_{i=1}^{n} c_{ij} z_i, \qquad j = 1, 2, 3$$

where the $c_{ij}$ are the cofactors of the $[A]$ matrix. Expanding the compact form about $j$ and summing about $i$ gives

$$x_1 = \frac{1}{|A|} [c_{11} z_1 + c_{21} z_2 + c_{31} z_3]$$

$$x_2 = \frac{1}{|A|} [c_{12} z_1 + c_{22} z_2 + c_{32} z_3]$$

$$x_3 = \frac{1}{|A|} [c_{13} z_1 + c_{23} z_2 + c_{33} z_3]$$

or in matrix form

$$\begin{Bmatrix} x_1 \\ x_2 \\ x_3 \end{Bmatrix} = \frac{1}{|A|} \begin{bmatrix} c_{11} & c_{21} & c_{31} \\ c_{12} & c_{22} & c_{32} \\ c_{13} & c_{23} & c_{33} \end{bmatrix} \begin{Bmatrix} z_1 \\ z_2 \\ z_3 \end{Bmatrix} \tag{5.5a}$$

or in compact matrix form

$$\{X\} = \frac{1}{|A|} [C]^t \{Z\} \tag{5.5b}$$

Comparison of equation (5.5) with equation (5.4) clearly shows that

$$[A]^{-1} = \frac{1}{|A|}[c_{ij}]^{t}, \qquad i = 1, \ldots, n, \quad j = 1, \ldots, n \tag{5.6}$$

Equation (5.6) simply states that the inverse of a matrix $[A]$ is equal to the transpose of the cofactors matrix divided by the determinant of $[A]$. In some textbooks, $[c_{ij}]^{t}$ is referred to as the "adjoint of $[A]$." Furthermore, if we are interested in solving for the $\{X\}$ vector in terms of the $\{Y\}$ vector rather than the $\{Z\}$ vector, then equation (5.5) can be written as

$$\begin{Bmatrix} x_1 \\ x_2 \\ x_3 \end{Bmatrix} = \frac{1}{|A|}\begin{bmatrix} c_{11} & c_{12} & c_{13} \\ c_{21} & c_{22} & c_{23} \\ c_{31} & c_{32} & c_{33} \end{bmatrix}^{t}\begin{bmatrix} b_{11} & b_{12} \\ b_{21} & b_{22} \\ b_{31} & b_{32} \end{bmatrix}\begin{Bmatrix} y_1 \\ y_2 \\ y_3 \end{Bmatrix}$$

or in compact matrix form

$$\{X\} = \frac{1}{|A|}[C]^{t}[B]\{Y\} = [A]^{-1}[B]\{Y\}$$

It is evident that the inverse approach to solving systems of linear algebraic equations is much neater than the methods outlined earlier in chapter 4. Unfortunately, it requires more effort. Note that the inverse approach involves dividing by the determinant of the coefficient matrix. Thus no inverse can be determined if $|A| = 0$. This is precisely the condition we placed on inverse calculations of a square matrix.

EXAMPLE 5.1

Solve the following set of equations using Cramer's method of cofactors:

$$2x_1 + x_2 = y_1$$

$$2x_1 + 3x_2 = y_2$$

Solution

Expressing the set in matrix form and then computing the cofactors yields

$$c_{11} = 3, \qquad c_{12} = -2$$

$$c_{21} = -1, \qquad c_{22} = 2$$

$$[C] = \begin{bmatrix} 3 & -2 \\ -1 & 2 \end{bmatrix}$$

The determinant of $[A]$ is calculated next from the cofactors just computed. Hence

$$|A| = (2)(3) - (2)(1) = 4$$

The inverse is then determined by using equation (5.6):

$$[A]^{-1} = \frac{1}{|A|}[c_{ij}]^{t} = \frac{1}{4}\begin{bmatrix} 3 & -1 \\ -2 & 2 \end{bmatrix}$$

The solution is given as follows:

$$\begin{Bmatrix} x_1 \\ x_2 \end{Bmatrix} = \frac{1}{4}\begin{bmatrix} 3 & -1 \\ -2 & 2 \end{bmatrix}\begin{Bmatrix} y_1 \\ y_2 \end{Bmatrix}$$

Thus

$$x_1 = \tfrac{3}{4}y_1 - \tfrac{1}{4}y_2, \qquad x_2 = -\tfrac{1}{2}y_1 + \tfrac{1}{2}y_2$$

E

EXAMPLE 5.2

Solve the following set of equations for $x_1$, $x_2$, and $x_3$ in terms of $y_1$ and $y_2$ using the inverse procedure:

$$\begin{aligned} x_1 + 2x_2 - 3x_3 &= 1 + 2y_1 + 3y_2 \\ 3x_1 - x_2 + 2x_3 &= 5 - y_1 + 2y_2 \\ 7x_1 + 4x_2 + x_3 &= 2 + 3y_1 - y_2 \end{aligned}$$

Solution

We begin by expressing the set in matrix form as follows:

$$\begin{bmatrix} 1 & 2 & -3 \\ 3 & -1 & 2 \\ 7 & 4 & 1 \end{bmatrix}\begin{Bmatrix} x_1 \\ x_2 \\ x_3 \end{Bmatrix} = \begin{bmatrix} 1 & 2 & 3 \\ 5 & -1 & 2 \\ 2 & 3 & -1 \end{bmatrix}\begin{Bmatrix} 1 \\ y_1 \\ y_2 \end{Bmatrix}$$

In compact matrix form, we have

$$[A]\{X\} = [B]\{Y\}$$

Solving for the $\{X\}$ vector gives

$$\{X\} = [A]^{-1}[B]\{Y\} \tag{5.7}$$

Therefore the problem is now reduced to determining the inverse of the $[A]$ matrix. Using Cramer's method we determine the cofactors as follows:

$$c_{11} = \begin{vmatrix} -1 & 2 \\ 4 & 1 \end{vmatrix} = -9$$

$$c_{21} = -\begin{vmatrix} 2 & -3 \\ 4 & 1 \end{vmatrix} = -14$$

$$c_{31} = \begin{vmatrix} 2 & -3 \\ -1 & 2 \end{vmatrix} = 1$$

$$c_{12} = -\begin{vmatrix} 3 & 2 \\ 7 & 1 \end{vmatrix} = 11$$

$$c_{22} = \begin{vmatrix} 1 & -3 \\ 7 & 1 \end{vmatrix} = 22$$

$$c_{32} = -\begin{vmatrix} 1 & -3 \\ 3 & 2 \end{vmatrix} = -11$$

$$c_{13} = \begin{vmatrix} 3 & -1 \\ 7 & 4 \end{vmatrix} = 19$$

$$c_{23} = -\begin{vmatrix} 1 & 2 \\ 7 & 4 \end{vmatrix} = 10$$

$$c_{33} = \begin{vmatrix} 1 & 2 \\ 3 & -1 \end{vmatrix} = -7$$

The cofactors matrix is then determined as

$$[C] = \begin{bmatrix} -9 & 11 & 19 \\ -14 & 22 & 10 \\ 1 & -11 & -7 \end{bmatrix}$$

The determinant of the $[A]$ matrix is calculated from the cofactors of a row or column. That is,

$$\begin{aligned} \det[A] &= \sum_{i=1}^{3} c_{ij}a_{ij}, \qquad j = 1 \\ &= c_{11}a_{11} + c_{21}a_{21} + c_{31}a_{31} \\ &= (-9)(1) + (-14)(3) + (1)(7) \\ &= -44 \end{aligned}$$

The inverse is given as

$$[A]^{-1} = \frac{1}{-44}\begin{bmatrix} -9 & -14 & 1 \\ 11 & 22 & -11 \\ 19 & 10 & -7 \end{bmatrix}$$

Substituting into equation (5.7) gives

$$\begin{Bmatrix} x_1 \\ x_2 \\ x_3 \end{Bmatrix} = \frac{1}{44}\begin{bmatrix} 9 & 14 & -1 \\ -11 & -22 & 11 \\ -19 & -10 & 7 \end{bmatrix}\begin{bmatrix} 1 & 2 & 3 \\ 5 & -1 & 2 \\ 2 & 3 & -1 \end{bmatrix}\begin{Bmatrix} 1 \\ y_1 \\ y_2 \end{Bmatrix}$$

which simplifies to

$$\begin{Bmatrix} x_1 \\ x_2 \\ x_3 \end{Bmatrix} = \frac{1}{44}\begin{bmatrix} 77 & 1 & 56 \\ -99 & 33 & -88 \\ -55 & -7 & -84 \end{bmatrix}\begin{Bmatrix} 1 \\ y_1 \\ y_2 \end{Bmatrix}$$

The solution for the unknown vector $\{X\}$ is now given as a linear combination of $y_1$ and $y_2$.

## 5.3 ELIMINATION METHOD

This method is an extension of the Gauss–Jordan method for solving systems of linear algebraic equations outlined in chapter 4. Consider the following $3 \times 3$ matrix:

$$[A] = \begin{bmatrix} a_{11} & a_{12} & a_{13} \\ a_{21} & a_{22} & a_{23} \\ a_{31} & a_{32} & a_{33} \end{bmatrix}$$

Suppose that the inverse of $[A]$ is given by a second matrix $[B]$. Hence

$$[A]^{-1} = [B] = \begin{bmatrix} b_{11} & b_{12} & b_{13} \\ b_{21} & b_{22} & b_{23} \\ b_{31} & b_{32} & b_{33} \end{bmatrix}$$

Then by definition we have

$$[A][B] = [B][A] = [I]$$

or more explicitly

$$\begin{bmatrix} a_{11} & a_{12} & a_{13} \\ a_{21} & a_{22} & a_{23} \\ a_{31} & a_{32} & a_{33} \end{bmatrix}\begin{bmatrix} b_{11} & b_{12} & b_{13} \\ b_{21} & b_{22} & b_{23} \\ b_{31} & b_{32} & b_{33} \end{bmatrix} = \begin{bmatrix} 1 & 0 & 0 \\ 0 & 1 & 0 \\ 0 & 0 & 1 \end{bmatrix}$$

Multiplication of $[A]$ by $[B]$ gives

$$\begin{aligned} a_{11}b_{11} + a_{12}b_{21} + a_{13}b_{31} &= 1 \\ a_{21}b_{11} + a_{22}b_{21} + a_{23}b_{31} &= 0 \\ a_{31}b_{11} + a_{32}b_{21} + a_{33}b_{31} &= 0 \end{aligned} \qquad \textbf{(5.8a)}$$

$$\begin{aligned} a_{11}b_{12} + a_{12}b_{22} + a_{13}b_{32} &= 0 \\ a_{21}b_{12} + a_{22}b_{22} + a_{23}b_{32} &= 1 \\ a_{31}b_{12} + a_{32}b_{22} + a_{33}b_{32} &= 0 \end{aligned} \tag{5.8b}$$

$$\begin{aligned} a_{11}b_{13} + a_{12}b_{23} + a_{13}b_{33} &= 0 \\ a_{21}b_{13} + a_{22}b_{23} + a_{23}b_{33} &= 0 \\ a_{31}b_{13} + a_{32}b_{23} + a_{33}b_{33} &= 1 \end{aligned} \tag{5.8c}$$

Note that equation (5.8) represents a set of nine equations in nine unknowns. The implication is that the inverse matrix requires $n$ times the computational effort necessary for solving a set of $n \times n$ equations! However, careful consideration of the coefficient matrix $[A]$ in each set reveals that it is exactly the same and only the designation of the unknowns $b_{11}, \ldots, b_{33}$ change. Therefore we may operate on the coefficient matrix directly using the Gauss–Jordan method:

$$[[A] \vdots [I]] = \left[\begin{array}{ccc|ccc} a_{11} & a_{12} & a_{13} & 1 & 0 & 0 \\ a_{21} & a_{22} & a_{23} & 0 & 1 & 0 \\ a_{31} & a_{32} & a_{33} & 0 & 0 & 1 \end{array}\right]$$

The object is then to reduce the augmented matrix to the following form:

$$[[I] \vdots [B]] = \left[\begin{array}{ccc|ccc} 1 & 0 & 0 & b_{11} & b_{12} & b_{13} \\ 0 & 1 & 0 & b_{21} & b_{22} & b_{23} \\ 0 & 0 & 1 & b_{31} & b_{32} & b_{33} \end{array}\right]$$

where $[B]$ is the inverse of $[A]$. This method of determining the inverse is summarized in one step for an $n \times n$ matrix:

$$[[A] \vdots [I]] \Rightarrow [[I] \vdots [B]] \tag{5.9}$$

Equation (5.9) states that given an $n \times n$ matrix $[A]$, an augmented matrix containing $[A]$ and its corresponding identity can be formed and that the inverse matrix is determined if the augmented matrix is transformed to a new augmented matrix in which the identity matrix replaces the $[A]$ matrix.

**E** EXAMPLE 5.3

Use the Gauss–Jordan elimination method to determine the solution of the following set of equations:

$$\begin{aligned} x_1 + 2x_2 &= y_1 \\ 2x_1 - x_2 &= y_2 \end{aligned}$$

Solution

$$\left[\begin{array}{cc|cc} 1 & 2 & 1 & 0 \\ 2 & -1 & 0 & 1 \end{array}\right] \begin{array}{c} \\ -2R_1 + R_2 \end{array} \Rightarrow \left[\begin{array}{cc|cc} 1 & 2 & 1 & 0 \\ 0 & -5 & -2 & 1 \end{array}\right] R_2/-5$$

$$\left[\begin{array}{cc|cc} 1 & 2 & 1 & 0 \\ 0 & 1 & \frac{2}{5} & -\frac{1}{5} \end{array}\right] \begin{array}{c} \\ -2R_2 + R_1 \end{array} \Rightarrow \left[\begin{array}{cc|cc} 1 & 0 & \frac{1}{5} & \frac{2}{5} \\ 0 & 1 & \frac{2}{5} & -\frac{1}{5} \end{array}\right]$$

The inverse is simply equal to the right-hand portion, namely,

$$[A]^{-1} = \frac{1}{5}\begin{bmatrix} 1 & 2 \\ 2 & -1 \end{bmatrix}$$

The solution is then given as follows:

$$\begin{Bmatrix} x_1 \\ x_2 \end{Bmatrix} = \frac{1}{5}\begin{bmatrix} 1 & 2 \\ 2 & -1 \end{bmatrix}\begin{Bmatrix} y_1 \\ y_2 \end{Bmatrix} = \begin{Bmatrix} \frac{1}{5}y_1 + \frac{2}{5}y_2 \\ \frac{2}{5}y_1 - \frac{1}{5}y_2 \end{Bmatrix}$$

## 5.4 REDUCING MATRIX METHOD

This method represents a slight modification of the reducing matrix method outlined in the previous chapter for solving systems of linear algebraic equations. The basis for the method can be best explained by considering the following set of equations:

$$\begin{bmatrix} a_{11} & a_{12} & a_{13} \\ a_{21} & a_{22} & a_{23} \\ a_{31} & a_{32} & a_{33} \end{bmatrix}\begin{Bmatrix} x_1 \\ x_2 \\ x_3 \end{Bmatrix} = \begin{Bmatrix} y_1 \\ y_2 \\ y_3 \end{Bmatrix} \tag{5.10}$$

The first column of the coefficient matrix in equation (5.10) can be reduced to that of its equivalent identity by simply multiplying both sides by the following reducing matrix:

$$[R_1] = \begin{bmatrix} 1/a_{11} & 0 & 0 \\ -a_{21}/a_{11} & 1 & 0 \\ -a_{31}/a_{11} & 0 & 1 \end{bmatrix}$$

Giving

$$\begin{bmatrix} 1 & a'_{12} & a'_{13} \\ 0 & a'_{22} & a'_{23} \\ 0 & a'_{32} & a'_{33} \end{bmatrix}\begin{Bmatrix} x_1 \\ x_2 \\ x_3 \end{Bmatrix} = [R_1]\begin{Bmatrix} y_1 \\ y_2 \\ y_3 \end{Bmatrix} \tag{5.11}$$

Now, reducing the second column of the new coefficient matrix to that of an identity is accomplished by multiplying equation (5.11) by the following reducing matrix:

$$[R_2] = \begin{bmatrix} 1 & -a'_{12}/a'_{22} & 0 \\ 0 & 1/a'_{22} & 0 \\ 0 & -a'_{31}/a'_{22} & 1 \end{bmatrix}$$

This gives

$$\begin{bmatrix} 1 & 0 & a''_{13} \\ 0 & 1 & a''_{23} \\ 0 & 0 & a''_{33} \end{bmatrix} \begin{Bmatrix} x_1 \\ x_2 \\ x_3 \end{Bmatrix} = [R_2][R_1] \begin{Bmatrix} y_1 \\ y_2 \\ y_3 \end{Bmatrix} \tag{5.12}$$

Finally, the third column is reduced by multiplying equation (5.12) by the following reducing matrix:

$$[R_3] = \begin{bmatrix} 1 & 0 & -a''_{13}/a''_{33} \\ 0 & 1 & -a''_{23}/a''_{33} \\ 0 & 0 & 1/a''_{33} \end{bmatrix}$$

which gives the solution for the $\{x\}$ vector as follows:

$$\begin{bmatrix} 1 & 0 & 0 \\ 0 & 1 & 0 \\ 0 & 0 & 1 \end{bmatrix} \begin{Bmatrix} x_1 \\ x_2 \\ x_3 \end{Bmatrix} = [R_3][R_2][R_1] \begin{Bmatrix} y_1 \\ y_2 \\ y_3 \end{Bmatrix}$$

Careful consideration of equation (5.10) clearly indicates that the inverse of the coefficient matrix is given by

$$[A]^{-1} = [R_3][R_2][R_1]$$

In general, for an $n \times n$ matrix the inverse is given by

$$[A]^{-1} = [R_n][R_{n-1}] \cdots [R_1] \tag{5.13}$$

where the coefficients of the reducing matrices are determined from the coefficient matrix. It should be pointed out that only partial matrix multiplication is necessary. This is an important consideration if the computational effort is to be reduced. Furthermore, if the pivot coefficient $a_{ii}$ at a given iteration is zero then the row in which that coefficient appears must be switched with another row appearing below it.

EXAMPLE 5.4

Solve the following set of equations using the reducing matrix method:

$$\begin{bmatrix} 1 & -1 & 1 \\ 2 & 1 & -3 \\ 1 & 5 & 2 \end{bmatrix} \begin{Bmatrix} x_1 \\ x_2 \\ x_3 \end{Bmatrix} = \begin{Bmatrix} y_1 \\ y_2 \\ y_3 \end{Bmatrix}$$

## Solution

Multiplying the coefficient matrix by the first reducing matrix gives

$$[R_1][A] = [A_1]$$

$$\begin{bmatrix} 1 & 0 & 0 \\ -2 & 1 & 0 \\ -1 & 0 & 1 \end{bmatrix} \begin{bmatrix} 1 & -1 & 1 \\ 2 & 1 & -3 \\ 1 & 5 & 2 \end{bmatrix} = \begin{bmatrix} 1 & -1 & 1 \\ 0 & 3 & -5 \\ 0 & 6 & 1 \end{bmatrix}$$

Note that the first row and first column of the new coefficient matrix $[A_1]$ are predetermined. Hence there is no need to carry out the multiplication for their coefficients. Now, multiplying $[A_1]$ by the second reducing matrix gives

$$[R_2][A_1] = [A_2]$$

$$\begin{bmatrix} 1 & \frac{1}{3} & 0 \\ 0 & \frac{1}{3} & 0 \\ 0 & -2 & 1 \end{bmatrix} \begin{bmatrix} 1 & -1 & 1 \\ 0 & 3 & -5 \\ 0 & 6 & 1 \end{bmatrix} = \begin{bmatrix} 1 & 0 & -\frac{2}{3} \\ 0 & 1 & -\frac{5}{3} \\ 0 & 0 & 11 \end{bmatrix}$$

Note that there is no need to multiply the $[A_2]$ matrix by the third reducing matrix. (Why?) Consequently, the third reducing matrix is determined from the $[A_2]$ matrix as

$$[R_3] = \begin{bmatrix} 1 & 0 & \frac{2}{33} \\ 0 & 1 & \frac{5}{33} \\ 0 & 0 & \frac{1}{11} \end{bmatrix}$$

The inverse is then determined by using equation (5.13) as follows:

$$[A]^{-1} = \begin{bmatrix} 1 & 0 & \frac{2}{11} \\ 0 & 1 & \frac{5}{33} \\ 0 & 0 & \frac{1}{11} \end{bmatrix} \begin{bmatrix} 1 & \frac{1}{3} & 0 \\ 0 & \frac{1}{3} & 0 \\ 0 & -2 & 1 \end{bmatrix} \begin{bmatrix} 1 & 0 & 0 \\ -2 & 1 & 0 \\ -1 & 0 & 1 \end{bmatrix}$$

$$= \begin{bmatrix} 1 & 0 & \frac{2}{33} \\ 0 & 1 & \frac{5}{33} \\ 0 & 0 & \frac{1}{11} \end{bmatrix} \begin{bmatrix} \frac{1}{3} & \frac{1}{3} & 0 \\ -\frac{2}{3} & \frac{1}{3} & 0 \\ 3 & -2 & 1 \end{bmatrix}$$

$$= \frac{1}{33} \begin{bmatrix} 17 & 7 & 2 \\ -7 & 1 & 5 \\ 9 & -6 & 3 \end{bmatrix}$$

Consequently, the solution to the set is now readily determined. That is,

$$\begin{Bmatrix} x_1 \\ x_2 \\ x_3 \end{Bmatrix} = \frac{1}{33} \begin{bmatrix} 17 & 7 & 2 \\ -7 & 1 & 5 \\ 9 & -6 & 3 \end{bmatrix} \begin{Bmatrix} y_1 \\ y_2 \\ y_3 \end{Bmatrix}$$

It is evident that a solution to the $\{X\}$ vector can be easily calculated for any $\{Y\}$ vector. The reader is reminded that Crout's method permits the factorization of a matrix $[A]$ into upper- and lower-triangular matrices (see section 4.7). Thus

$$[A] = [L][T]$$

Therefore, given a set of equations of the form

$$[A]\{X\} = \{Y\}$$

then

$$[L][T]\{X\} = \{Y\} \tag{5.14}$$

As a result, if we define

$$\{Z\} = [T]\{X\} \tag{5.15}$$

Then equation (5.15) can be substituted into (5.14) to give

$$[L]\{Z\} = \{Y\} \tag{5.16}$$

Consequently, solving equation (5.16) for the $\{Z\}$ vector by forward substitution then solving equation (5.15) for the $\{X\}$ vector provides a second procedure for using factorization in inverting a matrix.

## 5.5 PARTITIONING METHOD

The methods just outlined are fairly efficient in solving the inverse problem using hand-held calculators for small matrices; otherwise, a large computer must be used for solving large matrices. A problem arises in cases where one does not have access to a large computer.

This section should help you double the capacity of your personal computer to solve a system of equations and/or invert a matrix from $n \times n$ to $2n \times 2n$.

Assume that a given matrix is partitioned in such a way that the leading-diagonal submatrices are square matrices:

$$[A] = \begin{bmatrix} [A_{11}] & [A_{12}] \\ [A_{21}] & [A_{22}] \end{bmatrix} \tag{5.17}$$

Assume that the inverse of $[A]$ is given by matrix $[B]$, which is partitioned in exactly the same fashion as $[A]$:

$$[A]^{-1} = [B] = \begin{bmatrix} [B_{11}] & [B_{12}] \\ [B_{21}] & [B_{22}] \end{bmatrix} \tag{5.18}$$

Recall that

$$[A][B] = [I]$$

$$\begin{bmatrix} [A_{11}] & [A_{12}] \\ [A_{21}] & [A_{22}] \end{bmatrix} \begin{bmatrix} [B_{11}] & [B_{12}] \\ [B_{21}] & [B_{22}] \end{bmatrix} = \begin{bmatrix} [I] & [0] \\ [0] & [I] \end{bmatrix} \tag{5.19}$$

Multiplying matrix $[A]$ by $[B]$ gives

$$[A_{11}][B_{11}] + [A_{12}][B_{21}] = [I] \tag{5.20a}$$

$$[A_{21}][B_{11}] + [A_{22}][B_{21}] = [0] \tag{5.20b}$$

$$[A_{11}][B_{12}] + [A_{12}][B_{22}] = [0] \tag{5.20c}$$

$$[A_{21}][B_{12}] + [A_{22}][B_{22}] = [I] \tag{5.20d}$$

The object is to solve equations (5.20) for the unknown submatrices constituting the inverse matrix ($[B_{11}]$, $[B_{12}]$, $[B_{21}]$, and $[B_{22}]$). Therefore, multiply equation (5.20b) by $[A_{12}][A_{22}]^{-1}$:

$$[A_{12}][A_{22}]^{-1}[A_{21}][B_{11}] + [A_{12}][A_{22}]^{-1}[A_{22}][B_{21}] = [0]$$

or more simply

$$[A_{12}][A_{22}]^{-1}[A_{21}][B_{11}] + [A_{12}][B_{21}] = [0] \tag{5.21}$$

Subtract equation (5.21) from (5.20a). This yields

$$[A_{11}][B_{11}] + [A_{12}][B_{21}] - [[A_{12}][A_{22}]^{-1}[A_{21}][B_{11}] + [A_{12}][B_{21}]] = [I] - [0]$$

which simplifies to

$$[A_{11}][B_{11}] - [A_{12}][A_{22}]^{-1}[A_{21}][B_{11}] = [I]$$

Solving for $[B_{11}]$ we have

$$[B_{11}] = [[A_{11}] - [A_{12}][A_{22}]^{-1}[A_{21}]]^{-1} \tag{5.22a}$$

In a similar way, equation (5.20b) can be solved for $[B_{21}]$, giving

$$[B_{21}] = -[[A_{22}]^{-1}[A_{21}][B_{11}]] \tag{5.22b}$$

Similarly, equations (4.14) and (4.15) are solved for $[B_{12}]$ annd $[B_{22}]$, giving

$$[B_{22}] = [[A_{22}] - [A_{21}][A_{11}]^{-1}[A_{12}]]^{-1} \tag{5.22c}$$

$$[B_{12}] = -[[A_{11}]^{-1}[A_{12}][B_{22}]] \tag{5.22d}$$

The solutions of equation (4.22) constitute the inverse of matrix $[A]$.

**E** EXAMPLE 5.5

Invert the following matrix by partitioning:

$$[A] = \left[\begin{array}{ccc|ccc} 2 & 0 & 0 & -7 & 0 & 0 \\ 0 & 2 & 0 & 0 & -7 & 0 \\ 0 & 0 & 2 & 0 & 0 & -7 \\ \hline -3 & 0 & 0 & 9 & 0 & 0 \\ 0 & -3 & 0 & 0 & 9 & 0 \\ 0 & 0 & -3 & 0 & 0 & 9 \end{array}\right] = \begin{bmatrix} [A_{11}] & [A_{12}] \\ [A_{21}] & [A_{22}] \end{bmatrix}$$

First partition $[A]$ as shown above. Note that $[A]$ could have been partitioned differently. Then by inspection determine the following submatrices:

$$[A_{11}]^{-1} = \frac{1}{2}\begin{bmatrix} 1 & 0 & 0 \\ 0 & 1 & 0 \\ 0 & 0 & 1 \end{bmatrix} = \frac{1}{2}[I]$$

$$[A_{22}]^{-1} = \frac{1}{9}\begin{bmatrix} 1 & 0 & 0 \\ 0 & 1 & 0 \\ 0 & 0 & 1 \end{bmatrix} = \frac{1}{9}[I]$$

Solve for the inverse using equation (5.22a):

$$[B_{11}] = [2[I] - (-7)[I](\tfrac{1}{9})[I](-3)[I]]^{-1}$$

$$= (-3)\begin{bmatrix} 1 & 0 & 0 \\ 0 & 1 & 0 \\ 0 & 0 & 1 \end{bmatrix}$$

Using equation (5.22b) gives

$$[B_{21}] = -[(\tfrac{1}{9})[I](-3)[I](-3)[I]]$$

$$= (-1)\begin{bmatrix} 1 & 0 & 0 \\ 0 & 1 & 0 \\ 0 & 0 & 1 \end{bmatrix}$$

Similarly,

$$[B_{22}] = [(9)[I] - (3)[I](\tfrac{1}{2})[I](-7)[I]]^{-1}$$

$$= -\frac{2}{3}\begin{bmatrix} 1 & 0 & 0 \\ 0 & 1 & 0 \\ 0 & 0 & 1 \end{bmatrix}$$

and

$$[B_{12}] = -[(\tfrac{1}{2})[I](-7)[I](\tfrac{2}{3})[I]] = -\frac{7}{3}\begin{bmatrix} 1 & 0 & 0 \\ 0 & 1 & 0 \\ 0 & 0 & 1 \end{bmatrix}$$

The inverse is then given as

$$[A]^{-1} = \begin{bmatrix} -3 & 0 & 0 & -\frac{7}{3} & 0 & 0 \\ 0 & -3 & 0 & 0 & -\frac{7}{3} & 0 \\ 0 & 0 & -3 & 0 & 0 & -\frac{7}{3} \\ -1 & 0 & 0 & -\frac{2}{3} & 0 & 0 \\ 0 & -1 & 0 & 0 & -\frac{2}{3} & 0 \\ 0 & 0 & -1 & 0 & 0 & -\frac{2}{3} \end{bmatrix}$$

Generally speaking, the matrix in equation might be more difficult to partition than the one given in example 5.5. However, the same procedure can be employed.

## 5.6 MATRICES INVOLVING COMPLEX COEFFICIENTS

As is the case with linear algebraic equations, one may declare complex variables rather than real ones and use the real analysis outlined thus far for matrix inversion. A method in which the complex matrix in question is inverted by using real analysis without declaring complex variables is outlined herein. Recall that, given a complex matrix $[C]$, an equivalent matrix can be derived as given by equation (4.45), repeated here for convenience:

$$[C] = \begin{bmatrix} [a] & -[b] \\ [b] & [a] \end{bmatrix} \tag{5.23}$$

where $[a]$ is a submatrix containing the real parts of the complex coefficients and $[b]$ is a submatrix containing the imaginary parts of the complex coefficients in matrix $[C]$. It is evident that the square submatrices $[a]$ and $[b]$ are exactly the same size as matrix $[C]$. This problem is ready made for the matrix partitioning procedure outlined in section 5.5. Therefore, substituting into equations (5.22) yields the following submatrices, which constitute the inverse of the partitioned matrix given by equation (5.23):

$$[B_{11}] = [[a] + [b][a]^{-1}[b]]^{-1} \tag{5.24a}$$

$$[B_{21}] = -[[a]^{-1}[b][B_{11}]] \tag{5.24b}$$

$$[B_{22}] = [B_{11}] \tag{5.24c}$$

$$[B_{12}] = -[B_{21}] \tag{5.24d}$$

It is evident that a matrix involving complex coefficients is a special case of a partitioned real matrix and requires half the computational effort required by a general real partitioned matrix $[A]$. In fact, these results should have been expected and there should be no need for equations (5.24c) and (5.24d). This is because the inverse involves two unique submatrices which represent the real and imaginary parts of the matrix $[C]^{-1}$.

**E** EXAMPLE 5.6

Invert the following complex matrix:

$$[C] = \begin{bmatrix} 1 + J & 2 - J \\ 1 - J & 3 + 2J \end{bmatrix}$$

Solution

We begin by first identifying the submatrices $[a]$ and $[b]$:

$$[a] = \begin{bmatrix} 1 & 2 \\ 1 & 3 \end{bmatrix}, \qquad [b] = \begin{bmatrix} 1 & -1 \\ -1 & 2 \end{bmatrix}$$

Therefore the inverse of $[a]$ is readily determined as

$$[a]^{-1} = \begin{bmatrix} 3 & -2 \\ -1 & 1 \end{bmatrix}$$

Substituting into equation (5.24) yields

$$[B_{11}] = \left[\begin{bmatrix} 1 & 2 \\ 1 & 3 \end{bmatrix} + \begin{bmatrix} 1 & -1 \\ -1 & 2 \end{bmatrix}\begin{bmatrix} 3 & -2 \\ -1 & 1 \end{bmatrix}\begin{bmatrix} 1 & -1 \\ -1 & 2 \end{bmatrix}\right]^{-1}$$

$$= \frac{1}{8}\begin{bmatrix} 2 & 1 \\ 1 & 1 \end{bmatrix}$$

The $[B_{11}]$ matrix is now substituted into equation (5.24b) to give

$$[B_{21}] = -\frac{1}{8}\left[\begin{bmatrix} 3 & -2 \\ -1 & 1 \end{bmatrix}\begin{bmatrix} 1 & -1 \\ -1 & 2 \end{bmatrix}\begin{bmatrix} 2 & 1 \\ 1 & 1 \end{bmatrix}\right]$$

$$= -\frac{1}{8}\begin{bmatrix} 3 & -2 \\ -1 & 1 \end{bmatrix}$$

Consequently, the inverse of $[C]$ is given as

$$[C]^{-1} = \frac{1}{8}\begin{bmatrix} 2 - 3J & 1 + 2J \\ 1 + J & 1 - J \end{bmatrix}$$

Note that the real parts of the $[C]^{-1}$ are given by $[B_{11}]$ and the imaginary parts are given by $[B_{21}]$. This is because

$$\begin{bmatrix} [a] & -[b] \\ [b] & [a] \end{bmatrix}\begin{bmatrix} [B_{11}] & -[B_{21}] \\ [B_{21}] & [B_{11}] \end{bmatrix} = \begin{bmatrix} [I] & [0] \\ [0] & [I] \end{bmatrix}$$

and thus

$$[a][B_{11}] - [b][B_{21}] = [I] \tag{5.25a}$$

$$[b][B_{11}] + [a][B_{21}] = [0] \tag{5.25b}$$

$$-[a][B_{21}] - [b][B_{11}] = [0] \tag{5.25c}$$

$$-[b][B_{21}] + [a][B_{11}] = [I] \tag{5.25d}$$

Obviously only two of the four equations given by (5.25) are unique.

## 5.7 SPECIAL MATRICES

Often, the analysis of physical systems involves symmetrical matrices. The stiffness, damping, and mass matrices derived for the mechanical system discussed in chapter 2 are examples of symmetrical matrices. In addition, the solution of linear algebraic equations using Crout's and Gauss's methods involves triangular matrices. Consequently, one may find it necessary to establish schemes that handle the inversion problem of these unique matrices, especially when computer memory becomes an issue. This may happen when one is dealing with very large matrices.

### 5.7.1 Triangular Matrices

The method described in this section addresses the problem of inverting triangular matrices. The discussion is limited to upper-triangular matrices even though lower-triangular matrices can be inverted by using the same procedure. This is true since the transpose of an upper-triangular matrix is a lower-triangular matrix. Consider the following:

$$[A] = \begin{bmatrix} a_{11} & a_{12} & a_{13} \\ 0 & a_{22} & a_{23} \\ 0 & 0 & a_{33} \end{bmatrix}$$

If we define the inverse of $[A]$ as

$$[B] = \begin{bmatrix} b_{11} & b_{12} & b_{13} \\ b_{21} & b_{22} & b_{23} \\ b_{31} & b_{32} & b_{33} \end{bmatrix}$$

then the following is true:

$$\begin{bmatrix} a_{11} & a_{12} & a_{13} \\ 0 & a_{22} & a_{23} \\ 0 & 0 & a_{33} \end{bmatrix} \begin{bmatrix} b_{11} & b_{12} & b_{13} \\ b_{21} & b_{22} & b_{23} \\ b_{31} & b_{32} & b_{33} \end{bmatrix} = \begin{bmatrix} 1 & 0 & 0 \\ 0 & 1 & 0 \\ 0 & 0 & 1 \end{bmatrix}$$

Carrying out the product of $[A][B]$ then equating corresponding terms yields

$$\begin{aligned} a_{11}b_{11} + a_{12}b_{21} + a_{13}b_{31} &= 1 \\ a_{22}b_{21} + a_{23}b_{31} &= 0 \\ a_{33}b_{31} &= 0 \end{aligned} \tag{5.26}$$

$$\begin{aligned} a_{11}b_{12} + a_{12}b_{22} + a_{13}b_{32} &= 0 \\ a_{22}b_{22} + a_{23}b_{32} &= 1 \\ a_{33}b_{32} &= 0 \end{aligned} \tag{5.27}$$

$$\begin{aligned} a_{11}b_{13} + a_{12}b_{23} + a_{13}b_{33} &= 0 \\ a_{22}b_{23} + a_{23}b_{33} &= 0 \\ a_{33}b_{33} &= 1 \end{aligned} \tag{5.28}$$

Equations (5.26) can be solved for the unknowns $b_{11}$, $b_{21}$, and $b_{31}$ to give

$$b_{31} = 0, \qquad b_{21} = 0, \qquad b_{11} = \frac{1}{a_{11}}$$

Obviously, this solution holds as long as the diagonal elements $a_{ii} \neq 0$. Hence equation (5.27) is solved for the second column of the inverse matrix to give

$$b_{32} = 0, \qquad b_{22} = \frac{1}{a_{22}}, \qquad b_{12} = -\frac{a_{12}b_{22}}{a_{11}}$$

Finally, equations (5.28) are solved for the third column of the inverse matrix. Thus

$$b_{33} = \frac{1}{a_{33}}, \qquad b_{23} = -\frac{a_{23}b_{33}}{a_{22}}, \qquad b_{13} = -\frac{a_{12}b_{23} + a_{13}b_{33}}{a_{11}}$$

Clearly, the inverse of an upper-triangular matrix is an upper-triangular matrix as well. In general, for a triangular matrix of order $n$ the inverse is given simply as follows:

$$b_{ii} = \frac{1}{a_{ii}}, \qquad i = 1, \ldots, n \tag{5.29a}$$

$$b_{ij} = -\frac{1}{a_{ii}} \sum_{k=i+1}^{n} a_{ik}b_{kj}, \qquad i < j \tag{5.29b}$$

$$b_{ij} = 0, \qquad i > j \tag{5.29c}$$

Equations (5.29) provide a straightforward and efficient method for inverting triangular matrices irrespective of their size.

E

EXAMPLE 5.7

Solve the following set of linear algebraic equations for $\{y\}_1^t = \{1, 1, 1\}$, $\{y\}_2^t = \{2, -1, 1\}$, and $\{y\}_3^t = \{3, 4, 2\}$:

$$\begin{bmatrix} 2 & 3 & 1 \\ 0 & 7 & -2 \\ 0 & 0 & 3 \end{bmatrix} \begin{Bmatrix} x_1 \\ x_2 \\ x_3 \end{Bmatrix} = \begin{Bmatrix} y_1 \\ y_2 \\ y_3 \end{Bmatrix}$$

**Solution**

We begin by inverting the coefficient matrix using equation (5.29). Thus, from equation (5.29a), we have

$$b_{11} = \tfrac{1}{2}, \qquad b_{22} = \tfrac{1}{7}, \qquad b_{33} = \tfrac{1}{3}$$

Therefore, using equation (5.29b) we solve for the remaining unknowns:

$$b_{12} = -\frac{1}{a_{11}} \sum_{k=2}^{2} a_{1k} b_{k2} = -\frac{1}{a_{11}} a_{12} b_{22} = -\frac{3}{14}$$

$$b_{23} = -\frac{1}{a_{22}} \sum_{k=3}^{3} a_{2k} b_{k3} = -\frac{1}{a_{22}} a_{23} b_{33} = \frac{2}{21}$$

$$b_{13} = -\frac{1}{a_{11}} \sum_{k=2}^{3} a_{1k} b_{13} = -\frac{1}{a_{11}} (a_{12} b_{23} + a_{13} b_{33}) = -\frac{13}{42}$$

The inverse of the coefficient matrix is now given as

$$\frac{1}{42}\begin{bmatrix} 21 & -9 & -13 \\ 0 & 6 & 4 \\ 0 & 0 & 14 \end{bmatrix}$$

The solution is determined for the various $\{y\}$ vectors as follows:

$$\begin{Bmatrix} x_1 \\ x_2 \\ x_3 \end{Bmatrix} = \frac{1}{42}\begin{bmatrix} 21 & -9 & -13 \\ 0 & 6 & 4 \\ 0 & 0 & 14 \end{bmatrix}\left[\begin{array}{c|c|c} 1 & 2 & 3 \\ 1 & -1 & 4 \\ 1 & 1 & 2 \end{array}\right]$$

$$\begin{Bmatrix} x_1 \\ x_2 \\ x_3 \end{Bmatrix} = \frac{1}{42}\left[\begin{array}{c|c|c} -1 & 38 & 1 \\ 10 & -2 & 32 \\ 14 & 14 & 28 \end{array}\right]$$

Note that the dashed lines separate the solutions obtained for the three $\{X\}$ vectors corresponding to the three given $\{Y\}$ vectors.

### 5.7.2 Symmetrical Matrices

In most engineering applications including those using finite difference and finite element methods, the matrices involved are symmetrical. Consequently, one may use the methods outlined for the general matrix to invert these matrices. Alternatively, one may take advantage of their symmetry in reducing computational and storage requirements, especially when solving large sets of equations involving variable $\{b\}$ vectors. That is, given

$$[A]\{X\} = \{Y\} \tag{5.30a}$$

where $[A]$ is symmetrical, then using the square root method, we can rewrite equation (5.30a) as follows:

$$[u]^t[u]\{X\} = \{Y\} \qquad \textbf{(5.30b)}$$

Note that the $[u]$ matrix is an upper-triangular matrix. Therefore premultiplying equation (5.30b) by $[u]^{-1}[[u]^t]^{-1}$ yields

$$\{X\} = [u]^{-1}[[u]^t]^{-1}\{Y\} \qquad \textbf{(5.31)}$$

It is evident that the inverse of $[A]$ is given as

$$[A]^{-1} = [u]^{-1}[[u]^t]^{-1} = [u]^{-1}[[u]^{-1}]^t \qquad \textbf{(5.32)}$$

The implication is that by decomposing the symmetric $[A]$ matrix into $[u]^t[u]$ we can determine the inverse of $[A]$ by simply inverting an upper-triangular matrix. However, this is precisely the problem we have addressed in section 5.7.1. Furthermore, it is clear that the inverse of a symmetrical matrix is symmetric as well. This is true since any matrix multiplied by its transpose results in a symmetric matrix.

**E** EXAMPLE 5.8

Invert the following symmetrical matrix:

$$[A] = \begin{bmatrix} 4 & 2 & 4 \\ 2 & 10 & 5 \\ 4 & 5 & 6 \end{bmatrix}$$

Solution

Using the square root method we decompose the $[A]$ matrix into

$$[A] = \begin{bmatrix} 2 & 0 & 0 \\ 1 & 3 & 0 \\ 2 & 1 & 1 \end{bmatrix} \begin{bmatrix} 2 & 1 & 2 \\ 0 & 3 & 1 \\ 0 & 0 & 1 \end{bmatrix}$$

where

$$[u] = \begin{bmatrix} 2 & 1 & 2 \\ 0 & 3 & 1 \\ 0 & 0 & 1 \end{bmatrix}$$

The inverse of $[u]$ is determined by using equation (5.29) to give

$$[u]^{-1} = \begin{bmatrix} \frac{1}{2} & -\frac{1}{6} & -\frac{5}{6} \\ 0 & \frac{1}{3} & -\frac{1}{3} \\ 0 & 0 & 1 \end{bmatrix} = \frac{1}{6}\begin{bmatrix} 3 & -1 & -5 \\ 0 & 2 & -2 \\ 0 & 0 & 6 \end{bmatrix}$$

Using equation (5.31) permits the determination of the inverse of [A]. Hence

$$[A]^{-1} = \frac{1}{36}\begin{bmatrix} 3 & -1 & -5 \\ 0 & 2 & -2 \\ 0 & 0 & 6 \end{bmatrix}\begin{bmatrix} 3 & 0 & 0 \\ -1 & 2 & 0 \\ -5 & -2 & 6 \end{bmatrix}$$

$$= \frac{1}{36}\begin{bmatrix} 35 & 8 & -30 \\ 8 & 8 & -12 \\ -30 & -12 & 36 \end{bmatrix}$$

## Recommended Reading

*Numerical Methods for Scientists and Engineers*, R. W. Hamming, McGraw-Hill Book Company, New York, 1962.

*Elementary Numerical Analysis*, S. D. Conti, McGraw-Hill Book Company, New York, 1965.

*Numerical Algorithms—Origins and Applications*, W. B. Arden, and N. S. Kenneth, Addison-Wesley Publishing Company, Reading, Mass., 1970.

*Numerical Mathematics and Computers*, W. Cheney and D. Kincaid, Brooks/Cole Publishing Company, Monterey, Calif., 1980.

*Numerical Analysis: A Practical Approach*, M. J. Maron, Macmillan Publishing Co., Inc., New York, 1982.

P

## PROBLEMS

**5.1** Invert the following matrix using Cramer's method of cofactors:

$$[A] = \begin{bmatrix} \cos\theta & \sin\theta \\ -\sin\theta & \cos\theta \end{bmatrix}$$

**5.2** Use Cramer's method of cofactors to determine the inverse for the following matrix:

$$[A] = \begin{bmatrix} 2 & 1 & 5 \\ -3 & 4 & 1 \\ 9 & 6 & 1 \end{bmatrix}$$

**5.3** Solve the following system of linear algebraic equations using the elimination method for matrix inversion:

$$2x_1 + 3x_2 - x_3 = A_1$$

$$-x_1 + 2x_2 + 3x_3 = A_2$$

$$x_1 - x_2 + 7x_3 = A_3$$

**5.4** Rework problem 5.3, if $A_1 = y_1 + y_2$, $A_2 = y_1 - y_2$, $A_3 = -y_1 + y_2$. Express the solution for the $\{X\}$ vector in terms of $y_1$ and $y_2$.

**5.5** Solve the following system of equations for $x_1$, $x_2$, and $x_3$ using matrix inversion:

$$\begin{aligned} x_1 + 7x_2 - 8x_3 &= y_1 - 3y_2 + 2y_3 \\ 2x_1 - x_2 + x_3 &= 3y_1 + 4y_2 + y_3 \\ 8x_1 + x_2 + 4x_3 &= y_1 + y_2 \end{aligned}$$

**5.6** Rework problem 5.5 solving for $y_1$, $y_2$, and $y_3$ in terms of $x_1$, $x_2$, and $x_3$.

**5.7** Invert the following matrices using the reducing matrix method:

$$\textbf{(a)}\ [A] = \begin{bmatrix} 1 & 2 & 3 \\ 2 & 7 & -1 \\ 1 & -1 & 5 \end{bmatrix} \qquad \textbf{(b)}\ [B] = \begin{bmatrix} 2 & 7 & -1 \\ 1 & 2 & 3 \\ 1 & -1 & 5 \end{bmatrix}$$

**5.8** Invert the following matrix using the method of partitioning:

$$[A] = \left[\begin{array}{ccc|cc} 3 & 4 & 6 & 0 & 0 \\ -7 & 1 & 6 & 0 & 0 \\ 6 & -8 & 2 & 0 & 0 \\ \hline 1 & 2 & 3 & 4 & 5 \\ 6 & 7 & 8 & 9 & 2 \end{array}\right]$$

**5.9** Invert the following matrix with complex coefficients:

$$[A] = \begin{bmatrix} 3 + J & 1 - 2J & 2 + 2J \\ -1 + 2J & 1 + J & 5 - J \\ 2 + 3J & -2 + 7J & 1 + 3J \end{bmatrix}$$

**5.10** Invert the following triangular matrices using the method outlined in section 5.7.1:

$$\textbf{(a)}\ [A] = \begin{bmatrix} 2 & 5 & -3 \\ 0 & -1 & 6 \\ 0 & 0 & 4 \end{bmatrix} \qquad \textbf{(b)}\ [B] = \begin{bmatrix} 7 & 0 & 0 \\ 3 & 4 & 0 \\ 2 & 1 & -8 \end{bmatrix}$$

Show all calculations.

**5.11** Invert the following decomposed symmetrical matrices:

$$\textbf{(a)}\ [A] = \begin{bmatrix} 1 & 0 & 0 \\ 2 & 1 & 0 \\ 2 & 2 & 1 \end{bmatrix}\begin{bmatrix} 1 & 2 & 2 \\ 0 & 1 & 2 \\ 0 & 0 & 1 \end{bmatrix} \qquad \textbf{(b)}\ [B] = \begin{bmatrix} -1 & 0 & 0 \\ 5 & 3 & 0 \\ 1 & -7 & 4 \end{bmatrix}\begin{bmatrix} -1 & 5 & 1 \\ 0 & 3 & -7 \\ 0 & 0 & 4 \end{bmatrix}$$

Check your results.

**5.12** The following equations represent the generalized Hooke's law for a homogeneous, isotropic material under the most general stress conditions:

$$\varepsilon_x = \frac{\sigma_x}{E} - \frac{\nu\sigma_y}{E} - \frac{\nu\sigma_z}{E}$$

$$\varepsilon_y = -\frac{\nu\sigma_x}{E} + \frac{\sigma_y}{E} = \frac{\nu\sigma_z}{E}$$

$$\varepsilon_z = -\frac{\nu\sigma_x}{E} - \frac{\nu\sigma_y}{E} + \frac{\sigma_z}{E}$$

where $\varepsilon_x$, $\varepsilon_y$, and $\varepsilon_z$ are strains; $\sigma_x$, $\sigma_y$, and $\sigma_z$ are normal stresses; and $\nu$ is Poisson's ratio. Express the stress vector as a function of the strain vector using Cramer's method of cofactors.

# chapter 6

# Nonlinear Algebraic Equations

This 164-ft. (50-meters) high spiral minaret was built during the Abbasid period (A.D. 900). *Courtesy of the Iraqi government.*

"Cube roots and square roots occurred whenever volumes and areas had to be reckoned with. To help irrigate or drain the land, the Sumerians (3500 B.C.) prepared tables for cube roots and square roots of numbers."

# 6

## 6.1 INTRODUCTION

Nonlinear algebraic equations are defined as those which contain powers of variable(s) and/or transcendental functions. Such equations arise frequently in engineering, especially when one is dealing with optimization, differential equations, and eigenproblems. The object is then to find the so-called zeros of the equation. That is, given

$$y = f(x)$$

we seek values of $x$ such that $y$ is zero. These $x$ values are referred to as the roots of the function $f(x)$. The roots may be real, complex, or both, and their number may be finite or infinite, depending on the function in question.

No general algebraic method is available for solving all nonlinear algebraic equations. Hence different techniques with varying degrees of accuracy and rates of convergence are presented in this chapter. The degree of accuracy attained for a given root is directly related to the amount of computational effort and the method being used.

## 6.2 GRAPHICAL METHOD

This method is frequently used in establishing a range of values for the real roots of a given function. It provides no information relative to complex roots. Furthermore, it is generally used in conjunction with more accurate numerical techniques to find a starting value approximation of roots. The method is fairly simple and involves sketching the function(s) in the desired range. Unfortunately, this may require a significant amount of effort, especially when hand calculation is employed. For illustration purposes, two functions are plotted in figure 6.1. The real

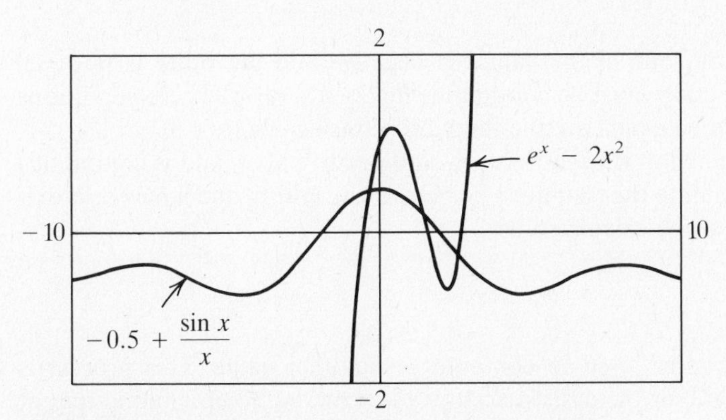


**Figure 6.1** Graphical representation of two functions.

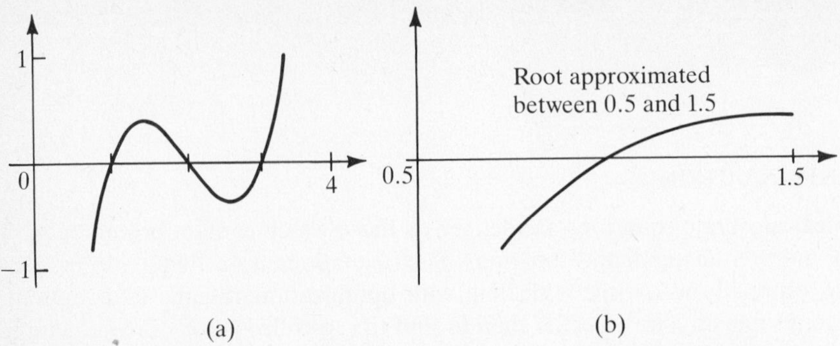


**Figure 6.2** Procedure used in approximating the real roots of the function $y = x^3 - 6x^2 + 11x - 6$.

roots of these functions correspond to the $x$-axis values where the functions are zero.

Clearly, varying degrees of accuracy can be attained by simply altering the scales of the ordinate and the abscissa. Figure 6.2 illustrates the process involved in determining a real root graphically.

The function illustrated in figure 6.2a has three real roots of 1, 2, and 3. By altering the scale, we are able to isolate the first of these roots (figure 6.2b); therefore one may repeat this process until desired results are attained.

## 6.3 INTERVAL-HALVING METHOD

This is one of the simplest iterative techniques for determining roots of a function. The basis for this method can be easily illustrated by considering the following function:

$$y = f(x)$$

Our object is to find an $x$ value for which $y$ is zero. Using this method, we begin by evaluating the function at two $x$ values, say $x_1$ and $x_2$, such that

$$f(x_1)f(x_2) < 0$$

The implication is that one of the values is negative and the other is positive. Furthermore, the function must be continuous for $x_1 \le x \le x_2$. These conditions can be easily satisfied by sketching the function. Consider figure 6.3.

Obviously, the function is negative at $x_1$ and positive at $x_2$ and is continuous for $x_1 \le x \le x_2$. Therefore the root must lie between $x_1$ and $x_2$ and a new approximation to the root can be calculated as

$$x_3 = \frac{x_1 + x_2}{2}$$

Clearly, $x_3$ and $x_1$ can be used to compute yet another value. This process is continued until $f(x) \approx 0$ or the desired accuracy is achieved. Keep in mind that at each iteration, the new $x$ value and one of the two previous values are used so that

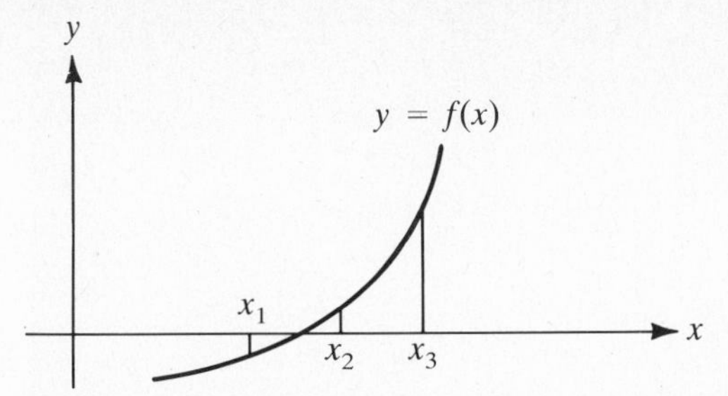


**Figure 6.3** Graphical illustration of interval-halving method.

continuity and functional products are satisfied. The convergence rate of this technique is rather slow. In fact, the interval in which the root $x_i$ lies is reduced by a factor of $2^{-k}$, where $k$ is the number of iterations.

**E**

EXAMPLE 6.1

Using the graphical method, the following function was found to have a real root between $x = 1$ and $x = 3$; approximate the root:

$$f(x) = x^3 - 5x^2 - 2x + 10$$

Solution

We begin by evaluating the function at the initial values

$$x_1 = 1: \qquad f(x_1) = 4$$

$$x_2 = 3: \qquad f(x_2) = -14$$

Obviously, $f(1)f(3) = (4)(-14) < 0$ and the root has a value between 1 and 3. Therefore a new value is approximated by

$$x_3 = \frac{1+3}{2} = 2: \qquad f(x_3) = -6$$

It is evident that the root is between $x_3$ and $x_1$, which must now be used to compute a new $x$ value. This is because $f(x_1)f(x_3) < 0$. Proceeding with the next five iterations gives

$$x_4 = \frac{x_1 + x_3}{2} = 1.50000: \qquad f(x_4) = -0.87500$$

$$x_5 = \frac{x_1 + x_4}{2} = 1.25000: \qquad f(x_5) = 1.64062$$

$$x_6 = \frac{x_5 + x_4}{2} = 1.37500: \qquad f(x_6) = 0.39648$$

$$x_7 = \frac{x_6 + x_4}{2} = 1.43750: \qquad f(x_7) = -0.23657$$

$$x_8 = \frac{x_7 + x_6}{2} = 1.40625: \qquad f(x_8) = 0.08071$$

It is evident that the functional values are approaching zero as the number of iterations is increased. After six iterations the approximated root of 1.40625 compares favorably with the exact value of $\sqrt{2}$.

**SUMMARY OF THE INTERVAL-HALVING METHOD**

| STEP | OPERATION | SYMBOL |
|---|---|---|
| 1 | Sketch the function in question. | |
| 2 | Establish $x_1$ and $x_2$ such that $f(x_1)$ and $f(x_2)$ are of opposite sign. | $f(x_1)f(x_2) < 0$ |
| 3 | Establish an error tolerance value for the function. | $T$ |
| 4 | Compute a new approximation for the root. | $x_3 = \frac{x_1 + x_2}{2}$ |
| 5 | Check tolerance. If $T \geq \|f(x_3)\|$, then use $x_3$ for the root; otherwise continue. | |
| 6 | If $f(x_3)$ is of opposite sign to $f(x_1)$, then set $x_2 = x_3$; otherwise set $x_1 = x_3$. | |
| 7 | Go to step 4. | |

## 6.4 FALSE-POSITION METHOD

The graphical and interval-halving techniques can be improved upon by using the false-position method, which is a very old technique. The method starts with two points at which the function has opposite sign; that is, we have to establish an interval in which there is a zero for the function, and we seek to decrease that interval. This is accomplished by extending a straight line through the two points to replace the function, and the zero of the straight line is used to approximate the zero of the function. This is shown graphically in figure 6.4.

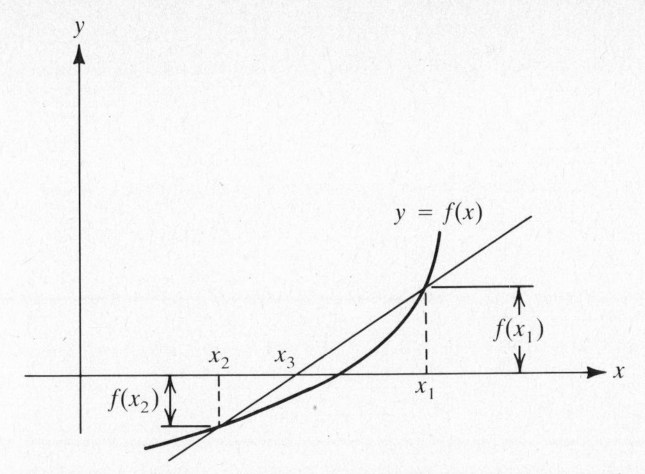


**Figure 6.4** Graphical illustration of the false-position method.

Given two points $(x_1, f(x_1))$ and $(x_2, f(x_2))$ such that

$$f(x_1)f(x_2) < 0$$

the equation of the line passing through them is readily determined:

$$\frac{f(x_1)}{x_1 - x_3} = \frac{f(x_2)}{x_3 - x_2}$$

Solving for $x_3$ gives the zero of the line. Thus

$$x_3 = \frac{x_2 f(x_1) - x_1 f(x_2)}{f(x_1) - f(x_2)} \tag{6.1}$$

It is evident that once $x_3$ is determined and $f(x_3)$ is evaluated a decision will have to be made as to which variable ($x_1$ or $x_2$) is replaced by $x_3$ in subsequent iterations. The main weakness of the false-position method is that it slowly converges from one side, as shown in figure 6.5.

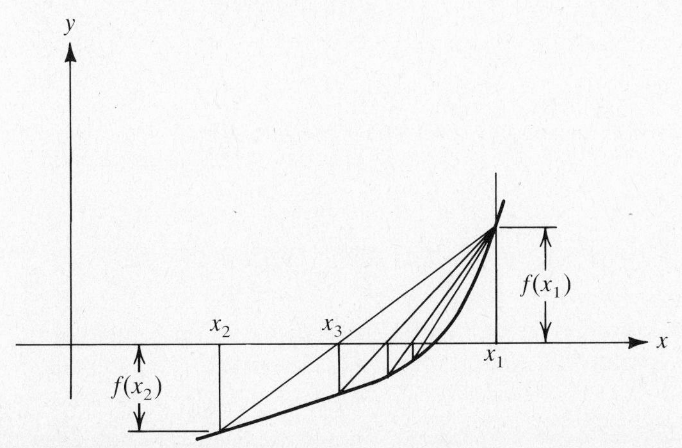


**Figure 6.5** Graphical illustration of convergence of the false-position method.

Note that corresponding to $x_3$ we have a negative functional value and that subsequent zeros of the straight lines all have negative functional values. Consider the following example.

E

EXAMPLE 6.2

The following function (see example 6.1) is known to have a root in the interval $1 \leq x \leq 3$:

$$f(x) = x^3 - 5x^2 - 2x + 10$$

Estimate the actual root.

Solution

We begin by letting $x_1 = 1$ and $x_2 = 3$ and evaluate the function at these points to give

$$x_1 = 1: \qquad f(x_1) = 4$$

$$x_2 = 3: \qquad f(x_2) = -14$$

Consequently, using equation (6.1), we determine a first approximation to the actual root. Hence

$$x_3 = \frac{(3)(4) - (1)(-14)}{4 - (-14)} = 1.44444: \qquad f(x_3) = -0.30727$$

Obviously, $x_3$ replaces $x_2$ since $f(x_3)$ is negative and $f(x_1)f(x_3) < 0$. Therefore, letting $x_2 = x_3$, we have

$$x_1 = 1: \qquad f(x_1) = 4$$

$$x_2 = 1.44444: \qquad f(x_2) = -0.30727$$

Using equation (6.1) once again gives a second approximation to the actual root. Thus

$$x_3 = \frac{(1.44444)(4) - (1)(-0.30727)}{4 - (-0.30727)} = 1.41273: \qquad f(x_3) = 0.01504$$

Clearly this method converges faster than the method of interval halving in that better accuracy is attained in just two iterations versus six for interval halving.

**SUMMARY OF THE FALSE-POSITION METHOD**

| STEP | OPERATION | SYMBOL |
|---|---|---|
| 1 | Establish an interval $x_1 \le x \le x_2$ such that $f(x_1)f(x_2) < 0$. This can be done graphically or by trial and error. | |
| 2 | Calculate the zero of the line passing through $(x_1, f(x_1))$ and $(x_2, f(x_2))$. | $x_3 = \dfrac{x_2 f(x_1) - x_1 f(x_2)}{f(x_1) - f(x_2)}$ |
| 3 | Establish a tolerance limit for the function. If the criterion is met, then end; otherwise continue. | $T \ge \lvert f(x_3)\rvert$ for $x_3 \approx$ root |
| 4 | Examine $f(x_1)f(x_3)$. If it is less than zero, then let $x_2 = x_3$ and go to step 2. Otherwise let $x_1 = x_3$. | $f(x_1)f(x_3) < 0, \quad x_2 = x_3$<br>$f(x_2)f(x_3) < 0, \quad x_1 = x_3$ |

## 6.5 NEWTON–RAPHSON FIRST METHOD

This is one of the more popular methods used for solving nonlinear algebraic equations. It is also known as the method of tangents. Furthermore, it converges faster than the methods described thus far. However, it suffers from the same limitations, in that an initial value close to the root must first be estimated and the derivative of the function must be evaluated. The basis for this method is that the actual root is estimated and the zero of the tangent to the function at that point is determined. Consider figure 6.6.

If a real root $x_1$ is to be assumed for the function, then one may easily compute the functional $f(x_1)$. Now, if we draw a line tangent to the curve at point $x_1$, then the tangent line intersects the $x$ axis at a point, say $x_2$, which is expected to be closer to the actual root than the assumed root $(x_1)$. The question is, can we find $x_2$? The answer is yes, if we realize that the slope of the tangent is equal to the first derivative of the function evaluated at $x = x_1$. That is,

$$f'(x_1) = \tan\theta \tag{6.2}$$

But the slope can also be determined from the following:

$$\tan\theta = \frac{f(x_1)}{x_1 - x_2} \tag{6.3}$$

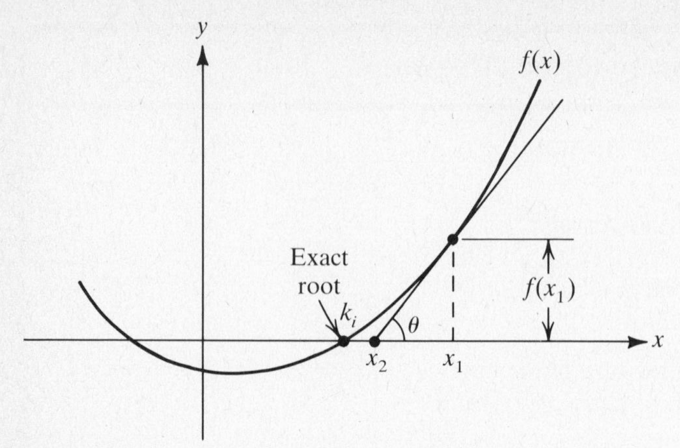


**Figure 6.6** Interpretation of Newton–Raphson first method.

Equating equation (6.2) to equation (6.3) gives

$$f'(x_1) = \frac{f(x_1)}{x_1 - x_2} \tag{6.4}$$

Solving for equation (6.4) for $x_2$ yields

$$x_2 = x_1 - \frac{f(x_1)}{f'(x_1)} \tag{6.5}$$

In general, equation (6.5) is given in the following form for the $k$th iteration:

$$x_{k+1} = x_k - \frac{f(x_k)}{f'(x_k)} \tag{6.6}$$

where

$x_{k+1}$ = approximate root after $k + 1$ iterations

$x_k$ = approximate root after $k$ iterations

$f(x_k)$ = functional value at $x_k$

$f'(x_k)$ = first-derivative value of the function at $x_k$

$k = 1, 2, \ldots$

Obviously, an initial approximation $x_1$ for the root must first be assumed or graphically determined for this method to work. It should be noted that this method is suitable for finding real as well as imaginary roots of polynomials. Unfortunately, it does not always converge to a root if $x_1$ is not sufficiently close to the actual root.

E

EXAMPLE 6.3

Determine a root of the following function:

$$f(x) = x^3 - 3.5^2 + 2x - 10$$

## Solution

Since the Newton–Raphson method requires that the value of the derivative be found, let us find the derivative first:

$$f'(x) = 3x^2 - 7x + 2$$

Now assume a root, say $x_1 = 3$. Then

$$x_1 = 3$$

$$f(x_1) = (3)^3 - 3.5(3)^2 + 2(3) - 10 = -8.5$$

$$f'(x_1) = 3(3)^2 - 7(3) + 2 = 8$$

Applying equation (6.6) and solving for $x_2$, we have

$$x_2 = x_1 - \frac{f(x_1)}{f'(x_1)}$$

$$= 3 - \frac{-8.5}{8}$$

$$= 4.0625$$

Using $x_2$ as the new estimate for the root and solving again yields

$$x_2 = 4.0625$$

$$f(x_2) = 7.4084$$

$$f'(x_2) = 23.074$$

Applying equation (6.6) and solving for $x_2$ gives

$$x_3 = x_2 - \frac{f(x_2)}{f'(x_2)}$$

$$= 4.0625 - \frac{7.4084}{23.074}$$

$$= 3.7414$$

Repeating this procedure would ultimately give the approximate root. After five iterations the root is approximated to be

$$x_6 = 3.691933958$$

and the functional value is reduced to

$$f(x_6) = -2.2 \times 10^{-9}$$

For most problems this degree of accuracy is more than adequate.

EXAMPLE 6.4

Develop an iterative procedure for evaluating the square root of a number using the Newton–Raphson method.

### Solution

This problem can be easily solved by noting that we seek to find a root to the function

$$f(x) = x^2 - N$$

where $x$ is the square root and $N$ is the number whose square root is to be found. Therefore, if $f(x) = 0$ then $x = \sqrt{N}$ is the exact root. Hence, assuming an initial estimate to the root, say $x = x_1$, and substituting into equation (6.6) gives

$$x_2 = x_1 - \frac{f(x_1)}{f'(x_1)}$$

where $f'(x) = 2x$, or more simply

$$x_2 = x_1 - (x_1^2 - N)/2x_1 = \tfrac{1}{2}(x_1 + N/x_1)$$

In general,

$$x_{k+1} = \tfrac{1}{2}(x_k + N/x_k), \qquad k = 1, \ldots, n \tag{6.7}$$

where $n$ is the number of iterations. For example, suppose we want the square root of $N = 13$. Assuming an initial guess of say 4, then substituting into equation (6.7) gives

$$k = 1: \qquad x_2 = \tfrac{1}{2}(4 + 13/4) = 3.625$$

$$k = 2: \qquad x_3 = \tfrac{1}{2}(3.625 + 13/3.625) = 3.6056035$$

After just two iterations the estimated value compares rather favorably with the exact value of 3.6055513. Furthermore, this technique can be used to calculate higher roots of a number.

EXAMPLE 6.5

Determine the complex roots of the following function by starting out with $x_1 = i$:

$$x^2 + x + 1 = 0$$

Solution

Clearly the complex roots of this function can be easily determined by using the quadratic formula. These are

$$x_{1,2} = \frac{-1 \pm \sqrt{1-4}}{2} = \frac{1}{2} \pm \frac{\sqrt{3}}{2} i$$

Applying Newton's method we have

$$x_1 = i$$

$$f(x_1) = (i)^2 + i + 1 = i$$

$$f'(x_1) = 2(i) + 1 = 2i + 1$$

$$x_2 = x_1 - \frac{f(x_1)}{f'(x_1)}$$

$$= i - \frac{i}{2i+1} \frac{2i-1}{2i-1}$$

$$= i - \frac{2i^2 - i}{4i^2 - 1}$$

$$= i - \frac{-2-i}{-4-1}$$

$$= i - \frac{2}{5} - \frac{i}{5} = -\frac{2}{5} + \frac{4}{5} i$$

Now, using $x_2$ as an initial approximate root and solving for $x_3$ should give better accuracy. However, one can see the degree of accuracy achieved after just one iteration. That is,

$$x_{\text{exact}} = -0.5 + 0.866i$$

$$x_2 = -0.4 + 0.8i$$

A criterion for convergence of the Newton–Raphson method can be established by rewriting equation (6.6) as follows:

$$y = x - \frac{f(x)}{f'(x)}$$

where $y$ is the improved estimate of the root. Consequently, if the method is to converge, then the absolute value of the rate of change of $y$ with respect to $x$ must be less than 1.0. That is,

$$\frac{dy}{dx} = \left| 1 - \frac{f'(x)f'(x) - f(x)f''(x)}{[f'(x)]^2} \right| < 1$$

or more simply

$$\left| \frac{f(x)f''(x)}{[f'(x)]^2} \right| < 1 \qquad \textbf{(6.8)}$$

Equation (6.8) represents a sufficient condition for convergence. It is evident that $f'(x)$ must not be zero. This is an important factor to consider when choosing the initial $x$ value.

The Newton–Raphson method suffers from four basic limitations. The first is that some functions are not easy to differentiate, in which case either the false-position or the modified method, discussed later, is used instead. The remaining limitations are illustrated graphically in figure 6.7.

Clearly, if the root is zero then $f'(x)$ slowly approaches zero. On the other hand, if the assumed root is taken in an interval containing a “local minimum,” then the method will oscillate. Finally, an interval containing an inflection point might cause problems, especially if the initial root approximation is not close to the exact root.

One way of addressing some of these limitations is to evaluate the function at the newly computed approximation; if this value is not smaller than the previous functional value, then the new estimate for the root is not accepted. Instead, compute the interval

$$x'_k = x_{k+1} - x_k = -\frac{f(x_k)}{f'(x_k)}$$

then estimate the new root as

$$x_{k+1} = x_k + \frac{x'_k}{2}$$

That is, we use half of the interval to estimate the new root. Furthermore, if this step does not reduce the functional value from the previous value then we repeat the halving process until it does.

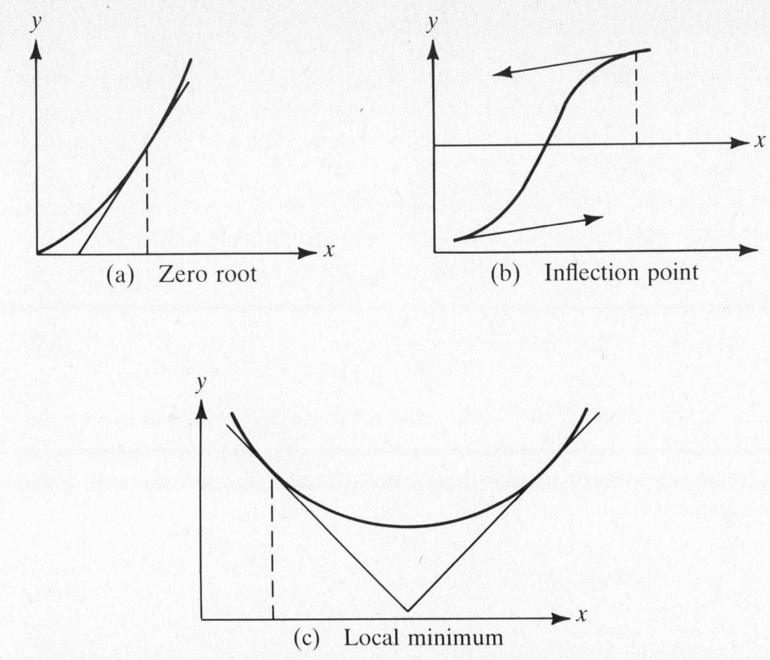


**Figure 6.7** Graphical illustration of the limitations associated with the Newton–Raphson first method.

**SUMMARY OF THE NEWTON–RAPHSON FIRST METHOD**

| STEP | OPERATION | SYMBOL |
|---|---|---|
| 1 | Sketch the function in question and determine an initial approximation for the root. | $x_k = x_1$ |
| 2 | Determine the first derivative of $f(x)$, then evaluate at $x = x_1$. If $f'(x_1) = 0$, then go to step 1. | |
| 3 | Establish tolerance value for the function (e.g., $T = 10^{-6}$). | $T \geq \lvert f(x_1) \rvert$ for $x_1 \approx$ root |
| 4 | Compute new approximation for the root. | $x_2 = x_1 - \dfrac{f(x_1)}{f'(x_1)}$ |
| 5 | Check tolerance. If $T \geq \lvert f(x_2) \rvert$, then end; otherwise set $x_1 = x_2$ and go to step 4. | |

## 6.6 NEWTON–RAPHSON SECOND METHOD

The method of tangents discussed in the preceding section does not always converge to a root. A second Newton–Raphson method achieves faster convergence and is more stable than all of the methods discussed so far. The basis for this method is illustrated in figure 6.8.

For a given function which varies continuously over a region where a root exists, a Taylor series can be written with respect to an initial $x_k$ value. The value of the function at a new point $x_{k+1}$ is therefore given as

$$f(x_{k+1}) = f(x_k) + f'(x_k)h + \frac{f''(x_k)h^2}{2} + \cdots \tag{6.9}$$

where $h = (x_{k+1} - x_k)$. Equation (6.9) represents a truncated Taylor series with the first three terms retained. Furthermore, it is assumed that the initial value $x_k$ is close enough to the exact root in question. Consequently, if equation (6.9) is to converge to a solution, then $f(x_{k+1})$ must go to zero. That is,

$$f(x_k) + f'(x_k)h + \frac{f''(x_k)}{2}h^2 = 0 \tag{6.10}$$

Recall that from the Newton–Raphson first method we have the relationship

$$x_{k+1} = x_k - \frac{f(x_k)}{f'(x_k)} \tag{6.11}$$

Thus

$$x_{k+1} - x_k = h = -\frac{f(x_k)}{f'(x_k)} \tag{6.12}$$

Rewriting equation (6.10) in the following form

$$\frac{1}{h} = -\frac{f'(x_k)}{f(x_k)} - \frac{1}{2}\frac{f''(x_k)}{f(x_k)}h \tag{6.13}$$

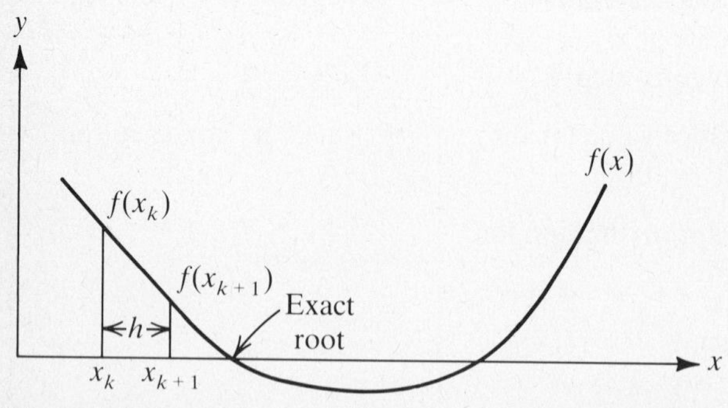


**Figure 6.8** Graphical interpretation of Newton–Raphson second method.

then substituting equation (6.12) into its right-hand side gives

$$\frac{1}{h} = -\frac{f'(x_k)}{f(x_k)} - \frac{1}{2}\frac{f''(x_k)}{f(x_k)}\left(-\frac{f(x_k)}{f'(x_k)}\right)$$

Simplifying, we obtain

$$\frac{1}{h} = -\frac{f'(x_k)}{f(x_k)} + \frac{1}{2}\frac{f''(x_k)}{f'(x_k)}$$

or more simply

$$h = \left(-\frac{f'(x_k)}{f(x_k)} + \frac{1}{2}\frac{f''(x_k)}{f'(x_k)}\right)^{-1} \tag{6.14}$$

But we know that $x_{k+1} - x_k = h$. Hence substituting into equation (6.14) gives

$$x_{k+1} - x_k = \left(-\frac{f'(x_k)}{f(x_k)} + \frac{1}{2}\frac{f''(x_k)}{f'(x_k)}\right)^{-1}$$

Solving for $x_{k+1}$ yields

$$x_{k+1} = x_k + \left(\frac{f''(x_k)}{2f'(x_k)} - \frac{f'(x_k)}{f(x_k)}\right)^{-1} \tag{6.15}$$

Equation (6.15) requires functional values for the first and second derivatives. This fact might present some complications when seeking roots of a function whose derivatives are not easy to calculate.

E

EXAMPLE 6.6

Find a root for $\tan x - x = 0$ using $x_1 = 5$. (Note that the Newton–Raphson method of tangents will not converge to a root for this choice of $x_1$.)

Solution

$$x_2 = x_1 + \left(-\frac{f'(x_1)}{f(x_1)} + \frac{1}{2}\frac{f''(x_1)}{f'(x_1)}\right)^{-1}$$

$$f(x) = \tan x - x = 0$$

$$f'(x) = \sec^2 x - 1 = \frac{2}{\cos^2 x} - 1 = \tan^2 x$$

$$f''(x) = 2\tan x \sec^2 x = 2\frac{\tan x}{\cos^2 x}$$

Using the initial assumed root, we have

$$x_1 = 5: \quad f(x_1) = -8.3805$$
$$f'(x_1) = 11.4279$$
$$f''(x_1) = -84.0252$$

Therefore the new approximation is given as

$$x_2 = 5 + \left( -\frac{11.4279}{-8.3805} + \frac{1}{2}\frac{-84.0252}{11.4279} \right)^{-1}$$

Continuing this process will yield

$$x_2 = 4.5676: \quad f(x_2) = 2.2908$$
$$f'(x_2) = 47.0382$$
$$f''(x_2) = 658.89$$

$$x_3 = 4.5676 + \left( -\frac{47.0382}{2.2908} + \frac{1}{2}\frac{(658.89)}{(47.0382)} \right)^{-1}$$
$$= 4.4936863: \quad f(x_3) = 5.6120 \times 10^{-3}$$
$$f'(x_3) = 20.243685$$
$$f''(x_3) = 191.16335$$

$$x_4 = 4.4936863 + \left( -\frac{20.243685}{5.6120 \times 10^{-3}} + \frac{1}{2}\frac{191.16335}{20.243685} \right)^{-1}$$
$$= 4.4934084: \quad f(x_4) = -3.504 \times 10^{-5}$$

Clearly this process can be continued to achieve greater accuracy.

**SUMMARY OF THE NEWTON–RAPHSON SECOND METHOD**

| STEP | OPERATION | SYMBOL |
|---|---|---|
| 1 | Sketch the function in question and determine an initial approximation for the root. | $x_k = x_1$ |
| 2 | Evaluate the first and second derivatives at $x_k = x_1$ and make sure they are not zero. | $f'(x_1) \neq 0$<br>$f''(x_1) \neq 0$ |

| | | |
|---|---|---|
| 3 | Establish a tolerance value for the function (e.g., $T = 10^{-6}$). | $T \geq \lvert f(x_1) \rvert$ |
| 4 | Compute a new approximation for the root. | $x_2 = x_1 + \left( \dfrac{f''(x_1)}{2f'(x_1)} - \dfrac{f'(x_1)}{f(x_1)} \right)^{-1}$ |
| 5 | Check tolerance. If $T \leq \lvert f(x_2) \rvert$, then end; otherwise set $x_1 = x_2$ and go to step 4. | |

## 6.7 MODIFIED NEWTON–RAPHSON METHODS

The disadvantage of using the Newton–Raphson methods is that the derivatives of the function in question must be known for the methods to apply. While this requirement might not be a problem for many functions, it may involve considerable effort for some. In addition, computer programs would have to be written such that the user would be required to input the derivatives of a given function. (It is interesting to note that many textbooks, including recent software manuals, claim that this is exactly the reason why their software is written using other less powerful methods.)

The authors totally disagree with the notion that finding the derivative of a function is a limitation, especially when dealing with approximate roots. Most students recall the basic definition for the first derivative of a function:

$$f'(x) = \lim_{\Delta x \to 0} \frac{f(x + \Delta x) - f(x)}{\Delta x} \tag{6.16}$$

Equation (6.16) is illustrated graphically in figure 6.9.

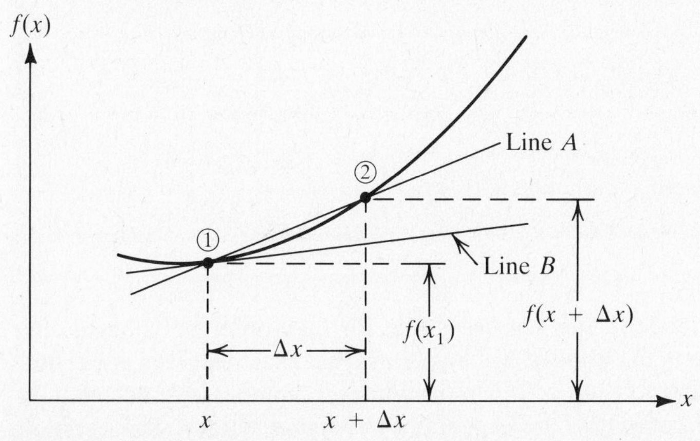


**Figure 6.9** Graphical illustration of the first derivative of a function.

Obviously, if $\Delta x$ is large, then the slope of the line passing through point 1 and point 2 is determined. As $\Delta x$ approaches zero the true slope at point 1 is approached. Therefore, if we are to evaluate the slope accurately, $\Delta x$ must be chosen to be very small; otherwise the derivative will be in error. The extent of the error depends not only on the magnitude of $\Delta x$ but also on the type of function being examined. This is true since the first derivative of a linear function is independent of the value of $\Delta x$ used.

E

EXAMPLE 6.7

Evaluate the first derivative of a first-order polynomial by selecting $\Delta x = h$ and by using the method of calculus at $x = x_0$.

Solution

$$f(x) = a + bx$$

Using a large $\Delta x$ one may use equation (6.16) as follows:

$$f'(x_0) = \lim_{\Delta x \to h} \frac{f(x_0 + h) - f(x_0)}{h}$$

where

$$f(x_0 + h) = a + b(x_0 + h)$$

$$f(x_0) = a + b(x_0)$$

$$f'(x_0) = \frac{a + bx_0 + bh - a - bx_0}{h}$$

$$= b$$

Using the method of calculus, we then have at $x = x_0$

$$f'(x_0) = b$$

Clearly, the values are exactly equal.

The point to be made in this respect is that for high-order polynomials the derivatives do depend on the value of $\Delta x$ chosen and that exact values may be computed only when $\Delta x$ approaches zero. Unfortunately, computers are not without limits. Most personal computers are capable of ten-digit accuracy. The question is then, how can one use a finite value of $\Delta x$? The answer to this question is addressed in detail in chapter 10. The student is asked to accept the following definitions for the first and second derivatives pending further discussion of the subject.

$$f'(x) = \frac{1}{60h}[-f(x - 3h) + 9f(x - 2h) - 45f(x - h) + 45f(x + h) - 9f(x + 2h) + f(x + 3h)] \quad \textbf{(6.17)}$$

$$f''(x) = \frac{1}{180h^2}[2f(x - 3h) - 27f(x - 2h) + 270f(x - h) - 490f(x) + 270f(x + h) - 27f(x + 2h) + 2f(x + 3h)] \quad \textbf{(6.18)}$$

where

$h$ = increment equivalent to $\Delta x$

$f(x - 3h)$ = functional value at a point $-3h$ from the point $x$ where the derivatives are to be evaluated

$f(x - 2h)$ = functional value at a point $-2h$ from the point $x$ where the derivatives are to be computed

⋮

$f(x + 3h)$ = functional value at a point $3h$ from the point $x$ where the derivatives are to be computed

These definitions enable us to use relatively large $\Delta x = h$ values without substantial error. The error magnitude is on the order of $h^6$. Equations (6.17) and (6.18) can be programmed on a computer so that the user need only input the function whose roots are to be found. That is, there is no need for determining the derivatives.

So far, we have avoided the question of what $h$ value should be used. After a careful review of a variety of functions, the authors have determined that using an $h$ value of 0.01–0.001 results in highly accurate results. The author's choice for $h$ is 0.01. This value should provide a greater degree of flexibility in analyzing most functions than the lower value of 0.001.

E

EXAMPLE 6.8

Evaluate the first and second derivatives for the following functions using the methods of calculus and equations (6.17) and (6.18); evaluate both functions at $x = 2$:

(a) $f(x) = 3x^3 - 2x^2 + 7x + 1$

(b) $f(x) = x \sin x$

Solution

(a) Using the methods of calculus:

$$f'(x) = 9x^2 - 4x + 7$$

$$f''(x) = 18x - 4$$

At $x = 2$:

$$f'(1) = 9(2)^2 - 4(2) + 7 = 35$$

$$f''(1) = 18(2) - 4 = 32$$

To use equations (6.17) and (6.18), we need to determine the various functional values. These are

$$f(x - 3h) = f(2 - 0.3) = f(1.7)$$

$$f(x - 2h) = f(2 - 0.2) = f(1.8)$$

$$f(x - h) = f(2 - 0.1) = f(1.9)$$

$$f(x) = f(2)$$

$$f(x + h) = f(2 + 0.1) = f(2.1)$$

$$f(x + 2h) = f(2 + 0.2) = f(2.2)$$

$$f(x + 3h) = f(2 + 0.3) = f(2.3)$$

Note that $h = 0.1$ was used:

$$f(1.7) = 3(1.7)^3 - 2(1.7)^2 + 7(1.7) + 1 = 21.859$$

$$f(1.8) = 3(1.8)^3 - 2(1.8)^2 + 7(1.8) + 1 = 24.616$$

$$f(1.9) = 3(1.9)^3 - 2(1.9)^2 + 7(1.9) + 1 = 27.657$$

$$f(2) = 3(2)^3 - 2(2)^2 + 7(2) + 1 = 31$$

$$f(2.1) = 3(2.1)^3 - 2(2.1)^2 + 7(2.1) + 1 = 34.663$$

$$f(2.2) = 3(2.2)^3 - 2(2.2)^2 + 7(2.2) + 1 = 38.664$$

$$f(2.3) = 3(2.3)^3 - 2(2.3)^2 + 7(2.3) + 1 = 43.021$$

The first derivative can now be evaluated by substituting the various values into equation (6.17):

$$f'(2) = \frac{1}{60(0.1)}[-21.859 + 9(24.616) - 45(27.657)$$

$$+ 45(34.663) - 9(38.664) + 43.021)]$$

$$= 35$$

The second derivative is determined by substituting the same functional values into equation 6.18:

$$f''(2) = \frac{1}{180(0.1)^2}[2(21.859) - 27(24.616) + 270(27.657)$$
$$-490(31) + 270(34.663) - 27(38.664) + 2(43.021)]$$
$$= 32$$

Note that equations (6.17) and (6.18) gave exact answers. One may argue that it is easier to find the derivatives directly. This is true if one ignores the fact that some functions might be more difficult to analyze than the function given in this example. In addition, this type of calculation can be easily handled on a computer with little effort. On the other hand, computers cannot easily find derivatives based on the methods of calculus.

(b) Once again applying equations (6.17) and (6.18) relative to the function $x \sin x$, we get

$$f(1.7) = 1.7 \sin 1.7 = 1.685830178$$
$$f(1.8) = 1.8 \sin 1.7 = 1.752925736$$
$$f(1.9) = 1.9 \sin 1.9 = 1.797970167$$
$$f(2) = 2 \sin 2 = 1.818594854$$
$$f(2.1) = 2.1 \sin 2.1 = 1.81273967$$
$$f(2.2) = 2.2 \sin 2.2 = 1.778692088$$
$$f(2.3) = 2.3 \sin 2.3 = 1.715121988$$

Applying equation (6.17) we determine the first derivative:

$$f'(2) = 0.07700371267$$

The exact value is

$$f'(2) = 0.07700375358$$

Applying equation (6.18) we determine the second derivative:

$$f''(2) = -2.650888556$$

The exact value is

$$f''(2) = -2.650888527$$

Once again, excellent results are obtained by using the approximate procedure. Further discussion of associated errors involved in approximate derivative expressions is given in chapter 10.

The first and second Newton–Raphson methods can now be modified rather easily so that no derivatives appear in the expressions. In particular, using the above approximations for the derivatives with $h = 0.1$ we can now express the first method as

$$x_{k+1} = x_k - \frac{f(x_k)}{F_1(x_k)} \tag{6.19}$$

where

$$F_1(x_k) = \tfrac{1}{6}[-f(x_k - 0.3) + 9f(x_k - 0.2) - 45f(x_k - 0.1) + 45f(x_k + 0.1) - 9f(x_k + 0.2) + f(x_k + 0.3)] \tag{6.20}$$

The second Newton–Raphson method may now be expressed as follows:

$$x_{k+1} = x_k + \left(\frac{F_2(x_k)}{2F_1(x_k)} - \frac{F_1(x_k)}{f(x_k)}\right)^{-1} \tag{6.21}$$

where $F_1(x_k)$ is given by equation (5.19) and $F_2(x_k)$ is given as

$$F_2(x_k) = \frac{1}{1.8}[2f(x_k - 0.3) - 27f(x_k - 0.2) + 270f(x_k - 0.1) - 490f(x_k) + 270f(x_k + 0.1) - 27f(x + 0.2) + 2f(x_k + 0.3)] \tag{6.22}$$

Equations (6.20)–(6.22) can be used to approximate roots to any arbitrary function without any need for determining the derivatives in a closed form.

## 6.8 LIN–BAIRSTOW METHOD FOR ROOTS OF POLYNOMIALS

The Lin–Bairstow method of quadratic factors is one of the most efficient for determining real and complex roots of polynomials with real coefficients. Polynomials do occur frequently in engineering and especially in eigenvalue problems. This method is based on the fact that for a given polynomial a quadratic factor can be extracted and the remainder term is forced to go to zero. Consider the following:

$$f(x) = a_n x^n + a_{n-1} x^{n-1} + \cdots + a_0 \tag{6.23}$$

Assume the following quadratic factor:

$$x^2 - yx - z \tag{6.24}$$

Divide equation (6.23) by (6.24) to get

$$f(x) = (x^2 - yx - z)(b_n x^{n-2} + b_{n-1} x^{n-3} + \cdots + b_2) + R_1 \tag{6.25}$$

where $R_1$ is a remainder having the following form:

$$R_1 = b_1(x - y) + b_0 \tag{6.26}$$

Equating equation (6.23) to equation (6.25) gives

$$a_n x^n + a_{n-1} x^{n-1} + \cdots + a_0 = (x^2 - yx - z)(b_n x^{n-2} + b_{n-1} x^{n-3} + \cdots + b_2) + b_1(x - y) + b_0 \tag{6.27}$$

Obviously, if the quadratic factor is exact, then $b_1$ and $b_0$ are equal to zero. However, this is not generally the case. Thus

$$\begin{aligned} a_n x^n &+ a_{n-1}x^{n-1} + \cdots + a_0 \\ &= b_n x^n + (b_{n-1} - yb_n)x^{n-1} + (b_{n-2} - yb_{n-1} - zb_n)x^{n-2} \\ &+ (b_{n-3} - yb_{n-2} - zb_{n-1})x^{n-3} + \cdots + (b_1 - yb_2 - zb_3)x \\ &+ (b_0 - b_1 y - zb_2) \end{aligned} \tag{6.28}$$

Equating terms of equal $x$ power gives

$$\begin{aligned} a_n &= b_n \\ a_{n-1} &= b_{n-1} - yb_n \\ a_{n-2} &= b_{n-2} - yb_{n-1} - zb_n \\ &\vdots \\ a_1 &= b_1 - yb_2 - zb_3 \\ a_0 &= b_0 - b_1 y - zb_2 \end{aligned} \tag{6.29}$$

Solving equations (6.29) for the $b$'s gives

$$\begin{aligned} b_n &= a_n \\ b_{n-1} &= a_{n-1} + yb_n \\ b_{n-2} &= a_{n-2} + yb_{n-1} + zb_n \\ &\vdots \\ b_1 &= a_1 + yb_2 + zb_3 \\ b_0 &= a_0 + yb_1 + zb_2 \end{aligned} \tag{6.30}$$

The question that we must address at this stage is, how can we force $b_1$ and $b_0$ to be zero? Fortunately, this can be accomplished iteratively as summarized below.

1. Assume $y$ and $z$ values, then evaluate all $b$'s except $b_1$ and $b_0$.
2. Set $b_1$ and $b_0$ to zero in the last two equations of (6.30), giving

   $$a_1 + yb_2 + zb_3 = 0$$

   $$a_0 + zb_2 = 0$$

   Then solve these equations for $y$ and $z$. Thus

   $$y = \frac{1}{b_2^2}(b_3 a_0 - b_2 a_1) \tag{6.31}$$

   $$z = \frac{1}{b_2^2}(-a_0 b_2) = -\frac{a_0}{b_2} \tag{6.32}$$

3. Use $y$ and $z$ from equations (6.31) and (6.32) to determine all $b$ values except $b_1$ and $b_0$. This is because the $b_0$ and $b_1$ equations were used in deriving equations (6.31) and (6.32).
4. Repeat this procedure until changes in $y$ and $z$ become insignificant.

The basis for this method may best be explained by considering the following third-order polynomial:

$$f(x) = a_3x^3 + a_2x^2 + a_1x + a_0$$

Obviously, the solution to this problem is started by first assuming a $y$ and a $z$ value. Thus let us assume

$$y = y_e, \qquad z = z_e$$

Hence

$$b_3 = a_3, \qquad b_2 = a_2 + a_3y_e$$

Substituting $b_3$ and $b_2$ into equations (6.31) and (6.32) gives

$$y = \frac{1}{(a_2 + a_3y_e)^2}[a_3a_0 - (a_2 + a_3y_e)a_1] \tag{6.33}$$

and

$$z = -\frac{a_0}{a_2 + a_3y_e} \tag{6.34}$$

Equations (6.33) and (6.34) provide a straightforward numerical solution to a cubic equation. That is, if $y_e = 0$, then the new estimates of $y$ and $z$ are given as follows:

$$y = \frac{1}{a_2^2}(a_3a_0 - a_2a_1) \tag{6.35}$$

$$z = -\frac{a_0}{a_2} \tag{6.36}$$

It is interesting to note that equations (6.33) and (6.35) are independent of the initial $z$ value used. This is true only for a cubic. For higher-order polynomials both $y$ and $z$ are dependent on the initial $y$ and $z$ values. Equation (6.33) clearly illustrates that we are dealing with a nonlinear equation of the third order. In addition, its solution is obtained iteratively. It is extremely important to note that $y_e$ must be selected in such a way that $a_2 \neq -a_3y_e$. Therefore, from now on, equations (6.35) and (6.36) are used for calculating rather than assuming the initial $y$ and $z$ values. This method of solution converges rather slowly, yet it provides more insight than other more popular methods.

**E** EXAMPLE 6.9

Determine approximate roots of the following polynomial:

$$f(x) = x^3 - 6x^2 + 11x - 6$$

Solution

$$y_e = \frac{1}{(-6)^2}[(1)(-6) - (-6)(11)] = 1.67$$

Thus

$$b_3 = 1, \qquad b_2 = -6 + (1)(1.67) = -4.33$$

Solving for new $y$ and $z$ values yields

$$y = \frac{1}{(-4.33)^2}[(1)(-6) - (-4.33)(11)] = 2.22$$

$$z = -\frac{(-6)}{(-6) + (1)(1.67)} = -1.38$$

Now these values are used to calculate new $b$ values:

$$b_3 = 1, \qquad b_2 = -6 + (1)(2.22) = -3.78$$

Solving for new $y$ and $z$ values gives

$$y = \frac{1}{(-3.78)^2}[(1)(-6) - (-3.78)(11)] = 2.49$$

$$z = -\frac{6}{3.78} = -1.59$$

After 11 iterations the following values are computed:

$$y = 3.0, \qquad z = -2$$

The quadratic factor is now given as follows:

$$x^2 - 3x + 2$$

The roots are computed as

$$x_{1,2} = \frac{3 \pm \sqrt{9 - 4 \times 2}}{2} = \frac{3 \pm 1}{2}$$

That is, $x_1 = 1$ and $x_2 = 2$. The third root is determined by noting that

$$(-1)(-2)(-x_3) = -6$$

Consequently, $x_3 = 3$.

Bairstow showed that $b_0$ and $b_1$ can be determined differently by expanding $b_1$ and $b_0$ as a Taylor series for functions of two variables in $y$ and $z$ as follows: Define

$$\Delta y = y_a - y \tag{6.37}$$

$$\Delta z = z_a - z \tag{6.38}$$

where

$\Delta y$ = increment of assumed $y$ value

$\Delta z$ = increment of assumed $z$ value

$y_a$ = adjusted $y$ value

$z_a$ = adjusted $z$ value

Therefore

$$b_1(y_a, z_a) = b_1(y, z) + \frac{\partial b_1}{\partial y}\Delta y + \frac{\partial b_1}{\partial z}\Delta z + \cdots \tag{6.39}$$

$$b_0(y_a, z_a) = b_0(y, z) + \frac{\partial b_0}{\partial y}\Delta y + \frac{\partial b_0}{\partial z}\Delta z + \cdots \tag{6.40}$$

Note that derivative terms higher than the first are ignored. This is possible if we assume that $\Delta y$ and $\Delta z$ are relatively small. The object is to force $b_1(y, z)$ and $b_0(y_a, z_a)$ to go to zero. Hence, rewriting equations (6.39) and (6.40) in matrix form, we have

$$\begin{bmatrix} \dfrac{\partial b_1}{\partial y} & \dfrac{\partial b_1}{\partial z} \\ \dfrac{\partial b_0}{\partial y} & \dfrac{\partial b_0}{\partial z} \end{bmatrix} \begin{Bmatrix} \Delta y \\ \Delta z \end{Bmatrix} = -\begin{Bmatrix} b_1 \\ b_0 \end{Bmatrix} \tag{6.41}$$

Equation (6.41) can then be solved for the unknown increments if the partial derivatives are known. Fortunately, these partials can be easily computed as follows:

$$\begin{aligned}
\frac{\partial b_n}{\partial y} &= \frac{\partial a_n}{\partial y} = 0 \\
\frac{\partial b_{n-1}}{\partial y} &= y\frac{\partial b_n}{\partial y} + b_n = b_n = D_n \\
\frac{\partial b_{n-2}}{\partial y} &= y\frac{\partial b_{n-1}}{\partial y} + b_{n-1} + z\frac{\partial b_n}{\partial y} = b_{n-1} + yD_n = D_{n-1} \\
\frac{\partial b_{n-3}}{\partial y} &= y\frac{\partial b_{n-2}}{\partial y} + b_{n-2} + z\frac{\partial b_{n-1}}{\partial y} = b_{n-2} + yD_{n-1} + zD_n = D_{n-2} \\
&\vdots \\
\frac{\partial b_1}{\partial y} &= y\frac{\partial b_2}{\partial y} + b_2 + z\frac{\partial b_3}{\partial y} = b_2 + yD_3 + zD_4 = D_2 \\
\frac{\partial b_0}{\partial y} &= y\frac{\partial b_1}{\partial y} + b_1 + z\frac{\partial b_2}{\partial y} = b_1 + yD_2 + zD_3 = D_1
\end{aligned} \tag{6.42a}$$

Similarly, the partials with respect to $z$ can be found. Thus

$$
\begin{aligned}
\frac{\partial b_n}{\partial z} &= \frac{\partial a_n}{\partial z} = 0 \\
\frac{\partial b_{n-1}}{\partial z} &= \frac{\partial a_{n-1}}{\partial z} + y\frac{\partial b_n}{\partial z} = 0 \\
\frac{\partial b_{n-2}}{\partial z} &= y\frac{\partial b_{n-1}}{\partial z} + z\frac{\partial b_n}{\partial z} + b_n = b_n = D_n \\
\frac{\partial b_{n-3}}{\partial z} &= y\frac{\partial b_{n-2}}{\partial z} + z\frac{\partial b_{n-1}}{\partial z} + b_{n-1} = b_{n-1} + yD_n = D_{n-1} \\
\frac{\partial b_{n-4}}{\partial z} &= y\frac{\partial b_{n-3}}{\partial z} + z\frac{\partial b_{n-2}}{\partial z} + b_{n-2} = b_{n-2} + yD_{n-1} + zD_n = D_{n-2} \\
&\vdots \\
\frac{\partial b_1}{\partial z} &= y\frac{\partial b_2}{\partial z} + z\frac{\partial b_3}{\partial z} + b_3 = b_3 + yD_4 + zD_5 = D_3 \\
\frac{\partial b_0}{\partial z} &= y\frac{\partial b_1}{\partial z} + z\frac{\partial b_2}{\partial z} + b_2 = b_2 + yD_3 + zD_4 = D_2
\end{aligned}
\tag{6.42b}
$$

Obviously, the $D$'s are related to the $b$'s in the same manner as the $b$'s are related to the $a$'s. Equation (5.41) can now be expressed in terms of $D$ values as follows:

$$
\begin{bmatrix} D_2 & D_3 \\ D_1 & D_2 \end{bmatrix} \begin{Bmatrix} \Delta y \\ \Delta z \end{Bmatrix} = -\begin{Bmatrix} b_1 \\ b_0 \end{Bmatrix} \tag{6.43}
$$

Consequently, equation (6.43) is used to determine changes in the assumed $y$ and $z$ values. The procedure is repeated until the changes in $\Delta y$ and $\Delta z$ approach zero. Note that a solution is possible if the determinant of the coefficient matrix in equation (6.43) is not zero, that is, if $D_2^2 = D_1 D_3$. To avoid such a problem, the reader is advised to exercise the utmost care in choosing proper $y$ and $z$ values initially. For example, using $y = \pi$ and $z = \sqrt{2}$ may eliminate that possibility.

**E** EXAMPLE 6.10

Determine the roots of the following polynomial:

$$x^3 - 6x^2 + 11x - 6 = 0$$

### Solution

Assume $y = 3$ and $z = -3$, then determine the $b$ values:

$$b_3 = a_3 = 1$$

$$b_2 = a_2 + yb_3 = -6 + 3(1) = -3$$

$$b_1 = a_1 + yb_2 + zb_3 = 11 + 3(-3) + (-3)(1) = -1$$

$$b_0 = a_0 + yb_1 + zb_2 = -6 + 3(-1) + (-3)(-3) = 0$$

Since $b_1$ and $b_0$ are not equal to zero, compute the $D$'s:

$$D_3 = b_3 = 1$$

$$D_2 = b_2 + yD_3 = -3 + (3)(1) = 0$$

$$D_1 = b_1 + yD_2 + zD_3 = -1 + (3)(0) + (-3)(1) = -4$$

Substitute into equation (6.43) and solve for the increments:

$$\begin{bmatrix} 0 & 1 \\ -4 & 0 \end{bmatrix} \begin{Bmatrix} \Delta y \\ \Delta z \end{Bmatrix} = -\begin{Bmatrix} -1 \\ 0 \end{Bmatrix}$$

$$\Delta y = 0, \qquad \Delta z = +1$$

Therefore the adjusted $y$ and $z$ values can now be computed:

$$y_a = y + y = 0 + 3 = 3$$

$$z_a = z + z = 1 - 3 = -2$$

Repeating the procedure once again by computing the $b$ values for $y = 3$ and $z = -2$, we get

$$b_3 = 1$$

$$b_2 = -6 + 3(1) = -3$$

$$b_1 = 11 + 3(-3) + (-2)(1) = 0$$

$$b_0 = -6 + 3(0) + (-2)(-3) = 0$$

Clearly both $b_1$ and $b_0$ are equal to zero. This means that the assumed $y$ and $z$ values are exact. Thus the quadratic factor is

$$x^2 - 3x + 2$$

This factor has the roots

$$x_1 = 1 \qquad \text{and} \qquad x_2 = 2$$

Now, one may easily use synthetic division to extract the third root:

$$
\begin{array}{r|l}
 & x - 3 \\
x^2 - 3x + 2 & x^3 - 6x^2 + 11x - 6 \\
 & x^3 - 3x^2 + 2x \\
\hline
 & -3x^2 + 9x - 6 \\
 & -3x^2 + 9x - 6 \\
\hline
 & 0
\end{array}
$$

Hence $x_3 = 3$ is a root. Alternatively, we could have found $x_3$ by noting that $(1)(2)(x_3) = 6$, in which case $x_3 = 3$ as before. Note that the $b$ values as well as the $D$ values can be computed by using synthetic division. This is not recommended since it becomes cumbersome when dealing with polynomials of high powers. The Bairstow method can only be applied when every term in a polynomial appears in the expression. Fortunately, this requirement can be met rather easily.

EXAMPLE 6.11

Transform the following polynomial so that the Lin–Bairstow method can be applied.

$$x^3 + x + 1 = 0$$

Solution

Assume the following transformation:

$$x = y + 1$$

Then, substitute into the equation to get

$$(y + 1)^3 + (y + 1) + 1 = 0$$

or more simply

$$y^3 + 3y^2 + 4y + 2 = 0$$

Consequently, one may solve for the roots $y_1$, $y_2$, and $y_3$, and then solve for $x_1$, $x_2$, and $x_3$. That is, $x_1 = y_1 + 1, \ldots$.

**SUMMARY OF THE LIN–BAIRSTOW METHOD**

| STEP | OPERATION | SYMBOL |
|---|---|---|
| 1 | Assume initial values for the coefficients of the quadratic factor. | $y = y_0$<br>$z = z_0$ |
| 2 | Compute the $b$'s using equation (6.30). | $b_n = a_n$<br>$b_{n-1} = a_{n-1} + yb_n$<br>$b_i = a_i + yb_{i+1} + zb_{i+2}$<br>$i = n-2,\ldots,0$ |
| 3 | If $b_1$ and $b_0$ are not zero, then compute the $D$'s using equation (6.42). | $D_n = b_n$<br>$D_{n-1} = b_{n-1} + yD_n$<br>$D_i = b_i + yD_{i+1} + zD_{i+2}$<br>$i = n-2,\ldots,0$ |
| 4 | Solve for the increments in $y_0$ and $z_0$, using equation (6.43). | $\begin{bmatrix} D_2 & D_3 \\ D_1 & D_2 \end{bmatrix}\begin{Bmatrix} \Delta y \\ \Delta z \end{Bmatrix} = -\begin{Bmatrix} b_1 \\ b_0 \end{Bmatrix}$ |
| 5 | Establish tolerance limits for $\Delta y$ and $\Delta z$. If $T$ is satisfied, then end; otherwise continue. | $T \geq \max\lvert\Delta y, \Delta z\rvert$ |
| 6 | Compute the adjusted $y$ and $z$ values, then let $y = y_a$ and $z = z_a$, then go to step 2. | $y_a = \Delta y + y$<br>$z_a = \Delta z + z$ |

## 6.9 NEWTON–RAPHSON METHOD FOR SYSTEMS OF EQUATIONS

A system of nonlinear algebraic equations may arise when one is dealing with problems involving optimization and numerical integration (Gauss quadratures). In fact, a single nonlinear algebraic equation may in some cases represent a system of nonlinear algebraic equations, and vice versa. For example, the equation

$$x^3 + A_1x^2 + A_2x + A_3 = 0$$

has three possible roots, say $x_1$, $x_2$, and $x_3$. Therefore

$$(x - x_1)(x - x_2)(x - x_3) = x^3 + A_1x^2 + A_2x + A_3$$

or more simply

$$\begin{aligned} x^3 - (x_1 + x_2 + x_3)x^2 + (x_1x_2 + x_1x_3 + x_2x_3)x - x_1x_2x_3 \\ = x^3 + A_1x^2 + A_2x + A_3 \end{aligned}$$

Equating terms of equal power yields three nonlinear algebraic equations in the three unknown roots. That is,

$$-(x_1 + x_2 + x_3) = A_1$$
$$x_1x_2 + x_1x_3 + x_2x_3 = A_2$$
$$-x_1x_2x_3 = A_3$$

These equations can be expressed in the following form:

$$f_1 = A_1 + x_1 + x_2 + x_3$$
$$f_2 = A_2 - x_1x_2 - x_1x_3 - x_2x_3$$
$$f_3 = A_3 + x_1x_2x_3$$

It is evident that solving a single nonlinear equation is easier than solving a system. However, solving a system provides approximations for all of the roots at once. In general, for a polynomial of the form

$$x^n + A_1x^{n-1} + A_2x^{n-2} + \cdots + A_n = 0 \tag{6.44}$$

the corresponding system of nonlinear algebraic equations that can be solved for the roots is

$$\begin{aligned} f_1 &= A_1 + \sum_{i=1}^{n} x_i \\ f_2 &= A_2 - \sum_{\substack{i=1\\j=1}}^{n} x_ix_j \\ f_3 &= A_3 + \sum_{\substack{i=1\\j=1\\k=1}}^{n} x_ix_jx_k \\ &\vdots \\ f_n &= A_n - (-1)^n(x_1x_2\cdots x_n) \end{aligned} \tag{6.45}$$

where $i \neq j \neq k$.

More generally, the system of equations may not be of the polynomial variety. Therefore a system of $n$ equations in $n$ unknowns is called nonlinear if one or more of the equations in the system is/are nonlinear. The object is to determine values for $x_1, \ldots, x_n$ such that the functionals $f_1, \ldots, f_n$ are approximately zero. One technique for solving this problem is the Newton–Raphson first method. Consider the following system of nonlinear algebraic equations, evaluated at $x_1, \ldots, x_n$:

$$\begin{aligned} f_1(x_1, x_2, \ldots, x_n) &= f_1(x_i) \\ f_2(x_1, x_2, \ldots, x_n) &= f_2(x_i) \\ &\vdots \\ f_n(x_1, x_2, \ldots, x_n) &= f_n(x_i) \end{aligned}$$

Using Taylor series, one can evaluate these functions at $x_1 + h_1, \ldots, x_n + h_n$ by expanding about the initial values $x_1, \ldots, x_n$. Thus

$$f_1(x_1 + h_1, x_2 + h_2, \ldots, x_n + h_n) = f_1(\bar{x}_i)$$
$$f_2(x_1 + h_1, x_2 + h_2, \ldots, x_n + h_n) = f_2(\bar{x}_i)$$
$$\vdots$$
$$f_n(x_1 + h_1, x_2 + h_2, \ldots, x_n + h_n) = f_n(\bar{x}_i)$$

$$\begin{aligned} f_1(\bar{x}_i) &= f_1(x_i) + \sum_{j=1}^{n} h_j \frac{\partial f_1(x_i)}{\partial x_j} \\ f_2(\bar{x}_i) &= f_2(x_i) + \sum_{j=1}^{n} h_j \frac{\partial f_2(x_i)}{\partial x_j} \\ &\vdots \\ f_n(\bar{x}_i) &= f_n(x_i) + \sum_{j=1}^{n} h_j \frac{\partial f_n(x_i)}{\partial x_j} \end{aligned} \tag{6.46}$$

Equations (6.46) represent a truncated Taylor series in which only the linear terms are retained, which is why this method is sometimes referred to as the "linear method." Therefore, forcing the functions $f_1(\bar{x}_i), \ldots, f_n(\bar{x}_i)$ to be zeros permits the derivation of an iterative formula for determining roots of nonlinear algebraic equations. That is,

$$\begin{aligned} 0 &= f_1(x_i) + \sum_{j=1}^{n} h_j \frac{\partial f_1(x_i)}{\partial x_j} \\ 0 &= f_2(x_i) + \sum_{j=1}^{n} h_j \frac{\partial f_2(x_i)}{\partial x_j} \\ &\vdots \\ 0 &= f_n(x_i) + \sum_{j=1}^{n} h_j \frac{\partial f_n(x_i)}{\partial x_j} \end{aligned}$$

Expanding, then using the matrix form yields

$$\begin{bmatrix} \dfrac{\partial f_1(x_i)}{\partial x_1} & \dfrac{\partial f_1(x_i)}{\partial x_2} & \cdots & \dfrac{\partial f_1(x_i)}{\partial x_n} \\ \dfrac{\partial f_2(x_i)}{\partial x_1} & \dfrac{\partial f_2(x_i)}{\partial x_2} & \cdots & \dfrac{\partial f_2(x_i)}{\partial x_n} \\ \vdots & \vdots & & \vdots \\ \dfrac{\partial f_n(x_i)}{\partial x_1} & \dfrac{\partial f_n(x_i)}{\partial x_2} & \cdots & \dfrac{\partial f_n(x_i)}{\partial x_n} \end{bmatrix} \begin{Bmatrix} h_1 \\ h_2 \\ \vdots \\ h_n \end{Bmatrix} = - \begin{Bmatrix} f_1(x_i) \\ f_2(x_i) \\ \vdots \\ f_n(x_i) \end{Bmatrix} \tag{6.47}$$

or more simply

$$[J]\{h\} = \{f\}$$

Here $\{h\}$ is the increments vector and $\{f\}$ is the function matrix; the square matrix of partial derivatives is denoted by $[J]$ and is called the Jacobian. Therefore, solving for the increments vector gives

$$\{h\} = -[J]^{-1}\{f\} \tag{6.48}$$

However, since $h_1 = \bar{x}_1 - x_1, h_2 = \bar{x}_2 - x_2, \ldots, h_n = \bar{x}_n - x_n$, substituting into equation (6.48) yields

$$\{\bar{x}\} = \{x\} - [J]^{-1}\{f\} \tag{6.49}$$

It is evident that equation (6.49) has the same form as that derived for a single linear algebraic equation. Therefore, using $k$ to designate the number of iterations performed, we can express equation (6.49) in the following general form:

$$\{x\}_{k+1} = \{x\}_k - [J]_k^{-1}\{f\}_k, \qquad k = 1, 2, \ldots \tag{6.50}$$

The solution procedure consists of making initial approximations for each of the $n$ variables, then evaluating the Jacobian and the $n$ equations so that

$$\det[J] \neq 0$$

after which a test for errors is made. This is accomplished in one of two ways. In the first

$$T_1 \geq \sqrt{h_1^2 + h_2^2 + \cdots + h_n^2} = \sqrt{\sum_{i=1}^{n} h_i^2} \tag{6.51}$$

where $T$ is an appropriate tolerance value.

The second test is to use the maximum $h$ value as a basis for accuracy. Thus

$$T_2 \geq \max|h_i|, \qquad i = 1, \ldots, n \tag{6.52}$$

Equation (6.52) is by far the most widely used tolerance test for convergence.

The reader is advised that since some problems may not converge, a limit on the maximum number of iterations should be included in any computer program pertaining to roots of nonlinear algebraic equations.

Thus far we have made no mention of how the initial approximation vector is chosen. Unfortunately, there are no general means by which an initial solution can be obtained. A knowledge of the physical problem being solved is one way of obtaining an initial estimate. On the other hand, for systems involving two equations, one may be able to get a good estimate of the real roots graphically. Finally, the assumed values must be such that $\det[J] = 0$. This does not guarantee convergence, but it does provide some guidance as to the appropriateness of one's initial approximations.

E

EXAMPLE 6.12

Estimate one set of roots for the following system of nonlinear algebraic equations using the Newton–Raphson method:

$$x_1^2 + x_2^2 = 18: \qquad f_1(x_1, x_2) = x_1^2 + x_2^2 - 18$$

$$x_1 - x_2 = 0: \qquad f_2(x_1, x_2) = x_1 - x_2$$

### Solution

Obviously this system has an exact solution of

$$x_1 = \pm 3, \qquad x_2 = \pm 3$$

However, let us see how the method just described is used to estimate these roots. We begin by estimating one set of roots, say $x_1 = 2$ and $x_2 = 2$, graphically to get

$$\begin{Bmatrix} x_1 \\ x_2 \end{Bmatrix} = \begin{Bmatrix} 2 \\ 2 \end{Bmatrix}_1$$

Then the Jacobian is determined and its inverse is evaluated. That is,

$$[J] = \begin{bmatrix} \dfrac{\partial f_1}{\partial x_1} & \dfrac{\partial f_1}{\partial x_2} \\ \dfrac{\partial f_2}{\partial x_1} & \dfrac{\partial f_2}{\partial x_2} \end{bmatrix} = \begin{bmatrix} 2x_1 & 2x_2 \\ 1 & -1 \end{bmatrix}$$

and

$$[J]^{-1} = \frac{1}{2(x_1 + x_2)} \begin{bmatrix} 1 & 2x_2 \\ 1 & -2x_1 \end{bmatrix}$$

Therefore, evaluating $[J]$ at the initial assumed values of $x_1 = 2$ and $x_2 = 2$ gives

$$[J]_1^{-1} = \frac{1}{8} \begin{bmatrix} 1 & 4 \\ 1 & -4 \end{bmatrix}_1$$

The functions are now evaluated to give

$$\begin{Bmatrix} f_1 \\ f_2 \end{Bmatrix}_1 = \begin{Bmatrix} -10 \\ 0 \end{Bmatrix}_1$$

Substituting into equation (6.50) gives

$$\begin{Bmatrix} x_1 \\ x_2 \end{Bmatrix}_2 = \begin{Bmatrix} 2 \\ 2 \end{Bmatrix}_1 - \frac{1}{8} \begin{bmatrix} 1 & 4 \\ 1 & -4 \end{bmatrix}_1 \begin{Bmatrix} -10 \\ 0 \end{Bmatrix}_1 = \begin{Bmatrix} 3.25 \\ 3.25 \end{Bmatrix}_2$$

The inverse of the Jacobian and the functions are now evaluated at $x_1 = 3.25$ and $x_2 = 3.25$; then, using equation (6.50), we get

$$\begin{Bmatrix} x_1 \\ x_2 \end{Bmatrix}_3 = \begin{Bmatrix} 3.25 \\ 3.25 \end{Bmatrix}_1 - \frac{1}{13}\begin{bmatrix} 1 & 2.5 \\ 1 & -2.5 \end{bmatrix}_2 \begin{Bmatrix} 3.125 \\ 0 \end{Bmatrix}_2 = \begin{Bmatrix} 3.0096 \\ 3.0096 \end{Bmatrix}_3$$

Clearly, in just two iterations, we have achieved an acceptable level of accuracy. Therefore we may proceed to estimate the rest of the roots by applying this method to yet another set of initial estimates for the roots.

E

EXAMPLE 6.13

Approximate all of the roots to the following nonlinear algebraic equation by first transforming it into its equivalent system of nonlinear algebraic equations:

$$f(x) = x^3 - 6x^2 + 11x - 6$$

### Solution

Since this is a polynomial equation, we may use equation (6.45) to determine its equivalent system. That is,

$$f_1(x_1, x_2, x_3) = -6 + x_1 + x_2 + x_3$$

$$f_2(x_1, x_2, x_3) = 11 - x_1x_2 - x_1x_3 - x_2x_3$$

$$f_3(x_1, x_2, x_3) = -6 + x_1x_2x_3$$

The corresponding Jacobian is evaluated next to give

$$[J]_1 = \begin{bmatrix} \frac{\partial f_1}{\partial x_1} & \frac{\partial f_1}{\partial x_2} & \frac{\partial f_1}{\partial x_3} \\ \frac{\partial f_2}{\partial x_1} & \frac{\partial f_2}{\partial x_2} & \frac{\partial f_2}{\partial x_3} \\ \frac{\partial f_3}{\partial x_1} & \frac{\partial f_3}{\partial x_2} & \frac{\partial f_3}{\partial x_3} \end{bmatrix} = \begin{bmatrix} 1 & 1 & 1 \\ -x_2 - x_3 & -x_1 - x_3 & -x_1 - x_2 \\ x_2x_3 & x_1x_3 & x_1x_2 \end{bmatrix}$$

Now an initial guess as to the approximate roots is made. Note here that $x_1 = x_2 = x_3 = 0$ is a bad guess, and so is any initial guess that assumes equal values for all of the roots. This is because the resulting determinant will be zero. Consequently, by plotting the original function, an estimate of the approximate roots can be made. Say

$$\begin{Bmatrix} x_1 \\ x_2 \\ x_3 \end{Bmatrix}_1 = \begin{Bmatrix} 0.5 \\ 1.5 \\ 2.5 \end{Bmatrix}_1$$

The actual roots are 1, 2, and 3. Therefore the Jacobian becomes

$$[J]_1 = \begin{bmatrix} 1 & 1 & 1 \\ -4 & -3 & -2 \\ 3.75 & 1.25 & 0.75 \end{bmatrix}$$

and

$$[J]_1^{-1} = \begin{bmatrix} 0.125 & 0.25 & 0.50 \\ -2.25 & -0.15 & -1.00 \\ 3.125 & 1.25 & 0.50 \end{bmatrix}_1$$

Evaluating the function $f_1, f_2$, and $f_3$ at $\{x\}_0$ then substituting into equation (6.50) yields

$$\begin{Bmatrix} x_1 \\ x_2 \\ x_3 \end{Bmatrix}_2 = \begin{Bmatrix} 0.5 \\ 1.5 \\ 2.5 \end{Bmatrix}_1 - \begin{bmatrix} 0.125 & 0.25 & 0.50 \\ -2.25 & -1.5 & -1 \\ 3.125 & 1.25 & 0.50 \end{bmatrix}_1 \begin{Bmatrix} -1.5 \\ 5.25 \\ 4.125 \end{Bmatrix}_1 = \begin{Bmatrix} 1.4375 \\ 1.8750 \\ 2.6875 \end{Bmatrix}_2$$

Now, using the new vector, we can approximate better values for the roots. That is,

$$\{x\}_2 = \begin{Bmatrix} 1.4375 \\ 1.8750 \\ 2.6875 \end{Bmatrix}_2$$

$$[J]_2 = \begin{bmatrix} 1 & 1 & 1 \\ -4.5625 & -4.1250 & -3.3125 \\ 5.0391 & 3.8633 & 2.6953 \end{bmatrix}_2$$

$$[J]_2^{-1} = \begin{bmatrix} 3.7788 & 2.6286 & 1.8286 \\ -9.8905 & -5.2748 & -2.8132 \\ 7.1117 & 2.6462 & 0.9846 \end{bmatrix}_2$$

and

$$\begin{Bmatrix} f_1 \\ f_2 \\ f_3 \end{Bmatrix}_2 = \begin{Bmatrix} 0 \\ -0.5976 \\ 1.2436 \end{Bmatrix}_2$$

Clearly, the functional values are much closer to zero than the initial assumed roots. Substituting into equation (6.50) and using $k = 1$ gives

$$\begin{Bmatrix} x_1 \\ x_2 \\ x_3 \end{Bmatrix}_3 = \begin{Bmatrix} 1.4375 \\ 1.8750 \\ 2.6875 \end{Bmatrix}_2 - [J]_2^{-1} \begin{Bmatrix} 0 \\ -0.5976 \\ 1.2436 \end{Bmatrix}_2 = \begin{Bmatrix} 0.7343 \\ 2.221 \\ 3.0439 \end{Bmatrix}_3$$

Repeating this procedure one more time for $k = 3$ gives

$$\begin{Bmatrix} x_1 \\ x_2 \\ x_3 \end{Bmatrix}_4 = \begin{Bmatrix} 0.7343 \\ 2.221 \\ 3.0438 \end{Bmatrix}_3 - \begin{bmatrix} 0.1569 & 0.2138 & 0.2912 \\ -4.0319 & -1.8153 & -0.8174 \\ 4.8749 & 1.6015 & 0.5261 \end{bmatrix} \begin{Bmatrix} 0 \\ 0.3735 \\ -1.0358 \end{Bmatrix}_3$$

Simplifying, we have

$$\begin{Bmatrix} x_1 \\ x_2 \\ x_3 \end{Bmatrix}_4 = \begin{Bmatrix} 0.9561 \\ 2.0523 \\ 2.9906 \end{Bmatrix}_4$$

Subsequent evaluations of the three functions at $\{x\}_3$ yields

$$\begin{Bmatrix} f_1 \\ f_2 \\ f_3 \end{Bmatrix}_4 = \begin{Bmatrix} -0.0010 \\ 0.0409 \\ -0.1318 \end{Bmatrix}_4$$

These functional values clearly indicate that the new approximate roots are fairly close to the exact ones. On the other hand, had we established a tolerance limit for the roots, we could easily check the changes in the $\{x\}$ vector at each iteration. For example, suppose

$$T_1 = 0.10 \overset{?}{\leq} \sqrt{h_1^2 + h_2^2 + h_3^2}$$

Then for $k = 1$, we have

$$\begin{aligned} 0.10 &\overset{?}{\leq} \sqrt{(1.4375 - 0.5)^2 + (1.875) - 1.5)^2 + (2.6875 - 2.5)^2} \\ &\overset{?}{\leq} 1.0269 \qquad \text{NO GOOD} \end{aligned}$$

Similarly for $k = 2$,

$$\begin{aligned} 0.10 &\overset{?}{\leq} \sqrt{(0.7343 - 1.4375)^2 + (2.221 - 1.875)^2 + (3.0439 - 2.6875)^2} \\ &\overset{?}{\leq} 0.8609 \qquad \text{NO GOOD} \end{aligned}$$

For the final iteration performed, $k = 3$,

$$\begin{aligned} 0.01 &\overset{?}{\leq} \sqrt{(0.9561 - 0.7343)^2 + (2.0523 - 1.875)^2 + (2.9906 - 3.0439)^2} \\ &\overset{?}{\leq} 0.0835 \qquad \text{GOOD} \end{aligned}$$

Clearly, the desired results have been achieved. However, this does not mean that greater accuracy is not possible. In fact, one may establish a different tolerance value and continue the iterative process until more accurate roots are obtained.

**SUMMARY OF THE NEWTON–RAPHSON METHOD FOR SYSTEMS OF EQUATIONS**

| STEP | OPERATION | SYMBOL |
|---|---|---|
| 1 | Make an initial guess for the roots of the systems, so that the determinant of the Jacobian is not zero. | $\{x\}_k = \{x\}_1$ |
| 2 | Establish a tolerance limit using the $T_1$ or $T_2$ criterion. | $T_1 \geq \sqrt{\sum_{i=1}^{n} h_i^2}$ <br> $T_2 \geq \max\lvert h_i\rvert$ |
| 3 | Evaluate the Jacobian at $\{x\}_k = \{x\}_1$, then invert $[J]_1$. | $[J]_1 = \begin{bmatrix} \frac{\partial f_1}{\partial x_1} & \cdots & \frac{\partial f_1}{\partial x_n} \\ \vdots & & \vdots \\ \frac{\partial f_n}{\partial x_1} & & \frac{\partial f_n}{\partial x_n} \end{bmatrix}$ <br> $\{f\}_1 = \begin{Bmatrix} f_1 \\ \vdots \\ f_n \end{Bmatrix}_1$ |
| 4 | Calculate the new approximation to the roots, then determine the increments vector $\{h\}$. | $\{x\}_2 = \{x\}_1 - [J]^{-1}\{f\}_1$ <br> $\{h\}_2 = \{x\}_2 - \{x\}_1$ |
| 5 | Check the tolerance limits. If the criterion is met, then end; otherwise let $\{x\}_1 = \{x\}_2$, then go to step 3. | |

## 6.10 PRACTICAL CONSIDERATIONS

Thus far, we have avoided questions relating to truncation and roundoff errors. In this section, these questions as well as problems pertaining to multiple roots are addressed.

### 6.10.1 Errors

It is evident that all of the methods outlined in this chapter involve making initial guesses as to the root. These estimates are arbitrary in that even when the function is sketched, one has to pick an approximate root from the infinite possibilities that exist in the interval where a root is located. Therefore one might expect that errors pertaining to inaccurately evaluating the function or its derivative may not in some cases be as critical as we might think. Errors pertaining to truncation are caused by the method itself in that numerical methods can usually be related to a truncated Taylor series. Truncation errors are generally reduced by repeated iterations.

Roundoff errors are attributed to computing devices (calculators, microcomputers, and mainframes). There is little that we can do about the fact that calculators can retain ten digits or less. As a result, scaling of coefficients may sometimes be a useful tool in providing approximate accuracy.

In fact, the most important error service is that pertaining to formulation of the mathematical model. That is, if one is to determine the coefficients of a nonlinear algebraic equation approximately, then the corresponding error associated with the roots may not be acceptable. Consider the following function:

$$f(x) = x^2 - 2x + 1$$

Clearly, the roots can be determined by means of the quadratic formula as

$$x_1 = 1, \qquad x_2 = 1$$

Suppose that the constant term is not known exactly to be 1.0, but instead has values of $1 \pm 0.10$. That is, it may be 0.9 or 1.1 or somewhere in between. Consequently, corresponding to the low and high value of the constant TERM, the solution is given as

$$\text{Constant} = 0.9: \qquad x_1 = 1.316, \qquad x_2 = 0.684$$

$$\text{Constant} = 1.1: \qquad x_1 = 0.5 + 0.316J, \quad x_2 = 0.5 - 0.316J$$

A change of $\pm 10\%$ in the constant term has resulted in significant changes in the roots; these are larger variations than are likely from roundoff or truncation. This problem is even more significant when one is dealing with a higher-order polynomial. This is precisely why polynomials are referred to as ill conditioned: small changes in their coefficients result in significant changes in their roots.

### 6.10.2 Root Multiplicity

A root is called a simple root if it is distinct; otherwise roots that are of the same order of magnitude are called multiple. The numerical methods illustrated thus far and in particular the Newton–Raphson methods converge more slowly when used for functions involving multiple roots. In fact, the rate of convergence is reduced from second to first order. That is, the interval where a root is located is reduced linearly, rather than quadratically. Consider the following function with multiple roots $x_1 = x_2 = 1$:

$$f(x) = x^2 - 2x + 1$$

Using the Newton–Raphson first method, we express the new approximation to the root as

$$x_{k+1} = x_k - \frac{x_k^2 - 2x_k + 1}{2x_k - 2}$$

or more simply

$$x_{k+1} = \tfrac{1}{2}(x_k + 1)$$

It is evident that by using an initial guess of $x_1 = 2$, a new approximation $x_2$ is calculated as

$$x_1 = 2, \qquad x_2 = 1.5$$

Furthermore,

$$x_2 = 1.5, \qquad x_3 = 1.25$$

Linear convergence implies that the interval between two successive approximations is reduced by the same factor. In this case

$$x_1 - x_2 = 0.5, \qquad x_2 - x_3 = 0.25$$

The interval is clearly cut in half at every iteration. On the other hand, if the function has two distinct roots, then the interval is reduced quadratically.

Generally, we may not know in advance whether the function in question has multiple roots. Therefore we need to identify such functions. It can be shown that a root $x_i$ is of multiplicity $m$ (meaning the function involves $m$ roots of values $x_i$) if and only if

$$f(x_i) = f^{(1)}(x_i) = \cdots = f^{(m)}(x_i) = 0$$

The implication is that not only would the function in question be zero at $x = x_i$, but so also would its derivatives up to the $m$th order. For example, the following cubic has multiplicity of order 3 at $x = a$:

$$f(x) = x^3 - 3ax^2 + 3a^2x - a^3$$

Therefore

$$f(a) = a^3 - 3a^3 + 3a^3 - a^3 = 0$$

$$f^{(1)}(a) = 3a^2 - 6a^2 + 3a^2 = 0$$

$$f^{(2)}(a) = 6a - 6a = 0$$

It is evident that, while the first two derivatives evaluated at $x = a$ (multiple root) as well as the functions are all zero, $f^{(3)}(a)$ is not zero. With this in mind, it might be appropriate to test the function in question and its derivatives, once an approximation to a root is determined wherever multiplicity is suspected.

## Recommended Reading

*Relaxation Methods*, D. N. Allen, McGraw-Hill Book Company, New York, 1954.

*Numerical Methods in Engineering*, Mario G. Salvadori and Melvin L. Baron, Prentice-Hall, Englewood Cliffs, N.J., 1961.

*Theory of Functions*, L. V. Toralballa, Merrill, 1963.

*Iterative Methods for the Solution of Equations*, J. F. Traub, Prentice-Hall, Englewood Cliffs, N.J., 1964.

*A First Course in Numerical Analysis*, Anthony Ralston, McGraw-Hill Book Company, New York, 1965.

*Brief Numerical Methods*, Wendell E. Grone, Prentice-Hall, Englewood Cliffs, N.J., 1966.

Solution of nonlinear equations. M. C. Y. Kuo, *IEEE Transactions on Computers*, **C-17** (1968).

*Introduction to Applied Numerical Analysis*, R. W. Hamming, McGraw-Hill Book Company, New York, 1971.

## PROBLEMS

**6.1** Estimate the largest positive root of the following function graphically:

$$f(x) = x^3 - 3x^2 + 6x - 4$$

Show at least one plot of the isolated largest root.

**6.2** Use the interval-halving method to approximate the largest positive root of the function given in problem 6.1. The root must be accurate to four significant figures.

**6.3** Approximate the roots of the functions

**(a)** $e^{-x} - 2 = 0$ **(b)** $e^{x} - 2 = 0$

Use the false-position method so that the computed root is accurate to three decimal places.

**6.4** Determine the points of intersection between the following two functions:

$$y = x^2 - 3x + 2 \qquad \text{and} \qquad y = 2\sin x$$

Then determine the area bounded between them using the methods of calculus.

**6.5** The following equation has three positive roots and one negative root:

$$x^4 - 4x^3 - 7x^2 + 34x - 24 = 0$$

Approximate one positive root and the negative root by starting at $x_0 = 5$ and $x_0 = -5$, respectively, using the Newton–Raphson first method of tangents. Perform at least five iterations for each of the roots. Determine the remaining roots using the quadratic formula on the deflated polynomial.

**6.6** The buckling load $P_{cr}$ for a column fixed at one end and hinged at the other is related to the positive roots of the following equation:

$$\tan x - x = 0$$

where $P_{cr} = (EI/L^2)x^2$ ($L$ is the length of the column, $E$ is Young's modulus of elasticity, and $I$ is the moment of inertia). Use the Newton–Raphson second method to estimate the least load other than zero that a given column can carry without buckling.

**6.7** Evaluate the second largest root of the function given in problem 6.6 by starting at $x_0 = 10$. Use five iterations.

**6.8** Approximate the roots of the following polynomial using the Lin–Bairstow method:

$$4x^4 + 2x^3 + 3x^2 + x + 1 = 0$$

Use two iterations only and $y = z = 0$.

**6.9** Approximate all roots of the following polynomial using the Lin–Bairstow method:

$$x^4 - 3x^3 + 10 = 0$$

(*Note:* You must first transform the equation.) Use two iterations only and $y = z = 1$ as your initial estimate.

**6.10** Solve the following system of equations using the Newton–Raphson method:

$$x + y + z = 11$$

$$xy + xz + yz = 31$$

$$xyz = 21$$

Assume $x_0 = 1.5$, $y_0 = 2.4$, and $z_0 = 6.5$ and use at least two iterations. Show that these roots are the roots of a cubic.

**6.11** The following equation has two unique roots. Find these roots and the root of multiplicity.

$$x^3 + 3x^2 - 9x + 5$$

Use $x_0 = 0.5$.

**6.12** The vertical stress increment ($\Delta\sigma_z$) due to a point load acting on the surface of linearly elastic medium is given by

$$\Delta\sigma_z = 3Pz^3/[2\pi(r^2 + z^2)^{1/2}]$$

where $P$ is the magnitude of the load, $r$ is the lateral distance, and $z$ is the depth of the point where the stress is to be calculated. Determine the depth $z$ at which the stress increment is 100 pounds/ft$^2$ if $P = 10{,}000$ pounds and $r = 2.0$ ft. Use the Newton–Raphson first method.

**6.13** Rework problem 6.12 with $P = 10{,}000$ pounds and $r = 5.0$ ft. Use the Newton–Raphson second method.

**6.14** The vertical stress increment due to an infinite strip loading is given as follows

$$\Delta\sigma_z = \frac{P}{\pi}[\alpha + \sin\alpha\cos(\alpha + 2\delta)]$$

Determine the value of the angle $\alpha$ if $P = 100$ pounds/ft, $\delta = \pi/4$, and $\Delta\sigma_z = 10$ pounds/ft$^2$. Use the Newton–Raphson second method.

# chapter 7

# Eigenproblems

The space shuttle. *Courtesy of N.A.S.A.*

"Eigen problems arise in the analysis and design of space vehicles, launching pads, and complex electrical devices."

# 7

## 7.1 INTRODUCTION

The analysis of many physical systems leads to sets of homogeneous linear algebraic equations (i.e., all elements of the column vector appearing on the right-hand side are zero). These may include the dynamic response of mechanical, structural, and electrical systems. In fact, the solution of ordinary, as well as partial differential equations often results in a homogeneous set of linear algebraic equations. Such equations have a so-called trivial solution in that a null vector is a solution. The eigenproblem solution technique permits the determination of a nontrivial solution to such a set of equations. Consider the following set of equations:

$$\begin{aligned} q_{11}x_1 + q_{12}x_2 + \cdots + q_{1n}x_n &= 0 \\ q_{21}x_1 + q_{22}x_2 + \cdots + q_{2n}x_n &= 0 \\ &\vdots \\ q_{n1}x_1 + q_{n2}x_2 + \cdots + q_{nn}x_n &= 0 \end{aligned} \tag{7.1}$$

The coefficients $q_{ij}$ may involve constants, variables, and/or functions. Obviously, the set has a trivial solution, namely

$$x_1 = x_2 = \cdots = x_n = 0$$

This solution is of no practical use to the engineer. Consequently, a nontrivial solution is required if physical problems are to be mathematically modeled and, thus, properly studied. This is precisely the essence of the eigenproblem.

Recall that in chapter 4 we saw that a set of linear algebraic equations can be solved by using Cramer's rule. For the set given by equation (7.1) we have

$$x_j = \frac{|Q_j|}{|Q|}, \qquad j = 1, 2, \ldots, n \tag{7.2}$$

Multiplying both sides of equation (7.2) by $|Q|$ gives

$$|Q|x_j = |Q_j|, \qquad j = 1, 2, \ldots, n$$

where $|Q_j|$ is the determinant of the $q_{ij}$ coefficient matrix whose $j$th column is replaced by the null vector, $|Q|$ is the determinant of the $q_{ij}$ coefficient matrix, and $x_j$ is the value of the $j$th variable. Clearly, the determinant of $|Q_j|$ is always zero. This is because the null vector will always appear as a column in the matrix $|Q_j|$. Thus

$$|Q|x_j = 0, \qquad j = 1, 2, \ldots, n \tag{7.3}$$

It is evident that if $x_j \neq 0$ then $|Q|$ must be forced to equal zero. This is indeed the condition that must be satisfied in eigenproblems if a nontrivial solution is to be attained. That is,

$$|Q| = 0 \tag{7.4}$$

is a sufficient condition to ensure that $x_j$ is not trivial. Obviously, the coefficients of $|Q|$ must contain a variable whose values can be determined in such a way that the resulting determinant has a value of zero. Furthermore, the expansion of the determinant of $|Q|$ results in a nonlinear algebraic equation in the unknown variable. This is called the characteristic equation of the system. Hence the roots of the equation are in fact the values that force $|Q|$ to equal zero. These roots are referred to as the "eigenvalues." Corresponding to every eigenvalue an "eigenvector" can be determined from equation (7.1). It is interesting to note that the determinant of a square matrix is zero when two of its rows are the same. The implication is that by forcing the determinant $|Q|$ to go to zero we are actually forcing the set of equations to become dependent. This concept can be easily verified graphically by considering the following set:

$$\begin{aligned} q_{11}x_1 + q_{12}x_2 &= 0 \\ q_{21}x_1 + q_{22}x_2 &= 0 \end{aligned} \tag{7.5}$$

Suppose now that we are interested in a nontrivial solution. We shall proceed by rewriting equation (7.5) in the following form:

$$x_1 = -\frac{q_{12}}{q_{11}} x_2 \tag{7.6a}$$

$$x_1 = -\frac{q_{22}}{q_{21}} x_2 \tag{7.6b}$$

Since the $q_{ij}$ coefficients involve variables, equations (7.6a) and (7.6b) must generally be expected to have different slopes. These equations are shown graphically in figure 7.1.

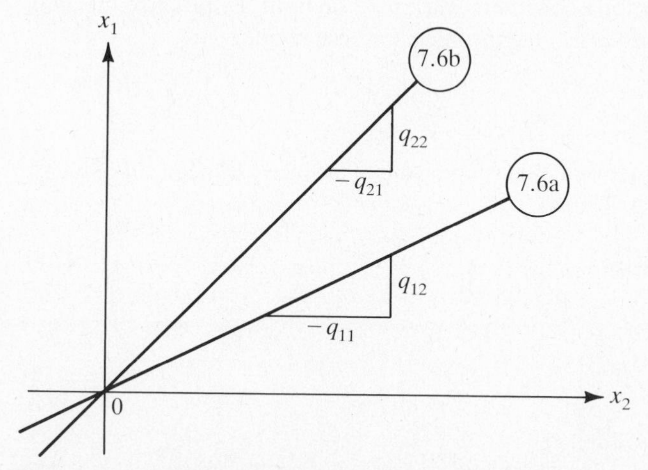


**Figure 7.1** Graphical illustration of eigenvalue problems.

Consideration of this figure shows that equations (7.6) intersect at the origin. Furthermore, the only way these two equations would intersect at points other than the origin is if they had the same slope. The implication is

$$-\frac{q_{12}}{q_{11}} = -\frac{q_{22}}{q_{21}}$$

Rearranging terms gives

$$q_{11}q_{22} - q_{12}q_{21} = 0 \tag{7.7}$$

Equation (7.7) is clearly the determinant of the matrix of coefficients of equation (7.5). This is exactly the same conclusion we reached earlier. Keep in mind that by forcing the determinant to equal zero, we no longer have a unique solution. That is to say we have infinite solutions to the problem at hand. This observation raises another question, simply, which one of these solutions should one select? The answer is that any one of them will do. In fact, the solution should not have been expected to be a unique one. This is because nothing was said about the system's initial and/or boundary conditions to begin with.

## 7.2 CHARACTERISTIC EQUATION DETERMINATION

So far, we have outlined the conditions involved in eigenproblems, but a reference was made to the fact that not all of the $q_{ij}$ coefficients are constants. The purpose of this section is to outline a method for expanding the determinant of the $[Q]$ matrix as is required by equation (7.4). For convenience, it is assumed that, in general, the $[Q]$ matrix is related to two other matrices as follows:

$$[Q] = [A] - [B] \tag{7.8}$$

where $[A]$ is a matrix containing only constant coefficients, while the coefficients of the $[B]$ matrix may include constants, variables, or both. In practice, we may encounter one of three classes of eigenproblems. These include

### Type I

Here the $[B]$ matrix in equation (7.8) is composed of elements some or all of which are functions of a variable. That is,

$$[Q_1]\{x\} = [[A] - [B]]\{x\} = \{0\} \tag{7.9}$$

where

$$[B] = \begin{bmatrix} f_{11}(\lambda) & \cdots & f_{1n}(\lambda) \\ \vdots & & \vdots \\ f_{n1}(\lambda) & \cdots & f_{nn}(\lambda) \end{bmatrix}$$

Clearly, $f_{11}(\lambda), \ldots, f_{nn}(\lambda)$ may take on any value including zero.

Type II

In this case, the matrix $[B]$ is equal to the product of the variable $\lambda$ and a matrix $[C]$. Thus

$$[Q_2]\{x\} = [[A] - \lambda[C]]\{x\} = \{0\} \tag{7.10}$$

where the coefficients of $[C]$ are all constants.

Type III

This is the simplest eigenproblem that one may encounter in practice. Here the $[B]$ matrix is equal to the product of the variable $\lambda$ and the identity matrix. That is,

$$[Q_3]\{x\} = [[A] - \lambda[I]]\{x\} = \{0\} \tag{7.11}$$

Note that in all three types, the $[A]$ matrix contains scalar coefficients. Furthermore, the eigenproblem given by (7.10) can be easily reduced to that given by equation (7.11) by simply premultiplying it by the inverse of the $[C]$ matrix. Hence

$$[[C]^{-1}[A] - \lambda[C]^{-1}[C]]\{x\} = \{0\}$$

or more simply

$$[[D] - \lambda[I]]\{x\} = \{0\} \tag{7.12}$$

where

$$[D] = [C]^{-1}[A]$$

For some applications the form given by equation (7.12) might be more appropriate to use than that given by equation (7.10).

### 7.2.1 Direct Determinant Expansion

The determinant of each of the $[Q]$ matrices corresponding to equations (7.9)–(7.11) results in nonlinear algebraic equations in $\lambda$. These are referred to as the characteristic equations of the physical system being analyzed. Obviously, the method outlined in chapter 2 for determinant expansion can be directly used. For problems involving determinants of order two or three, these methods are appropriate. Unfortunately, for larger determinants the direct methods of expansion are highly inefficient and are not recommended. The difficulty lies in the fact that these determinants involve a variable ($\lambda$) rather than constants. Consider the following:

$$\begin{bmatrix} 2-\lambda & -1 & 0 \\ -1 & 3-\lambda & -1 \\ 0 & -1 & 2-\lambda \end{bmatrix} \begin{Bmatrix} x_1 \\ x_2 \\ x_3 \end{Bmatrix} = \begin{Bmatrix} 0 \\ 0 \\ 0 \end{Bmatrix}$$

Obviously, a nontrivial solution can be obtained by forcing the determinant of the coefficient matrix to zero. Using the method of cofactors gives

$$\det[Q] = (2-\lambda)\begin{vmatrix} 3-\lambda & -1 \\ -1 & 2-\lambda \end{vmatrix} - (-1)\begin{vmatrix} -1 & 0 \\ -1 & 2-\lambda \end{vmatrix} + (0)\begin{vmatrix} -1 & 0 \\ 3-\lambda & -1 \end{vmatrix}$$

$$\det[Q] = \lambda^3 - 7\lambda^2 + 14\lambda - 8 = 0$$

Clearly this is the characteristic equation whose roots are the eigenvalues of the system under study. Note that, while this is a polynomial, in general the characteristic equation may involve transcendental functions. Furthermore, the method of pivotal condensation is equally applicable in expanding the determinant of $[Q]$. In fact, even the upper-triangular method could have been used for that same purpose. These applications do not lend themselves to computer applications and are cumbersome for hand calculations.

### 7.2.2 Indirect Determinant Expansion

This method is based on the fact that the determinant containing the variable $\lambda$ can be treated as a polynomial having the same order as the determinant itself. That is, if the determinant is an $n \times n$, then the order of the characteristic equation is a polynomial also of order $n$. It is evident that this technique is not suitable for eigenproblems of type I. Consequently, while limited to type II and type III eigenproblems, the method has the advantage of being easy to program. In particular, given the determinant

$$\det[Q(\lambda)] = \begin{vmatrix} q_{11} & q_{12} & \cdots & q_{1n} \\ q_{21} & q_{22} & \cdots & q_{2n} \\ \vdots & \vdots & & \vdots \\ q_{n1} & q_{n2} & \cdots & q_{nn} \end{vmatrix} = f(\lambda) \tag{7.13}$$

since the $q_{ij}$ coefficients are generally functions of $\lambda$,

$$f(\lambda) = a_n\lambda^n + a_{n-1}\lambda^{n-1} + \cdots + a_0$$

or more simply

$$f(\lambda) = \sum_{i=0}^{n} a_i\lambda^i \tag{7.14}$$

Now, by selecting $n+1$ arbitrary values for $\lambda$ (say, $\lambda_0, \lambda_1, \ldots, \lambda_n$) the corresponding determinants can be easily evaluated by using equation (7.13). However, these determinants represent the left-hand sides of equation (7.14). That is,

$$\begin{aligned} \lambda_0: &\quad f(\lambda_0) = a_0 + a_1\lambda_0 + \cdots + a_n\lambda_0^n \\ \lambda_1: &\quad f(\lambda_1) = a_0 + a_1\lambda_1 + \cdots + a_n\lambda_1^n \\ \vdots \quad &\quad \vdots \\ \lambda_n: &\quad f(\lambda_n) = a_0 + a_1\lambda_n + \cdots + a_n\lambda_n^n \end{aligned}$$

These equations can be expressed in matrix form as follows:

$$\begin{bmatrix} 1 & \lambda_0 & \cdots & \lambda_0^n \\ 1 & \lambda_1 & \cdots & \lambda_1^n \\ \vdots & \vdots & & \vdots \\ 1 & \lambda_n & \cdots & \lambda_n^n \end{bmatrix} \begin{Bmatrix} a_0 \\ a_1 \\ \vdots \\ a_n \end{Bmatrix} = \begin{Bmatrix} f(\lambda_0) \\ f(\lambda_1) \\ \vdots \\ f(\lambda_n) \end{Bmatrix} \tag{7.15}$$

Obviously, the unknowns $a_0, \ldots, a_n$ can be easily determined. Consequently, the characteristic equation is given by equation (7.14). The arbitrary $\lambda$ values chosen should be comparable in magnitude to the diagonal elements in the $[Q]$ matrix; otherwise equation (7.15) may become unstable.

E

EXAMPLE 7.1

Determine the characteristic equation corresponding to the following matrix using the indirect procedure:

$$|Q| = \begin{vmatrix} 2-\lambda & -3 \\ -1 & 3-\lambda \end{vmatrix}$$

### Solution

First, the characteristic equation can be easily evaluated by using the direct method of determinant expansion to give

$$\lambda^2 - 5\lambda + 3 = 0$$

(Unfortunately, it is not always this simple, especially when the determinant is of large order.) Alternatively, let us proceed by applying the indirect method. We begin by assuming three arbitrary values for $\lambda$, say

$$\lambda_0 = 0, \qquad \lambda_1 = 2, \qquad \lambda_2 = 3$$

Then evaluating the three corresponding determinants as follows yields

$$\lambda_0 = 0: \qquad f(0) = \begin{vmatrix} 2 & -3 \\ -1 & 3 \end{vmatrix} = 3$$

$$\lambda_1 = 2: \qquad f(2) = \begin{vmatrix} 0 & -3 \\ -1 & 1 \end{vmatrix} = -3$$

$$\lambda_2 = 3: \qquad f(3) = \begin{vmatrix} -1 & -3 \\ -1 & 0 \end{vmatrix} = -3$$

These determinants correspond to three linear algebraic equations in the unknown characteristic equation coefficients. That is,

$$f(\lambda_0) = a_0 + a_1\lambda_0 + a_2\lambda_0^2$$
$$f(\lambda_1) = a_0 + a_1\lambda_1 + a_2\lambda_1^2$$
$$f(\lambda_2) = a_0 + a_1\lambda_2 + a_2\lambda_2^2$$

Substituting the determinant values and the assumed $\lambda$ values gives

$$3 = a_0$$
$$-3 = a_0 + 2a_1 + 4a_2$$
$$-3 = a_0 + 3a_1 + 9a_2$$

Solving for the unknowns yields

$$a_0 = 3, \qquad a_1 = -5, \qquad a_2 = 1$$

The characteristic equation is then given by

$$f(\lambda) = \lambda^2 - 5\lambda + 3$$

This of course, is exactly the same equation we have derived using the direct method of expansion.

E

EXAMPLE 7.2

Determine the characteristic equation of the following matrix using the indirect method of determinant expansion:

$$\det[Q] = \begin{vmatrix} 5-\lambda & -4 & -2 & 0 \\ -4 & 5-\lambda & -2 & -1 \\ -3 & -4 & 6-\lambda & -4 \\ 0 & -3 & -1 & 7-\lambda \end{vmatrix}$$

Solution

Clearly the characteristic equation is a polynomial of the fourth order; thus we need to assume five $\lambda$ values. Hence

$$\lambda_0 = 0, \qquad \lambda_1 = 5, \qquad \lambda_2 = 6, \qquad \lambda_3 = 10, \qquad \lambda_4 = 12$$

are initially assumed. We proceed by establishing the corresponding linear algebraic equations in the unknown coefficients of the characteristic equation. That is,

$$f(\lambda_0) = a_4\lambda_1^4 + a_3\lambda_1^3 + a_2\lambda_1^2 + a_1\lambda_1 + a_0$$

$$f(\lambda_1) = a_4\lambda_2^4 + a_3\lambda_2^3 + a_2\lambda_2^2 + a_1\lambda_2 + a_0$$

$$f(\lambda_2) = a_4\lambda_3 + a_3\lambda_3 + a_2\lambda_3 + a_1\lambda_3 + a_0$$

$$f(\lambda_3) = a_4\lambda_4 + a_3\lambda_4 + a_2\lambda_4 + a_1\lambda_4 + a_0$$

$$f(\lambda_4) = a_4\lambda_5 + a_3\lambda_5 + a_3\lambda_5 + a_1\lambda_5 + a_0$$

where the determinants of $f(\lambda_i)$ are calculated easily as

$$f(0) = -860, \qquad f(5) = -170, \qquad f(6) = -44$$

$$f(10) = 20, \qquad f(12) = 628$$

Substituting into the above set of equations and solving for the unknowns gives

$$a_0 = -806, \qquad a_1 = -212, \qquad a_2 = 160, \qquad a_3 = -23, \qquad a_4 = 1.0$$

Consequently the characteristic equation is given by

$$f(\lambda) = \lambda^4 - 23\lambda^3 + 160\lambda^2 - 212\lambda - 860 = 0$$

Using different assumed values of $\lambda$ yields the same characteristic equation. For example, letting

$$\lambda_0 = 0, \qquad \lambda_1 = 10, \qquad \lambda_2 = 20, \qquad \lambda_3 = 30, \qquad \lambda_4 = 4$$

yields the following determinants:

$$f(0) = -860, \qquad f(10) = 20, \qquad f(20) = 34{,}900,$$

$$f(30) = 325{,}780, \qquad f(40) = 1{,}334{,}660$$

Substituting into the corresponding set of linear algebraic equations yields

$$-860 = a_0$$

$$20 = a_0 + (10)a_1 + (10)^2a_2 + (10)^3a_3 + (10)^4a_4$$

$$34{,}900 = a_0 + (20)a_1 + (20)^2a_2 + (20)^3a_3 + (20)^4a_4$$

$$325{,}780 = a_0 + (30)a_1 + (30)^2a_2 + (30)^3a_3 + (30)^4a_4$$

$$1{,}334{,}660 = a_0 + (40)a_1 + (40)^2a_2 + (40)^4a_4 + (40)^4a_4$$

Solving for the unknown coefficients yields the same characteristic equation derived earlier.

**SUMMARY OF THE INDIRECT DETERMINANT EXPANSION METHOD**

| STEP | OPERATION | SYMBOL |
|---|---|---|
| 1 | Form the $[Q]$ matrix for the eigenproblem and make sure it is not of the first type. | $[Q] = [A] - \lambda[B]$ |
| 2 | Assume $n + 1$ values for $\lambda$ and evaluate the corresponding determinants of the $[Q]$ matrix. | $\lambda_i = \lambda_0, \ldots, \lambda_n,$ <br> $i = 1, \ldots, n$ <br> $f(\lambda_i) = \lvert Q \rvert$ |
| 3 | Form an equivalent set of linear algebraic equations to the order of $n + 1$. | $f(\lambda_0) = a_0 + a_1\lambda_0 + \cdots + a_n\lambda_0^n$ <br> $f(\lambda_1) = a_0 + a_1\lambda_1 + \cdots + a_n\lambda_1^n$ <br> $\vdots$ <br> $f(\lambda_n) = a_0 + a_1\lambda_n + \cdots + a_n\lambda_n^n$ |
| 4 | Determine the unknown coefficients of the characteristic equation by solving the set of equations formed in step 3. | $f(\lambda) = a_0 + a_1\lambda + \cdots + a_n\lambda^n$ |

## 7.3 EIGENVALUES AND EIGENVECTORS

In this section the problem of evaluating eigenvalues and their corresponding vectors are addressed. In addition, the interpretation of some common physical eigenproblems is covered. Consider, for example, the following system of homogeneous linear algebraic equations:

$$\begin{bmatrix} 2-\lambda & -1 & 0 \\ -1 & 3-\lambda & -1 \\ 0 & -1 & 2-\lambda \end{bmatrix} \begin{Bmatrix} x_1 \\ x_2 \\ x_3 \end{Bmatrix} = \begin{Bmatrix} 0 \\ 0 \\ 0 \end{Bmatrix} \tag{7.16}$$

It is evident that a nontrivial solution to the $\{x\}$ vector can be found by setting the determinant of the coefficient matrix to zero. Therefore

$$f(\lambda) = \lambda^3 - 7\lambda^2 + 14\lambda - 8 = 0 \tag{7.17}$$

Note that this problem was addressed earlier in section 7.2.1. The object is now to find $\lambda$ values such that the determinant of the coefficient matrix is zero. That is, we need to find the roots of the polynomial given by equation (7.17). These roots are

$\lambda_1 = 1$ (first eigenvalue)

$$\lambda_2 = 2 \qquad \text{(second eigenvalue)}$$

$$\lambda_3 = 4 \qquad \text{(third eigenvalue)}$$

The implication is that any one of these three eigenvalues forces the determinant of the coefficient matrix of equation (7.16) to be zero. Furthermore, there are three nontrivial solutions to the $\{x\}$ vectors, each corresponding to a $\lambda$ value.

### First Eigenvector

This is determined by substituting $\lambda = 1$ into equation (7.16). That is,

$$\begin{bmatrix} 1 & -1 & 0 \\ -1 & 2 & -1 \\ 0 & -1 & 1 \end{bmatrix} \begin{Bmatrix} x_1 \\ x_2 \\ x_3 \end{Bmatrix} = \begin{Bmatrix} 0 \\ 0 \\ 0 \end{Bmatrix}$$

The unknown $\{x\}$ vector is determined by using the Gauss–Jordan elimination method. Hence, forming the augmented matrix gives

$$\left[\begin{array}{ccc|c} 1 & -1 & 0 & 0 \\ -1 & 2 & -1 & 0 \\ 0 & -1 & 1 & 0 \end{array}\right] \quad R_1 + R_2 \quad \Rightarrow \quad \left[\begin{array}{ccc|c} 1 & -1 & 0 & 0 \\ 0 & 1 & -1 & 0 \\ 0 & -1 & 1 & 0 \end{array}\right] \begin{matrix} R_2 + R_1 \\ \\ R_2 + R_3 \end{matrix}$$

which simplifies to the following augmented matrix:

$$\left[\begin{array}{ccc|c} 1 & 0 & -1 & 0 \\ 0 & 1 & -1 & 0 \\ 0 & 0 & 0 & 0 \end{array}\right]$$

Note that the last row is reduced to one with zero coefficients, which indicates that the determinant is indeed equal to zero. Consequently, the set of three equations is now reduced to two equations in the three unknowns $x_1$, $x_2$, and $x_3$. That is,

$$\begin{bmatrix} 1 & 0 & -1 \\ 0 & 1 & -1 \end{bmatrix} \begin{Bmatrix} x_1 \\ x_2 \\ x_3 \end{Bmatrix} = \begin{Bmatrix} 0 \\ 0 \\ 0 \end{Bmatrix}$$

or more simply

$$x_1 - x_3 = 0, \qquad x_2 - x_3 = 0$$

However, this set has infinite solutions, one of which is the trivial solution. Consequently, solving for $x_1$ and $x_2$ in terms of $x_3$ gives

$$x_1 = x_3, \qquad x_2 = x_3, \qquad x_3 = x_3$$

in vector form

$$\{x\}_1 = \begin{Bmatrix} x_3 \\ x_3 \\ x_3 \end{Bmatrix} = x_3 \begin{Bmatrix} 1 \\ 1 \\ 1 \end{Bmatrix}_1 \qquad \text{(first vector)}$$

Note that the first eigenvector represents a solution to the set of linear algebraic equations given by (7.16) for $\lambda_1 = 1$.

### Second Eigenvector

This vector is determined by substituting $\lambda = 2$ into equations (7.16) and solving for the $\{x\}$ vector. Hence

$$\begin{bmatrix} 0 & -1 & 0 \\ -1 & 1 & -1 \\ 0 & -1 & 0 \end{bmatrix} \begin{Bmatrix} x_1 \\ x_2 \\ x_3 \end{Bmatrix} = \begin{Bmatrix} 0 \\ 0 \\ 0 \end{Bmatrix}$$

It is evident that the first and third rows are identical, which once again implies that the determinant of the coefficient matrix is zero. The solution is then given as

$$\{x\}_2 = x_3 \begin{Bmatrix} 1 \\ 0 \\ -1 \end{Bmatrix} \qquad \text{(second eigenvector)}$$

Once again, choosing any value for $x_3$ gives a solution to equations (7.16) when $\lambda = 2$.

### Third Eigenvector

Substituting $\lambda = 4$ into equations (7.16) permits the determination of the final vector. Hence

$$\begin{bmatrix} -2 & -1 & 0 \\ -1 & -1 & -1 \\ 0 & -1 & -2 \end{bmatrix} \begin{Bmatrix} x_1 \\ x_2 \\ x_3 \end{Bmatrix} = \begin{Bmatrix} 0 \\ 0 \\ 0 \end{Bmatrix} \tag{7.18}$$

Using the Gauss–Jordan elimination method, we reduce this set to

$$\begin{bmatrix} 1 & 0 & -1 \\ 0 & 1 & 2 \\ 0 & 0 & 0 \end{bmatrix} \begin{Bmatrix} x_1 \\ x_2 \\ x_3 \end{Bmatrix} = \begin{Bmatrix} 0 \\ 0 \\ 0 \end{Bmatrix}$$

Therefore the solution is given as

$$\{x\}_3 = x_3 \begin{Bmatrix} -1 \\ -2 \\ 1 \end{Bmatrix}$$

Normally, the eigenvectors are determined by simply ignoring the last equation, assuming a constant value for the $n$th variable, and then solving for the remaining unknowns. For example, the third vector could have been determined by solving

$$-2x_1 - x_2 = 0, \qquad -x_1 - x_2 = x_3$$

These are the first two equations appearing in the set given by equations (7.18). Now, if $x_3 = 1$, then $x_1 = -1$ and $x_2 = 2$, which corresponds to the third vector.

In order to have some understanding as to the meaning of the eigenproblem and its physical interpretation, the following applications are provided. The reader may ignore the physics of these applications and simply consider the governing equations and the methods used.

### 7.3.1 Column Buckling

An engineering system, or one of its parts, is perceived to have failed when it ceases to perform its intended purpose. Therefore a column must be designed in such a way that limiting values of deformations are observed. An important mode of deformation is called buckling, which occurs when a system fails to carry the load it is intended to support. This mode of failure is manifested by significant geometrical shape changes. Columns buckle because in practice no column can be assumed to be perfect in the sense that loads are applied through its center and its geometry conforms to an ideal geometrical shape. This problem was formulated in chapter 3. Let us begin by considering the column shown in figure 7.2.

The column is considered to have a uniform cross section and a constant material property ($E$ is Young's modulus, $I$ the moment of inertia). Recall that the governing equation for determining the critical buckling load $P$ was given earlier by equation (3.25), repeated here for convenience as

$$x = A_1 \sin(ky) + A_2 \cos(ky) + A_3 \frac{y}{L} + A_4 \tag{7.19}$$

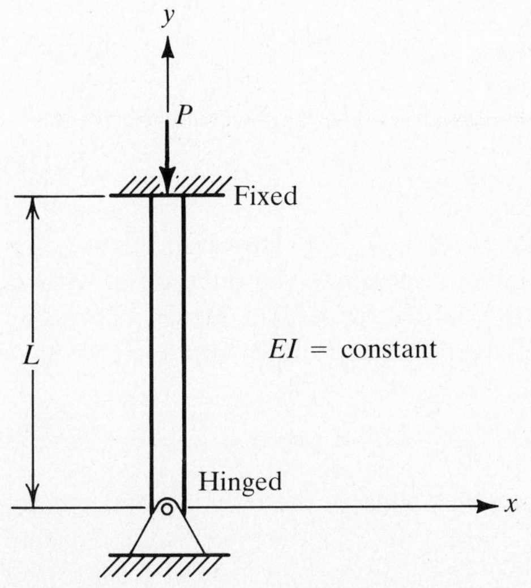


**Figure 7.2** A column hinged at one end and fixed at the other end.

The boundary conditions at the two ends of the column are given as follows.

At $y = 0$:

$$(1)\quad x = 0 \qquad (2)\quad \frac{d^2x}{dy^2} = 0$$

At $y = L$:

$$(3)\quad x = 0 \qquad (4)\quad \frac{dx}{dy} = 0$$

These four boundary conditions permit the determination of the four unknowns $A_1$, $A_2$, $A_3$, and $A_4$ in equation (7.19). However, since the boundary conditions involve the first and second derivatives, let us proceed by first determining the following:

$$\frac{dx}{dy} = A_1 k \cos(ky) - A_2 k \sin(ky) + \frac{A_3}{L} \tag{7.20}$$

$$\frac{d^2x}{dy^2} = -A_1 k^2 \sin(ky) - A_2 k^2 \cos(ky) \tag{7.21}$$

Applying the boundary conditions to equations (7.19)–(7.21) yields four linear algebraic equations in the four unknowns. That is,

$$(1)\ 0 = 0 + A_2 + 0 + A_4$$

$$(2)\ 0 = 0 - A_2 k^2$$

$$(3)\ 0 = A_1 \sin(kL) + A_2 \cos(kL) + A_3 + A_4$$

$$(4)\ 0 = A_1 k \cos(kL) - A_2 \sin(kL) + \frac{A_3}{L}$$

Letting $\lambda = kL$ and noting that $A_2 = A_4 = 0$ reduces the set of $4 \times 4$ equations to a set of $2 \times 2$. Thus

$$\begin{aligned} A_1 \sin\lambda + A_3 &= 0 \\ A_1 \lambda \cos\lambda + A_3 &= 0 \end{aligned} \tag{7.22a}$$

Clearly a trivial solution is in order. That is, $A_1 = A_3 = 0$. However, this solution is meaningless in that it does not provide the answer to the question of what is the minimum load necessary to cause the column to buckle. Hence, expressing equation (7.22a) in matrix form and applying the eigenproblem approach, we have

$$\begin{bmatrix} \sin\lambda & 1 \\ \lambda\cos\lambda & 1 \end{bmatrix} \begin{Bmatrix} A_1 \\ A_3 \end{Bmatrix} = \begin{Bmatrix} 0 \\ 0 \end{Bmatrix} \tag{7.22b}$$

It is evident that the coefficient matrix corresponds to the most general type of eigenproblem. Therefore

$$\det[Q] = \begin{vmatrix} \sin\lambda & 1 \\ \lambda\cos\lambda & 1 \end{vmatrix} = \tan\lambda - \lambda = 0 \tag{7.23}$$

The solution of equation (7.23) should provide a nontrivial solution. It is interesting to note that the most obvious solution is $\lambda = 0$. Do not be alarmed, however: this is a nonlinear equation with an infinite number of solutions. Note that the Newton–Raphson first method of tangents does not converge in this case. The Newton–Raphson second method is used to find the first root. The solution to the first nonzero root is given directly.

### First Eigenvalue

$$\lambda_1 = 4.494 = kL$$

Recall that

$$P = k^2 EI$$

The buckling load is then given as follows:

$$P = \left(\frac{4.5}{L}\right)^2 EI = 20.2\frac{EI}{L^2} \tag{7.24}$$

Obviously, there are other eigenvalues and correspondingly equal numbers of buckling loads. Let's investigate the mode of deflection for the column. That is to say, determine the first eigenvector. Substitution of $\lambda_1$ into equation (7.22b) gives the following:

$$\begin{aligned} \begin{bmatrix} \sin(4.494) & 1 \\ (4.494)\cos(4.494) & 1 \end{bmatrix} \begin{Bmatrix} A_1 \\ A_3 \end{Bmatrix} &= \begin{Bmatrix} 0 \\ 0 \end{Bmatrix} \\ \begin{bmatrix} -0.976 & 1 \\ -0.976 & 1 \end{bmatrix} \begin{Bmatrix} A_1 \\ A_3 \end{Bmatrix} &= \begin{Bmatrix} 0 \\ 0 \end{Bmatrix} \end{aligned} \tag{7.25}$$

Note that we now have a single equation in two unknowns. This implies that we have an infinite number of solutions. Therefore

$$\begin{aligned} -0.976A_1 + A_3 &= 0 \\ A_3 &= +0.976A_1 \end{aligned} \tag{7.26}$$

The physical meaning of (7.26) is that only relative values are possible. This is expected, since we really have said nothing about the physical properties of the column material ($EI$), that is, the cross-sectional properties, length, and material type.

### First Eigenvector

$$\{V\}_1^t = \{A_1, 0.976A_1\}$$

Substituting into equation (7.19) and recalling that $A_2 = A_4 = 0$ gives

$$x = c_1[\sin(ky) + 0.976y/L] \tag{7.27a}$$

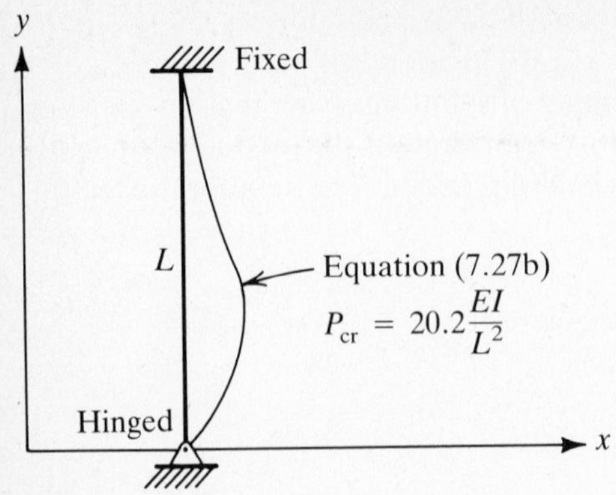


**Figure 7.3** First buckling mode of a column hinged at one end and fixed at the other.

Now substituting $\lambda_1 = KL = 4.494$ into equation (7.27a) yields

$$x = c_1[\sin(4.494y/L) + 0.976y/L] \tag{7.27b}$$

Figure 7.3 shows a plot of the first mode corresponding to the first eigenvector.

Note that buckling loads are inversely proportional to the square of column length. That is, long columns are expected to buckle more easily than short ones. Furthermore, while it is possible to compute other eigenvalues (buckling loads) these are greater in magnitude than the first one computed. Consequently, they are generally ignored because failure is set once the first buckling load is reached. Note that end conditions play an important role in determining buckling loads. However, the same solution technique is used to determine their values.

### 7.3.2 Mechanical Vibrations

The dynamic response of a mass–dashpot–spring assembly involves an eigenproblem of the second or third type. That is, the resulting characteristic equation is a polynomial equivalent in order of magnitude to the number of degrees of freedom in which the system can be allowed to displace. Such problems occur frequently in civil and mechanical engineering. Dynamic analysis involving earthquake effects on structures and machine vibrations are two examples of situations in which eigenproblems arise. The corresponding mathematical models describing the behavior of such systems was treated earlier in section 3.3.2. Moreover, the effects of damping are generally ignored when the dynamic parameters involved are evaluated. This is done partly because it is easier to handle undamped systems and partly because damping can be included in determining actual response, once the eigenproblem is solved. Consider the undamped system given in figure 7.4.

Obviously, this system represents an initial-value problem in that the dynamic response is due to an initial displacement and/or initial velocity. The mathematical model describing this system's behavior is of the following form (see chapter 3):

$$\begin{bmatrix} 3k & -2k \\ -2k & 2k \end{bmatrix} \begin{Bmatrix} x_1 \\ x_2 \end{Bmatrix} + \begin{bmatrix} m & 0 \\ 0 & 2m \end{bmatrix} \begin{Bmatrix} \ddot{x}_1 \\ \ddot{x}_2 \end{Bmatrix} = \begin{Bmatrix} 0 \\ 0 \end{Bmatrix} \tag{7.28}$$

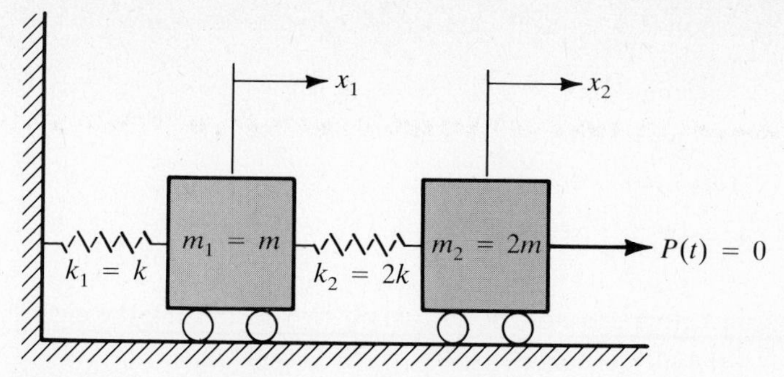


**Figure 7.4** A two-degrees-of-freedom mass–spring mechanical system.

It can be shown that the response of a system experiencing free vibration is sinusoidal and of the form

$$x_1 = A_1 \sin(wt + \theta)$$

$$x_2 = A_2 \sin(wt + \theta)$$

where $A_1$ and $A_2$ are the amplitudes of the displacements for masses $m$ and $2m$, respectively; $w$ is the natural circular frequency of the system; and $\theta$ is its phase angle. Therefore, calculating the corresponding accelerations

$$\ddot{x}_1 = -A_1 w^2 \sin(wt + \theta)$$

$$\ddot{x}_2 = -A_2 w^2 \sin(wt + \theta)$$

and substituting them and the displacement expressions into equation (7.25) we obtain

$$k\begin{bmatrix} 3 & -2 \\ -2 & 2 \end{bmatrix}\begin{Bmatrix} A_1 \\ A_2 \end{Bmatrix} - mw^2\begin{bmatrix} 1 & 0 \\ 0 & 2 \end{bmatrix}\begin{Bmatrix} A_1 \\ A_2 \end{Bmatrix} = \begin{Bmatrix} 0 \\ 0 \end{Bmatrix}$$

Letting $\lambda = mw^2/k$ and simplifying gives the following set of homogeneous linear algebraic equations:

$$\begin{bmatrix} 3-\lambda & -2 \\ -2 & 2-2\lambda \end{bmatrix}\begin{Bmatrix} A_1 \\ A_2 \end{Bmatrix} = \begin{Bmatrix} 0 \\ 0 \end{Bmatrix} \qquad \textbf{(7.29)}$$

It is evident that equation (7.29) represents an eigenproblem. Therefore, forcing the determinant of the coefficient matrix to equal zero yields the characteristic equation

$$\det[Q] = \lambda^2 - 4\lambda + 1 = 0$$

Using the quadratic formula, one can easily determine the roots (eigenvalues). Thus

$$\lambda_1 = 2 - \sqrt{3} = 0.2679 \Rightarrow \text{first eigenvalue}$$

$$\lambda_2 = 2 + \sqrt{3} = 3.7320 \Rightarrow \text{second eigenvalue}$$

Consequently, the corresponding natural circular frequencies are calculated as follows:

$$\lambda_1 = \frac{m}{k} w_1^2 = 0.2679 \Rightarrow w_1 = 0.518 \sqrt{\frac{k}{m}}$$

$$\lambda_2 = \frac{m}{k} w_2^2 = 3.7320 \Rightarrow w_2 = 1.932 \sqrt{\frac{k}{m}}$$

The corresponding eigenvectors can now be calculated by substituting the eigenvalue into equation (7.29) and solving for $A_1$ and $A_2$, we have

### First Eigenvector

This is determined by substituting $\lambda_1 = 0.2679$ into equation (7.29) and solving for the $\{A\}$ vector. Hence

$$\begin{bmatrix} 2.7321 & -2 \\ -2 & 1.4642 \end{bmatrix} \begin{Bmatrix} A_1 \\ A_2 \end{Bmatrix} = \begin{Bmatrix} 0 \\ 0 \end{Bmatrix}$$

Obviously, the two linear algebraic equations are identical. Therefore using one of the two equations gives

$$2.7321 A_1 - 2A_2 = 0 \Rightarrow A_1 = 0.7320 A_2$$

The eigenvector is determined by assuming an arbitrary value for $A_2$ then solving for $A_1$. Hence letting $A_2 = 1$ gives $A_1 = 0.7320$. Consequently

$$\{V\}_1 = \begin{Bmatrix} 0.7320 \\ 1 \end{Bmatrix}$$

is the first eigenvector.

### Second Eigenvector

This is determined by substituting $\lambda_2 = 3.7320$ into equation (7.28) and solving for the $\{A\}$ vector. Thus

$$\begin{bmatrix} -0.732 & -2 \\ -2 & -5.464 \end{bmatrix} \begin{Bmatrix} A_1 \\ A_2 \end{Bmatrix} = \begin{Bmatrix} 0 \\ 0 \end{Bmatrix}$$

Once again, using any of the two equations gives

$$A_1 = -2.732 A_2$$

Assuming $A_2 = 1$, this gives $A_1 = -2.732$. Therefore

$$\{V\}_2 = \begin{Bmatrix} -2.732 \\ 1 \end{Bmatrix}$$

is the second eigenvector. Note that in engineering these vectors are normalized so that in part, relative values can be compared. This can be accomplished as follows: Given a vector

$$\{V\} = \begin{Bmatrix} a_1 \\ \vdots \\ a_n \end{Bmatrix}$$

the normalized vector is given simply by

$$\{\hat{V}\} = \frac{1}{\sqrt{a_1^2 + \cdots + a_n^2}} \begin{Bmatrix} a_1 \\ \vdots \\ a_n \end{Bmatrix} \tag{7.30}$$

For the problem at hand, the first and second vectors are given directly as follows:

$$\{\hat{V}\}_1 = \begin{Bmatrix} 0.5907 \\ 0.8069 \end{Bmatrix} \Rightarrow \text{first normalized vector}$$

and

$$\{V\}_2 = \begin{Bmatrix} -0.939 \\ 0.3437 \end{Bmatrix} \Rightarrow \text{second normalized vector}$$

Note that there are other normalizing schemes in addition to the one given by equation (7.30). Consideration of the two vectors provides some insight into the so-called modes (vectors) of vibration of the two masses. These are shown graphically in figure 7.5.

Note when using the first vector that the relative amplitude of the displacement is given by the first eigenvector. This represents the first mode of vibration in which the masses displace in the same direction. Similarly, the second eigenvector provides information relative to the second mode of vibration, in which the masses move in opposite direction.

The total response (displacement) of the two masses given in figure 7.4 will depend upon the initial conditions imposed on them, that is, on their initial displacements and initial velocities. Furthermore, the solution will involve both of the eigenvalues and eigenvectors. Therefore

$$\begin{Bmatrix} x_1 \\ x_2 \end{Bmatrix} = A\{\hat{V}\}_1 \sin(w_1 t + \theta) + B\{\hat{V}\}_2 \sin(w_2 t + \theta_2)$$

where $A$ and $B$ are arbitrary constants pertaining to the amplitude of the displacements and $\theta_1$ and $\theta_2$ are the phase angles. These four unknowns can be easily determined from the four initial conditions. Consequently,

$$\begin{Bmatrix} x_1 \\ x_2 \end{Bmatrix} = A\begin{Bmatrix} 0.5907 \\ 0.8069 \end{Bmatrix} \sin\left[0.518\sqrt{\frac{k}{m}}t + \theta_1\right] + B\begin{Bmatrix} -0.9391 \\ 0.3437 \end{Bmatrix} \sin\left[1.932\sqrt{\frac{k}{m}}t + \theta_2\right] \tag{7.31}$$

Note that at time $t = 0$ the initial conditions are given by

(1) $x_1(t) = x_1(0)$ (2) $x_2(t) = x_2(0)$

(3) $\dot{x}_1(t) = x_1(0)$ (4) $\dot{x}_2(0) = x_2(0)$

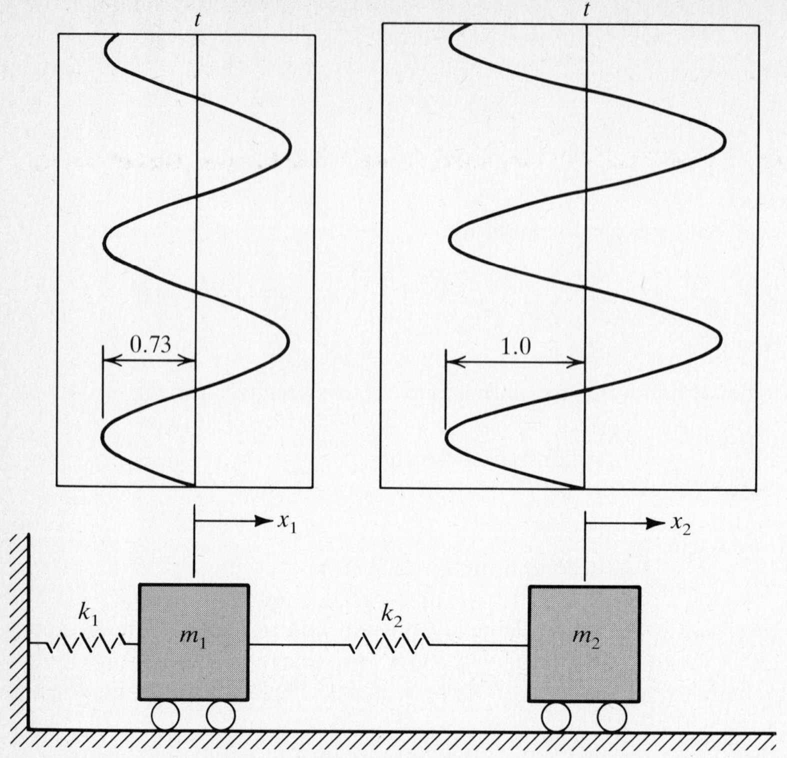


(a) First natural frequency and corresponding modes (first eigenvector)

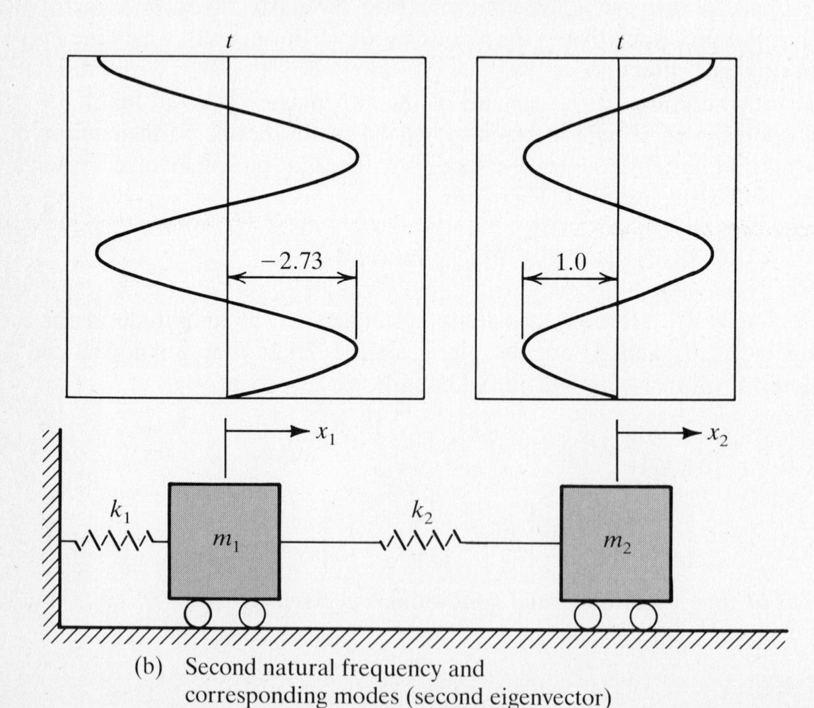


(b) Second natural frequency and corresponding modes (second eigenvector)

**Figure 7.5** Modes of vibration corresponding to the two natural circular frequencies $w_1$ and $w_2$.

These values permit the evaluation of the four unknowns. In addition, once the displacement functions are determined the velocities and accelerations of the two masses can be calculated by using equation (7.31). In fact, the forces in the spring as well as the inertial forces for each mass can be calculated rather easily and at any given time as follows:

$$F_{s1} = kx_1 = \text{force in spring 1}$$

$$F_{s2} = 2kx_2 = \text{force in spring 2}$$

$$F_{I1} = m\ddot{x}_1 = \text{inertial force for mass 1}$$

$$F_{I2} = 2m\ddot{x}_2 = \text{inertial force for mass 2}$$

where $x_1$, $x_2$, $x_1$, and $x_2$ are obtained from the expression given by equation (7.31) and its second derivative.

### 7.3.3 Electrical Circuits

The solution of problems pertaining to electrical engineering circuits often involves solving an eigenproblem. Such problems arise when dealing with direct current circuits as well as alternating current circuits. For AC circuits, the eigenproblem relates primarily to establishing part of the total circuit response due to initial conditions. On the other hand, for DC circuits, the eigenproblem relates to establishing the total response represented by the initial conditions. The implication is that as long as the input voltage is constant or zero in a given circuit, then an eigenproblem will have to be solved.

Consider the following DC circuit, then determine the resonance frequency of the system and the corresponding eigenvectors:

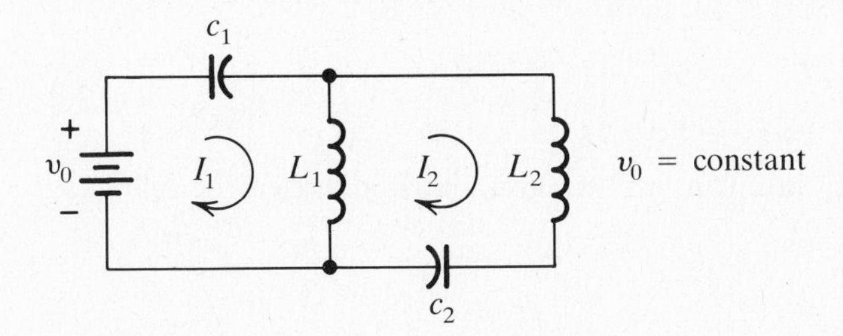


The voltages across capacitor, inductor, and resistor are given, respectively, as follows:

$$v_c = \frac{1}{c}\int I\,dt$$

$$v_L = L\frac{dI}{dt}$$

$$v_R = IR$$

Using Kirchhoff's law, one can easily develop the following differential equations for loop 1 and loop 2:

Loop 1:

$$-\frac{1}{c_1}\int_{-\infty}^{t} I_1\,dt - L_1\frac{d}{dt}(I_1 - I_2) = 0$$

Differentiating with respect to time eliminates the integral and the constant voltage from the expression to give

$$-\frac{1}{c_1}I_1 - L_1\frac{d^2I_1}{dt^2} + L_1\frac{d^2I_2}{dt^2} = 0$$

Rearranging terms yields

$$-L_1\frac{d^2I_1}{dt^2} + L_1\frac{d^2I_2}{dt^2} - \frac{I_1}{c_1} = 0 \tag{7.32}$$

Loop 2:

$$L_1\frac{d}{dt}(I_1 - I_2) - L_2\frac{dI_2}{dt} - \frac{1}{c_2}\int_{-\infty}^{t} I_2\,dt = 0$$

Differentiating with respect to time, we obtain

$$L_1\frac{d^2I_1}{dt^2} - L_1\frac{d^2I_2}{dt^2} - L_2\frac{d^2I_2}{dt^2} - \frac{I_2}{c_2} = 0$$

Rearranging terms yields

$$L_1\frac{d^2I_1}{dt^2} - (L_1 + L_2)\frac{d^2I_2}{dt^2} - \frac{I_2}{c_2} = 0 \tag{7.33}$$

Expressing equations (7.32) and (7.33) in matrix form gives

$$\begin{bmatrix} 1/c_1 & 0 \\ 0 & 1/c_2 \end{bmatrix}\begin{Bmatrix} I_1 \\ I_2 \end{Bmatrix} + \begin{bmatrix} L_1 & -L_1 \\ -L_1 & L_1 + L_2 \end{bmatrix}\begin{Bmatrix} \ddot{I}_1 \\ \ddot{I}_2 \end{Bmatrix} = \begin{Bmatrix} 0 \\ 0 \end{Bmatrix} \tag{7.34}$$

Now, if we assume that the current in each loop is sinusoidal then the following equations are obtained:

$$I_1 = A_1 \sin(wt + \theta)$$

$$\ddot{I}_1 = -A_1 w^2 \sin(wt + \theta)$$

$$I_2 = A_2 \sin(wt + \theta)$$

$$\ddot{I}_2 = -A_2 w^2 \sin(wt + \theta)$$

where

$A_1, A_2$ = current amplitudes in loop 1 and loop 2, respectively

$I_1, I_2$ = current in loop 1 and loop 2, respectively

$\theta$ = phase angle

$w$ = natural frequency

Substituting the currents and their second derivatives into equations (7.34) and simplifying gives

$$\begin{bmatrix} 1/c_1 & 0 \\ 0 & 1/c_2 \end{bmatrix} \begin{Bmatrix} A_1 \\ A_2 \end{Bmatrix} - w^2 \begin{bmatrix} L_1 & -L_1 \\ -L_1 & L_1 + L_2 \end{bmatrix} \begin{Bmatrix} A_1 \\ A_2 \end{Bmatrix} = \begin{Bmatrix} 0 \\ 0 \end{Bmatrix} \tag{7.35}$$

Clearly this is an eigenvalue problem. Let's make the following substitutions:

$$\zeta = \frac{c_1}{c_2}, \qquad \beta = \frac{L_1 + L_2}{L_1}, \qquad \lambda = w^2 L_1 c_1$$

$$\begin{bmatrix} 1 & 0 \\ 0 & \zeta \end{bmatrix} \begin{Bmatrix} A_1 \\ A_2 \end{Bmatrix} - \lambda \begin{bmatrix} 1 & -1 \\ -1 & \beta \end{bmatrix} \begin{Bmatrix} A_1 \\ A_2 \end{Bmatrix} = \begin{Bmatrix} 0 \\ 0 \end{Bmatrix}$$

which simplifies to

$$\begin{bmatrix} 1 - \lambda & -\lambda \\ -\lambda & \zeta - \lambda\beta \end{bmatrix} \begin{Bmatrix} A_1 \\ A_2 \end{Bmatrix} = \begin{Bmatrix} 0 \\ 0 \end{Bmatrix} \tag{7.36}$$

Therefore, forcing the determinant of the coefficient matrix to equal zero gives

$$(\beta - 1)\lambda^2 - (\zeta + \beta)\lambda + \zeta = 0 \tag{7.37}$$

Equation (7.37) is the characteristic equation of the circuit. Solving for the roots using the quadratic formula yields

$$\lambda_{1,2} = \frac{(\beta + \zeta) \pm \sqrt{(\zeta + \beta)^2 - 4(\beta - 1)}}{2(\beta - 1)} \tag{7.38}$$

Consider the special case in which

$$c_1 = c_2 = c, \qquad L_1 = L_2 = L$$

Therefore

$$\zeta = \frac{c_1}{c_2} = \frac{c}{c} = 1, \qquad \beta = \frac{L_1 + L_2}{L_1} = \frac{L + L}{L} = 2$$

Substituting into (7.38) permits the evaluation of the eigenvalues. That is,

$$\lambda_{1,2} = \frac{3 \pm \sqrt{(3)^2 - 4(2 - 1)(1)}}{2(2 - 1)}$$

$$\lambda_{1,2} = \frac{3 \pm \sqrt{5}}{2}$$

$$\lambda_1 = \frac{3 - \sqrt{5}}{2} = 0.382 \qquad \text{(first eigenvalue)}$$

$$\lambda_2 = \frac{3 + \sqrt{5}}{2} = 2.618 \qquad \text{(second eigenvalue)}$$

The natural circular frequencies of the system are

$$\lambda_1 = 0.382 = w_1^2 Lc \Rightarrow w_1 = 0.618/\sqrt{Lc}$$

$$\lambda_2 = 2.618 = w_2^2 Lc \Rightarrow w_2 = 1.618/\sqrt{Lc}$$

The corresponding eigenvectors are determined by substituting the eigenvalues into equation (7.36). That is, substituting $\lambda_1 = 0.382$ gives the first eigenvector:

$$\begin{bmatrix} 1 - 0.382 & -0.382 \\ -0.382 & 1 - 2(0.382) \end{bmatrix} \begin{Bmatrix} A_1 \\ A_2 \end{Bmatrix} = \begin{Bmatrix} 0 \\ 0 \end{Bmatrix}$$

$$\begin{bmatrix} 1 & -0.618 \\ 1 & -0.618 \end{bmatrix} \begin{Bmatrix} A_1 \\ A_2 \end{Bmatrix} = \begin{Bmatrix} 0 \\ 0 \end{Bmatrix}$$

Clearly the first and second rows are the same. The implication is that it makes no difference which one we use and that there are an infinite number of possible solutions. Hence

$$\{V\}_1 = \begin{Bmatrix} 0.618 \\ 1 \end{Bmatrix} A_2 \qquad \text{(first eigenvector)}$$

$$\{\hat{V}\}_1 = \begin{Bmatrix} 0.526 \\ 0.851 \end{Bmatrix} \qquad \text{(normalized first vector)}$$

The meaning of the first eigenvector is best explained by considering figure 7.6a, which shows the relative amplitude of the currents in loop 1 and loop 2 when the circuit is operating at its first (smallest) frequency.

The second eigenvector is found by substituting the second eigenvalue into equation (7.37) and solving for $A_1$ and $A_2$:

$$\begin{bmatrix} 1 - 2.618 & -2.618 \\ -2.618 & 1 - 2(2.618) \end{bmatrix} \begin{Bmatrix} A_1 \\ A_2 \end{Bmatrix} = \begin{Bmatrix} 0 \\ 0 \end{Bmatrix}$$

$$\{V\}_2 = \begin{Bmatrix} 1.618 \\ -1 \end{Bmatrix} A_2 \qquad \text{(second eigenvector)}$$

$$\{\hat{V}\}_2 = \begin{Bmatrix} 0.851 \\ -0.526 \end{Bmatrix} \qquad \text{(normalized second eigenvector)}$$

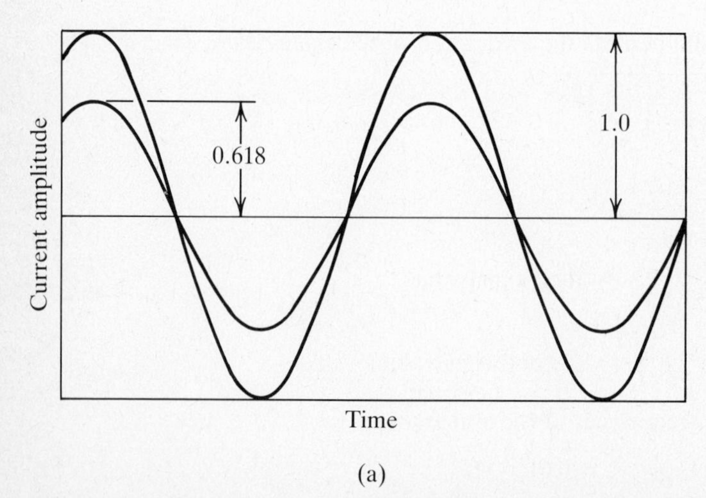


(a)

**Figure 7.6a** First eigenvector corresponding to the first natural frequency.

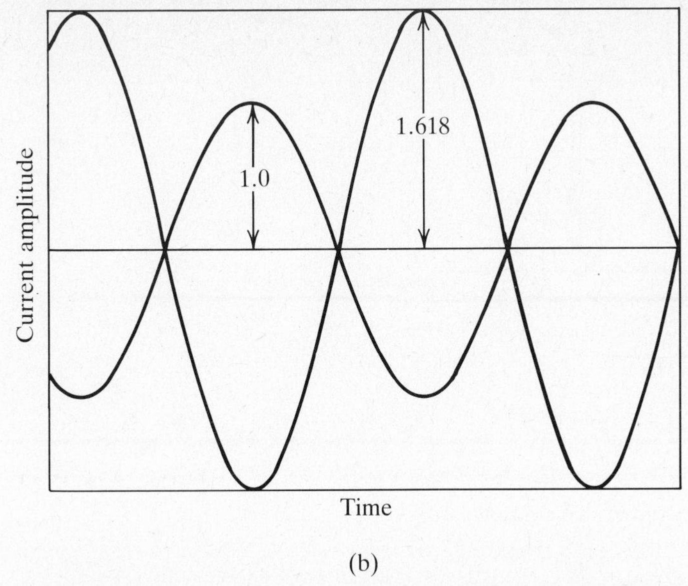


(b)

**Figure 7.6b** Second eigenvector corresponding to the second natural frequency.

Figure 7.6b shows the amplitudes of the currents when the circuit is operating at its second frequency.

The currents in each of the two loops are now determined in terms of the eigenvalues and their corresponding vectors to give

$$\begin{Bmatrix} I_1 \\ I_2 \end{Bmatrix} = A\{V\}_1 \sin(w_1 t + \theta_1) + B\{V\}_2 \sin(w_2 y + \theta_2) \tag{7.39}$$

Note that there are four unknowns, $A$, $B$, $\theta_1$, and $\theta_2$, which can be easily determined from initial conditions. These are, at time $t = 0$,

(1) $I_1(t) = I_1(0)$ (2) $I_1(t) = I_1(0)$

(3) $I_2(t) = I_2(0)$ (4) $I_2(t) = I_2(0)$

The derivative of the currents at time $t = 0$ can be evaluated by measuring the voltages across the inductances in loops 1 and 2. That is,

$$V_1 = L_1 I_1(0), \qquad I_1(0) = 1/L_1$$

$$V_2 = L_2 I_2(0), \qquad I_2(0) = 2/L_2$$

Consequently, the unknowns are evaluated to give the currents in each of the loops as a function of time [equation (7.39)].

### 7.3.4 Eigenproblems and Dynamic Response

In dynamic analysis, we are often faced with the problem of determining the model response of a given system and the frequency of that response. It turns out that the eigenvalues are related to the frequency and that the eigenvectors define the

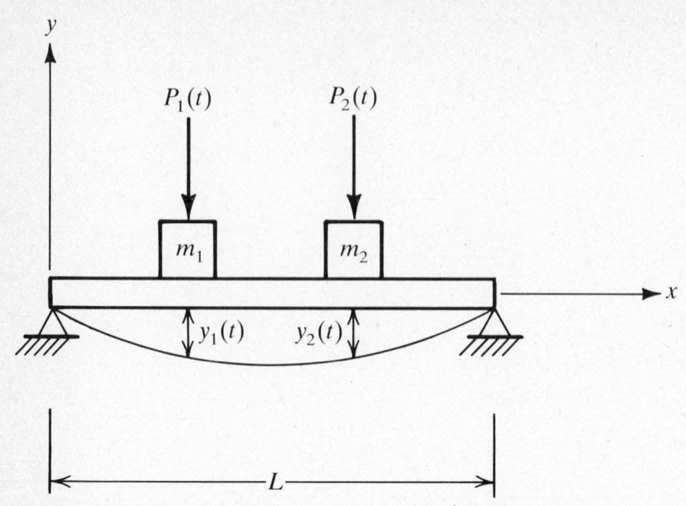


**Figure 7.7** Beam subjected to two dynamic loads.

possible deflected shapes of a structural system. Consider the beam shown in figure 7.7. Let's assume that the mass of the beam is small as compared to the masses $m_1$ and $m_2$.

Obviously, if the beam is to be displaced vertically as shown in the figure and then released, then one would expect it to vibrate. The question that we need to answer is, how much does it vibrate and in what way? These questions can be answered by considering the following principles.

*D'Alembert's principle.* A dynamic system may be treated as a static system if we consider the inertiaı forces as real forces. What is an inertial force? It is the negative of the mass times acceleration. Thus it represents the resistance of an object to acceleration.

$$\sum F = 0 \qquad \text{(static)}$$

$$m\ddot{y} = \sum F \qquad \text{(dynamic)} \qquad \sum F + (-m\ddot{x}) = 0$$

We presume that the beam is linearly elastic. Therefore the principle of superposition applies. That is, the total effect is equal to the sum of the individual effects:

$$\begin{aligned} y_1 &= (-m_1\ddot{y}_1 + P_1)d_{11} + (-m_2\ddot{y}_2 + P_2)d_{12} \\ y_2 &= (-m_1\ddot{y}_1 + P_1)d_{21} + (-m_2\ddot{y}_2 + P_2)d_{22} \end{aligned} \qquad \textbf{(7.40)}$$

where $d_{ij}$ is the displacement in coordinate $i$ due to a unit force applied in coordinate $j$.

Let's start the solution process by first assuming $P_1(t) = P_2(t) = 0$. This means that there are no external forces acting on the beam. This situation is often referred to as "free vibration." It can be shown that the response of $m_1$ and $m_2$ can be

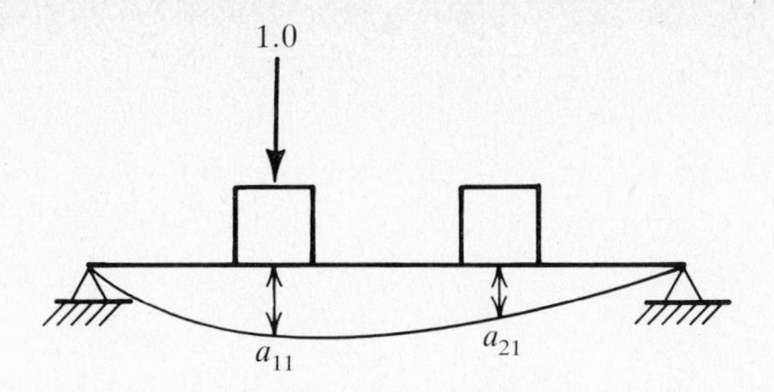


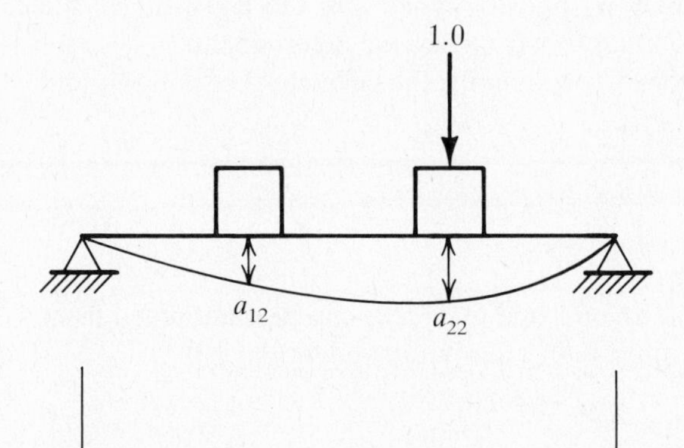


**Figure 7.8** Beam deflection due to point static loads $P_1$ and $P_2$.

precisely described by the following sinusoidal functions:

$$y_1 = a_1 \sin(wt + \theta) \qquad \text{and} \qquad y_2 = a_2 \sin(wt + \theta)$$

Therefore taking the derivatives twice with respect to time gives

$$\ddot{y}_1 = -w^2 a_1 \sin(wt + \theta) \qquad \text{and} \qquad \ddot{y}_2 = -w^2 a_2 \sin(wt + \theta)$$

Substituting in equation (7.40) gives

$$a_1 \sin(wt + \theta) = m_1 w^2 a_1 \sin(wt + \theta)\, d_{11} + m_2 w^2 a_2 \sin(wt + \theta)\, d_{12}$$

$$a_2 \sin(wt + \theta) = m_1 w^2 a_1 \sin(wt + \theta)\, d_{21} + m_2 w^2 a_2 \sin(wt + \theta)\, d_{22}$$

Simplifying and rearranging gives

$$\begin{aligned} (m_1 w^2 d_{11} - 1)a_1 + (m_2 w^2 d_{12})a_2 &= 0 \\ (m_1 w^2 d_{21})a_1 + (m_2 w^2 d_{22} - 1)a_2 &= 0 \end{aligned} \qquad \textbf{(7.41)}$$

Equation (7.41) is called the mode equation.

Once again a nontrivial solution for $a_1$ and $a_2$ is possible only if the determinant is zero. This is an eigenvalue problem of type I:

$$\begin{vmatrix} m_1 w^2 d_{11} - 1 & m_2 w^2 d_{12} \\ m_1 w^2 d_{12} & m_2 w^2 d_{22} - 1 \end{vmatrix} = 0 \qquad \textbf{(7.42)}$$

Expanding equation (7.42) gives

$$m_1 m_2 (w^2)^2 d_{11} d_{22} - m_1 w^2 d_{11} - m_2 w^2 d_{22} + 1 - m_1 m_2 (w^2)^2 d_{12} d_{21} = 0$$

$$(m_1 m_2 d_{11} d_{22} - m_1 m_2 d_{12} d_{21})(w^2)^2 + (-m_1 d_{11} - m_2 d_{22})(w^2) + 1 = 0$$

$$c_1 (w^2)^2 + c_2 (w^2) + 1 = 0$$

$$(w^2) = \frac{-c_2 \pm \sqrt{c_2^2 - 4c_1}}{2c_1}$$

Clearly, we have two roots, namely $w_1$ and $w_2$. These are the eigenvalues. We obtain $a_1 = a_{11}$ and $a_{21}$ corresponding to $w_1$; for $w_2$, we determine $a_1 = a_{12}$ and $a_2 = a_{22}$. These are the eigenvectors. Consequently, the deflections of points 1 and 2 are

$$\begin{aligned} y_1 &= a_{11} \sin(w_1 t + \theta_1) + a_{12} \sin(w_2 t + \theta_2) \\ y_2 &= a_{21} \sin(w_1 t + \theta_1) + a_{22} \sin(w_2 t + \theta_2) \end{aligned} \tag{7.43}$$

Here we have four unknowns, $a_{11}$, $a_{22}$, $\theta_1$, and $\theta_2$. These can be determined from the initial conditions: $y_1 = y_1(0)$, $\dot{y}_1 = \dot{y}_1(0)$, $y_2 = y_2(0)$, and $y_2 = y_2(0)$.

## E EXAMPLE 7.3

Determine the frequencies of vibrations and the corresponding modal shapes of the beam shown below.

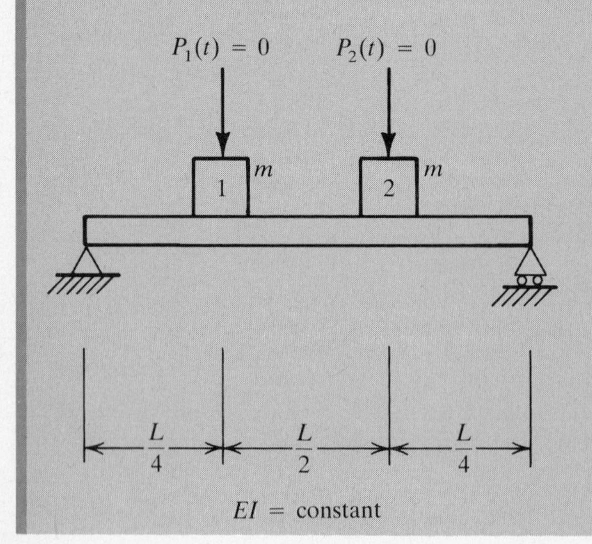

## Solution

The $d_{ij}$ are determined by using elastic deflection formulas:

$$
\begin{aligned}
d_{11} &= 9L^3/768EI \\
d_{12} &= 7L^3/768EI \\
d_{21} &= 7L^3/768EI \\
d_{22} &= 9L^3/768EI
\end{aligned}
\tag{7.44}
$$

For simplicity, let's assume that

$$\alpha = \frac{mL^3}{768EI} \qquad \text{and} \qquad \lambda = \frac{1}{w^2} \tag{7.45}$$

Substituting equations (7.45) into (7.44) gives

$$
\begin{aligned}
d_{11} &= 9\alpha/m \\
d_{12} &= 7\alpha/m \\
d_{21} &= 7\alpha/m \\
d_{22} &= 9\alpha/m
\end{aligned}
$$

Noting that $m_1 = m_2$ then substituting into equation (7.41) gives

$$\begin{bmatrix} 9\alpha - \lambda & 7\alpha \\ 7\alpha & 9\alpha - \lambda \end{bmatrix} \begin{Bmatrix} a_1 \\ a_2 \end{Bmatrix} = \begin{Bmatrix} 0 \\ 0 \end{Bmatrix} \tag{7.46}$$

Clearly this is an eigenproblem. Therefore, forcing the determinant of the coefficients to zero yields the characteristic equation:

$$\lambda^2 - 18\alpha\lambda + 32\alpha^2 = 0$$

Solving for the $\lambda$'s gives

$$\lambda_1 = 16\alpha, \qquad \lambda_2 = 2\alpha$$

and

$$\lambda_1 = \frac{1}{w_1^2} \Rightarrow w_1 = \left(\frac{48EI}{mL^3}\right)^{1/2}$$

Now, solving for $w_2$ we get

$$w_2 = \sqrt{8}\left(\frac{48EI}{mL^3}\right)^{1/2}$$

These values represent the so-called natural circular frequencies of vibration. In dynamic analysis the smallest frequencies are of the most concern. This is why they are arranged in the order given above.

The corresponding first eigenvector can be easily determined by substituting $\lambda_1 = 16$ into equation (7.46) to give

$$\begin{bmatrix} 9\alpha - 16\alpha & 7\alpha \\ 7\alpha & 9\alpha - 16\alpha \end{bmatrix} \begin{Bmatrix} a_{11} \\ a_{21} \end{Bmatrix} = \begin{Bmatrix} 0 \\ 0 \end{Bmatrix}$$

or simply

$$\begin{bmatrix} -7 & 7 \\ 7 & -7 \end{bmatrix} \begin{Bmatrix} a_{11} \\ a_{21} \end{Bmatrix} = \begin{Bmatrix} 0 \\ 0 \end{Bmatrix}$$

Clearly it makes no difference which of the two equations we use to determine $a_{11}$ and $a_{21}$, since they are identical. Therefore

$$-7a_{11} + 7a_{21} = 0$$

$$a_{11} = a_{21}$$

The first eigenvector is arbitrarily defined as $\{V\}_1$ and is given as

$$\{V\}_1 = \begin{Bmatrix} a_{11} \\ a_{11} \end{Bmatrix} = a_{11} = \begin{Bmatrix} 1 \\ 1 \end{Bmatrix}$$

The interpretation of the first vector is shown in figure 7.9, which assumes $a_{11} = 1$. Thus the first eigenvector is now given as

$$\{V\}_1 = \begin{Bmatrix} 1 \\ 1 \end{Bmatrix}$$

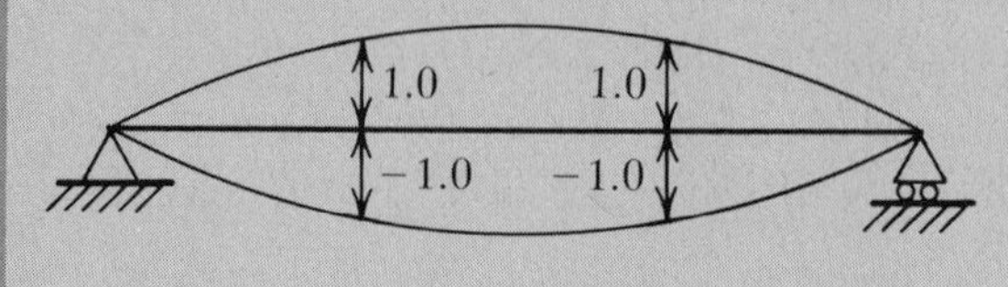


$$w_1 = \sqrt{\frac{48EI}{mL^3}}$$

**Figure 7.9** First mode of vibration.

The significance of this is that the beam will deflect in the fashion shown in the figure when it is operating at a frequency $w_1$. Similarly, the second eigenvector is determined by substituting $w_2$ into equation (7.46). Hence

$$\begin{bmatrix} 9\alpha - 2\alpha & 7\alpha \\ 7\alpha & 9\alpha - 2\alpha \end{bmatrix} \begin{Bmatrix} a_{12} \\ a_{22} \end{Bmatrix} = \begin{Bmatrix} 0 \\ 0 \end{Bmatrix}$$

Once again either equation may be used to define the second eigenvector. That is,

$$7a_{12} + 7a_{22} = 0$$

$$a_{12} = -a_{22}$$

The second natural mode shape is now arbitrarily defined as follows:

$$\{V\}_2 = \begin{Bmatrix} a_{12} \\ -a_{12} \end{Bmatrix} = \begin{Bmatrix} 1 \\ 1 \end{Bmatrix}$$

A schematic representation of the second vector is shown in figure 7.10.

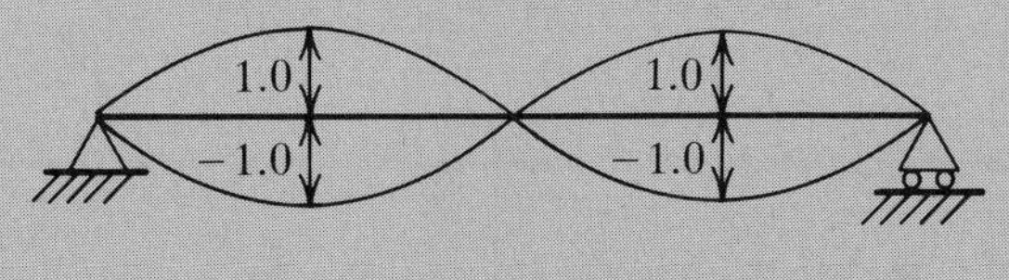


$$w_2 = \sqrt{8} w_1$$

**Figure 7.10** Second mode of vibration.

Clearly the eigenproblem we have just solved provided us with a wealth of information as to the dynamic behavior of a structural member. This type of analysis is called multidegree analysis because there is more than one degree of freedom.

So far we have said nothing about how one can determine the actual response as a function of time. Instead, we have chosen to assume arbitrary values for the displacement. Let's now see how we can fix these arbitrary values.

Since the beam motion is not caused by any external forces $[P_1(t) = P_2(t) = 0]$, it must be caused by initial conditions. These may include initial displacements of mass 1 and mass 2 by, say, $y_1(0)$ and $y_2(0)$, respectively; in addition, the initial velocities of the two masses may not be zero at time zero. Thus we could possibly have $y_1(0)$ and $y_2(0)$ as initial velocities at nodes 1 and 2. Recall that the motion of each node is completely described by equation (7.43). These equations are repeated here for convenience:

$$y_1 = a_{11} \sin(w_1 t + \theta_1) + a_{12} \sin(w_2 t + \theta_2)$$

$$y_2 = a_{21} \sin(w_1 t + \theta_1) + a_{22} \sin(w_2 t + \theta_2)$$

Therefore, substituting $y_1 = y_1(0)$ and $y_2 = y_2(0)$ at $t = 0$, we then get

$$y_1(0) = a_{11} \sin \theta_1 + a_{12} \sin \theta_2 \tag{7.47a}$$

$$y_2(0) = a_{21} \sin \theta_1 + a_{22} \sin \theta_2 \tag{7.47b}$$

But $a_{11} = a_{21}$ and $a_{12} = -a_{22}$; substituting these values into (7.47) gives

$$y_1(0) = a_{11} \sin \theta_1 - a_{22} \sin \theta_2 \tag{7.48a}$$

$$y_2(0) = a_{11} \sin \theta_1 + a_{22} \sin \theta_2 \tag{7.48b}$$

Clearly we still have four unknowns, namely, $a_{11}$, $a_{22}$, $\theta_1$, and $\theta_2$, but only two equations. Fortunately the other two equations can be readily formulated from initial velocities. Hence

$$\dot{y}_1 = a_{11} w_1 \cos(w_1 t + \theta_1) + a_{12} w_2 \cos(w_2 t + \theta_2)$$

$$\dot{y}_2 = a_{21} w_1 \cos(w_1 t + \theta_1) + a_{12} w_2 \cos(w_2 t + \theta_2)$$

so at $t = 0$, $\dot{y}_1 = \dot{y}_1(0)$ and $\dot{y}_2 = \dot{y}_2(0)$. Thus

$$\dot{y}_1(0) = a_{11}w_1 \cos\theta_1 - a_{22}w_2 \cos\theta_2 \tag{7.49a}$$

$$\dot{y}_2(0) = a_{11}w_1 \cos\theta_1 + a_{22}w_2 \cos\theta_2 \tag{7.49b}$$

Now, equations (7.48) and (7.49) can be solved for the unknowns as follows. Adding equation (7.48a) to (7.48b) gives

$$y_1(0) + y_2(0) = 2a_{11} \sin\theta_1 \tag{7.50a}$$

adding equations (7.49a) to (7.49b) gives

$$\dot{y}_1(0) + \dot{y}_1(0) = 2a_{11}w_1 \cos\theta_1 \tag{7.50b}$$

Now, dividing equation (7.50a) by (7.50b) gives

$$\frac{y_1(0) + y_2(0)}{\dot{y}_1(0) + \dot{y}_2(0)} = \frac{\sin\theta_1}{w_1 \cos\theta_1} = \frac{\tan\theta_1}{w_1}$$

Solving for $\theta_1$ in terms of initial conditions and the first natural circular frequency gives

$$\theta_1 = \tan^{-1}\left(\frac{y_1(0) + y_2(0)}{\dot{y}_1(0) + \dot{y}_2(0)}(w_1)\right)$$

Now, substituting equation $\theta_1$ into equation (7.49a) and solving for $a_{11}$, we get

$$a_{11} = \frac{y_1(0) + y_2(0)}{2 \sin\theta_1}$$

The values of $\theta_2$ and $a_{22}$ are determined similarly. Therefore the response of the beam in terms of initial conditions is now determined for $t \geq 0$.

## 7.4 VECTOR ITERATION TECHNIQUES

In the previous section, we discussed classical methods for evaluating the eigenvalues and eigenvectors for different systems. It is evident that these methods become impractical as the matrices involved become large. Consequently, iterative techniques are used for that purpose. The solution of some problems occurring in the various fields of engineering may involve matrices exceeding 300 × 300 in size. For example, the Saturn V rocket used in the Apollo space program involved solving models of 30, 120, 300, and 400 degrees of freedom. The analysis of the dynamic response of a typical high-rise building may exceed 100 degrees of freedom. The point to be made is that conventional methods of solving eigenproblems may not well be suited for tackling a number of practical problems. Therefore more practical techniques are needed.

### 7.4.1 Largest Eigenvalue

This method permits the determination of the largest eigenvalue as well as its corresponding vector. The so-called "power" technique is an iterative method that can be applied to eigenproblems of types II and III, but not type I. The solution involves assuming a trial vector which is refined by using the property matrices for a given system. Consider the eigenproblem most often encountered in engineering,

$$[[A] - \lambda[B]]\{X\} = \{0\}$$

This can be expressed in the following form:

$$[A]\{X\} = \lambda[B]\{X\}$$

Premultiplying by the inverse of the $[B]$ matrix and simplifying yields

$$[B]^{-1}[A]\{X\} = \lambda\{X\} \tag{7.51a}$$

Assuming $[C] = [B]^{-1}[A]$ gives

$$[C]\{X\} = \lambda\{X\} \tag{7.51b}$$

Note that, for the special case in which the $[B]$ matrix is the identity matrix, $[C] = [A]$. In general, equations (7.51) can be expressed in an iterative form as follows:

$$[C]\{X\}_i = \{X\}_{i+1} = \lambda\{X\}_{i+1} \tag{7.52}$$

The implication is that if we assume a trial vector $\{X\}_0 \neq \{0\}$ then a new vector $\{\bar{X}\}_1$ can be calculated. The new vector is normalized by factoring its largest coefficient. This coefficient is then taken as a first approximation to the largest eigenvalue and the resulting vector $\{X\}_1$ represents the first approximation to the corresponding eigenvector. This process is continued by substituting the new eigenvector and determining a second approximation, etc., until the vector produces itself. That is,

$$\begin{aligned}
[C]\{X\}_0 &= \{\bar{X}\}_1 = \lambda_1\{X\}_1 \\
[C]\{X\}_1 &= \{\bar{X}\}_2 = \lambda_2\{X\}_2 \\
&\vdots \\
[C]\{X\}_{k-1} &= \{\bar{X}\}_k = \lambda\{X\}_k
\end{aligned}$$

Note that the normalized vector coefficients will always be less than or equal to one. Furthermore, the assumed vector is generally taken as a unit vector ($\{X\}_0 = \{1\}_0$).

E

EXAMPLE 7.4

The designs of many engineering systems generally involve the application of a so-called maximum-stress theory of failure. This theory is based on the assumption that the maximum principal stress acting on a body determines failure. Consider the cubic element shown below and the forces acting on it, then determine the major principal stress.

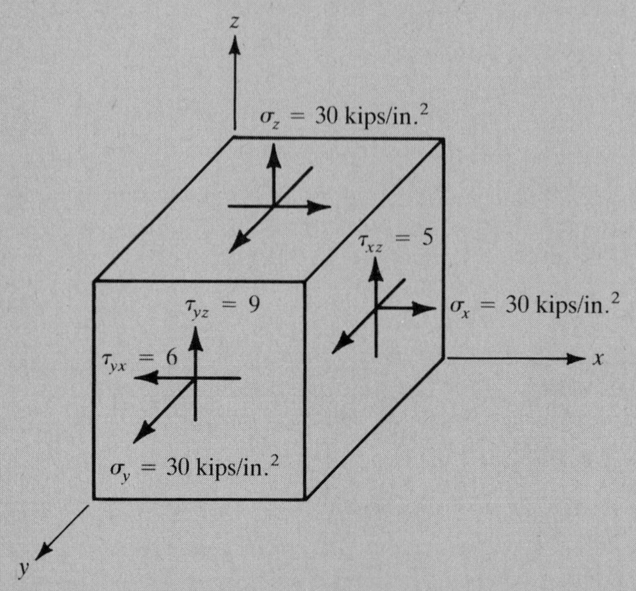


**Figure 7.11** Stresses acting on a cubic element.

## Solution

This is an eigenvalue problem of the following form:

$$\left[\begin{bmatrix} \sigma_x & \tau_{yx} & \tau_{zx} \\ \tau_{xy} & \sigma_y & \tau_{zy} \\ \tau_{xz} & \tau_{yz} & \sigma_z \end{bmatrix} - \sigma \begin{bmatrix} 1 & 0 & 0 \\ 0 & 1 & 0 \\ 0 & 0 & 1 \end{bmatrix}\right] \begin{Bmatrix} l \\ m \\ n \end{Bmatrix} = \begin{Bmatrix} 0 \\ 0 \\ 0 \end{Bmatrix} \tag{7.53}$$

Note here that $\tau_{yx} = \tau_{xy}$, $\tau_{xz} = \tau_{zx}$, and $\tau_{yz} = \tau_{zy}$. Thus substituting these values into equation (7.53) and rearranging gives

$$\begin{bmatrix} 30 & 6 & 5 \\ 6 & 30 & 9 \\ 5 & 9 & 30 \end{bmatrix} \begin{Bmatrix} l \\ m \\ n \end{Bmatrix} = \sigma \begin{bmatrix} 1 & 0 & 0 \\ 0 & 1 & 0 \\ 0 & 0 & 1 \end{bmatrix} \begin{Bmatrix} l \\ m \\ n \end{Bmatrix} \tag{7.54}$$

or more simply

$$\begin{bmatrix} 30 & 6 & 5 \\ 6 & 30 & 9 \\ 5 & 9 & 30 \end{bmatrix} \begin{Bmatrix} l \\ m \\ n \end{Bmatrix}_i = \sigma \begin{Bmatrix} l \\ m \\ n \end{Bmatrix}_{i+1} \tag{7.55}$$

Now assuming an initial vector $\{X\}_0^t = \{0, 0, 1\}$, solving for a new one yields

$i = 0$

$$\begin{bmatrix} 30 & 6 & 5 \\ 6 & 30 & 9 \\ 5 & 9 & 30 \end{bmatrix} \begin{Bmatrix} 0 \\ 0 \\ 1 \end{Bmatrix}_0 = \begin{Bmatrix} 5 \\ 9 \\ 30 \end{Bmatrix}_1 = 30 \begin{Bmatrix} 0.166 \\ 0.300 \\ 1.000 \end{Bmatrix}_1$$

$i = 1$

$$\begin{bmatrix} 30 & 6 & 5 \\ 6 & 30 & 9 \\ 5 & 9 & 30 \end{bmatrix} \begin{Bmatrix} 0.166 \\ 0.300 \\ 1.000 \end{Bmatrix}_1 = \begin{Bmatrix} 11.780 \\ 18.996 \\ 33.530 \end{Bmatrix}_2 = 33.53 \begin{Bmatrix} 0.351 \\ 0.566 \\ 1.000 \end{Bmatrix}_2$$

$i = 2$

$$\begin{bmatrix} 30 & 6 & 5 \\ 6 & 30 & 9 \\ 5 & 9 & 30 \end{bmatrix} \begin{Bmatrix} 0.351 \\ 0.566 \\ 1.000 \end{Bmatrix}_2 = \begin{Bmatrix} 18.926 \\ 29.886 \\ 36.849 \end{Bmatrix}_3 = 36.849 \begin{Bmatrix} 0.514 \\ 0.811 \\ 1.000 \end{Bmatrix}_3$$

Obviously, the vector is still changing significantly with each iteration. Therefore we must continue.

$i = 3$

$$\begin{bmatrix} 30 & 6 & 5 \\ 6 & 30 & 9 \\ 5 & 9 & 30 \end{bmatrix} \begin{Bmatrix} 0.514 \\ 0.811 \\ 1.000 \end{Bmatrix}_3 = \begin{Bmatrix} 25.286 \\ 36.414 \\ 39.869 \end{Bmatrix}_4 = 39.869 \begin{Bmatrix} 0.634 \\ 0.913 \\ 1.000 \end{Bmatrix}_4$$

$i = 4$

$$\begin{bmatrix} 30 & 6 & 5 \\ 6 & 30 & 9 \\ 5 & 9 & 30 \end{bmatrix} \begin{Bmatrix} 0.634 \\ 0.913 \\ 1.000 \end{Bmatrix}_4 = \begin{Bmatrix} 29.505 \\ 40.194 \\ 41.387 \end{Bmatrix}_5 = 41.387 \begin{Bmatrix} 0.713 \\ 0.971 \\ 1.000 \end{Bmatrix}_5$$

$i = 5$

$$\begin{bmatrix} 30 & 6 & 5 \\ 6 & 30 & 9 \\ 5 & 9 & 30 \end{bmatrix} \begin{Bmatrix} 0.713 \\ 0.971 \\ 1.000 \end{Bmatrix}_5 = \begin{Bmatrix} 32.22 \\ 42.41 \\ 42.30 \end{Bmatrix}_6 = 42.41 \begin{Bmatrix} 0.760 \\ 1.000 \\ 0.998 \end{Bmatrix}_6$$

$i = 6$

$$\begin{bmatrix} 30 & 6 & 5 \\ 6 & 30 & 9 \\ 5 & 9 & 30 \end{bmatrix} \begin{Bmatrix} 0.760 \\ 1.000 \\ 0.998 \end{Bmatrix}_6 = \begin{Bmatrix} 33.79 \\ 43.54 \\ 42.74 \end{Bmatrix}_7 = 43.54 \begin{Bmatrix} 0.776 \\ 1.000 \\ 0.982 \end{Bmatrix}_7$$

$i = 7$

$$\begin{bmatrix} 30 & 6 & 5 \\ 6 & 30 & 9 \\ 5 & 9 & 30 \end{bmatrix} \begin{Bmatrix} 0.776 \\ 1.000 \\ 0.982 \end{Bmatrix}_7 = \begin{Bmatrix} 34.19 \\ 43.49 \\ 42.34 \end{Bmatrix}_8 = 43.49 \begin{Bmatrix} 0.786 \\ 1.000 \\ 0.974 \end{Bmatrix}_8$$

We can clearly see that the values are not changing significantly, and so the process is now terminated. The largest eigenvalue is determined easily to be

$$\sigma_1 = 43.49$$

The corresponding eigenvector is

$$\{X\}_1 = \begin{Bmatrix} 0.800 \\ 1.000 \\ 0.965 \end{Bmatrix}$$

The student may wonder about the physical meaning of the eigenvalue. In this problem, it represents the maximum normal stress acting on the element. The eigenvector, on the other hand, is the directional cosine vector that determines the position of the plane on which the maximum principal stress acts. Thus

$$\{X\}_1 = \begin{Bmatrix} l \\ m \\ n \end{Bmatrix} = \begin{Bmatrix} \cos\theta_1 \\ \cos\theta_2 \\ \cos\theta_3 \end{Bmatrix}$$

where the angles are defined in figure 7.12.

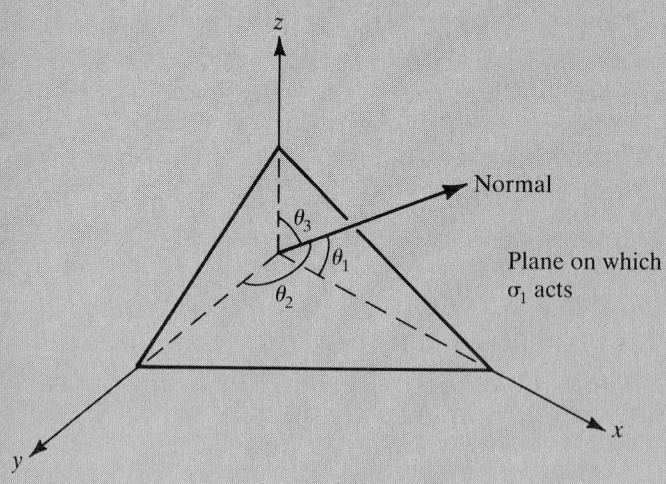


**Figure 7.12** Plane of maximum stress.

At first one might question the meaning of the eigenvector since we know that

$$\cos^2\theta_1 + \cos^2\theta_2 + \cos^2\theta_2 = 1 \quad \textbf{(7.56)}$$

and yet the values of the cosines as given by the eigenvector would not satisfy equation (7.56). The student should not be alarmed because the eigenvector does not give the exact values; instead it gives the ratios of the cosine values: geometrical considerations will ultimately determine the actual magnitudes.

EXAMPLE 7.5

Determine the actual directional cosines vector for the previous example.

Solution

Let's first substitute the eigenvalue of $\sigma_1 = 43.49$ into equation (7.54). Thus

$$\begin{bmatrix} 30 & 6 & 5 \\ 6 & 30 & 9 \\ 5 & 9 & 30 \end{bmatrix} \begin{Bmatrix} l \\ m \\ n \end{Bmatrix} = 43.49 \begin{bmatrix} 1 & 0 & 0 \\ 0 & 1 & 0 \\ 0 & 0 & 1 \end{bmatrix} \begin{Bmatrix} l \\ m \\ n \end{Bmatrix}$$

Combining terms gives

$$\begin{bmatrix} 30-43.49 & 6 & 5 \\ 6 & 30-43.49 & 9 \\ 5 & 9 & 30-43.49 \end{bmatrix} \begin{Bmatrix} l \\ m \\ n \end{Bmatrix} = \begin{Bmatrix} 0 \\ 0 \\ 0 \end{Bmatrix}$$

Simplifying, we obtain

$$\begin{bmatrix} -13.49 & 6 & 5 \\ 6 & -13.49 & 9 \\ 5 & 9 & -13.49 \end{bmatrix} \begin{Bmatrix} l \\ m \\ n \end{Bmatrix} = \begin{Bmatrix} 0 \\ 0 \\ 0 \end{Bmatrix}$$

Solving this system of equations using Gauss's elimination method yields

$$\left[\begin{array}{ccc|c} 1 & -0.445 & -0.371 & 0 \\ 0 & -10.810 & 11.226 & 0 \\ 0 & 11.225 & -11.623 & 0 \end{array}\right] \begin{array}{l} R_1/-13.49 = R_1' \\ -6R_1' + R_2 \\ -5R_1' + R_3 \end{array}$$

$$\left[\begin{array}{ccc|c} 1 & -0.445 & -0.371 & 0 \\ 0 & 1 & -1.038 & 0 \\ 0 & 1 & -1.036 & 0 \end{array}\right] \begin{array}{l} R_2/-10.81 \\ R_3/11.255 \\ \end{array}$$

Clearly the second and third rows are identical. This is expected because the eigenvalue problem does just that. We now have two equations in three unknowns. That is,

$$l - 0.445m - 0.371n = 0 \quad \textbf{(7.57a)}$$

$$m - 1.038n = 0 \quad \textbf{(7.57b)}$$

Now recall that the eigenvector in example 7.4 is

$$\{X\}_1 = \begin{Bmatrix} 0.80 \\ 1.00 \\ 0.965 \end{Bmatrix}$$

This vector should satisfy equations (7.57). Thus

$$0.8 - 0.445(1) - 0.371(0.965) = 0$$

$$1(1) - 1.038(0.965) = 0$$

The reason we have said that this vector is arbitrary is that it represents one of an infinite number of solutions. That is, any scalar multiple of this solution is also a solution. For example,

$$\{X\}_1 = \begin{Bmatrix} 1.60 \\ 2.00 \\ 1.93 \end{Bmatrix}$$

represents another solution. Consequently, a unique solution is possible only if we have three equations in three unknowns. A unique solution to this problem is definitely possible because we do have a third equation, namely,

$$l^2 + m^2 + n^2 = 1 \quad \textbf{(7.58)}$$

Solving equation (7.57b) for $m$ gives

$$m = 1.038n \quad \textbf{(7.59a)}$$

Substituting equation (7.59a) into equation (7.57a) and solving for $l$ gives

$$l = 0.828n \quad \textbf{(7.59b)}$$

Now substituting equations (7.59) into equation (7.58) yields the value of $n$. That is,

$$(0.828n)^2 + (1.038n)^2 + n^2 = 1$$

$$n = \pm 0.602$$

Hence

$$m = \pm 0.602(1.038) = \pm 0.625$$

$$l = \pm 0.602(0.828) = \pm 0.4963$$

These values are the true values and are not arbitrary. In vector form,

$$\{X\}_1 = \begin{Bmatrix} 0.498 \\ 0.625 \\ 0.602 \end{Bmatrix} = 0.625 \begin{Bmatrix} 0.797 \\ 1.000 \\ 0.963 \end{Bmatrix}$$

Compare this vector with the original one of

$$= \begin{Bmatrix} 0.800 \\ 1.000 \\ 0.965 \end{Bmatrix}$$

Obviously, they are of the same relative magnitude.

**SUMMARY OF VECTOR ITERATIONS—LARGEST EIGENVALUE**

| STEP | OPERATION | SYMBOL |
|---|---|---|
| 1 | Formulate the physical problem. | $[A]\{X\} = \lambda[B]\{X\}$ |
| 2 | Invert the $[B]$ matrix, then determine matrix $[C]$. | $[C] = [B]^{-1}[A]$ |
| 3 | Assume a trial vector that is not equal to the null vector. | $\{X\}_i = \{X\}_0 \neq \{0\}_0$ |
| 4 | Compute a new approximation to the true vector corresponding to the largest eigenvalue. | $\{\bar{X}\}_1 = [C]\{X\}_0$ |
| 5 | Factor out the largest coefficient in the $\{\bar{X}\}_1$ vector computed in step 4 to get $\{X\}_1$. | $\{\bar{X}\}_1 = \lambda_1\{X\}_1$ |
| 6 | Check the tolerance and if the criterion is satisfied, then end; otherwise, let $\{X\}_0 = \{X\}_1$ and go to step 4. | $T \geq \lvert\{X\}_1 - \{X\}_0\rvert$ |

### 7.4.2 Smallest Eigenvalue

In the previous section we dealt with problems in which the largest eigenvalue is of most interest. There are many situations in which the engineer will be interested in finding the smallest eigenvalue. This problem may arise when one is dealing with frictional materials such as soils. The basic criterion used in determining the state of stress for such materials is based on the so-called Mohr failure criteria in which both the maximum principal stress $\sigma_1$ and minor principal stress $\sigma_3$ are needed.

The solution procedure is a little more involved than that presented for finding the largest eigenvalue. Fortunately, it is just as straightforward. It can be accomplished by considering the basic eigenvalue equation. That is,

$$[[A] - \lambda[B]]\{X\} = \{0\} \tag{7.60a}$$

or

$$[A]\{X\} = \lambda[B]\{X\} \tag{7.60b}$$

Premultiplying equation (7.60b) by the inverse of $[A]$ gives

$$[A]^{-1}[A]\{X\} = [A]^{-1}\lambda[B]\{X\}$$

Simplifying, we obtain

$$[I]\{X\} = [A]^{-1}\lambda[B]\{X\}$$

Dividing both sides by $\lambda$ gives in general the following iterative formula:

$$[[A]^{-1}[B]]\{X\}_i = \frac{1}{\lambda}\{X\}_{i+1}, \qquad i = 1, \ldots, k \tag{7.61a}$$

For the special case in which the $[B]$ matrix is the identity matrix, we get

$$[A]^{-1}\{X\}_i = \frac{1}{\lambda}\{X\}_{i+1} \tag{7.61b}$$

The solution procedure is initiated by first assuming a vector $\{X\}_i$, improving the solution by getting $\{X\}_{i+1}$, and so on until $\{X\}_{k-1}$ is approximately equal to $\{X\}_k$. Consider the following example.

E

EXAMPLE 7.6

Given the force configuration of example 7.4, determine the smallest eigenvalue.

$$\begin{bmatrix} 30 & 6 & 5 \\ 6 & 30 & 9 \\ 5 & 9 & 30 \end{bmatrix} \begin{Bmatrix} l \\ m \\ n \end{Bmatrix} = \sigma \begin{Bmatrix} l \\ m \\ n \end{Bmatrix}$$

Solution

$$[A]\{X\} = \sigma\{X\}$$

First we must find the inverse of matrix $[A]$. Thus

$$[A]^{-1} = \frac{1}{23{,}280}\begin{bmatrix} 819 & -135 & -96 \\ -135 & 875 & -240 \\ -96 & -240 & 864 \end{bmatrix}$$

Now, using equation (7.61) and assuming a trial vector of the form

$$\{X\}_0 = \begin{Bmatrix} 1 \\ 0 \\ 0 \end{Bmatrix}_0$$

and solving for new vectors, we have

$i = 0$

$$\frac{1}{23{,}280}\begin{bmatrix} 819 & -135 & -96 \\ -135 & 875 & -240 \\ -96 & -240 & 864 \end{bmatrix}\begin{Bmatrix} 1 \\ 0 \\ 0 \end{Bmatrix}_0 = \begin{Bmatrix} 0.0352 \\ -0.0058 \\ -0.0041 \end{Bmatrix}_1$$

$$= 0.035\begin{Bmatrix} 1.000 \\ -0.165 \\ -0.117 \end{Bmatrix}_1$$

$i = 1$

$$\frac{1}{23{,}280}\begin{bmatrix} 819 & -135 & -96 \\ -135 & 875 & -240 \\ -96 & -240 & 864 \end{bmatrix}\begin{Bmatrix} 1.000 \\ -0.165 \\ -0.117 \end{Bmatrix}_1 = \begin{Bmatrix} 0.0366 \\ -0.0108 \\ -0.0068 \end{Bmatrix}_2$$

$$= 0.0366\begin{Bmatrix} 1.000 \\ -0.295 \\ -0.186 \end{Bmatrix}_2$$

$i = 2$

$$\frac{1}{23{,}280}\begin{bmatrix} 819 & -135 & -96 \\ -135 & 875 & -240 \\ -96 & -240 & 864 \end{bmatrix}\begin{Bmatrix} 1.000 \\ -0.295 \\ -0.186 \end{Bmatrix}_2 = \begin{Bmatrix} 0.0377 \\ -0.0150 \\ -0.0080 \end{Bmatrix}_3$$

$$= 0.0377\begin{Bmatrix} 1.000 \\ -0.398 \\ -0.212 \end{Bmatrix}_3$$

This process may be continued until the desired accuracy is achieved. After 30 iterations the eigenvalue converges to 21.148 and the eigenvector to $\{0.187 - 10.881\}^t$.

Thus far, we have dealt with methods for finding the maximum and minimum eigenvalues. The question is, can we determine intermediate values? Fortunately, the answer is yes. However, there is no unique method that can handle the various types of eigenvalue problems; instead there are different methods for different situations. A complete treatment of these methods is beyond the scope of this text. Instead, a procedure for finding intermediate values of problems having symmetrical coefficient matrices is outlined in the next section.

**SUMMARY OF VECTOR ITERATION—SMALLEST EIGENVALUE**

| STEP | OPERATION | SYMBOL |
|---|---|---|
| 1 | Formulate the physical problem. | $[A]\{X\} = \lambda[B]\{X\}$ |
| 2 | Invert the $[A]$ matrix, then determine a matrix $[D]$. | $[D] = [A]^{-1}[B]$ |
| 3 | Assume a trial vector that is not equal to the null vector. | $\{X\}_i = \{X\}_0 \neq \{0\}_0$ |
| 4 | Compute a new approximation to the true vector corresponding to the smallest eigenvalue. | $\{\bar{X}\}_1 = [D]\{X\}_0$ |
| 5 | Factor out the largest coefficient in the $\{\bar{X}\}_1$ vector computed in step 4 to get $\{X\}_1$. Note that the smallest eigenvalue is now greater than or equal to the factored value. | $\{\bar{X}\}_1 = \dfrac{1}{\lambda_1}\{X\}_1$ |
| 6 | Check the tolerance and if the criterion is satisfied, then end; otherwise, let $\{X\}_0 = \{X\}_1$ and go to step 4. | $T \geq \lvert\{X\}_1 - \{X\}_0\rvert$ |

### 7.4.3 Intermediate Eigenvalues

Once the largest eigenvalue is determined, it is possible to find the next largest eigenvalue by transforming the original coefficient matrix to one possessing only the remaining eigenvalues. This procedure is called *matrix deflation* and it is applicable to *symmetrical coefficient matrices only*. Hence, using the procedure outlined earlier for finding the largest eigenvalue $\lambda_1$ and its corresponding vector $\{X\}_1$, one determines matrix $[A]$ by invoking the orthogonality condition to give

$$[A_i] = [A] - \lambda_i\{\hat{X}\}_i\{\hat{X}\}_i^t, \qquad i = 1, \ldots, n \tag{7.62}$$

where $\{\hat{X}\}_i$ is the normalized eigenvector $\{X\}_i$ and $i$ is an index pertaining to the eigenvalue being factored out. It is evident that this process can be continued until all of the eigenvalues have been extracted. Although this technique shows promise, it does have a significant drawback. That is, at each iteration performed in deflating the original coefficient matrix, any errors in the computed eigenvalues and eigenvectors will be passed on to the next eigenvectors. This could result in serious inaccuracies, especially when dealing with large eigenproblems. This is precisely why this method is generally used for small eigenproblems. For large problems, it is recommended that methods outlined in the next section be used instead.

E

EXAMPLE 7.7

Determine the remaining eigenvalue of the problem given in example 7.4. Recall that the largest eigenvalue and its corresponding vector were determined earlier. That is,

$$\sigma_1 = 43.48 \qquad \text{(eigenvalue)}$$

$$\{X\}_1 = \begin{Bmatrix} 0.800 \\ 1.000 \\ 0.965 \end{Bmatrix} \qquad \text{(eigenvector)}$$

Solution

The normalized eigenvector is obtained easily as follows:

$$\{\hat{X}\}_1 = \frac{1}{(0.8^2 + 1.0^2 + 0.965^2)^{1/2}} \begin{Bmatrix} 0.8 \\ 1.00 \\ 0.965 \end{Bmatrix}$$

Simplifying yields

$$\{\hat{X}\}_1 = \begin{Bmatrix} 0.499 \\ 0.624 \\ 0.602 \end{Bmatrix}$$

Substituting the normalized eigenvector, the matrix $[A]$, and the eigenvalue into equation (7.62) gives

$$[A_1] = \begin{bmatrix} -30 & 6 & 5 \\ -6 & 30 & 9 \\ 5 & 9 & 30 \end{bmatrix} - 43.48 \begin{Bmatrix} 0.499 \\ 0.624 \\ 0.602 \end{Bmatrix} \{0.499 \quad 0.624 \quad 0.602\}$$

$$[A_1] = \begin{bmatrix} 19.17 & -7.54 & -8.06 \\ -7.54 & 13.07 & -7.33 \\ -8.06 & -7.33 & 14.24 \end{bmatrix}$$

The largest eigenvalue corresponding to the matrix $[A_0]$ is now determined. Assume a trial vector $\{X\}_0^t = \{1, 0, 0\}$; a new vector $\{X\}_1$ is found as follows:

$$\begin{bmatrix} 19.17 & -7.54 & -8.06 \\ -7.54 & 13.07 & -7.33 \\ -8.06 & -7.33 & 14.24 \end{bmatrix} \begin{Bmatrix} 1 \\ 0 \\ 0 \end{Bmatrix}_0 = \begin{Bmatrix} 19.17 \\ -7.54 \\ -8.06 \end{Bmatrix}_1 = 19.17 \begin{Bmatrix} 1.000 \\ -0.393 \\ -0.42 \end{Bmatrix}_1$$

and

$i = 1$

$$\begin{bmatrix} 19.17 & -7.54 & -8.06 \\ -7.54 & 13.07 & -7.33 \\ -8.06 & -7.33 & 14.24 \end{bmatrix} \begin{Bmatrix} 1.000 \\ -0.393 \\ -0.420 \end{Bmatrix}_1 = \begin{Bmatrix} 25.52 \\ -9.50 \\ -11.16 \end{Bmatrix}_2$$

$$= 25.52 \begin{Bmatrix} 1.000 \\ -0.376 \\ -0.438 \end{Bmatrix}_2$$

$i = 2$

$$\begin{bmatrix} 19.17 & -7.54 & -8.06 \\ -7.54 & 13.07 & -7.33 \\ -8.06 & -7.33 & 14.24 \end{bmatrix} \begin{Bmatrix} 1.000 \\ -0.376 \\ -0.438 \end{Bmatrix}_2 = \begin{Bmatrix} 25.54 \\ -9.24 \\ -11.54 \end{Bmatrix}_3$$

$$= 25.54 \begin{Bmatrix} 1.000 \\ -0.362 \\ -0.452 \end{Bmatrix}_3$$

Clearly the change in the eigenvalue is very small and thus the procedure may be terminated. Consequently, the eigenvalue of 25.54 represents the largest value corresponding to the deflated matrix $[A_1]$ and it corresponds to the intermediate eigenvalue of the original matrix $[A]$.

| SUMMARY OF VECTOR ITERATION—INTERMEDIATE EIGENVALUES | | |
|---|---|---|
| STEP | OPERATION | SYMBOL |
| 1 | Compute the largest eigenvalue and its corresponding vector. | $\lambda_1, \{X\}_1$ |
| 2 | Normalize the eigenvector $\{X\}_1$. | $\{\hat{X}\}_1 = \dfrac{1}{(a_1^2 + \cdots + a_n^2)^{1/2}} \begin{Bmatrix} a_1 \\ \vdots \\ a_n \end{Bmatrix}$ |
| 3 | Deflate the original coefficient $[A]$. | $[A_1] = [A] - \lambda_1\{\hat{X}\}_1\{\hat{X}\}_1^t$ |
| 4 | Use the method outlined in section 7.4.1 to calculate the largest eigenvalue of the deflated $[A_1]$ matrix. | |
| 5 | Repeat steps 2–4 until all eigenvalues are determined. | |

## 7.5 POLYNOMIAL ITERATION METHOD

This technique of determining the eigenvalues and their corresponding vectors is implicit in the sense that the polynomial (characteristic equation) need not be known in advance. Instead, one may use the techniques outlined in sections 7.4.1 and 7.4.2 to calculate the range of eigenvalues to be expected for a given problem, then search for intermediate values. In fact, the errors associated with the largest and the smallest eigenvalues can be reduced by using this technique. Furthermore, intermediate eigenvalues can be refined as well. Consequently, this method may be looked at as a tool for improving the accuracy of eigenvalues evaluated by using the vector iteration techniques. Consider the following eigenproblem:

$$[Q]\{X\} = \{0\}$$

where

$$[Q] = [A] - \lambda[B]$$

It is evident that the determinant of the $[Q]$ matrix gives rise to a polynomial in $\lambda$. That is,

$$\det[Q] = f(\lambda) = 0$$

Suppose that an approximation, say $\lambda_k$, to one of the roots (largest or smallest) has been determined; then

$$\det[Q] = f(\lambda_k) = 0$$

The implication is that since $\lambda_k$ is only an approximation the function $f(\lambda_k)$ will not be zero. The object is to refine $\lambda_k$ so that $f(\lambda_k)$ is close to zero. This can be accomplished by using the Newton–Raphson first method for finding roots of nonlinear algebraic equations. That is,

$$\lambda_{k+1} = \lambda_k - \frac{f(\lambda_k)}{f'(\lambda_k)} \tag{7.63}$$

It is evident that the functional value of the characteristic equation can be evaluated by simply evaluating the determinant of the $[Q]$ matrix at $\lambda = \lambda_k$. Unfortunately, the derivative of the characteristic equation can only be approximated since the characteristic equation is not known. This is illustrated graphically in figure 7.13.

Note that the slope given by line $B$ is approximated by line $A$. That is,

$$f'(\lambda_k) = \frac{f(\lambda_{k-1}) - f(\lambda_k)}{\lambda_{k-1} - \lambda_k}$$

Substituting into equation (7.63) and then simplifying permits the elimination of the derivative from the eigenvalue approximation formula. Hence

$$\lambda_{k+1} = \lambda_k - f(\lambda_k)\frac{\lambda_{k-1} - \lambda_k}{f(\lambda_{k-1}) - f(\lambda_k)} \tag{7.64}$$

The value of $\lambda_{k+1}$ represents an improved estimate of the approximate $\lambda_k$ eigenvalue. Consequently, one may substitute the improved $\lambda_{k+1}$ value to compute yet another approximation until the desired accuracy is achieved. This method is highly accurate and is strongly recommended provided a computer is used to carry out the computations.

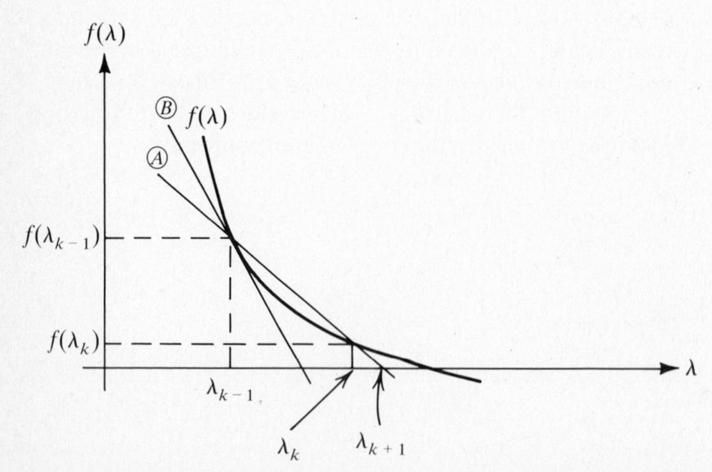


**Figure 7.13** Eigenvalue approximation of a characteristic equation.

E

EXAMPLE 7.8

Approximate the largest eigenvalue of the problem given in example 7.4 using the estimated value of $43 < \sigma < 44$.

Solution

Using equation (7.64) for $k = 1$ yields

$$\lambda_2 = \lambda_1 - f(\lambda_1)\left[\frac{\lambda_0 - \lambda_1}{f(\lambda_0) - f(\lambda_1)}\right]$$

Note here that the characteristic equation $f(\lambda)$ is known implicitly to be equal to the determinant of the $[Q]$ matrix. Thus letting $\lambda = \sigma$ gives

$$f(\lambda) = \begin{bmatrix} 30-\lambda & 6 & 5 \\ 6 & 30-\lambda & 9 \\ 5 & 9 & 30-\lambda \end{bmatrix}$$

Now assuming

$$\lambda_0 = 43, \qquad \lambda_1 = 44$$

permits the determination of the functional values and ultimately the improved eigenvalue. That is,

$$f(\lambda_0) = \begin{vmatrix} -13 & 6 & 5 \\ 6 & -13 & 9 \\ 5 & 9 & -13 \end{vmatrix} = 189$$

$$f(\lambda_1) = \begin{vmatrix} -14 & 6 & 5 \\ 6 & -14 & 9 \\ 5 & 9 & -14 \end{vmatrix} = -216$$

Solving for the new approximation gives

$$\lambda_2 = 44 - (-216)\frac{43 - 44}{189 + 216}$$

$$\lambda_2 = 43.4667$$

This values compares with the exact value of 43.49 after just one iteration. Consequently, repeating this procedure for $k = 2$ and taking

$$\lambda_1 = 44: \qquad f(\lambda_1) = -216$$

$$\lambda_2 = 43.4667: \qquad f(\lambda_2) = 10.072$$

yields the second approximation to the root to give

$$\lambda_3 = 43.4667 - (10.072)\left|\frac{44 - 43.4667}{-216 + 10.072}\right|$$

$$= 43.4928$$

Obviously, this value is extremely close to the exact value. In fact, the value of the characteristic equation as given by $f(\lambda)$ at $\lambda = \lambda_3$ is equal to $-0.4629$. Clearly, the computed value can be improved further and the original matrix is deflated until all of the eigenvalues have been determined.

**SUMMARY OF POLYNOMIAL ITERATION**

| STEP | OPERATION | SYMBOL |
|---|---|---|
| 1 | Evaluate the largest eigenvalue using the vector iteration technique. | $\lambda_0$ = largest value |
| 2 | Assume a second eigenvalue close to the approximate value evaluated in step 1. | $\lambda_1 = \lambda_0 \pm \Delta\lambda$ |
| 3 | Compute the improved eigenvalue using equation (7.63). | $\lambda_2 = \lambda_1 - f(\lambda_1)\dfrac{\lambda_0 - \lambda_1}{f(\lambda_0) - f(\lambda_1)}$ where $f(\lambda) = \lvert Q\rvert$ |
| 4 | Establish a tolerance limit on the function being considered, then check if the criterion is satisfied, then end; otherwise let $\lambda_0 = \lambda_1, \lambda_1 = \lambda_2$. | $T \geq \lVert f(\lambda_1)\rvert - \lvert f(\lambda_0)\rVert$ |
| 5 | Go to step 3. | |

## 7.6 TRANSFORMATION METHODS

These methods involve the transformation of the original $[Q]$ matrix into a new matrix having the same eigenvalues but a simpler form. Clearly, the best form for the transformed matrix would be a diagonal one. Unfortunately, many of the methods covered in this section do not achieve this form. Furthermore, these techniques are generally limited to symmetrical matrices with real coefficients.

### 7.6.1 Jacobi Method

The basic Jacobi method was proposed over a century ago for solving the so-called standard eigenproblems. These eigenproblems involve symmetric $[A]$ matrices and identity $[B]$ matrices. That is,

$$[[A] - \lambda[I]]\{x\} = \{0\}$$

This form represents the simplest eigenproblem that one might expect to encounter in practice. Fortunately, many engineering systems result in such a form.

The Jacobi method permits the transformation of a symmetrical $[A]$ matrix into a diagonal one having the same eigenvalues as the original $[A]$ matrix. This is accomplished by eliminating each off-diagonal element in a systematic fashion. Unfortunately, the method requires an infinite number of iterations to produce the diagonal form. This is because the reduction of a given element to zero in a matrix will most likely introduce a nonzero element into a previous zero coefficient. Hence the method can be viewed as an iterative procedure that can approach a diagonal form using a finite number of steps. The implication is that the off-diagonal coefficients will be close to zero rather than exactly equal to zero. This technique is simple to use and provides answers to eigenproblems involving positive, negative, and/or even zero eigenvalues. The basis for the method is outlined as follows. Given

$$[A]\{X\} = \lambda\{X\} \tag{7.65}$$

where $[A]$ is a symmetrical matrix of order $n \times n$, let us suppose that the solution to equation (7.65) produces the eigenvalues $\lambda_1, \ldots, \lambda_n$ and their corresponding vectors $\{V_1\}, \ldots, \{V_n\}$; then the following equalities hold:

$$[A]\{V_1\} = \lambda_1\{V_1\}$$
$$[A]\{V_2\} = \lambda_2\{V_2\}$$
$$\vdots$$
$$[A]\{V_n\} = \lambda_n\{V_n\}$$

In compact form this gives

$$[A][V] = [V][\lambda] \tag{7.66}$$

where

$$[V] = [\{V_1\} \cdots \{V_n\}]$$

$$[\lambda] = \begin{bmatrix} \lambda_1 & & 0 \\ & \ddots & \\ 0 & & \lambda_n \end{bmatrix}$$

It can be shown that since $[A]$ is symmetrical, then its eigenvectors are orthogonal. The implication is that

$$[V]^t = [V]^{-1}$$

Consequently premultiplying equation (7.66) by $[V]^t$ gives

$$[V]^t[A][V] = [\lambda] \tag{7.67}$$

That is, if $[A]$ is operated on as indicated by equation (7.67), then the eigenvalues are readily determined as the diagonal coefficients of the transformed matrix. In practice, the $[V]$ matrix is constructed iteratively. The basic computational scheme is to reduce the original $[A]$ matrix to a diagonal form using real-plane rotation matrices as follows:

$$[A_1] = [A]$$
$$[A_2] = [P_1]^t[A_1][P_1]$$
$$[A_3] = [P_2]^t[A_2][P_2] = [P_2]^t[P_1]^t[A_1][P_1][P_2]$$
$$\vdots$$
$$[A_k] = [P_{k-1}]^t \cdots [P_1]^t[A][P_1] \cdots [P_{k-1}]$$

Consequently as $k \to \infty$ then

$$[A_k] \to [\lambda]$$
$$[P_1] \cdots [P_{k-1}] \to [V]$$

The matrix $[P_k]$ is a plane rotation matrix which is constructed in such a way that an off-diagonal coefficient in $[A_k]$ is reduced to zero. Hence, if the $a_{rs}$ coefficient is to be reduced to zero, then $[P_k]$ is given as

$$[P_k] = \begin{bmatrix} 1 & & & & \\ & \ddots & & & \\ & & \cos\theta & -\sin\theta & \\ & & \sin\theta & \cos\theta & \\ & & & & \ddots & \\ & & & & & 1 \end{bmatrix} \begin{matrix} \\ \\ r \\ s \\ \\ \\ \end{matrix} \qquad \text{(columns } r, s\text{)} \tag{7.68}$$

It is evident that the $[P_k]$ matrix is an orthogonal matrix, where the various elements are determined as follows:

$$P_{rr} = P_{ss} = \cos\theta$$
$$P_{rs} = -P_{sr} = \sin\theta$$
$$P_{ii} = 1 \qquad \text{for} \quad i \neq r, s$$
$$P_{ij} = 0 \qquad \text{(elsewhere)}$$

The value of $\theta$ is selected in such a way that the $a_{rs}$ coefficient in the $[A_k]$ matrix is reduced to zero. That is,

$$\tan 2\theta = \frac{2a_{rs}}{a_{rr} - a_{ss}} \tag{7.69}$$

Theoretically there are an infinite number of $\theta$ values corresponding to the infinite $[A_k]$ matrices. However, as $\theta$ approaches zero the $[P_k]$ matrices become the identity matrix and no further transformations are required.

E

EXAMPLE 7.9

Show that equation (7.69) is true for a 2 × 2 symmetrical matrix.

Solution

Suppose that the matrix to be transformed is of the following form:

$$[A] = \begin{bmatrix} a_{11} & a_{12} \\ a_{12} & a_{22} \end{bmatrix} = [A_1]$$

Then to reduce $[A]$ to zero implies that $r = 1$ and $s = 2$. The corresponding $[P]$ matrix is then given as follows:

$$[P_1] = \begin{bmatrix} \cos\theta & -\sin\theta \\ \sin\theta & \cos\theta \end{bmatrix}$$

The corresponding $[A_2]$ matrix is now constructed to give

$$[A_2] = [P_1]^t[A_1][P_1]$$

$$\begin{bmatrix} a'_{11} & a'_{12} \\ a'_{12} & a'_{22} \end{bmatrix} = \begin{bmatrix} \cos\theta & \sin\theta \\ -\sin\theta & \cos\theta \end{bmatrix}\begin{bmatrix} a_{11} & a_{12} \\ a_{12} & a_{22} \end{bmatrix}\begin{bmatrix} \cos\theta & -\sin\theta \\ \sin\theta & \cos\theta \end{bmatrix}$$

Since our task is to reduce $a'_{12}$ to zero, carrying out the multiplication on the right-hand side and using matrix equality gives

$$a'_{12} = 0 = -\sin\theta(\cos\theta)a_{11} + (\cos^2\theta)a_{12} - a_{12}\sin^2\theta + \cos\theta(\sin\theta)a_{22}$$

Simplifying and rearranging gives

$$\frac{\sin\theta\cos\theta}{\cos^2\theta - \sin^2\theta} = \frac{a_{12}}{a_{11} - a_{22}}$$

$$\frac{\frac{1}{2}\sin 2\theta}{\cos 2\theta} = \frac{a_{12}}{a_{11} - a_{22}}$$

or more simply

$$\tan 2\theta = \frac{2a_{12}}{a_{11} - a_{22}}$$

Clearly this is exactly equivalent to equation (7.69) for $a_{rs} = a_{12}$. In fact a 2 × 2 matrix requires a single iteration for the $[A]$ matrix to become a diagonal one.

E

EXAMPLE 7.10

Approximate the eigenvalues and their corresponding vectors for the problem analyzed earlier in example 7.4.

Solution

This is a standard eigenproblem in that $[A]$ is symmetrical and $[B]$ is the identity matrix. Therefore the Jacobi transformation is applicable. Recall that

$$[A] = \begin{bmatrix} 30 & 6 & 5 \\ 6 & 30 & 9 \\ 5 & 9 & 30 \end{bmatrix} = [A_1]$$

Obviously our task is to reduce the off-diagonal coefficients to zero. We begin by reducing $A_{12} = 6$ to zero. Thus

$r = 1, s = 2$:

$$\tan 2\theta = \frac{2(6)}{30 - 30} \Rightarrow \theta = 45^\circ$$

$$[P] = \begin{bmatrix} 0.7071 & -0.7071 & 0 \\ 0.7071 & 0.7071 & 0 \\ 0 & 0 & 1 \end{bmatrix}$$

$$[A_3] = [P_2]^t[A_2][P_2] = \begin{bmatrix} 43.44 & 1.691 & 0 \\ 1.691 & 23.99 & 2.271 \\ 0 & 2.271 & 22.71 \end{bmatrix}$$

Note that as a result of reducing $a_{13}$ to zero, we have introduced an $a_{12} = 1.691$ into the previous zero coefficient at $a_{12}$. To complete the first sweep, we now reduce $a_{23} = 2.271$ to zero. Hence

$r = 2, s = 3$:

$$\tan 2\theta = \frac{2(2.271)}{23.99 - 22.71} \Rightarrow \theta = 37.10^\circ$$

$$[P_3] = \begin{bmatrix} 1 & 0 & 0 \\ 0 & 0.7976 & -0.6032 \\ 0 & 0.6032 & -0.7979 \end{bmatrix}$$

$$[A_4] = [P_3]^t[A_3][P_3] = \begin{bmatrix} 43.44 & 1.349 & -1.020 \\ 1.349 & 25.72 & 0 \\ -1.020 & 0 & 21.00 \end{bmatrix}$$

It is evident that, even though the off-diagonal elements are not all equal to zero, the diagonal elements compare rather well with the eigenvalues computed earlier. We now attempt a second sweep through the off-diagonal elements by first eliminating $a_{12} = 1.349$. Thus

$r = 1$, $s = 2$:

$$\tan 2\theta = \frac{2(1.349)}{43.44 - 25.72} \Rightarrow \theta = 4.328^\circ$$

$$[P_4] = \begin{bmatrix} 0.9971 & -0.0755 & 0 \\ 0.0755 & 0.9971 & 0 \\ 0 & 0 & 1 \end{bmatrix}$$

$$[A_5] = [P_4]^t[A_4][P_4] = \begin{bmatrix} 43.53 & 0 & -1.017 \\ 0 & 25.61 & 0.0765 \\ -1.017 & 0.0765 & 21.00 \end{bmatrix}$$

reducing $a_{13} = -1.017$ to zero gives

$r = 1$, $s = 3$:

$$\tan 2\theta = \frac{2(-1.017)}{43.53 - 21.00} \Rightarrow \theta = -2.579^\circ$$

$$[P_5] = \begin{bmatrix} 0.9990 & 0 & 0.0450 \\ 0 & 1 & 0 \\ -0.0450 & 0 & 0.999 \end{bmatrix}$$

$$[A_6] = [P_5]^t[A_5][P_5] = \begin{bmatrix} 43.58 & -0.003 & 0 \\ -0.003 & 25.61 & 0.076 \\ 0 & 0.076 & 20.95 \end{bmatrix}$$

To complete the second sweep, we now eliminate $a_{23} = 0.076$.

$r = 2$, $s = 3$:

$$\tan 2\theta = \frac{2(0.0765)}{25.61 - 20.95} \Rightarrow \theta = 0.940^\circ$$

$$[P_6] = \begin{bmatrix} 1 & 0 & 0 \\ 0 & 0.9999 & -0.0164 \\ 0 & 0.0164 & 0.9999 \end{bmatrix}$$

$$[A_7] = [P_6]^t[A_6][P_6] = \begin{bmatrix} 43.58 & -0.003 & 0 \\ -0.003 & 25.62 & 0 \\ 0 & 0 & 20.95 \end{bmatrix}$$

Obviously, the off-diagonal coefficients are close to zero and a third sweep should not be necessary. Therefore the eigenvalues are

$$\sigma_1 = 43.58, \qquad \sigma_2 = 25.62, \qquad \sigma_3 = 20.95$$

The corresponding vectors are determined as follows:

$$[V] = [P_1][P_2][P_3][P_4][P_5][P_6]$$

$$= \begin{bmatrix} 0.517 & -0.863 & 0.122 \\ 0.626 & 0.260 & -0.738 \\ 0.603 & 0.448 & 0.664 \end{bmatrix}$$

Note that the first, second, and third columns are the eigenvectors corresponding to the eigenvalues 43.48, 25.62, and 20.95, respectively. Furthermore, these eigenvectors are normalized. Why?

**SUMMARY OF THE JACOBI METHOD**

| STEP | OPERATION | SYMBOL |
|---|---|---|
| 1 | Given a symmetric $[A]$ matrix, select an off-diagonal coefficient $a_{rs} \neq 0$. | $[A] = \begin{bmatrix} a_{11} & a_{12} & \cdots & a_{1n} \\ a_{12} & a_{22} & \cdots & a_{2n} \\ \vdots & \vdots & & \vdots \\ a_{1n} & a_{2n} & & a_{nn} \end{bmatrix} = [A_1]$ |
| 2 | Calculate the angle $\theta$. | $\theta = \frac{1}{2}\tan[2a_{rs}/(a_{rr} - a_{ss})]$ |
| 3 | Construct the corresponding rotation matrix $[P_1]$. | $[P_1] = \begin{bmatrix} 1 & & & & \\ & \ddots & & & \\ & & \cos\theta & -\sin\theta & \\ & & \sin\theta & \cos\theta & \\ & & & & \ddots \\ & & & & & 1 \end{bmatrix}$ (columns $r$, $s$; rows $r$, $s$) |
| 4 | Compute a new matrix $[A_2]$. | $[A_2] = [P_1]^t[A_1][P_1]$ |
| 5 | Check all off-diagonal coefficients if they are not close to zero, then let $[A_1] = [A_2]$ and go to step 1; otherwise end. | $a_{ij} \geq T$, $i \neq j$, $i = 1, \ldots, n$, $j = 1, \ldots, n$, $T$ = tolerance |

### 7.6.2 Householder Method

This solution technique permits the transformation of a symmetrical matrix into a tridiagonal form having the same eigenvalues. It uses a more complex transformation than the Jacobi method to reduce whole rows and columns of off-diagonal coefficients to zero. However, it is significantly more efficient than the Jacobi technique in that it required a finite number of iterations. The tridiagonal form can then be used to construct a sequence of polynomials that can be solved for the eigenvalues. In fact, the Jacobi as well as other techniques can be used to calculate the eigenvalues, once the tridiagonal form is achieved. The Householder method requires the construction of reflection matrices and $n - 2$ transformations of the form.

$$[A_1] = [A]$$

$$[A_2] = [P_1]^t[A_1][P_1]$$

$$\vdots$$

$$[A_{n-2}] = [P_{n-3}]^t[A_{n-3}][P_{n-3}]$$

where $n$ is the order of the $[A]$ matrix under consideration. These equations can be summarized more simply as follows:

$$[A_{kh}] = [P_k]^t[A_k][P_k], \qquad k = 1, \ldots, n-2 \tag{7.70}$$

The $[P_k]$ matrices are the Householder transformation matrices and can be constructed from the following:

$$[P_k] = [I] - S_k\{W_k\}\{W_k\}^t \tag{7.71a}$$

$$S_k = \frac{2}{\{W_k\}^t\{W_k\}} \tag{7.71b}$$

The coefficients of the $\{W_k\}$ vector are defined in terms of the $[A]$ matrix coefficients as follows:

$$W_{ik} = \begin{cases} 0 & \text{for} \quad i = 1, 2, \ldots, k \\ a_{ik} & \text{for} \quad i = k+2, \ldots, n \end{cases}$$

$$W_{k+1,k} = a_{k+1,k} \pm \sqrt{\sum_{i=k+1}^{n} a_{ik}^2}$$

The sign preceding the square root is taken to be the same as that for the coefficient $a_{k+1,k}$. Consequently, once a tridiagonal form has been achieved, it then becomes necessary to calculate the eigenvalues. This is accomplished by expanding the following determinant:

$$\det[T] = \begin{vmatrix} A_1 - \lambda & B_2 & & & \\ B_2 & A_2 - \lambda & B_3 & & \\ & \ddots & & \ddots & \\ & & \ddots & & B_n \\ & & & B_n & A_n - \lambda \end{vmatrix} = 0$$

where $[T]$ is the tridiagonal form of the original $[A]$ matrix. The determinant $\det[T]$ is given by the following polynomials:

$$f_0(\lambda) = 1$$

$$f_1(\lambda) = A_1 - \lambda$$

$$f_m(\lambda) = (A_m - \lambda)f_{m-1}(\lambda) - B_m^2 f_{m-2}(\lambda) \quad \textbf{(7.72)}$$

That is, using the quadratic formula $(m - 2)$, one is able to identify approximations for two of the roots; therefore, for $m = 3$ a third root is approximated; etc. This information is quite valuable when using the Newton–Raphson methods for root approximations of nonlinear algebraic equations.

EXAMPLE 7.11

Reduce the $[A]$ matrix given in example 7.4 to a tridiagonal form using the Householder technique, then approximate the eigenvalues.

### Solution

We begin our solution by letting

$$[A_1] = \begin{bmatrix} 30 & 6 & 5 \\ 6 & 30 & 9 \\ 5 & 9 & 30 \end{bmatrix}$$

Consequently, for $k = 1$ we construct the $\{W_1\}$ vector as follows:

$$W_{11} = 0$$

$$W_{31} = a_{31} = 5$$

$$W_{21} = a_{21} \pm \sqrt{a_{21}^2 + a_{31}^2}$$
$$= 6 \pm \sqrt{6^2 + 5^2}$$

Note that since $a_{12}$ has a positive coefficient the positive sign must be used. That is,

$$W_{12} = 13.81$$

Therefore the $\{W_1\}$ vector is now determined to be

$$\{W_1\} = \begin{Bmatrix} 0 \\ 13.81 \\ 5 \end{Bmatrix}$$

$$S_1 = \frac{2}{0^2 + 13.81^2 + 5^2} = 0.0093$$

The reflection matrix needed for the first iteration is determined by using equation (7.71a) to give

$$[P_1] = \begin{bmatrix} 1 & 0 & 0 \\ 0 & 1 & 0 \\ 0 & 0 & 1 \end{bmatrix} - 0.0093 \begin{Bmatrix} 0 \\ 13.81 \\ 5 \end{Bmatrix} \{0 \quad 13.81 \quad 5\}$$

$$= \begin{bmatrix} 1 & 0 & 0 \\ 0 & -0.7682 & -0.6402 \\ 0 & -0.6402 & 0.7682 \end{bmatrix}$$

The tridiagonal matrix is given by equation (7.70). That is,

$$[A_2] = [P_1]^t[A_1][P_1] = \begin{bmatrix} 30 & -7.810 & 0 \\ -7.810 & 38.85 & -1.622 \\ 0 & -1.622 & 21.15 \end{bmatrix}$$

The transformed tridiagonal matrix as given by $[A_2]$ can be operated on to determine the eigenvalues using equation (7.72):

$$\begin{aligned} f_2(\lambda) &= (A_2 - \lambda)f_1(\lambda) - B_2^2 f_0(\lambda) \\ &= (38.85 - \lambda)(30 - \lambda) - (-7.81)^2(1) \\ &= \lambda^2 - 68.85\lambda + 1104.50 \end{aligned}$$

and the roots are given as

$$\lambda_1 = 43.40, \qquad \lambda_2 = 25.45$$

Clearly these roots are close to the values computed earlier. Hence for $m = 3$ we have

$$f_2(\lambda) = (A_3 - \lambda)f_2(\lambda) - (-1.622)^2 f_1(\lambda)$$

$$f_3(\lambda) = (21.15 - \lambda)(\lambda^2 - 68.85\lambda + 1104.5) - 2.63(30 - \lambda)$$

The third root is approximated as $\lambda_3 = 21.15$; better approximations to all of the roots are achieved by using $f_3(\lambda)$ with starting roots of $\lambda_1$, $\lambda_2$, and $\lambda_3$.

**E**

EXAMPLE 7.12

Reduce the following matrix to tridiagonal form using the Householder technique.

$$[A_1] = \begin{bmatrix} 7 & 1 & 2 & 1 \\ 1 & 8 & 1 & -1 \\ 2 & 1 & 3 & 1 \\ 1 & -1 & 1 & 2 \end{bmatrix}$$

Solution

Since the matrix is of order 4 × 4, two iterations are required in order to reduce the first and second columns to the appropriate tridiagonal form. Thus

For $k = 1$ (first column):

$$\{W_1\} = \begin{Bmatrix} 0 \\ 3.449 \\ 2 \\ 1 \end{Bmatrix}$$

$$S_1 = \frac{2}{0^2 + 3.449^2 + 2^2 + 1^2} = 0.11835$$

$$[P_1] = \begin{bmatrix} 1 & 0 & 0 & 0 \\ 0 & -0.4072 & -0.8164 & -0.4082 \\ 0 & -0.8164 & 0.5266 & -0.2367 \\ 0 & -0.4082 & -0.236 & 0.8816 \end{bmatrix}$$

$$[A_2] = [P_1]^t[A_1][P_1] = \begin{bmatrix} 7 & -2.448 & 0 & 0 \\ -2.448 & 4.658 & 1.564 & 1.189 \\ 0 & 1.564 & 4.780 & 2.996 \\ 0 & 1.189 & 2.996 & 3.551 \end{bmatrix}$$

In order to reduce the second column to a tridiagonal form, a second iteration is required. Thus

For $k = 2$ (second column):

$$\{W_2\} = \begin{Bmatrix} 0 \\ 0 \\ 3.5286 \\ 1.1890 \end{Bmatrix}$$

$$S_2 = \frac{2}{3.5286^2 + 1.189^2} = 0.14424$$

$$[P_2] = \begin{bmatrix} 1 & 0 & 0 & 0 \\ 0 & 1 & 0 & 0 \\ 0 & 0 & -0.7960 & -0.6052 \\ 0 & 0 & -0.6052 & 0.7960 \end{bmatrix}$$

$$[A_3] = [P_2]^t[A_2][P_2] = \begin{bmatrix} 7 & -2.448 & 0 & 0 \\ -2.448 & 4.658 & -1.464 & 0 \\ 0 & -1.964 & 7.216 & -2.089 \\ 0 & 0 & -2.089 & 1.114 \end{bmatrix}$$

Clearly the $[A_3]$ matrix is a tridiagonal one and has the same eigenvalues as the original $[A]$ matrix. Furthermore, once a coefficient is reduced to zero, it remains so in subsequent iterations.

**SUMMARY OF THE HOUSEHOLDER METHOD**

| STEP | OPERATION | SYMBOL |
|---|---|---|
| 1 | Reduce the first column of an $n \times n$ symmetrical matrix to that of a tridiagonal matrix $[A_2]$. Note $[A_1] = [A]$. | $[A_2] = [P_k]^t[A_1][P_k]$<br>$[P_k] = [I] - S_k\{W_k\}\{W_k\}^t$<br>$S_k = 2/\{W_k\}^t\{W_k\}$ |
| 2 | Check the remaining columns of $[A_2]$ and if they are not of tridiagonal form then let $[A_1] = [A_2]$ and $k = 2$. Go to step 1. | |
| 3 | Repeat step 2 for $k = 3, \ldots, n-1$. | $[A_2] = \begin{bmatrix} A_1 & \ddots & B_2 & \ddots & \\ B_2 & \ddots & A_2 & \ddots & B_n \\ & & B_n & & A_n \end{bmatrix}$ |
| 4 | Solve for the eigenvalues. | $f(\lambda) = \lvert [A_2] - \lambda[I] \rvert = 0$ |
| 5 | Solve for the eigenvectors. | |

## 7.7 FUNCTIONS OF A MATRIX

The numerical solution of ordinary and partial differential equations often involves the evaluation of functions of matrices. These functions may include raising a matrix to a power, raising a scalar to a matrix power, and other functions. In this section, two methods for evaluating such functions are examined.

### 7.7.1 Caley–Hamilton Method

This is based on the theory that every $n \times n$ matrix satisfies its own characteristic equation. For a given matrix $[A]$, the characteristic equation is determined by using the following:

$$f(\lambda) = \lvert [A] - \lambda[I] \rvert = 0$$

This is called a standard eigenproblem. In general, for an $n \times n$ matrix we have

$$f(\lambda) = a_0 + a_1\lambda + \cdots + a_{n-1}\lambda^{n-1} + \lambda^n = 0 \tag{7.73}$$

Note that the coefficient of the $\lambda^n$ term is equal to one. Now substituting the $[A]$ matrix for $\lambda$ in equation (7.73) gives

$$f([A]) = a_0[I] + a_1[A] + \cdots + a_{n-1}[A]^{n-1} + [A]^n = [0]$$

Therefore solving for $[A]^n$ yields

$$[A]^n = -a_0[I] - a_1[A] - \cdots - a_{n-1}[A]^{n-1} \tag{7.74a}$$

The implication is that the matrix $[A]^n$ can be evaluated in terms of the $[A]$ matrices of powers 1 to $n - 1$. Consequently, the results can be extended by multiplying equation (7.74a) by matrix $[A]$ to give

$$[A]^{n+1} = -a_0[A] - a_1[A]^2 - \cdots - a_{n-1}[A]^n \tag{7.74b}$$

However, since $[A]^n$ is given by equation (7.74a), the matrix $[A]^{n+1}$ can still be evaluated by using the $[A]$ matrices of powers 1 to $n - 1$. In general, then, raising a matrix to the $P$ power can be expressed as follows:

$$[A]^P = b_0[I] + b_1[A] + \cdots + b_{n-1}[A]^{n-1} \tag{7.75}$$

where $b_0, b_1, \ldots, b_{n-1}$ are functions of $a_0, a_1, \ldots, a_{n-1}$. Equation (7.75) is rather significant in that it provides an efficient means by which a matrix can be raised to any power by simply evaluating matrices of powers of $n - 1$ and less.

The $b$'s are determined rather easily by noting that

$$f(\lambda) = b_0 + b_1 + \cdots + b_{n-1}\lambda^{n-1} \tag{7.76}$$

It is evident that once the eigenvalues $\lambda_1, \ldots, \lambda_n$ are computed—assuming that they are distinct—a set of linear algebraic equations can then be formed in the unknowns $b_0, \ldots, b_{n-1}$. That is,

$$\begin{aligned} f(\lambda_1) &= b_0 + b_1\lambda_1 + \cdots + b_{n-1}\lambda_1^{n-1} \\ f(\lambda_2) &= b_0 + b_1\lambda_2 + \cdots + b_{n-1}\lambda_2^{n-1} \\ &\vdots \\ f(\lambda_n) &= b_0 + b_1\lambda_n + \cdots + b_{n-1}\lambda_n^{n-1} \end{aligned} \tag{7.77}$$

Hence, knowing the $b$'s, one can then write

$$f([A]) = \sum_{i=0}^{n-1} b_i[A]^i \tag{7.78}$$

Note that if the eigenvalues are not distinct, then the set of equations given by (7.77) are linearly dependent. In this case, a set of linearly independent equations is formed as follows: given an $[A]$ matrix of order $n \times n$ with $N$ distinct eigenvalues $\lambda_1, \lambda_2, \ldots, \lambda_n$ and an eigenvalue $\lambda_j$ of multiplicity $m$, then

$$f(\lambda_k) = \sum_{i=0}^{n-1} b_i\lambda_k^i, \qquad k = 1, \ldots, N$$

and

$$\frac{d^j}{d\lambda^j} f(\lambda) = \frac{d^j}{d\lambda^j} \sum_{i=0}^{n-1} b_i \lambda^i, \qquad j = 1, \ldots, m-1 \tag{7.79}$$

**E** EXAMPLE 7.13

Determine $f([A])$, where

$$f([A]) = [A]^P, \qquad [A] = \begin{bmatrix} 2 & 1 \\ 1 & 2 \end{bmatrix}$$

Solution

We begin by evaluating the characteristic equation and its eigenvalues as follows:

$$f(\lambda) = \left\| \begin{bmatrix} 2 & 1 \\ 1 & 2 \end{bmatrix} - \lambda \begin{bmatrix} 1 & 0 \\ 0 & 1 \end{bmatrix} \right\| = 0$$

$$f(\lambda) = \lambda^2 - 4\lambda + 3$$

Therefore the eigenvalues are

$$\lambda_1 = 1, \qquad \lambda_2 = 3$$

and using equation (7.77) we have

$$f(\lambda_1) = (1)^P = b_0 + (1)b_1$$

$$f(\lambda_2) = (3)^P = b_0 + (3)b_1$$

Solving for the unknowns $b_0$ and $b_1$ gives

$$b_0 = \tfrac{1}{2}(3 - 3^P), \qquad b_1 = \tfrac{1}{2}(3^P - 1)$$

Therefore $[A]^P$ is found by using equation (7.78). That is,

$$f([A]) = b_0[I] + b_1[A] = [A]^P$$

$$[A]^P = \tfrac{1}{2}(3 - 3^P)\begin{bmatrix} 1 & 0 \\ 0 & 1 \end{bmatrix} + \tfrac{1}{2}(3^P - 1)\begin{bmatrix} 2 & 1 \\ 1 & 2 \end{bmatrix}$$

$$[A]^P = \frac{1}{2}\begin{bmatrix} 3^P + 1 & 3^P - 1 \\ 3^P - 1 & 3^P + 1 \end{bmatrix}$$

It is evident that raising a matrix to any power has been reduced to raising scalars to that same power.

**E** EXAMPLE 7.14

Determine $f([A])$, where

$$f([A]) = [A]^P, \qquad [A] = \begin{bmatrix} 2 & 0 \\ 1 & 2 \end{bmatrix}$$

### Solution

The characteristic equation corresponding to the $[A]$ matrix is given as

$$f(\lambda) = \left\| \begin{bmatrix} 2 & 0 \\ 1 & 2 \end{bmatrix} - \lambda \begin{bmatrix} 1 & 0 \\ 0 & 1 \end{bmatrix} \right\| = 0$$

$$f(\lambda) = \lambda^2 - 4\lambda + 4$$

The eigenvalues are

$$\lambda_1 = 2, \qquad \lambda_2 = 2$$

Since these are repeated roots, the independent set of linear algebraic equations needed for determining $b_0$ and $b_1$ is formed as follows:

$$f(\lambda_1) = 2^P = b_0 + 2b_1$$

Now, using equation (7.79) we have

$$\frac{d}{d\lambda}(2)^P = \frac{d}{d\lambda}(b_0 + b_1\lambda)$$

or more simply, the second equation is given as

$$P2^{P-1} = b_1$$

Hence solving for $b_0$ yields

$$b_0 = 2^P(1 - P)$$

Then

$$[A]^P = b_0 \begin{bmatrix} 1 & 0 \\ 0 & 1 \end{bmatrix} + b_1 \begin{bmatrix} 2 & 0 \\ 1 & 2 \end{bmatrix}$$

$$= 2^P \begin{bmatrix} 1 & 0 \\ \frac{1}{2}P & 1 \end{bmatrix}$$

The solution technique presented here is not limited to that of raising a matrix to a power. In fact, any function of a matrix can be handled by implementing the technique just outlined.

### 7.7.2 Spectral Matrix Decomposition Method

This method of determining functions of a matrix is often faster than the method presented in section 7.7.1. It is based on the fact that an $n \times n$ matrix can be expressed as a linear combination of constituent matrices that are of a simpler form than the $[A]$ matrix. That is,

$$[A] = \lambda_1[C_1] + \lambda_2[C_2] + \cdots + \lambda_n[C_n]$$
$$= \sum_{i=1}^{n} \lambda_i[C_i] \tag{7.80}$$

where $[C_i]$, $i = 1, \ldots, n$, are the constituent matrices corresponding to the $[A]$ matrix. These have the following properties:

$$[C_i][C_j] = 0, \qquad i \neq j$$
$$[C_i][C_j] = [C_i], \qquad i = j$$
$$\sum_{i=1}^{n} [C_i] = [I]$$

and

$$[C_i][A] = [A][C_i] = \lambda_i[C_i], \qquad i = 1, \ldots, n$$

Using equation (7.79) we have

$$f([A]) = \sum_{i=1}^{n} f(\lambda_i)[C_i] \tag{7.81}$$

Therefore, if

$$f([A]) = [A]^P$$

then substituting into equation (7.81) yields

$$[A]^P = \lambda_1^P[C_1] + \lambda_2^P[C_2] + \cdots + \lambda_n^P[C_n] \tag{7.82}$$

Hence taking $P = 0, \ldots, n-1$ yields a set of matrix equations in $[C_i]$. These are independent for distinct eigenvalues.

**E**

EXAMPLE 7.15

Rework example 7.13 using the spectral decomposition method.

Solution

The eigenvalues were determined earlier to be

$$\lambda_1 = 1, \qquad \lambda_2 = 3$$

Therefore using equation (7.82) for $P = 0, 1$ gives

$$[I] = [C_1] + [C_2]$$
$$[A] = \lambda_1[C_1] + \lambda_2[C_2]$$

These equations can be expressed as follows:

$$\begin{Bmatrix} [I] \\ [A] \end{Bmatrix} = \begin{bmatrix} 1 & 1 \\ 1 & 3 \end{bmatrix} \begin{Bmatrix} [C_1] \\ [C_2] \end{Bmatrix}$$

Upon matrix inversion this yields the unknown constituent matrices:

$$[C_1] = \frac{1}{2}\begin{bmatrix} 1 & -1 \\ -1 & 1 \end{bmatrix}, \qquad [C_2] = \frac{1}{2}\begin{bmatrix} 1 & 1 \\ 1 & 1 \end{bmatrix}$$

Therefore using equation (7.81) and noting that $f([A]) = [A]^P$ gives

$$[A]^P = \lambda_1^P[C_1] + \lambda_2^P[C_2]$$
$$= (1)^P\begin{bmatrix} \frac{1}{2} & -\frac{1}{2} \\ -\frac{1}{2} & \frac{1}{2} \end{bmatrix} + (3)^P\begin{bmatrix} \frac{1}{2} & \frac{1}{2} \\ \frac{1}{2} & \frac{1}{2} \end{bmatrix}$$
$$= \frac{1}{2}\begin{bmatrix} 3^P + 1 & 3^P - 1 \\ 3^P - 1 & 3^P + 1 \end{bmatrix}$$

Obviously, this gives the same result we obtained earlier.

## 7.8 STATIC CONDENSATION

There are situations in which engineers must model systems in which the number of eigenvalues is less than the order of the $[A]$ matrix. While such problems can be handled by using the conventional methods discussed thus far, these are not efficient for such problems. Instead, one can take advantage of the special form these problems possess to more effectively evaluate the eigenvalues. The basic idea is to reduce the original eigenproblem to one having the same order as the number of eigenvalues. For example, given an eigenproblem of the form

$$[A]\{X\} = \lambda[B]\{X\}$$

where

$$\begin{bmatrix} [A_{11}] & [A_{12}] \\ [A_{21}] & [A_{22}] \end{bmatrix} \begin{Bmatrix} \{X\}_1 \\ \{X\}_2 \end{Bmatrix} = \lambda \begin{bmatrix} [I] & [0] \\ [0] & [0] \end{bmatrix} \begin{Bmatrix} \{X\}_1 \\ \{X\}_2 \end{Bmatrix}$$

we can easily reduce the problem by solving for $\{X\}_1$ and $\{X\}_2$ provided $[A_{11}]$ and $[A_{22}]$ are square matrices that are not singular. Thus

$$[A_{11}]\{X\}_1 + [A_{12}]\{X\}_2 = \lambda[I] \quad \textbf{(7.83a)}$$

$$[A_{21}]\{X\}_1 + [A_{22}]\{X\}_2 = [0] \quad \textbf{(7.83b)}$$

Solving equation (7.83b) for $\{X\}_2$ gives

$$\{X\}_2 = -[A_{22}]^{-1}[A_{21}]\{X\}_1 \tag{7.84}$$

Substituting equation (7.84) into equation (7.83a) yields the following reduced eigenproblem:

$$[[A_{11}] - [A_{12}][A_{22}]^{-1}[A_{21}]]\{X\}_1 = \lambda[I] \tag{7.85}$$

or more simply

$$[A]_r\{X\}_1 = \lambda[I] \tag{7.86}$$

It is evident that once the eigenproblem given by equation (7.85) is solved, the complete solution for the eigenvectors is given by equation (7.84).

EXAMPLE 7.16

Evaluate the characteristic equation for the following eigenproblem using the method of static condensation:

$$\begin{bmatrix} 2 & -1 & 0 \\ -1 & 2 & -1 \\ 0 & -1 & 2 \end{bmatrix} \begin{Bmatrix} x_1 \\ x_2 \\ x_3 \end{Bmatrix} = \lambda \begin{bmatrix} 1 & 0 & 0 \\ 0 & 1 & 0 \\ 0 & 0 & 0 \end{bmatrix} \begin{Bmatrix} x_1 \\ x_2 \\ x_3 \end{Bmatrix}$$

### Solution

Obviously, a solution is possible by expanding the corresponding 3 × 3 determinant. On the other hand, since this problem involves two eigenvalues only (why?) we shall proceed by first reducing it to an equivalent 2 × 2 problem. That is,

$$\begin{bmatrix} \begin{bmatrix} 2 & -1 \\ -1 & 2 \end{bmatrix} & \begin{Bmatrix} 0 \\ -1 \end{Bmatrix} \\ \{0 \quad -1\} & [2] \end{bmatrix} \begin{Bmatrix} \begin{Bmatrix} x_1 \\ x_2 \end{Bmatrix} \\ \{x_3\} \end{Bmatrix} = \lambda \begin{bmatrix} \begin{bmatrix} 1 & 0 \\ 0 & 1 \end{bmatrix} & \begin{Bmatrix} 0 \\ 0 \end{Bmatrix} \\ \{0 \quad 0\} & \{0\} \end{bmatrix} \begin{Bmatrix} \begin{Bmatrix} x_1 \\ x_2 \end{Bmatrix} \\ \{x_3\} \end{Bmatrix}$$

Therefore, using equation (7.86), we have

$$[A]_r = \begin{bmatrix} 2 & -1 \\ -1 & 2 \end{bmatrix} - \begin{Bmatrix} 0 \\ 1 \end{Bmatrix} [2]^{-1} \{0 \quad -1\}$$

$$= \begin{bmatrix} 2 & -1 \\ -1 & 1.5 \end{bmatrix}$$

The reduced eigenproblem is now given by equation (7.85) as follows:

$$\begin{bmatrix} 2 & -1 \\ -1 & 1.5 \end{bmatrix}\begin{Bmatrix} x_1 \\ x_2 \end{Bmatrix} = \lambda\begin{bmatrix} 1 & 0 \\ 0 & 1 \end{bmatrix}\begin{Bmatrix} x_1 \\ x_2 \end{Bmatrix}$$

and the characteristic equation is given as

$$\lambda^2 - 3.5\lambda + 2 = 0$$

This is precisely the same equation we would have gotten had we solved the original $3 \times 3$ eigenproblem directly.

## Recommended Reading


Dynamic analysis of structural systems using component modes. W. C. Hurty, *A.I.A.A. Journal*, **3** (1965).

*A First Course in Numerical Analysis*, A. Ralson, McGraw-Hill Book Company, New York, 1965.

*The Algebraic Eigenvalue Problem*, J. H. Wilkinson, Clarendon Press, Oxford, 1965.

Vibration analysis by dynamic partitioning. R. L. Goldmann, *A.I.A.A. Journal*, **7** (1969).

*Numerical Algorithms—Origins and Applications.* Bruce W. Arden and Kenneth N. Astill, Addison-Wesley Publishing Company, Reading, Mass., 1970.

An economical method for determining the smallest eigenvalues of large linear systems. G. C. Wright and G. A. Miles, *International Journal for Numerical Methods in Engineering*, **3** (1971).

*Numerical Computation*, P. W. William, Harper and Row, New York, 1972.

*Numerical Methods in Finite Elements Analysis*, Klaus-Jurgen Bathe and Edward L. Wilson, Prentice-Hall, Inc., Englewood Cliffs, N.J., 1976.

*A Practical Guide to Computer Methods for Engineers*, Terry E. Shoup, Prentice-Hall, Inc., Englewood Cliffs, N.J., 1979.


## P PROBLEMS

**7.1** Evaluate the following determinants using the direct method of expansion.

**(a)** $\det[A] = \begin{vmatrix} 2-\lambda & 3 \\ 8 & 2 \end{vmatrix}$

**(b)** $\det[B] = \begin{vmatrix} 2-\lambda & 3 \\ 8 & 2-\lambda \end{vmatrix}$

(c) $\det[C] = \begin{vmatrix} 3-\lambda & -2 & 0 \\ -2 & 5-\lambda & -2 \\ 0 & -2 & 2-\lambda \end{vmatrix}$

**7.2** Rework problem 7.1 using the indirect determinant expansion method.

**7.3** Evaluate the eigenvalues and their corresponding vectors for the characteristic equations obtained from problem 7.1.

**7.4** Evaluate the first two buckling loads and corresponding buckling modes for a column fixed at both ends. Assume $EI$ constant and column length $L$.

**7.5** Evaluate the natural frequencies of vibration and the corresponding modes for the mechanical system shown below.

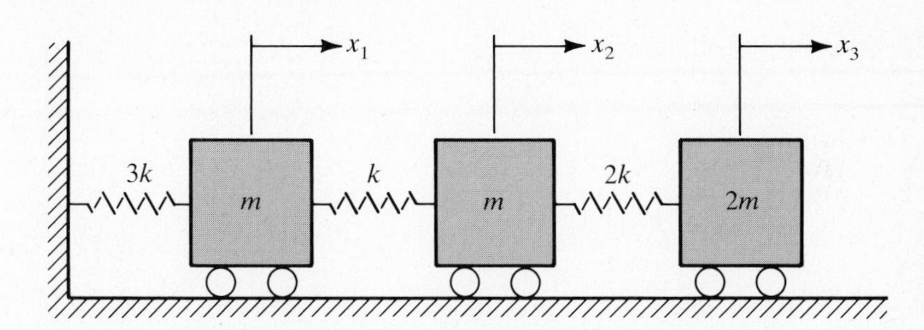


Normalize the eigenvector so that the largest entry is 1.0.

**7.6** Evaluate the natural frequencies and the corresponding modes for the electrical circuit shown below.

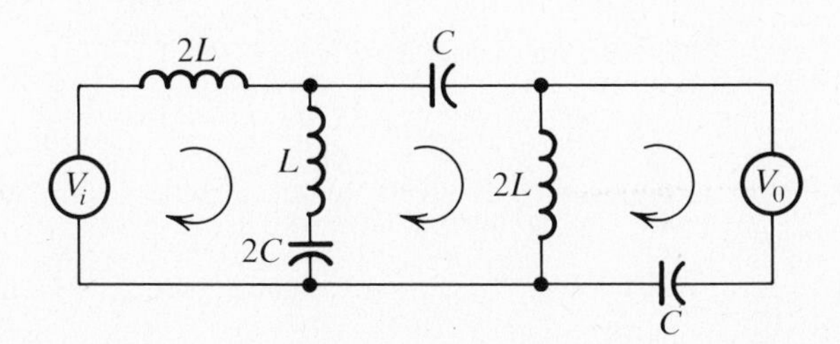


**7.7** Use the vector iteration technique to find the largest eigenvalues for the following:

(a) $[A] = \begin{bmatrix} 1 & -2 & 0 \\ -2 & 8 & -2 \\ 0 & -2 & 5 \end{bmatrix}$ (b) $[B] = \begin{bmatrix} 5 & -2 & 2 \\ -2 & 8 & -2 \\ 2 & -2 & 5 \end{bmatrix}$

Assume a unit vector initially and perform at least five iterations.

**7.8** Evaluate the smallest eigenvalues of the matrices given in problem 7.7.

**7.9** Evaluate the intermediate eigenvalue for the matrices given in problem 7.7.

**7.10** Use the Jacobi method to evaluate the eigenvalues and the normalized eigenvectors of the matrices given in problem 7.7.

**7.11** Reduce the following matrix to a tridiagonal form using the Householder transformation.

$$[A] = \begin{bmatrix} 5 & -3 & -2 \\ -3 & 7 & -1 \\ -2 & -1 & 8 \end{bmatrix}$$

**7.12** Evaluate $[A]^{10}$ using the Caley–Hamilton method for $[A]$ given in problem 7.11.

**7.13** Rework problem 7.12 using the spectral matrix decomposition approach.

**7.14** Solve the following set using the static condensation method:

$$4x_1 + 2x_2 - x_3 = \lambda x_1$$

$$2x_1 + 5x_2 + 3x_3 = \lambda x_2$$

$$-x_1 + 4x_2 + 7x_3 = 0$$

# chapter 8

# Interpolation

A view of the space shuttle sailing in space. *Courtesy of N.A.S.A.*

"Interpolating functions form the basis for many numerical techniques."

## 8.1 INTRODUCTION

Interpolation is a rational process generally used in estimating a missing functional value by taking a weighted average of known functional values at neighboring points. This brief dictionary definition falls short of describing the versatility and the many uses of interpolation techniques in engineering practice. In dealing with tabular quantities, three basic problems are encountered. First, given a mathematical relationship in tabular form, one may wish to extend its range beyond that given by the original data. Second, one may wish to approximate a functional value between two data points. Third, one may wish to approximate an independent variable corresponding to a given functional value. These problems and others occur quite frequently in engineering practice, especially when one is dealing with field and/or laboratory experimental data. Furthermore, interpolation techniques play a significant role in the numerical solution of ordinary and partial differential equations.

## 8.2 INTERPOLATING POLYNOMIALS FOR EVEN INTERVALS

Suppose we are given a set of $n$ data points relating a dependent variable $f(x)$ to an independent variable $x$ as follows:

| $x_i$ | $x_0$ | $x_1$ | $\cdots$ | $x_n$ |
|---|---|---|---|---|
| $f(x_i)$ | $f(x_0)$ | $f(x_1)$ | $\cdots$ | $f(x_n)$ |

Generally, the base points $x_0, \ldots, x_n$ are arbitrary; however, let us assume that the interval between two adjacent points is fixed. Furthermore, assume that the data are organized in such a way that $x_0 < x_1 < \cdots < x_n$. Hence, if we designate the interval as $h$, then

$$h = x_1 - x_0 = x_2 - x_1 = \cdots = x_n - x_{n-1}$$

In general,

$$h = x_{i+1} - x_i, \qquad i = 0, \ldots, n$$

The task at hand is to determine a continuous function that passes through the $n$ data points. It is evident that a variety of functions can be assumed. That is, one can assume a function with as many unknown coefficients as the number of data points ($n$). Probably the most common form for the function is that of a polynomial. This is because any function can be expressed as a Taylor series. Hence

$$f(x) \approx a_0 + a_1 x + \cdots + a_n x^n \tag{8.1}$$

Note that in general, we have no idea as to the form of the true function. Therefore we shall always assume that the derived function $p(x)$ is only an approximation to the true function $f(x)$. Consequently, a set of $n + 1$ linear algebraic equations in the unknown coefficients $a_0, \ldots, a_n$ is formed as follows:

$$\begin{aligned} f(x_0) &= a_0 + a_1 x_0 + \cdots + a_n x_0^n \\ f(x_1) &= a_0 + a_1 x_1 + \cdots + a_n x_1^n \\ &\vdots \\ f(x_n) &= a_0 + a_1 x_n + \cdots + a_n x_n^n \end{aligned} \tag{8.2}$$

These equations can be solved for the unknowns, which are then substituted into equation (8.1) to yield the desired interpolating function.

### 8.2.1 Forward Interpolation

Consider, for example, the situation of three equally spaced base points 0, $h$, and $2h$ with corresponding functional values of $f_0$, $f_1$, and $f_2$ as shown in figure 8.1.

Obviously, a second-order interpolating polynomial can be determined so that it passes through the three data points. Such a polynomial is referred to as a forward interpolating function because the base point $x_0 = 0$ is related to points forward of it. Thus, using equation (8.1) for $n = 2$, we have

$$f(x) \approx a_0 + a_1 x + a_2 x^2 \tag{8.3}$$

Substituting the three data points into equation (8.3) yields three linear algebraic equations as follows:

$$\begin{aligned} f_0 &= a_0 \\ f_1 &= a_0 + ha_1 + h^2 a_2 \\ f_2 &= a_0 + 2ha_1 + 4h^2 a_2 \end{aligned}$$

Note that $f_0 = f(x_0)$, $f_1(x_1)$, and $f_2 = f(x_2)$. Consequently, expressing the set in matrix form gives

$$\begin{Bmatrix} f_0 \\ f_1 \\ f_2 \end{Bmatrix} = \begin{bmatrix} 1 & 0 & 0 \\ 1 & 1 & 1 \\ 1 & 2 & 4 \end{bmatrix} \begin{Bmatrix} a_0 \\ a_1 h \\ a_2 h^2 \end{Bmatrix}$$

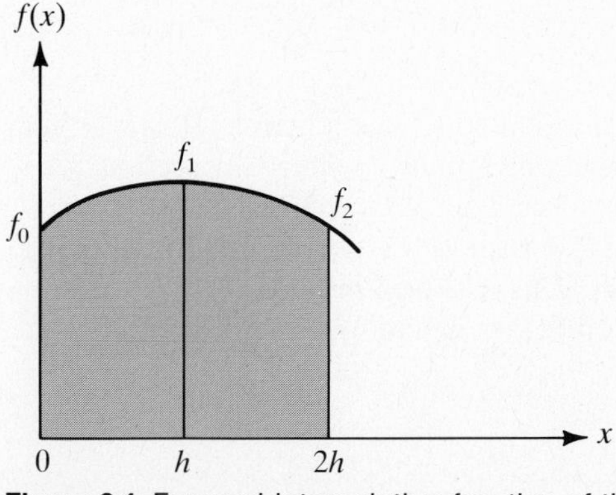


**Figure 8.1** Forward interpolating function of the second order.

The unknowns are readily determined by matrix inversion to give

$$\begin{Bmatrix} a_0 \\ a_1 h \\ a_2 h^2 \end{Bmatrix} = \frac{1}{2}\begin{bmatrix} 2 & 0 & 0 \\ -3 & 4 & -1 \\ 1 & -2 & 4 \end{bmatrix}\begin{Bmatrix} f_0 \\ f_1 \\ f_2 \end{Bmatrix}$$

or more simply

$$a_0 = f_0$$

$$a_1 = \frac{1}{2h}(-3f_0 + 4f_1 - f_2)$$

$$a_2 = \frac{1}{2h^2}(f_0 - 2f_1 + f_2)$$

These values can now be substituted into the assumed second-order polynomial equation (8.3) to get the interpolating function:

$$f(x) = f_0 + \frac{1}{2h}(-3f_0 + 4f_1 - f_2)x + \frac{1}{2h^2}(f_0 - 2f_1 + f_2)x^2 \qquad \textbf{(8.4)}$$

Equation (8.4) is called a "forward interpolation formula." It can be used to approximate functional values at any $x$ value. Clearly this function passes through the three data points used in its derivation. That is,

$$\text{at } x = 0: \qquad f(0) = f_0$$

$$\text{at } x = h: \qquad f(h) = f_1$$

$$\text{at } x = 2h: \qquad f(2h) = f_2$$

Furthermore, since the function is continuous, it can be used to approximate functional values at other $x$ values such that $0 \le x \le 2h$. This is called interpolation. In addition, it can be used to estimate functional values at $x > 2h$ and $x < 0$. This is referred to as extrapolation. In fact, the function can also be integrated or differentiated. Obviously, such results may be in doubt if the true function is not a second-order polynomial.

### 8.2.2 Backward Interpolation

Obviously, polynomial interpolation is not limited to one function. However, for clarity let us reconsider the second-order polynomial assumed in section 8.2.1 with one modification. That is, given three base points $0, -h, -2h$ with the corresponding functional values $f_0, f_{-1}, f_{-2}$ we wish to determine an interpolating function that passes through the data set. This is shown in figure 8.2.

Once again, substituting the data points into equation (8.3) gives

$$f_0 = a_0$$

$$f_{-1} = a_0 - ha_1 + h^2 a_2$$

$$f_{-2} = a_0 - 2ha_1 + 4h^2 a_2$$

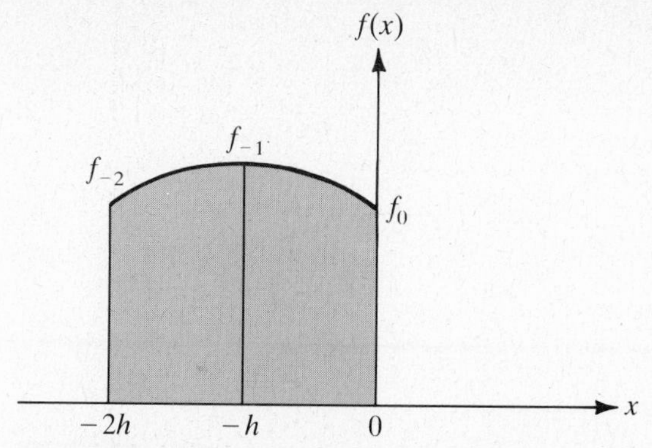

**Figure 8.2** Backward interpolating function of the second order.

Solving for the unknowns and then substituting into equation (8.3) yields the "backward interpolating formula" for a second-order polynomial. Thus

$$\begin{Bmatrix} a_0 \\ a_1 h \\ a_2 h^2 \end{Bmatrix} = \frac{1}{2} \begin{bmatrix} 2 & 0 & 0 \\ 3 & -4 & 1 \\ 1 & -2 & 1 \end{bmatrix} \begin{Bmatrix} f_0 \\ f_{-1} \\ f_{-2} \end{Bmatrix}$$

and

$$f(x) = f_0 + \frac{1}{2h}(f_{-2} - 4f_{-1} + 3f_0)x + \frac{1}{2h^2}(f_{-2} - 2f_{-1} + f_0)x^2 \tag{8.5}$$

It is evident that equation (8.5) gives the exact functional values $f_0$, $f_{-1}$, and $f_{-2}$ at $x = 0$, $-h$, and $-2h$, respectively.

### 8.2.3 Central Interpolation

In this section a second-order interpolating function is derived by using the base points $-h$, 0, $h$ and their corresponding functional values $f_{-1}$, $f_0$, and $f_1$, respectively. This is shown graphically in figure 8.3.

Obviously, central interpolation is possible for even-order polynomials only. This is because an odd-order polynomial possesses an even number of unknown coefficients, thus making it impossible to have a central base point. Substituting the data points into equation (8.3) gives

$$f_{-1} = a_0 - ha_1 + h^2 a_2$$

$$f_0 = a_0$$

$$f_1 = a_0 + ha_1 + h^2 a_2$$

Solving for the unknowns using matrix inversion gives

$$\begin{Bmatrix} a_0 \\ a_1 h \\ a_2 h^2 \end{Bmatrix} = \frac{1}{2} \begin{bmatrix} 0 & 2 & 0 \\ -1 & 0 & 1 \\ 1 & -2 & 1 \end{bmatrix} \begin{Bmatrix} f_{-1} \\ f_0 \\ f_1 \end{Bmatrix}$$

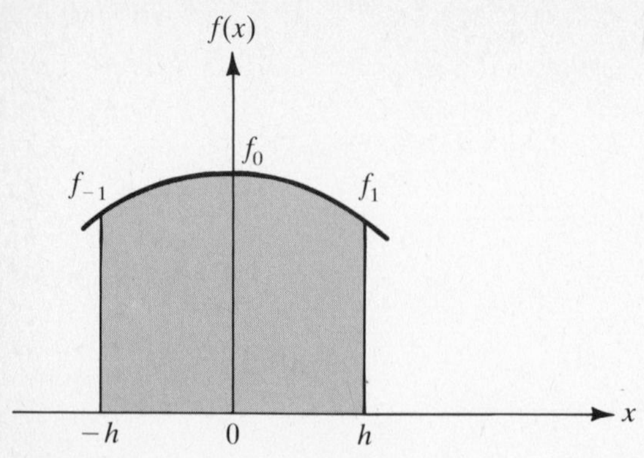


**Figure 8.3** Central interpolating function of the second order.

Substituting $a_0$, $a_1$, and $a_2$ into equation (8.3) gives the "central interpolating formula." That is,

$$f(x) = f_0 + \frac{1}{2h}(-f_{-1} + f_1)x + \frac{1}{2h^2}(f_{-1} - 2f_0 + f_1)x^2 \tag{8.6}$$

Interpolation functions of higher order can be derived by using the procedure just described. Furthermore, the arrangement of the various terms in the interpolation equations can take on different forms. For example, collection of terms can be made with respect to the known functional values. This can be achieved rather easily by expressing the assumed interpolating polynomial in matrix form. Hence for the second-order function given by equation (8.3) we have

$$f(x) = \{1 \quad x \quad x^2\}\begin{Bmatrix} a_0 \\ a_1 \\ a_2 \end{Bmatrix} \tag{8.7}$$

Therefore the forward interpolating function can be easily determined by noting that

$$\begin{Bmatrix} a_0 \\ a_1 \\ a_2 \end{Bmatrix} = \frac{1}{2}\begin{bmatrix} 2 & 0 & 0 \\ -3/h & 4/h & -1/h \\ 1/h^2 & -2/h^2 & 1/h^2 \end{bmatrix}\begin{Bmatrix} f_0 \\ f_1 \\ f_2 \end{Bmatrix}$$

and substituting into equation (8.7) gives

$$\begin{aligned} f(x) &= \frac{1}{2}\{1 \quad x \quad x^2\}\begin{bmatrix} 2 & 0 & 0 \\ -3/h & 4/h & -1/h \\ 1/h^2 & -2/h^2 & 1/h^2 \end{bmatrix}\begin{Bmatrix} f_0 \\ f_1 \\ f_2 \end{Bmatrix} \\ &= \frac{1}{2}\left(2 - \frac{3x}{h} + \frac{x^2}{h^2}\right)f_0 + \left(2\frac{x}{h} - \frac{x^2}{h^2}\right)f_1 + \frac{1}{2}\left(\frac{x}{h} + \frac{x^2}{h^2}\right)f_2 \end{aligned} \tag{8.8}$$

It is evident that equation (8.8) is exactly equivalent to equation (8.4). In fact, it could have been derived by operating directly on equation (8.4). Furthermore, the coefficients of the functional values are second-order polynomials. Therefore one may express the interpolating formula more conveniently as follows:

$$f(x) = N_0 f_0 + N_1 f_1 + N_2 f_2 = \sum_{i=0}^{2} N_i f_i \tag{8.9a}$$

where

$$N_0 = \frac{1}{2}\left(2 - \frac{3x}{h} + \frac{x^2}{h^2}\right) = N_0(x) \tag{8.9b}$$

$$N_1 = \frac{2x}{h} - \frac{x^2}{h^2} = N_1(x) \tag{8.9c}$$

$$N_2 = \frac{1}{2}\left(-\frac{x}{h} + \frac{x^2}{h^2}\right) = N_x(x) \tag{8.9d}$$

The functions $N_0$, $N_1$, and $N_2$ are called "shape functions," weighting functions which have unique properties. These properties are

at $x = 0$: $\quad N_0 = 1, \quad N_1 = 0, \quad N_2 = 0$

at $x = h$: $\quad N_0 = 0, \quad N_1 = 1, \quad N_2 = 0$

at $x = 2h$: $\quad N_0 = 0, \quad N_1 = 0, \quad N_2 = 1$

Furthermore, for any $x$ value

$$N_0 + N_1 + N_2 = 1$$

This is precisely why they are considered weight functions: for any $x$ value the interpolated functional value is determined as a weighted average of the functional values at $x = 0$, $h$, $2h$. The shape functions for the forward interpolating formula given by equation (8.9) are shown graphically in figure 8.4.

In general, for a polynomial of the $n$th order, the corresponding interpolating function is given as

$$f(x) = N_0 f_0 + N_1 f_1 + \cdots + N_n f_n = \sum_{i=0}^{n} N_i f_i \tag{8.10a}$$

and the shape function's properties are

$$\sum_{i=1}^{n} N_i = 1 \tag{8.10b}$$

$$N_i(x_j) = 0 \quad \text{for} \quad i \neq j \tag{8.10c}$$

$$= 1 \quad \text{for} \quad i = j \tag{8.10d}$$

Note that the shape functions have the same order as the assumed polynomial. Furthermore, backward as well as central interpolating formulas can be derived

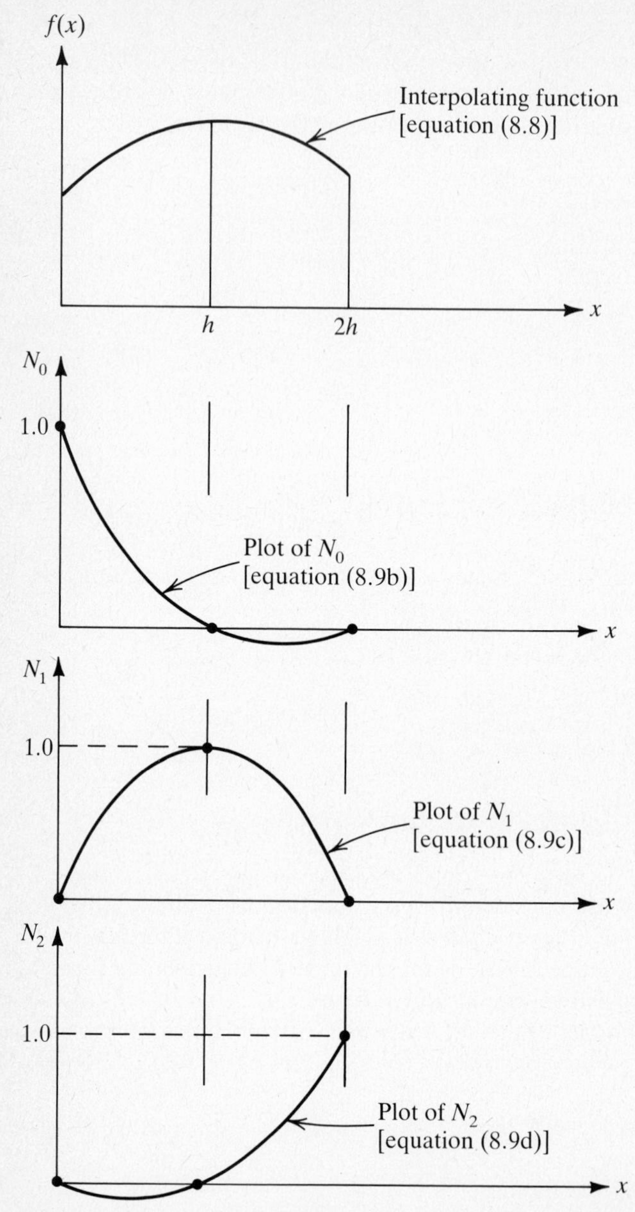


**Figure 8.4** Shape functions for a second-order forward interpolating function.

with shape functions having the same properties as those given by equation (8.10). Interpolating formulas expressed in terms of shape functions play an important role in the finite element methods and provide the basis for interpolation through unevenly spaced base points.

E

## EXAMPLE 8.1

Given the following data, approximate the functional value at $x = 5$ using the forward, backward, and central interpolating formulas derived earlier.

| $x$ | 2 | 4 | 6 |
|---|---|---|---|
| $f(x)$ | 8 | 2 | 8 |

### Solution

Note that if the number of data points exceeds three, then a higher-order interpolating function must be derived if accurate results are to be achieved.

*Forward Interpolation*

We begin by first transforming the data set so that it corresponds to equation (8.4). That is, $x_0 = 0$, $x_1 = h$, and $x_2 = 2h$. This can be accomplished by subtracting 2.0 from each base point to give

| | GIVEN DATA | | TRANSFORMED DATA | |
|---|---|---|---|---|
| $i$ | $x_i$ | $f(x_i) = f_i$ | $(x_i)_{tr}$ | $f(x_i) = f_i$ |
| 0 | 2 | 8 | 0 | 8 |
| 1 | 4 | 2 | 2 | 2 |
| 2 | 6 | 8 | 4 | 8 |
| | 5 | ? | 3 | ? |

Substituting into equation (8.4) and noting that $h = 2$ gives

$$f(x) = 8 - 6x_{tr} + 1.5x_{tr}^2$$

Therefore, at $x_{tr} = 3$ ($x = 5$) we have

$$f(3) = 3.5$$

Note that the interpolating polynomial can be easily expressed in terms of the original data by substituting $x_{tr} = x - 2$ into the expression to give

$$f(x) = 26 - 12x + 1.5x^2$$

*Backward Interpolation*

Transforming the data in such a way that $x_6 = 0$, $x_{-1} = -h$, and $x_{-2} = -2h$ permits the application of equation (8.5). This is accomplished by subtracting

6.0 from each $x$ value to give

| $i$ | GIVEN DATA $x_i$ | $f(x_i) = f_i$ | TRANSFORMED DATA $(x_i)_{tr}$ | $f(x_i) = f_i$ |
|---|---|---|---|---|
| −2 | 2 | 8 | −4 | 8 |
| −1 | 4 | 2 | −2 | 2 |
| 0 | 6 | 8 | 0 | 8 |
| | 5 | ? | −1 | ? |

Substituting the transformed data into equation (8.5) and noting that $h = 2$ gives

$$f(x) = 8 + 6x_{tr} + 1.5x_{tr}^2$$

Therefore at $x_{tr} = -1$ ($x = 5$) we have

$$f(-1) = 3.5$$

This is exactly the same answer we have determined earlier using forward interpolation. In fact, the result should have been expected because the second-order polynomials pass through the set of three given data points. Consequently, for the central interpolating formula we have

*Central Interpolation*

The transformation is accomplished by subtracting 4 from every $x$ value so that $x_0 = 0$, $x_{-1} = -h$, and $x_1 = h$. Thus

| $i$ | GIVEN DATA $x_i$ | $f(x_i) = f_i$ | TRANSFORMED DATA $(x_i)_{tr}$ | $f(x_i) = f_i$ |
|---|---|---|---|---|
| −1 | 2 | 8 | −2 | 8 |
| 0 | 4 | 2 | 0 | 2 |
| 1 | 6 | 8 | 2 | 8 |
| | 5 | ? | 1 | ? |

Substituting the transformed values into equation (8.6) and noting that $h = 2$ gives

$$f(x) = 2 + 1.5x_{tr}^2$$

Hence at $x_{tr} = 1$ ($x = 5$) we have

$$f(1) = 3.5$$

Note that the exact same results are obtained using shape functions. That is, using equation (8.9) for the forward interpolation formula

$$f(x) = 8N_0 + 2N_1 + 8N_2$$

and at $x_{tr} = 3$ we obtain for the shape functions

$$N_0(3) = -\tfrac{1}{8}, \qquad N_1(3) = \tfrac{3}{4}, \qquad N_2(3) = \tfrac{3}{8}$$

and

$$f(3) = -1 + \tfrac{3}{2} + 3 = 3.5$$

Clearly the sum of the shape functions is 1.0 as it should be.

## 8.3 DIFFERENCE OPERATORS AND DIFFERENCE TABLES

The basic idea of determining a polynomial which passes through a set of data points is relatively simple and straightforward. Unfortunately, the methods described in section 8.2 tend to become cumbersome and do at times obscure some useful relationships, especially when the order of the assumed polynomial is large. In fact, as the number of data points increases, the resulting set of linear algebraic equations in the unknown coefficients tends to become unstable. Furthermore, solving simultaneous linear algebraic equations is computationally more complex than forming interpolating functions using the methods presented in this section. Hence interpolations for equally spaced base points are considered.

### 8.3.1 Forward Differences

Suppose that a table relating a dependent variable $f(x)$ to an independent variable $x$ is given as

| $x_i$ | $x_0$ | $x_i$ | $\cdots$ | $x_n$ |
|---|---|---|---|---|
| $f(x_i)$ | $f_0$ | $f_i$ | $\cdots$ | $f_n$ |

and that the base points are equally spaced and are arranged in increasing order. That is,

$$h = x_{i+1} - x_i, \qquad i = 0, \ldots, n$$

and

$$x_0 < x_1 < \cdots < x_n$$

Forward differences for the various known functional values can be established as follows:

$$\begin{aligned}
\Delta f_0 &= f_1 - f_0 \\
\Delta f_1 &= f_2 - f_1 \\
&\vdots \\
\Delta f_n &= f_{n+1} - f_n
\end{aligned}$$

These are first forward differences and $\Delta$ is referred to as the forward difference operator. In general,

$$\Delta f_i = f_{i+1} - f_i, \qquad i = 0, \ldots, n \tag{8.11}$$

Furthermore, higher difference expressions can be readily defined by using equation (8.11). For example, the second forward differences are given as

$$\begin{aligned} \Delta^2 f_0 &= \Delta(f_1 - f_0) \\ \Delta^2 f_1 &= \Delta(f_2 - f_1) \\ &\vdots \\ \Delta^2 f_n &= \Delta(f_{n+1} - f_n) \end{aligned}$$

Therefore the second forward differences can be expressed more conveniently as

$$\Delta^2 f_i = \Delta(f_{i+1} - f_i) \tag{8.12}$$

In general, for the $n$th forward difference we have

$$\Delta^n f_i = \Delta^{n-1}(f_{i+1} - f_i) \tag{8.13}$$

Equation (8.13) can be used to develop forward difference expressions of any order. These differences are then used to interpolate between functional values. For illustration purposes, let us develop the forward differences for a fourth-order polynomial. Using equation (8.13) we have for any node $i$

First forward difference ($n = 1$):

$$\Delta f_i = f_{i+1} - f_i \tag{8.14}$$

Second forward difference ($n = 2$):

$$\begin{aligned} \Delta^2 f_i &= \Delta(f_{i+1} - f_i) = \Delta f_{i+1} - \Delta f_i \\ &= (f_{i+2} - f_{i+1}) - (f_{i+1} - f_i) \\ &= f_{i+2} - 2f_{i+1} + f_i \end{aligned} \tag{8.15}$$

Third forward difference ($n = 3$):

$$\begin{aligned} \Delta^3 f_i &= \Delta^2(f_{i+1} - f_i) = \Delta^2 f_{i+1} - \Delta^2 f_i \\ &= (f_{i+3} - 2f_{i+2} + f_{i+1}) - (f_{i+2} - 2f_{i+1} + f_i) \\ &= f_{i+3} - 3f_{i+2} + 3f_{i+1} - f_i \end{aligned} \tag{8.16}$$

Fourth forward difference ($n = 4$):

$$\begin{aligned} \Delta^4 f_i &= \Delta^3(f_{i+1} - f_i) = \Delta^3 f_{i+1} - \Delta^3 f_i \\ &= (f_{i+4} - 3f_{i+3} + 3f_{i+2} - f_{i+1}) - (f_{i+3} - 3f_{i+2} + 3f_{i+1} - f_i) \\ &= f_{i+4} - 4f_{i+3} + 5f_{i+2} - 4f_{i+1} + f_i \end{aligned} \tag{8.17}$$

It is evident that the process of determining differences can be continued for any $n$ value. Note, however, that the coefficients appear to take on a precise pattern. Indeed, these coefficients are the binomial coefficients and can be evaluated as follows:

$$b_{n,k} = (-1)^k \frac{n!}{k!(n-k)!} \tag{8.18}$$

where $b_{n,k}$ is the $k$th binomial coefficient for an $n$th-order difference. Therefore the $n$th forward difference is given directly as

$$\Delta^n f_i = \sum_{k=0}^{n} (-1)^k \frac{n!}{k!(n-k)!} f_{i-k+n} \tag{8.19}$$

Consequently, equation (8.17) can now be developed directly by using the expression given by equation (8.19) without the need for defining $\Delta f_i$, $\Delta^2 f_i$, or $\Delta^3 f_i$. That is, for $n = 4$,

$$\Delta^4 f_i = \sum_{k=0}^{4} (-1)^k \frac{4!}{k!(4-k)!} f_{i-k+4}$$

$$= \frac{4!}{0!4!} f_{i+4} - \frac{4!}{1!3!} f_{i+3} + \frac{4!}{2!2!} f_{i+2} - \frac{4!}{3!1!} f_{i+1} + \frac{4!}{4!0!} f_i$$

$$= f_{i+4} - 4f_{i+3} + 6f_{i+2} - 4f_{i+1} + f_i$$

This expression is obviously the same as that given by equation (8.17). Furthermore, for clarity and in hand calculations, differences are normally arranged as shown in table 8.1.

Such tables are called "difference tables." Interpolation using difference tables was once an important topic. However, recent advances in computer hardware have minimized their usefulness as well as their use. Note that differences, on the other hand, are extremely useful in the development of interpolating formulas (see section 8.4). For the purposes of the present discussion, differences are especially useful when the function as given by the data behaves like a polynomial. Consider,

Table 8.1 FORWARD DIFFERENCE TABLE FOR A FOURTH-ORDER INTERPOLATING POLYNOMIAL

| $x_i$ | $f_i$ | $\Delta f_i$ | $\Delta^2 f_i$ | $\Delta^3 f_i$ | $\Delta^4 f_i$ |
|---|---|---|---|---|---|
| $x_0$ | $f_0$ | | | | |
| | | $\Delta f_0$ | | | |
| $x_1$ | $f_1$ | | $\Delta^2 f_0$ | | |
| | | $\Delta f_1$ | | $\Delta^3 f_0$ | |
| $x_2$ | $f_2$ | | $\Delta^2 f_1$ | | $\Delta^4 f_0$ |
| | | $\Delta f_2$ | | $\Delta^3 f_1$ | |
| $x_3$ | $f_3$ | | $\Delta^2 f_2$ | | |
| | | $\Delta f_3$ | | | |
| $x_4$ | $f_4$ | | | | |

for example, the following function:

$$f(x) = x^2 + x + 1$$

Using a uniform interval of $h = 1$, we construct the corresponding difference table:

| $x_i$ | $f_i$ | $\Delta f_i$ | $\Delta^2 f_i$ | $\Delta^3 f_i$ |
|---|---|---|---|---|
| $x$ | $x^2 + x + 1$ | | | |
| | | $2x + 2$ | | |
| $x + 1$ | $x^2 + 3x + 3$ | | 2 | |
| | | $2x + 4$ | | 0 |
| $x + 2$ | $x^2 + 5x + 7$ | | 2 | |
| | | $2x + 6$ | | 0 |
| $x + 3$ | $x^2 + 7x + 13$ | | 2 | |
| | | $2x + 8$ | | |
| $x + 4$ | $x^2 + 9x + 21$ | | | |
| $\vdots$ | $\vdots$ | $\vdots$ | $\vdots$ | $\vdots$ |

It is evident that the second differences are constants and that the third differences are zeros. The implications are that the second differences are related to the second derivative and that the third differences are related to the third derivative. This is because the second and third derivatives of a second-order polynomial are indeed constant and zero, respectively.

### 8.3.2 Backward Differences

In the backward difference it is assumed that the difference in question is referred to the greater of the considered $x$ values in a given tabular data, that is, the first backward difference

$$\nabla f_i = f_i - f_{i-1}, \qquad i = 0, \ldots, n$$

where $\nabla$ is the backward difference operator. In general, the $n$th backward difference is given as

$$\nabla^n f_i = \nabla^{n-1}(f_i - f_{i-1}) \tag{8.20}$$

Consequently, the first four backward differences can be evaluated in the same way as the four forward differences were evaluated in the preceding section. Obviously, higher differences can be computed as well. Thus for the $i$th data point we have from equation (8.20)

First backward difference ($n = 1$):

$$\nabla f_i = f_i - f_{i-1} \tag{8.21}$$

Second backward difference ($n = 2$):

$$\begin{aligned}\nabla^2 f_i &= \nabla(f_i - f_{i-1}) = \nabla f_i - \nabla f_{i-1}\\ &= (f_i - f_{i-1}) - (f_{i-1} - f_{i-2})\\ &= f_i - 2f_{i-1} + f_{i-2}\end{aligned} \tag{8.22}$$

Third backward difference ($n = 3$):

$$\begin{aligned}\nabla^3 f_i &= \nabla^2(f_i - f_{i-1}) = \nabla^2 f_i - \nabla^2 f_{i-1}\\ &= (f_i - 2f_{i-1} + f_{i-2}) - (f_{i-1} - 2f_{i-2} + f_{i-3})\\ &= f_i - 3f_{i-1} + 3f_{i-2} - f_{i-3}\end{aligned} \tag{8.23}$$

Fourth backward difference ($n - 4$):

$$\begin{aligned}\nabla^4 f_i &= \nabla^3(f_i - f_{i-1}) = \nabla^3 f_i - \nabla^3 f_{i-1}\\ &= (f_i - 3f_{i-1} + 3f_{i-2} - f_{i-3}) - (f_{i-1} - 3f_{i-2} + 3f_{i-3} - f_{i-4})\\ &= f_i - 4f_{i-1} + 6f_{i-2} - 4f_{i-3} + f_{i-4}\end{aligned} \tag{8.24}$$

It is interesting to note that the backward difference expressions involve the same coefficients as those derived for the forward differences. Hence equation (8.18) can be used in their evaluation. In general, the $n$th difference can now be written as follows:

$$\nabla^n f_i = \sum_{k=0}^{n} (-1)^k \frac{n!}{k!(n-k)!} f_{i-k} \tag{8.25}$$

Equation (8.25) can be used in determining any difference expression directly. Consequently, a backward difference table is constructed as follows:

Table 8.2 BACKWARD DIFFERENCE TABLE FOR A FOURTH-ORDER INTERPOLATING FUNCTION

| $x_i$ | $f_i$ | $\nabla f_i$ | $\nabla^2 f_i$ | $\nabla^3 f_i$ | $\nabla^4 f_i$ |
|---|---|---|---|---|---|
| $x_0$ | $f_0$ | | | | |
| | | $\nabla f_1$ | | | |
| $x_1$ | $f_1$ | | $\nabla^2 f_2$ | | |
| | | $\nabla f_2$ | | $\nabla^3 f_3$ | |
| $x_2$ | $f_2$ | | $\nabla^2 f_3$ | | $\nabla^4 f_4$ |
| | | $\nabla f_3$ | | $\nabla^3 f_4$ | |
| $x_3$ | $f_3$ | | $\nabla^2 f_4$ | | |
| | | $\nabla f_4$ | | | |
| $x_4$ | $f_4$ | | | | |

Note that additional data points can be included in table 8.2.

### 8.3.3 Central Differences

Central differences are used rather extensively in the development of numerical methods for solving differential equations. Central difference expressions contain the same number of function values forward of the reference location as backward from it. That is, the first central difference for the $i$th data point is given as

$$\delta f_{i+1/2} = f_{i+1} - f_i, \qquad i = 0, \ldots, n$$

where $\delta$ is the central difference operator. In general, the $n$th central difference is given by

$$\delta^n f_{i+1/2} = \delta^{n-1}(f_{i+1} - f_i) \tag{8.26}$$

It is evident that the right-hand side of equation (8.26) is of the same form as that defined for the forward difference. Consequently, the difference expression coefficients are expected to be the same. In fact, these are also equivalent to the backward difference expression coefficients. Therefore the $n$th central difference is given directly as follows:

$$\delta^n f_{i+1/2} = \sum_{k=0}^{n} (-1)^k \frac{n!}{k!(n-k)!} f_{i-k+n} \tag{8.27}$$

Consequently, a central difference table can be developed readily. Obviously, additional data points can be included and their corresponding differences computed in the same fashion as shown in table 8.3.

It is interesting to note that the $n$th derivatives and the $n$th differences are related. For example, the first derivative is defined as

$$\frac{df(x)}{dx} = \lim_{h \to 0} \frac{f(x+h) - f(x)}{h}$$

Table 8.3 CENTRAL DIFFERENCE TABLE FOR A FOURTH-ORDER INTERPOLATING POLYNOMIAL

| $x_i$ | $f_i$ | $\delta f_i$ | $\delta^2 f_i$ | $\delta^3 f_i$ | $\delta^4 f_i$ |
|---|---|---|---|---|---|
| $x_0$ | $f_0$ | | | | |
| | | $\delta_{1/2}$ | | | |
| $x_1$ | $f_1$ | | $\delta_1^2$ | | |
| | | $\delta_{3/2}$ | | $\delta_{3/2}^3$ | |
| $x_2$ | $f_2$ | | $\delta_2^2$ | | $\delta_2^4$ |
| | | $\delta_{5/2}$ | | $\delta_{5/2}^3$ | |
| $x_3$ | $f_3$ | | $\delta_3^2$ | | |
| | | $\delta_{7/2}$ | | | |
| $x_4$ | $f_4$ | | | | |

However, the forward difference is given by

$$\Delta f(x) = f(x + h) - f(x)$$

Consequently,

$$\frac{df(x)}{dx} = \lim_{h \to 0} \frac{\Delta f(x)}{h}$$

But, since the function is general and may not be known when one is dealing with tabulated data, one may then write

$$\frac{df(x)}{dx} \approx \frac{\Delta f(x)}{h}$$

and use the first difference expression for approximating the derivative, provided the $n$th differences are nearly constants. It is extremely important to note that derivatives evaluated by using difference operators may involve significant errors, especially when one is dealing with data whose $(n + 1)$th differences are not constants. This situation may arise when the given tabulated data involve experimental error. Generally, the $n$th derivative may be approximated by the relationship

$$\frac{d^n f(x)}{dx^n} = f^{(n)}(x) = \frac{\Delta^n f(x)}{h^n}$$

This expression represents the forward approximation to the $n$th derivative. Moreover, similar expressions can be developed in terms of the backward and central operators.

### 8.3.4 Relationships between Difference Operators

It is evident from the preceding discussion that the three difference operators $\Delta$, $\nabla$, and $\delta$ refer to a particular position containing the functional values at that position and at those forward of it. Consequently, it would seem appropriate to introduce a new operator in order to simplify and relate the various operators. The shift operator $E$ is used for this purpose and is defined as follows:

$$f(x + h) = Ef(x)$$

or more simply for the $i$th data point, we have

$$f_{i+1} = Ef_i, \qquad i = 0, \ldots, n \tag{8.28}$$

Consequently, the shift operator can now be related to the forward, backward, and central operators as follows:

#### Forward Operator

Let us begin by considering the first forward difference. Recall that

$$\Delta f_i = f_{i+1} - f_i \tag{8.29}$$

Substituting equation (8.28) into equation (8.29) and simplifying yields

$$\Delta f_i = Ef_i - f_i = (E - 1)f_i$$

from which we determine

$$\Delta = E - 1$$

For the second forward difference we have

$$\begin{aligned}\Delta^2 f_i &= f_{i+2} - 2f_{i+1} + f_i \\ &= Ef_{i+1} - 2Ef_i + f_i \\ &= E(Ef_i) - 2Ef_i + f_i \\ &= (E^2 - 2E + 1)f_i \\ &= (E - 1)^2 f_i\end{aligned}$$

and in general for the $n$th forward difference the relationship is given by

$$\Delta^n f_i = (E - 1)^n f_i \tag{8.30a}$$

or more simply

$$\Delta^n = (E - 1)^n \tag{8.30b}$$

### Backward Operator

Recall that the first backward difference is given by

$$\nabla f_i = f_i - f_{i-1} \tag{8.31}$$

Equation (8.28) can be rewritten as follows:

$$f_i = Ef_{i-1}, \qquad f_{i-1} = \frac{1}{E} f_i$$

Substituting into equation (8.31) and simplifying yields

$$\nabla f_i = f_i - \frac{1}{E} f_i = \left(1 - \frac{1}{E}\right) f_i$$

or more simply

$$\nabla = \frac{E - 1}{E}$$

Therefore the relationship between the forward and backward operators is readily established by noting that $\Delta = E - 1$ to give

$$\nabla = \frac{\Delta}{\Delta + 1} \tag{8.32}$$

Consequently the $n$th differences are related as

$$\nabla^n = \frac{(E - 1)^n}{E^n} = \frac{\Delta^n}{(\Delta + 1)^n} \tag{8.33}$$

### Central Operator

The first central difference was defined earlier as

$$\delta f_{i+1/2} = f_{i+1} - f_i$$

Therefore, using the shift operator notation, we have

$$\delta E^{1/2} f_i = E f_i - f_i$$

which gives

$$\delta = E^{1/2} - E^{-1/2}$$

This can in turn be expressed as

$$\delta = (1 + \Delta)^{1/2} - (1 + \Delta)^{-1/2}$$

which simplifies to

$$\delta = \frac{\Delta}{(1 + \Delta)^{1/2}}$$

Consequently for the $n$th difference

$$\delta^n = \frac{\Delta^n}{(1 + \Delta)^{n/2}} \tag{8.34}$$

It is evident that the central difference operator can also be related to the backward difference operator to give for the $n$th difference

$$\delta^n = \frac{\nabla^n}{(1 - \nabla)^{n/2}} \tag{8.35}$$

Clearly the difference and shift operators behave in the same manner as algebraic symbols. In fact, the distributive, associative, and commutative laws of algebra apply to the operators. Furthermore, the purpose of developing the various relationships is to simplify the derivation of interpolating formulas. This topic is covered in the following section.

E

EXAMPLE 8.2

Determine whether the following data are those of a polynomial and if so, determine the order of that polynomial.

| $x_i$ | 2 | 3 | 4 | 5 | 6 | 7 | 8 | 9 | 10 |
|---|---|---|---|---|---|---|---|---|---|
| $f_i$ | −16 | −27 | 0 | 125 | 432 | 1029 | 2048 | 3645 | 6000 |

## Solution

Using the forward difference expressions given by equation (8.13), we establish the following difference table:

| $i$ | $x_i$ | $f_i$ | $\Delta f_i$ | $\Delta^2 f_i$ | $\Delta^3 f_i$ | $\Delta^4 f_i$ | $\Delta^5 f_i$ |
|---|---|---|---|---|---|---|---|
| 0 | 2 | −16 | | | | | |
| | | | −11 | | | | |
| 1 | 3 | −27 | | 38 | | | |
| | | | 27 | | 60 | | |
| 2 | 4 | 0 | | 98 | | 24 | |
| | | | 125 | | 84 | | 0 |
| 3 | 5 | 125 | | 182 | | 24 | |
| | | | 307 | | 108 | | 0 |
| 4 | 6 | 432 | | 290 | | 24 | |
| | | | 597 | | 132 | | 0 |
| 5 | 7 | 1029 | | 422 | | 24 | |
| | | | 1019 | | 156 | | 0 |
| 6 | 8 | 2048 | | 578 | | 24 | |
| | | | 1597 | | 180 | | |
| 7 | 9 | 3645 | | 758 | | | |
| | | | 2355 | | | | |
| 8 | 10 | 6000 | | | | | |

Clearly the fourth differences are constants and the fifth are zeros. Consequently the data are those of a fourth-order polynomial. This is because the derivative of a fourth-order polynomial is a constant. In fact, the data were derived from the function

$$f = x^4 - 4x^3$$

Note that the backward as well as the central differences provide the same results. Furthermore, the fourth-order differences can be computed directly by using equation (8.17). This gives for $i = 0$ $(x = 2)$

$$\begin{aligned}\Delta^4 f_0 &= f_4 - 4f_3 + 6f_2 - 4f_1 + f_0 \\ &= 432 - 4(125) + 6(0) - 4(-27) + (-16) \\ &= 24\end{aligned}$$

It is evident that other differences for $i = 1, 2, 3, 4, 5$ yield precisely the values of $\Delta^4 f_i = 24$ computed in the difference table.

## 8.4 DIFFERENCES AND INTERPOLATING POLYNOMIALS

When the $n$th differences corresponding to a set of $n + 1$ data points are constants or nearly so, the set can be replaced by a polynomial. This polynomial is the interpolating function which passes through the $n + 1$ points. In this section, direct methods of evaluating interpolation functions are outlined. These methods will not require the solution of sets of linear algebraic equations as was described in section 8.2. Instead, the development of interpolating formulas is made possible by using the difference operators and their interrelationships. Recall that

$$f(x + h) = Ef(x)$$

Therefore the functional value at $x + 2h$ is given as

$$f(x + 2h) = Ef(x + h) = E^2 f(x)$$

In general at $x + h$ the functional is related to the shift operator as follows:

$$f(x + \alpha h) = E^\alpha f(x) \tag{8.36}$$

Recall that the shift operator $E$ is related to the forward, backward, and central difference operators as given by equations (8.30), (8.33), and (8.34), respectively. Therefore, in terms of forward differences equation (8.36) is given as follows for the $i$th point:

$$f(x + \alpha h) = (1 + \Delta)^\alpha f(x_i)$$

Now, expanding $(1 + \Delta)^\alpha$ by the binomial formula yields Newton's well-known forward difference interpolating formula. That is,

$$f(x_i + \alpha h) = \left(1 + \alpha\Delta + \frac{\alpha(\alpha - 1)}{2!}\Delta^2 + \frac{\alpha(\alpha - 1)(\alpha - 2)}{3!}\Delta^3 + \cdots + \frac{\alpha(\alpha - 1)\cdots(\alpha - n + 1)}{n!}\Delta^2\right) f(x_i) \tag{8.37}$$

Equation (8.37) is represented graphically in figure 8.5.

Clearly, the value of $\alpha$ can be conveniently expressed in terms of two adjacent base points. That is,

$$\alpha = \frac{x - x_i}{h} \tag{8.38}$$

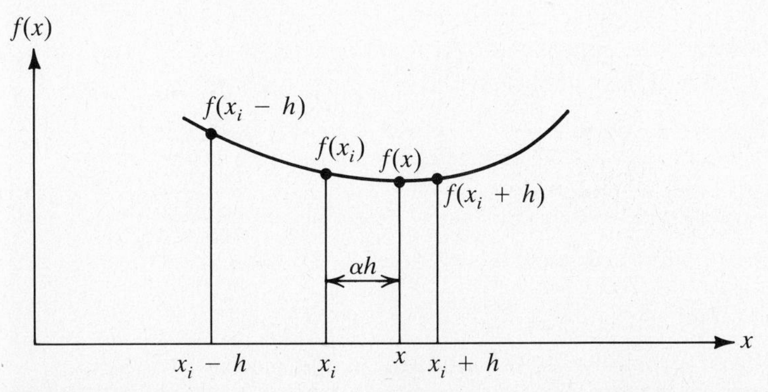


**Figure 8.5** Graphical illustration of Newton's forward interpolating formula.

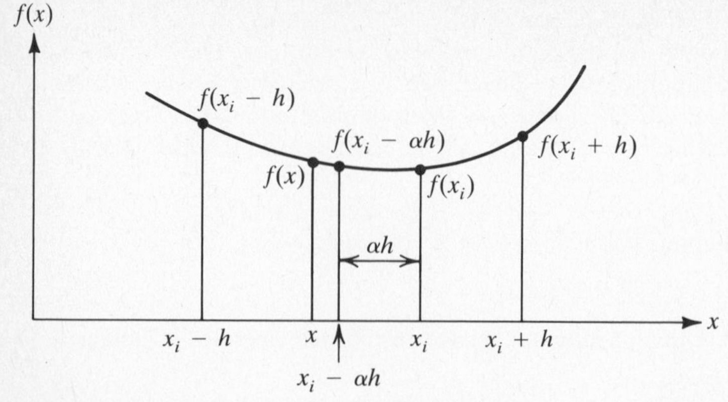


**Figure 8.6** Graphical illustration of Newton's backward interpolating formula.

Similar interpolating formulas can be easily developed by using the procedure just outlined. Consequently, Newton's backward interpolating formula is given as

$$f(x + \alpha h) = Ef(x) = (1 - \nabla)^{-\alpha} f(x)$$

Once again, expanding the term $(1 - \nabla)^{-\alpha}$ using the binomial formula gives for the $i$th point

$$f(x_i + \alpha h) = \left(1 + \alpha\nabla + \frac{\alpha(\alpha+1)}{2!}\nabla^2 \frac{\alpha(\alpha+1)(\alpha+2)}{3!}\nabla^3 + \frac{\alpha(\alpha+1)\cdots(\alpha+n-1)}{n!}\nabla^n\right) f(x_i) \quad \textbf{(8.39)}$$

Equation (8.39) is illustrated graphically in figure 8.6.

It is evident that other interpolating formulas can be developed by using the difference operators and the binomial formula.

**E** EXAMPLE 8.3

Given the following data, approximate the functional value at $x = 5.65$ using Newton's forward and backward interpolating formulas.

| $x_i$ | 0 | 1 | 2 | 3 | 4 | 5 | 6 | 7 |
|---|---|---|---|---|---|---|---|---|
| $f(x_i)$ | −2 | −1 | 6 | 25 | 62 | 123 | 214 | 341 |

Solution

We proceed by first establishing the following difference table for forward interpolation.

| $i$ | $x_i$ | $f(x_i)$ | $\Delta f_i$ | $\Delta^2 f_i$ | $\Delta^3 f_i$ | $\Delta^4 f_i$ |
|---|---|---|---|---|---|---|
| 0 | 0 | −2 | | | | |
| | | | 1 | | | |
| 1 | 1 | −1 | | 6 | | |
| | | | 7 | | 6 | |
| 2 | 2 | 6 | | 12 | | 0 |
| | | | 19 | | 6 | |
| 3 | 3 | 25 | | 18 | | 0 |
| | | | 37 | | 6 | |
| 4 | 4 | 62 | | 24 | | 0 |
| | | | 61 | | 6 | |
| 5 | 5 | 123 | | 30 | | 0 |
| | | | 91 | | 6 | |
| 6 | 6 | 214 | | 36 | | |
| | | | 127 | | | |
| 7 | 7 | 341 | | | | |

Clearly the data are those of a third-order polynomial. Hence using equation (8.37) gives

$$f(x_i + \alpha h) = \left(1 + \alpha\Delta + \frac{\alpha(\alpha - 1)}{2}\Delta^2 + \frac{\alpha(\alpha - 1)(\alpha - 2)}{6}\Delta^3\right) f(x_i)$$

Note that $\Delta^4$ and the higher operators are not included because they are all equal to zero. Therefore, letting $x = 5.65$ and $x_i = x_0 = 0$ and noting that $h = 1.0$ permits the evaluation of $\alpha$. Thus

$$\alpha = \frac{5.65 - 0}{1} = 5.65$$

Furthermore, corresponding to $x = 0$, we have

$$f_0 = -2, \qquad \Delta f_0 = 1, \qquad \Delta^2 f_0 = 6, \qquad \Delta^3 f_0 = 6$$

Therefore

$$f(5.65) = -2 + 5.65(1) + \frac{5.65(4.65)}{2}(6) + \frac{5.65(4.65)(3.65)}{6}(6)$$

$$= 178.362125$$

This is the exact value calculated from the function $f(x) = x^3 - 2$ from which the data were determined. Note that a different $x$ value could have been selected for approximating the interpolated functional value. For example, taking $x_i = x_3 = 3$ and $x = 5.65$ gives

$$\alpha = \frac{5.65 - 3}{1} = 2.65$$

and the forward differences

$$f_3 = 25, \qquad \Delta f_3 = 37, \qquad \Delta^2 f_3 = 24, \qquad \Delta^3 f_3 = 6$$

Hence

$$f(5.65) = 25 + 2.65(37) + \frac{2.65(1.65)}{2}(24) + \frac{2.65(1.65)(0.65)}{6}(6)$$

$$= 178.362125$$

Now using the backward difference operators, we use the forward difference table to give for $x = 5.65$ and $x_i = x_7 = 7$

$$f_7 = 341, \qquad \nabla f_7 = 127, \qquad \nabla^2 f_7 = 36, \qquad \nabla^3 f_7 = 6$$

Hence using equation (8.39) yields

$$f(5.65) = 341 + (-1.35)(127) + \frac{(-1.35)(-0.35)}{2}(36)$$

$$+ \frac{(-1.35)(-0.35)(0.65)}{6}(6)$$

$$= 178.362125$$

Clearly this is the exact value computed by means of the forward difference interpolating formula. Note that in general experimental data may not behave exactly as a polynomial. As a result the $n$th differences may not be constants. In such cases the number of terms used in the Newton interpolating formula must be increased in order to achieve desired accuracies.

## 8.5 INTERPOLATING POLYNOMIALS FOR UNEVEN INTERVALS

When the base points in a given functional relationship are not equally spaced, the interpolation problem becomes more difficult to solve. The basis for this assertion lies in the fact that the interpolating polynomial coefficients will depend not only on the functional values, but also on the base points in question.

The Lagrange interpolation method provides a direct approach for determining interpolated values, regardless of the base-point spacing. The method is based on the fact that an interpolating function can be expressed as a linear combination of separate polynomials in the independent variable $x$. This was shown to be the case in section 8.2 [see equation (8.10)]. Consequently, an interpolating polynomial is given as

$$f(x) = L_0(x)f(x_0) + L_1(x)f(x_1) + \cdots + L_n(x)f(x_n)$$

or more simply

$$f(x) = \sum_{i=0}^{n} L_i(x_j)f(x_i), \qquad j = 0, 1, \ldots, n \tag{8.40a}$$

The $L_i(x_j)$ functions have the following special properties:

$$L_i(x_j) = \begin{cases} 0 & \text{for } i \neq j \\ 1 & \text{for } i = j \end{cases} \qquad \textbf{(8.40b)} \quad \textbf{(8.40c)}$$

$$\sum_{i=0}^{n} L_i(x) = 1 \qquad \textbf{(8.40d)}$$

Equation (8.40b) clearly states that the base points $x_0, \ldots, x_n$ are roots of the function $L_i(x_j)$. Therefore

$$L_i(x_j) = C_j(x - x_0)(x - x_1) \cdots (x - x_n) \qquad \textbf{(8.41)}$$

where $C_j$ is a proportionality constant. Note that the $(x - x_j)$ term is not part of the expression given by equation (8.41) because $L_j(x_j) = 1$ and not zero. Now applying equation (8.40c) gives for $x = x_j$

$$C_j(x_j - x_0)(x_j - x_1) \cdots (x_j - x_n) = 1$$

Solving for the $C_j$ value yields

$$C_j = \frac{1}{(x_j - x_0)(x_j - x_1) \cdots (x_j - x_n)} \qquad \textbf{(8.42)}$$

Therefore, substituting equation (8.42) into equation (8.41) permits the determination of the $L_i(x_j)$ functions. That is,

$$L_i(x_j) = \frac{(x - x_0)(x - x_1) \cdots (x - x_n)}{(x_j - x_0)(x_j - x_1) \cdots (x_j - x_n)} \qquad \textbf{(8.43)}$$

In an abbreviated form, we have

$$L_i(x_j) = \prod_{\substack{j=0 \\ j \neq i}}^{n} \frac{x - x_j}{x_i - x_j}, \qquad i = 0, 1, \ldots, n \qquad \textbf{(8.44)}$$

The interpolating function for uneven base-point intervals can now be readily evaluated by substituting equation (8.44) into equation (8.40a) to give

$$f(x) = \sum_{i=0}^{n} \prod_{\substack{j=0 \\ j \neq i}}^{n} \frac{x - x_j}{x_i - x_j} f(x_i) \qquad \textbf{(8.45)}$$

The expression given in equation (8.45) is known as the Lagrange interpolation formula for unevenly spaced base points. This powerful formula permits the determination of an interpolating function coefficient without the need for solving a set of linear algebraic equations as was required by the method outlined in section 8.2. In fact, for an evenly spaced base point it gives precisely the same results as those obtained earlier. Furthermore, in using the Lagrange interpolation scheme there is no need to construct difference tables.

E

EXAMPLE 8.4

Develop a second-order forward interpolating polynomial for an equally spaced base point using equation (8.45).

Solution

Since the notation given by equation (8.45) involves $x_0, \ldots, x_n$, we must first transform the base points and the functional values so that they are consistent with the definition of the forward interpolating polynomial. That is,

| EQUATION (8.45) | | FORWARD NOTATION | |
|---|---|---|---|
| $x_i$ | $f(x_i)$ | $x_i$ | $f(x_i)$ |
| $x_0$ | $f(x_0)$ | $x_0 = 0$ | $f(x_0)$ |
| $x_1$ | $f(x_1)$ | $x_1 = h$ | $f(x_1)$ |
| $x_2$ | $f(x_2)$ | $x_2 = 2h$ | $f(x_2)$ |

Therefore using equation (8.45) for $i = 0, 1, 2$ and $j = 0, 1, 2$ yields

$$f(x) = \frac{(x - x_1)(x - x_2)}{(x_0 - x_1)(x_0 - x_2)} f(x_0) + \frac{(x - x_0)(x - x_2)}{(x_1 - x_0)(x_1 - x_2)} f(x_1) + \frac{(x - x_0)(x - x_1)}{(x_2 - x_0)(x_2 - x_1)} f(x_2)$$

Substituting the transformed $x$ values and simplifying gives

$$f(x) = \frac{1}{2}\left(2 - 3\frac{x}{h} + \frac{x^2}{h^2}\right) f(x_0) + \left(\frac{2x}{h} - \frac{x^2}{h^2}\right) f(x_1) + \frac{1}{2}\left(\frac{x^2}{h^2} - \frac{x}{h}\right) f(x_2)$$

This expression is precisely the same as that given by equation (8.8).

EXAMPLE 8.5

Approximate the function value at $x = 2$ for the following data.

| $x_i$ | 0 | 3 | 4 | 7 |
|---|---|---|---|---|
| $f(x_i)$ | 2 | 8 | 9 | 6 |

Solution

Obviously, a third-order polynomial can be determined so that it passes through the four data points. However, let us interpolate directly using equation (8.45).

$$f(2) = \frac{(2-3)(2-4)(2-7)}{(0-3)(0-4)(0-7)}(2) + \frac{(2-0)(2-4)(2-7)}{(3-0)(3-4)(3-7)}(8)$$

$$+\frac{(2-0)(2-3)(2-7)}{(4-0)(4-3)(4-7)}(9) + \frac{(2-0)(2-3)(2-4)}{(7-0)(7-3)(7-4)}(6)$$

$$= \frac{10}{84}(2) + \frac{20}{12}(8) + \left(-\frac{10}{12}\right)(9) + \frac{4}{84}(6)$$

$$= \frac{534}{84} = 6.3571429$$

Note that the sum of the weight function $\Sigma L_i(x_j)$ is equal to 1.0 as it should be.

## 8.6 INTERPOLATION ERRORS

The errors associated with the interpolated values depend on many factors. These include truncation, rounding off, inaccurate data, and human errors. Truncation errors occur when a terminated series rather than an infinite one is used. Therefore the magnitude of such errors corresponds to the sum of the terms not included in the computations. Unfortunately, there is no general rule that can be implemented for the determination of such errors of magnitude unless infinite data points are given or the higher differences are nearly constant. Consequently, for the special case in which the $(n-1)$th difference terms are included in the interpolation and the $n$th differences are constants, the truncation error can be approximated. That is, for Newton's forward interpolation formula we have

$$R(\xi) = \frac{h^n f(\xi)^{(n)}}{n!}\alpha(\alpha-1)\cdots(\alpha-n-1) \quad \textbf{(8.46)}$$

where $R(\xi)$ is the truncation error term and $f^n(\xi)$ is the $n$th derivative, which is assumed to be constant over the range $x_0 \ll \xi < x_{n-1}$. Obviously, the function is not known; as a result the derivative must be approximated from the differences as follows:

$$f(\xi)^{(n)} = \frac{\Delta^n f(\xi)}{h^n} \quad \textbf{(8.47)}$$

Substituting equation (8.47) into equation (8.46) yields the desired approximation:

$$R(\xi) = \frac{\Delta^n f(\xi)}{n!}[\alpha(\alpha-1)\cdots(\alpha-n+1)] \quad \textbf{(8.48)}$$

Table 8.4 ERROR PROPAGATION IN AN ARBITRARY DATA SET

| $i$ | $x_i$ | $f(x_i)$ | $\Delta E_i$ | $\Delta^2 f_i$ | $\Delta^3 f_i$ |
|---|---|---|---|---|---|
| 0 | $x_0$ | $E_0$ | | | |
| | | | $E_1 - E_0$ | | |
| 1 | $x_1$ | $E_1$ | | $E_2 - 2E_1 + E_0$ | |
| | | | $E_2 - E_1$ | | $E_3 - 3E_2 + 3E_1 - E_0$ |
| 2 | $x_2$ | $E_2$ | | $E_3 - 2E_2 + E_1$ | |
| | | | $E_3 - E_2$ | | $E_4 - 3E_3 + 3E_2 - E_1$ |
| 3 | $x_3$ | $E_3$ | | $E_4 - 2E_3 + E_2$ | |
| | | | $E_4 - E_3$ | | $E_5 - 3E_4 + 3E_3 - E_2$ |
| 4 | $x_4$ | $E_4$ | | $E_5 - 2E_4 + E_3$ | |
| | | | $E_5 - E_4$ | | |
| 5 | $x_5$ | $E_5$ | | | |

Note that by taking $\xi = x_0$ and $\xi = x_{n-1}$, we can easily establish an upper and a lower bound for the truncation error.

For Newton's backward difference interpolating formula, the truncation error is approximated in the same manner to give

$$R(\xi) = \frac{\Delta^n f(\xi)}{n!}[\alpha(\alpha + 1) \cdots (\alpha + n - 1)] \tag{8.49}$$

The rounding off of numbers in the process of determining differences could have significant effect on the interpolated value. This is because the $n$th differences may oscillate when in fact they are constants. This situation is described in table 8.4.

Note here that the rounding off errors $E_i$ $(i = 0, \ldots, 5)$ propagate in the same manner as the original functional data values would. Clearly, if the roundoff errors

Table 8.5 SPECIAL CASE OF ERROR PROPAGATION IN A DATA SET

| $i$ | $x_i$ | $f(x_i)$ | $\Delta E_i$ | $\Delta^2 E_i$ | $\Delta^3 E_i$ |
|---|---|---|---|---|---|
| 0 | $x_0$ | 0 | | | |
| | | | 0 | | |
| 1 | $x_1$ | 0 | | $E$ | |
| | | | $E$ | | $-3E$ |
| 2 | $x_2$ | $E$ | | $-2E$ | |
| | | | $-E$ | | $3E$ |
| 3 | $x_3$ | 0 | | $E$ | |
| | | | 0 | | |
| 4 | $x_4$ | 0 | | | |

are of exactly the same magnitude, then all of the differences will be zero and the error will be equivalent to errors associated with the functional values. Experimental and/or human errors may for some problems be associated with one or more functional values. The errors in the computed differences will then depend on the magnitude of the error, its location in the difference table, and the number of data points where the error occurs. For illustration purposes, suppose that an error of magnitude $E$ is introduced in the $i = 2$ location of table 8.4. Then the data error propagates as shown in table 8.5. It is interesting to note that the sum of all errors for any difference column is zero. Similar expressions for different error locations can therefore be developed.

## 8.7 INVERSE INTERPOLATION

Thus far, we have addressed the question of how to interpolate for a functional value corresponding to a given independent variable $x$. Suppose that we now reverse the question so that we seek to determine an $x$ value corresponding to a given functional value. Then the problem becomes inverse interpolation. Since for all practical purposes the tabulated functional values are not equally spaced, Lagrange's interpolation formula can be directly applied by switching the dependent and independent variables in equation (8.45). This yields

$$x = \sum_{i=0}^{n} \sum_{\substack{j=0 \\ j \neq i}}^{n} \frac{f(x) - f(x_j)}{f(x_i) - f(x_j)} x_i \tag{8.50}$$

For cases in which the functional values are equally spaced alternative iterative procedures can be developed by using Newton's forward and backward interpolating formulas. Such methods are not included in this text because they are limited to unrealistic situations. Furthermore, equation (8.50) is more general and gives the proper interpolated value directly without the need for iterations. Unfortunately, the method in some instances gives poor results. The reason for this is that the $x$ variable expressed as a function of the tabulated functional values may not be well approximated with a polynomial.

## 8.8 CUBIC SPLINES

A spline is a flexible drafting device that can be constrained to pass smoothly through a set of plotted data points. Spline functions are a recent mathematical tool which is an adaptation of this idea. Although spline functions of varying orders can be developed, cubic splines are by far the most widely used in engineering practice. These are a connected group of third-order interpolating polynomials arranged so that adjacent functions are forced to join with continuous first and second derivatives. Consequently, cubic splines do not have the inherent wiggle problems associated with high-order interpolating polynomials. For illustration purposes, consider the set of data points given below:

| $x_i$ | −3 | −2 | −1 | 0 | 1 | 2 | 3 |
|---|---|---|---|---|---|---|---|
| $f(x_i)$ | 0.6 | 0.2 | 0.6 | 3 | 0.6 | 0.2 | 0.6 |

It is evident that a sixth-order interpolating polynomial can be fitted through the seven data points. Therefore, if we assume

$$f(x) = a_0 + a_1x + a_2x^2 + a_3x^3 + a_4x^4 + a_5x^5 + a_6x^6$$

we determine the unknowns by simply substituting the data points into the assumed polynomials and solving the resulting set of linear algebraic equations. This gives

$$f(x) = \tfrac{1}{150}(450 - 481x^2 + 130x^4 - 9x^6) \quad \textbf{(8.51)}$$

Clearly the interpolating polynomial given by equation (8.51) passes through every data point. In order to appreciate the need for cubic splines and to have a better understanding of the wiggle effect associated with polynomial interpolation, the data and equation (8.51) are shown graphically in figure 8.7.

It is interesting to note that while none of the functional values given in the data set is negative, the derived interpolating polynomial suggests that negative functional values do exist within the range of the given $x$ values. Hence a better indication of the probable behavior might be given by the dashed curve shown in figure 8.7. Therefore we seek a method that will eliminate the wiggle problem, but can still fit local irregularities. The method of cubic spline interpolation does precisely that.

The mathematical basis for the cubic spline method is illustrated in figure 8.8. Consequently, given a set of $n$ data points $(x_i, f_i)$ where $i = 1, \ldots, n$, we wish to pass a set of cubics through the $n$ data points that correspond to the drafting spline. This can be accomplished by forcing the slopes and curvatures to be the same for each pair of cubics that join at each point. Therefore, for the $(i-1)$th interval, the polynomial is given as follows:

$$f(x) = A_{i-1} + B_{i-1}(x - x_{i-1}) + C_{i-1}(x - x_{i-1})^2 + D_{i-1}(x - x_{i-1})^3 \quad \textbf{(8.52)}$$

where

$$x_{i-1} \leq x \leq x_i$$

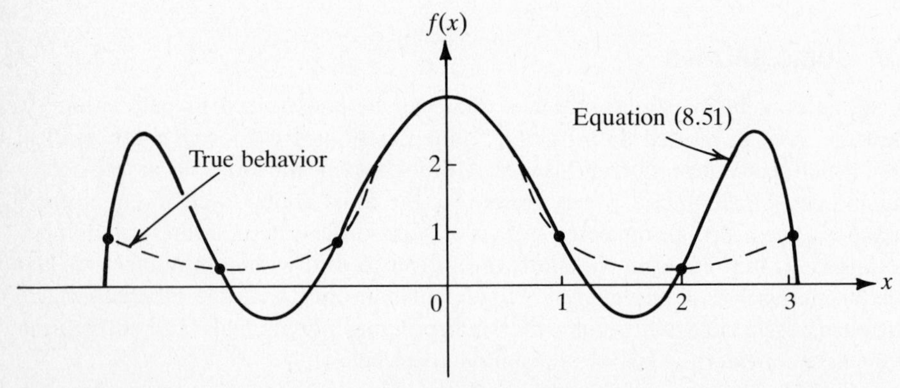


**Figure 8.7** Graphical illustration of the wiggle effect associated with polynomial interpolation.

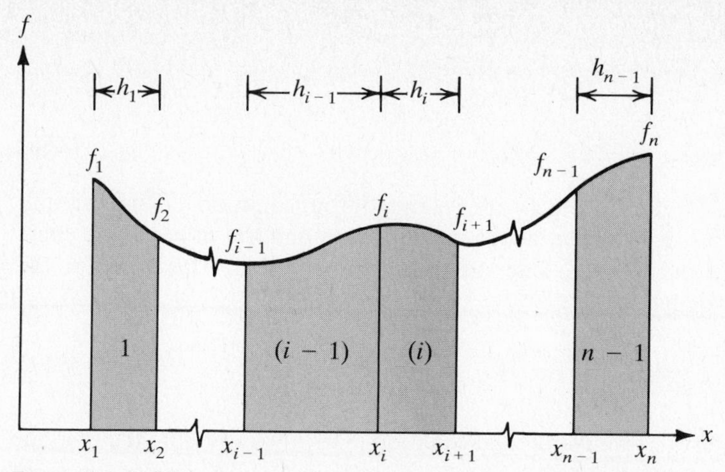


**Figure 8.8** Cubic spline interpolation for a set of $n$ data points.

Since this cubic fits at the end points $x_{i-1}$ and $x_i$ of the $(i-1)$th interval,

$$f(x_{i-1}) = f_{i-1} = A_{i-1} \tag{8.53}$$

$$f(x_i) = f_i = A_{i-1} + B_{i-1}h_{i-1} + C_{i-1}h_{i-1}^2 + D_{i-1}h_{i-1}^3 \tag{8.54}$$

Note that $h_{i-1} = x_i - x_{i-1}$. Consequently, the first and second derivatives of equation (8.52) are readily determined as follows:

$$\frac{df}{dx} = B_{i-1} + 2C_{i-1}(x - x_{i-1}) + 3D_{i-1}(x - x_{i-1})^2 \tag{8.55}$$

$$\frac{d^2f}{dx^2} = 2C_{i-1} + 6D_{i-1}(x - x_{i-1}) \tag{8.56}$$

In order to simplify the mathematical expressions, we denote the second derivative by $M$. We now fix the curvatures at nodes $i - 1$ and $i$ using equation (8.56). That is,

$$M_{i-1} = 2C_{i-1} \tag{8.57}$$

$$M_i = 2C_{i-1} + 6D_{i-1}h_{i-1} \tag{8.58}$$

Equations (8.53), (8.54), (8.57), and (8.58) can now be solved for the unknowns $A_{i-1}$, $B_{i-1}$, $C_{i-1}$, and $D_{i-1}$ to give

$$A_{i-1} = f_{i-1} \tag{8.59a}$$

$$B_{i-1} = \frac{1}{h_{i-1}}(f_i - f_{i-1}) - \frac{h_{i-1}}{6}(M_i + 2M_{i-1}) \tag{8.59b}$$

$$C_{i-1} = \tfrac{1}{2}M_{i-1} \tag{8.59c}$$

$$D_{i-1} = \frac{1}{6h_{i-1}}(M_i - M_{i-1}) \tag{8.59d}$$

It is evident that similar expressions for the remaining intervals can be readily developed by using equation (8.59). For example, the cubic spline for the $i$th interval is given as

$$f(x) = A_i + B_i(x - x_i) + C_i(x - x_i)^2 + D_i(x - x_i)^3 \tag{8.60}$$

and the unknowns $A_i$, $B_i$, $C_i$, and $D_i$ are evaluated from equation (8.59) by simply incrementing $i$ by $i + 1$. We now invoke the condition that the slopes of the cubic splines for the $i$th and $(i - 1)$th intervals be equal at $x = x_i$. Therefore, for the $(i - 1)$th interval, we have

$$\left.\frac{df}{dx}\right|_{x=x_i} = B_{i-1} + 2C_{i-1}h_{i-1} + 3D_{i-1}h_{i-1}^2 \tag{8.61}$$

and for the $i$th interval, we differentiate equation (8.60) then evaluate the first derivative at $x = x_i$ to get

$$\left.\frac{df}{dx}\right|_{x=x_i} = B_i$$

Note that $B_i$ is easily determined from equation (8.59b) by simply incrementing $i$ by $+1$ to give

$$\left.\frac{df}{dx}\right|_{x=x_i} = B_i = \frac{1}{h_i}(f_{i+1} - f_i) - \frac{h_i}{6}(M_{i+1} + 2M_i) \tag{8.62}$$

Now, equating equations (8.61) and (8.62) and substituting for $B_{i-1}$, $C_{i-1}$, and $D_{i-1}$ we obtain the following simplified expression:

$$h_{i-1}M_{i-1} + 2(h_{i-1} + h_i)M_i + h_iM_{i+1} = \frac{6}{h_{i-1}}f_{i-1} - 6\left(\frac{1}{h_{i-1}} + \frac{1}{h_i}\right)f_i + \frac{6}{h_i}f_{i+1} \tag{8.63}$$

Equation (8.63) is a linear algebraic equation in the unknown second derivatives $M_{i-1}$, $M_i$, and $M_{i+1}$. Furthermore, it can be applied at nodes $i = 2, \ldots, n - 1$ but not at nodes 1 and $n$. Consequently, this gives $n - 2$ equations in $n$ unknowns. The implication is that the second derivatives at $x = x_1$ and $x = x_n$ must be assumed in order to solve for the remaining $n - 2$ unknown second derivatives. Generally, there are three basic alternatives. These are

1. Assume $M_1 = M_n = 0$. This forces the splines for both ends to approach linearity at their extremities. Cubic splines derived in this manner are referred to as "natural splines" and are widely used in practice.

2. Assume $M_1 = M_2$ and $M_{n-1} = M_n$. This forces the splines at both ends to approach parabolas at their extremities.

3. Approximate $M_1$ and $M_n$ using interpolating polynomials of the third order. This can be easily accomplished by assuming

   $$f = a_0 + a_1x + a_2x^2 + a_3x^3$$

Then the second derivative is given as

$$\frac{d^2f}{dx^2} = 2a_2 + 6a_3x$$

Therefore at $x = x_1$ we have

$$M_1 = 2a_2 + 6a_3x_1 \tag{8.64a}$$

The unknown coefficients $(a_2, a_3)$ are calculated from the first four data points as follows:

$$\begin{bmatrix} 1 & x_1 & x_1^2 & x_1^3 \\ 1 & x_2 & x_2^2 & x_2^3 \\ 1 & x_3 & x_3^2 & x_3^3 \\ 1 & x_4 & x_4^2 & x_4^3 \end{bmatrix} \begin{Bmatrix} a_0 \\ a_1 \\ a_2 \\ a_3 \end{Bmatrix} = \begin{Bmatrix} f_1 \\ f_2 \\ f_3 \\ f_4 \end{Bmatrix} \tag{8.64b}$$

The second derivative at $x = x_n$ is defined in a similar fashion to give

$$M_n = 2b_2 + 6b_3x_n \tag{8.65a}$$

where the unknowns $b_2$ and $b_3$ are given by

$$\begin{bmatrix} 1 & x_{n-3} & x_{n-3}^2 & x_{n-3}^3 \\ 1 & x_{n-2} & x_{n-2}^2 & x_{n-2}^3 \\ 1 & x_{n-2} & x_{n-1}^2 & x_{n-1}^3 \\ 1 & x_n & x_n^2 & x_n^3 \end{bmatrix} \begin{Bmatrix} b_0 \\ b_1 \\ b_2 \\ b_3 \end{Bmatrix} = \begin{Bmatrix} f_{n-3} \\ f_{n-2} \\ f_{n-1} \\ f_n \end{Bmatrix} \tag{8.65b}$$

It is evident that in order to determine the second derivatives at the two ends, we must solve two sets of $4 \times 4$ linear algebraic equations. This technique permits the determination of the best spline fit possible. It is interesting that many authors do not seem to think that the second derivatives at the two ends can be computed—instead they resort to approximations in terms of the second derivatives at neighboring points.

The expressions given by equations (8.64) and (8.65) are greatly simplified if the base points are equally spaced. That is, if $h = x_{i+1} - x_i$ for $i = 1, \ldots, n-1$. For this special case the second derivatives are given by the following simple expressions:

$$M_1 = \frac{1}{h^2}(2f_i - 5f_2 + 4f_3 - f_4) \tag{8.66}$$

$$M_n = \frac{1}{h^2}(-f_{n-3} + 4f_{n-2} - 5f_{n-1} + 2f_n) \tag{8.67}$$

Obviously, these expressions are derived from equations (8.64) and (8.65) respectively.

E

EXAMPLE 8.6

Given the following data, derive cubic splines for the six intervals assuming $M_1 = M_7 = 0$.

| $i$ | 1 | 2 | 3 | 4 | 5 | 6 | 7 |
|---|---|---|---|---|---|---|---|
| $x_i$ | −3 | −2 | −1 | 0 | 1 | 2 | 3 |
| $f_i$ | 0.6 | 0.2 | 0.6 | 3 | 0.6 | 0.2 | 0.6 |

## Solution

Obviously, this set of data points is exactly the same as that shown graphically in figure 8.7. Moreover, the base points are equally spaced. That is,

$$h_1 = h_2 = \cdots = h_5 = 1.0$$

Consequently, equation (8.63) simplifies to the following:

$$M_{i-1} + 4M_i + M_{i+1} = 6f_{i-1} - 12f_i + 6f_{i+1} \qquad \textbf{(8.68)}$$

Equation (8.68) is now applied at the interior nodes $i = 2, 3, 4, 5$, and 6. This gives the following set of linear algebraic equations:

$$i = 2: \quad M_1 + 4M_2 + M_3 = 6f_1 - 12f_2 + 6f_3$$

$$i = 3: \quad M_2 + 4M_3 + M_4 = 6f_2 - 12f_3 + 6f_4$$

$$i = 4: \quad M_3 + 4M_4 + M_5 = 6f_3 - 12f_4 + 6f_5$$

$$i = 5: \quad M_4 + 4M_5 + M_6 = 6f_4 - 12f_5 + 6f_6$$

$$i = 6: \quad M_5 + 4M_6 + M_7 = 6f_5 - 12f_6 + 6f_7$$

Note that the right-hand side is given in the table and that since $M_1 = M_7 = 0$ the set is reduced to a set of five equations in five unknowns. That is,

$$\begin{bmatrix} 4 & 1 & 0 & 0 & 0 \\ 1 & 4 & 1 & 0 & 0 \\ 0 & 1 & 4 & 1 & 0 \\ 0 & 0 & 1 & 4 & 1 \\ 0 & 0 & 0 & 1 & 4 \end{bmatrix} \begin{Bmatrix} M_2 \\ M_3 \\ M_4 \\ M_5 \\ M_6 \end{Bmatrix} = \begin{Bmatrix} 4.8 \\ 12 \\ -28.8 \\ 12 \\ 4.8 \end{Bmatrix}$$

Solving for the unknowns yields

$$M_2 = -0.1846, \qquad M_3 = 5.5385, \qquad M_4 = -9.9692$$

$$M_5 = 5.5386, \qquad M_6 = -0.1846$$

These values can now be substituted into equation (8.59) to evaluate the coefficients of the cubic splines. That is, for $i = 2\ (-3 \le x \le -2)$, we have from equation (8.52)

$$f = A_1 + B_1(x - x_1) + C_1(x - x_1)^2 + D_1(x - x_1)^3$$

and the constants are determined from equation (8.59) as follows.

$$A_1 = f_1 = 0.6$$

$$B_1 = (f_2 - f_1) - \tfrac{1}{6}(M_2 + 2M_1) = -0.3692$$

$$C_1 = \tfrac{1}{2}M_1 = \tfrac{1}{2}(0) = 0$$

$$D_1 = \tfrac{1}{6}(M_2 - M_1) = \tfrac{1}{6}(-0.1846) = -0.0308$$

Hence

$$f = 0.6 - 0.3692(x + 3) - 0.0308(x + 3)^3$$

Clearly this spline function passes through $(-3, 0.6)$ and $(-2, 0.2)$. The remaining splines can be determined in a similar fashion, and are summarized as follows:

| EQUATION | INTERVAL | FUNCTION |
|---|---|---|
| $(A)$ | $-3 \le x \le -2$ | $0.6 - 0.3692(x + 3) - 0.0308(x + 3)^3$ |
| $(B)$ | $-2 \le x \le -1$ | $0.2 - 0.4616(x + 2) - 0.0923(x + 2)^2 + 0.9538(x + 2)^3$ |
| $(C)$ | $-1 \le x \le 0$ | $0.6 + 2.2154(x + 1) + 2.7693(x + 1)^2 - 2.5846(x + 1)^3$ |
| $(D)$ | $0 \le x \le 1$ | $3 - 4.9846x^2 + 2.5846x^3$ |
| $(E)$ | $1 \le x \le 2$ | $0.6 - 2.2154(x - 1) + 2.7693(x - 1)^2 - 0.9538(x - 1)^3$ |
| $(F)$ | $2 \le x \le 3$ | $0.2 + 0.4615(x - 2) - 0.0923(x - 2)^2 + 0.0308(x - 2)^3$ |

These functions are shown graphically, along with the sixth-order interpolating function derived earlier, in figure 8.9.

It is evident that the cubic splines are a better representation of the true behavior than that given by equation (8.51).

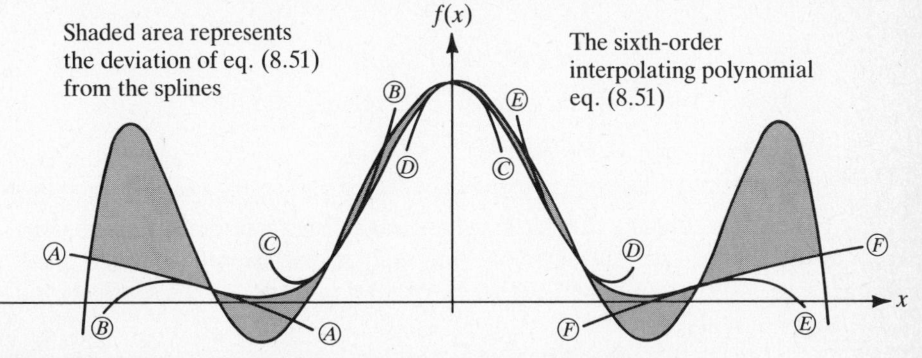


**Figure 8.9** Graphical illustration of the natural cubic splines derived for example 8.6.

E

EXAMPLE 8.7

Determine the second derivatives at the end points in the previous example using equations (8.66) and (8.67), then evaluate the remaining second derivatives at the interior data points.

Solution

Since the base points are equally spaced and $h = 1$, using equation (8.66) we have

$$M_1 = \frac{1}{1^2}[2(0.6) - 5(0.2) + 4(0.6) - 3] = -0.4$$

and from equation (8.67) the second derivative at the other end ($x = 3$) is given as

$$M_7 = \frac{1}{h^2}(-f_4 + 4f_5 - 5f_6 + 2f_7)$$

$$= \frac{1}{1^2}[-3 + 4(0.6) - 5(0.2) + 2(0.6)] = -0.4$$

Note that these values are not even close to zero as was assumed in example 8.6. Furthermore, owing to the symmetrical nature of the functional values, the second derivatives at $x_1$ and $x_7$ turned out to be equal. The remaining unknown derivatives at the interior nodes can be computed by using equation (8.63). Thus

$$\begin{bmatrix} 4 & 1 & 0 & 0 & 0 \\ 1 & 4 & 1 & 0 & 0 \\ 0 & 1 & 4 & 1 & 0 \\ 0 & 0 & 1 & 4 & 1 \\ 0 & 0 & 0 & 1 & 4 \end{bmatrix} \begin{Bmatrix} M_2 \\ M_3 \\ M_4 \\ M_5 \\ M_6 \end{Bmatrix} = \begin{Bmatrix} 4.8 + 0.4 \\ 12 \\ -28.8 \\ 12 \\ 4.8 + 0.4 \end{Bmatrix}$$

Solving for the unknowns gives

$$M_2 = -0.0769, \qquad M_3 = 5.5077, \qquad M_4 = -9.9538$$

$$M_5 = 5.5077, \qquad M_6 = -0.0769$$

Comparison of these values with those obtained in example 8.6 clearly indicates that only $M_2$ and $M_6$ have been significantly affected. This may not always be the case. However, one may deduce that assumptions made relative to end conditions will have their greatest influence on the shapes of the splines at the two ends.

The authors wish to alert the reader that although equation (8.63) is valid irrespective of the $h$ intervals used, problems may arise when these intervals are not of the same order of magnitude. This situation is encountered when the $x$ values are extremely small and the $y$ values are extremely large. In such cases transformation of the original date may be necessary.

## Recommended Reading

*The Theory of Splines and Their Application*, H. J. Ahlberg, E. N. Nilson, and J. L. Walsh, Academic Press, New York, 1967.

*Modern Methods of Engineering Computation*, R. L. Ketter and S. P. Prawel, Jr., McGraw-Hill Book Company, New York, 1969.

*Numerical Mathematics and Computing*, Ward Cheney and David Kincaid, Brooks/Cole Publishing Co., Monterey, Calif., 1980.

*Applied Numerical Analysis*, Curtis F. Gerald and Patrick O. Wheatley, Addison-Wesley, Reading, Mass., 1984.

## PROBLEMS

**8.1** The following table was obtained from a polynomial function.

| $x$ | 0 | 1 | 2 | 3 | 4 | 5 | 6 | 7 |
|---|---|---|---|---|---|---|---|---|
| $y$ | 0 | 0 | 14 | 78 | 252 | 620 | 1290 | 2394 |

Determine the order of the polynomial using a difference table.

**8.2** Evaluate the polynomial in problem 8.1.

**8.3** Develop a third-order interpolating polynomial to pass through the following sets of points:

**(a)**

| $x$ | 0 | 2 | 5 | 7 |
|---|---|---|---|---|
| $y$ | 1 | 3 | $-3$ | 3 |

**(b)**

| $x$ | 1 | 2 | 3 | 4 |
|---|---|---|---|---|
| $y$ | 1 | 5 | 4 | 1 |

What are the shape functions?

**8.4** Derive a fourth-order interpolating polynomial formula using
**(a)** forward interpolation,
**(b)** backward interpolation,
**(c)** central interpolation.

**8.5** Given the following data, approximate the functional value at $x = 3.75$ using Newton's forward interpolating formula.

| $x$ | $-1$ | 0 | 1 | 2 | 3 | 4 |
|---|---|---|---|---|---|---|
| $y$ | 2 | $-1$ | 1 | 3 | 5 | 2 |

**8.6** Rework problem 8.5 using Newton's backward interpolating formula.

**8.7** Use the Lagrange interpolation formula to approximate the functional value at $x = 3.5$ for the following data set:

| $x$ | 0 | 3 | 7 | 8 |
|---|---|---|---|---|
| $y$ | 2 | 4 | 19 | 28 |

**8.8** Plot the shape functions for a fourth-order polynomial passing through the following data set:

| $x$ | $-1$ | 2 | 4 | 5 | 6 |
|---|---|---|---|---|---|
| $y$ | 2 | 7 | 10 | 3 | 0 |

What is unique about their behavior?

**8.9** Develop a set of cubic splines for the data set given in problem 8.8 assuming
**(a)** the second derivatives at the ends are both equal to zero;
**(b)** the second derivatives at $x = -1$ and $x = 6$ are equal to those at $x = 2$ and $x = 5$, respectively;
**(c)** approximate the second derivatives at $x = -1$ and $x = 6$ using interpolating polynomials of the third order.

**8.10** Replace the following function by a set of cubic splines in the range $4 \leq x \leq 20$ at $h = 4$:

$$f(x) = 2\log(1 + \tfrac{6}{10}x)$$

Assume that the second derivatives at the ends are both zero.

**8.11** Fit a fourth-order interpolating polynomial through the data set obtained for problem 8.10 then plot the resulting function and the cubic splines obtained earlier.

# chapter 9

# Curve Fitting

A photograph of the moon surface. *Courtesy of N.A.S.A.*

"Engineers are problem solvers regardless of the challenges."

# 9

## 9.1 INTRODUCTION

The formulation of mathematical models of many physical systems is a basic first step in the process of evaluating their behavior. Unfortunately, such formulations may become too complex or may not even be possible. Consequently, empirical functional relationships are often developed to describe system behavior using experimental data.

In fitting data with an approximating function, there are two basic approaches. The first approach involves passing an assumed function (preferably a polynomial) through every data point. The interpolating polynomials discussed in chapter 8 have this special property. However, such approaches suffer from three significant limitations. These are

1. Unexpectedly large deviations from a smooth curve (wiggle).
2. An $n$th-order polynomial for $n + 1$ data points is two complex for large $n$ values.
3. Experimental data are subject to errors and to pass a polynomial through every point is inappropriate.

Obviously, the first two of these limitations can be eliminated by using cubic splines (see section 8.8). Unfortunately, there is little that can be done to address the third of these deficiencies. Alternatively, we may resort to the second approach. This involves assuming a function which best describes the shape of the curve representing the given data set, and having it pass as close as possible, but not necessarily through every data point. This is precisely what we will attempt to do in this chapter.

## 9.2 INTRODUCTION TO THE METHOD OF LEAST SQUARES

This method of evaluating empirical formulas was developed over a century ago and has been in use for many years. Like the method of cubic splines, it attempts to fit a simple function through a set of data points without the wiggle problem associated with high-order polynomials. Unlike the cubic splines technique, it presupposes that the derived functional relationship does not necessarily pass through every data point. The procedure involves approximating a function such that the sum of the squares of the differences between the approximating function and the actual functional values given by the data is a minimum. The basis for the method is represented graphically in figure 9.1.

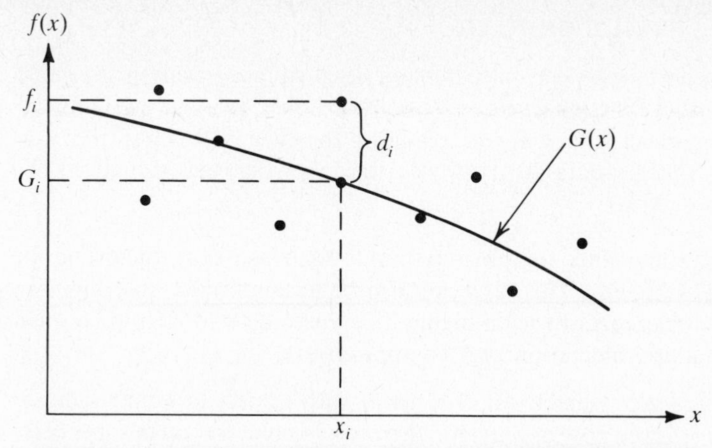


**Figure 9.1** Graphical interpretation of the least-squares method.

Here, given a set of $N$ data points, we wish to fit an approximating function of the form $G(x)$ having the following form:

$$G(x) = a_1 g_1(x) + a_2 g_2(x) + \cdots + a_m g_m(x) \tag{9.1}$$

where $m \leq N$ and $g_1(x), \ldots, g_m(x)$ are assumed functions of the independent variable $x$. The problem is then to evaluate the regression coefficients $a_1, \ldots, a_m$. The method of least squares suggests that these can be easily calculated by minimizing a deviation function $D$ defined as follows:

$$D = \sum_{i=1}^{N} [f(x_i) - G(x_i)]^2 \tag{9.2}$$

Note that since one can fit different $G$ functions through the data set then $a_1, \ldots, a_m$ may be looked at as variables. Therefore, using the method of calculus we can minimize the function given by equation (9.2) as shown below:

$$\begin{aligned} \frac{\partial D}{\partial a_1} &= 0 \\ \frac{\partial D}{\partial a_2} &= 0 \\ &\\ \frac{\partial D}{\partial a_m} &= 0 \end{aligned} \tag{9.3}$$

The set of $m \times m$ linear algebraic equations given by equations (9.3) can therefore be solved for the unknowns $a_1, \ldots, a_m$. These coefficients are then substituted into equation (9.1) to give the desired approximating function.

## 9.3 WHAT TYPE OF FUNCTION TO FIT

One of the most important aspects of regression (least squares) analysis is related to the question of what functional relationship $G(x)$ should be assumed. Unfortunately, there are no clear criteria for determining the type of function that can best represent an arbitrary data set. However, the following rules should be observed whenever possible:

1. Plot the data and look for obvious trends such as linear, quadratic, or higher-order behavior. This task can also be accomplished by examining the first few differences to see if higher differences tend to zero. If so, then a polynomial approximation may be appropriate.
2. See if the data are symmetrical. Symmetry with respect to $f$ may indicate polynomials of even powers only. Furthermore, symmetry can in some cases be achieved by transforming the data.
3. Consider periodicity. Trigonometric functions may be possible.
4. Consider plotting the data on a semilog and/or log-log scale. This may provide information on whether logarithmic or exponential functions can be assumed.
5. Break the data set into groups and consider the possibility of assuming different functions for the data subsets.

These rules are helpful in the absence of more suitable criteria, but should be used with discretion. Perhaps the most important rule is that of common sense. That is, if the given data are related to a particular physical phenomenon for which a functional relationship is known, then it makes no sense to assume something different. For example, when examining the relationship between stress and strain for a steel specimen, we know that the initial portion of this relationship is linear and fully described by Hooke's law. Therefore, if the experimental data indicate nonlinear behavior initially, then the data are in error and a linear function must be assumed.

## 9.4 LINEAR REGRESSION

In order to fully understand the significance of the method of least squares and be able to interpret the deviation function given by equation (9.2) let us consider figure 9.2.

Given a set of $N$ data points, a linear function is assumed to exist between the dependent variable $f$ and the independent variable $x$. That is,

$$G(x) = a_1 + a_2 x \tag{9.4}$$

where $g_1(x) = 1$ and $g_2(x) = x$. It is evident that many straight lines can be fitted through any two of the data points. Such an arbitrary procedure will undoubtedly be biased and significantly dependent on individual preference. In fact, there are

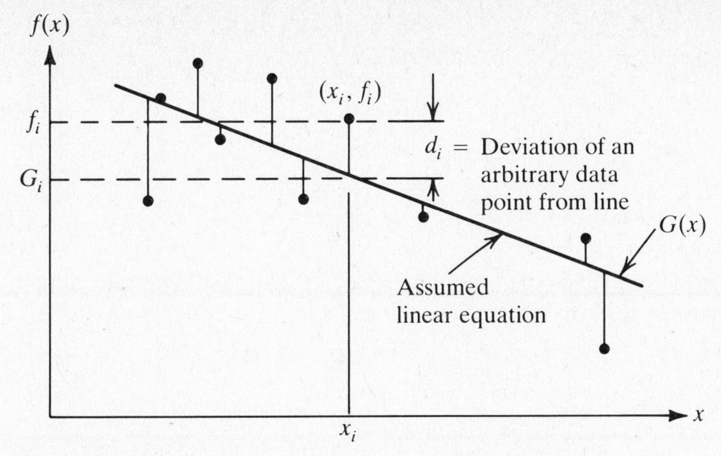


**Figure 9.2** Schematic representation of linear regression.

$S$ linear equations that can be determined where

$$S = \frac{N!}{2!(N-2)!}$$

where $N$ is the number of data points. The implication is that there are $S$ different sets of values for $a_1$ and $a_2$ that can be evaluated. This implies that $a_1$ and $a_2$ are indeed variables and must be fixed in such a way that the derived function is the best possible solution to the problem. Such solutions will have the smallest deviations $d_i$ from the actual data. The set of deviations is given by

$$\begin{aligned}
d_1 &= f_1 - G(x_1) = f_1 - a_1 - a_2x_1 \\
d_2 &= f_2 - G(x_2) = f_2 - a_1 - a_2x_2 \\
&\vdots \\
d_i &= f_i - G(x_i) = f_i - a_1 - a_2x_i \\
&\vdots \\
d_N &= f_N - G(x_N) = f_N - a_1 - a_2x_N
\end{aligned}$$

The object is to minimize the sum of all of the deviations. Unfortunately, this is not possible because the negative deviations will reduce the positive ones. However, if the sum of the squares of the deviations is determined then a true estimate of the total deviation is possible. This is precisely what equation (9.2) states. Consequently for an assumed linear approximation we have

$$d^2 = D = \sum_{i=1}^{N} [f(x_i) - G(x_i)]^2 = \sum_{i=1}^{N} (f_i - a_1 - a_2x_i)^2$$

Note that $f(x_i) = f_i$ and $G(x_i) = a_1 + a_2 x_i$. Differentiating with respect to $a_1$ and $a_2$ then setting the resulting expressions to zero gives

$$\frac{\partial D}{\partial a_1} = 0 = 2 \sum_{i=1}^{N} (f_i - a_1 - a_2 x_i)(-1)$$

$$\frac{\partial D}{\partial a_2} = 0 = 2 \sum_{i=1}^{N} (f_i - a_1 - a_2 x_i)(-x_i)$$

Simplifying and rearranging yields the following set of 2 × 2 linear algebraic equations in the unknowns $a_1$ and $a_2$:

$$\sum_{i=1}^{N} f_i = a_1 N + a_2 \sum_{i=1}^{N} x_i \tag{9.5a}$$

$$\sum_{i=1}^{N} f_i x_i = a_1 \sum_{i=1}^{N} x_i + a_2 \sum_{i=1}^{N} x_i^2 \tag{9.5b}$$

Equation (9.5) can therefore be solved to give

$$a_1 = \frac{\sum f_i \sum x_i^2 - \sum x_i \sum f_i x_i}{N \sum x_i^2 - (\sum x_i)^2} \tag{9.6a}$$

$$a_2 = \frac{N \sum f_i x_i - \sum x_i \sum f_i}{N \sum x_i^2 - (\sum x_i)^2} \tag{9.6b}$$

Note that for simplicity the limits have been omitted from the sums in equation (9.6). It is evident that the assumed linear functional relationship is now completely defined in terms of the given data set.

EXAMPLE 9.1

Use linear regression to fit the following data set:

| $x_i$ | 0 | 1 | 2 | 3 | 5 |
|---|---|---|---|---|---|
| $f_i$ | 0 | 1.4 | 2.2 | 3.5 | 4.4 |

Solution

Assuming a linear function such that $g_1(x) = 1$ and $g_2(x) = x$ gives $G(x) = a_1 + a_2 x$. Then $a_1$ and $a_2$ are given directly by equation (9.6). The solution to regression problems is often simplified by establishing a table for the various sums as shown below.

| $i$ | $x_i$ | $f_i$ | $x_i f_i$ | $x_i^2$ |
|---|---|---|---|---|
| 1 | 0 | 0 | 0 | 0 |
| 2 | 1 | 1.4 | 1.4 | 1 |
| 3 | 2 | 2.2 | 4.4 | 4 |
| 4 | 3 | 3.5 | 10.5 | 9 |
| 5 | 5 | 4.4 | 22.0 | 25 |
| $\sum$ | 11 | 11.50 | 38.3 | 39 |

Substituting the sums into equations (9.6) and noting that $n = 5$ yields

$$a_1 = \frac{(11.5)(39) - (11)(38.3)}{(5)(39) - (11)^2} = \frac{27.2}{74} = 0.368$$

$$a_2 = \frac{(5)(38.3) - (11)(11.50)}{74} = 0.878$$

Consequently, the assumed linear relationship that can best fit the data is established as

$$G(x) = 0.368 + 0.878x$$

Clearly, $G(x)$ replaces the tabulated functional relationship given by $f(x)$. The original data along with the approximating polynomials are shown graphically in figure 9.3.

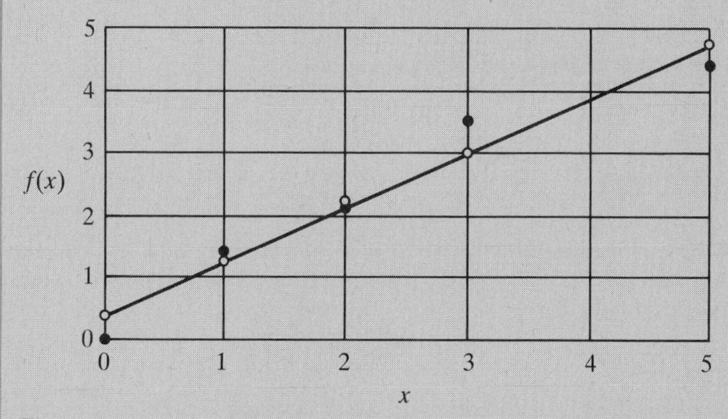


**Figure 9.3** A plot of given functional values versus estimated regression values.

## 9.5 LINEARIZATION

Often, the behavior exhibited by a given data set may appear to be nonlinear. However, in some cases the assumed nonlinear approximating function can be transformed into a linear one by directly operating on the data. This of course is advantageous in that it allows us to use equations (9.6) to achieve a solution. For illustration purposes, suppose that given data $(x_i, f_i)$ exhibit behavior that can be best described by the following function:

$$G(x) = a_1 x^{a_2} \tag{9.7}$$

We can easily transform equation (9.7) to an equivalent linear form by simply taking the natural logarithm of both sides (obviously, taking $\log_{10}$ will work just as well). Thus

$$\ln G = \ln a_1 + a_2 \ln x$$

Now making the following substitutions into equation (9.7),

$$\bar{G} = \ln G \tag{9.8a}$$

$$\bar{a}_1 = \ln a_1 \tag{9.8b}$$

$$\bar{x} = \ln x \tag{9.8c}$$

$$\bar{f} = \ln f \tag{9.8d}$$

yields the equivalent linear functional relationship to that given by equation (9.7). That is,

$$\bar{G} = \bar{a}_1 + a_2 \bar{x} \tag{9.8e}$$

Comparison of equation (9.8e) to equation (9.4) clearly shows that they are equivalent. Consequently, their solution must be equivalent as well. Hence, using equation (9.6), we have

$$\bar{a}_1 = \frac{\sum \bar{f}_i \sum \bar{x}_i^2 - \sum \bar{x}_i \sum \bar{f}_i \bar{x}_i}{N \sum \bar{x}_i^2 - (\sum \bar{x}_i)^2} \tag{9.9a}$$

$$a_2 = \frac{N \sum \bar{f}_i \bar{x}_i - \sum \bar{x}_i \sum \bar{f}_i}{N \sum \bar{x}_i^2 - (\sum \bar{x}_i)^2} \tag{9.9b}$$

Then, once $\bar{a}_1$ and $a_2$ are determined, equation (9.8b) is solved for $a_1$ and then substituted into equation (9.7). It is evident that similar transformations are possible for functions of the following form:

$$G(x) = a_1 e^{a_2 x}$$

For cases in which the given data involve zeros or negative numbers, one must transform the data to an equivalent set in which all values are positive. This may be accomplished by adding an arbitrary scalar and introducing a set of new variables.

In some applications the assumed functional relationship may be given in the form

$$G(x) = a_1 x^{a_2} + c$$

where $c$ is a known constant. Such problems can easily be handled by subtracting the constant $c$ from both sides to give

$$G(x) - c = a_1 x^{a_2}$$

Then introducing the transformation

$$T(x) = G(x) - c$$

yields the modified form

$$T(x) = a_1 x^{a_2}$$

Obviously, this equation is of the same form as equation (9.7) and can be transformed in a similar manner.

EXAMPLE 9.2

The following data were plotted and shown to exhibit behavior that can be best described by the function $G = a_1 x^{a_2}$. Evaluate the unknowns $a_1$ and $a_2$.

| $x_i$ | 1 | 2 | 3 | 4 |
|---|---|---|---|---|
| $f_i$ | 1 | 4 | 8 | 14 |

Solution

Obviously, the assumed function is of the same form as that given by equation (9.7). That is, the solution is readily given by equation (9.9). Hence we establish the following table for the various sums:

| $i$ | $x_i$ | $f_i$ | $x_i$ | $f_i$ | $x_i f_i$ | $x_i^2$ |
|---|---|---|---|---|---|---|
| 1 | 1 | 1 | 0 | 0 | 0 | 0 |
| 2 | 2 | 4 | 0.6931 | 1.3863 | 0.9608 | 0.4804 |
| 3 | 3 | 8 | 1.0986 | 2.0794 | 2.2844 | 1.2069 |
| 4 | 4 | 14 | 1.3863 | 2.6390 | 3.6585 | 1.9218 |
| $\Sigma$ | | | 3.1780 | 6.1047 | 6.9037 | 3.6091 |

Substituting into equation (9.9) gives

$$\bar{a}_1 = \frac{(6.1047)(3.6091) - (3.1780)(6.9037)}{(4)(3.6091) - (3.1780)^2} = 0.0213$$

$$a_2 = \frac{(4)(6.9037) - (3.1780)(6.1047)}{(4)(3.6091) - (3.1780)^2} = 1.8941$$

Since $\bar{a}_1 = \ln a_1$ this implies that $a_1 = e^{\bar{a}_1} = 1.0215$. Substituting $a_1$ and $a_2$ into the assumed functional approximation yields the desired expression. That is,

$$G = 1.0215x^{1.8941}$$

This regression equation gives reasonable results for the given tabulated values.

## 9.6 NONLINEAR REGRESSION

In the event that the assumed functional approximation is nonlinear and linearization is not possible, the regression coefficients must be established by using the deviation function. For example, suppose that the data $(x_i, f_i)$ suggest a second-order polynomial of the form

$$G(x) = a_1 + a_2 x + a_3 x^2 \tag{9.10}$$

where $g_1(x) = 1$, $g_2(x) = x$, and $g_3(x) = x^2$. The corresponding deviation function that must be minimized with respect to the unknown regression coefficients is given by equation (9.2) as follows:

$$D = \sum_{i=1}^{N} (f_i - a_1 - a_2 x_i - a_3 x_i^2)^2$$

Consequently, differentiating with respect to $a_1$, $a_2$, and $a_3$ yields the following three linear algebraic equations:

$$\frac{\partial D}{\partial a_1} = 0 = 2 \sum_{i=1}^{N} (f_i - a_1 - a_2 x_i - a_3 x_i^2)(-1)$$

$$\frac{\partial D}{\partial a_2} = 0 = 2 \sum_{i=1}^{N} (f_i - a_1 - a_2 x_i - a_3 x_i^2)(-x_i)$$

$$\frac{\partial D}{\partial a_3} = 0 = 2 \sum_{i=1}^{N} (f_i - a_1 - a_2 x_i - a_3 x_i^2)(-x_i^2)$$

Simplifying and rearranging yields the following more convenient form:

$$\sum_{i=1}^{N} f_i = a_1 N + a_2 \sum_{i=1}^{N} x_i + a_3 \sum_{i=1}^{N} x_i^2 \tag{9.11a}$$

$$\sum_{i=1}^{N} f_i x_i = a_1 \sum_{i=1}^{N} x_i + a_2 \sum_{i=1}^{N} x_i^2 + a_3 \sum_{i=1}^{N} x_i^3 \tag{9.11b}$$

$$\sum_{i=1}^{N} f_i x_i^2 = a_1 \sum_{i=1}^{N} x_i^2 + a_2 \sum_{i=1}^{N} x_i^3 + a_3 \sum_{i=1}^{N} x_i^4 \tag{9.11c}$$

Consequently, solving equation (9.11) for the unknowns $a_1$, $a_2$, and $a_3$, then substituting into equation (9.10) yields the desired approximating regression polynomial. It is evident that the procedure just outlined can be easily extended to high-order polynomials or even transcendental functions. Unfortunately, the resulting set of linear algebraic equations in the unknown regression coefficients tends to become unstable as the order of the assumed polynomial increases. Consequently, other techniques should be used instead, such as that of orthogonal polynomials presented in section 9.8.

EXAMPLE 9.3

The following data represent the elevation in feet ($f$) versus horizontal distance in 100 ft ($x$) for a hilly road section. Derive a second-order approximating polynomial.

| $x_i$ | 0 | 1 | 2 | 3 | 4 | 5 | 6 | 7 | 8 |
|---|---|---|---|---|---|---|---|---|---|
| $f_i$ | 4 | 5 | 10 | 17 | 21 | 16 | 11 | 3 | 1 |

Solution

A plot of the given data set indeed reveals that a second-order polynomial may provide a good estimate of the relationship between $f$ and $x$. Thus

$$G(x) = a_1 + a_2 x + a_3 x^2 \tag{9.12}$$

Obviously, equation (9.11) can be readily used to solve for the unknown regression coefficients $a_1$, $a_2$, and $a_3$. The various sums are given below:

$$\sum_{i=1}^{9} x_i = 13, \qquad \sum_{i=1}^{9} x_i^2 = 204$$

$$\sum_{i=1}^{9} x_i^3 = 1296, \qquad \sum_{i=1}^{9} x_i^4 = 8772$$

$$\sum_{i=1}^{9} f_i = 88, \qquad \sum_{i=1}^{9} f_i x_i = 335$$

$$\sum_{i=1}^{9} f_i x_i^2 = 1541, \qquad N = 9$$

Consequently, solving equation (9.11) yields

$$a_1 = 1.29, \qquad a_2 = 7.97, \qquad a_3 = -1.03$$

The approximating polynomial function is then given by equation (9.12). That is,

$$G(x) = 1.29 + 7.97x - 1.03x^2$$

Clearly, the function $G(x)$ is an approximation to the dependent variable $f$. The given data values as well as the least-squares fit are shown graphically in figure 9.4.

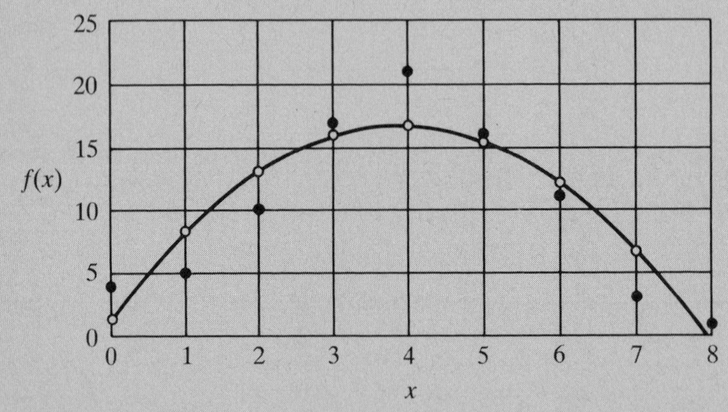


**Figure 9.4** Graphical representation of the given data and the corresponding second-order polynomial approximation.

## 9.7 MULTIPLE REGRESSION

Many experiments relating to the behavior of physical systems involve more than one independent variable. For example, the cost of building a house depends on many variables such as labor, volume, inflation, etc. Consequently, the dependent variable may be expressed as a linear combination of the independent variables, whose coefficients are determined by using regression analysis. For illustration consider a problem in which the data relating a dependent variable $f_i$ to two independent variables $(x_i, y_i)$ have been accumulated and an approximating function of the following form has been assumed:

$$G(x) = a_1 + a_2x + a_3y \tag{9.13}$$

The question we need to address is what are the best possible values for the regression coefficients $a_1$, $a_2$, and $a_3$ such that the function $G(x)$ can best represent the given data? Obviously, this problem can be easily handled by using the deviation function given by equation (9.2). That is,

$$D = \sum_{i=1}^{N} (f_i - a_1 - a_2 x - a_3 y)^2$$

Therefore, differentiating with respect to $a_1$, $a_2$, and $a_3$ yields the following set of linear algebraic equations:

$$\sum_{i=1}^{N} f_i = a_1 N + a_2 \sum_{i=1}^{N} x_i + a_3 \sum_{i=1}^{N} y_i \quad \textbf{(9.14a)}$$

$$\sum_{i=1}^{N} f_i x_i = a_1 \sum_{i=1}^{N} x_i + a_2 \sum_{i=1}^{N} x_i^2 + a_3 \sum_{i=1}^{N} y_i x_i \quad \textbf{(9.14b)}$$

$$\sum_{i=1}^{N} f_i y_i = a_1 \sum_{i=1}^{N} y_i + a_2 \sum_{i=1}^{N} x_i y_i + a_3 \sum_{i=1}^{N} y_i^2 \quad \textbf{(9.14c)}$$

The set of equations given by (9.14) can once again be solved for the regression coefficients, which are then substituted into equation (9.13) in order to fully describe the assumed functional approximation.

E

EXAMPLE 9.4

The following experimental data relate the current $I$ to time $t$ and sheet resistance $R$ in a device. Develop a functional approximation of the form $\bar{I} = a_1 + a_2 R + a_3 t$:

| | | | | | | |
|---|---|---|---|---|---|---|
| $I$ | 5.3 | 7.8 | 7.6 | 9.7 | 10.5 | 12.6 |
| $R$ | 66 | 85 | 70 | 140 | 95 | 125 |
| $t$ | 1.5 | 2.6 | 0.6 | 1.3 | 2.7 | 1.6 |

Solution

The regression equations whose solution yields the unknown regression coefficients are of the same form as those given by equation (9.14). That is,

$$\sum_{i=1}^{N} I_i = a_1 N + a_2 \sum_{i=1}^{N} R_i + a_3 \sum_{i=1}^{N} t_i$$

$$\sum_{i=1}^{N} I_i R_i = a_1 \sum_{i=1}^{N} R_i + a_2 \sum_{i=1}^{N} R_i^2 + a_3 \sum_{i=1}^{N} t_i R_i$$

$$\sum_{i=1}^{N} I_i t_i = a_1 \sum_{i=1}^{N} t_i + a_2 \sum_{i=1}^{N} R_i t_i + a_3 \sum_{i=1}^{N} t_i^2$$

where the sums are given as

$$\sum_{i=1}^{n} I_i = 53.5, \qquad \sum_{i=1}^{n} t_i = 10.3$$

$$\sum_{i=1}^{n} R_i = 581, \qquad \sum_{i=1}^{n} t_i R_i = 1000.5$$

$$\sum_{i=1}^{n} t_i^2 = 20.91, \qquad \sum_{i=1}^{n} R_i^2 = 60.731$$

$$\sum_{i=1}^{n} I_i t_i = 93.91, \qquad \sum_{i=1}^{n} I_i R_i = 5475.3$$

Substituting and solving for $a_1$, $a_2$, and $a_3$ yields

$$a_1 = 1.581, \qquad a_2 = 0.066, \qquad a_3 = 0.577$$

Therefore the approximating function is given by

$$\bar{I} = 0.577 + 0.066R + 1.581t$$

Note that multiple regression may involve independent variables that are nonlinear. Such problems are handled easily by minimizing the particular deviation function with respect to the various regression coefficients.

## 9.8 ORTHOGONAL POLYNOMIALS FOR EQUAL INTERVALS

The least-squares method for approximating a polynomial suffers from a rather significant deficiency. That is, the resulting regression equations for a high-order assumed polynomial are ill conditioned. Consequently, the regression coefficients cannot be determined accurately even when double precision is used. For polynomial curve fitting with evenly spaced base points, it is possible to develop an alternative approach in which the resulting regression equations are diagonal. Hence the regression coefficients can be evaluated without any difficulty. This approach involves selecting a set of orthogonal polynomials $P_0(x), \ldots, P_m(x)$ with the following properties:

$$\sum_{i=1}^{N} P_k(x_i) P_L(x_i) = 0 \quad \text{for} \quad k \neq L \tag{9.15a}$$

The assumed polynomial is expressed as a linear combination of the $m + 1$ orthogonal polynomials as follows:

$$G(x) = A_0 P_0(x) + A_1 P_1(x) + \cdots + A_m P_m(x) \tag{9.15b}$$

The corresponding deviation function for $N$ data points $(x_i, f_i)$ is then given by equation (9.2) and is repeated here for convenience.

$$D = \sum_{i=1}^{N} [f(x_i) - G(x_i)]^2$$

substituting equation (9.15b) for $G(x_i)$ yields

$$D = \sum_{i=1}^{N} [f(x_i) - A_0 P_0(x_i) - A_1 P_1(x_i) - \cdots - A_m P_m(x_1)]^2$$

Differentiating the deviation function with respect to the regression coefficients yields the necessary set of linear algebraic equations. That is

$$\sum_{i=1}^{N} f(x_i)P_0(x_i) = A_0 P_0^2(x_i) + A_1 \sum_{i=1}^{N} P_1(x_i) + \cdots + A_m \sum_{i=1}^{N} P_m(x_i)P_1(x_i)$$

$$\sum_{i=1}^{N} f(x_i)P_1(x_i) = A_0 \sum_{i=1}^{N} P_0(x_i)P_1(x_i) + A_1 \sum_{i=1}^{N} P_1^2(x_i) + \cdots + A_m \sum_{i=1}^{N} P_m(x_i)P_1(x_i)$$

$$\sum_{i=1}^{N} f(x_i)P_m(x_i) = A_0 \sum_{i=1}^{N} P_0(x_i)P_m(x_i) + a_1 \sum_{i=1}^{N} P_1(x_i)P_m(x_i) + \cdots + A_m \sum_{i=1}^{N} P_m^2(x_i)$$

Now using the special property of the orthogonal polynomials given by equation (9.15b) permits the simplification of the regression equation to the following:

$$\sum_{i=1}^{N} f(x_i)P_0(x_i) = A_0 \sum_{i=1}^{N} P_0^2(x_i)$$

$$\sum_{i=1}^{N} f(x_i)P_1(x_i) = A_1 \sum_{i=1}^{N} P_1^2(x_i)$$

$$\vdots$$

$$\sum_{i=1}^{N} f(x_i)P_m(x_i) = A_m \sum_{i=1}^{N} P_m^2(x_i)$$

Hence the solution is easily determined as follows:

$$A_0 = \frac{\sum_{i=1}^{N} f(x_i)P_0(x_i)}{\sum_{i=1}^{N} P_0^2(x_i)}$$

$$A_1 = \frac{\sum_{i=1}^{N} f(x_i)P_1(x_i)}{\sum_{i=1}^{N} P_1^2(x_i)} \tag{9.16}$$

$$\vdots$$

$$A_m = \frac{\sum_{i=1}^{N} f(x_i)P_m(x_i)}{\sum_{i=1}^{N} P_m^2(x_i)}$$

The only remaining question we need to address is how can the orthogonal polynomials be determined? A set of polynomials for which the conditions specified by equation (9.15) are met are the so-called gram polynomials. These are defined as follows:

$$P_m(\alpha) = \sum_{r=0}^{m} (-1)^r \frac{1}{r!r!} \frac{(m+r)!}{(m-r)!} \frac{(N-r-1)!}{(N-1)!} \frac{\alpha!}{(\alpha-r)!} \tag{9.17}$$

where

$$\alpha = \frac{x - x_0}{h}$$

Here $N$ is the number of equally spaced base points, $h$ is the $x$ interval, $x_0$ is the first $x$ data value, and $m$ is the order of the polynomial in question. It is evident that corresponding to each data value there is an $\alpha$ value. In fact, the $\alpha$ values will be $0, 1, \ldots, N-1$. Consequently, equation (9.17) is generally expressed in terms of $\alpha$. For illustration purposes the first three orthogonal polynomials corresponding to $N$ data points are given as follows:

$$P_0 = \sum_{r=0}^{0} (-1)^0 \frac{1}{0!0!} \frac{0!}{0!} \frac{(N-1)!}{(N-1)!} \frac{\alpha!}{\alpha!} = 1$$

$$P_1 = \sum_{r=0}^{1} (-1)^r \frac{1}{r!r!} \frac{(1+r)!}{(1-r)!} \frac{(N-r-1)!}{(N-1)!} \frac{\alpha!}{(\alpha-r)!} = 1 - \frac{2\alpha}{N-1}$$

$$P_2 = \sum_{r=0}^{2} (-1)^r \frac{1}{r!r!} \frac{(2+r)!}{(2-r)!} \frac{(N-r-1)!}{(N-1)!} \frac{\alpha!}{(\alpha-r)!}$$

$$= 1 - \frac{6\alpha}{(N-1)} + \frac{6\alpha(\alpha-1)}{(N-1)(N-2)}$$

Obviously, polynomials of higher order can be easily derived in a similar manner.

**E** EXAMPLE 9.5

Fit a second-order polynomial through the data set given below using orthogonal polynomials.

| $x_i$ | 0 | 0.5 | 1 | 1.5 | 2 | 2.5 | 3 |
|---|---|---|---|---|---|---|---|
| $f(x_i)$ | 1 | 3 | 6 | 10 | 18 | 24 | 35 |

Solution

Since the base points are equally spaced, the method of orthogonal polynomials can be implemented. Hence assuming the following polynomial approximation for the function $f$ gives

$$G(x) = A_0P_0(x) + A_1P_1(x) + A_2P_2(x)$$

The polynomials $P_0$, $P_1$, and $P_2$ are determined for $m = 2$ and $N = 7$ by using equation (9.17) as follows:

$$P_m(\alpha) = \sum_{r=0}^{2} (-1)^r \frac{1}{r!r!} \frac{(2+r)!}{(2-r)!} \frac{(6-r)!}{6!} \frac{\alpha!}{(\alpha - r)!}$$

Therefore, for $m = 0, 1, 2$ we have

$$P_0(\alpha) = 1$$

$$P_1(\alpha) = 1 - \frac{\alpha}{3}$$

$$P_2(\alpha) = 1 - \alpha + \frac{\alpha(\alpha - 1)}{5}$$

Note that $\alpha = (x - x_0)/h$, where $h = 0.50$ and $x_0 = 0$. Consequently, the polynomials are indeed functions of the independent variable $x$. The regression coefficients $A_0$, $A_1$, and $A_2$ are determined by using equation (9.16). For $m = 2$ and $N = 7$ we have

$$A_0 = \frac{\sum_{i=1}^{7} f(x_i)P_0(x_i)}{\sum_{i=1}^{7} P_0^2(x_i)}$$

$$A_1 = \frac{\sum_{i=1}^{7} f(x_i)P_1(x_i)}{\sum_{i=1}^{7} P_1^2(x_i)}$$

$$A_2 = \frac{\sum_{i=1}^{7} f(x_i)P_2(x_i)}{\sum_{i=1}^{7} P_2^2(x_i)}$$

The various sums are evaluated conveniently as follows:

| $x$ | $f$ | $\alpha$ | $P_0$ | $P_1$ | $P_2$ | $fP_0$ | $fP_1$ | $fP_2$ |
|---|---|---|---|---|---|---|---|---|
| 0 | 1 | 0 | 1 | 1 | 1 | 1.0000 | 1.0000 | 1.0000 |
| 0.5 | 3 | 1 | 1 | 0.6667 | 0 | 3.0000 | 2.0000 | 0 |
| 1 | 6 | 2 | 1 | 0.3333 | −0.6 | 6.0000 | 2.0000 | −3.6000 |
| 1.5 | 10 | 3 | 1 | 0 | −0.8 | 10.0000 | 0 | −8.0000 |
| 2 | 18 | 4 | 1 | −0.3333 | −0.6 | 18.0000 | −6.0000 | −10.0000 |
| 2.5 | 24 | 5 | 1 | −0.6667 | 0 | 24.0000 | −16.0000 | 0 |
| 3 | 35 | 6 | 1 | −1 | 1 | 35.0000 | −35.0000 | 35.0000 |
| $\Sigma$ | | | | | | 97.0 | −52.0 | 13.60 |

In addition, the sums of the squares of the orthogonal functions are

$$\sum_{i=1}^{7} P_0^2 = 1^2 + 1^2 + 1^2 + 1^2 + 1^2 + 1^2 + 1^2 = 7.0$$

$$\sum_{i=1}^{7} P_1^2 = (1)^2 + (0.6667)^2 + (0.333)^2 + (0)^2 + (-0.3333)^2 + (-0.6667)^2 + (-1)^2 = 3.1111$$

$$\sum_{i=1}^{7} P_2^2 = (1)^2 + (0)^2 + (-0.6)^2 + (-0.8)^2 + (-0.6)^2 + (0)^2 + (1)^2 = 3.36$$

Solving for the regression coefficients yields

$$A_0 = \frac{97.0}{7} = 13.857$$

$$A_1 = \frac{-52.0}{3.1111} = -16.714$$

$$A_2 = \frac{13.60}{3.36} = 4.048$$

Hence the polynomial approximation is given in terms of $\alpha$ as

$$G(x) = 13.857(1) - 16.714\left(1 - \frac{\alpha}{3}\right) + 4.048\left(1 - \alpha + \frac{\alpha(\alpha - 1)}{5}\right)$$

which simplifies to

$$G(x) = 1.191 + 0.7137\alpha + 0.8096\alpha^2$$

Now, substituting $\alpha = (x - 0)/0.5 = 2x$ gives

$$G(x) = 1.191 + 1.4274x + 3.2384x^2$$

Obviously, $G(x)$ represents a least-squares approximation of $f(x)$. That is,

| $x$ | 0 | 0.5 | 1.0 | 1.5 | 2.0 | 2.5 | 3.0 |
|---|---|---|---|---|---|---|---|
| $f(x)$ | 1 | 3 | 6 | 10 | 18 | 24 | 35 |
| $G(x)$ | 1.19 | 2.71 | 5.86 | 10.62 | 17.00 | 25.00 | 34.62 |

It is evident that the approximate values are reasonable when compared with the given functional values. This method can be easily extended to higher-order polynomials by simply following the steps outlined in this example.

## 9.9 GOODNESS OF FUNCTIONAL APPROXIMATIONS

Thus far we have purposely avoided the question of how good the assumed functional approximation is in describing the tabulated values. In a qualitative sense, one may be able to address this particular concern by simply plotting the data. Unfortunately, such a procedure suffers from major limitations in that the number of independent variables cannot exceed two. Furthermore, for a given data set, one is not able to relatively compare different assumed functional approximations. In this section two statistical measures are introduced that permit the determination of the goodness of fit in a quantitative sense.

### 9.9.1 Coefficients of Multiple Determination

For a given data set, one may wish to try different functional approximations and then choose one that is most suitable. The coefficient of multiple determination provides a quantitative measure of the correlation between the dependent and the independent variable(s) for a given functional approximation. The basis for this statistical measure can be easily demonstrated by considering figure 9.5.

It is evident that if the deviations of the given data points $(x_i, f_i)$ from the functional approximation $G_i$ are large, then the assumed approximation is not suitable. On the other hand, if the deviations are small, then the assumed functional approximation is appropriate. This concept can be expressed mathematically by relating these deviations to the average functional value $\bar{f}$ as follows:

$$R^2 = \frac{\sum_{i=1}^{N} (G_i - \bar{f})^2}{\sum_{i=1}^{N} (f_i - \bar{f})^2} \tag{9.18}$$

where $G_i = G(x_i)$, $f_i = f(x_i)$, $\bar{f} = \bar{G}$, and $R^2$ is the coefficient of multiple determination. Obviously, if the deviation of the functional values given by the data is small then $f_i$ will be approximately equal to $G_i$ and $R^2$ is approximately equal to unity. Otherwise $R^2$ approaches zero. In general, functions may involve one or more independent variables. For the special case in which the assumed relationship

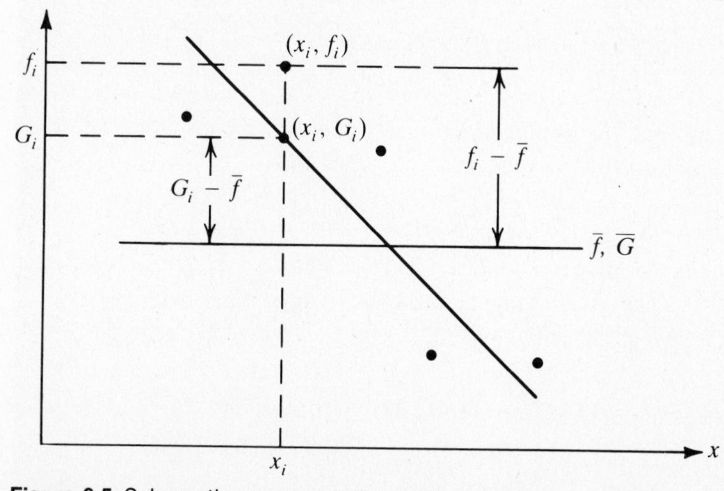


**Figure 9.5** Schematic representation of the coefficient of multiple determination.

between $f_i$ and $x_i$ is linear ($G(x) = a_1 + a_2 x$), equation (9.18) simplifies to

$$R^2 = \frac{\left(N \sum_{i=1}^{N} x_i f_i - \left(\sum_{i=1}^{N} f_i\right)^2\right)^2}{\left(N \sum_{i=1}^{N} x_i^2 - \left(\sum_{i=1}^{N} x_i\right)^2\right)\left(N \sum_{i=1}^{N} f_i^2 - \left(\sum_{i=1}^{N} f_i\right)^2\right)} \tag{9.19}$$

Clearly, equation (9.18) is more useful in that it is applicable regardless of the assumed functional approximation. The following rules pertaining to the $R^2$ value are included:

$R^2 = 1$ Perfect correlation

$R^2 = 0$ No correlation

In practice the value of $R^2$ ranges between these two limiting values (0 and 1) for any functional approximation.

**E**

EXAMPLE 9.6

Determine the value of $R^2$ for example 9.5.

Solution

Using the data, we begin by first calculating the average functional value $\bar{f}$ as follows:

$$\bar{f} = \frac{\sum_{i=1}^{N} f_i}{N} = \frac{1 + 3 + 6 + 10 + 18 + 24 + 35}{7} = 13.86$$

The approximate and the actual functional values were computed earlier. Hence, using equation (9.18) we have

$$\begin{aligned}\sum_{i=1}^{7} (G_i - \bar{f})^2 &= (1.19 - 13.86)^2 + (2.71 - 13.86)^2 + (5.86 - 13.86)^2 \\ &\quad + (10.62 - 13.86)^2 + (17 - 13.86)^2 \\ &\quad + (25 - 13.86)^2 + (34.62 - 13.86)^2 \\ &= 924.29\end{aligned}$$

$$\begin{aligned}\sum_{i=1}^{7} (f_i - \bar{f})^2 &= (1 - 13.86)^2 + (3 - 13.86)^2 + (6 - 13.86)^2 \\ &\quad + (10 - 13.86)^2 + (18 - 13.86)^2 + (24 - 13.86)^2 \\ &\quad + (35 - 13.86)^2 \\ &= 926.86\end{aligned}$$

Therefore

$$R^2 = \frac{924.29}{926.86} = 0.9972$$

Obviously the $R^2$ value indicates an excellent correlation between the dependent and the independent variables for the assumed approximation.

### 9.9.2 Standard Error of the Estimate

The coefficient of multiple determination suffers from a rather significant limitation. That is, as the number of data points approaches the number of regression coefficients the $R^2$ value approaches unity. The implication is that a linear approximation assumed for a set of two data points is expected to have an $R^2$ equal to 1.0 (why?). Consequently, we tend to be less certain of how appropriate a given functional approximation is regardless of the $R^2$ value if the number of data points is small. The standard error of the estimate provides a second measure for a given functional approximation in that it permits us to evaluate the relative certainty we can attach to derived functions. In that sense, it is similar to the standard deviation of random data. Consider figure 9.6.

The standard error $\sigma_e$ of the estimated functional relationship is given by

$$\sigma_e = \sqrt{\frac{\sum_{i=1}^{N} (f_i - G_i)}{N - M}} \tag{9.20}$$

where $N$ is the total number of data points and $M$ is the number of regression coefficients. It is evident that the smaller $\sigma_e$ is the more certain we can be of the

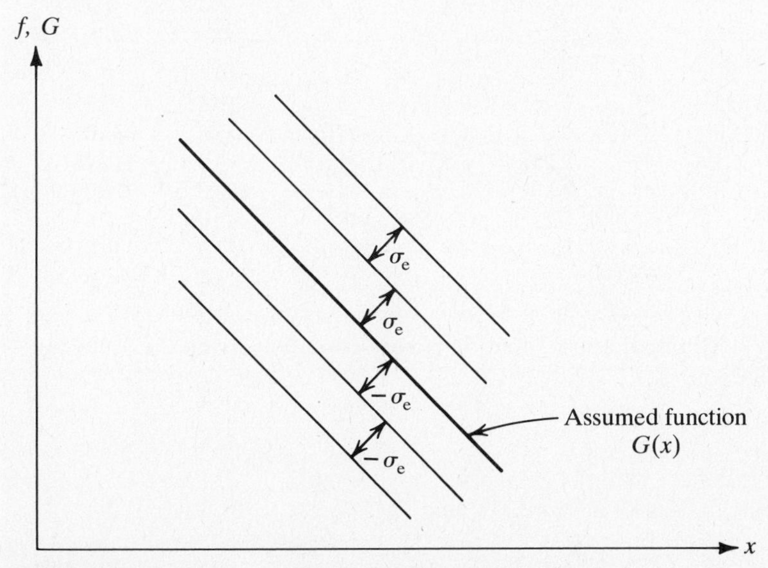


**Figure 9.6** Schematic representation of standard error.

derived approximation. This is because 68.3% of the observations can be expected to be within $\pm 2\sigma_e$ and 99.9% within $\pm 3\sigma_e$. That is,

$$G_i \pm 3\sigma_e \Rightarrow 99.9\% \text{ certainty}$$

$$G_i \pm 2\sigma_e \Rightarrow 68.3\% \text{ certainty}$$

Consequently, the smaller the $\sigma_e$ value and the higher the $R^2$ value, the better the assumed functional approximation.

E

EXAMPLE 9.7

A fighter aircraft lands on an aircraft carrier and is brought to rest in 3.0 s. An accelerometer attached to the airplane provides the acceleration record given below. Determine the coefficients of multiple determination and the standard errors for a linear and a second-order approximating polynomial relating the acceleration $a$ to time $t$.

| $t$ (s) | 0 | 0.5 | 1.0 | 1.5 | 2.0 | 2.5 | 3.0 |
|---|---|---|---|---|---|---|---|
| $a$ (ft/s$^2$) | 0 | 21.0 | 35.0 | 49.0 | 54.0 | 48.0 | 61.0 |

Solution

The regression coefficients can be easily computed by first establishing the following table.

| $i$ | $t_i$ | $a_i$ | $t_i^2$ | $t_i^3$ | $t_i^4$ | $a_i t_i$ | $a_i t_i^2$ |
|---|---|---|---|---|---|---|---|
| 1 | 0 | 0 | 0 | 0 | 0 | 0 | 0 |
| 2 | 0.5 | 21 | 0.25 | 0.125 | 10.5 | 10.5 | 5.25 |
| 3 | 1.0 | 35 | 1.0 | 1.0 | 35.0 | 35.0 | 35.0 |
| 4 | 1.5 | 49 | 2.25 | 3.375 | 73.5 | 73.5 | 110.25 |
| 5 | 2.0 | 54 | 4.0 | 8.0 | 108.0 | 108.0 | 216.0 |
| 6 | 2.5 | 58 | 6.25 | 15.625 | 145.0 | 145.0 | 362.5 |
| 7 | 3.0 | 61 | 9.0 | 27.0 | 183.0 | 183.0 | 549.0 |
| $\sum$ | 10.5 | 278 | 22.75 | 55.125 | 142.188 | 555.0 | 1278.0 |

The linear polynomial approximation is now given by solving the following set of equations:

$$278 = 7a_1 + 10.5a_2$$

$$555 = 10.5a_1 + 22.75a_2$$

This gives $a_1 = 10.143$, $a_2 = 19.714$, and

$$G_L = 10.143 + 19.714t \tag{9.21a}$$

Similarly, the second-order polynomial approximation is determined by solving the following:

$$278 = 7a_1 + 10.5a_2 + 22.75a_3$$

$$555 = 10.5a_1 + 22.75a_2 + 55.125a_3$$

$$1278 = 22.75a_1 + 55.125a_2 + 142.188a_3$$

which yield $a_1 = 0.7390$, $a_2 = 42.284$, $a_3 = -7.5230$, and

$$G_s = 0.7390 + 42.284t - 7.523t^2 \tag{9.21b}$$

Note that $G_L$ is the linear approximation and $G_s$ is the second-order approximation of the acceleration. Therefore, for the linear case and using equations (9.18) and (9.20),

$$R_L^2 = \frac{\sum_{i=1}^{7} (G_{Li} - \bar{a})^2}{\sum_{i=1}^{7} (a_i - \bar{a})^2}$$

$$\sigma_{eL} = \sqrt{\frac{\sum_{i=1}^{7} (a_i - G_{Li})^2}{7 - 2}}$$

Note that $M = 2$ and $\bar{a} = \frac{278}{7} = 39.714$. The various sums are established by using equation (9.21a) as follows:

| $i$ | $a_i$ | $G_{Li}$ | $(G_{Li} - \bar{a})^2$ | $(a_i - \bar{a})^2$ | $(a_i - G_{Li})^2$ |
|---|---|---|---|---|---|
| 1 | 0 | 10.143 | 874.21 | 1577.2 | 102.88 |
| 2 | 21 | 20.000 | 388.48 | 350.21 | 1.0 |
| 3 | 35 | 29.857 | 97.16 | 22.22 | 26.45 |
| 4 | 49 | 39.714 | 0 | 86.23 | 86.23 |
| 5 | 54 | 49.571 | 97.16 | 204.09 | 19.62 |
| 6 | 58 | 59.428 | 388.80 | 334.38 | 2.04 |
| 7 | 61 | 69.285 | 874.68 | 453.09 | 68.69 |
| $\Sigma$ | | | 2720.57 | 3027.42 | 306.91 |

Therefore

$$R_L^2 = \frac{2720.57}{3027.42} = 0.899$$

$$\sigma_{eL} = \frac{306.91}{5} = 7.835$$

It is evident that while the $R_L^2$ value seems reasonable, $\sigma_{eL}$ indicates a significant flaw in the linear fit. That is, for us to have 99.9% confidence in the linear approximation,

$$G = G_L \pm 3\sigma_{eL} = (10.143 \pm 23.505) + 19.714t$$

Thus at any time $t$ we may expect our results to vary significantly. For example, at $t = 0$ the approximate acceleration value is given as

$$-13.362 \leq G \leq 33.648$$

For the second-order polynomial approximation, we have

$$R_s^2 = \frac{\sum_{i=1}^{7} (G_{si} - \bar{a})^2}{\sum_{i=1}^{7} (a_i - \bar{a})^2}$$

$$\sigma_{es} = \sqrt{\frac{\sum_{i=1}^{7} (a_i - G_{si})^2}{7 - 3}}$$

Note that $M = 3$ since there are three regression coefficients. Furthermore, the denominator in $R_s^2$ remains the same as that computed for the linear approximation. Therefore the remaining sums are calculated from the following table.

| $i$ | $a_i$ | $G_{si}$ | $(G_{si} - \bar{a})^2$ | $(a_i - G_{si})^2$ |
|---|---|---|---|---|
| 1 | 0 | 0.739 | 1519.05 | 0.55 |
| 2 | 21 | 21.881 | 318.02 | 0.78 |
| 3 | 35 | 35.500 | 17.76 | 0.25 |
| 4 | 49 | 47.238 | 56.61 | 3.10 |
| 5 | 54 | 55.215 | 240.28 | 1.48 |
| 6 | 58 | 59.430 | 388.72 | 2.04 |
| 7 | 61 | 59.884 | 406.83 | 1.25 |
| $\Sigma$ | | | 2947.27 | 9.45 |

Consequently,

$$R_s^2 = \frac{2947.27}{3027.42} = 0.974$$

$$\sigma_{es} = \frac{9.45}{4} = 1.537$$

These values indicate that a second-order polynomial approximation is significantly better than the linear one, in that not only the $R_s^2$ is high but also the standard error is a lot lower. Hence for 99.9% confidence we have

$$G = G_s \pm 3\sigma_e = (0.739 \pm 4.611) + 42.284t - 7.523t^2$$

The implication is that the expected functional values will vary within $\pm 4.611$. This is substantially less than the $\pm 23.505$ calculated with the linear approximation.

## Recommended Reading

*Brief Numerical Methods*, Wendell E. Grove, Prentice-Hall, Inc., Englewood Cliffs, N.J., 1966.

*Mathematical Methods in the Physical Sciences*, Merle C. Potter, Prentice-Hall, Inc., Englewood Cliffs, N.J., 1978.

*Numerical Methods for Scientists and Engineers*, R. W. Hamming, McGraw-Hill Book Company, New York, 1973.

*A First Course in Numerical Analysis*, Anthony Ralston and P. Rabinowitz, McGraw-Hill Book Company, New York, 1978.

*Finite Mathematics with Business Applications*, John G. Kemeny et al., Prentice-Hall, Inc., Englewood Cliffs, N.J., 1972.

*Numerical Algorithms—Origins and Applications*, Bruce W. Arden and K. N. Astill, Addison-Wesley, Reading, Mass., 1970.

## PROBLEMS

**9.1** Derive the regression equations needed for fitting the following third-order polynomial:

$$f(x) = ax^3 + bx^2 + cx + d$$

**9.2** Derive the regression equations for the following function:

$$f(x) = a(\tan x)^b$$

**9.3** Derive the regression equations corresponding to the following function:

$$f(x) = ax^3 + b \sin x + c$$

**9.4** Given the table

| $x$ | 0 | 1 | 2 | −1 | −2 |
|---|---|---|---|---|---|
| $y$ | 1 | 3 | 11 | 5 | 7 |

fit a first-order polynomial through the data set.

**9.5** Fit a second-order polynomial through the data set given in problem 4.

**9.6** Fit a third-order polynomial through the data set given in problem 4.

**9.7** Determine the correlation coefficient for problem 4.

**9.8** Determine the correlation coefficient for problem 5.

**9.9** Determine the correlation coefficient for problem 6.

**9.10** Develop the regression equations for the following function:

$$y = az^2 + bx + cw^{1/2}$$

where $y$, $z$, $x$, and $w$ are variables and $a$, $b$, and $c$ are the regression coefficients.

**9.11** In geotechnical engineering, the so-called liquid limit is determined by using the following function:

$$w = a \log N + b$$

where $w$ is the water content and $N$ is the blow count. Determine $a$ and $b$ for the following data obtained in such a test:

| $w$ | 15 | 24.6 | 31.8 |
|---|---|---|---|
| $N$ | 50 | 32 | 20 |

Compute the correlation coefficient then determine whether the test is good or bad.

**9.12** In many experimental tests relating to the tensile test a transducer is used to relate deformation to voltage change in mV. Determine the Young's modulus of elasticity for the following hypothetical data:

| Voltage $V$ | 0 | 1 | 2 | 3 | 4 |
|---|---|---|---|---|---|
| Strain $\varepsilon$ | 0 | $\frac{1}{2}$ | 1 | $1\frac{1}{3}$ | $1\frac{3}{4}$ |

Note that stress is related to voltage as follows:

$$\sigma = 2 \times 10^6 V \qquad \text{(psi)}$$

**9.13** The resistance of a given material is determined as the ratio of voltage to current. Given the following data, determine the resistance:

| Current $I$ | 1 | 2 | 3.5 | 6 |
|---|---|---|---|---|
| Voltage $V$ | 2 | 3.5 | 7 | 10 |

**9.14** The cost of many structures can be estimated by relating cost to area, height, location, price index, and type of structure. In many cases, a rule of thumb is used instead, relating the cost in dollars to volume in cubic feet. Determine such a rule of thumb using the following data.

| Cost ($) | 10,000 | 25,000 | 50,000 | 60,000 |
|---|---|---|---|---|
| Volume (ft$^3$) | 1250 | 2000 | 5000 | 45,000 |
| Price index | 1.0 | 1.08 | 1.05 | 1.2 |

Develop a linear regression equation that would relate cost to volume and price index.

**9.15** Use orthogonal polynomials to fit a fourth-order polynomial through the following data set:

| $x$ | 4 | 5 | 6 | 7 | 8 | 9 | 10 | 11 |
|---|---|---|---|---|---|---|---|---|
| $y$ | 1 | 7 | 3 | $-1$ | $-4$ | $-10$ | 0 | 4 |

Evaluate the regression coefficient and the standard error of the estimate.

# chapter 10

# Numerical Differentiation

The space shuttle thundering into space. *Courtesy of N.A.S.A.*

"When traditional mathematical methods fail, engineers learn to seek alternatives."

# 10

## 10.1 INTRODUCTION

The solution of many engineering problems involves mathematical models expressed in terms of differential equations. Such equations are classified according to the number of independent variables involved. If only one independent variable exists, the differential equation is called "ordinary." If the differential equation is expressed in terms of more than one independent variable, then it is called "partial." Furthermore, partial differential equations may or may not involve mixed partial derivatives. Unfortunately, very few differential equations can be solved by using the methods of calculus. For this reason, the topic of numerical differentiation is extremely important in engineering problem solving. Unlike closed-form solutions (solutions obtained by using calculus methods), numerical solutions are not limited by the complexity of the differential equation, or by the physical model it represents. In the recent past, numerical methods were for the most part ignored because of the amount of number crunching involved. Fortunately, this is no longer a problem owing to recent advances in computer hardware and software.

While some still make the claim that numerical methods of solving differential equations are inaccurate, they neglect to mention that the so-called exact solutions are generally constrained by various assumptions and simplifications. The accuracy obtained when using numerical methods is limited only by the power of the computer being used. The advantage of using numerical methods lies in their versatility and simplicity. Thus it is now possible to tackle the most complex of engineering problems for which solutions were nowhere in sight even a few years ago. This chapter lays the foundations for the numerical techniques to be used in solving differential equations.

## 10.2 REVIEW OF TAYLOR SERIES

The basis for many numerical methods can often be traced back to Taylor series. This series is based on the assumption that any function can be expressed as a polynomial having an infinite number of terms. Consider the arbitrary function $f(x)$ given in figure 10.1. Furthermore, assume that $f(x)$ is continuous and has a known value $f(x_0)$ at $x = x_0$. If the functional value at a point $x = x_0 + h$ is derived, then we should be able to approximate $f(x_0 + h)$ by considering the slope of the function $f^{(1)}(x)$ evaluated at $x = x_0$. Therefore the slope of the tangent line is easily evaluated in terms of the functional values $f(x_0)$ and $G(x_0 + h)$ to give

$$f^{(1)}(x_0) = \frac{G(x_0 + h) - f(x_0)}{h}$$

Solving for $G(x_0 + h)$ yields

$$G(x_0 + h) = hf^{(1)}(x_0) + f(x_0) \tag{10.1}$$

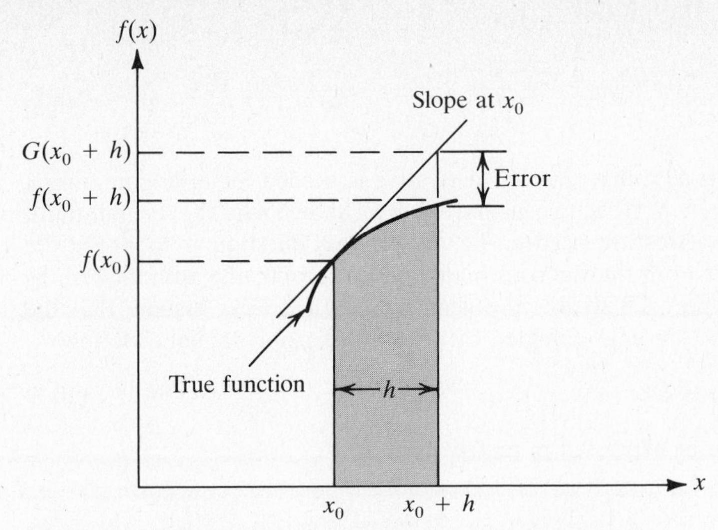


**Figure 10.1** Graphical illustration of functional approximation.

Obviously $G(x_0 + h)$ is not exactly equal to $f(x_0 + h)$; however, it can be used to approximate its value. That is,

$$f(x_0 + h) = G(x_0 + h) \pm \text{ERROR} \tag{10.2}$$

The reason that a $\pm$ sign is used is because if the true function is concave upward then the error is negative. Substituting equation (10.1) into equation (10.2) yields the following approximation:

$$f(x_0 + h) = f(x_0) + hf^{(1)}(x_0) \pm \text{ERROR} \tag{10.3}$$

Note that here $f^{(1)}(x_0)$ is the first derivative of the function evaluated at $x = x_0$.

The Taylor series simply states that $f(x_0 + h)$ is an infinite series containing higher derivative terms. That is, the error term can be reduced to zero if the second, third, . . . , and infinite derivative are known at $x_0$. Hence

$$f(x_0 + h) = f(x_0) + hf^{(1)}(x_0) + \frac{h^2}{2!} f^{(2)}(x_0) + \cdots + \frac{h^n}{n!} f^{(n)}(x_0) + R_{n+1} \tag{10.4a}$$

where

$$R_{n+1} = \frac{h^{n+1}}{(n+1)!} f^{(n+1)}(s), \qquad x_0 \leq s \leq x_0 + h \tag{10.4b}$$

Equation (10.4) is a Taylor series and equation (10.3) is then a truncated Taylor series. The term $R_{n+1}$ is called the remainder term of the series. In general,

$$f(x_0 \pm h) = \sum_{n=0}^{\infty} \frac{(\pm h)^n}{n!} f^{(n)}(x_0) \tag{10.5}$$

Note that $f^{(n)}(x_0)$ is the $n$th derivative evaluated at $x = x_0$. The so-called Maclaurin series is a special case of the Taylor series, in that $x$ is evaluated at $h$. This means that the expansion is made with respect to the origin. Consequently,

$$f(\pm h) = \sum_{n=0}^{\infty} (\pm h)^n \frac{f^n(0)}{n!} \tag{10.6}$$

Thus far we have shown that the first derivative is needed for approximating a function at point $x_0 + h$. However, we have not yet shown why the second, third, ..., and $n$th derivatives are needed. To answer this question is to prove the Taylor series. Aside from the rigorous mathematical proofs, the authors wish to outline a unique approach to this problem, namely matrices. Assume that the function of figure 10.1 is approximated by a third-order polynomial as follows:

$$f(x) = a_0 + a_1 x + a_2 x^2 + a_3 x^3 \tag{10.7}$$

Equation (10.7) can be expressed in matrix form as

$$f(x) = \{1 \quad x \quad x^2 \quad x^3\} \begin{Bmatrix} a_0 \\ a_1 \\ a_2 \\ a_3 \end{Bmatrix} \tag{10.8}$$

Now define the derivatives of $f(x)$:

$$f^{(1)}(x) = a_1 + 2a_2 x + 3a_3 x^2 \tag{10.9a}$$

$$f^{(2)}(x) = 2a_2 + 6a_3 x \tag{10.9b}$$

$$f^{(3)}(x) = 6a_3 \tag{10.9c}$$

Equation (10.8) involves the unknowns $a_0$, $a_1$, $a_2$, and $a_3$. Consequently, four equations are needed. These equations can be easily obtained by setting $x = x_0$ in equation (10.9). Thus

$$f(x_0) = a_0 + a_1 x_0 + a_2 x_0^2 + a_3 x_0^3$$

$$f^{(1)}(x_0) = a_1 + 2a_2 x_0 + 3a_3 x_0^2$$

$$f^{(2)}(x_0) = 2a_2 + 6a_3 x_0$$

$$f^{(3)}(x_0) = 6a_3$$

These equations are now expressed in matrix form:

$$\begin{Bmatrix} f(x_0) \\ f^{(1)}(x_0) \\ f^{(2)}(x_0) \\ f^{(3)}(x_0) \end{Bmatrix} = \begin{bmatrix} 1 & x_0 & x_0^2 & x_0^3 \\ 0 & 1 & 2x_0 & 3x_0^2 \\ 0 & 0 & 2 & 6x_0 \\ 0 & 0 & 0 & 6 \end{bmatrix} \begin{Bmatrix} a_0 \\ a_1 \\ a_2 \\ a_3 \end{Bmatrix}$$

or in compact matrix form

$$\{F\} = [x]\{a\}$$

Note that the $[\bar{x}]$ matrix is an upper-triangular matrix; therefore its inverse must be an upper-triangular matrix as well. Hence

$$\{a\} = [\bar{x}]^{-1}\{F\} \tag{10.10}$$

Therefore evaluating $|\bar{x}|^{-1}$ in equation (10.10) yields

$$\{a\} = \begin{bmatrix} 1 & -x_0 & x_0^2/2 & -x_0^3/6 \\ 0 & 1 & -x_0 & x_0^2/2 \\ 0 & 0 & \frac{1}{2} & -x_0/2 \\ 0 & 0 & 0 & \frac{1}{6} \end{bmatrix} \begin{Bmatrix} f(x_0) \\ f^{(1)}(x_0) \\ f^{(2)}(x_0) \\ f^{(3)}(x_0) \end{Bmatrix} \tag{10.11}$$

Substituting equation (10.11) into equation (10.8) gives

$$f(x) = \{1 \quad x \quad x^2 \quad x^3\} \begin{bmatrix} 1 & -x_0 & x_0^2/2 & -x_0^3/6 \\ 0 & 1 & -x_0 & x_0^2/2 \\ 0 & 0 & -\frac{1}{2} & -x_0/2 \\ 0 & 0 & & \frac{1}{6} \end{bmatrix} \begin{Bmatrix} f(x_0) \\ f^{(1)}(x_0) \\ f^{(2)}(x_0) \\ f^{(3)}(x_0) \end{Bmatrix}$$

Simplifying, we have

$$f(x) = \{1 \quad x - x_0 \quad \tfrac{1}{2}x_0^2 - xx_0 + \tfrac{1}{2}x^2 \quad -\tfrac{1}{6}x_0^3 + \tfrac{1}{2}x_0^2 x - \tfrac{1}{2}x_0 x^2 + \tfrac{1}{6}x^3\}$$

$$\times \begin{Bmatrix} f(x_0) \\ f^{(1)}(x_0) \\ f^{(2)}(x_0) \\ f^{(3)}(x_0) \end{Bmatrix} \tag{10.12}$$

If we define

$$A_0 = 1 \tag{10.13}$$

$$A_1 = x - x_0 \tag{10.14}$$

$$A_2 = \tfrac{1}{2}x_0^2 - xx_0 + \tfrac{1}{2}x^2 \tag{10.15}$$

$$A_3 = -\tfrac{1}{6}x_0^3 + \tfrac{1}{2}x_0^2 - \tfrac{1}{2}x_0 x^2 + \tfrac{1}{6}x^3 \tag{10.16}$$

equation (10.12) can now be expressed in terms of the variables $A_0$, $A_1$, $A_2$, and $A_3$:

$$f(x) = A_0 f(x_0) + A_1 f^{(1)}(x_0) + A_2 f^{(2)}(x_0) + A_3 f^{(3)}(x_0) \tag{10.17}$$

It should be emphasized that while the derivatives are evaluated at $x_0$, the values of the $A$ coefficients are dependent on both $x_0$ and $x$. Therefore, if $x = x_0 + h$, then the following $A$ values are computed:

$$A = 1, \qquad A_1 = h, \qquad A_2 = \tfrac{1}{2}h^2, \qquad A_3 = \tfrac{1}{6}h^3$$

Substituting these values into equation (10.17) gives

$$f(x) = f(x_0 + h) = f(x_0) + hf^{(1)}(x_0) + \frac{h^2}{2} f^{(2)}(x_0) + \frac{h^3}{6} f^{(3)}(x_0) \tag{10.18}$$

Equation (10.18) is a truncated Taylor series. This can be easily verified by comparing it with equation (10.5). For the special case in which $x_0 = 0$, the coefficients $A_0, \ldots, A_3$ remain exactly the same. However, the functional values change to give the so-called "Maclaurin's series" as follows:

$$f(h) = f(0) + hf^{(1)}(0) + \frac{h^2}{2} f^{(2)}(0) + \frac{h^3}{6} f^{(3)}(0)$$

In general, a range for the error term can be estimated as

$$R_{n+1} = \frac{h^{n+1}}{(n+1)!} f^{(n+1)}(s) \tag{10.19}$$

where

$$x_0 \leq s \leq x_0 + h \tag{10.20}$$

Therefore one may calculate the error term corresponding to $x = x_0$ and $x = x_0 + h$. Consider the following example.

**E** EXAMPLE 10.1

Determine the square root of 13 using the value of the square root of 9.

### Solution

The function we need to expand in Taylor series is of the form

$$f(x) = \sqrt{x}$$

where

$$x_0 = 9$$

$$x = x_0 + h = 13$$

and

$$h = 13 - 9 = 4$$

In Taylor series equation (10.26) is expanded as follows:

$$f(x) = f(x_0) + \frac{hf^{(1)}(x_0)}{1!} + \frac{h^2}{2!} f^{(2)}(x_0) + \cdots$$

Let us use five terms of the series. Hence the first, second, third, and fourth derivatives must be evaluated at $x_0 = 9$:

$$f(x) = \sqrt{x} = x^{1/2}$$

$$f^{(1)}(x) = \frac{1}{2}x^{-1/2}$$

$$f^{(2)}(x) = -\frac{1}{4}x^{-3/2}$$

$$f^{(3)}(x) = \frac{3}{8}x^{-5/2}$$

$$f^{(4)}(x) = -\frac{15}{16}x^{-7/2}$$

Therefore

$$f(x_0) = f(9) = (9)^{1/2} = 3$$

$$f^{(1)}(x_0) = f^{(1)}(9) = \frac{1}{2}(9)^{-1/2} = \frac{1}{6}$$

$$f^{(2)}(x_0) = f^{(2)}(9) = -\frac{1}{4}(9)^{-3/2} = -\frac{1}{108}$$

$$f^{(3)}(x_0) = f^{(3)}(9) = \frac{3}{8}(9)^{-5/2} = \frac{3}{1944}$$

$$f^{(4)}(x_0) = f^{(4)}(9) = -\frac{15}{16}(9)^{-7/2} = -\frac{15}{34{,}992}$$

Substituting into the assumed truncated Taylor series gives

$$f(x) = f(x_0) + hf^{(1)}(x_0) + \frac{h^2}{2}f^{(2)}(x_0) + \frac{h^3}{6}f^{(3)}(x_0) + \frac{h^4}{24}f^{(4)}(x_0) + \text{ERROR}$$

$$f(13) = 3 + (4)\frac{1}{6} + \frac{(4)^2}{2}\left(-\frac{1}{108}\right) + \frac{(4)^3}{6}\frac{3}{1944} + \frac{(4)^4}{24}\left(-\frac{15}{34{,}992}\right) + \text{ERROR}$$

$$= 3.60448 + \text{ERROR}$$

Since $n = 4$, the error term is now given as

$$R_5 = \frac{h^5}{5!}f^{(5)}(s) = \frac{(4)^5}{120}f^{(5)}(s) \qquad \textbf{(10.21)}$$

where

$$f^{(5)}(s) = \frac{105}{32}s^{-9/2} \qquad \textbf{(10.22)}$$

Substituting equation (10.22) into equation (10.21) gives

$$R_5 = \frac{1024}{120} \times \frac{105}{32} s^{-9/2} = 28s^{-9/2}$$

The error term given by equation (10.34) is now evaluated at $s = 9$ and $s = 13$:

$$s = 9: \qquad R_s = 28(9)^{-9/2} = 1.4225 \times 10^{-3}$$

$$s = 13: \qquad R_s = 28(13)^{-9/2} = 0.2719 \times 10^{-3}$$

The error term is given as a range of

$$0.2719 \times 10^{-3} \leq R_5 \leq 1.4225 \times 10^{-3}$$

The exact value for the error is $1.0702 \times 10^{-3}$, which clearly falls within the established range. The reader may question the need for going through all of these steps to find the square root of a number. However, this example is meant to illustrate a very important principle, namely errors associated with the Taylor series.

## 10.3 NUMERICAL DIFFERENTIATION OF FUNCTIONS

In this section, various methods are presented which permit the development of derivative approximation formulas. The basis for these methods lies in the fact that if a function can be reasonably approximated by an interpolating polynomial, then it is assumed that its derivatives can also be approximated by taking the derivatives of the approximating polynomial. Obviously, this assumption presents a serious problem in that interpolating polynomials show significant deviations from a smooth curve. This deviation is referred to as the "wiggle effect" and is illustrated in figure 10.2.

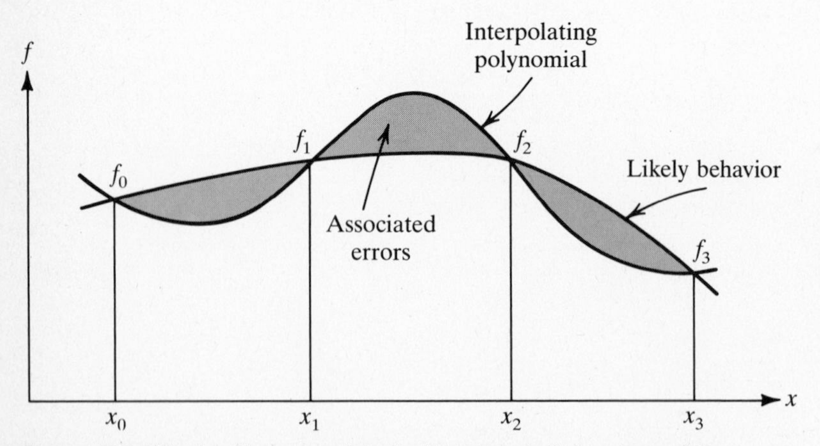


**Figure 10.2** Wiggle effect associated with interpolating polynomials.

Note here that, even though the interpolating polynomial may adequately describe the function, its derivatives may be totally different from those of the actual function. Fortunately, many of the engineering applications involve functions which can be approximated by polynomials of reasonable order. Furthermore, one can reduce the errors associated with the wiggle effect by simply reducing base point intervals. This of course is possible only if the function is known explicitly. For practical considerations, only derivative expressions for equally spaced points are discussed.

### 10.3.1 Interpolating Polynomial Methods

In chapter 8, it was pointed out that for a given equally spaced data set, one can develop three different types of polynomials that can be forced to pass through each data point. These include forward, backward, and central interpolating formulas. With that in mind, it is feasible to approximate derivatives of varying order using each of these interpolating formulas. However, since the derivation of the derivative approximation is similar in all three cases, only the first forward derivative approximation is presented here.

Recall that the forward interpolating polynomial approximation was given earlier by equation (8.37). It is repeated here for convenience:

$$f(x_i + \alpha h) = \left[1 + \alpha\Delta + \frac{\alpha(\alpha - 1)}{2!}\Delta^2 + \frac{\alpha(\alpha - 1)(\alpha - 2)}{3!}\Delta^3 + \frac{\alpha(\alpha - 1)(\alpha - 2)(\alpha - 3)}{4!}\Delta^4 + \cdots\right] f(x_i) \tag{10.23}$$

For simplicity let us make the following substitutions:

$$x_{i+1} = x_i + \alpha h$$

$$f(x_i) = f_i$$

$$f(x_{i+1}) = f_{i+1}$$

Consequently, equation (10.23) is given more simply by the expression

$$f_{i+1} = f_i + \alpha\,\Delta f_i + \frac{1}{2}(\alpha^2 - \alpha)\,\Delta^2 f_i + \frac{1}{6}(\alpha^3 - 3\alpha^2 + 2\alpha)\,\Delta^3 f_i + \frac{1}{24}(\alpha^4 - 6\alpha^3 + 11\alpha^2 - 6\alpha)\,\Delta^4 f_i + \cdots \tag{10.24}$$

Equation (10.24) can now be differentiated with respect to $x$ by using the chain rule. That is,

$$f_{i+1}^{(1)} = \frac{df_{i+1}}{dx} = \frac{df_{i+1}}{d\alpha}\frac{d\alpha}{dx} = \frac{1}{h}\frac{df_{i+1}}{d\alpha}$$

Therefore the first forward derivative at $x = x_i + \alpha h$ is given as

$$f_{i+1}^{(1)} = \frac{1}{h}\left(\Delta f_i + \frac{1}{2}(2\alpha - 1)\Delta^2 f_i + \frac{1}{6}(3\alpha^2 - 6\alpha + 2)\Delta^3 f_i + \frac{1}{24}(4\alpha^3 - 18\alpha^2 + 22\alpha - 6)\Delta^4 f_i + \cdots\right)$$

However, since we are interested in expressing the derivative at $x = x_i$, letting $\alpha = 0$ yields the following simplified expression:

$$f_i^{(1)} = \frac{1}{h}\left(\Delta f_i - \frac{1}{2}\Delta^2 f_i + \frac{1}{3}\Delta^3 f_i - \frac{1}{4}\Delta^4 f_i + \cdots \pm \frac{1}{n}\Delta^n f_i\right) \qquad \textbf{(10.25a)}$$

or more compactly,

$$f_i^{(1)} = \frac{1}{h}\sum_{k=1}^{n}(-1)^{k+1}\frac{\Delta^k f_i}{k} \qquad \textbf{(10.25b)}$$

where $n$ is the order of the forward interpolating polynomial being used for estimating the first derivative. In addition, for a given $n$ value, equation (10.25) can be expressed directly in terms of the functional values by using equation (8.19). For example, we can express the first forward derivative using first- and second-order interpolating polynomials, respectively, as follows:

$$n = 1 \rightarrow f_i^{(1)} \cong \frac{1}{h}\Delta f_i \cong \frac{1}{h}(-f_i + f_{i+1})$$

$$\begin{aligned} n = 2 \rightarrow f_i^{(2)} &\cong \frac{1}{h}\left(\Delta f_i - \frac{1}{2}\Delta^2 f_i\right) \\ &\cong \frac{1}{h}\left((f_{i+1} - f_i) - \frac{1}{2}(f_{i+12} - 2f_{i+1} + f_i)\right) \\ &\cong \frac{1}{h}(-3f_i + 4f_{i+1} - f_{i+1}) \end{aligned}$$

Obviously, similar expressions can be readily derived for a higher-order interpolating polynomial. This procedure can be easily employed to derive derivative approximations by using backward and central interpolating polynomials. Furthermore, for the speccial case in which an $n$th-order forward interpolating polynomial is used to evaluate the $n$th derivative, the following expression is used:

$$f_i^{(n)} = \frac{\Delta^n f_i}{h^n} \qquad \textbf{(10.26)}$$

Note that when using finite-order interpolating polynomials the derivative can only be determined approximately. Consequently, an estimate of the errors involved can be obtained by considering the remainder term. Therefore, for the first forward derivative, the error is given as

$$E_n = \frac{(-1)^n}{n+1} h^n f^{(n+1)}(s), \qquad x_i \le s \le x_{i+1}$$

It is evident that for a given interval $h$ and polynomial order $n$ the function $f^{(n+1)}(s)$ may or may not be zero. Consequently, for the error to be minimized, the interval must be reduced to an appropriate value.

### 10.3.2 Taylor Series Method

The Taylor series provides a straightforward procedure for approximating derivatives. This assertion is based on the fact that any function including polynomials can be expressed as an infinite series in terms of its derivatives and base point interval. Consequently, one may write a Taylor series for each pivotal point in a data set and solve for the derivative in terms of the functional values. For illustration purposes, consider figure 10.3.

Suppose that it is required to develop forward, backward, and central derivative expressions at $x = x_i$, using a second-order approximating polynomial. Note that the central derivative can only be defined for an even-order polynomial, since it requires equal numbers of pivotal points on either side of $x_i$. Moreover, $f_{i-2} = f(x_{i-2}), \ldots, f_{i+2} = f(x_{i+2})$ and $x_{i-2} = x_i - 2h, \ldots, x_{i+2} = x_i + 2h$. Therefore the first forward derivative is determined by expanding a Taylor series about points $x_{i+1}$ and $x_{i+2}$ to give

$$f_{i+1} = f_i + hf_i^{(1)} + \frac{h^2}{2!} f_i^{(2)} \pm \text{ERROR}, \qquad x = x_{i+1} \tag{10.27}$$

$$f_{i+2} = f_i + (2h)f_i^{(1)} + \frac{(2h)^2}{2!} f_i^{(2)} \pm \text{ERROR}, \qquad x = x_{i+2} \tag{10.28}$$

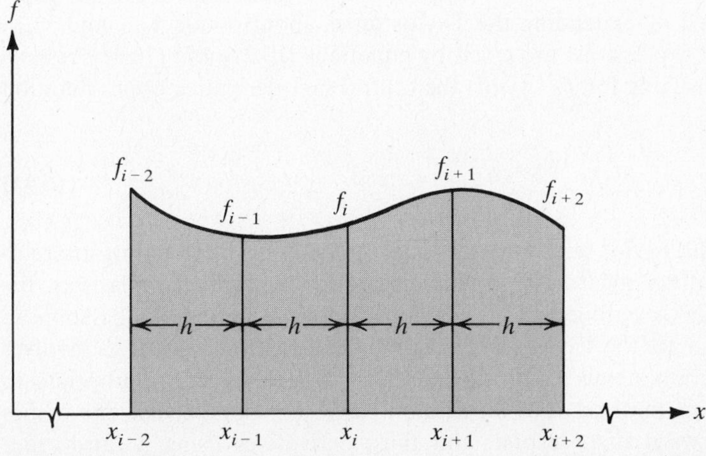


**Figure 10.3** Graphical representation of Taylor series derivative approximation.

These two equations can be easily solved for the first derivative $f_i^{(1)}$ by eliminating $f_i^{(2)}$. Hence multiplying equation (10.27) by $-4$ then adding to equation (10.28) yields

$$-4f_{i+1} + f_{i+2} = -3f_i - 2hf_i^{(1)}$$

Solving for $f_i^{(1)}$ gives the first forward derivative approximation using an order-two polynomial:

$$f_i^{(1)} \cong \frac{1}{2h}(-3f_i + 4f_{i+1} - f_{i+2}) \tag{10.29}$$

Equation (10.29) is precisely the same expression we derived earlier using interpolation methods. It is evident that the backward first derivative can be approximated by expanding the Taylor series about points $x_{i-2}$ and $x_{i-1}$ to give

$$f_{i-2} = f_i - 2hf_i^{(1)} + \frac{(-2h)^2}{2!}f_i^{(2)} \pm \text{ERROR}, \qquad x = x_{i-2} \tag{10.30}$$

$$f_{i-1} = f_i - hf_i^{(1)} + \frac{(-h)^2}{2!}f_i^{(2)} \pm \text{ERROR}, \qquad x = x_{i-1} \tag{10.31}$$

Once again, eliminating $f_i^{(2)}$ from equations (10.30) and (10.31) then solving for the remaining unknown $f_i^{(1)}$ yields the backward first derivative. That is,

$$f_i^{(1)} \cong \frac{1}{2h}(f_{i-2} - 4f_{i-1} + 3f_i) \tag{10.32}$$

Note that expressions for the second derivative can be readily determined by eliminating the first derivative from equations (10.27) and (10.28) for the forward second derivative and from (10.30) and (10.31) for the backward second derivative.

Finally, the central first derivative approximation using a second-order polynomial is evaluated by expanding the Taylor series about points $x_{i+1}$ and $x_{i-1}$. In this case, these expressions are given by equations (10.27) and (10.31), respectively. Therefore, solving for $f_i^{(1)}$ yields the central first derivative approximation as follows:

$$f_i^{(1)} = \frac{1}{2h}(-f_{i-1} + f_{i+1}) \tag{10.33}$$

It is evident that the Taylor series provides a direct method for evaluating approximations of derivatives regardless of their order and type. By comparison, the interpolation methods outlined in the previous section are more cumbersome to use. Furthermore, since the Taylor series method of evaluating derivatives involves the solution of simultaneous algebraic equations, it would seem appropriate to formalize it by using matrices. For illustration purposes, suppose that one wishes to obtain the forward first, second, and third derivatives using a third-order polynomial. This implies that four equally spaced base points must be specified as shown in figure 10.4.

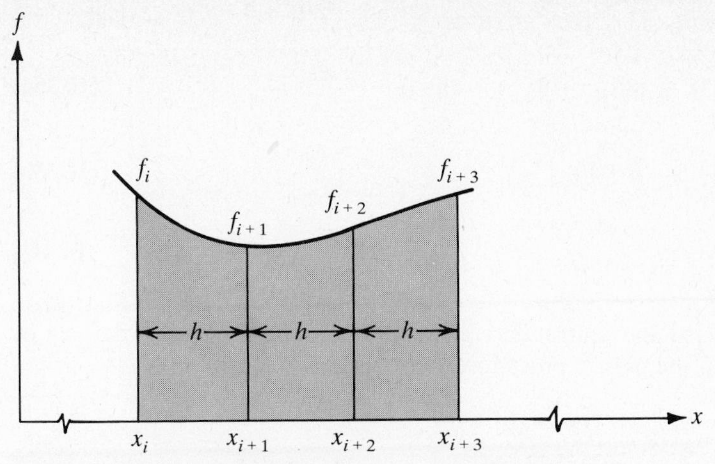


**Figure 10.4** Forward derivatives for a third-order polynomial.

The solution of this problem can be accomplished by expanding the Taylor series as follows:

$$f_{i+1} = f_i + hf_i^{(1)} + \frac{h^2}{2} f_i^{(2)} + \frac{h^3}{6} f_i^{(3)}, \qquad x = x_{i+1}$$

$$f_{i+2} = f_i + 2hf_i^{(1)} + \frac{(2h)^2}{2} f_i^{(2)} + \frac{(2h)^3}{6} f_i^{(3)}, \qquad x = x_{i+2}$$

$$f_{i+3} = f_i + 3hf_i^{(1)} + \frac{(3h)^2}{2} f_i^{(2)} f_i^{(2)} + \frac{(3h)^3}{6} f_i^{(3)}, \qquad x = x_{i+3}$$

These equations can be easily expressed in matrix form:

$$\begin{bmatrix} 1 & \frac{1}{2} & \frac{1}{6} \\ 2 & 2 & \frac{4}{3} \\ 3 & \frac{9}{2} & \frac{9}{2} \end{bmatrix} \begin{Bmatrix} hf_i^{(1)} \\ h^2 f_i^{(2)} \\ h^3 f_i^{(3)} \end{Bmatrix} \cong \begin{bmatrix} -1 & 1 & 0 & 0 \\ -1 & 0 & 1 & 0 \\ -1 & 0 & 0 & 1 \end{bmatrix} \begin{Bmatrix} f_i \\ f_{i+1} \\ f_{i+2} \\ f_{i+3} \end{Bmatrix} \tag{10.34}$$

Solving for the unknown derivatives by premultiplying equation (10.34) by the inverse of the square coefficient matrix appearing on the left-hand side yields

$$\begin{Bmatrix} hf_i^{(1)} \\ h^2 f_i^{(2)} \\ h^3 f_i^{(3)} \end{Bmatrix} \cong \begin{bmatrix} 3 & -\frac{3}{2} & \frac{1}{3} \\ -5 & 4 & -1 \\ 3 & -3 & 1 \end{bmatrix} \begin{bmatrix} -1 & 1 & 0 & 0 \\ -1 & 0 & 1 & 0 \\ -1 & 0 & 0 & 1 \end{bmatrix} \begin{Bmatrix} f_i \\ f_{i+1} \\ f_{i+2} \\ f_{i+3} \end{Bmatrix}$$

or more simply

$$\begin{Bmatrix} hf_i^{(1)} \\ h^2 f_i^{(2)} \\ h^3 f_i^{(3)} \end{Bmatrix} \cong \begin{bmatrix} -\frac{11}{6} & 3 & -\frac{3}{2} & \frac{1}{3} \\ 2 & -5 & 4 & -1 \\ -1 & 3 & -3 & 1 \end{bmatrix} \begin{Bmatrix} f_i \\ f_{i+1} \\ f_{i+2} \\ f_{i+3} \end{Bmatrix}$$

The forward derivatives are then given as

$$f_i^{(1)} \cong \frac{1}{6h}(-11f_i + 18f_{i+1} - 9f_{i+2} + 2f_{i+3}) \tag{10.35a}$$

$$f_i^{(2)} \cong \frac{1}{h^2}(2f_i - 5f_{i+1} + 4f_{i+2} - f_{i+3}) \tag{10.35b}$$

$$f_i^{(3)} \cong \frac{1}{h^3}(-f_i + 3f_{i+1} - 3f_{i+2} + f_{i+3}) \tag{10.35c}$$

Obviously, backward and central derivative approximations of any order can be evaluated by using the matrix procedure just outlined.

EXAMPLE 10.2

Use equations (10.29), (10.32), and (10.33) to approximate the first derivative of the following function at $x = 2$:

$$f(x) = e^{x-2}$$

Take $h = 0.1$.

Solution

Clearly the function in question is not a polynomial. However, an estimate of its first derivative can be made if we assume that it can be approximated by a polynomial (in this case of second order) in the range at which the derivative is required. Therefore, using the forward approximation we have

$$f_i^{(1)} = \frac{1}{2h}(-3f_i + 4f_{i+1} - f_{i+2})$$

where $h = 0.1$ (given) and the functional values are evaluated as follows:

$$f_i = f(2) = e^{2-2} = 1.0$$

$$f_{i+1} = f(2.1) = e^{2.1-2} = 1.1051709$$

$$f_{i+2} = f(2.2) = e^{2.2-2} = 1.2214027$$

The forward first derivative approximation is now readily evaluated as

$$f^{(1)}(2) = \frac{1}{2(0.1)}[-3(1) + 4(1.1051709) - 1.2214027]$$

$$f^{(1)}(2) = 0.9964045$$

Similarly, the backward first derivative is approximated by using equation (10.32) to give

$$f^{(1)}(2) = \frac{1}{0.2}[f(1.8) - 4f(1.9) + 3f(1.0)]$$

$$f^{(1)}(2) = \frac{1}{0.2}[0.8187308 - 4(0.9048374) + 3(1)]$$

$$f^{(1)}(2) = 0.9969058$$

Finally, the central first derivative approximation is given by equation (10.33) as

$$f^{(1)}(2) = \frac{1}{0.2}[-f(1.9) + f(2.1)]$$

$$f^{(1)}(2) = \frac{1}{0.2}(-0.9048374 + 1.1051709)$$

$$f^{(1)}(2) = 1.0016675$$

These values compare with the exact value of 1.0. It is evident that while better accuracy can be achieved by choosing a smaller $h$ value, the central derivative approximation is more accurate than the backward or forward approximation.

### 10.3.3 Undetermined Coefficients Method

This method of developing approximate derivative expressions is based on the fact that the derivative of a function is defined as the rate of change of the function relative to changes in the independent variable. Consequently, we should expect that derivative expressions can be developed as a linear combination of functional values. Furthermore, the fact that derivatives are dependent on the shape of the function, regardless of its position, makes it possible to simplify the development of this technique. This is accomplished by shifting the function relative to the $x$ axis. Suppose we want to find the forward first and second derivatives using a second-order polynomial. Then we can proceed by first assuming the following polynomial approximation:

$$f(x) = a_0 + a_1 x + a_2 x^2 \qquad \textbf{(10.36)}$$

Clearly the coefficients $a_0$, $a_1$, and $a_2$ are unknowns that must be determined if we are to evaluate the derivatives. Furthermore, higher-order polynomials may be assumed for the same purpose. Consequently, the derivatives are evaluated directly from equation (10.36) as follows:

$$f^{(1)}(x) = a_1 + 2a_2 x$$

$$f^{(2)}(x) = 2a_2$$

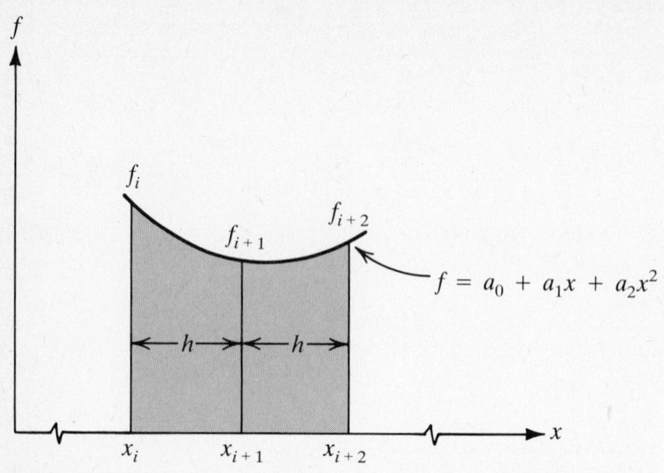


**Figure 10.5** Forward derivative approximations using a second-order polynomial.

Obviously our solution is greatly simplified if we are to evaluate these derivatives at $x_i = 0$. Hence

$$f_i^{(1)} = a_1 \tag{10.37}$$

$$f_i^{(2)} = 2a_2 \tag{10.38}$$

The remaining task is to evaluate the constants $a_1$ and $a_2$. This is accomplished readily by considering figure 10.5.

Using equation (10.36), we can develop three equations in the unknowns $a_0$, $a_1$, and $a_2$ as follows:

$$x_i = 0 \Rightarrow f_i = a_0$$

$$x_{i+1} = h \Rightarrow f_{i+1} = a_0 + ha_1 + h^2a_2$$

$$x_{i+2} = 2h \Rightarrow f_{i+2} = a_0 + 2ha_1 + 4h^2a_2$$

These algebraic equations can now be solved for the unknowns by matrix methods to give

$$\begin{Bmatrix} a_0 \\ a_1h \\ a_2h^2 \end{Bmatrix} = \frac{1}{2}\begin{bmatrix} 2 & 0 & 0 \\ -3 & 4 & -1 \\ 1 & -2 & 1 \end{bmatrix}\begin{Bmatrix} f_i \\ f_{i+1} \\ f_{i+2} \end{Bmatrix}$$

Then the first and second forward derivative approximations are determined from equations (10.37) and (10.38), respectively. That is,

$$f_i^{(1)} = \frac{1}{2h}(-3f_i + 4f_{i+1} - f_{i+2}) \tag{10.39}$$

$$f_i^{(2)} = \frac{1}{h^2}(f_i - 2f_{i+1} + f_{i+2}) \tag{10.40}$$

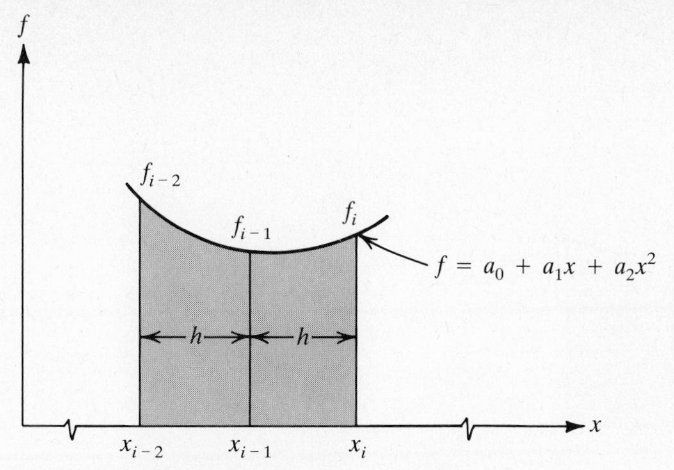


**Figure 10.6** Backward derivative approximations using a second-order polynomial.

The backward first and second derivative approximations are evaluated in similar manner. Consider figure 10.6.

Once again, taking $x_i = 0$ we evaluate the unknowns $a_0$, $a_1$, and $a_2$ in equation (10.36) in terms of the backward functional values. Thus

$$\begin{aligned} f_{i-2} &= a_0 - 2ha_1 + 4h^2a_2 \\ f_{i-1} &= a_0 - ha_1 + h^2a_2 \\ f_i &= a_0 \end{aligned}$$

Solving for the unknowns using the matrix inversion procedure, we have

$$\begin{Bmatrix} a_0 \\ a_1h \\ a_2h^2 \end{Bmatrix} = \frac{1}{2}\begin{bmatrix} 0 & 0 & 1 \\ 1 & -4 & 3 \\ 1 & -2 & 1 \end{bmatrix}\begin{Bmatrix} f_{i-2} \\ f_{i-1} \\ f_i \end{Bmatrix}$$

Substituting $a_1$ and $a_2$ into equations (10.37) and (10.38) yields the first and second backward derivative approximations, respectively. Thus

$$f_i^{(1)} = \frac{1}{2h}(f_{i-2} - 4f_{i-1} + f_i) \tag{10.41}$$

$$f_i^{(2)} = \frac{1}{h^2}(f_{i-2} - 2f_{i-1} + f_i) \tag{10.42}$$

Finally, the central first and second derivative approximations are developed by evaluating the unknowns $a_0$, $a_1$, and $a_2$ in terms of the proper functional values shown in figure 10.7 and assuming $x_i = 0$.

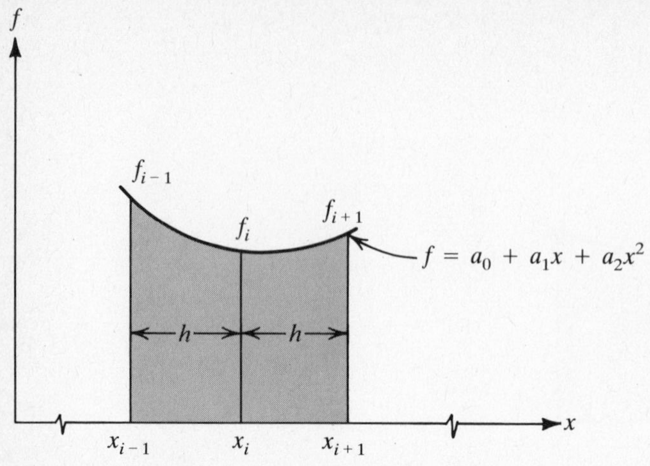


**Figure 10.7** Central derivative approximations using a second-order polynomial.

Therefore, using equation (10.36), we have

$$x_{i-1} = -h \Rightarrow f_{i-1} = a_0 - ha_1 + h^2a_2$$

$$x_i = 0 \Rightarrow f_i = a_0$$

$$x_{i+1} = h \Rightarrow f_{i+1} = a_0 + ha_1 + h^2a_2$$

Solving for the unknown coefficients yields

$$\begin{Bmatrix} a_0 \\ a_1h \\ a_2h^2 \end{Bmatrix} = \frac{1}{2}\begin{bmatrix} 0 & 2 & 0 \\ -1 & 0 & 1 \\ 1 & -2 & 1 \end{bmatrix}\begin{Bmatrix} f_{i-1} \\ f_i \\ f_{i+1} \end{Bmatrix}$$

The first and second central derivatives are now given by equations (10.37) and (10.38), respectively. Hence

$$f_i^{(1)} \cong \frac{1}{2h}(-f_{i-1} + f_{i+1}) \tag{10.43}$$

$$f_i^{(2)} \cong \frac{1}{h^2}(f_{i-1} - 2f_i + f_{i+1}) \tag{10.44}$$

Clearly $x_i$ can be taken at any point. Similar expressions for higher-order derivative approximations can be developed in a similar manner.

**E** EXAMPLE 10.3

Derive the central derivative approximations up to the fourth derivative using a fourth-order polynomial.

Solution

The approximating polynomial is of the form

$$f(x) = a_0 + a_1x + a_2x^2 + a_3x^3 + a_4x^4 \tag{10.45}$$

The derivatives are given as follows:

$$f^{(1)}(x) = a_1 + 2a_2x + 3a_3x^2 + 4a_4x^3$$

$$f^{(2)}(x) \cong 2a_2 + 6a_3x + 12a_4x^2$$

$$f^{(3)}(x) \cong 6a_3 + 24a_4x$$

$$f^{(4)}(x) \cong 24a_4$$

These derivatives are greatly simplified if they are evaluated at $x_i = 0$. Thus

$$f_i^{(1)} \cong a_1 \tag{10.46}$$

$$f_i^{(2)} \cong 2a_2 \tag{10.47}$$

$$f_i^{(3)} \cong 6a_3 \tag{10.48}$$

$$f_i^{(4)} \cong 24a_4 \tag{10.49}$$

The unknown coefficients are determined in terms of the functional value and the $h$ interval. Consider figure 10.8.

Using equation (10.45), we develop the following set of linear algebraic equations in the unknowns $a_0, \ldots, a_4$:

$$\begin{Bmatrix} f_{i-2} \\ f_{i-1} \\ f_i \\ f_{i+1} \\ f_{i+2} \end{Bmatrix} = \begin{bmatrix} 1 & -2 & 4 & -8 & 16 \\ 1 & -1 & 1 & -1 & 1 \\ 1 & 0 & 0 & 0 & 0 \\ 1 & 1 & 1 & 1 & 1 \\ 1 & 2 & 4 & 8 & 16 \end{bmatrix} \begin{Bmatrix} a_0 \\ a_1h \\ a_2h^2 \\ a_3h^3 \\ a_4h^4 \end{Bmatrix}$$

Solving for the unknowns then substituting into equations (10.46)–(10.49) yields the first through fourth central derivative approximations:

$$f_i^{(1)} = \frac{1}{12h}(f_{i-2} - 8f_{i-1} + 8f_{i+1} - f_{i+2})$$

$$f_i^{(2)} = \frac{1}{12h^2}(-f_{i-2} + 16f_{i-1} - 30f_i + 16f_{i+1} - f_{i+2})$$

$$f_i^{(3)} = \frac{1}{2h^3}(-f_{i-2} + 2f_{i-1} - 2f_{i+1} + f_{i+2})$$

$$f_i^{(4)} = \frac{1}{h^4}(f_{i-2} - 4f_{i-1} + 6f_i - 4f_{i+1} + f_{i+2})$$

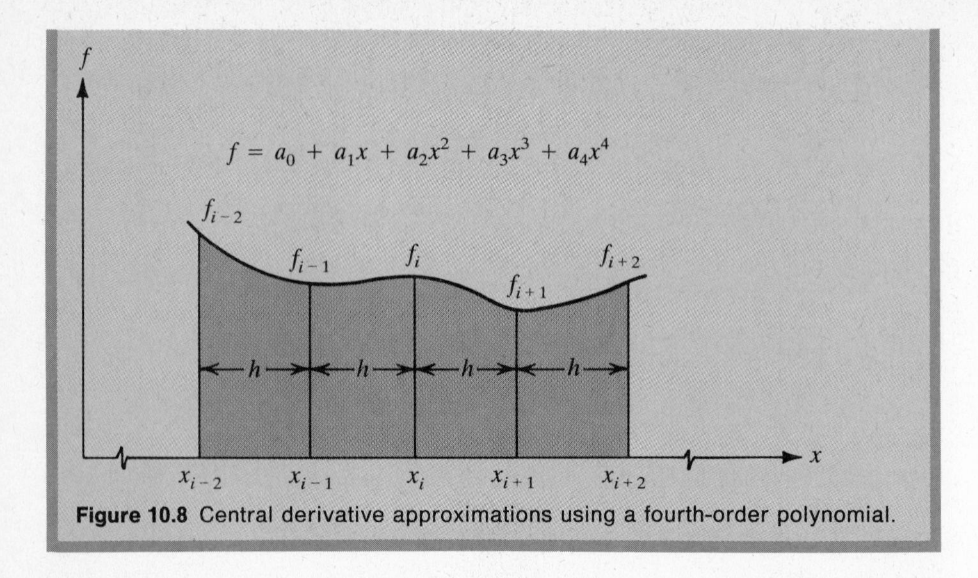

**Figure 10.8** Central derivative approximations using a fourth-order polynomial.

The reader should be aware that, regardless of the technique being employed in developing derivative approximations, the results are always the same. The implication is that the above approximations can also be obtained by using the interpolation method or the Taylor series method.

### 10.3.4 Errors

One question we have purposely avoided thus far is that of errors associated with derivative approximations. Nor have we addressed the question of what $h$ value to use. In this section, we shall see that these two important questions are not unrelated. Generally, errors associated with various numerical derivative expressions may be reduced by simply choosing a small interval $h$. In fact, as $h$ approaches zero, the difference expressions give exact values for the derivative. This observation is consistent with the basic definition of a derivative using the methods of calculus. That is,

$$f_i^{(1)} = \lim_{h \to 0} \left( \frac{f_{i+1} - f_i}{h} \right)$$

However, this approach is not realistic in that computers have a limited ability to retain a large number of significant figures. Consequently, for all practical purposes, the interval $h$ must be finite. This can be accomplished by using high-order polynomials.

While any one of the three methods outlined in this section can be employed in estimating errors associated with difference expressions, the Taylor series method provides the most direct approach. Therefore let us begin by investigating errors associated with the derivative approximations relative to the forward, backward, and central expressions. Hence, for the forward first and second derivative approximation and using figure 10.5 we write the Taylor series for the forward points. That is,

$$f_{i+1} = f_i + hf_i^{(1)} + \frac{h^2}{2} f_i^{(2)} + \frac{h^3}{6} f_i^{(3)} + \frac{h^4}{24} f_i^{(4)} \tag{10.50}$$

$$f_{i+2} = f_i + 2hf_i^{(1)} + 2h^2 f_i^{(2)} + \frac{4h^3}{3} f_i^{(3)} + \frac{2h^4}{3} f_i^{(4)} \tag{10.51}$$

Note that we have purposely included the third and fourth derivative terms in these expressions. This is so we can study the effect of truncating higher derivative values on the derived first and second derivative approximations. Obviously, we could have included more of the higher derivatives, but that is not necessary in illustrating the concept. Therefore solving equations (10.50) and (10.51) for $f_i^{(1)}$ and $f_i^{(2)}$ using the matrix approach gives

$$\begin{bmatrix} 1 & \frac{1}{2} \\ 2 & 2 \end{bmatrix} \begin{Bmatrix} hf_i^{(1)} \\ h^2 f_i^{(2)} \end{Bmatrix} \cong \begin{bmatrix} -1 & 1 & 0 & -\frac{1}{6} & -\frac{1}{24} \\ -1 & 0 & 1 & -\frac{4}{3} & -\frac{2}{3} \end{bmatrix} \begin{Bmatrix} f_i \\ f_{i+1} \\ f_{i+2} \\ h^3 f_i^{(3)} \\ h^4 f_i^{(4)} \end{Bmatrix}$$

Consequently, premultiplying by the inverse of the coefficient matrix on the left-hand side of the above equation yields the solution

$$\begin{Bmatrix} hf_i^{(1)} \\ h^2 f_i^{(2)} \end{Bmatrix} \cong \begin{bmatrix} -\frac{3}{2} & 2 & -\frac{1}{2} & \frac{1}{3} & \frac{1}{4} \\ 1 & -2 & 1 & -1 & -\frac{7}{12} \end{bmatrix} \begin{Bmatrix} f_i \\ f_{i+1} \\ f_{i+2} \\ h^3 f_i^{(3)} \\ h^4 f_i^{(4)} \end{Bmatrix}$$

The first and second forward derivatives are now given as follows:

$$f_i^{(1)} = \frac{1}{2h}(-3f_i + 4f_{i+1} - f_{i+2}) + \frac{h^2}{3} f_i^{(3)} + \frac{h^3}{4} f_i^{(4)} \tag{10.52}$$

$$f_i^{(2)} = \frac{1}{h^2}(f_i - 2f_{i+1} + f_{i+2}) - hf_i^{(3)} - \frac{7h^2}{12} f_i^{(4)} \tag{10.53}$$

Note that for a small $h$ interval, the error terms may be approximated by the term involving the third derivative. It is customary to express these errors in terms of the power of $h$. That is, in equation (10.52), the error is assumed to be of order $h^2$ and is expressed $(O)h^2$. Therefore equations (10.52) and (10.53) are normally given as

$$f_i^{(1)} = \frac{1}{2h}(-3f_i + 4f_{i+1} - f_{i+2}) + (O)h^2$$

$$f_i^{(2)} = \frac{1}{h^2}(f_i - 2f_{i+1} + f_{i+2}) + (O)h$$

The implication is that $f_i^{(1)}$ involves less error than $f_i^{(2)}$ for the same value of $h$. Hence, from now on we shall compare difference formulas in terms of their order of error.

The orders of error for the backward difference approximations are determined by considering figure 10.6 and expanding the Taylor series about the backward points $x_{i-2}$ and $x_{i-1}$ to give

$$f_{i-2} = f_i - 2hf_i^{(1)} + 2h^2f_i^{(1)} + 2h^2f_i^{(2)} - \frac{4h^3}{3}f_i^{(3)} + \frac{2h^4}{3}f_i^{(4)}$$

$$f_{i-1} = f_i - hf_i^{(1)} + \frac{h^2}{2}f_i^{(2)} - \frac{h^3}{6}f_i^{(3)} + \frac{h^4}{24}f_i^{(4)}$$

Solving for the backward first and second derivatives using the matrix procedure yields

$$f_i^{(1)} = \frac{1}{2h}(3f_{i-2} - 4f_{i-1} + f_i) + \frac{h^2}{3}f_i^{(3)} - \frac{h^3}{3}f_i^{(4)} \tag{10.54}$$

$$f_i^{(2)} = \frac{1}{h^2}(f_{i-2} - 2f_{i-1} + f_i) + hf_i^{(3)} - \frac{7h^2}{12}f_i^{(4)} \tag{10.55}$$

Clearly the orders of error for the backward derivatives are exactly the same as those derived for the forward approximations. In fact, the coefficients are exactly the same as if we disregarded the signs. This is not the case with the central derivative approximations. Hence, using figure 10.7 we evaluate the following central difference approximations:

$$f_i^{(1)} = \frac{1}{2h}(-f_{i-1} + f_{i+1}) - \frac{h^2}{6}f_i^{(3)} + 0 \tag{10.56}$$

$$f_i^{(2)} = \frac{1}{h^2}(f_{i-1} - 2f_i + f_{i+1}) + 0 - \frac{h^2}{12}f_i^{(4)} \tag{10.57}$$

It is interesting to note that the fourth derivative term in equation (10.56) and the third derivative term in equation (10.57) are both zero. Consequently, the orders of error for the first and second derivatives are $(O)h^2$ and $(O)h^2$, respectively. Consideration of the error terms clearly indicates that the central difference approximations are more accurate than the backward and forward approximations. That is, the first central derivative as given by equation (10.56) is $\frac{1}{6}h^2f_i^{(3)}$. This compares with $\frac{1}{3}h^2f_i^{(3)}$ for the forward and backward approximations. Hence, while the order of error is the same, the actual coefficients are different. This is indeed the conclusion we reached earlier in example 10.2.

Furthermore, when dealing with ordinary and partial differential equations, it is essential to use difference expressions having the same order of error. This will help eliminate the instability problems caused by error propagation.

Knowing that central differences are more accurate, the reader may question the need for backward and forward difference expressions. While central differences are used extensively in solving complex engineering problems, the forward and backward differences play a rather important role in solving initial-value problems.

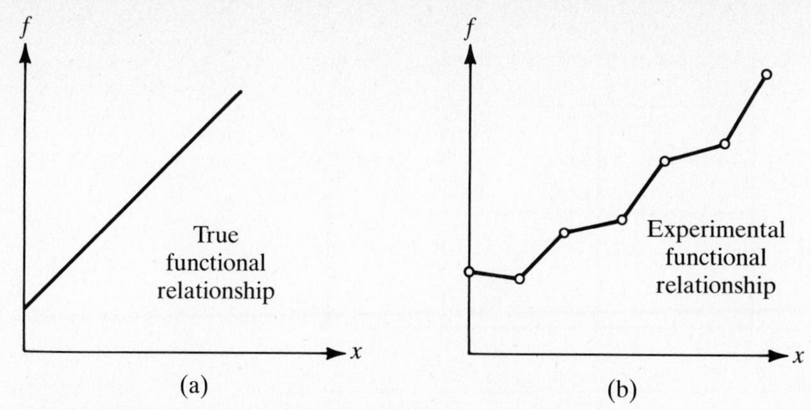


**Figure 10.9** Graphical illustration of experimental data errors.

## 10.4 NUMERICAL DIFFERENTIATION OF DATA

In the preceding section, we outlined methods for approximating the derivatives of a function. In some cases, it may be necessary to approximate derivatives of functions known only at some discrete points. Experimental data represent a case in point. Obviously, when dealing with continuous functions, we are able to evaluate these at equally spaced points. This is not the case with tabulated functions. Furthermore, experimental data with equally spaced base points involve errors. Consequently, one would expect the derivatives to be in error as well. In order to understand the magnitude of experimental error and its effect on the approximated derivative values, consider figure 10.9.

It is evident that if the true relationship between the dependent variable $f$ and the independent variable $x$ is linear as shown in figure 10.9a, then it would be unwise to assume that the data given by figure 10.9b are likely to yield appropriate derivative approximations. This is because derivative approximations involve passing an interpolating polynomial through a given set of points. These polynomials do not possess the same derivatives nor the same functional values as the true function. This situation occurs even if the wiggle problem associated with polynomials is eliminated by using cubic splines or low-order polynomials. The implication is that errors associated with experimental data may be significantly amplified when derivative approximations are determined using the methods described in section 10.3. Therefore derivatives of experimental data should be avoided whenever possible. The alternative is to use the method of least squares in approximating the true relationship, then differentiate the resulting function. Obviously, this is possible only if we know exactly the true nature of the functional relationship between $f$ and $x$.

## 10.5 MIXED DERIVATIVES

Under certain circumstances, it may be necessary to approximate the derivatives of a function involving more than one independent variable. These are referred to as partial derivatives. Obviously, these may include mixed partial derivatives of a function with respect to two or more independent variables. Such problems

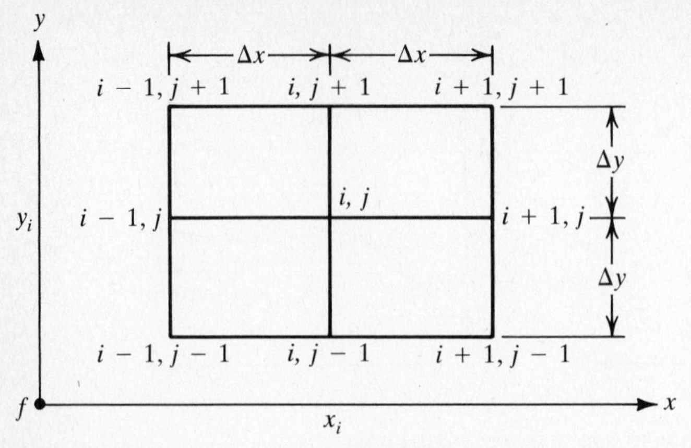


**Figure 10.10** Cartesian representation of mixed derivative grid.

arise when one is dealing with mathematical models involving partial differential equations. For example, suppose we are interested in approximating the mixed partial derivative $\partial^2 f/\partial x\, \partial y$. Clearly we must first specify the type of derivative and the order of the approximating polynomial to be used. Therefore, using central derivative approximations and a second-order polynomial, we develop the grid shown in figure 10.10.

The mixed partial derivative is now defined at node $(i, j)$ so that equal numbers of nodes exist at either side in both the $x$ and $y$ directions. Hence

$$\left.\frac{\partial^2 f}{\partial x\, \partial y}\right|_{x_i, y_i} = \frac{\partial}{\partial x}\left(\frac{\partial f}{\partial y}\right) \cong \frac{\partial}{\partial x}\left(\frac{1}{2\,\Delta y}(-f_{i,j-1} + f_{i,j+1}\right)$$

Since $\Delta y$ is constant, it can be factored out. This gives

$$\left.\frac{\partial^2 f}{\partial x\, \partial y}\right|_{x_i, y_i} \cong \frac{1}{2\,\Delta y}\left(-\frac{\partial f_{i,j-1}}{\partial x} + \frac{\partial f_{i,j+1}}{\partial x}\right)$$

Note that $f_{i,j-1}$ and $f_{i,j+1}$ are functions of the variable $x$. Therefore, using the central difference expression, we have

$$\left.\frac{\partial^2 f}{\partial x\, \partial y}\right|_{x_i, y_i}$$

$$\cong \frac{1}{2\,\Delta y}\left(-\frac{1}{2\,\Delta x}(-f_{i-1,j-1} + f_{i+1,j-1}) + \frac{1}{2\,\Delta x}(-f_{i-1,j+1} + f_{i+1,j+1})\right) \tag{10.58}$$

Simplifying equation (10.58) yields the desired finite difference approximation:

$$\left.\frac{\partial^2 f}{\partial x\, \partial y}\right|_{x_i, y_i} \cong \frac{1}{4\,\Delta x\,\Delta y}(f_{i-1,j-1} - f_{i+1,j-1} - f_{i-1,j+1} + f_{i+1,j+1}) \tag{10.59}$$

Mixed derivatives using forward as well as backward approximations can be developed in similar fashion. In addition, higher mixed derivatives can also be evaluated by using appropriate difference expressions.

E

EXAMPLE 10.4

Approximate the mixed partial derivative $\partial^4 f/\partial x^2\,\partial y^2$ at $x = 1$, $y = 1$ for the function

$$f(x, y) = x^3 y^3$$

Use central differences and a second-order polynomial approximation.

Solution

Assuming a constant interval $h$ in both the $x$ and $y$ directions, we define the derivative at $(x_i, y_i)$ as follows:

$$\left.\frac{\partial^4 f}{\partial x^2\,\partial y^2}\right|_{x_i, y_i} = \frac{\partial^2}{\partial x^2}\left(\frac{\partial^2 f}{\partial y^2}\right)$$

Using figure 10.10 and the central difference approximation for the second derivative in the $y$ direction gives

$$\begin{aligned}\left.\frac{\partial^4 f}{\partial x^2\,\partial y^2}\right|_{x_i, y_i} &\cong \frac{\partial^2}{\partial x^2}\left[\frac{1}{h^2}(f_{i,j-1} - 2f_{ij} + f_{i,j+1})\right] \\ &\cong \frac{1}{h^2}\left(\frac{\partial^2 f_{i,j-1}}{\partial x^2} - 2\frac{\partial^2 f_{ij}}{\partial x^2} + \frac{\partial^2 f_{i,j+1}}{\partial x^2}\right) \\ &\cong \frac{1}{h^2}\left[\frac{1}{h^2}(f_{i-1,j-1} - 2f_{i,j-1} + f_{i+1,j-1})\right. \\ &\quad - \frac{2}{h^2}(f_{i-1,j} - 2f_{i,j} + f_{i+1,j}) \\ &\quad \left. + \frac{1}{h^2}(f_{i-1,j+1} - 2f_{i,j+1} + f_{i+1,j+1})\right]\end{aligned}$$

Simplifying yields

$$\begin{aligned}\left.\frac{\partial^4 f}{\partial x^4\,\partial y^2}\right|_{x_i, y_i} &\cong \frac{1}{h^4}[f_{i-1,j-1} - 2f_{i,j-1} + f_{i+1,j-1} - 2f_{i-1,j} \\ &\quad + 4f_{i,j} - 2f_{i+1,j} + f_{i-1,j+1} - 2f_{i,j+1} + f_{i+1,j+1}]\end{aligned} \tag{10.60}$$

In order to evaluate the mixed partial derivative given by equation (10.60), we need to specify the $h$ value. While there is no golden rule that can be followed, let us assume that $h = 0.1$, then evaluate the various functional values. Hence

$$f_{i-1,j-1} = f(0.9, 0.9) = 0.531441$$

$$f_{i,j-1} = f(1.0, 0.9) = 0.729000$$

$$f_{i+1,j-1} = f(1.1, 0.9) = 0.970299$$

$$f_{i-1,j} = f(0.9, 1.0) = 0.729000$$

$$f_{i,j} = f(1.0, 1.0) = 1.00000$$

$$f_{i+1,j} = f(1.1, 1.0) = 1.33100$$

$$f_{i-1,j+1} = f(0.9, 1.1) = 0.970299$$

$$f_{i,j+1} = f(1.0, 1.1) = 1.33100$$

$$f_{i+1,j-1} = f(1.1, 1.1) = 1.77156$$

Substituting into equation (10.60) yields

$$\left.\frac{\partial^4 f}{\partial x^2\, \partial y^2}\right|_{x_i, y_i} = \frac{1}{(0.1)^4}[0.531441 - 2(0.729) + 0.970299 - 2(0.729) + 4 - 2(1.331) + 0.970299 - 2(1.331) + 1.77156]$$

$$= 36.000$$

Obviously this is the exact value of the mixed partial derivative. The reader should be aware that the difference expressions derived thus far are extremely useful in solving ordinary and partial differential equations.

## 10.6 SPECIAL DERIVATIVE APPROXIMATIONS

The solution of many engineering problems involving differential equations may in some cases require us to evaluate derivative expressions in which the $h$ interval is not constant. Consequently, we must develop different approximations for the derivatives from those outlined thus far. Consider figure 10.11.

Furthermore, suppose we are interested in developing central difference expressions for the first and second derivatives using a second-order polynomial. Consequently, using Taylor series, we can easily expand about points $x_{i-1}$ and $x_{i+\alpha}$ to get

$$f_{i+1} = f_i - hf_i^{(1)} + \frac{h^2}{2} f_i^{(2)} + \cdots$$

$$f_{i+\alpha} = f_i + \alpha h f_i^{(1)} + \frac{(\alpha h)^2}{2} f_i^{(2)} + \cdots$$

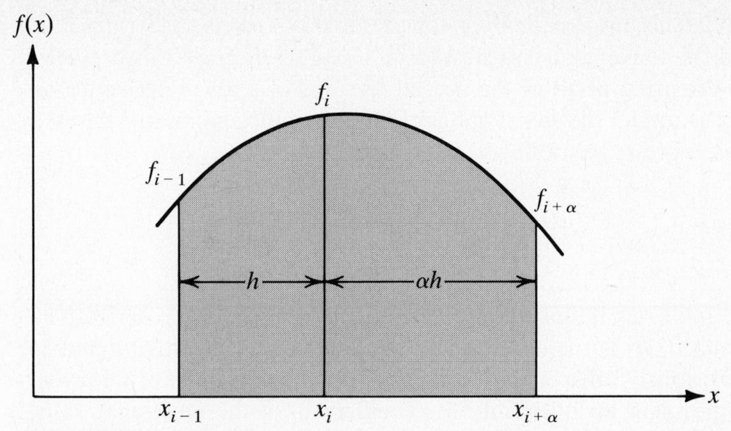


**Figure 10.11** Derivative approximations for unevenly spaced base points using a second-order polynomial.

These two equations can be easily solved for the central derivatives as follows:

$$\begin{bmatrix} -1 & \frac{1}{2} \\ \alpha & \alpha^2/2 \end{bmatrix} \begin{Bmatrix} hf_i^{(1)} \\ h^2 f_i^{(2)} \end{Bmatrix} = \begin{bmatrix} 1 & -1 & 0 \\ 0 & -1 & 1 \end{bmatrix} \begin{Bmatrix} f_{i-1} \\ f_i \\ f_{i+\alpha} \end{Bmatrix}$$

$$\begin{Bmatrix} hf_i^{(1)} \\ h^2 f_i^{(2)} \end{Bmatrix} = \frac{2}{\alpha + \alpha^2} \begin{bmatrix} -\alpha^2/2 & \frac{1}{2} \\ \alpha & 1 \end{bmatrix} \begin{bmatrix} 1 & -1 & 0 \\ 0 & -1 & 1 \end{bmatrix} \begin{Bmatrix} f_{i-1} \\ f_i \\ f_{i+\alpha} \end{Bmatrix}$$

Therefore

$$f_i^{(1)} = \frac{1}{h}\left( -\frac{\alpha^2}{\alpha + \alpha^2} f_{i-1} + \frac{\alpha^2 - 1}{\alpha + \alpha^2} f_i + \frac{1}{\alpha + \alpha^2} f_{i+\alpha} \right) \tag{10.61}$$

$$f_i^{(2)} = \frac{1}{h^2}\left( \frac{2\alpha}{\alpha + \alpha^2} f_{i-1} - \frac{2}{\alpha} f_i + \frac{2}{\alpha + \alpha^2} f_{i+\alpha} \right) \tag{10.62}$$

For $\alpha = 1$ these equations give the central difference expressions derived for equal base point spacing. It is evident that higher derivatives can be approximated by using forward and central differences. However, it is apparent that such expressions can become extremely complex and therefore difficult to use. Fortunately, equations (10.61) and (10.62) can be used effectively in solving large numbers of practical problems. The implication is that there is little need for additional difference expressions beyond those given here. Otherwise, the finite difference procedure is not recommended. Instead, the methods of finite elements should be employed; they are discussed in more detail later in this text.

## 10.7 STENCIL REPRESENTATION OF DERIVATIVES

The finite difference expressions developed thus far can be greatly simplified if we focus on their important features. Clearly the significant features are the direction, functional values, coefficients, and order of error. Consequently, for equally spaced base points the difference approximations for the first and second derivatives obtained earlier can be conveniently expressed in so-called "stencil" form. Such a form permits the retention of all of the special features of a given derivative approximation. For example, the first forward derivative approximation given by equation (10.52) is expressed as follows:

$$f^{(1)} = \frac{1}{2h}\left(\textcircled{\textcircled{-3}} \overset{h}{—} \textcircled{4} \overset{h}{—} \textcircled{-1}\right) + (O)h^2$$

The double circle indicates the position of the base point where the derivative is being evaluated and $(O)h^2$ is the order of the error associated with the difference approximation. Obviously, adjacent circles are separated by the interval $h$. Furthermore, such an expression includes only the coefficients of the functional values present in the expression. It is interesting to note that the sum of the coefficients enclosed by the circles is zero. This is always the case with derivative approximations, even when dealing with unevenly spaced base points. This observation can be very useful in checking the accuracy of the derived derivative approximation.

The second forward derivative given by equation (10.53) can thus be expressed as

$$f^{(2)} = \frac{1}{h^2}\left(\textcircled{\textcircled{1}} \overset{h}{—} \textcircled{-2} \overset{h}{—} \textcircled{1}\right) + (O)h$$

Note that once again the sum of the coefficients is zero. Similarly, the corresponding backward derivatives given by equations (10.54) and (10.55) are expressed as

$$f^{(1)} = \frac{1}{2h}\left(\textcircled{3} \overset{h}{—} \textcircled{-4} \overset{h}{—} \textcircled{\textcircled{1}}\right) + (O)h^2$$

$$f^{(2)} = \frac{1}{h^2}\left(\textcircled{1} \overset{h}{—} \textcircled{-2} \overset{h}{—} \textcircled{\textcircled{1}}\right) + (O)h$$

Similarly, the central derivative expressions given by equations (10.56) and (10.57) are expressed in stencil form as

$$f^{(1)} = \frac{1}{2h}\left(\textcircled{-1} \overset{h}{—} \textcircled{\textcircled{0}} \overset{h}{—} \textcircled{1}\right) + (O)h^2$$

$$f^{(2)} = \frac{1}{h^2}\left(\textcircled{1} \overset{h}{—} \textcircled{\textcircled{-2}} \overset{h}{—} \textcircled{1}\right) + (O)h^2$$

Note that the position of the double circle is shifted for the various difference expressions to reflect the type of derivative approximation being used.

Additional derivative approximation formulas expressed in stencil form along with their corresponding errors are given in tables 10.1–10.4.

Table 10.1 FINITE DIFFERENCE APPROXIMATIONS OF THE FIRST DERIVATIVE FOR EQUALLY SPACED BASE POINTS

| STENCILS FOR $f^{(1)}(x) = df(x)/dx$ | ERROR | POLYNOMIAL | INTERPOLATION |
|---|---|---|---|
| $\frac{1}{2h}$ (−1 — [0] — 1) | $(O)h^2$ | Second | Central |
| $\frac{1}{2h}$ ([−3] — 4 — −1) | $(O)h^2$ | Second | Forward |
| $\frac{1}{2h}$ (1 — −4 — [3]) | $(O)h^2$ | Second | Backward |
| $\frac{1}{6h}$ ([−11] — 18 — −9 — 2) | $(O)h^3$ | Third | Forward |
| $\frac{1}{6h}$ (−2 — 9 — −18 — [11]) | $(O)h^3$ | Third | Backward |
| $\frac{1}{12h}$ (1 — −8 — [0] — 8 — −1) | $(O)h^4$ | Fourth | Central |
| $\frac{1}{12h}$ ([−25] — 48 — −36 — 16 — −3) | $(O)h^4$ | Fourth | Forward |

(*continued*)

Table 10.1 (*Continued*)

| STENCILS FOR $f^{(1)}(x) = df(x)/dx$ | ERROR | POLYNOMIAL | INTERPOLATION |
|---|---|---|---|
| $\frac{1}{12h}$ (3 — −16 — 36 — −48 — 25) | $(O)h^4$ | Fourth | Backward |
| $\frac{1}{60h}$ (−137 — 300 — −300 — 200 — −75 — 12) | $(O)h^5$ | Fifth | Forward |
| $\frac{1}{60h}$ (−12 — 75 — −200 — 500 — −300 — 137) | $(O)h^5$ | Fifth | Backward |
| $\frac{1}{60h}$ (−1 — 9 — −45 — 0 — 45 — −9 — 1) | $(O)h^6$ | Sixth | Central |
| $\frac{1}{60h}$ (−147 — 360 — −450 — 400 — −225 — 72 — −10) | $(O)h^6$ | Sixth | Forward |
| $\frac{1}{60h}$ (10 — −72 — 225 — −400 — 450 — −360 — 147) | $(O)h^6$ | Sixth | Backward |

Table 10.2 FINITE DIFFERENCE APPROXIMATIONS OF THE SECOND DERIVATIVE FOR EQUALLY SPACED BASE POINTS

| STENCILS FOR $f^{(2)}(x) = d^2f(x)/dx^2$ | ERROR | POLYNOMIAL | INTERPOLATION |
|---|---|---|---|
| $\frac{1}{h^2}\left(1 — \boxed{-2} — 1\right)$ | $(O)h^2$ | Second | Central |
| $\frac{1}{h^2}\left(\boxed{1} — -2 — 1\right)$ | $(O)h$ | Second | Forward |
| $\frac{1}{h^2}\left(1 — -2 — \boxed{1}\right)$ | $(O)h$ | Second | Backward |
| $\frac{1}{h^2}\left(\boxed{2} — -5 — 4 — -1\right)$ | $(O)h^2$ | Third | Forward |
| $\frac{1}{h^2}\left(-1 — 4 — -5 — \boxed{2}\right)$ | $(O)h^2$ | Third | Backward |
| $\frac{1}{12h^2}\left(-1 — 16 — \boxed{-30} — 16 — -1\right)$ | $(O)h^4$ | Fourth | Central |
| $\frac{1}{12h^2}\left(\boxed{35} — -104 — 114 — -56 — 11\right)$ | $(O)h^3$ | Fourth | Forward |

*(continued)*

Table 10.2 (*Continued*)

| STENCILS FOR $f^{(2)}(x) = d^2f(x)/dx^2$ | ERROR | POLYNOMIAL | INTERPOLATION |
|---|---|---|---|
| $\frac{1}{12h^2}$ (11 — −56 — 114 — −104 — 35) | $(O)h^3$ | Fourth | Backward |
| $\frac{1}{60h^2}$ (225 — −770 — 1070 — −780 — 305 — −50) | $(O)h^4$ | Fifth | Forward |
| $\frac{1}{60h^2}$ (−50 — 350 — −780 — 1070 — −770 — 225) | $(O)h^4$ | Fifth | Backward |
| $\frac{1}{180h^2}$ (2 — −27 — 270 — −490 — 270 — −27 — 2) | $(O)h^6$ | Sixth | Central |
| $\frac{1}{180h^2}$ (812 — −3132 — 5265 — −5080 — 2970 — −972 — 137) | $(O)h^5$ | Sixth | Forward |
| $\frac{1}{180h^2}$ (137 — −972 — 2970 — −5080 — 5265 — −3132 — 812) | $(O)h^5$ | Sixth | Backward |

Table 10.3 FINITE DIFFERENCE APPROXIMATIONS OF THE THIRD DERIVATIVE FOR EQUALLY SPACED BASE POINTS

| STENCILS FOR $f^{(3)}(x) = d^3f(x)/dx^3$ | ERROR | POLYNOMIAL | INTERPOLATION |
|---|---|---|---|
| $\frac{1}{h^3}$ ( **−1** — 3 — −3 — 1 ) | $(O)h$ | Third | Forward |
| $\frac{1}{h^3}$ ( −1 — 3 — −3 — **1** ) | $(O)h$ | Third | Backward |
| $\frac{1}{2h^3}$ ( −1 — 2 — **0** — −2 — 1 ) | $(O)h^2$ | Fourth | Central |
| $\frac{1}{2h^3}$ ( **−5** — 18 — −24 — 14 — −3 ) | $(O)h^2$ | Fourth | Forward |
| $\frac{1}{2h^3}$ ( 3 — −14 — 24 — −18 — **5** ) | $(O)h^2$ | Fourth | Backward |
| $\frac{1}{4h^3}$ ( **−17** — 71 — −118 — 98 — −41 — 7 ) | $(O)h^3$ | Fifth | Forward |
| $\frac{1}{4h^3}$ ( −7 — 41 — −98 — 118 — −71 — **17** ) | $(O)h^3$ | Fifth | Backward |
| $\frac{1}{8h^3}$ ( 1 — −8 — 13 — **0** — −13 — 8 — −1 ) | $(O)h^4$ | Sixth | Central |
| $\frac{1}{8h^3}$ ( **−49** — 232 — −461 — 496 — −307 — 104 — −15 ) | $(O)h^4$ | Sixth | Forward |
| $\frac{1}{8h^3}$ ( 15 — −104 — 307 — −496 — 461 — −232 — **49** ) | $(O)h^4$ | Sixth | Backward |

Table 10.4 FINITE DIFFERENCE APPROXIMATIONS OF THE FOURTH DERIVATIVE FOR EQUALLY SPACED BASE POINTS

| STENCILS FOR $f^{(4)}(x) = d^4f(x)/dx^4$ | ERROR | POLYNOMIAL | INTERPOLATION |
|---|---|---|---|
| $\frac{1}{h^4}$ (1 — −4 — **6** — −4 — 1) | $(O)h^2$ | Fourth | Central |
| $\frac{1}{h^4}$ (**1** — −4 — 6 — −4 — 1) | $(O)h$ | Fourth | Forward |
| $\frac{1}{h^4}$ (1 — −4 — 6 — −4 — **1**) | $(O)h$ | Fourth | Backward |
| $\frac{1}{h^4}$ (3 — −14 — 26 — −24 — 11 — −2) | $(O)h^2$ | Fifth | Forward |
| $\frac{1}{h^4}$ (−2 — 11 — −24 — 26 — −14 — 3) | $(O)h^2$ | Fifth | Backward |
| $\frac{1}{6h^4}$ (−1 — 12 — −39 — **56** — −39 — 12 — −1) | $(O)h^4$ | Sixth | Central |
| $\frac{1}{6h^4}$ (**35** — −186 — 411 — −484 — 321 — −114 — 17) | $(O)h^3$ | Sixth | Forward |
| $\frac{1}{6h^4}$ (17 — −114 — 321 — −484 — 411 — −186 — **35**) | $(O)h^3$ | Sixth | Backward |

## 10.8 STENCIL REPRESENTATION OF MIXED DERIVATIVES

The concept of stencil representation of derivatives involving one indepenent variable can be extended to mixed and partial derivatives. The basic idea is to keep proper direction in the physical sense of the various derivatives in question. For illustration purposes, let us consider the mixed partial derivative $\partial^2 f/\partial x\,\partial y$ given by equation (10.59). A careful look at the various functional values relative to figure 10.10 clearly shows that a two-dimensional stencil can be used to represent the derivative. That is,

$$\frac{\partial^2 f}{\partial x\,\partial y} = \frac{1}{4\,\Delta x\,\Delta y}\left\{\begin{matrix} -1 & 0 & 1 \\ 0 & 0 & 0 \\ 1 & 0 & -1 \end{matrix}\right\} \qquad \text{(spacing } \Delta x \text{ horizontally, } \Delta y \text{ vertically)} \tag{10.63}$$

Once again, adjacent circles along the $x$ axis are spaced by the interval $\Delta x$ and along the $y$ axis by $\Delta y$. Note that the sum of all coefficients appearing inside the circles is zero as it should be. Obviously, the stencil representation of derivatives is extremely useful and easier to use than the difference algebraic form; this is especially true in the case of mixed derivatives. Furthermore, the equivalent stencil form for the mixed partial derivative given by equation (10.60) is expressed simply as follows:

$$\frac{\partial^4 f}{\partial x^2\,\partial y^2} = \frac{1}{h^4}\left\{\begin{matrix} 1 & -2 & 1 \\ -2 & \textcircled{4} & -2 \\ 1 & -2 & 1 \end{matrix}\right\} \qquad \text{(spacing } h \text{ horizontally and vertically)} \tag{10.64}$$

Clearly the double circle represents the point $(x_i, y_i)$ at which the derivative is being evaluated. For this special case adjacent circles are spaced by the interval $h$. It is evident that other mixed as well as partial derivative approximations can be easily expressed in the more desirable stencil form.

It is interesting to note that equations (10.63) and (10.64) can be derived symbolically by using vector multiplication. That is, we form vectors in the coefficients of derivatives in the $x$ and $y$ directions separately, then perform the multiplication. For example, maintaining the proper positive directions of the coordinates

$x$ and $y$, we have

$$\frac{\partial^2 f}{\partial x\,\partial y} = \frac{\partial}{\partial y}\left(\frac{\partial f}{\partial x}\right) = \frac{1}{2\,\Delta y}\begin{Bmatrix} 1 \\ 0 \\ -1 \end{Bmatrix}\{-1 \quad 0 \quad 1\}\frac{1}{2\,\Delta x}$$

$$= \frac{1}{4\,\Delta x\,\Delta y}\begin{bmatrix} -1 & 0 & 1 \\ 0 & 0 & 0 \\ 1 & 0 & -1 \end{bmatrix}$$

Therefore, knowing that the derivative is to be evaluated at the center point $(x_i, y_i)$, we have

$$\frac{\partial^2 f}{\partial x\,\partial y} = \frac{1}{4\,\Delta y\,\Delta x}\begin{Bmatrix} (-1) & (0) & (1) \\ (0) & (0) & (0) \\ (1) & (0) & (-1) \end{Bmatrix}$$

Similarly, equation (10.64) can be derived easily for $\Delta x = \Delta y = h$ as follows:

$$\frac{\partial^4 f}{\partial x^2\,\partial y^2} = \frac{1}{h^2}\begin{Bmatrix} 1 \\ -2 \\ 1 \end{Bmatrix}\{1 \quad -2 \quad 1\}\frac{1}{h^2}$$

$$= \frac{1}{h^4}\begin{bmatrix} 1 & -2 & 1 \\ -2 & 4 & -2 \\ 1 & -2 & 1 \end{bmatrix}$$

Obviously, these coefficients correspond precisely to those given in equation (10.64).

## Recommended Reading

*A Practical Guide to Computer Methods for Engineers*, Terry E. Shoup, Prentice-Hall, Inc., Englewood Cliffs, N.J., 1979.

*Numerical Mathematics and Computing*, Cheney, Ward, and Kincaid, Brooks/Cole Publishing Co., Monterey, Calif., 1980.

*Applied Numerical Analysis*, Curtis F. Gerald and P. O. Wheatley, Addison-Wesley, Reading, Mass., 1984.

*Numerical Analysis—A Practical Approach*, Melvin J. Maron, Macmillan, New York, 1982.

## P PROBLEMS

**10.1** Approximate the square root of 19.0 using Taylor series and retaining the first five terms of the series. Expand about $x_0 = 16$.

**10.2** Use Taylor series to develop the first, second, and third forward difference approximations. Retain the first five derivative terms in the series.

**10.3** Rework problem 10.2 retaining the first six derivative terms in the series.

**10.4** Use the method of undetermined coefficients to develop backward derivative approximations up to the second order. Employ a fifth-order polynomial.

**10.5** Evaluate the first, second, third, and fourth derivatives of the following polynomial using the most accurate stencils given in tables 10.1–10.4 and $h = 0.1$:

$$f(x) = x^4 - 3x^3 + 4x^2 + 10$$

Evaluate at $x = 1.0$. Then compare your results with the exact values.

**10.6** Rework problem 10.5 using the least accurate central derivative stencil; however, use $h = 0.01$.

**10.7** Approximate the first and second derivatives of the following expression using an $(O)h^4$ stencil and $h = 0.05$:

$$f(x) = \frac{x^2}{\sin x} \ln x$$

Evaluate at $x = 10$.

**10.8** Rework problem 10.7 using $h = 0.01$.

**10.9** In chapter 6, the modified Newton–Raphson methods were modified so that the derivatives needed for evaluating the roots were approximated using difference expressions. Develop these approximations for the first and second derivatives.

**10.10** Use the difference tables given by tables 10.1–10.4 to approximate the following mixed partial derivatives:

(a) $\dfrac{\partial^2 f}{\partial x\, \partial y}$ (b) $\dfrac{\partial^3 f}{\partial x^2\, \partial y}$ (c) $\dfrac{\partial^3 f}{\partial x\, \partial y^2}$ (d) $\dfrac{\partial^4 f}{\partial x^2\, \partial y^2}$ (e) $\dfrac{\partial^4 f}{\partial x^3\, \partial y}$

Use central difference approximations with error of $(O)h^2$.

# chapter 11

# Numerical Integration

An astronaut swimming in space. *Courtesy of N.A.S.A.*

"Numerical integration has become a valuable tool in solving complex engineering problems. It represents the basis for many numerical schemes including the finite element methods."

# 11

## 11.1 INTRODUCTION

The solution of many engineering problems may involve evaluating one or more integrals. These integrals cannot always be evaluated analytically. For such cases as well as problems involving experimental data, numerical methods are used to arrive at a solution. The strategy in developing the so-called integration formulas is to use interpolating functions. That is, a polynomial is passed through a set of points, and the resulting function is integrated analytically.

Unlike numerical differentiation, numerical integration is stable and more accurate for a given interpolating polynomial. This fact is illustrated graphically in figure 11.1.

Here a function is assumed to be continuous in the interval

$$x_1 \leq x \leq x_3$$

A polynomial is used to approximate the actual function. The area under the polynomial and the $x$ axis is clearly less than the actual area for

$$x_1 \leq x \leq x_2$$

and exceeds the actual area for

$$x_2 \leq x \leq x_3$$

Therefore the sum of the error associated with the integral for $x = x_1$ to $x = x_3$ is reduced. Evidently the associated error is determined by the order of the interpolating function used. Hence a first-order function is less accurate than a second, etc.

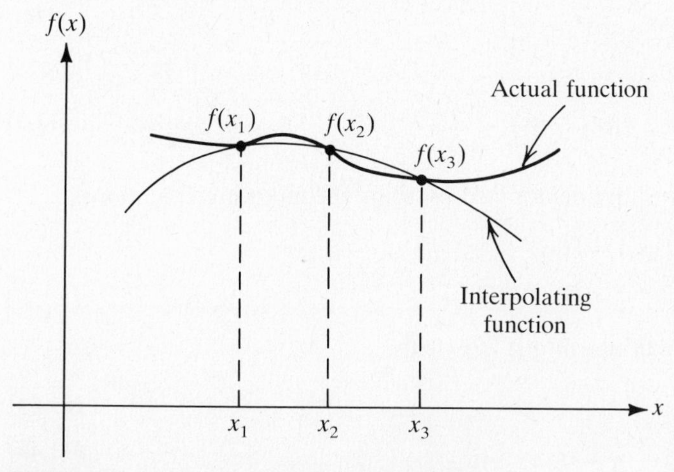


**Figure 11.1** Numerical integration of a function.

## 11.2 TRAPEZOIDAL RULE OF INTEGRATION

This is one of the simplest integration methods available. It applies for equally spaced base points only. Consider the integral

$$I = \int_{x_1}^{x_2} f(x)\,dx$$

The value of $I$ can be determined analytically provided that the function $f(x)$ is of relatively simple form and continuous. Generally the evaluation of definite integrals by conventional analytical methods is difficult or even impossible. In addition, integration may be required for cases in which $f(x)$ is known at discrete points. Consequently, numerical methods of integration represent a natural alternative whenever conventional methods fail to yield a solution.

Consider the arbitrary function $f(x)$ as shown in figure 11.2. Suppose that the area bounded by the function and the $x$ axis from $x_1$ to $x_2$ is required. If we now assume that the integral is approximately equal to the area enclosed by the linear function and the $x$ axis, then a general rule of integration can be easily developed as follows:

$$I \simeq \int_{x_1}^{x_2} F(x)\,dx \simeq \int_{x_1}^{x_2} f(x)\,dx \tag{11.1}$$

It is evident that the simplest form for $F(x)$ is a first-order polynomial. Thus

$$F(x) = a_0 + a_1 x \tag{11.2}$$

Substituting equation (11.2) into equation (11.1) then integrating gives

$$\begin{aligned} I &= \int_{x_1}^{x_2} (a_0 + a_1 x)\,dx \\ &= a_0 x + a_1 \frac{x^2}{2}\bigg|_{x_1}^{x_2} \\ &= a_0 x_2 + a_1 \frac{x_2^2}{2} - a_0 x_1 - a_1 \frac{x_1^2}{2} \\ &= a_0(x_2 - x_1) + \frac{1}{2} a_1 (x_2^2 - x_1^2) \end{aligned} \tag{11.3}$$

The constants $a_0$ and $a_1$ are determined by using the following conditions:

At $x = x_1$: $F(x_1) = f(x_1)$

At $x = x_2$: $F(x_2) = f(x_2)$

Substituting these conditions into (11.2) yields

$$f(x_1) = a_0 + a_1 x_1 \tag{11.4a}$$

$$f(x_2) = a_0 + a_1 x_2 \tag{11.4b}$$

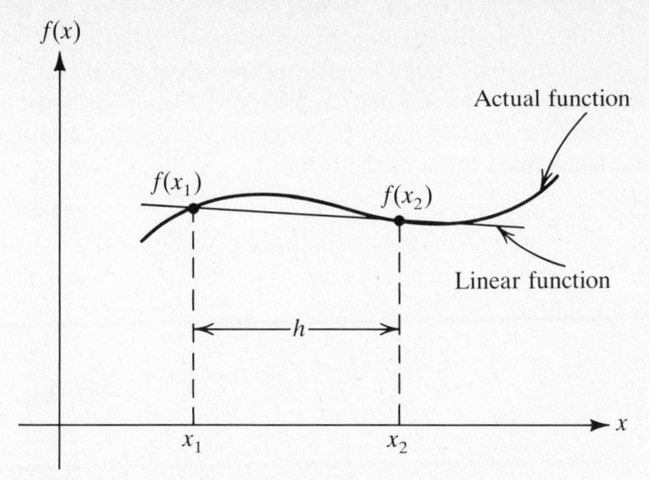


**Figure 11.2** Integral evaluation using the trapezoidal rule.

Solving equations (11.4) for the unknowns $a_0$ and $a_1$ gives

$$a_0 = \frac{1}{x_2 - x_1}[x_2 f(x_1) - x_1 f(x_2)] \tag{11.5a}$$

$$a_1 = \frac{1}{x_2 - x_1}[-f(x_1) + f(x_2)] \tag{11.5b}$$

Substituting equations (11.5) into equation (11.3) then simplifying yields the so-called trapezoidal rule of integration:

$$I = \tfrac{1}{2}(x_2 - x_1)[f(x_1) + f(x_2)] \tag{11.6}$$

This equation states that from the functional values at $x_1$ and $x_2$ the value of the integral can be determined rather easily. Note that

$$x_2 - x_1 = h$$

Hence

$$I = \tfrac{1}{2}h[f(x_1) + f(x_2)] \tag{11.7}$$

It is clear that equation (11.7) could have been derived in an easier way had we selected $x_1 = 0$ and $x_2 = h$. Since $h$ is independent of the particular values of $x_1$ and $x_2$, it is rather dependent on the difference between them.

Consideration of figure 11.2 clearly shows that better accuracy can be obtained if more intermediate base points are used between $x = x_1$ and $x = x_2$. Therefore equation (11.7) can be written in a global form

$$I = \tfrac{1}{2}h[f(x_1) + 2f(x_1 + h) + 2f(x_1 + 2h) + \cdots + f(x_2)] \tag{11.8}$$

where $h$ is the increment between two adjacent base points.

The trapezoidal rule of integration derives its name from the fact that a given area is approximated by a trapezoid or a number of trapezoids. Equation (11.7) can be derived somewhat differently by noting that the resulting expression must be given exact results for constant as well as linear functions. Consequently, if the desired integration formula is assumed to be of the form

$$I = a_0 f(x_1) + a_1 f(x_2) \tag{11.9}$$

then for $x_1 = 0$ and $x_2 = h$

$f(x) = 1$:

$$\begin{aligned} \int_0^h 1\,dx &= a_0 f(x_1) + a_1 f(x_2) \\ h &= a_0(1) + a_1(1) \end{aligned} \tag{11.10a}$$

$f(x) = x$:

$$\begin{aligned} \int_0^h x\,dx &= a_0 f(x_1) + a_1 f(x_2) \\ \tfrac{1}{2}h^2 &= a_0(0) + a_1(h) \end{aligned} \tag{11.10b}$$

Solving equations (11.10) for $a_0$ and $a_1$ gives

$$a_1 = \tfrac{1}{2}h, \qquad a_0 = \tfrac{1}{2}h$$

Substituting into equation (11.9) yields the desired expression:

$$I = \tfrac{1}{2}h[f(x_1) + f(x_2)] \tag{11.11}$$

Obviously, equation (11.11) is exactly the same equation as was derived earlier [equation (11.7)]. This solution procedure is referred to as the method of undetermined coefficients.

EXAMPLE 11.1

Evaluate the following integral using the trapezoidal rule and $h = 0.1$.

$$I = \int_1^{1.6} e^{x^2}\,dx$$

Solution

It should be evident that the smaller $h$ is the higher the accuracy will be. Hence

$$I = \tfrac{1}{2}h[f(1) + 2f(1.1) + 2f(1.2) + 2f(1.3) + 2f(1.4) + 2f(1.5) + f(1.6)] \tag{11.12}$$

where

$$f(x) = e^{x^2}$$

$$f(1.0) = 2.7182818$$

$$f(1.1) = 3.3534846$$

$$f(1.2) = 4.2206958$$

$$f(1.3) = 5.4194807$$

$$f(1.4) = 7.0993271$$

$$f(1.5) = 9.4877357$$

$$f(1.6) = 12.935817$$

Substituting these values into equation (11.12) gives

$$I = \frac{0.1}{2} 74.815547 = 3.740773$$

EXAMPLE 11.2

Given the following experimental data, determine the approximate area enclosed by a curve passing through the data and the $x$ axis:

| $x$ | 1 | 2 | 3 | 5 |
|---|---|---|---|---|
| $y$ | 2 | 4 | 5 | 7 |

## Solution

It is evident that no information is provided as to the function in question; however, this is not an obstacle to determining the answer. Note that the base points ($x$ values) are not equally spaced; hence the trapezoidal rule must be applied as many times as the number of $h$ values in a given data set. In this case the integral is evaluated in two steps:

$$I_1 = \tfrac{1}{2}h_1[f(1) + 2f(2) + f(3)]$$

$$I_2 = \tfrac{1}{2}h_2[f(3) + f(5)]$$

where $h_1 = 1$ and $h_2 = 2$. Thus

$$I = I_1 + I_2 = \tfrac{1}{2}[2 + 2(4) + 5] + \tfrac{2}{2}(5 + 7)$$
$$= 19.5$$

Note that the trapezoidal rule of integration involves no restrictions relative to the number of base points involved. This is not the case with more elaborate methods yet to be discussed. This is one reason why it is one of the favorite numerical integration methods used in engineering.

# E

EXAMPLE 11.3

The deformation of the axially loaded member shown below is completely defined by the differential equation

$$\frac{du}{dx} = \frac{N(x)}{A(x)E(x)} \tag{11.13}$$

where

$u$ = axial deformation

$N(x)$ = axial force applied

$E(x)$ = Young's modulus of elasticity

$A(x)$ = cross-sectional area of member

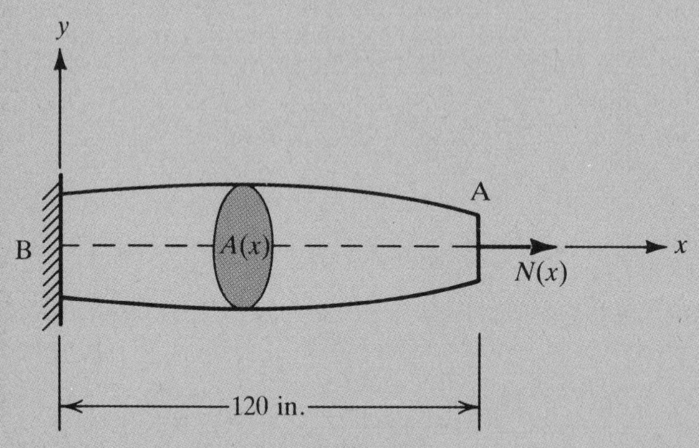


Determine the relative deformation of point A with respect to B if

$N(x) = 10$ kips

$L = 120$ in.

$A(x) = -\frac{1}{3600}x^2 + \frac{1}{30}x + 1$ in.$^2$

$E(x) = 30 \times 10^3$ ksi

Use $h = 10$ in.

Solution

This problem can be easily solved by integrating the differential equation (11.13) pertaining to axial deformation. Thus

$$\int_{u_B}^{u_A} du = \int_0^{120} \frac{N(x)}{E(x)A(x)}\,dx$$

$$u_A - u_B = \int_0^{120} \frac{10\,dx}{30 \times 10^3(-\frac{1}{3600}x^2 + \frac{1}{30}x + 1)} \tag{11.14}$$

$$u_{A/B} = \int_0^{120} \frac{dx}{-\frac{30}{36}x^2 + 100x + 3000}$$

Obviously the methods of calculus are not well suited for solving equation (11.14). Therefore a numerical procedure must be employed. Thus

$$h = 10 \text{ in.}$$

$$f(x) = (-\tfrac{30}{36}x^2 + 100x + 3000)^{-1}$$

$$\begin{aligned} u_{A/B} = \tfrac{1}{2}h[&f(0) + 2f(10) + 2f(20) + 2f(30) + 2f(40) + 2f(50) + 2f(60) \\ &+ 2f(70) + 2f(80) + 2f(90) + 2f(100) + 2f(110) + f(120)] \end{aligned} \tag{11.15}$$

where

$f(0) = 3.33333 \times 10^{-4}$ $\quad$ $f(70) = 1.69014 \times 10^{-4}$

$f(10) = 2.55319 \times 10^{-4}$ $\quad$ $f(80) = 1.76471 \times 10^{-4}$

$f(20) = 2.14285 \times 10^{-4}$ $\quad$ $f(90) = 1.90476 \times 10^{-4}$

$f(30) = 1.90476 \times 10^{-4}$ $\quad$ $f(100) = 2.14285 \times 10^{-4}$

$f(40) = 1.76471 \times 10^{-4}$ $\quad$ $f(110) = 2.55319 \times 10^{-4}$

$f(50) = 1.69014 \times 10^{-4}$ $\quad$ $f(120) = 3.33333 \times 10^{-4}$

$f(60) = 1.66667 \times 10^{-4}$

Substituting into equation (11.15) gives the relative deformation:

$$\begin{aligned} u_{A/B} &= \tfrac{10}{2}(50.2226 \times 10^{-4}) \\ &= 0.0251113 \text{ in.} \end{aligned}$$

It is interesting to note that most engineering mechanics textbooks avoid dealing with axial deformation of members of variable cross section partly because of the difficulty in evaluating integrals similar to that of equation (11.14).

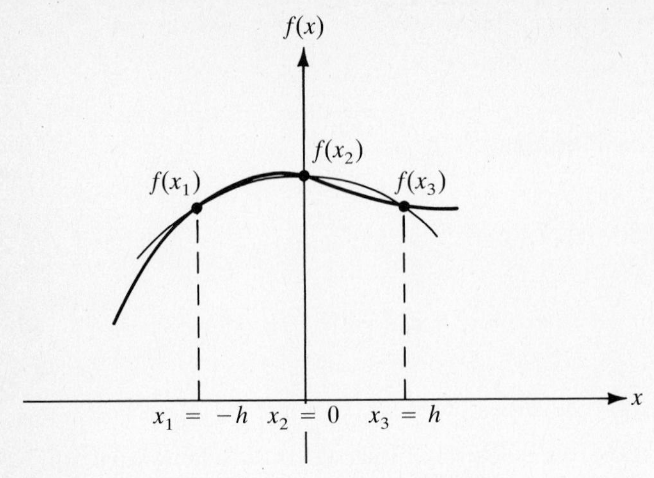


**Figure 11.3** Integration using Simpson's rule.

## 11.3 SIMPSON'S $\frac{1}{3}$ RULE OF INTEGRATION METHOD

This method of integration is more accurate than the trapezoidal rule in that it assumes a second-order interpolating function to approximate a given function. The *disadvantage of Simpson's rule* is that it can only be applied to an even number of increments. Consider figure 11.3.

The value of the integral for $x = -h$ to $x = h$ is approximated as follows:

$$I \simeq \int_{-h}^{h} f(x)\,dx \simeq \int_{-h}^{h} F(x)\,dx \tag{11.16}$$

where

$$F(x) = a_0 + a_1 x + a_2 x^2$$

Integration of the right-hand side of equation (11.16) gives

$$\begin{aligned} I &= \int_{-h}^{h} (a_0 + a_1 x + a_2 x^2)\,dx \\ &= a_0 x + \frac{a_1 x^2}{2} + \frac{a_2 x^3}{3}\bigg|_{-h}^{h} \\ &= 2ha_0 + \frac{2h^3}{3} a_2 \end{aligned} \tag{11.17}$$

In matrix form

$$I = \left\{ 2h \quad 0 \quad \frac{2h}{3} \right\} \left\{ \begin{matrix} a_0 \\ a_1 h \\ a_2 h^2 \end{matrix} \right\} \tag{11.18}$$

or in compact matrix form

$$I = \{c\}\{a\} \tag{11.19}$$

The $\{a\}$ vector can be determined by using the three pivot points given in figure 11.3. That is,

$$f(x_1) = a_0 - ha_1 + h^2 a_2 \tag{11.20a}$$

$$f(x_2) = a_0 + (0)a_1 + (0)^2 a_2 \tag{11.20b}$$

$$f(x_3) = a_0 + ha_1 + h^2 a_2 \tag{11.20c}$$

Expressing equations (11.20) in matrix form gives

$$\begin{Bmatrix} f(x_1) \\ f(x_2) \\ f(x_3) \end{Bmatrix} = \begin{bmatrix} 1 & -1 & 1 \\ 1 & 0 & 0 \\ 1 & 1 & 1 \end{bmatrix} \begin{Bmatrix} a_0 \\ a_1 h \\ a_2 h^2 \end{Bmatrix} \tag{11.21}$$

$$\begin{Bmatrix} a_0 \\ a_1 h \\ a_2 h^2 \end{Bmatrix} = \begin{bmatrix} 1 & -1 & 1 \\ 1 & 0 & 0 \\ 1 & 1 & 1 \end{bmatrix}^{-1} \begin{Bmatrix} f(x_1) \\ f(x_2) \\ f(x_3) \end{Bmatrix} \tag{11.22}$$

$$\begin{Bmatrix} a_0 \\ a_1 h \\ a_2 h^2 \end{Bmatrix} = \begin{bmatrix} 0 & 1 & 0 \\ -\frac{1}{2} & 0 & \frac{1}{2} \\ \frac{1}{2} & -1 & \frac{1}{2} \end{bmatrix} \begin{Bmatrix} f(x_1) \\ f(x_2) \\ f(x_3) \end{Bmatrix} \tag{11.23}$$

Substituting equation (11.23) into equation (11.18) gives

$$I = \begin{Bmatrix} 2h & 0 & \dfrac{2h}{3} \end{Bmatrix} \begin{bmatrix} 0 & 1 & 0 \\ -\frac{1}{2} & 0 & \frac{1}{2} \\ \frac{1}{2} & -1 & \frac{1}{2} \end{bmatrix} \begin{Bmatrix} f(x_1) \\ f(x_2) \\ f(x_3) \end{Bmatrix}$$

$$= \begin{Bmatrix} \dfrac{h}{3} & \dfrac{4h}{3} & \dfrac{h}{3} \end{Bmatrix} \begin{Bmatrix} f(x_1) \\ f(x_2) \\ f(x_3) \end{Bmatrix}$$

$$= \frac{h}{3}[f(x_1) + 4f(x_2) + f(x_3)] \tag{11.24}$$

Equation (11.24) is the so-called Simpson's $\frac{1}{3}$ rule of integration. Once again one could use the method of undetermined coefficients to derive Simpson's $\frac{1}{3}$ rule. It is left to the student to show that indeed this is true. The authors wish to emphasize that this method of solution is more appropriate in that it provides additional insight into matrices and their uses.

In general, equation (11.24) may be expressed for $n$ base points as follows:

$$I = \frac{h}{3}[f(x_1) + 4f(x_1 + h) + 2f(x_1 + 2h) + 4f(x_1 + 3h) + \cdots + f(x_n)] \tag{11.25}$$

where

$x_1$ = lower limit of integration

$x_n$ = upper limit of integration

$h$ = increment

$n$ = number of base points

Note that $n$ must be an odd number. This guarantees an even number of increments.

EXAMPLE 11.4

Solve example 11.3 using $h = 30$ in.

Solution

$$u_{A/B} = \frac{h}{3}[f(0) + 4f(30) + 2f(60) + 4f(90) + f(120)]$$

$$= 0.025238$$

This value is close to that obtained in example 11.3 despite the fact we have only used 5 points rather than 12!

## 11.4 DEVELOPMENT OF SPECIAL INTEGRATION FORMULAS

It should be apparent by now that a polynomial of the $n$th order may be assumed for approximating a set of data points or a function whose integral is desired. In essence there is no limit to the number of integration formulas that can be developed. Unfortunately, as was illustrated with Simpson's rule, higher accuracy is attained at the expense of constraints imposed on the number of increments used.

In this section, integration formulas are developed for cases in which the range of integration is bounded by fewer base points than the number of base points through which the interpolating function passes. That is, the shape of the approximating function is fixed in such a way that the actual function is approximated beyond the limits of integration. Consider figure 11.4.

Here we are interested in developing an integration formula that applies for $x = -h$ to $x = h$ yet using more base points than those specified by the limits

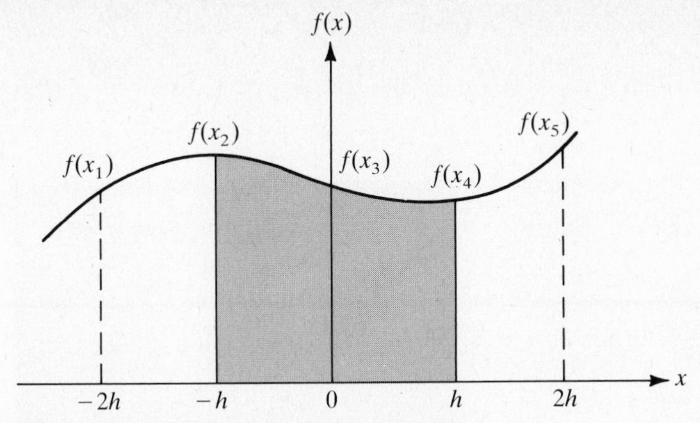


**Figure 11.4** Special integration techniques.

of integration. Assuming a fourth-order interpolating function of the form

$$F(x) \simeq f(x) = a_0 + a_1 x + a_2 x^2 + a_3 x^3 + a_4 x^4 \tag{11.26}$$

we have

$$\begin{aligned} I &\simeq \int_{-h}^{h} F(x)\,dx = \int_{-h}^{h} (a_0 + a_1 x + a_2 x^2 + a_3 x^3 + a_4 x^4)\,dx \\ &\simeq a_0 x + \frac{a_1 x^2}{2} + \frac{a_2 x^3}{3} + \frac{a_3 x^4}{4} + \frac{a_4 x^5}{5} \Big|_{-h}^{h} \\ &\simeq 2ha_0 + \tfrac{2}{3}h^3 a_2 + \tfrac{2}{5}h^5 a_4 \end{aligned} \tag{11.27}$$

The constants $a_0, \ldots, a_4$ appearing in equation (11.27) are determined readily as follows:

$$\begin{Bmatrix} f(x_1) \\ f(x_2) \\ f(x_3) \\ f(x_4) \\ f(x_5) \end{Bmatrix} = \begin{bmatrix} 1 & -2 & 4 & -8 & 16 \\ 1 & -1 & 1 & -1 & 1 \\ 1 & 0 & 0 & 0 & 0 \\ 1 & 1 & 1 & 1 & 1 \\ 1 & 2 & 4 & 8 & 16 \end{bmatrix} \begin{Bmatrix} a_0 \\ a_1 h \\ a_2 h^2 \\ a_3 h^3 \\ a_4 h^4 \end{Bmatrix}$$

Solving for the $\{a\}$ vector using the inverse procedure gives

$$a_0 = f(x_3)$$

$$a_2 = \frac{1}{24h^2}\,[-f(x_1) + 16f(x_2) - 30f(x_3) + 16f(x_4) - f(x_5)]$$

$$a_4 = \frac{1}{24h^4}\,[f(x_1) - 4f(x_2) + 6f(x_3) - 4f(x_4) + f(x_5)]$$

Substituting into equation (11.27) then simplifying gives

$$I \simeq \frac{h}{90}[-f(x_1) + 34f(x_2) + 114f(x_3) + 34f(x_4) - f(x_5)] \quad \textbf{(11.28)}$$

EXAMPLE 11.5

Determine the value of the integral given below using equation (11.28) and $h = 0.1$:

$$I = \int_0^{0.2} e^{10x^2}\, dx$$

### Solution

A schematic representation of the function may help visualize the method. Thus

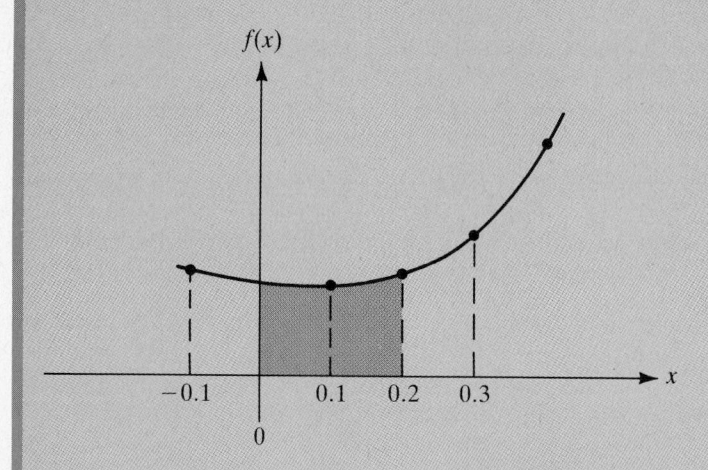


Clearly the functional values at each of the base points must be determined. That is,

$$f(-0.1) = \exp(-0.1)^2(10) = 1.1051709$$

$$f(0) = \exp(0)(10) = 1.000$$

$$f(0.1) = \exp(0.1)^2(10) = 1.1051709$$

$$f(0.2) = \exp(0.2)^2(10) = 1.4918247$$

$$f(0.3) = \exp(0.3)^2(10) = 2.4596031$$

The integral is now evaluated by using equation (11.28):

$$I = \frac{h}{90}[-f(-0.1) + 34f(0) + 114f(0.1) + 34f(0.2) - f(0.3)]$$

$$= \frac{0.1}{90} 207.14675$$

$$= 0.23016$$

This value compares with 0.23041 obtained from Simpson's rule and 0.2351 from the trapezoidal rule of integration. Obviously equation (11.28) gives the more accurate answer.

## 11.5 INTEGRATION OF UNEVENLY SPACED BASE POINTS

When dealing with experimental data one may be faced with a situation in which base points are not equally spaced. That is,

$$x_1 - x_2 \neq x_2 - x_3 \neq \cdots \neq x_{n-1} - x_n$$

Consider the simple case illustrated in figure 11.5.

It is evident that a second-order polynomial is needed for interpolating through the three points:

$$f(x) = a_0 + a_1x + a_2x^2 \tag{11.29}$$

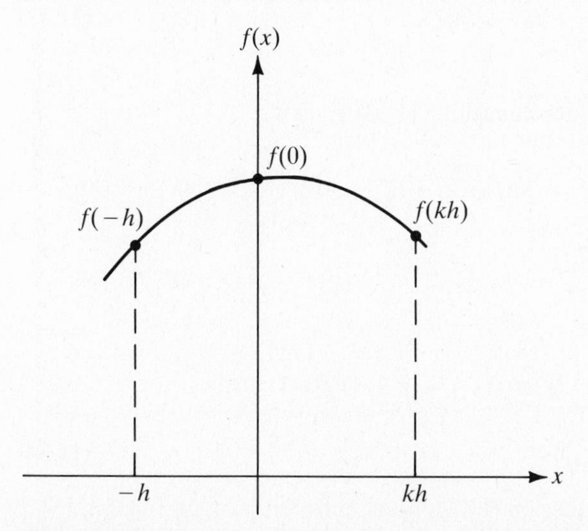


**Figure 11.5** Integration of unevenly spaced base points.

Substituting the pivotal points $(-h, f(-h))$, $(0, f(0))$, and $(kh, f(kh))$ gives

$$f(-h) = a_0 - a_1 h + a_2 h^2 \tag{11.30a}$$

$$f(0) = a_0 + a_1(0) + a_2(0)^2 \tag{11.30b}$$

$$f(kh) = a_0 + a_1(kh) + a_2(kh)^2 \tag{11.30c}$$

or in matrix form

$$\begin{Bmatrix} f(-h) \\ f(0) \\ f(kh) \end{Bmatrix} = \begin{bmatrix} 1 & -1 & 1 \\ 1 & 0 & 0 \\ 1 & k & k^2 \end{bmatrix} \begin{Bmatrix} a_0 \\ a_1 h \\ a_2 h^2 \end{Bmatrix} \tag{11.31}$$

Solving for the unknown vector containing $a_0$, $a_1$, and $a_2$ gives

$$\begin{Bmatrix} a_0 \\ a_1 h \\ a_2 h^2 \end{Bmatrix} = \frac{1}{k + k^2} \begin{bmatrix} 0 & k + k^2 & 0 \\ -k^2 & -1 + k^2 & 1 \\ k & -k - 1 & 1 \end{bmatrix} \begin{Bmatrix} f(-h) \\ f(0) \\ f(kh) \end{Bmatrix} \tag{11.32}$$

The integral is evaluated as follows:

$$\begin{aligned} I &= \int_{-h}^{kh} (a_0 + a_1 x + a_2 x^2)\, dx \\ &= a_0 x + a_1 \frac{x^2}{2} + a_2 \frac{x^3}{3} \bigg|_{-h}^{kh} \\ &= a_0 h\,(k + 1) + \frac{a_1 h^2}{2}(k^2 - 1) + \frac{a_2 h^3}{3}(k^3 + 1) \\ &= \frac{h}{6} \{6(k + 1) \quad 3(k^2 - 1) \quad 2(k^3 + 1)\} \begin{Bmatrix} a_0 \\ a_1 h \\ a_2 h^2 \end{Bmatrix} \end{aligned} \tag{11.33}$$

Substituting equation (11.32) into equation (11.33) gives

$$I = \frac{h}{6(k + k^2)} [(-k^4 + 3k^2 + 2k) f(-h) + (k^4 + 4k^3 + 6k^2 + 4k + 1) f(0) + (2k^3 + 3k^2 - 1) f(kh)]$$

or more simply

$$I = \frac{h}{6(k + k^2)} [(-k^4 + 3k^2 + 2k) f(x_1) + (k + 1)^4 f(x_2) + (2k^3 + 3k^2 - 1) f(x_3)] \tag{11.34}$$

It is interesting to note that equation (11.34) reduces to Simpson's $\frac{1}{3}$ rule for $k = 1$ as it should. It also shows the difficulty involved in dealing with unevenly spaced points.

E

EXAMPLE 11.6

Given the following unevenly spaced data, determine the area enclosed by the points and the $x$ axis.

| $f(x)$ | 1 | 7 | 9 |
|---|---|---|---|
| $x$ | 5 | 13 | 15 |

Solution

$$h = 13 - 5 = 8$$

$$kh = 15 - 13 = 2$$

$$k = \tfrac{2}{8} = \tfrac{1}{4}$$

Substituting into equation (11.34) gives

$$I = \frac{8}{6(0.25 + 0.0625)}[0.68359f(5) + 2.44141f(13) - 0.78125f(15)]$$

$$= 45.862782$$

Using the trapezoidal rule of integration twice gives

$$I = \frac{h_1}{2}[f(5) + f(13)] + \frac{h_2}{2}[f(13) + f(15)]$$

$$= \tfrac{8}{2}(1 + 7) + \tfrac{2}{2}(7 + 9)$$

$$= 48$$

In many cases one may use regression analysis to develop a function or a group of functions that can be integrated by using the method of calculus. Consider the following example.

E

EXAMPLE 11.7

Determine the area enclosed by the following data and the $x$ axis using regression analysis.

| $x$ | 1 | 3 | 4 | 6 |
|---|---|---|---|---|
| $y$ | 1 | 2 | 5 | 10 |

### Solution

Let's assume that a second-order polynomial can adequately represent the data set. That is,

$$y = ax^2 + bx + c \tag{11.35}$$

The constants $a$, $b$, and $c$ can be easily determined by using the following linear algebraic equations:

$$\sum x_i^2 y_i = a \sum x_i^4 + b \sum x_i^3 + c \sum x_i^2 \tag{11.36a}$$

$$\sum x_i y_i = a \sum x_i^3 + b \sum x_i^2 + c \sum x_i \tag{11.36b}$$

$$\sum y_i = a \sum x_i^2 + b \sum x_i + cn \tag{11.36c}$$

where $n = 5$ and

$$\sum x_i^2 y_i = 589, \qquad \sum y_i = 18$$

$$\sum x_i y_i = 97, \qquad \sum x_i^2 = 75$$

$$\sum x_i = 15, \qquad \sum x_i^4 = 2739$$

Substituting into equations (11.36) and solving for the unknowns gives

$$a = 0.133, \qquad b = 0.48, \qquad c = 0.20$$

The regression equation is now given as

$$y = 0.133x^2 + 0.48x + 0.20$$

Integrating this simple function from $x = 1$ to $x = 6$ gives

$$I = \int_1^6 (0.133x^2 + 0.48x + 0.2)\,dx$$

$$= 0.133\frac{x^3}{3} + 0.48\frac{x^2}{2} + 0.2x\Big|_1^6$$

$$= 18.93$$

This value compares with 21.5 obtained by means of the trapezoidal rule of integration.

## 11.6 STENCIL REPRESENTATION OF INTEGRATION FORMULAS

It is evident that there is no limit to the number of integration formulas that one could develop. In this section, a few more integration formulas are provided along with their associated errors. The order of error is determined from Taylor series. For convenience, these formulas are given in stencil form (Table 11.1). Note that double circles are used for points falling within the range of integration.

Consideration of the various formulas reveals that the sum of all numbers inside the circles multiplied by the coefficient yields the number of increments. For

Table 11.1 STENCILS FOR INTEGRATION FORMULAS USING EQUALLY SPACED BASE POINTS

| INTEGRATION FORMULA FOR $I = \int f(x)\,dx$ | LOCAL ERROR | GLOBAL ERROR | POLYNOMIAL |
|---|---|---|---|
| $\frac{h}{2}$(1—1) | $(O)h^3$ | $(O)h^2$ | First |
| $\frac{h}{3}$(1—4—1) | $(O)h^5$ | $(O)h^4$ | Second |
| $\frac{h}{3}$(−1—13—13—−1) | $(O)h^5$ | $(O)h^4$ | Third |
| $\frac{3h}{8}$(1—3—3—1) | $(O)h^5$ | $(O)h^4$ | Third |
| $\frac{h}{90}$(−1—34—114—34—−1) | $(O)h^7$ | $(O)h^6$ | Fourth |
| $\frac{2h}{45}$(7—32—12—32—7) | $(O)h^7$ | $(O)h^6$ | Fourth |
| $\frac{h}{1440}$(11—−93—802—802—−93—11) | $(O)h^7$ | $(O)h^6$ | Fifth |
| $\frac{3h}{160}$(−1—23—58—58—23—−1) | $(O)h^7$ | $(O)h^6$ | Fifth |
| $\frac{5h}{288}$(19—75—50—50—75—19) | $(O)h^7$ | $(O)h^6$ | Fifth |
| $\frac{h}{3780}$(5—−72—1503—4688—1503—−72—5) | $(O)h^9$ | $(O)h^8$ | Sixth |
| $\frac{2h}{945}$(−4—171—612—332—612—171—−4) | $(O)h^9$ | $(O)h^8$ | Sixth |

example, the first line gives the trapezoidal rule of integration. Now, if you sum the numbers then multiply by the factor corresponding to the formula, then

$$(1 + 1)\frac{h}{2} = h$$

The second line represents Simpson's rule. Here we have

$$(1 + 4 + 1)\frac{h}{3} = 2h$$

Obviously Simpson's rule applies to a minimum of two increments. These observations are helpful in checking newly derived formulas.

## 11.7 ROMBERG'S INTEGRATION

When dealing with integrals involving continuous functions one may increase the accuracy of the calculated value of the integral by considering error terms. The trapezoidal rule of integration is used for this purpose because of its flexibility, that is, its applicability regardless of the number of base points involved. Generally, given an integral of the form

$$I = \int_A^B f(x)\,dx$$

the trapezoidal rule can be used to approximate $I$ as

$$I \simeq \frac{h}{2}[f(A) + 2f(A + h) + \cdots + f(B)]$$

where

$$h = (B - A)/(\text{Number of increments})$$

Obviously, the calculated $I$ value is dependent on the choice of $h$ used. Thus one may write in general that the calculated value is a function of $h$ and an improved $I$ value as follows:

$$I = I_0 + a_0h^2 + a_1h^4 + a_2h^6 + \cdots \tag{11.37}$$

where $I$ is the integral value calculated by using intervals of $h$; $I_0$ is the improved integral value; and $a_0, a_1, \ldots,$ are constants to be determined by using different $h$ values. Romberg assumed the following simple form of equation (11.37):

$$I = I_0 + a_0h^2 \tag{11.38}$$

The basis for his assumption was the fact that the trapezoidal rule has an associated global error of order $O(h^2)$. Equation (11.38) contains two unknowns, namely $I_0$ and $a_0$, which can be easily determined by assuming two different $h$ values and calculating the corresponding $I$ values. Thus

$$\text{if} \quad h = h_1 \quad \text{gives} \quad I = I_1$$

$$\text{and} \quad h = h_2 \quad \text{gives} \quad I = I_2$$

then substituting these values into equation (11.38) yields

$$I_1 = I_0 + a_0 h_1^2 \tag{11.39a}$$

$$I_2 = I_0 + a_0 h_2^2 \tag{11.39b}$$

Solving equation (11.39) for the unknown $I_0$ value, we obtain

$$I_0 = \frac{h_2^2 I_1 - h_1^2 I_2}{h_2^2 - h_1^2} \tag{11.40}$$

This is the so-called Romberg integration formula. Clearly one may repeatedly reduce the $h$ value and reuse equation (11.40) until the desired accuracy is achieved. Therefore one may rewrite the improved integral value in the following more general form:

$$I_0 = \frac{h_{i+1}^2 I_i - h_i^2 I_{i+1}}{h_{i+1}^2 - h_i^2} \tag{11.41}$$

where $i$ is the number of integrations performed. Equation (11.40) can be simplified if we are to assume a definite relationship between the intervals used. Suppose that

$$h_i = 2h_{i+1}$$

Then substituting into (11.41) and simplifying gives

$$I_{0i} = \tfrac{1}{3}(4I_{i+1} - I_i) \tag{11.42}$$

Consider the following example.

EXAMPLE 11.8

Evaluate the following integral using $h = 4, 2, 1$:

$$I = \int_0^4 e^x \, dx$$

Solution

The exact value of the integral is easily determined as follows:

$$I = e^x \Big|_0^4 = 53.59815$$

Taking $h_1 = 4$ we evaluate the integral using the trapezoidal rule:

$$I_1 = \frac{h_1}{2}[f(0) + f(4)]$$

$$= \tfrac{4}{2}(e^0 + e^4) = 111.1963$$

Now for $h_2 = 2$ the integral is evaluated once more:

$$I_2 = \frac{h_2}{2}[f(0) + 2f(2) + f(4)]$$

$$= \tfrac{2}{2}(e^0 + 2e^2 + e^4) = 70.376262$$

Obviously, both of these values are in error. Now using equation (11.41) we get

$$I_{01} = \tfrac{1}{3}[4(70.376262) - 111.1963]$$
$$= 56.769583$$

This value compares much better with the exact value than either one of the previous two values. Hence, if we now calculate $I_3$ corresponding to $h_3 = 1$, we get

$$I_3 = \frac{h_3}{2}[f(0) + 2f(1) + 2f(2) + 2f(3) + f(4)]$$

$$= 57.99195$$

Substituting into equation (11.42) yields

$$I_{02} = \tfrac{1}{3}[4(57.99195) - 70.376262]$$
$$= 53.863846$$

This procedure may be continued until there is no more change in the calculated $I_{0i}$ value. The accuracy of the calculated integral value can be improved even further if we are to include an additional term of $(O)h^4$ in the function assumed for calculating the improved integral value as follows:

$$I = I_0 + a_0h^2 + a_1h^4 \qquad \textbf{(11.43)}$$

Once again, we can solve for the $I_0$ value by selecting three different $h$ values. That is,

$h = h_1$ gives $I = I_1$

$h = h_2$ gives $I = I_2$

$h = h_3$ gives $I = I_3$

Substituting these values into equation (11.43) gives

$$I_1 = I_0 + a_0h_1^2 + a_1h_1^4 \qquad \textbf{(11.44a)}$$

$$I_2 = I_0 + a_0h_2^2 + a_1h_2^4 \qquad \textbf{(11.44b)}$$

$$I_3 = I_0 + a_0h_3^2 + a_1h_3^4 \qquad \textbf{(11.44c)}$$

These equations are difficult to solve in their present form. Therefore, if we assume the following relationships between the $h$ values:

$$h_1 = 2h_2 = 4h_3$$

then substituting into equations (11.44) we get

$$\begin{Bmatrix} I_1 \\ I_2 \\ I_3 \end{Bmatrix} = \begin{bmatrix} 1 & 1 & 1 \\ 1 & \frac{1}{4} & \frac{1}{16} \\ 1 & \frac{1}{16} & \frac{1}{256} \end{bmatrix} \begin{Bmatrix} I_0 \\ a_0 h_1^2 \\ a_1 h_1^4 \end{Bmatrix} \tag{11.45}$$

Equation (11.45) can now be solved for $I_0$ to give

$$I_0 = \tfrac{1}{45}(I_1 - 20I_2 + 64I_3) \tag{11.46}$$

This equation gives much better approximations for the integral of an arbitrary function.

E

EXAMPLE 11.9

Rework example 11.8 using the newly derived expression given by equation (11.46).

Solution

In the previous example the following calculations were made:

$$h_1 = 4 \qquad I_1 = 111.1963$$

$$h_2 = 2 \qquad I_2 = 70.376262$$

$$h_3 = 1 \qquad I_3 = 57.99195$$

Substituting the three integral values into equation (11.46) gives

$$\begin{aligned} I_0 &= \tfrac{1}{45}[111.1963 - 20(70.376262) + 64(57.99195)] \\ &= 53.67013 \end{aligned}$$

This value is closer to the exact value than that obtained by using Romberg's formula for the same number of integral evaluations. *The Romberg formula derived earlier is important in finite element and finite difference analysis.* That is, one may solve a given problem by assuming rather large increments, then extrapolate for the more exact solutions. While a complete treatment of this important subject is beyond the scope of this text, we emphasize that these formulas are applicable to other problems besides the evaluation of integrals.

## 11.8 GAUSS QUADRATURE FORMULAS

Unlike all of the methods covered thus far, these formulas are determined without specifying pivotal points. That is, given

$$I = \int_A^B f(x)\,dx \tag{11.47}$$

the integral is approximated by a function of the form

$$I = a_1 f(x_1) + a_2 f(x_2) + \cdots + a_n f(x_n) \tag{11.48}$$

Obviously, the simplest integration formula that can be assumed is one that involves a single term. Hence

$$I_1 = a_1 f(x_1) \tag{11.49}$$

Since equation (11.49) involves two unknowns $(a_1, x_1)$, the integral given by (11.47) must hold for $f(x) = 1$ and $f(x) = x$. Therefore

$f(x) = 1$:

$$\int_A^B 1\,dx = B - A = a_1(1) \tag{11.50}$$

$f(x) = x$:

$$\int_A^B x\,dx = \frac{B^2 - A^2}{2} = a_1 x_1 \tag{11.51}$$

Solving for $a_1$ and $x_1$ gives

$$a_1 = B - A$$

$$x_1 = \frac{B + A}{2}$$

Substituting $a_1$ and $x_1$ into equation (11.49) gives the simplest Gauss quadrature formula:

$$I = (B - A) f\left(\frac{B + A}{2}\right) \tag{11.52}$$

Equation (11.52) is illustrated in figure 11.6.

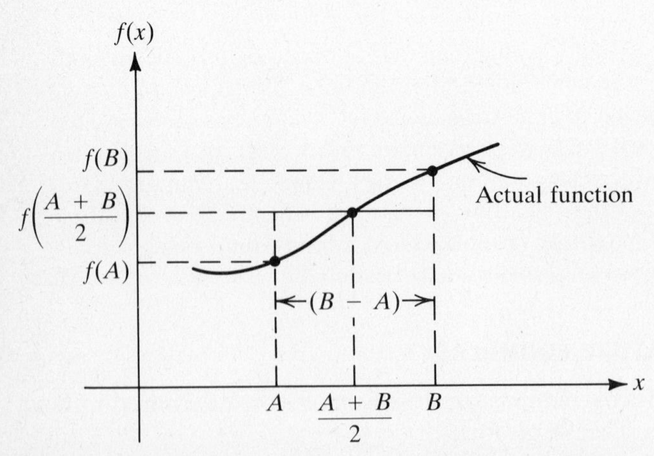


**Figure 11.6** Interpretation of a one-term Gauss quadrature.

The integral is evaluated without regard to the actual functional values at $x = A$ or $x = B$. Instead the functional value at a point representing the average of the sum of $A$ and $B$ is used.

Gauss quadrature formulas involving two and higher terms result in sets of nonlinear algebraic equations. It is for this reason that the limits of integration are normalized as follows:

$$I = \int_A^B f(x)\,dx \tag{11.53}$$

The integral is first transformed to a simpler form by linearly relating $x$ to a new variable $r$:

$$x = c_0 + c_1 r \tag{11.54}$$

The constants $c_0$ and $c_1$ are determined by assuming new limits of integration. That is,

For $x = A$ let $r = -1$

For $x = B$ let $r = +1$

Substituting these values into equation (11.54) gives

$$A = c_0 - c_1, \qquad B = c_0 + c_1$$

Solving for $c_0$ and $c_1$ gives

$$c_0 = \tfrac{1}{2}(B + A), \qquad c_1 = \tfrac{1}{2}(B - A)$$

Substituting these values into equation (11.54) yields the desired transformation function:

$$x = \tfrac{1}{2}(B + A) + \tfrac{1}{2}(B - A)r \tag{11.55}$$

The derivative of $x$ is now defined in terms of the derivative of $r$ as

$$dx = \tfrac{1}{2}(B - A)\,dr \tag{11.56}$$

Substituting equations (11.55) and (11.56) into equation (11.53) gives

$$I = \int_A^B f(x)\,dx = \frac{1}{2}(B - A)\int_{-1}^{1} f\left[\frac{1}{2}(B + A) + \frac{1}{2}(B - A)r\right]dr \tag{11.57}$$

This equation can be best explained by considering the following example.

**E** EXAMPLE 11.10

Evaluate the following integral using the transformation given by equation (11.57):

$$I = \int_0^1 (e^x + 1)\,dx \tag{11.58}$$

### Solution

The solution of this problem is easily determined without the need for transformation. However, this is not the point. Hence

$$I = e^x + x \Big|_0^1 = e$$

Now using equation (11.57) we obtain

$$I = \frac{1}{2}(1-0)\int_{-1}^{1} f\left[\frac{1}{2}(1+0) + \frac{1}{2}(1-0)r\right] dr$$

$$= \frac{1}{2}\int_{-1}^{1} f\left(\frac{1}{2} + \frac{1}{2}r\right) dr$$

The function $f(\frac{1}{2} + \frac{1}{2}r)$ is determined by substituting $x = \frac{1}{2} + \frac{1}{2}r$ into the function given by equation (11.58). Thus

$$I = \frac{1}{2}\int_{-1}^{1} (e^{(1+r)/2} + 1)\, dr$$

$$I = \frac{1}{2}\int_{-1}^{1} (e^{1/2}e^{r/2} + 1)\, dr$$

$$= \tfrac{1}{2}(2e^{1/2}e^{r/2} + r)\Big|_{-1}^{1}$$

$$= \tfrac{1}{2}[(2e^{1/2}e^{1/2} + 1) - (2e^{1/2}e^{-1/2} - 1)]$$

$$= \tfrac{1}{2}[(2e + 1) - (2 - 1)]$$

$$= e$$

It is evident that this type of integration is more cumbersome than the direct procedure; however, the advantage will become apparent shortly.

Suppose that a Gauss quadrature formula having two terms is needed. That is,

$$I = \int_{-1}^{1} f(r)\, dr = a_1 f(r_1) + a_2 f(r_2) \tag{11.59}$$

While the conventional Gauss quadrature approach assumes four unknowns $a_1$, $a_2$, $r_1$, and $r_2$, the authors contend that this problem can be solved by using three unknowns. This is done by using the same interval $H$ in the positive and negative $r$ directions rather than assuming two distinct values. This assumption is extremely important in that it simplifies the solution of the resulting nonlinear equations. Consider figure 11.7.

Now let us assume that the integral of an arbitrary function is given by the equation.

$$I = a_1 f(-H) + a_2 f(H) \tag{11.60}$$

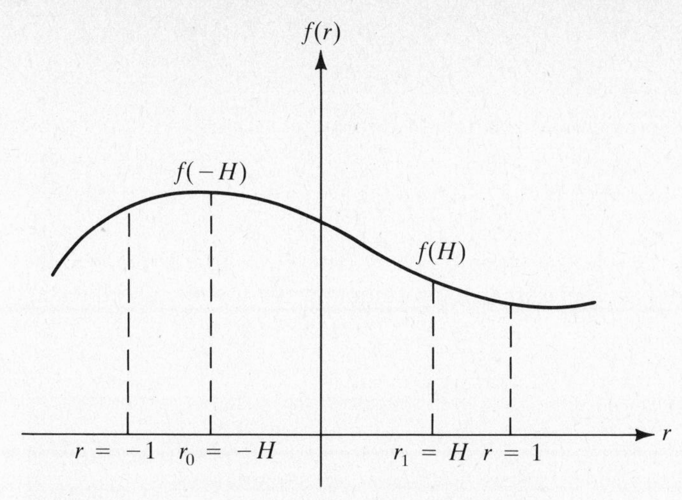


**Figure 11.7** Gauss quadrature involving two terms.

Note that $H$ is not known. Hence there are three unknowns. Thus the integral must hold for $f(r) = 1$, $f(r) = r$, and $f(r) = r^2$:

$f(r) = 1$:

$$\int_{-1}^{1} 1\, dr = 2 = a_1 + a_2 \tag{11.61}$$

$f(r) = r$:

$$\int_{-1}^{1} r\, dr = 0 = a_1(-H) + a_2(H) \tag{11.62}$$

$f(r) = r^2$:

$$\int_{-1}^{1} r^2\, dr = \frac{2}{3} = a_1(H^2) + a_2(H^2) \tag{11.63}$$

Note that if the limits of integration were $A$ and $B$, then a more complex set of equations would have resulted. Rewriting equations (11.61), (11.62), and (11.63) gives

$$a_1 + a_2 = 2 \tag{11.64}$$

$$-a_1 + a_2 = 0 \tag{11.65}$$

$$a_1 + a_2 = \frac{2}{3H^2} \tag{11.66}$$

Clearly the left-hand sides of equations (11.64) and (11.66) are exactly the same. Therefore their right-hand sides must be equal. That is,

$$2 = \frac{2}{3H^2} \quad \text{or} \quad H = \frac{1}{\sqrt{3}}$$

Now equations (11.64) and (11.65) are solved for $a_1$ and $a_2$ to give

$$a_1 = 1, \qquad a_2 = 1$$

Consequently, the two-term Gauss quadrature formula is given as

$$I = \int_{-1}^{1} f(r)\,dr = f\left(\frac{1}{-\sqrt{3}}\right) + f\left(\frac{1}{\sqrt{3}}\right) \tag{11.67}$$

What this formula says is that the integral of any function whose limits of integration are $-1$ and $+1$ is approximated by summing the functional values at $r = -1/\sqrt{3}$ and $r = 1/\sqrt{3}$.

**E** EXAMPLE 11.11

Evaluate the following integral using equation (11.67):

$$I = \int_{-1}^{1} x^2\,dx$$

Solution

The exact value is calculated rather easily:

$$I = \left.\frac{x^3}{3}\right|_{-1}^{1} = \frac{2}{3}$$

Using equations (11.67) we obtain

$$f(x) = x^2$$

$$f\left(\frac{1}{-\sqrt{3}}\right) = \frac{1}{3}$$

$$f\left(\frac{1}{\sqrt{3}}\right) = \frac{1}{3}$$

and

$$I = \frac{2}{3}$$

In general the limits of integration are not $-1$ and $+1$; therefore the function must be transformed so that its limits of integration are equal to $-1$ and $+1$. Equation (11.57) may be evaluated in general as follows:

$$I = \int_{A}^{B} f(x)\,dx = (B - A)[0.5f(x_1) + 0.5f(x_2)] \tag{11.68a}$$

where

$$x_1 = \frac{1}{2}(B + A) - \frac{1}{2}(B - A)\frac{1}{\sqrt{3}}$$

$$= 0.2113248654B + 0.7886751346A \qquad \textbf{(11.68b)}$$

$$x_2 = \frac{1}{2}(B + A) + \frac{1}{2}(B - A)\frac{1}{\sqrt{3}}$$

$$= 0.7886751346B + 0.2113248654A \qquad \textbf{(11.68c)}$$

Equation (11.68) can now be used to approximate the integral of any function for any arbitrary limits $A$ and $B$.

EXAMPLE 11.12

Evaluate the following integral using equation (11.68).

$$I = \int_0^3 x^2\, dx$$

### Solution

The exact value is $I = 9$. Using equation (11.68) we must first calculate $x_0$ and $x_1$. Thus

$$x_1 = 0.2113248654(3) + 0.7886751346(0) = 0.6339745962$$

$$x_2 = 0.7886751346(3) + 0.2113248654(0) = 2.366025404$$

The functional values at $x = x_1$ and $x = x_2$ are computed next:

$$f(x_1) = x_1^2 = 0.4019237886$$

$$f(x_2) = x_2^2 = 5.598076212$$

Substituting into equation (11.68) gives

$$I = \tfrac{1}{2}(3 - 0)(0.4019237886 + 5.598076212)$$
$$= 9.0$$

Obviously, this is the exact answer. Equation (11.68) gives exact answers for polynomials up to the third degree.

Gauss quadrature formulas involving three or more terms can be developed by using the procedure outlined earlier. Thus, for a three-term Gauss quadrature,

$$I = \int_{-1}^{1} f(r)\, dr = a_1 f(-H) + a_2 f(0) + a_3 f(H) \qquad \textbf{(11.69)}$$

Hence

$f(r) = 1$:

$$\int_{-1}^{1} 1\,dr = 2 = a_1 + a_2 + a_3 \tag{11.70}$$

$f(r) = r$:

$$\int_{-1}^{1} r\,dr = 0 = -a_1 H + a_2(0) + a_3 H \tag{11.71}$$

$f(r) = r^2$

$$\int_{-1}^{1} r^2\,dr = \frac{2}{3} = a_1 H^2 + a_2(0) + a_3 H^2 \tag{11.72}$$

$f(r) = r^3$:

$$\int_{-1}^{1} r^3\,dr = 0 = -a_1 H^3 + a_2(0) + a_3 H^3 \tag{11.73}$$

$f(r) = r^4$:

$$\int_{-1}^{1} r^4\,dr = \frac{2}{5} = a_1 H^4 + a_2(0) + a_3 H^4 \tag{11.74}$$

Dividing equation (11.72) by $H^2$ and equation (11.74) by $H^4$ then equating their right-hand sides gives

$$\frac{2}{3H^2} = \frac{2}{5H^4}, \qquad \text{or} \qquad H = \sqrt{\tfrac{3}{5}} = \sqrt{0.6}$$

Equations (11.70) and (11.72) can now be rewritten in matrix form as follows:

$$\begin{bmatrix} 1 & 1 & 1 \\ -1 & 0 & 1 \\ 1 & 0 & 1 \end{bmatrix} \begin{Bmatrix} a_1 \\ a_2 \\ a_3 \end{Bmatrix} = \begin{Bmatrix} 2 \\ 0 \\ 10/9 \end{Bmatrix}$$

Solving for the unknown $a$ values gives

$$a_1 = \tfrac{5}{9}, \qquad a_2 = \tfrac{8}{9}, \qquad a_3 = \tfrac{5}{9}$$

The integration formula is now given as

$$I = \int_{-1}^{1} f(r)\,dr = \frac{5}{9} f(-\sqrt{0.6}) + \frac{8}{9} f(0) + \frac{5}{9} f(\sqrt{0.6})$$

or in general

$$I = \int_{A}^{B} f(x)\,dx = (B - A)\left[\frac{5}{18} f(x_1) + \frac{4}{9} f(x_2) + \frac{5}{18} f(x_3)\right] \tag{11.75a}$$

where

$$x_1 = 0.1127016654B + 0.8872983345A \tag{11.75b}$$

$$x_2 = 0.5B + 0.5A \tag{11.75c}$$

$$x_3 = 0.8872983345B + 0.1127016654A \tag{11.75d}$$

Equation (11.75) gives exact answers to integrals involving polynomials up to the fifth degree.

**E**

EXAMPLE 11.13

Evaluate the following integral using the three-term Gauss quadrature given by equation (11.75):

$$I = \int_0^4 e^x$$

Solution

This integral was evaluated earlier, in example 11.8. Its exact truncated value is 53.59815. Let us proceed by calculating $x_0$, $x_1$, and $x_2$ as given by equations (11.75). Thus

$$x_1 = 0.1127016654(4) = 0.450806616$$

$$x_2 = 0.5(4) = 2.0$$

$$x_3 = 0.8872983345(4) = 3.549193338$$

Evaluating the function at the various $x$ values gives

$$f(x_1) = e^{0.450806616} = 1.569577722$$

$$f(x_2) = e^{2.0} = 7.389056099$$

$$f(x_3) = e^{3.54919338} = 34.78524769$$

Now substituting these values into equation (11.75a) gives the value of the integral in question:

$$\begin{aligned} I &= (4 - 0)[\tfrac{5}{18}(1.569577722) + \tfrac{4}{9}(7.389056099) + \tfrac{5}{18}(34.78524769)] \\ &= 53.5303502 \end{aligned}$$

This value compares rather well with the exact value.

Gauss quadrature equations containing four, five, and six terms are provided in table 11.2. These integration formulas were derived by using the same procedure outlined in this section.

The four-term, five-term, and six-term integration formulas give exact integral values for polynomials of order seven, nine, and eleven, respectively. Note that

Table 11.2 GAUSS QUADRATURE EQUATIONS OF THE FOURTH, FIFTH, AND SIXTH ORDER

| NUMBER OF TERMS | INTEGRATION FORMULA FOR $I = \int_A^B f(x)\,dx$ |
|---|---|
| 4 | $I = (B - A)\{0.1739274226[f(x_1) + f(x_2)] + 0.3260725775[f(x_3) + f(x_4)]\}$ |
| | $x_1 = 0.0694318442B + 0.9305681558A$ |
| | $x_2 = 0.9305681558B + 0.0694318442A$ |
| | $x_3 = 0.3300094782B + 0.6699905218A$ |
| | $x_4 = 0.6699905218B + 0.3300094782A$ |
| 5 | $I = (B - A)\{0.1184634425[f(x_1) + f(x_2)] + 0.2844444444f(x_3)$ |
| | $+ 0.2393143353[f(x_4) + f(x_5)]\}$ |
| | $x_1 = 0.04691007705B + 0.953089923A$ |
| | $x_2 = 0.953089923B + 0.04691007705A$ |
| | $x_3 = 0.5B + 0.5A$ |
| | $x_4 = 0.230765345B + 0.769234655A$ |
| | $x_5 = 0.769234655B + 0.230765345A$ |
| 6 | $I = (B - A)\{0.0856622462[f(x_1) + f(x_2)] + 0.1803807865[f(x_3) + f(x_4)]$ |
| | $+ 0.2339569673[f(x_5) + f(x_6)]\}$ |
| | $x_1 = 0.0337652429B + 0.9662347571A$ |
| | $x_2 = 0.9662347571B + 0.0337652429A$ |
| | $x_3 = 0.1693953068B + 0.8306046932A$ |
| | $x_4 = 0.8306046932B + 0.1693953068A$ |
| | $x_5 = 0.380690407B + 0.619309593A$ |
| | $x_6 = 0.619309593B + 0.380690407A$ |

by specifying the limits of integration $A$ and $B$, one can easily determine the $x$ values $x_1, x_2, \ldots$, and thus evaluate the functional values. The integral is then determined by substituting the functional values into the proper formula.

**E**

EXAMPLE 11.14

Rework example 11.10 using the six-term formula.

### Solution

Evaluating the $x$ values gives

$$x_1 = 0.0337652429(4) + 0 = 0.1350609716$$
$$x_2 = 0.9662347571(4) + 0 = 3.864939028$$
$$x_3 = 0.1693953068(4) + 0 = 0.6775812272$$
$$x_4 = 0.8306046932(4) + 0 = 3.322418773$$
$$x_5 = 0.380690407(4) + 0 = 1.522761628$$
$$x_6 = 0.619309593(4) + 0 = 2.477238372$$

Now evaluating the corresponding functional values yields

$$f(x_1) = e^{x_1} = 1.144606571$$
$$f(x_2) = e^{x_2} = 47.70036396$$
$$f(x_3) = e^{x_3} = 1.96910914$$
$$f(x_4) = e^{x_4} = 27.72733565$$
$$f(x_5) = e^{x_5} = 4.584869432$$
$$f(x_6) = e^{x_6} = 11.90833258$$

The integral is determined by substituting these values into the six-term formula. Thus

$$\begin{aligned} I &= (4 - 0)\{0.0856622462[(1.144606571) + (47.70036396)] \\ &\quad + 0.1803807865[(1.96910914) + (27.72733565)] \\ &\quad + 0.2339569673[(4.584869432) + (11.90833258)]\} \\ &= 53.59814992 \end{aligned}$$

This value compares extremely well with the exact value. The error is 0.00000011.

EXAMPLE 11.15

In example 11.3 the deformation of an axially loaded member is given as

$$u_{A/B} = \int_0^{120} \frac{dx}{-\frac{30}{36}x^2 + 100x + 3000}$$

Evaluate the deformation $u_{A/B}$ using two-term Gauss quadratures.

### Solution

The solution of this problem is obtained by first determining the pivotal points corresponding to the two terms. Using equations (11.68b) and (11.68c), we have

$$x_1 = 0.2113248654(120) + 0 = 25.35898385$$

$$x_2 = 0.7886751346(120) + 0 = 94.64101615$$

Therefore

$$f(x_1) = [-\tfrac{30}{36}(25.35898385)^2 + 100(25.35898385) + 3000]^{-1}$$
$$= 0.0002$$

$$f(x_2) = [-\tfrac{30}{36}(94.64101615)^2 + 100(94.64101615) + 3000]^{-1}$$
$$= 0.0002$$

The deformation is calculated by using equation (11.68a):

$$u_{A/B} = (120 - 0)(0.0002 + 0.0002)(0.5)$$
$$= 0.024$$

**E**

EXAMPLE 11.16

The torsion of circular shafts is fully defined by the following expression:

$$\frac{d\phi}{dZ} = \frac{T(Z)}{G(Z)J(Z)} \tag{11.76}$$

where

$\phi$ = angle of twist

$G(Z)$ = shear modulus of shaft material

$J(Z)$ = polar moment of inertia of the cross section

$T(Z)$ = applied torque

For a shaft having the following configuration, determine the angle of twist using the two-term Gauss formula:

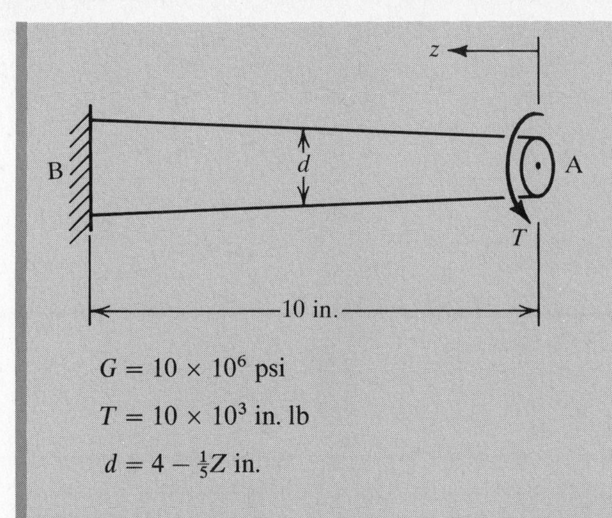


$G = 10 \times 10^6$ psi

$T = 10 \times 10^3$ in. lb

$d = 4 - \frac{1}{5}Z$ in.

## Solution

Rearranging equation (11.76) as

$$d\phi = \frac{T(Z)}{G(Z)J(Z)}\,dZ$$

then integrating, we have

$$\int_{\phi_A}^{\phi_B} d\phi = \int_0^{10} \frac{T}{GJ(Z)}\,dZ$$

$$\phi_A - \phi_B = \int_0^{10} \frac{T}{GJ(Z)}\,dZ$$

Note that both $T$ and $G$ are constant but not $J(Z)$:

$$\phi_{A/B} = \int_0^{10} \frac{T}{GJ(Z)}\,dZ \tag{11.77}$$

The polar moment of inertia is given in terms of the diameter as follows:

$$J(Z) = \frac{\pi}{32}d^4 = \frac{\pi}{32}\left(4 - \frac{Z}{5}\right)^4$$

Substituting the various values into equation (11.77) yields

$$\phi_{A/B} = \int_0^{10} \frac{10 \times 10^3\,dZ}{(10 \times 10^6)(\pi/32)(4 - Z/5)^4}$$

Simplifying, we obtain

$$\phi_{A/B} = \int_0^{10} \frac{0.01018591636}{(4 - Z/5)^4}\, dZ$$

Using equation (11.68b) and (11.68c) we get

$$x_1 = 0.2113248654(10) + 0 = 2.113248654$$

$$x_2 = 0.78867513468(10) + 0 = 7.8867513468$$

The corresponding functional values are determined next:

$$f(x_1) = 6.21948242 \times 10^{-5}$$

$$f(x_2) = 2.956906513 \times 10^{-4}$$

Substituting into equation (11.68a) gives the angle of twist as follows:

$$\begin{aligned}\phi_{A/B} &= (10 - 0)(6.21948242 \times 10^{-5} + 2.956906513 \times 10^{-4})(0.5) \\ &= 1.789427378 \times 10^{-3} \text{ radians}\end{aligned}$$

## 11.9 DOUBLE INTEGRATION

There are situations where a double or a triple integral may occur, in which case a modification of the procedure outlined thus far must be introduced. Consider the following double integral:

$$I = \int_{y_0}^{y_n} \int_{x_0}^{x_m} f(x, y)\, dx\, dy \tag{11.78}$$

Integrating with respect to $x$ using the trapezoidal rule gives

$$I = \int_{y_0}^{y_1} \frac{\Delta x}{2} [f(x_0, y) + 2f(x_1, y) + \cdots + f(x_m, y)]\, dy \tag{11.79}$$

Now integrating with respect to $y$ using the trapezoidal rule once again yields the desired expression for double integration. Hence

$$\begin{aligned}I = \frac{\Delta y}{2}\Bigg(&\frac{\Delta x}{2}[f(x_0, y_0) + 2f(x_1, y_0) + \cdots + f(x_m, y_0)] \\ &+ 2\frac{\Delta x}{2}[f(x_0, y_1) + 2f(x_1, y_1) + \cdots + f(x_m, y_1)] \\ &+ \cdots + 2\frac{\Delta x}{2}[f(x_0, y_n) + 2f(x_1, y_n) + \cdots + f(x_m, y_n)]\Bigg)\end{aligned} \tag{11.80}$$

For convenience equation (11.80) is expressed in matrix form as

$$I = \frac{\Delta x\,\Delta y}{4}\{1 \quad 2 \quad \cdots \quad 2 \quad 1\}$$

$$\times \begin{bmatrix} f(x_0, y_0) & f(x_0, y_1) & \cdots & f(x_0, y_n) \\ f(x_1, y_0) & f(x_1, y_1) & \cdots & f(x_1, y_n) \\ \vdots & \vdots & & \vdots \\ f(x_m, y_0) & f(x_m, y_1) & \cdots & f(x_m, y_n) \end{bmatrix} \begin{Bmatrix} 1 \\ 2 \\ \vdots \\ 2 \\ 1 \end{Bmatrix} \tag{11.81a}$$

In compact matrix form,

$$I = \frac{\Delta x\,\Delta y}{4}\{a\}_{1\times m}[f]_{m\times n}\{b\}_{n\times 1} \tag{11.81b}$$

A similar expression for Simpson's rule of integration is given by

$$I = \frac{\Delta x\,\Delta y}{9}\{1 \quad 4 \quad 2 \quad \cdots \quad 2 \quad 4 \quad 1\}$$

$$\times \begin{bmatrix} f(x_0, y_0) & f(x_0, y_1) & \cdots & f(x_0, y_n) \\ f(x_1, y_0) & f(x_1, y_1) & \cdots & f(x_1, y_n) \\ \vdots & \vdots & & \vdots \\ f(x_m, y_0) & f(x_m, y_1) & \cdots & f(x_m, y_n) \end{bmatrix} \begin{Bmatrix} 1 \\ 4 \\ 2 \\ \vdots \\ 2 \\ 4 \\ 1 \end{Bmatrix} \tag{11.82a}$$

in compact matrix form,

$$I = \frac{\Delta x\,\Delta y}{9}\{c\}_{1\times m}[f]_{m\times n}\{d\}_{n\times 1} \tag{11.82b}$$

It is recommended that the increments in the $x$ direction be taken equal to the increment in the $y$ direction. Doing so should reduce the errors involved. Note that some authors apply the trapezoidal rule in one direction and Simpson's rule in the other direction. This practice is not recommended and should be avoided.

**E** EXAMPLE 11.17

Evaluate the following integral using the trapezoidal rule of integration and $\Delta x = \Delta y = 1$:

$$I = \int_0^3 \int_0^2 xy\,dx\,dy$$

### Solution

The exact answer to this problem is readily obtained by direct integration.

$$I = \int_0^3 \frac{x^2}{2} y \bigg|_0^2 dy = \int_0^3 2y\,dy$$

$$= 2\frac{y^2}{2}\bigg|_0^3 = 9$$

Using equation (11.81) we must first determine the following functional values:

$$f(0,0) = (0)(0) = 0, \qquad f(0,2) = (0)(2) = 0$$

$$f(1,0) = (1)(0) = 0, \qquad f(1,2) = (1)(2) = 2$$

$$f(2,0) = (2)(0) = 0, \qquad f(2,2) = (2)(2) = 4$$

$$f(0,1) = (0)(1) = 0, \qquad f(0,3) = (0)(3) = 0$$

$$f(1,1) = (1)(1) = 1, \qquad f(1,3) = (1)(3) = 3$$

$$f(2,1) = (2)(1) = 2, \qquad f(2,3) = (2)(3) = 6$$

Substituting into equation (11.81) gives

$$I = \frac{(1)(1)}{4}\{1 \quad 2 \quad 1\}\begin{bmatrix} 0 & 0 & 0 & 0 \\ 0 & 1 & 2 & 3 \\ 0 & 2 & 4 & 6 \end{bmatrix}\begin{Bmatrix} 1 \\ 2 \\ 2 \\ 1 \end{Bmatrix}$$

$$= \frac{1}{4}\{0 \quad 4 \quad 8 \quad 12\}\begin{Bmatrix} 1 \\ 2 \\ 2 \\ 1 \end{Bmatrix}$$

$$= \tfrac{36}{4} = 9$$

Obviously, the answer is equal to the exact value.

## Recommended Reading

*The Finite Element Method*, O. C. Zienkiewicz, McGraw-Hill Book Company, New York, 1977. (Gaussian quadrature formulas up to the tenth order are provided on p. 198.)

*Numerical Algorithms—Origins and Applications*, Bruce W. Arder and K. N. Astill, Addison-Wesley, Reading, Mass., 1970.

*Numerical Analysis—A Practical Approach*, Melvin J. Maron, MacMillan Publishing Co., London, 1982.

*A Practical Guide to Computer Methods for Engineers*, Terry E. Shoup, Prentice-Hall, Inc., Englewood Cliffs, N.J., 1979.

## PROBLEMS

**11.1** Use the trapezoidal rule of integration to determine the value of the following integrals. Assume $h = 0.1$.

**(a)** $I = \int_{0.5}^{1} e^x \, dx$ **(b)** $I = \int_{0.5}^{1} e^{x^2} \, dx$

**(c)** $I = \int_{0}^{1} xe^x \, dx$ **(d)** $I = \int_{1}^{2} x \log x \, dx$

**11.2** Redo problem 1 using $h = 0.05$.

**11.3** Use Simpson's $\frac{1}{3}$ rule to evaluate the following integrals. Assume $h = 1$.

**(a)** $I = \int_{1}^{3} x^2 \, dx$ **(b)** $I = \int_{1}^{5} (x^3 + x^2 - 5) \, dx$

**(c)** $I = \int_{0}^{2} x \sin x \, dx$ **(d)** $I = \int_{1}^{5} \frac{x}{\sin x} dx$

Check the accuracy of (a) and (b) using direct integration.

**11.4** Develop an integration formula using the following interpolating function:

$$y = a_0 + a_1 x^2$$

**11.5** Develop Simpson's $\frac{3}{8}$ rule then check your result with that given in table 11.1. Note that a third-order interpolating function is used for this purpose.

**11.6** Use regression analysis to fit a second order through the following data set then evaluate the integral $I$.

| $x$ | 0 | $\pm 1$ | $\pm 3$ | $\pm 7$ |
|---|---|---|---|---|
| $f(x)$ | 1 | 4 | 18 | 50 |

and

$$I = \int_{-7}^{7} f(x) \, dx$$

Assume a second-order polynomial.

**11.7** Use Romberg's equation to evaluate the following integral to three decimal places. Assume $h_i = 2h_{i+1}$.

$$I = \int_{1}^{2} e^{x^2} \, dx$$

**11.8** Use equation (11.46) to solve problem 7.

**11.9** Evaluate the following integral using the two-term Gauss formula.

$$I = \int_{0}^{0.1} x^2 e^{x^3} \, dx$$

Compare your answer with the exact value.

**11.10** Evaluate the integral of problem 9 using (a) the three-term Gauss formula, (b) the four-term Gauss formula. Compare your answers with the exact value.

**11.11** Determine the axial deformation for a member having the following characteristics:

$N = \frac{1}{12}x^2$ kips/in.

$L = 12$ ft

$A = -\frac{1}{4000}x^2 + \frac{1}{60}x + 2$ in.$^2$

$E = 30 \times 10^3$ ksi

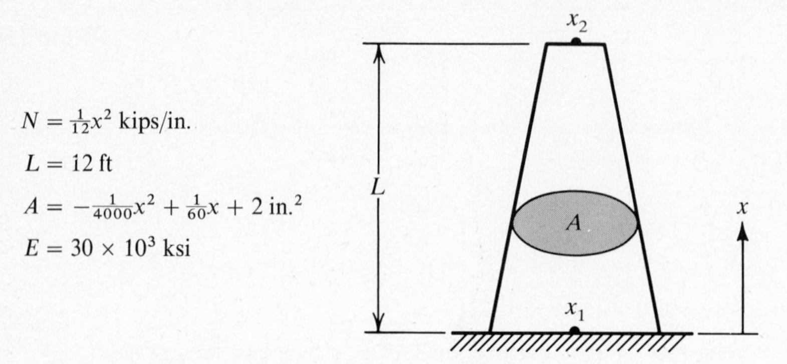


Use the two-term Gauss formula, knowing that

$$u = \int_{x_1}^{x_2} \frac{N}{EA}\,dx$$

**11.12** The settlement of structures is often calculated using the following equation.

$$\Delta H = \int_{Z_1}^{Z_2} \frac{C_c}{1 + e_0} \log\left(1 + \frac{\Delta\sigma}{P_0}\right) dZ$$

Evaluate the settlement for a building placed on a soil layer with the following characteristics:

$\Delta\sigma = 1000$ psf

$P_0 = 60Z$ psf

$C_c = 0.2 + 0.01Z$

$e_0 = 0.7 + 0.01Z$

$Z_1 = 4$ ft

$Z_2 = 18$ ft

Use the trapezoidal rule of integration with $h = 1.0$ ft.

**11.13** Rework problem 12 using the two-term Gauss formula.

**11.14** Evaluate the following double integral using $\Delta x = \Delta y = 0.5$ and the trapezoidal rule of integration.

$$I = \int_0^2 \int_0^1 (x^2 y + 5)\,dx\,dy$$

**11.15** Rework problem 14 using Simpson's $\frac{1}{3}$ rule of integration.

**11.16** Evaluate the following integral using the trapezoidal rule and $h = \Delta x = \Delta y = 0.2$:

$$I = \int_0^1 \int_0^2 (y^3 + x^2 y^2)\,dx\,dy$$

Compare your answer with the exact value.

# chapter 12

# Numerical Solution of Ordinary Differential Equations

A Sumerian statue dating back to 3500 B.C.
*Courtesy of the Iraqi government.*

"Numerical and classical solution of mathematical models go hand in hand."

# 12

## 12.1 INTRODUCTION

A differential equation is any equation containing one or more derivative terms. An ordinary differential equation is one involving a single independent variable. These are classified according to the order of derivative terms involved and their power. Hence, if only first derivatives are involved, then it is referred to as a first-order ordinary differential equation. On the other hand, if the highest derivative is of second order then it is called a second-order differential equation, etc. Furthermore, if any of the derivative terms is present as a multiple, then the equation is called nonlinear, otherwise linear. For example,

$$C_0(x) + \sum_{i=1}^{n} C_i(x) \frac{d^i y}{dx^i} = 0$$

is a linear ordinary differential equation, while

$$C_0(x) + \sum_{i=1}^{n} C_i(x) \left( \frac{d^i y}{dx^i} \right)^m = 0$$

is a nonlinear ordinary equation for $m \neq 1$. Furthermore, if the coefficient $C_0(x)$ is zero, the equation is called homogeneous, otherwise nonhomogeneous.

Ordinary differential equations of all types occur quite frequently in engineering practice, since most physical laws are expressed in terms of derivatives. For example, the velocity of a particle is given as the rate of change of distance relative to time. In addition, the force acting on a mass is related to the mass and acceleration by Newton's second law.

The analytical solution of ordinary differential equations, as well as partial differential equations, is called the "closed-form solution." This solution requires that the constants of integrations be evaluated by using prescribed values of the independent variable(s). Therefore an ordinary differential equation of order $n$ requires that $n$ conditions be specified. In practice, these conditions are defined in two different general ways. First, if the differential equation in question is used to describe the behavior of a physical system, such as deflection of a beam, temperature distribution in a region, etc., then the specific conditions at the boundaries are called boundary conditions and the equation is called a boundary-value problem. For example, the beam shown in figure 12.1 has a known deflection of zero at its boundaries $x = 0$ and $x = L$.

In some cases, the specific behavior of a system is known at a particular time; for example, the deflection of the beam at $x = a$ is shown at time $t = 0$ to be equal to $y_0$ and that we are interested in determining the response for $t > 0$. These conditions are called initial conditions, and the differential equation used for response determination is known as an "initial-value problem."

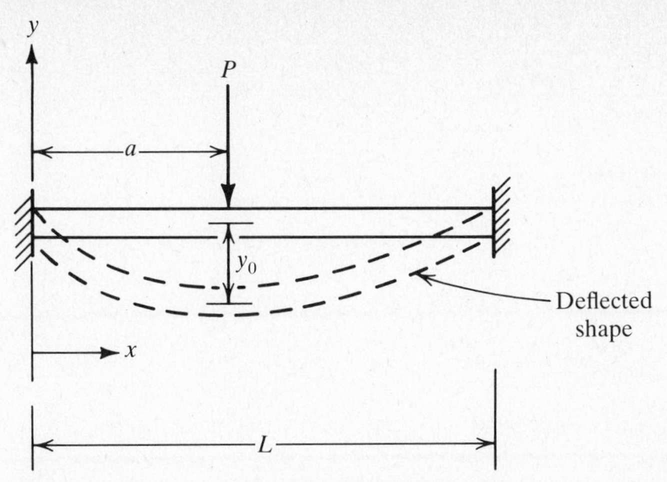


**Figure 12.1** Graphical illustration of initial and boundary value problems.

In practice, only a few differential equations at best can be solved analytically in closed form. Solutions of most practical engineering problems involving differential equations require the use of numerical methods.

## 12.2 TAYLOR SERIES METHOD

The basis for many numerical solution techniques dealing with ordinary differential equations can be traced back to the Taylor series. This method is relatively easy to use for certain types of problems in which the derivatives can be readily determined. Therefore, for problems involving a single first-order differential equation, this method applies regardless of whether the equation in question is a boundary-value or an initial-value problem. In fact, it can also be used for high-order ordinary differential equations as well. Unfortunately, it is not well suited for approximating solutions when a wide range of values of the independent variable are needed. The following example is chosen to illustrate the solution technique. Suppose we are given

$$\frac{dy}{dx} - \frac{1}{2}y = 0, \qquad y(0) = 1 \tag{12.1}$$

The closed-form solution is easily determined by using direct integration. That is,

$$y = e^{x/2} \tag{12.2}$$

It is evident that the solution given by equation (12.2) satisfies the condition that $y = 1$ when $x = 0$. The numerical solution can now be compared with the exact one by first expanding the independent variable $y$ about the initial condition $x_0 = 0$ to get

$$y(x) = y(0) + hy^{(1)}(0) + \frac{h^2}{2}y^{(2)}(0) + \frac{h^3}{6}y^{(3)}(0) + \frac{h^4}{24}y^{(4)}(0) + \cdots \tag{12.3}$$

Table 12.1 TAYLOR SERIES SOLUTION VERSUS EXACT SOLUTION

| $x$ | $y$ (TAYLOR) | $y$ (EXACT) |
|---|---|---|
| 0 | 1.0000 | 1.0000000 |
| 0.1 | 1.0512711 | 1.0512711 |
| 0.2 | 1.1051708 | 1.1051709 |
| 0.3 | 1.1618336 | 1.1618342 |
| 0.4 | 1.2214000 | 1.2214027 |
| 0.5 | 1.2840170 | 1.2840254 |
| 0.6 | 1.3498375 | 1.3498588 |
| 0.7 | 1.4190211 | 1.4190675 |
| 0.8 | 1.4917333 | 1.4918247 |
| 0.9 | 1.5681461 | 1.5683122 |
| 1.0 | 1.6484375 | 1.6487212 |

Note that equation (12.3) is consistent with the definition of a Taylor series if we realize that $x = x_0 + h$, where $h$ is the interval between $x$ and $x_0$. Obviously this is a truncated Taylor series. Therefore, the degree of accuracy attained is directly dependent on the number of terms used in finding the solution. The derivatives appearing in equation (12.3) can be easily evaluated from equation (12.1) as follows:

$$y^{(1)}(x) = \tfrac{1}{2}y, \qquad y^{(1)}(0) = (\tfrac{1}{2})(1) = \tfrac{1}{2} \tag{12.4a}$$

$$y^{(2)}(x) = \tfrac{1}{2}y^{(1)}, \qquad y^{(2)}(0) = (\tfrac{1}{2})(\tfrac{1}{2}) = \tfrac{1}{4} \tag{12.4b}$$

$$y^{(3)}(x) = \tfrac{1}{2}y^{(2)}, \qquad y^{(3)}(0) = (\tfrac{1}{2})(\tfrac{1}{4}) = \tfrac{1}{8} \tag{12.4c}$$

$$y^{(4)}(x) = \tfrac{1}{2}y^{(3)}, \qquad y^{(4)}(0) = (\tfrac{1}{2})(\tfrac{1}{8}) = \tfrac{1}{16} \tag{12.4d}$$

Substituting equation (12.4) into equation (12.3) and noting that $y(0) = 1$ and $x = h$ gives

$$y(x) = 1 + \tfrac{1}{2}x + \tfrac{1}{8}x^2 + \tfrac{1}{48}x^3 + \tfrac{1}{384}x^4 + \cdots \tag{12.5}$$

Equation (12.5) represents the Taylor series solution to the problem at hand. A comparison of the numerical solution and the analytical solution is given in table 12.1 for $0 \le x \le 1$.

It is evident that the higher the $x$ value is, the lesser the accuracy of the numerical solution becomes. This is expected, since only a few terms of the Taylor series are used. To illustrate this point graphically, a comparison is made between the analytical solution and the Taylor series solutions involving two terms, three terms, four terms, and five terms in figure 12.2.

It is evident that the number of terms to be retained in the solution is a matter of judgment. Generally, the rule is to truncate the series when the contributions of the succeeding terms become insignificant. Unfortunately, this is not possible without including a large number of terms to begin with. Finally, since for all practical purposes the computation is done on a computer the derivative terms cannot be evaluated directly. Consequently, numerical approximations of the de-

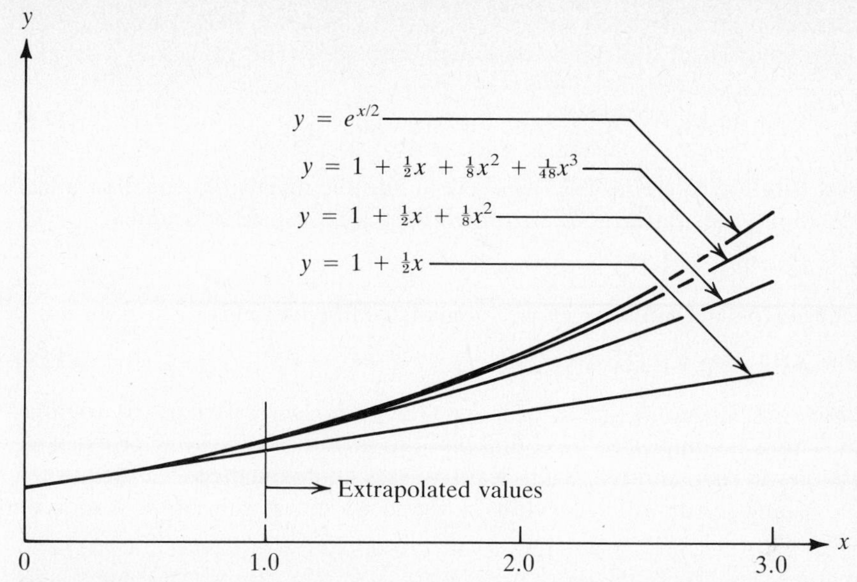


**Figure 12.2** Graphical illustration of truncation errors associated with Taylor series solutions.

rivatives must be used, in which case a second source of errors is introduced. For this and other reasons to be discussed later, this technique is not widely used.

## 12.3 LEAST-SQUARES METHOD

This method of solution requires that the sum of the square of error over the entire range of the independent variable be reduced to zero. Unlike the Taylor series method, where the number of terms is infinite, a polynomial having a finite number of terms is assumed.

The solution procedure can be illustrated by using the same example problem discussed in the previous section. That is,

$$\frac{dy}{dx} - \frac{1}{2}y = 0, \qquad y(0) = 1$$

The solution begins by assuming a polynomial of a finite degree, say, second order:

$$y = a_0 + a_1x + a_2x^2 \tag{12.6}$$

Obviously, equation (12.6) can be forced to satisfy the condition given by $y = 1$ when $x = 0$ as follows:

$$y(0) = 1 = a_0 + a_1(0) + a_2(0)^2$$

Solving for $a_0$ gives

$$a_0 = 1$$

Therefore equation (12.6) is rewritten in the following revised form:

$$y = 1 + a_1x + a_2x^2 \tag{12.7}$$

The task before us is to determine the unknowns $a_1$ and $a_2$. This is accomplished by taking the derivative of the independent variable $y$ with respect to $x$ to give

$$\frac{dy}{dx} = a_1 + 2a_2x \tag{12.8}$$

Then substituting equations (12.7) and (12.8) into the differential equation under consideration yields the error $E$ associated with the assumed polynomial:

$$E = a_1 + 2a_2x - \tfrac{1}{2}(1 + a_1x + a_2x^2)$$

Collecting terms in the unknown polynomial coefficients yields

$$E = a_1(1 - \tfrac{1}{2}x) + a_2(2x - \tfrac{1}{2}x^2) - \tfrac{1}{2} \tag{12.9}$$

Since the error function given by equation (12.9) is not zero, we must try to minimize it. This is accomplished by noting that this is precisely the same problem we faced earlier in chapter 9 relative to least-squares approximations. Consequently, by establishing the deviation function as the mean square sum of the errors, we have

$$D = \int_x E^2\,dx$$

The conditions ensuring that $D$ be a minimum are easily established by taking the partial derivatives of $D$ relative to $a_1$ and $a_2$ to give

$$\frac{\partial D}{\partial a_1} = 0 = \int_x 2E\frac{\partial E}{\partial a_1}\,dx \tag{12.10}$$

$$\frac{\partial D}{\partial a_2} = 0 = \int_x 2E\frac{\partial E}{\partial a_2}\,dx \tag{12.11}$$

The partial derivatives of the errors appearing in equations (12.10) and (12.11) are readily evaluated by using equation (12.9). Thus

$$\frac{\partial E}{\partial a_1} = 1 - \frac{x}{2} \tag{12.12}$$

$$\frac{\partial E}{\partial a_2} = 2x - \frac{x^2}{2} \tag{12.13}$$

Equations (12.9), (12.12), and (12.13) are now substituted into equations (12.10) and (12.11). This gives two linear algebraic equations in the two unknowns $a_1$ and $a_2$. Note that the limits of integration are established by considering the range of values for which the solution is to be applied. For illustration, suppose $0 \le x \le 1$; then

$$\int_0^1 \left[a_1\left(1 - \frac{x}{2}\right) + a_2\left(2x - \frac{x^2}{2}\right) - \frac{1}{2}\right]\left(1 - \frac{x}{2}\right)dx$$

$$\int_0^1 \left[a_1\left(1 - \frac{x}{2}\right) + a_2\left(2x - \frac{x^2}{2}\right) - \frac{1}{2}\right]\left(2x - \frac{x^2}{2}\right)dx$$

Table 12.2 LEAST-SQUARES VERSUS EXACT SOLUTIONS

| $x$ | $y$ (LEAST SQUARES) | $y$ (EXACT) |
|---|---|---|
| 0 | 1.0000000 | 1.0000000 |
| 0.1 | 1.0503274 | 1.0512711 |
| 0.2 | 1.1038848 | 1.1051709 |
| 0.3 | 1.1606722 | 1.1618342 |
| 0.4 | 1.2206896 | 1.2214027 |
| 0.5 | 1.2839371 | 1.2840254 |
| 0.6 | 1.3504146 | 1.3498588 |
| 0.7 | 1.4201222 | 1.4190675 |
| 0.8 | 1.4930598 | 1.4918247 |
| 0.9 | 1.5692274 | 1.5683122 |
| 1.0 | 1.6586250 | 1.6487212 |

Integrating then evaluating at the limits gives

$$\tfrac{7}{12}a_1 + \tfrac{9}{16}a_2 = \tfrac{3}{8}$$

$$\tfrac{9}{16}a_1 + \tfrac{53}{60}a_2 = \tfrac{5}{16}$$

Solving for the unknowns, we have

$$a_1 = 0.4871235, \qquad a_2 = 0.1615015$$

Consequently, the solution is given by equation (12.7) as

$$y = 1 + 0.4871235x + 0.1615015x^2 \tag{12.14}$$

Obviously, this solution is applicable for $x = 0$ to $x = 1$ only. A comparison of the least-squares approximation and the analytical solution is given in table 12.2. It is interesting to note that the approximate solution values are sometimes less and in some cases greater than the corresponding exact values. This result should of course have been expected, since the method of least squares minimizes the sum of deviations on either side of the assumed approximation. Furthermore, since the numerical integration is more accurate than numerical differentiation, then we could easily program this technique on a computer. The implication is that this technique is better suited for solving practical problems than the Taylor series method.

## 12.4 GALERKIN METHOD

This method of solving ordinary differential equations is based on the fact that the errors associated with a given solution are orthogonal over the range in question to a set of linearly independent weighting functions $M_i$. That is,

$$M_i = \int_x W_i E\, dx = 0, \qquad i = 1, \ldots, n \tag{12.15}$$

where $E$ is the error function, the $W_i$ are the weighting functions, and the $M_i$ are the $n$ functions to be minimized. It is evident that equation (12.15) is similar in principle to the deviation function used in the least-squares method. However, the individual weighting functions are normally taken as the individual terms of the assumed functional approximation. Clearly this makes it much easier to evaluate the various integrals involved in evaluating the approximate solution. For illustration, let us consider the problem given by equation (12.1) repeated here for convenience:

$$\frac{dy}{dx} - \frac{1}{2}y = 0, \qquad y(0) = 1$$

If the assumed solution is taken as a second-order polynomial, then

$$y = 1 + a_1x + a_2x^2$$

Consequently, the weighting functions $W_1$ and $W_2$ are given simply by the individual terms appearing in the assumed approximation. That is,

$$W_1 = x, \qquad W_2 = x^2$$

Note that the error function $E$ is given by equation (12.9). Therefore, substituting into equation (12.15) for $i = 1, 2$ yields the following linear algebraic equations:

$$M_1 = \int_0^1 \left[ a_1\left(1 - \frac{x}{2}\right) + a_2\left(2x - \frac{x^2}{2}\right) - \frac{1}{2} \right] x\,dx = 0$$

$$M_2 = \int_0^1 \left[ a_1\left(1 - \frac{x}{2}\right) + a_2\left(2x - \frac{x^2}{2}\right) - \frac{1}{2} \right] x^2\,dx = 0$$

integrating yields

$$8a_1 + 13a_2 = 6$$

$$25a_1 + 48a_2 = 20$$

Solving for the unknowns gives

$$a_1 = 0.4745763, \qquad a_2 = 0.1694915$$

Substituting into the assumed second-order polynomial approximation yields the desired result. Thus

$$y = 1 + 0.4745763x + 0.1694915x^2 \tag{12.16}$$

Note that equation (12.16) is applicable for $x = 0$ to $x = 1.0$ (why?). Better accuracy can be attained by using a higher-order polynomial. Once again, numerical integration techniques are well suited for approximating the $M_i$ integrals.

The solution given by equation (12.16) is now compared with the exact solution obtained earlier as shown in table 12.3. Clearly the approximations compare reasonably well with the exact values.

Table 12.3 GALERKIN VERSUS EXACT SOLUTIONS

| $x$ | $y$ (GALERKIN) | $y$ (EXACT) |
|---|---|---|
| 0 | 1.0000000 | 1.0000000 |
| 0.1 | 1.0491525 | 1.0512711 |
| 0.2 | 1.1016949 | 1.1051709 |
| 0.3 | 1.1576271 | 1.1618342 |
| 0.4 | 1.2169492 | 1.2214027 |
| 0.5 | 1.2796610 | 1.2840254 |
| 0.6 | 1.3457627 | 1.3498588 |
| 0.7 | 1.4152542 | 1.4190675 |
| 0.8 | 1.4881356 | 1.4918247 |
| 0.9 | 1.5644068 | 1.5683122 |
| 1.0 | 1.6440678 | 1.6487212 |

## 12.5 EULER AND MODIFIED EULER METHODS

Thus far we have introduced several techniques for solving ordinary differential equations of the first order. It is evident that these techniques involve approximating the solution by assuming an appropriate polynomial for the range of values in question. Hence, given

$$\frac{dy}{dx} = f(x, y) \tag{12.17}$$

we seek to determine the approximate functional relationship that relates $y$ to $x$. Obviously, the function $f(x, y)$ in a given differential equation may be a simple expression or a complicated one. Therefore it would seem reasonable to assume that the derivatives may also be difficult to evaluate even when one is using numerical approximations. This is especially true in the case of the Taylor series method. In fact, this is one of the basic criticisms we have made of the Taylor series method. However, one thing we do know is that the error associated with a given approximation is directly related to the interval used, $h$. That is, the smaller $h$ is, the greater the accuracy of the approximation will be.

The basic Euler method presupposes that the interval is small enough that a truncated Taylor series involving the first derivative term only can be used in approximating the solution. That is,

$$y(x_0 + h) = y(x_0) + hy^{(1)}(x_0) \tag{12.18}$$

Equation (12.18) states that the solution at $x_0 + h$ is obtained by using the initial condition $y(x_0)$ and the first derivative $y^{(1)}(x_0)$, which is determined from equation (12.17). Furthermore, for a given interval $h$ the error term is of order two. Hence, rewriting equation (12.18) in subscripted form, we have

$$y_{i+1} = y_i + hy_i^{(1)} + (O)h^2 \tag{12.19}$$

The various terms appear in figure 12.3. Note that $(O)h^2$ is the error order.

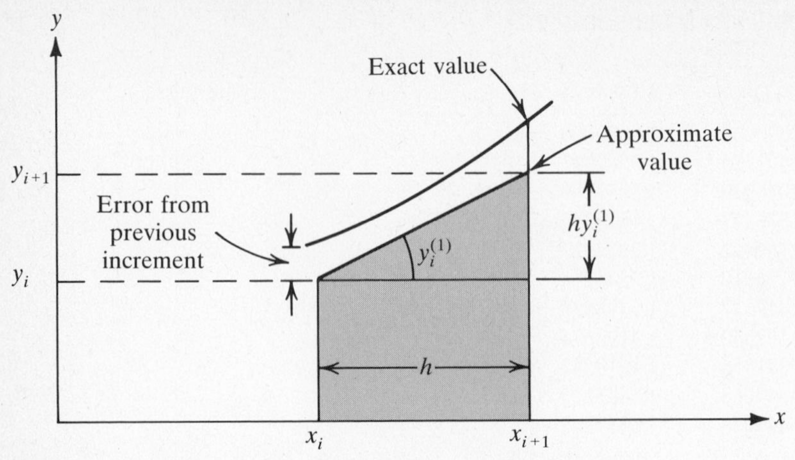


**Figure 12.3** Graphical illustration of Euler's basic method.

It is evident that as the interval is reduced, the approximated value approaches the exact value. However, while the error associated with each increment (local error) is of order $(O)h^2$, the error accumulated over many intervals (global error) is of order $(O)h$. For this reason, this method is highly inaccurate and should be avoided whenever possible. The solution given by equation (12.19) can be expressed more directly by noting that $y_i^{(1)} = f(x_i, y_i)$. Hence

$$y_{i+1} = y_i + hf(x_i, y_i) + (O)h^2 \tag{12.20}$$

Note that the local error $(O)h^2$ may be positive or negative, depending on the nature of the relationship. That is, if the actual relationship is concave upward, as shown in figure 12.3, then the error term is positive, otherwise it is negative.

The modified Euler method uses a truncated Taylor series in which the first and second derivative terms are retained. That is,

$$y_{i+1} = y_i + hy_i^{(1)} + \frac{h^2}{2} y_i^{(2)} \tag{12.21}$$

Clearly the local error term is now of order three and the global error is of order two. Consequently, the modified Euler method should be more accurate than the basic Euler method for a given interval. Furthermore, the second derivative term $y_i^{(2)}$ can be eliminated from equation (12.21) by using the following forward derivative approximation:

$$y_i^{(2)} = \frac{y_{i+1}^{(1)} - y_i^{(1)}}{h} \tag{12.22}$$

Substituting equation (12.22) into (12.21) yields the modified Euler formula:

$$y_{i+1} = y_i + hy_i^{(1)} + \frac{h^2}{2} \frac{y_{i+1}^{(1)} - y_i^{(1)}}{h} + (O)h^3$$

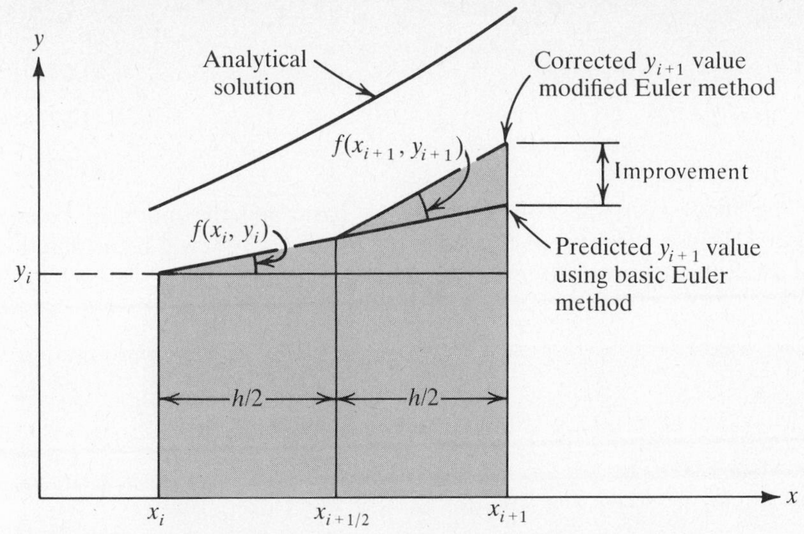


**Figure 12.4** Graphical illustration of the modified Euler method.

Simplifying and rearranging gives

$$y_{i+1} = y_i + \frac{h}{2}(y_i^{(1)} + y_{i+1}^{(1)}) + (O)h^3 \tag{12.23}$$

where $y_{i+1}$ is the approximated functional value and $y_i$ is the initial condition (given). The derivatives are then evaluated as follows:

$$y_i^{(1)} = f(x_i, y_i) \tag{12.24}$$

$$y_{i+1}^{(1)} = f(x_{i+1}, y_{i+1}) \tag{12.25}$$

Obviously, the derivative term at $x_i$ can be easily evaluated by using equation (12.24). However, the derivative at $x_{i+1}$ cannot be readily calculated since it is given in terms of $y_{i+1}$ as indicated by equation (12.25). Fortunately, this difficulty can be surmounted by approximating the functional value $y_{i+1}$ using the basic Euler method as given by equation (12.20). The approximated value is then used to calculate an improved value using equation (12.23). In essence, the modified Euler method is a simple predictor–corrector technique. The method is illustrated graphically in figure 12.4.

It is interesting to note that while the basic Euler method assumes a linear approximation for the true function over the interval $h$, the modified method assumes two linear approximations for that purpose.

The Euler methods are normally expressed more conveniently as follows:

### Basic Euler Method

$$y_{i+1} = y_i + k_1 + (O)h^2 \tag{12.26a}$$

$$k_1 = hf(x_i, y_i) \tag{12.26b}$$

Modified Euler Method

$$y_{i+1} = y_i + \tfrac{1}{2}(k_1 + k_2) + (O)h^3 \quad \textbf{(12.27a)}$$

$$k_1 = hf(x_i, y_i) \quad \textbf{(12.27b)}$$

$$k_2 = hf(x_{i+1}, y_i + k_1) \quad \textbf{(12.27c)}$$

Note that the factor $k_1$ is the same in both the basic and the modified Euler methods. Furthermore, the form of equations (12.26) and (12.27) will be helpful in illustrating the Runge–Kutta method, covered in the next section.

EXAMPLE 12.1

Solve the following differential equation for $0 \leq x \leq 1$ using $h = 0.1$:

$$\frac{dy}{dx} - \frac{1}{2}y = 0, \qquad y(0) = 1$$

Use the basic Euler method.

### Solution

This differential equation is the same one we solved in the preceding sections. Therefore, using equation (12.20) for $i = 0$, we have

$$y_1 = y_0 + hf(x_0, y_0) = 1 + 0.1f(0, 1)$$

Note that we could have used equations (12.26) just as well. The functional value $f(0, 1)$ is evaluated as follows:

$$f(x_0, y_0) = \tfrac{1}{2}y_0 = \tfrac{1}{2}(1) = \tfrac{1}{2}$$

Therefore the approximate functional value at $x_1 = h = 0.1$ is determined as

$$y_1 = 1 + 0.1(\tfrac{1}{2}) = 1.05$$

Furthermore, for $i = 1$, we have

$$y_2 = y_1 + hf(x_1, y_1) = 1.05 + 0.1f(0.1, 1.05)$$

where

$$f(0.1, 1.05) = \tfrac{1}{2}(1.05) = 0.525$$

Consequently, the approximated functional value is calculated at $x_2 = 2h = 0.2$ to be

$$y_2 = 1.05 + 0.1(0.525) = 1.1025$$

This procedure is repeated for $i = 2, \ldots, 9$; a summary of the results is given in table 12.4.

Table 12.4 BASIC EULER METHOD VERSUS EXACT SOLUTIONS

| $i$ | $x_i$ | $y_i$ | $f(x_i, y_i)$ | $y_{i+1}$ (EULER) | $y_{i+1}$ (EXACT) |
|---|---|---|---|---|---|
| 0 | 0 | 1.0000000 | 0.5000000 | 1.0500000 | 1.0512711 |
| 1 | 0.1 | 1.0500000 | 0.5250000 | 1.1025000 | 1.1051709 |
| 2 | 0.2 | 1.1025000 | 0.5512500 | 1.1576250 | 1.1618834 |
| 3 | 0.3 | 1.1576250 | 0.5788125 | 1.2155063 | 1.2214027 |
| 4 | 0.4 | 1.2155063 | 0.6077531 | 1.2762816 | 1.2840254 |
| 5 | 0.5 | 1.2762816 | 0.6381408 | 1.3400956 | 1.3498588 |
| 6 | 0.6 | 1.3400956 | 0.6700478 | 1.4071004 | 1.4190675 |
| 7 | 0.7 | 1.4071004 | 0.7035502 | 1.4774554 | 1.4918247 |
| 8 | 0.8 | 1.4774554 | 0.7387277 | 1.5513282 | 1.5683122 |
| 9 | 0.9 | 1.5513282 | 0.7756641 | 1.6288946 | 1.6487212 |

The approximate value indicates that in the interval $0 \le x \le 1$ the true functional relationships is concave upward (why?). Better accuracy can be achieved by reducing the interval $h$.

EXAMPLE 12.2

Rework example 12.1 using the modified Euler method.

### Solution

We begin by computing the $k_1$ factor using equation (12.27) for the first interval $i = 0$:

$$k_1 = hf(x_0, y_0) = h(\tfrac{1}{2}y_0)$$
$$= 0.1(\tfrac{1}{2}) = 0.05$$

Then using equation (12.27c) we calculate $k_2$:

$$k_2 = hf(x_1, y_0 + k) = h\left(\frac{y_0 + k_1}{2}\right)$$

$$= 0.1\left(\frac{1 + 0.05}{2}\right) = 0.0525$$

The functional approximation at $x_1 = 0.1$ ($i = 1$) is now given directly by equation (12.27a). That is,

$$y_1 = y_0 + \tfrac{1}{2}(k_1 + k_2)$$
$$= 1 + \tfrac{1}{2}(0.05 + 0.0525) = 1.05125$$

Therefore, at $x_2 = 0.2$, we have

$$k_1 = 0.1\left(\frac{1.05125}{2}\right) = 0.0525625$$

$$k_2 = 0.1\left(\frac{1.051255 + 0.0525625}{2}\right) = 0.0551909$$

$$y_2 = 1.05125 + \tfrac{1}{2}(0.0525625 + 0.0551909) = 1.1051267$$

This procedure is applied for $i = 2,\ldots,9$ to give the functional approximations shown in table 12.5.

Table 12.5 MODIFIED EULER METHOD VERSUS EXACT SOLUTIONS

| $i$ | $x_i$ | $y_i$ | $k_1$ | $k_2$ | $y_{i+1}$ (MODIFIED EULER) | $y_{i+1}$ (EXACT) |
|---|---|---|---|---|---|---|
| 0 | 0 | 1.0000000 | 0.0500000 | 0.0525000 | 1.0512500 | 1.0512711 |
| 1 | 0.1 | 1.0512500 | 0.0525625 | 0.0551909 | 1.1051267 | 1.1051709 |
| 2 | 0.2 | 1.1051267 | 0.0525634 | 0.0580192 | 1.1617644 | 1.1618834 |
| 3 | 0.3 | 1.1617644 | 0.0580882 | 0.0699263 | 1.2213048 | 1.2214027 |
| 4 | 0.4 | 1.2213048 | 0.0610652 | 0.0641185 | 1.2838967 | 1.2840254 |
| 5 | 0.5 | 1.2838967 | 0.0641948 | 0.0674045 | 1.3513013 | 1.3498588 |
| 6 | 0.6 | 1.3513013 | 0.0675651 | 0.0709433 | 1.4222446 | 1.4190675 |
| 7 | 0.7 | 1.4222446 | 0.0711122 | 0.0746678 | 1.4969124 | 1.4918247 |
| 8 | 0.8 | 1.4969124 | 0.0748456 | 0.0785879 | 1.5736292 | 1.5683122 |
| 9 | 0.9 | 1.5736292 | 0.0786815 | 0.0826155 | 1.6562447 | 1.6487212 |

The results clearly indicate that the modified Euler method gives better accuracy for the same $h$ interval when compared with the basic Euler method.

## 12.6 RUNGE–KUTTA METHODS

In the preceding section, we showed that the accuracy of the numerical solutions or ordinary differential equations can be improved by including a greater number of terms in the truncated Taylor series. The Runge–Kutta methods employ this strategy in realizing the approximated functional values efficiently and accurately. The basis for the Runge–Kutta procedure can be easily illustrated by reconsidering the modified Euler method. Recall that in approximating the second derivative in terms of the first derivatives at $x_i$ and $x_{i+1}$ it was necessary that we make an additional internal functional evaluation at $x_i + h/2$. Consequently, if higher derivative terms are included, then an additional number of internal functional evaluations will be needed. Obviously an unlimited number of formulas can be developed to correspond to different truncation strategies. For this reason, we may think of the Runge–Kutta methods as a set of formulas having different orders of error. Furthermore, any level of accuracy is possible: we are limited only by our desire to limit the computational effort involved.

In order to illustrate how the Runge–Kutta methods are developed, we show the derivation of the second-order method. Recall that from the modified Euler method we have

$$y_{i+1} = y_i + \frac{h}{2}(k_1 + k_2) \tag{12.28}$$

as the solution to a first-order ordinary differential equation of the form

$$\frac{dy}{dx} = f(x, y)$$

The Runge–Kutta method assumes a similar approximation to that given by equation (12.28),

$$y_{i+1} = y_i + ak_1 + bk_2 \tag{12.29a}$$

where

$$k_1 = hf(x_i, y_i) \tag{12.29b}$$

$$k_2 = hf(x_i + Ah, y_i + Bk_1) \tag{12.29c}$$

Substituting equation (12.29b) into (12.29c) yields the following expression for $k_2$:

$$k_2 = hf[x_i + Ah, y_i + Bhf(x_i, y_i)] \tag{12.30}$$

Equation (12.30) may be expressed in terms of the given functional $f(x_i, y_i)$ and its derivatives by using a Taylor series in two variables to give

$$k_2 = h\left(f(x_i, y_i) + Ah\frac{\partial f(x_i, y_i)}{\partial x} + Bhf(x_i, y_i)\frac{\partial f(x_i, y_i)}{\partial y}\right) \tag{12.31a}$$

This expression involves only the first term of the series. If we now adopt the notation $f = f(x_i, y_i)$, $f_x = \partial f/\partial x$, and $f_y = \partial f/\partial y$, then equation (12.31a) is given more conveniently as

$$k_2 = hf + Ah^2 f_x + Bh^2 ff_y \tag{12.31b}$$

Furthermore, equation (12.29c) is now given more simply as

$$k_1 = hf \tag{12.32}$$

Substituting equations (12.31) and (12.32) into equation (12.29a), then simplifying, yields the second-order Runge–Kutta formula. That is,

$$y_{i+1} = y_i + (a + b)hf + Abh^2 f_x + bBh^2 ff_y \tag{12.33}$$

It is evident that equation (12.33) involves the four unknowns $a$, $b$, $A$, and $B$. Consequently, in order to determine their values, we invoke the condition that equation (12.33) must correspond to a second-order Taylor series. That is,

$$y_{i+1} = y_i + hy_i^{(1)} + \frac{h^2}{2} y_i^{(2)} \tag{12.34}$$

The derivatives $y_1^{(1)}$ and $y_i^{(2)}$ are readily evaluated as follows:

$$\begin{aligned} y_1^{(1)} &= f(x_i, y_i) = f \\ y_i^{(2)} &= \left.\frac{d^2y}{dx^2}\right|_i = \left.\frac{\partial f}{\partial x} + \frac{\partial f}{\partial y}\frac{dy}{dx}\right|_i \end{aligned} \tag{12.35}$$

or more simply

$$y_i^{(2)} = f_x + f_y f \tag{12.36}$$

Substituting equations (12.35) and (12.36) into equation (12.34) gives

$$y_{i+1} = y_i + hf + \frac{h^2}{2} f_x + \frac{h^2}{2} ff_y \tag{12.37}$$

Equating equation (12.33) to equation (12.37) permits the evaluation of the unknowns as follows:

$$y_i + (a+b)hf + Abh^2 f_x + bBh^2 ff_y = y_i + hf + \frac{h^2}{2} f_x + \frac{h^2}{2} ff_y \tag{12.38}$$

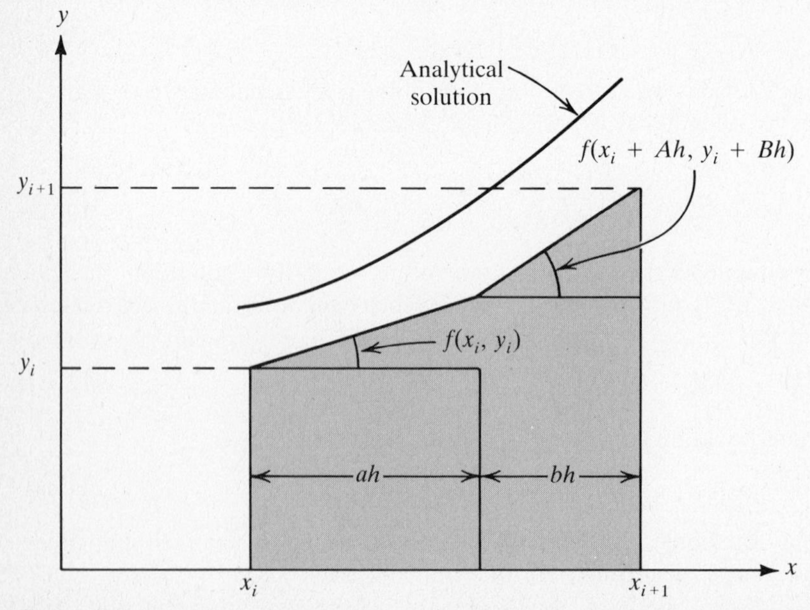


**Figure 12.5** Graphical representation of the second-order Runge–Kutta method.

Consequently, the following equations are determined by equating the individual $f$, $f_x$, and $ff_y$ coefficients to give

$$a + b = 1 \tag{12.39a}$$

$$Ab = \tfrac{1}{2} \tag{12.39b}$$

$$Bb = \tfrac{1}{2} \tag{12.39c}$$

Note that we now have three equations in four unknowns. Fortunately, this poses no problem in that one of the unknowns can be chosen arbitrarily. This is quite interesting, for it implies that we could choose the spacing of functional evaluation or the weighting factors. This concept is shown graphically in figure 12.5.

Therefore, if the weighting factor $a$ is chosen as $(\frac{1}{2})$, then, using equations (12.39), we determine the remaining unknowns to be $b = \frac{1}{2}$, $A = 1$, and $B = 1$. This situation corresponds precisely to the modified Euler method.

One of the most widely used methods of solving ordinary differential equations numerically is the fourth-order Runge–Kutta method. It involves the following equations given here without derivation:

$$y_{i+1} = y_i + \tfrac{1}{6}(k_1 + 2k_2 + 2k_3 + k_4) \tag{12.40a}$$

$$k_1 = hf(x_i, y_i) \tag{12.40b}$$

$$k_2 = hf\left(x_i + \frac{h}{2}, y_i + \frac{k_1}{2}\right) \tag{12.40c}$$

$$k_3 = hf\left(x_i + \frac{h}{2}, y_i + \frac{k_2}{2}\right) \tag{12.40d}$$

$$k_4 = hf(x_i + h, y_i + k_3) \tag{12.40e}$$

While the global error magnitude associated with the modified Euler method is of order $(O)h^2$, the global error associated with the fourth-order Runge–Kutta method is of order $(O)h^4$. Consequently, greater accuracy can be achieved with fewer steps by using the fourth-order method.

It is evident that while they are based on the Taylor series, the Runge–Kutta methods do not involve the evaluations of derivatives.

**E** EXAMPLE 12.3

Solve the following differential equations, using the fourth-order Runge–Kutta method:

$$\frac{dy}{dx} = \frac{1}{2}y, \qquad y(0) = 1$$

use $h = 1$.

Solution

Once again, this is the same problem solved earlier by Euler methods. Hence,

$$f(x, y) = y/2$$

The $k$ factors can easily be determined by using equations (12.40). Thus

$$k_1 = hf(x_0, y_0) = (1)f(0, 1)$$

$$= (1)\left(\frac{1}{2}\right) = \frac{1}{2}$$

$$k_2 = hf\left(x_0 + \frac{h}{2}, y_0 + \frac{k_1}{2}\right) = (1)f\left(\frac{1}{2}, \frac{5}{4}\right)$$

$$= \frac{\frac{5}{4}}{2} = \frac{5}{8}$$

$$k_3 = hf\left(x_0 + \frac{h}{2}, y_0 + \frac{k_2}{2}\right) = (1)f\left(\frac{1}{2}, \frac{21}{16}\right)$$

$$= \frac{\frac{21}{16}}{2} = \frac{21}{32}$$

$$k_4 = hf(x_0 + h, y_0 + k_3) = (1)f\left(1, \frac{53}{32}\right)$$

$$= \frac{\frac{53}{32}}{2} = \frac{53}{64}$$

Now substituting the various $k$ values into equation (12.40a) gives the desired solution:

$$y = y(1) = y_0 + \tfrac{1}{6}k_1 + 2k_2 + 2k_3 + k_4$$

$$y(1) = 1 + \tfrac{1}{6}\tfrac{1}{2} + 2\tfrac{5}{8} + 2\tfrac{21}{32} + \tfrac{53}{64}$$

$$y(1) = 1.6484375$$

This compares with the exact value of

$$y(1) = 1.6487212$$

Note that even though $h = 1$, the calculated value is a better approximation than that obtained by using the Euler method with $h = 0.1$.

## 12.7 PREDICTOR–CORRECTOR METHODS

The methods discussed in the preceding sections are called single-step methods because they use information obtained from the previous step. In essence, we used the initial values to compute new functional values which were then used as the new initial value to continue the solution. The implication is that they are self-starting and that the step size ($h$ interval) can be changed between iterations. In this respect, they are ideal for starting the solution and for providing additional information relative to functional values. Therefore it would seem logical to use these values along with the finite difference expressions discussed in chapter 10 to continue the solution. The predictor–corrector methods do precisely that. However, since difference expressions are generally derived for equal interval spacings, the predictor–corrector methods are generally developed for a fixed step size. Furthermore, they are not self-starting. Therefore methods such as the Runge–Kutta are used to begin the solution. In fact, the modified Euler method is one of the most basic predictor–corrector methods, in that a predicted $y_{i+1}$ value is calculated by using the basic Euler method, then corrected by using the modified scheme.

### 12.7.1 Adams Methods

This method qualifies as one of the simplest predictor methods available. However, it is not a predictor–corrector method. It is included here in order to lay the foundations for other techniques covered later in this section.

The basis for the Adams method is that given a first-order differential equation in the form

$$\frac{dy}{dx} = f(x, y) \tag{12.41}$$

we can integrate directly if the function $f(x, y)$ is replaced by a polynomial of suitable order. This is accomplished by forcing the approximating polynomial to fit through a set of past points. Hence, if we use two past points, the approximating polynomial will be of first order. Alternatively, if we use three past points, it will be a quadratic. The implication is that the more past points we use, the higher the order of the approximating polynomial will be and the better the accuracy.

The first Adams method approximates $f(x, y)$ as a second-order polynomial in $x$ for three past points. The resulting approximation is then integrated over the interval $x_i$ to $x_{i+1}$ to extrapolate for the new $y_{i+1}$ value. Thus using equation (12.41) after rearranging gives

$$\int_{y_i}^{y_{i+1}} dy = \int_{x_i}^{x_{i+1}} f(x, y)\, dx \tag{12.42a}$$

However, the function $f(x, y)$ is approximated as follows:

$$f(x, y) = f(x) = a_0 + a_1 x + a_2 x^2$$

Therefore equation (12.42a) can now be expressed in terms of the polynomial approximation to give

$$\int_{y_i}^{y_{i+1}} dy = \int_{x_i}^{x_{i+1}} (a_0 + a_1 x + a_2 x^2)\, dx \tag{12.42b}$$

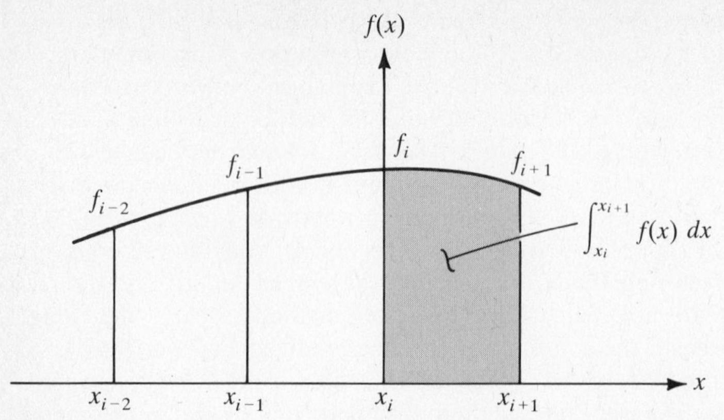


**Figure 12.6** Graphical interpretation of second-order Adams method.

Equation (12.42b) is illustrated graphically in figure 12.6. Note that the value of the integral (shaded area) is not affected by the positions of the base points. Hence, taking the limits of integration $x_i = 0$ and $x_{i+1} = h$, we can easily evaluate equation (12.42) to give

$$y_{i+1} - y_i = a_0 x + a_1 \frac{x^2}{2} + a_2 \frac{x^3}{3}\bigg|_0^h$$

$$y_{i+1} = y_i + a_0 h + a_1 \frac{h^2}{2} + a_2 \frac{h^3}{3} \tag{12.43}$$

It is evident that equation (12.43) involves three unknowns, namely, $a_0$, $a_1$, and $a_2$. These are readily evaluated in terms of the past functional values $f_{i-2}$, $f_{i-1}$, and $f_i$ as follows:

$$f_{i-2} = a_0 - 2ha_1 + 4h^2 a_2$$

$$f_{i-1} = a_0 - ha_1 + h^2 a_2$$

$$f_i = a_0$$

Solving for the unknowns using matrix inversion yields

$$\begin{Bmatrix} a_0 \\ a_1 h \\ a_2 h^2 \end{Bmatrix} = \frac{1}{2}\begin{bmatrix} 0 & 0 & 2 \\ 1 & -4 & 3 \\ 1 & -2 & 1 \end{bmatrix}\begin{Bmatrix} f_{i-2} \\ f_{i-1} \\ f_i \end{Bmatrix} \tag{12.44}$$

Equation (12.43) is now expressed more simply as follows:

$$y_{i+1} = y_i + \frac{h}{6}\{6 \quad 3 \quad 2\}\begin{Bmatrix} a_0 \\ a_1 h \\ a_2 h^2 \end{Bmatrix} \tag{12.45}$$

Substituting equation (12.44) into equation (12.45) yields the second-order Adams method. Hence

$$y_{i+1} = y_i + \frac{h}{6}\{6 \quad 3 \quad 2\}\frac{1}{2}\begin{bmatrix} 0 & 0 & 2 \\ 1 & -4 & 3 \\ 1 & -2 & 1 \end{bmatrix}\begin{Bmatrix} f_{i-2} \\ f_{i-1} \\ f_i \end{Bmatrix}$$

Simplifying by first multiplying the square matrix by the row vector gives

$$y_{i+1} = y_i + \frac{h}{12}(5f_{i-2} - 16f_{i-1} + 23f_i) \tag{12.46}$$

Clearly equation (12.46) uses a weighted average of three past functional values to extrapolate for a new $y$ value. Furthermore, the local error associated with equation (12.46) is of order $(O)h^4$. This can be easily verified by using forward interpolation instead of the method of undetermined coefficients.

The second Adams method achieves higher accuracy by assuming a third-order polynomial in approximating the function $f(x, y)$. Hence four past points are needed. Therefore

$$f(x, y) = f(x) = a_0 + a_1x + a_2x^2 + a_3x^3$$

Then substituting into equation (12.42) gives

$$\int_{y_i}^{y_{i+1}} dy = \int_{x_i}^{x_{i+1}} (a_0 + a_1x + a_2x^2 + a_3x^3)\,dx \tag{12.47}$$

Equation (12.47) is shown graphically in figure 12.7.

Once again, taking the limits of integration, such that $x_i = 0$ and $x_{i+1} = h$, then integrating equation (12.47), we get

$$y_{i+1} = y_i + a_0h + a_2\frac{h^2}{2} + a_2\frac{h^3}{3} + a_3\frac{h^4}{4}$$

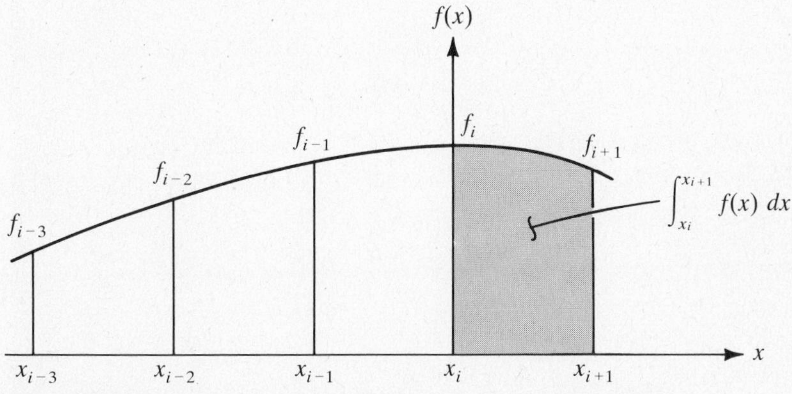


**Figure 12.7** Graphical interpretation of third-order Adams method.

or more simply

$$y_{i+1} = y_i + \frac{h}{24}\{24 \quad 12 \quad 8 \quad 6\}\begin{Bmatrix} a \\ a_1 h \\ a_2 h^2 \\ a_3 h^3 \end{Bmatrix} \tag{12.48}$$

The unknown vector in equation (12.48) is evaluated in terms of the functional values $f_{i-3}$, $f_{i-2}$, $f_{i-1}$, and $f_i$ by using the method of undetermined coefficients. That is,

$$\begin{Bmatrix} a_0 \\ a_1 h \\ a_2 h^2 \\ a_3 h^3 \end{Bmatrix} = \frac{1}{6}\begin{bmatrix} 0 & 0 & 0 & 6 \\ -2 & 9 & -18 & 11 \\ -3 & 12 & -15 & 6 \\ -1 & 3 & -3 & 1 \end{bmatrix}\begin{Bmatrix} f_{i-3} \\ f_{i-2} \\ f_{i-1} \\ f_i \end{Bmatrix} \tag{12.49}$$

Substituting (12.49) into (12.48) yields the desired third-order Adams formula:

$$y_{i+1} = y_i + \frac{h}{24}(-9f_{i-3} + 37f_{i-2} - 59f_{i-1} + 55f_i) \tag{12.50}$$

The local error associated with equation (12.50) is of order $(O)h^5$. It is evident that fourth- and higher-order approximating polynomials can be assumed to develop other formulas.

EXAMPLE 12.4

Approximate the $y$ value at $x = 0.4$ of the following differential equation using second- and third-order Adams methods:

$$\frac{dy}{dx} = \frac{1}{2}y, \qquad y(0) = 1.0$$

Solution

This is the same problem we solved in examples 12.1 and 12.2. Consequently, using the information given in table 12.5 relative to past points, we have

| $i$ | $x_i$ | $y_i$ | $f(x_i, y_i) = y_i/2$ |
|---|---|---|---|
| 0 | 0 | 1.0000000 | 0.5000000 |
| 1 | 0.1 | 1.0512500 | 0.5256250 |
| 2 | 0.2 | 1.1051267 | 0.5525634 |
| 3 | 0.3 | 1.1617644 | 0.5808822 |
| 4 | 0.4 | ? | ? |

Noting that $h = 0.1$ and using equation (12.46), we approximate the $y$ value at $x = 0.4$ as follows:

$$y_4 = y_3 + \frac{h}{12}[5f(x_1, y_1) - 16f(x_2, y_2) + 23f(x_3, y_3)]$$

$$= 1.1617644 + \frac{0.1}{12}[5(0.5256250 - 16(0.5525634) + 23(0.4808822)]$$

$$= 1.2213261$$

This value compares reasonably well with the exact value $y(0.4) = 1.2214027$. Furthermore, greater accuracy can be attained by using the fourth-order Runge–Kutta method in establishing past functional values. Hence, using equation (12.50), we have

$$y_4 = 1.1617644 + \frac{0.1}{24}[-9(0.5) + 37(0.525625)$$

$$- 59(0.5525634 + 55(0.5808822)]$$

$$= 1.2213286$$

Clearly this value is more accurate than the one we obtained using the second-order method.

### 12.7.2 Milne Method

This method qualifies as a true predictor–corrector method in that it first predicts a value for $y_{i+1}$ by extrapolating the derivative values. Unlike the Adams method, here the predicted value is corrected before going on to the subsequent step. Like the Adams method it requires that past values be calculated by using another scheme such as the Runge–Kutta. The Milne method assumes that four equally spaced starting $y$ values are known as $x_{i-3}$, $x_{i-2}$, $x_{i-1}$, and $x_i$ whereby $f(x, y)$ is replaced with a second-order approximating polynomial. The limits of integration are now taken at $x_{i-3}$ to $x_i$. Therefore equation (12.42b) now becomes

$$\int_{y_{i-3}}^{y_i} dy = \int_{x_{i-3}}^{x_i} (a_0 + a_1 x + a_2 x^2)\, dx \tag{12.51}$$

Note that while a second-order polynomial was assumed for the first Adams method, the limits of integration are different. Equation (12.51) is illustrated graphically in figure 12.8.

It is evident that the second-order approximation is first fitted through the three points corresponding to $x_{i-2}$, $x_{i-1}$, and $x_i$, then integrated from $x_{i-3}$ to $x_{i+1}$ to give the extrapolated (predicted) $y$ values. Hence using equation (12.51) with

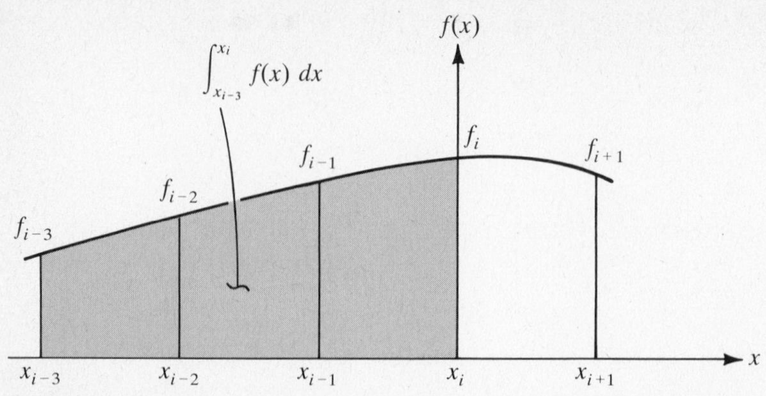


**Figure 12.8** Graphical illustration of the Milne method of determining the predicted *y* value.

$x_{i-3} = -3h$ and $x_{i+1} = h$ gives

$$y_{i+1} = y_{i-3} + \int_{-3h}^{h} (a_0 + a_1 x + a_2 x^2)\,dx$$

$$= y_{i-3} + \frac{h}{3}\{12 \quad -12 \quad 28\}\begin{Bmatrix} a_0 \\ a_1 h \\ a_2 h^2 \end{Bmatrix} \tag{12.52}$$

The unknown vector on the right-hand side of equation (12.52) is evaluated by using the functional values at $x_{i-2}$, $x_{i-1}$, and $x_i$. However, this was already done for the second-order Adams method (why?). Therefore substituting equation (12.44) into (12.52) yields the Milne formula for predicting the $y_{i+1}$ value. That is,

$$y_{i+1,P} = y_{i-3} + \frac{h}{3}\{12 \quad -12 \quad 28\}\frac{1}{2}\begin{bmatrix} 0 & 0 & 2 \\ 1 & -4 & 3 \\ 1 & -2 & 1 \end{bmatrix}\begin{Bmatrix} f_{i-2} \\ f_{i-1} \\ f_i \end{Bmatrix}$$

which simplifies to

$$y_{i+1,P} = y_{i-3} + \frac{4h}{3}(2f_{i-2} - f_{i-1} + 2f_i) \tag{12.53}$$

With the predicted $y_{i+1,P}$ value the function $f(x_{i+1}, y_{i+1})$ can be calculated with reasonable accuracy. The predicted value is then corrected by forcing a second-order polynomial to pass through points corresponding to $x_{i-1}$, $x_i$, and $x_{i+1}$ then integrating. Hence,

$$\int_{y_{i-1}}^{y_{i+1}} dy = \int_{x_{i-1}}^{x_{i+1}} (a_0 + a_1 x + a_2 x^2)\,dx \tag{12.54}$$

Equation (12.54) is illustrated graphically in figure 12.9.

Consequently, if we take the limits of integration to be $x_{i-1} = -h$ and $x_{i+1} = h$, then integrate equation (12.54), we obtain

$$y_{i+1,c} - y_{i-1} = \int_{-h}^{h} (a_0 + a_1 x + a_2 x^2)\,dx$$

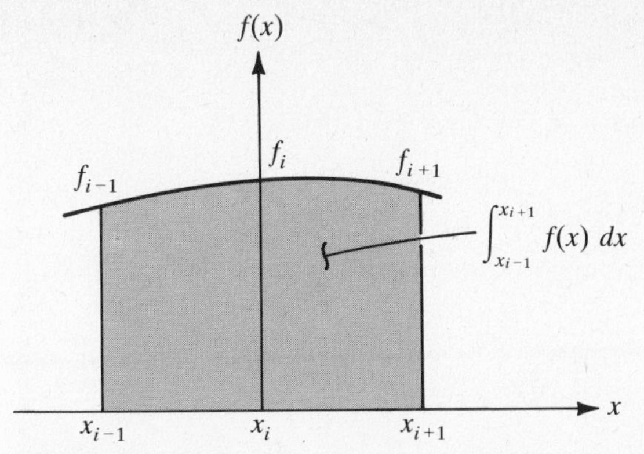


**Figure 12.9** Graphical interpretation of Milne corrector formula.

Note that the integral is exactly equivalent to Simpson's $\frac{1}{3}$ rule. Therefore we can write directly

$$y_{i+1,c} = y_{i-1} + \frac{h}{3}(f_{i-1} + 4f_i + f_{i+1}) \tag{12.55}$$

Equation (12.55) is the Milne corrector formula. Note that $f_{i+1}$ is estimated by using the predicted $y_{i+1,P}$ value. Furthermore, the local errors associated with the predicted value [equation (12.53)] and the corrected value [equation (12.55)] are of order $(O)h^5$. However, the predicted value is 28 times the corrected value.

Despite its simplicity and general accuracy, the Milne method is subject to an instability problem in some cases. The instability problem relates primarily to differential equations whose true solution is of an exponential form. Obviously, since for all practical purposes we have no idea what the true function is, the Milne method is not recommended.

EXAMPLE 12.5

Rework example 12.4 using the Milne predictor–corrector formulas.

Solution

The predicted $y$ value at $x = 4$ and $i = 3$ is given by equation (12.53). Thus

$$y_{4,P} = y_0 + \frac{4(0.1)}{3}[2f(x_1, y_1) - f(x_2, y_2) + 2f(x_3, y_3)]$$

Therefore, using the past values given in example 12.4, we have

$$y_{4,P} = 1 + \frac{0.4}{3}[2(0.5256250) - (0.5525634) + 2(0.5808822)]$$

$$= 1.2213935$$

The derivative at $x = 0.4$ can now be approximated by using the predicted value to give

$$\left.\frac{dy}{dx}\right|_{x_4} = f(x_4, y_4) = \frac{1}{2}(y_{4,P}) = 0.6106967$$

Therefore the corrected $y_4$ is obtained by using equation (12.55):

$$y_{4,c} = y_2 + \frac{0.1}{3}[f(x_2, y_2) + 4f(x_3, y_3) + f(x_4, y_4)]$$

Substituting the various functional values, we have

$$y_{4,c} = 1.1051267 + \frac{0.1}{3}[0.5525634 + 4(0.5808822) + 0.6106967]$$

$$= 1.221353$$

It is interesting to note that the predicted value is closer to the exact value of 1.2214027 than the corrected value. This is precisely the kind of stability problem associated with the Milne's method that we alluded to earlier.

### 12.7.3 Adams–Moulton Method

As the name implies this method employs the third-order Adams formulas to predict a $y$ value, then corrects it by using a second formula. The corrector formula is developed by approximating the function $f(x, y)$ with a third-order polynomial that passes through four points corresponding to $x_{i-2}$, $x_{i-1}$, $x_i$, and $x_{i+1}$. The predictor formula [equation (12.50)] is given in terms of functional values at $x_{i-3}$, $x_{i-2}$, $x_{i-1}$, and $x_i$. That is,

$$y_{i+1,P} = y_i + \frac{h}{24}(-9f_{i-3} + 37f_{i-2} - 59f_{i-1} + 55f_i) \qquad \textbf{(12.56)}$$

The corrector formula is then given by moving the integration one step forward:

$$\int_{y_i}^{y_{i+1}} dy = \int_{x_i}^{x_{i+1}} (a_0 + a_1 x + a_2 x^2 + a_3 x^3)\, dx \qquad \textbf{(12.57)}$$

Equation (12.57) is illustrated graphically in figure 12.10.

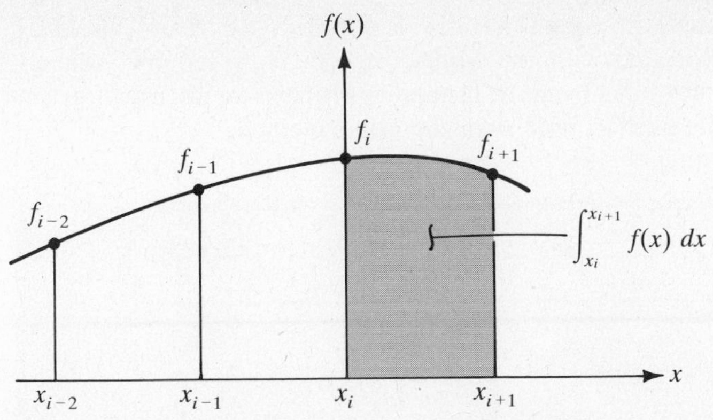


**Figure 12.10** Graphical interpretation of Adams–Moulton corrector formula.

Therefore, using the limits $x_i = 0$ and $x_{i+1} = h$, we develop the corrector formula in terms of the unknowns $a_0$, $a_1$, $a_2$, and $a_3$ as follows:

$$y_{i+1,c} = y_i + ha_0 + \frac{h^2}{2}a_1 + \frac{h^3}{3}a_2 + \frac{h^4}{4}a_3$$

or more simply

$$y_{i+1,c} = y_i + \frac{h}{12}\{12 \quad 6 \quad 4 \quad 3\}\begin{Bmatrix} a_0 \\ a_1 h \\ a_2 h^2 \\ a_3 h^3 \end{Bmatrix} \tag{12.58}$$

The unknown vector in equation (12.58) is easily determined from the function values at $x_{i-2}$, $x_{i-1}$, $x_i$, and $x_{i+1}$. That is,

$$\begin{Bmatrix} f_{i-2} \\ f_{i-1} \\ f_i \\ f_{i+1} \end{Bmatrix} = \begin{bmatrix} 1 & -2 & 4 & -8 \\ 1 & -1 & 1 & -1 \\ 1 & 0 & 0 & 0 \\ 1 & 0 & 1 & 1 \end{bmatrix}\begin{Bmatrix} a_0 \\ a_1 h \\ a_2 h^2 \\ a_3 h^3 \end{Bmatrix}$$

Solving for the unknown vector using matrix inversion then substituting into equation (12.58) gives the Adams–Moulton corrector formula:

$$y_{i+1,c} = y_i + \frac{h}{12}\{12 \quad 6 \quad 4 \quad 3\}\frac{1}{6}\begin{bmatrix} 0 & 0 & 6 & 0 \\ 1 & -6 & 3 & 2 \\ 0 & 3 & -6 & 3 \\ -1 & 3 & -3 & 1 \end{bmatrix}\begin{Bmatrix} f_{i-2} \\ f_{i-1} \\ f_i \\ f_{i+1} \end{Bmatrix}$$

Upon simplification, we have

$$y_{i+1,c} = y_i + \frac{h}{24}(f_{i-2} - 5f_{i-1} + 19f_i + 9f_{i+1}) \tag{12.59}$$

The local error associated with equation (12.59) is of order $(O)h^5$. This is precisely the order of error associated with the Milne method. However, the Adams–Moulton method does not suffer from the instability problem we discussed earlier. For this reason it is more widely used than the Milne method.

EXAMPLE 12.6

Rework example 12.4 using the Adams–Moulton method.

Solution

The predicted value at $x = 0.4$ is given by equation (12.56). Hence, using the values given in table 12.5, we have

$$y_{4,P} = y_3 + \frac{0.1}{24}(-9f_0 + 37f_1 - 59f_2 + 55f_3)$$

$$= 1.2213286$$

The corrected value is obtained by first evaluating $f(x_4, y_4)$, then substituting into equation (12.59). That is,

$$f_4 = f(x_4, y_{4,P}) = \tfrac{1}{2}(1.2213286) = 0.6106643$$

and

$$y_{4,c} = y_3 + \frac{0.1}{24}(f_1 - 5f_2 + 19f_3 + 9f_4)$$

$$= 1.1617644 + \frac{0.1}{24}[0.525625 - 5(0.5525634)$$

$$+ 19(0.5808822) + (0.6106643)]$$

$$= 1.2213292$$

The corrected value is clearly more accurate than the predicted value when compared with the exact value of $y_4 = 1.2214027$.

### 12.7.4 Step Size and Errors

The predictor–corrector methods are efficient techniques when compared with other methods such as Taylor series. In fact, the Adams–Moulton method is twice as efficient as the fourth-order Runge–Kutta method. Unfortunately, they all suffer from one rather important limitation. That is, the step size $h$ cannot be altered

Table 12.6 COMPARISON OF PREDICTOR–CORRECTOR METHODS

| METHOD | LOCAL ERROR | GLOBAL ERROR | EQUATION |
|---|---|---|---|
| Second-order Adams | $(O)h^4$ | $(O)h^3$ | (12.46) |
| Third-order Adams | $(O)h^5$ | $(O)h^4$ | (12.50) |
| Milne | $(O)h^5$ | $(O)h^4$ | (12.53), (12.55) |
| Adams–Moulton | $(O)h^5$ | $(O)h^4$ | (12.56), (12.59) |

without additional cost in terms of computer time. Obviously, the one-step methods, such as the Runge–Kutta methods, do not suffer from this limitation. The need for changing step size arises when the predicted and corrected values agree to as many significant figures as the desired accuracy. In this case, it will be appropriate to save computer time by increasing step size. However, in order to accomplish this, we need to develop additional predictor–corrector methods to bridge the two intervals. The alternative is to use a one-step method to establish yet another set of functional values whereby a predictor–corrector method is employed to continue the calculation for the new $h$ value. Furthermore, the past functional values needed to start the predictor–corrector methods should be calculated with utmost care. The fourth-order Runge–Kutta method is highly recommended for this purpose. The user should select an appropriate $h$ interval, then recalculate the functional values by using $h/2$ then compare corresponding functional values. Hence, if these do not vary significantly, then an interval $zh$ should be employed and the procedure should be repeated until significant differences in the functional values are observed, in which case the previous interval should be used in calculating the past values.

The local errors associated with the predictor–corrector methods can be estimated by using the forward interpolating formula. Table 12.6 provides a comparison of the various predictor–corrector methods outlined in this section.

Recall that the local error pertains to errors associated with each computational step, whereas global errors reflect errors accumulated over many steps. It is evident that the error order of $(O)h^5$ is not indicative of the exact error term. The implication is that different degrees of accuracy are possible using formulas with the same order of error. Furthermore, since the local error for the fourth-order Runge–Kutta method is of order $(O)h^5$, it is highly recommended in establishing past functional values. This is because it is always better to use expressions having the same order of error. Otherwise, error propagation may render the solution useless regardless of the precision of the predictor–corrector method being used.

The total error associated with numerical solutions is the sum of the truncation and roundoff errors. That is, although the truncation error is reduced by reducing the step size, such action may result in a corresponding increase in the roundoff errors. This is because computers can retain only a finite number of significant figures. Consequently, the net effect may in fact be an increase in the total error as the step size is reduced. This concept is shown graphically in figure 12.11. Clearly, the ideal step size for a given problem varies with the particular application and method being used.

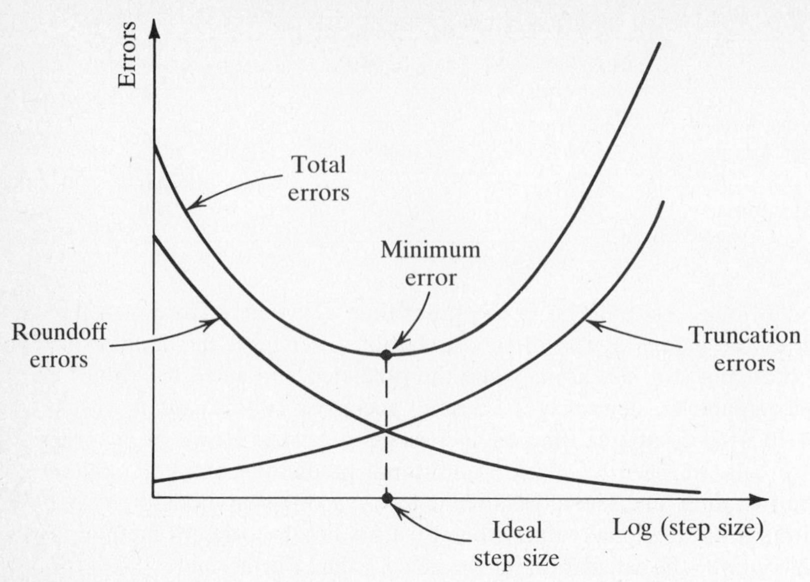


**Figure 12.11** Graphical illustration of total errors associated with numerical solutions.

## 12.8 STIFF EQUATIONS

When dealing with ordinary differential equations, it is possible that some do not lend themselves to solution by any of the methods presented thus far. Such equations are referred to as "stiff equations." These equations arise when the first-order time constant is widely different in magnitude when compared with the $h$ interval. The time constant is defined as the time required for the transient portion of the solution to decay by a factor of $1/e$, where $e$ is the base of the natural log. While it is possible to address this problem by altering the step size, this approach has two disadvantages. The first is that computers have limited capacity to retain a large number of significant figures. The second is that small $h$ intervals are costly when measured in computer time.

Stiff equations occur frequently in modeling electrical circuits. Furthermore, significant efforts have been and continue to be directed toward the development of more efficient techniques. One such technique was proposed by Gear. The method uses a predictor–corrector scheme with error of order $(O)h^6$. The basis for the method is that both $y$ values and derivative values are corrected at each step.

## 12.9 HIGH-ORDER EQUATIONS AND SYSTEMS OF FIRST ORDER

In the preceding sections, we have treated the problem of solving first-order differential equations numerically. Most differential equations pertaining to physical models are of higher order. This fact is supported by the many mathematical models presented in chapter 3. Consider, for example, the dynamic response model pertaining to a single-degree-of-freedom system shown in figure 12.12.

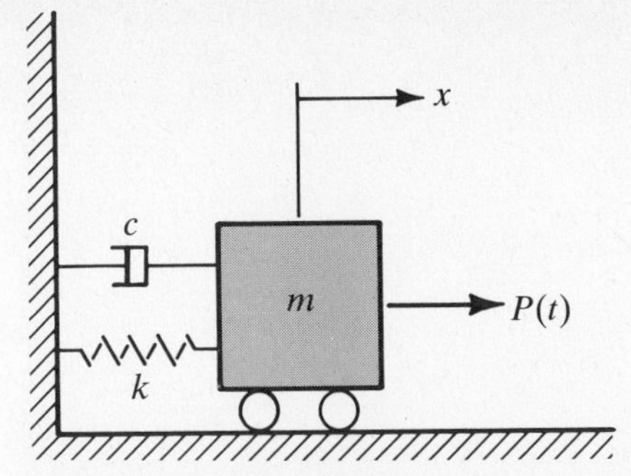


**Figure 12.12** Single-degree-of-freedom mass–dashpot–spring system.

This model represents a mass attached to a spring and a dashpot (shock absorber) in which the mass is subjected to an external force. The object is to find the displacement function $x(t)$ caused by the external force $P(t)$ and initial conditions. The initial conditions represent the initial velocity and displacement of the mass at time $t = 0$. We refer to such a system as a single-degree-of-freedom system because the mass $m$, the spring stiffness $k$, the damping coefficient $c$, and the external force $P(t)$ are concentrated at one location. The mathematical model pertaining to such a system was developed earlier and is repeated here for convenience:

$$m\frac{d^2x}{dt^2} + c\frac{dx}{dt} + kx = P(t) \tag{12.60}$$

It is evident that two or more such ordinary differential equations would have to be solved for higher degrees of freedom. While it is possible to develop special methods to facilitate the direct solution of certain types of higher-order ordinary differential equations, it is also possible to use the methods described before. This can be accomplished by reducing the high-order equations to sets of first-order ones. Consequently, if we specify the initial conditions for equation (12.60) at time $t = t_0$ to be

$$x_0 = x(t_0) = \text{initial displacement}$$

$$x_0^{(1)} = x^{(1)}(t_0) = \left.\frac{dx}{dt}\right|_{t=t_0} = \text{initial velocity}$$

then it is possible to reduce equation (12.60) to a set involving two first-order equations by introducing a new variable $y$ such that

$$\frac{dx}{dt} = y, \qquad x(t_0) = x_0 \tag{12.61a}$$

$$\frac{dy}{dt} = \frac{1}{m}[P(t) - cy - kx], \qquad y(t_0) = y_0 = x^{(1)}(0) \tag{12.61b}$$

Equations (12.61) are first-order differential equations that can be solved by the methods presented earlier. This concept of reducing an $n$th-order ordinary differential equation to an equivalent set of $n$ first-order equations can be extended to a set of higher order.

## EXAMPLE 12.7

Use Taylor series to solve the following second-order ordinary differential equation at $t = 0.1$:

$$4\frac{d^2x}{dt^2} + 3\frac{dx}{dt} + 16x = 0$$

with the initial conditions

$$x(0) = 1, \qquad \dot{x}(0) = 1$$

### Solution

Since the external force is zero, the response is caused by the initial conditions only. Hence assuming

$$\frac{dx}{dt} = y, \qquad x(0) = 1 = x_0 \tag{12.62}$$

the second order is expressed as

$$\frac{dy}{dt} = -4x - \frac{3}{4}y, \qquad y(0) = 1 = y_0 \tag{12.63}$$

The solution is now attained by observing that equations (12.62) and (12.63) are two first-order ordinary differential equations which can be expanded as Taylor series. Hence, taking $t_0 + h = t$ and $t_0 = 0$ gives

$$x(t) = x_0 + tx_0^{(1)} + \frac{t^2}{2}x_0^{(2)} + \frac{t^3}{6}x_0^{(3)} + \cdots \tag{12.64}$$

$$y(t) = y_0 + ty_0^{(1)} + \frac{t^2}{2}y_0^{(2)} + \frac{t^3}{6}y_0^{(3)} + \cdots \tag{12.65}$$

Obviously, we could have included additional terms in these expressions if high accuracy were desired. In any case, equations (12.64) and (12.65) involve derivatives at $t = 0$ which can be evaluated as follows:

$$x^{(1)} = y, \qquad x_0^{(1)} = 1$$

$$y^{(1)} = -4x - \tfrac{3}{4}y, \qquad y_0^{(1)} = -4(1) - \tfrac{3}{4}(1) = -\tfrac{19}{4}$$

$$x^{(2)} = y^{(1)}, \qquad x_0^{(2)} = -\tfrac{19}{4}$$

$$y^{(2)} = -4x^{(1)} - \tfrac{3}{4}y^{(1)}, \qquad y_0^{(2)} = -4(1) - \tfrac{3}{4}(-\tfrac{19}{4}) = -\tfrac{7}{16}$$

$$x^{(3)} = y^{(2)}, \qquad x_0^{(3)} = -\tfrac{7}{16}$$

$$y^{(3)} = -4x^{(2)} = \tfrac{3}{4}y^{(2)}, \qquad y_0^{(3)} = -4(-\tfrac{19}{4}) - \tfrac{3}{4}(-\tfrac{7}{16}) = \tfrac{1247}{64}$$

Substituting these values into equations (12.64) and (12.65) yields the desired solution for the displacement and velocity functions in terms of time $t$ (note $h = t$):

$$x(t) = 1 + t - \tfrac{19}{8}t^2 - \tfrac{7}{96}t^3 + \cdots$$

$$y(t) = 1 - \tfrac{19}{4}t - \tfrac{7}{32}t^2 + \tfrac{1247}{384}t^3 + \cdots$$

Therefore at $t = 0.1$ we have

$$x(0.1) = 1 + 0.1 - \tfrac{19}{8}(0.1)^2 - \tfrac{7}{96}(0.1)^3 = 1.0761771$$

$$y(0.1) = 1 - \tfrac{19}{4}(0.1) - \tfrac{7}{32}(0.1)^2 + \tfrac{1247}{384}(0.1)^3 = 0.5260599$$

These values compare with the exact values of $x(0.1) = 1.0794013$ and $y(0.1) = 0.5827277$. Note that while the displacements are reasonably close the velocity $y(0.1)$ is not so accurate. This is caused by truncation, which is more critical when evaluating the derivative $y = dx/dt$.

EXAMPLE 12.8

Rework example 12.7 using the fourth-order Runge–Kutta method.

Solution

Using the same transformation applied in example 12.7, we have

$$\frac{dx}{dt} = y = G(t, y), \qquad x(0) = 1 = x_0 \tag{12.66}$$

$$\frac{dy}{dt} = -4x - \tfrac{3}{4}y = f(x, y), \qquad y(0) = 1 = y_0 \tag{12.67}$$

The solution to equations (12.66) and (12.67) is obtained by using a modified version of equation (12.40) as shown below:

| $dx/dt = G(t, x, y)$ | $dy/dt = f(t, x, y)$ |
|---|---|
| $x_{i+1} = x_i + \frac{1}{6}(r_1 + 2r_2 + 2r_3 + r_4)$ | $y_{i+1} = y_i + \frac{1}{6}(k_1 + 2k_2 + 2k_3 + k_4)$ |
| $r_1 = hG(t_i, x_i, y_i)$ | $k_1 = hf(t_i, x_i, y_i)$ |
| $r_2 = hG(t_i + \frac{1}{2}h, x_i + \frac{1}{2}r_1, y_i + \frac{1}{2}k_1)$ | $k_2 = hf(t_i + \frac{1}{2}h, x_i + \frac{1}{2}r_1, y_i + \frac{1}{2}k_1)$ |
| $r_3 = hG(t_i + \frac{1}{2}h, x_i + \frac{1}{2}r_2, y_i + \frac{1}{2}k_2)$ | $k_3 = hf(t_i + \frac{1}{2}h, x_i + \frac{1}{2}r_2, y_i + \frac{1}{2}k_2)$ |
| $r_4 = hG(t_i + h, x_i + r_3, y_i + k_3)$ | $k_4 = hf(t_i + h, x_i + r_3, y_i + k_3)$ |

Consequently, the solution for the problem at hand is summarized as follows, for $h = 0.1$, $t_0 = 0$, $x_0 = 1$, and $y_0 = 1$:

| $G(t, x, y) = y$ | $f(t, x, y) = -4x - \frac{3}{4}y$ |
|---|---|
| $r_1 = 0.1(1) = 0.1$ | $k_1 = 0.1[-4(1) - \frac{3}{4}(1)] = -0.475$ |
| $r_2 = 0.07625$ | $k_2 = -0.4724375$ |
| $r_3 = 0.0763781$ | $k_3 = -0.4725592$ |
| $r_4 = 0.0527440$ | $k_4 = -0.4678278$ |

Therefore

$$\begin{aligned} x(0.1) &= 1 + \tfrac{1}{6}[0.1 + 2(0.07625) + 2(0.0763781) + 0.0527440] \\ &= 1.0763306 \end{aligned}$$

$$\begin{aligned} y(0.1) &= 1 + \tfrac{1}{6}[-0.475 - 2(0.4724375) - 2(0.4725592) - 0.4678278] \\ &= 0.5278631 \end{aligned}$$

Clearly these values are closer to the exact solution than those obtained by means of the truncated Taylor series method. However, once again the velocity $y(0.1)$ is less accurate than the displacement. Better accuracy can be achieved by using a smaller $h$ interval.

## 12.10 INITIAL-VALUE PROBLEMS INVOLVING HIGH-ORDER EQUATIONS

This section deals with specific applications pertaining to ordinary differential equations of the second order in which time-dependent responses are needed. These are referred to as initial-value problems because the response is known only at a given time. A variety of initial-value problems of practical importance occur in practice for which a closed-form solution is difficult if not impossible to determine. While the numerical methods outlined earlier in this chapter are appropriate for solving these types of problems, they are not as effective as the direct numerical integration procedures yet to be outlined. Figure 12.13 describes various initial-value problems encountered in engineering practice.

It is obvious that the three models described in figure 12.13 involve second-order differential equations that are extremely similar. The difference is in the variable names. Consequently, we need to find a way of solving only one of these models.

### 12.10.1 Finite Difference Methods

These methods involve replacing the derivatives in a given ordinary differential equation with equivalent difference approximations. It follows that any convenient finite difference can be used in approximating the solution. Theoretically, a large number of methods could be employed depending on the order of error and

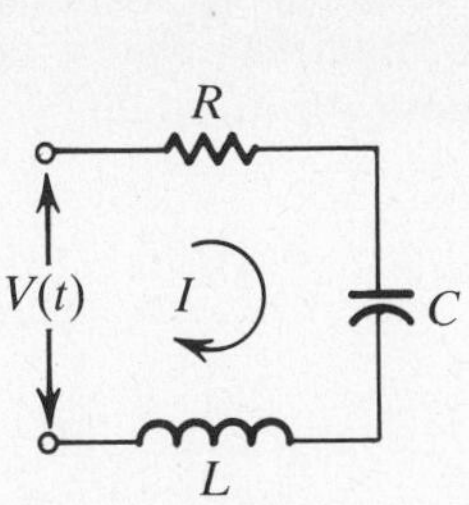


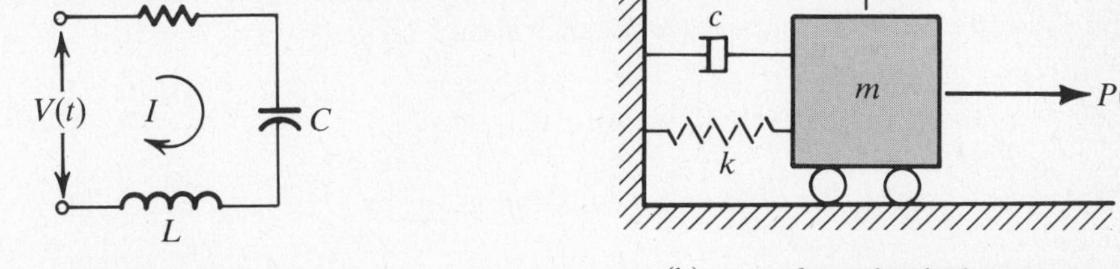


(a) *RCL* electrical circuit:

$$L\frac{d^2I}{dt^2} + R\frac{dI}{dt} + \frac{1}{C}I = \frac{dv}{dt}$$

(b) *m-c-k* mechanical system:

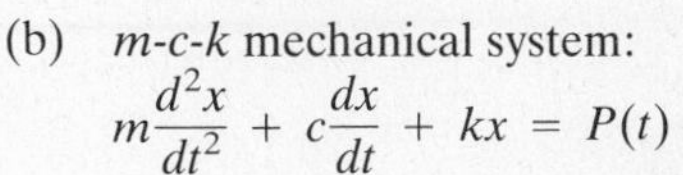

$$m\frac{d^2x}{dt^2} + c\frac{dx}{dt} + kx = P(t)$$

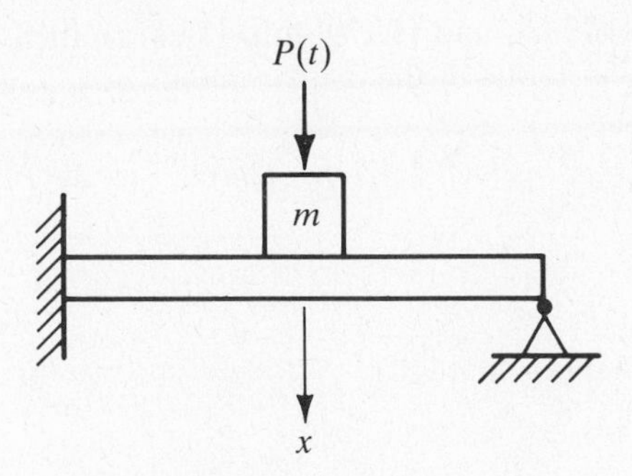


(c) Vibrating beam:

$$m\frac{d^2y}{dt^2} + c\frac{dy}{dt} + ky = P(t)$$

**Figure 12.13** Initial-value problems.

on whether forward, backward, or central differences are used. However, from the practical standpoint the central difference approximation provides better accuracy for a given interpolating function. For this reason, it is by far the most widely used technique in solving high-order differential equations.

For illustration purposes, only the *m-c-k* system is analyzed in this section. Consider the model

$$m\ddot{x} + c\dot{x} + kx = P(t) \qquad \textbf{(12.68a)}$$

with the initial conditions

$$x(t_0) = x_0 \qquad \textbf{(12.68b)}$$

$$\dot{x}(t_0) = \dot{x}_0 \qquad \textbf{(12.68c)}$$

Note that, for simplicity, we have used the notation $\dot{x} = dx/dt$ and $\ddot{x} = d^2x/dt^2$. Consequently, the second derivative in equation (12.68a) may be approximated by using a central difference formula with error of order $(O)h^2$ to give

$$\ddot{x}_i = \frac{1}{h^2}(x_{i-1} - 2x_i + x_{i+1}) + (O)h^2 \qquad \textbf{(12.69)}$$

The increment $h$ is now equal to $t$ (the time increment). Furthermore, the first derivative is approximated by using a central difference formula having the same error order as that of 12.69. This is necessary if we are to limit the instability problems associated with numerical techniques. Hence

$$\dot{x}_i = \frac{1}{2h}(-x_{i-1} + x_{i+1}) + (O)h^2 \tag{12.70}$$

At any given time $t_i$, the differential equation given by (12.68a) is in equilibrium; hence

$$m\ddot{x}_i + c\dot{x}_i + kx_i = P_i \tag{12.71a}$$

$$x(t_i) = x_i, \qquad \dot{x}(t_i) = \dot{x}_i \tag{12.71b}$$

Substituting equations (12.69) and (12.70) into (12.71a), then solving for the displacement at time $t_{i+1}$ in terms of the displacements at $t_{i-1}$ and $t_i$ yields

$$x_{i+1} = \frac{1}{2m + 2ch}[(4m - 2kh^2)x_i + (ch - 2m)x_{i-1} + 2h^2P_i]$$

or more simply

$$x_{i+1} = \frac{1}{\alpha_0}(\alpha_1 x_i + \alpha_2 x_{i-1} + \alpha_3 P_i) \tag{12.72a}$$

where

$$\alpha_0 = 2m + 2ch \tag{12.72b}$$

$$\alpha_1 = 4m - 2kh^2 \tag{12.72c}$$

$$\alpha_2 = ch - 2m \tag{12.72d}$$

$$\alpha_3 = 2h^2 \tag{12.72e}$$

It is interesting to note that equation (12.72a) is similar to a predictor equation, in that two previous values for $x_{i-1}$ and $x_i$ must be calculated before we can solve for $x_{i+1}$. The implication is that this procedure is not self-starting. Therefore, given the initial conditions $(x_i, \dot{x}_i)$ at time $t = t_i$, we need to approximate $x_{i-1}$. This is accomplished by expanding the displacement function $x(t)$ using Taylor series. Thus

$$x_{i-1} = x_i - h\dot{x}_i + \frac{h^2}{2}\ddot{x}_i \tag{12.73}$$

The acceleration at time $t_i$ is approximated from equation (12.71a) as

$$\ddot{x}_i = \frac{1}{m}(P_i - c\dot{x}_i - kx_i) \tag{12.74}$$

Substituting equation (12.74) into (12.73) and simplifying yields the unknown $x_{i-1}$. That is,

$$x_{i-1} = \frac{h^2}{2m}P_i + \left(1 - \frac{kh^2}{2m}\right)x_i - \left(h + \frac{ch^2}{2m}\right)\dot{x}_i \tag{12.75}$$

Thus equation (12.75) can be used to start the solution, then equation (12.72) is used for its continuation.

It is evident that using higher-order finite difference approximations for solving initial-value problems is possible. However, such an approach would introduce additional past points, making it impossible to solve for the succeeding displacement value without using a starting procedure such as the Runge–Kutta method. Furthermore, the technique illustrated here involves errors of order $(O)h^2$, which dictates that the solution interval $h$ must be kept reasonably small. It is apparent that this technique of solving high-order differential equations is extremely simple and efficient when compared with the Taylor series or Runge–Kutta methods. Furthermore, for a given step size, this method involves computations of the $\alpha$ coefficients only once. In fact, the $\alpha$ values given by equation (12.72) are changed only if the step size is changed.

EXAMPLE 12.9

Solve the following differential equation for $0 \le t \le 0.5$ using a time increment $h = 0.1$:

$$4\frac{d^2x}{dt^2} + 3\frac{dx}{dt} + 16x = 0, \qquad x_0 = 1, \quad \dot{x}_0 = 1$$

Solution

These are the same differential equations we solved earlier in example 12.8. However, in order to show the versatility and efficiency of the finite difference method, more steps are added. Hence, using equation (12.71) for $h = 0.1$, $m = 4$, $c = 3$, $k = 16$, and $P = 0$, we determine the solution in terms of two past displacements:

$$\alpha_0 = 2(4) = 3(0.1) = 8.3$$

$$\alpha_1 = 4(4) - 2(16)(0.1)^2 = 15.68$$

$$\alpha_2 = 3(0.1) - 2(4) = -7.7$$

$$\alpha_3 = 2(0.1)^2 = 0.02$$

and

$$x_{i+1} = \frac{1}{8.3}(15.68x_i - 7.7x_{i-1}) \tag{12.76}$$

Note that the displacement function given by equation (12.76) is expressed in terms of the given initial displacement $x_i$ and the past displacement $x_{i-1}$, which can be determined from initial conditions. Therefore, using equation (12.75) for $i = 0$ gives

$$x_{-1} = 0 + \left(1 - \frac{16(0.1)^2}{2(4)}\right)x_0 - \left(0.1 + \frac{3(0.1)^2}{2(4)}\right)\dot{x}_0$$

$$= 0.8762500$$

The solution is now given directly by equation (12.76) for any number of time increments. That is,

$t = 0.1$ ($i = 0$):

$$x_1 = \frac{1}{8.3}(15.68x_0 - 7.7x_{-1}) = 1.0762500$$

$t = 0.2$ ($i = 1$):

$$x_2 = \frac{1}{8.3}(15.68x_1 - 7.7x_0) = 1.105494$$

$t = 0.3$ ($i = 2$):

$$x_3 = \frac{1}{8.3}(15.68x_2 - 7.7x_1) = 1.0900025$$

$t = 0.4$ ($i = 3$):

$$x_4 = \frac{1}{8.3}(15.68x_3 - 7.7x_2) = 1.0336066$$

$t = 0.5$ ($i = 4$):

$$x_5 = \frac{1}{8.3}(15.68x_4 - 7.7x_3) = 0.94143768$$

The displacement at $t = 0.1$ compares rather well with the Runge–Kutta value of 1.0763306 obtained in example 12.8. However, the amount of computation involved in the central difference method is significantly less. In fact, the displacement for subsequent time increments involves nothing more than substituting past values. Furthermore, the velocity and acceleration at any time increment can be readily calculated by using equations (12.69) and (12.70), respectively. It is evident that better accuracy is made possible by using smaller step size ($h$ interval). It can be shown that the finite difference procedure is stable for $h \leq T/10$ where $T$ is the natural period, which is determined as $T = 2\pi\sqrt{m/k}$. However, this does not mean that the solution will be accurate. Instead, we should use this recommended value as a guide in the selection of the $h$ value. Consequently, we may first choose $h = T/10$, then halve the interval until the calculated values at any time $t$ are not significantly different or until desired accuracies are achieved.

### 12.10.2 Trapezoidal Rule Method

This is the second of a series of direct methods that can be developed for solving high-order ordinary differential equations. These direct methods are self-starting in that only initial conditions are needed to start the solution. The basis for the trapezoidal rule is that high-order derivatives can be expressed in terms of lower-order ones directly. For illustration purposes, let us consider the mechanical model analyzed in the preceding section. That is,

$$m\ddot{x} + c\dot{x} + kx = P, \qquad x(t_0) = x_0, \qquad \dot{x}(t_0) = \dot{x}_0 \tag{12.77}$$

Note that while equation (12.77) is of second order, in general the procedure to be outlined is equally applicable to higher-order ordinary differential equations. In any case, we proceed by expressing the second derivative $\ddot{x}$ as follows:

$$\ddot{x} = \frac{d\dot{x}}{dt}$$

Integrating over the interval $t_i$ to $t_{i+1}$ gives

$$\int_{\dot{x}_i}^{\dot{x}_{i+1}} d\dot{x} = \int_{t_i}^{t_{i+1}} \ddot{x}\,dt$$

$$\dot{x}_{i+1} = x_i + \int_{t_i}^{t_{i+1}} \ddot{x}\,dt \tag{12.78}$$

The implication is that the first derivative at time $t_{i+1}$ is given in terms of the velocity at time $t_i$ and the integral of the second derivative over the same time interval. The integral in equation (12.78) is illustrated graphically in figure 12.14.

It is evident that the shaded area can be approximated by using the trapezoidal rule of integration. Therefore equation (12.78) is given in a more useful form as

$$\dot{x}_{i+1} = \dot{x}_i + \frac{h}{2}(\ddot{x}_i + \ddot{x}_{i+1}) \tag{12.79}$$

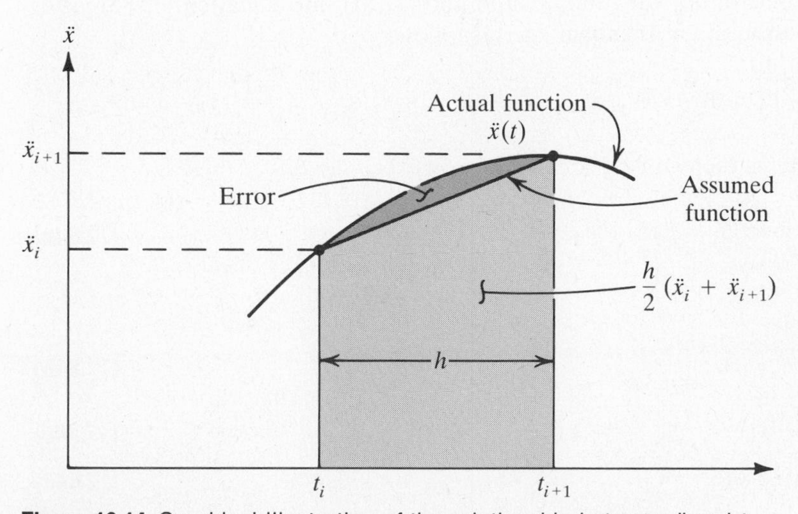


**Figure 12.14** Graphical illustration of the relationship between $\ddot{x}$ and $t$.

Similarly, the displacement at time $t_{i+1}$ is estimated to give

$$x_{i+1} = x_i + \frac{h}{2}(\dot{x}_i + \dot{x}_{i+1}) \tag{12.80}$$

The unknown $\dot{x}_{i+1}$ appearing in equation (12.80) can be easily eliminated by replacing it with equation (12.79). Hence

$$x_{i+1} = x_i + \frac{h}{2}\left(\dot{x}_i + \dot{x}_i + \frac{h}{2}\ddot{x}_i + \frac{h}{2}\ddot{x}_{i+1}\right)$$

Simplifying yields

$$x_{i+1} = x_i + h\dot{x}_i + \frac{h^2}{4}(\ddot{x}_i + \ddot{x}_{i+1}) \tag{12.81}$$

Note that since the trapezoidal rule was used in the derivation of equation (12.81), the associated local error order is $(O)h^2$. Furthermore, at a given time $t_i$ the initial conditions $x_i$ and $\dot{x}_i$ are known. Consequently the only remaining unknowns are $\ddot{x}_i$ and $\ddot{x}_{i+1}$. However, $\ddot{x}_i$ is related to the initial conditions and can be easily evaluated by using equation (12.77) at time $t = t_i$. Hence

$$m\ddot{x}_i + c\dot{x}_i + kx_i = P_i$$

solving for $\ddot{x}_i$ yields

$$\ddot{x}_i = \frac{1}{m}(P_i - kx_i - c\dot{x}_i) \tag{12.82}$$

The remaining unknown $x_{i+1}$ is evaluated by using the equilibrium equation at time $t = t_{i+1}$. That is,

$$m\ddot{x}_{i+1} + c\dot{x}_{i+1} + kx_{i+1} = P_{i+1} \tag{12.83}$$

Therefore, substituting equations (12.79) and (12.81) into equation (12.83) yields a single equation in an unknown $x_{i+1}$ as follows:

$$m\ddot{x}_{i+1} + c\left(\dot{x}_i + \frac{h}{2}\ddot{x}_i + \frac{h}{2}\ddot{x}_{i+1}\right) + k\left(x_i + h_2\dot{x}_i + \frac{h^2}{4}\ddot{x}_i + \frac{h^2}{4}\ddot{x}_{i+1}\right) = P_{i+1}$$

Collecting terms then solving for the unknown $\ddot{x}_{i+1}$ gives

$$\ddot{x}_{i+1} = \frac{1}{\beta_0}(\beta_1\ddot{x}_i + \beta_2\dot{x}_i + \beta_3 x_i + P_{i+1}) \tag{12.84a}$$

where

$$\beta_0 = m + \frac{ch}{2} + \frac{kh^2}{4} \tag{12.84b}$$

$$\beta_1 = m - \beta_0 \tag{12.84c}$$

$$\beta_2 = -c - kh \tag{12.84d}$$

$$\beta_3 = -k \tag{12.84e}$$

The method just outlined is stable as long as the time interval $h$ is chosen such that

$$h \leq \frac{T}{10} \leq \frac{\pi}{5}\sqrt{\frac{m}{k}}$$

Better accuracy may be obtained by using $h \leq \frac{1}{20}T$. The solution to a second-order differential equation is now summarized as follows:

*STEP 1:* Determine $\ddot{x}_i$ using equation (12.82).

*STEP 2:* Determine $\beta_0$, $\beta_1$, $\beta_2$, and $\beta_3$ in terms of $m$, $c$, $k$, and $h$ using equations (12.84b)–(12.84e). Then substitute into equation (12.84a).

*STEP 3:* Evaluate the given loading function $P(t)$ at time $t_{i+1}$ then calculate the $\ddot{x}_{i+1}$ at $t_{i+1}$ from equation (12.84a).

*STEP 4:* Determine $\dot{x}_{i+1}$ at $t_{i+1}$ using equation (12.79).

*STEP 5:* Determine $x_{i+1}$ at time $t_{i+1}$ using equation (12.80).

*STEP 6:* Repeat steps 3–5 for next time increment.

EXAMPLE 12.10

Rework example 12.9 using the trapezoidal rule method.

Solution

We begin by noting that initial conditions are $x_0 = 1$ and $\dot{x}_0 = 1$, then solve for the initial acceleration $\ddot{x}_0$ using equation (12.82). Thus

$$\ddot{x}_0 = \tfrac{1}{4}[-3(1) - 16(1)] = -4.75$$

We now solve for the $\beta$ values using $m = 4$, $c = 3$, $k = 16$, and $h = 0.1$:

$$\beta_0 = 4 + \frac{3(0.1)}{2} + \frac{16(0.1)^2}{4} = 4.19$$

$$\beta_1 = 4 - 4.19 = -0.19$$

$$\beta_2 = -3 - 16(0.1) = -4.6$$

$$\beta_3 = -16$$

Substituting into equation (12.84a) yields $x_{i+1}$ as follows:

$$\ddot{x}_{i+1} = \frac{1}{4.19}(-0.19\ddot{x}_i - 4.6\dot{x}_i - 16x_i) \qquad \textbf{(12.85a)}$$

Consequently, using equations (12.79) and (12.80), we have

$$\dot{x}_{i+1} = \dot{x}_i + \frac{h}{2}(\ddot{x}_i + \ddot{x}_{i+1}) \quad \textbf{(12.85b)}$$

$$x_{i+1} = x_i + \frac{h}{2}(\dot{x}_i + \dot{x}_{i+1}) \quad \textbf{(12.85c)}$$

The solution is now summarized using the initial conditions at time $t = 0$ ($i = 0$) and equations (12.85).

$t = 0.1$:

$$\ddot{x}_1 = \frac{1}{4.19}[-0.19(-4.75) - 4.6(1) - 16(1)] = -4.701074$$

$$\dot{x}_1 = 1 + \frac{0.1}{2}(-4.75 - 4.701074) = 0.5274463$$

$$x_1 = 1 + \frac{0.1}{2}(1 + 0.5274463) = 1.0753723$$

$t = 0.2$:

$$\ddot{x}_2 = \frac{1}{4.19}(-0.19x_1 - 4.6x_1 - 16x_1) = -4.476135$$

$$\dot{x}_2 = \dot{x}_1 + \frac{0.1}{2}(x_1 + x_2) = 0.0685858$$

$$x_2 = x_1 + \frac{0.1}{2}(\dot{x}_1 + \dot{x}_2) = 1.1061739$$

$t = 0.3$:

$$\ddot{x}_3 = \frac{1}{4.19}(-0.19\ddot{x}_2 - 4.6\dot{x}_2 - 16x_2) = -4.096374$$

$$\dot{x}_3 = \dot{x}_2 + \frac{0.1}{2}(\ddot{x}_2 + \ddot{x}_3) = -0.3600397$$

$$x_3 = x_2 + \frac{0.1}{2}(\dot{x}_2 + \dot{x}_3) = 1.091601$$

$t = 0.4$:

$$\ddot{x}_4 = \frac{1}{4.19}(-0.19\ddot{x}_3 - 4.6\dot{x}_3 - 16x_2) = -3.587381$$

$$\dot{x}_4 = \dot{x}_3 + \frac{0.1}{2}(\ddot{x}_3 + \ddot{x}_4) = -0.7442275$$

$$x_4 = x_3 + \frac{0.1}{2}(\dot{x}_3 + \dot{x}_4) = 1.036388$$

$t = 0.5$:

$$\ddot{x}_5 = \frac{1}{4.19}(-0.19\ddot{x}_4 - 4.6\dot{x}_4 - 16x_4) = -2.9778418$$

$$\dot{x}_5 = \dot{x}_3 + \frac{0.1}{2}(\ddot{x}_4 + \ddot{x}_5) = -1.072488$$

$$x_5 = x_4 + \frac{0.1}{2}(\dot{x}_4 + \dot{x}_5) = 0.9455521$$

The displacement values compare reasonably well, considering the size of the step used ($h = 0.1$), with the exact solution given by

$$x = \sqrt{1.25}\cos(2t - 0.46364761)$$

E

EXAMPLE 12.11

Develop the direct integration scheme using the trapezoidal rule for the electrical circuit shown in figure 12.13a.

Solution

The solution can be obtained directly by simply comparing the governing differential equations of the mechanical system with the electrical system. That is,

$$\text{Mechanical:} \quad m\ddot{x} + c\dot{x} + kx = P$$

$$\text{Electrical:} \quad L\ddot{I} + R\dot{I} + \frac{I}{c} = \dot{v}$$

It is evident that

$$m \equiv L$$

$$c \equiv R$$

$$k \equiv 1/c$$

$$P(t) \equiv \dot{v}$$

$$\ddot{x} \equiv \ddot{I}$$

$$\dot{x} \equiv \dot{I}$$

$$x \equiv I$$

Substituting into equations (12.82), (12.84), (12.79), and (12.80) yields the desired formulas for the electrical circuit, respectively:

$$\ddot{I}_i = \frac{1}{L}\left(\dot{v}_i - \frac{1}{c}I_i - R\dot{I}_i\right) \tag{12.86}$$

$$\ddot{I}_{i+1} = \frac{1}{\gamma_0}(\gamma_1\ddot{I}_i + \gamma_2\dot{I}_i + \gamma_3 I_i + \dot{v}_{i+1}) \tag{12.87a}$$

$$\gamma_0 = L + \frac{Rh}{2} + \frac{h^2}{4c} \tag{12.87b}$$

$$\gamma_1 = L - \gamma_0 \tag{12.87c}$$

$$\gamma_2 = -R - \frac{h}{c} \tag{12.87d}$$

$$\gamma_3 = -\frac{1}{c} \tag{12.87e}$$

$$\dot{I}_{i+1} = \dot{I}_i + \frac{h}{2}(\ddot{I}_i + \ddot{I}_{i+1}) \tag{12.88}$$

and

$$I_{i+1} = I_i + \frac{h}{2}(\dot{I}_i + \dot{I}_{i+1}) \tag{12.89}$$

Obviously, similar transformations are possible for other second-order differential equations. In fact, higher-order equations can be solved by using the technique outlined in this section. The solution given by equations (12.86)–(12.89) is stable for

$$h \leq \tfrac{1}{5}\pi\sqrt{LC}$$

This is obviously the same condition we applied relative to the mechanical system.

**E**

EXAMPLE 12.12

Determine the current for the following circuit at $t = 0.1$ if it is subjected to an initial current of $I_0 = 1$ A at $t = 0$.

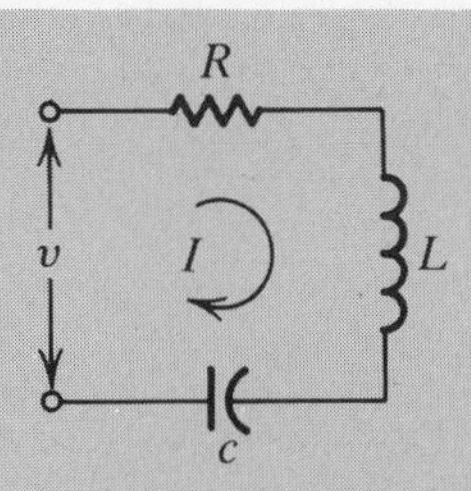


where

$$R = 1.8\ \Omega$$
$$L = 0.081\ \text{H}$$
$$c = 0.001\ \text{F}$$
$$v = 10\ \text{V}$$

## Solution

Obviously this problem can be easily solved analytically. However, let us see how the numerical solution predicts the behavior of the given circuit by comparing our results with the exact values.

In order to have some knowledge of electrical circuits, let us begin by defining the damping ratio $\alpha$ as follows:

$$\alpha = \frac{1}{2} R \sqrt{\frac{c}{L}} \tag{12.90}$$

This equation provides a way of determining whether a given circuit is damped, underdamped, critically damped, or overdamped. Thus

| | |
|---|---|
| $\alpha = 0$: | undamped circuit |
| $0 < \alpha < 1$: | underdamped circuit |
| $\alpha = 1$: | critically damped circuit |
| $\alpha > 1$: | overdamped circuit |

For our problem

$$\alpha = \frac{1}{2} \times 1.8 \sqrt{\frac{0.001}{0.081}} = 0.1$$

Obviously, the circuit in question is underdamped. Consequently, assuming

$$h \leq \frac{\pi}{10} \sqrt{Lc}$$

gives

$$h \leq 2.8274 \times 10^{-3}\ \text{s}$$

Therefore, taking $h = 2 \times 10^{-3}$ permits the determination of the system response. Thus we have the equations developed in example 12.11:

$$\gamma_0 = 0.081 + \frac{1.8(2 \times 10^{-3})}{2} + \frac{(2 \times 10^{-3})^2}{4(0.001)}$$

$$= 0.0838$$

$$\gamma_1 = -\left(\frac{1.8(2 \times 10^{-3})}{2} + \frac{(2 \times 10^{-3})^2}{4(0.001)}\right)$$

$$= -0.0028$$

$$\gamma_2 = -\left(1.8 + \frac{2 \times 10^{-3}}{0.001}\right)$$

$$= -3.8$$

$$\gamma_3 = -\frac{1}{0.001}$$

$$= -1000$$

Substituting these values into equation (12.87a) yields

$$\ddot{I}_{i+1} = \frac{-1}{8.38}(0.28\ddot{I}_i + 380\dot{I}_i + 100000I_i)$$

The initial second derivative is estimated for $\dot{I}(0) = 0$, $\dot{v}(0) = 0$, and $I(0) = 1$ by using equation (12.86):

$$\ddot{I}(0) = \frac{1}{L}\left(\dot{v}(0) - R\dot{I}(0) - \frac{1}{c}I(0)\right)$$

$$= \frac{1}{0.081}\left(0 - 1.8(0) - \frac{1}{0.001}(1)\right)$$

$$= -\tfrac{1}{81} \times 10^6$$

For $i = 0$, $t = 2 \times 10^{-3}$ s:

$$\ddot{I}_1 = -\frac{1}{8.38}\left[0.28\left(\frac{-1}{81} \times 10^6\right) + 380(0) + 10^5(1)\right]$$

$$= -11520.66944$$

Substituting $\ddot{I}_1$ into equation (12.88) allows us to compute $\dot{I}_1$:

$$\dot{I}_1 = \dot{I}_0 + \frac{h}{2}(\ddot{I}_0 + \ddot{I}_1)$$

$$= 0 + \frac{2 \times 10^{-3}}{2}\left(-\frac{1}{81} \times 10^6 - 11520.66944\right)$$

$$= -23.86634845$$

The current is computed next from equation (12.88):

$$I_1 = I_0 + \frac{h}{2}(\dot{I}_0 + \dot{I}_1)$$

$$= 1 + \frac{2 \times 10^{-3}}{2}(0 - 23.86634845)$$

$$= 0.9761336516$$

For $i = 2$, $t = 4 \times 10^{-3}$ s:

$$\ddot{I}_2 = -\frac{1}{8.38}(0.28\ddot{I}_1 + 380\dot{I}_1 + 100000I_1)$$

$$\dot{I}_2 = \dot{I}_1 + (\ddot{I}_1 + \ddot{I}_2)\tfrac{1}{1000}$$

$$I_2 = I_1 + (\dot{I}_1 + \dot{I}_2)\tfrac{1}{1000}$$

Obviously, $\ddot{I}_1$, $\dot{I}_1$, and $I_1$ have already been computed. Therefore one may proceed by computing as many values as necessary. It is evident that hand calculation is not realistic; therefore a computer program was used to evaluate the response up to $t = 0.1$ s. The results are given in table 12.7.

Table 12.7 COMPARISON OF EMPIRICAL AND EXACT CURRENTS FOR EXAMPLE 12.12

| INCREMENT | TIME (s) | CALCULATED $I$ | EXACT $I$ |
|---|---|---|---|
| 0 | 0 | 1.0 | 1.0 |
| 1 | 0.002 | 0.97613 | 0.97577 |
| 2 | 0.004 | 0.90670 | 0.90566 |
| 3 | 0.006 | 0.79697 | 0.79503 |
| 4 | 0.008 | 0.65391 | 0.65096 |
| 5 | 0.01 | 0.48578 | 0.48185 |
| 6 | 0.012 | 0.30169 | 0.29690 |
| 7 | 0.014 | 0.11111 | 0.10572 |
| 8 | 0.016 | −0.07659 | −0.08224 |
| 9 | 0.018 | −0.25257 | −0.25807 |
| 10 | 0.020 | −0.40893 | −0.41384 |
| 11 | 0.022 | −0.53906 | −0.54294 |
| 12 | 0.024 | −0.63787 | −0.64031 |
| 13 | 0.026 | −0.70198 | −0.70264 |
| 14 | 0.028 | −0.72983 | −0.72846 |
| 15 | 0.030 | −0.72165 | −0.71812 |
| 16 | 0.032 | −0.67938 | −0.67369 |
| 17 | 0.034 | −0.60649 | −0.59878 |
| 18 | 0.036 | −0.50778 | −0.49832 |
| 19 | 0.038 | −0.38908 | −0.37827 |
| 20 | 0.040 | −0.25690 | −0.24523 |
| 21 | 0.042 | −0.11814 | −0.10618 |
| 22 | 0.044 | 0.02030 | 0.03193 |

Table 12.7 *(Continued)*

| INCREMENT | TIME (s) | CALCULATED $I$ | EXACT $I$ |
|---|---|---|---|
| 23 | 0.046 | 0.15182 | 0.16250 |
| 24 | 0.048 | 0.27044 | 0.27958 |
| 25 | 0.050 | 0.37107 | 0.37812 |
| 26 | 0.052 | 0.44965 | 0.45419 |
| 27 | 0.054 | 0.50340 | 0.50510 |
| 28 | 0.056 | 0.53089 | 0.52951 |
| 29 | 0.058 | 0.53170 | 0.52738 |
| 30 | 0.060 | 0.50718 | 0.49998 |
| 31 | 0.062 | 0.45950 | 0.44971 |
| 32 | 0.064 | 0.39194 | 0.38000 |
| 33 | 0.066 | 0.30857 | 0.29504 |
| 34 | 0.068 | 0.21405 | 0.19958 |
| 35 | 0.070 | 0.11338 | 0.09867 |
| 36 | 0.072 | 0.01618 | 0.00259 |
| 37 | 0.074 | −0.08632 | −0.09934 |
| 38 | 0.076 | −0.17594 | −0.18210 |
| 39 | 0.078 | −0.25331 | −0.26204 |
| 40 | 0.080 | −0.31526 | −0.32112 |
| 41 | 0.082 | −0.35950 | −0.36219 |
| 42 | 0.084 | −0.38469 | −0.38404 |
| 43 | 0.086 | −0.39042 | −0.38648 |
| 44 | 0.088 | −0.37728 | −0.37022 |
| 45 | 0.090 | −0.34669 | −0.33685 |
| 46 | 0.092 | −0.30087 | −0.28873 |
| 47 | 0.094 | −0.24265 | −0.22882 |
| 48 | 0.096 | −0.17536 | −0.16050 |
| 49 | 0.098 | −0.10258 | −0.08744 |
| 50 | 0.10 | −0.02803 | −0.01334 |

*Analytical solution*

$$I_{\text{exact}} = e^{-\alpha wt}\left(\frac{\alpha I(0)}{1-\alpha^2}\sin(w_D t) + \cos w_D t\right)$$

$$\alpha = 0.10$$

$$I(0) = 1$$

$$w_D = (1-\alpha^2)\sqrt{\frac{1}{Lc}} = 110.55$$

$$w = \frac{1}{\sqrt{Lc}} = 111.11$$

While it is obvious that there are errors involved with the numerical procedures, desired accuracies can be achieved by reducing the time interval. The advantage of the numerical solution besides its adaptability to generalized $v$ is that it is not dependent on whether the system is undamped, underdamped, critically damped, or overdamped. This is not the case with analytical solutions. The results for this underdamped system are given in figure 12.15.

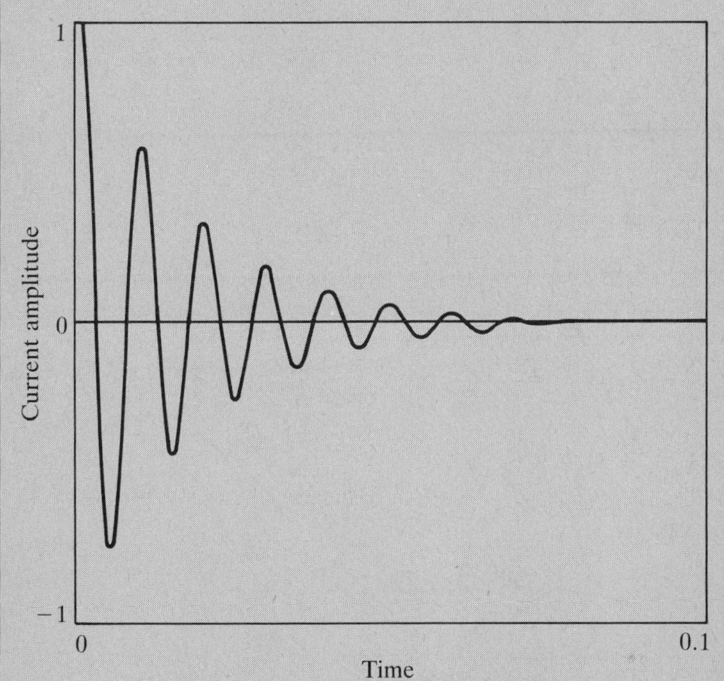


**Figure 12.15** Plot of current versus time for example 12.12.

E

EXAMPLE 12.13

Rework example 12.12 if the resistor is replaced by one with $R = 18\ \Omega$.

### Solution

For illustration purposes, let us investigate the damping ratio for the circuit:

$$\alpha = \frac{1}{2}(18)\sqrt{\frac{0.001}{0.081}} = 1.0$$

Therefore the circuit is critically damped. This means that the response $I$ will not oscillate. The solution procedure does not need to be altered. The solution is summarized in table 12.8. A summary of preliminary calculations is given below:

$\gamma_0 = 0.1 \qquad \gamma_2 = -20.0$

$\gamma_1 = -0.019 \qquad \gamma_3 = -1000$

Note that only the $a$'s change, but not the general equations used in the solution.

Table 12.8 COMPARISON OF EMPIRICAL AND EXACT CURRENT FOR EXAMPLE 12.13

| INCREMENT | TIME (s) | CALCULATED $I$ | EXACT $I$ |
|---|---|---|---|
| 0 | 0 | 1 | 1 |
| 1 | 0.002 | 0.98000 | 0.97868 |
| 2 | 0.004 | 0.92800 | 0.92615 |
| 3 | 0.006 | 0.85760 | 0.85570 |
| 4 | 0.008 | 0.77824 | 0.77655 |
| 5 | 0.010 | 0.69632 | 0.69496 |
| 6 | 0.012 | 0.61604 | 0.61506 |
| 7 | 0.014 | 0.54002 | 0.53941 |
| 8 | 0.016 | 0.46976 | 0.46948 |
| 9 | 0.018 | 0.40600 | 0.40601 |
| 10 | 0.020 | 0.34897 | 0.34919 |
| 11 | 0.022 | 0.29850 | 0.29889 |
| 12 | 0.024 | 0.25426 | 0.25477 |
| 13 | 0.026 | 0.21578 | 0.21637 |
| 14 | 0.028 | 0.18252 | 0.18316 |
| 15 | 0.030 | 0.15393 | 0.15459 |
| 16 | 0.032 | 0.12948 | 0.13013 |
| 17 | 0.034 | 0.10865 | 0.10928 |
| 18 | 0.036 | 0.09097 | 0.09159 |
| 19 | 0.038 | 0.07602 | 0.07659 |
| 20 | 0.040 | 0.06341 | 0.06394 |
| 21 | 0.042 | 0.05280 | 0.05327 |
| 22 | 0.044 | 0.04390 | 0.04434 |
| 23 | 0.046 | 0.03645 | 0.03685 |
| 24 | 0.048 | 0.03022 | 0.03058 |
| 25 | 0.050 | 0.02503 | 0.02534 |

*Analytical solution*

$$I_{\text{exact}} = [I(0) + wt]e^{-wt}$$

$$w = \sqrt{\frac{1}{Lc}} = \sqrt{\frac{1}{0.081 \times 0.001}} = \frac{1000}{9} \text{ rad/s}$$

$$I(0) = 1.0$$

A plot of the current versus time is shown in figure 12.16.

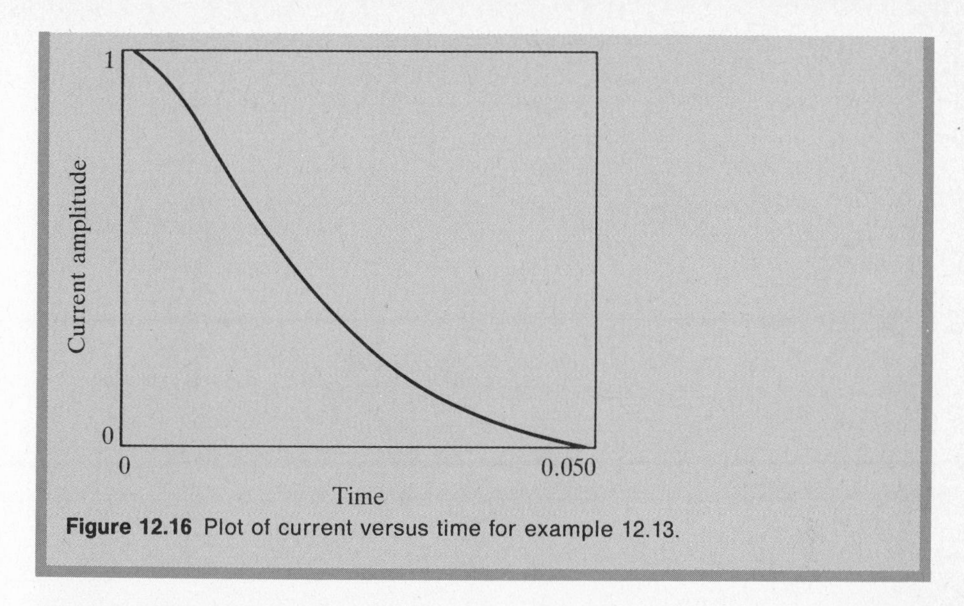

**Figure 12.16** Plot of current versus time for example 12.13.

**E**

EXAMPLE 12.14

Determine the displacement for the following *m-c-k* system at $t = 2.5$ s. Assume $\ddot{x}(0) = \dot{x}(0) = x(0) = 0$

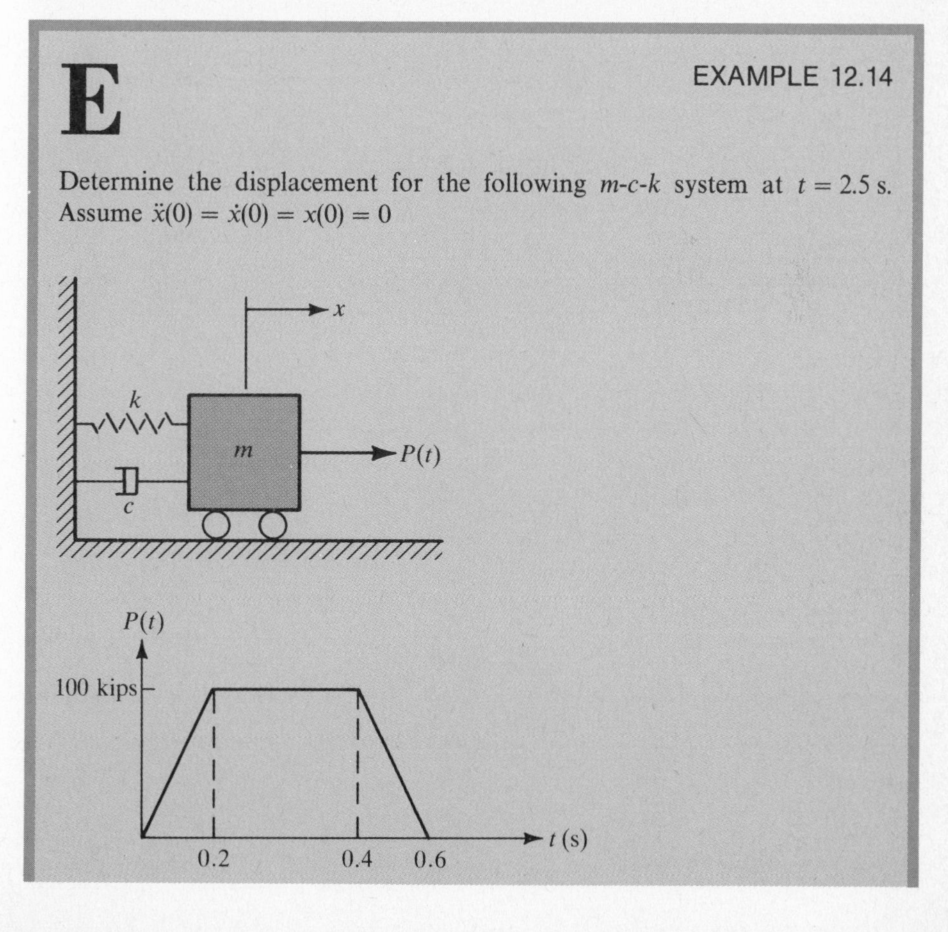

where

$$k = 27 \text{ kips/in.}$$

$$m = 3 \text{ kips s}^2\text{/in.}$$

$$c = 0.9 \text{ kips in./s}$$

## Solution

In order to gain a basic understanding of mechanical system behavior, let us begin by studying the damping characteristics of such systems. The damping ratio is defined in precisely the same way as that used for electrical circuits. However, to distinguish between the two models, we shall use a different variable name:

$$\xi = \frac{c}{2\sqrt{mk}}$$

Hence

| | |
|---|---|
| $\xi = 0$: | undamped system |
| $0 < \xi < 1$: | underdamped system |
| $\xi = 1$: | critically damped system |
| $\xi > 1$: | overdamped system |

For our present example problem

$$\xi = \frac{0.9}{2\sqrt{(27)(3)}} = 0.05$$

Therefore the system is underdamped. The implication is that the response is oscillatory in nature. Now calculate an appropriate time interval:

$$h \le \frac{\pi}{10}\sqrt{\frac{m}{k}} = 0.10471 \text{ s}$$

Take $h = 0.05$ s. The $\beta$ values can now be computed by using equation (12.84):

$$\beta_0 = 3 + \frac{(0.9)(0.05)}{2} + \frac{(27)(0.05)^2}{4} = 3.039375$$

$$\beta_1 = -\left(\frac{(0.9)(0.05)}{2} + \frac{(27)(0.05)^2}{4}\right) = -0.039375$$

$$\beta_2 = -[0.9 + (27)(0.05)] = -2.25$$

$$\beta_3 = -27$$

Substituting these values into equations (12.84a), (12.79), and (12.80) and noting that the external force is not zero gives

$$\ddot{x}_{i+1} = \frac{1}{3.039375}(-0.039375\ddot{x}_1 - 2.25\dot{x}_i - 27x_i + P_{i+1})$$

$$\dot{x}_{i+1} = \dot{x}_i + \frac{0.05}{2}(\ddot{x}_i + \ddot{x}_{i+1})$$

$$x_{i+1} = x_i + \frac{0.05}{2}(\dot{x}_i + \dot{x}_{i+1})$$

For $i = 0$, $t = 0.05$:

$$P(0.05) = 25 \text{ kips}$$

$$\ddot{x}_1 = \frac{1}{3.03937}(-0 - 0 - 0 + 25)$$

$$= 8.2253888 \text{ in./s}^2$$

$$\dot{x}_1 = 0 + \frac{0.05}{2}(0 + 8.2253888)$$

$$= 0.20563 \text{ in./s}$$

$$x_1 = 0 + \frac{0.05}{2}(0 + 0.20563)$$

$$= 5.1408 \times 10^{-3} \text{ in.}$$

For $i = 1$, $t = 0.10$:

$$P(0.1) = 50 \text{ kips}$$

$$\ddot{x}_2 = \frac{1}{3.039375}(-0.039375\ddot{x}_1 - 2.25\dot{x}_1 - 27x_1 + P_2)$$

$$\dot{x}_2 = \dot{x}_1 + \frac{0.05}{2}(\ddot{x}_1 + \ddot{x}_2)$$

$$x_2 = x_1 + \frac{0.05}{2}(\dot{x}_1 + \dot{x}_2)$$

It is evident that a computer is needed to carry out the computation. Table 12.9 gives the displacement versus time. Figure 12.17 provides a graphical illustration of the displacement of the *m-c-k* system with time.

At this stage it should be emphasized that structures subjected to wind loads, earthquakes, and blasts are often analyzed in terms of the *m-c-k* system.

Table 12.9 SUMMARY OF DISPLACEMENT VERSUS TIME FOR EXAMPLE 12.14

| | INCREMENT | TIME (s) | DISPLACEMENT CALCULATED | INCREMENT | TIME (s) | DISPLACEMENT CALCULATED |
|---|---|---|---|---|---|---|
| Forced vibration $[P(t) \neq 0]$ | 0 | 0 | 0 | 26 | 1.30 | 0.56041 |
| | 1 | 0.05 | 0.00514 | 27 | 1.35 | 0.032760 |
| | 2 | 0.10 | 0.03065 | 28 | 1.40 | −0.48781 |
| | 3 | 0.15 | 0.09623 | 29 | 1.45 | −0.98983 |
| | 4 | 0.20 | 0.22040 | 30 | 1.50 | −1.46244 |
| | 5 | 0.25 | 0.41495 | 31 | 1.55 | −1.89557 |
| | 6 | 0.30 | 0.67965 | 32 | 1.60 | −2.28019 |
| | 7 | 0.35 | 1.00759 | 33 | 1.65 | −2.60848 |
| | 8 | 0.40 | 1.39055 | 34 | 1.70 | −2.87398 |
| | 9 | 0.45 | 1.81408 | 35 | 1.75 | −3.07172 |
| | 10 | 0.50 | 2.25274 | 36 | 1.80 | −3.19832 |
| | 11 | 0.55 | 2.67599 | 37 | 1.85 | −3.25201 |
| | 12 | 0.60 | 3.05412 | 38 | 1.90 | −3.23268 |
| Free vibration $[P(t) = 0]$ | 13 | 0.65 | 3.36396 | 39 | 1.95 | −3.14185 |
| | 14 | 0.70 | 3.59450 | 40 | 2.00 | −2.98260 |
| | 15 | 0.75 | 3.741810 | 41 | 2.05 | −2.75943 |
| | 16 | 0.80 | 3.80383 | 42 | 2.10 | −2.47831 |
| | 17 | 0.85 | 3.78045 | 43 | 2.15 | −2.14631 |
| | 18 | 0.90 | 3.67347 | 44 | 2.20 | −1.77155 |
| | 19 | 0.95 | 3.48649 | 45 | 2.25 | −1.36301 |
| | 20 | 1.00 | 3.22484 | 46 | 2.30 | −0.93023 |
| | 21 | 1.05 | 2.89545 | 47 | 2.35 | −0.48321 |
| | 22 | 1.10 | 2.50663 | 48 | 2.40 | −0.03208 |
| | 23 | 1.15 | 2.06790 | 49 | 2.45 | 0.41308 |
| | 24 | 1.20 | 1.58974 | 50 | 2.50 | 0.84249 |
| | 25 | 1.25 | 1.08336 | | | |

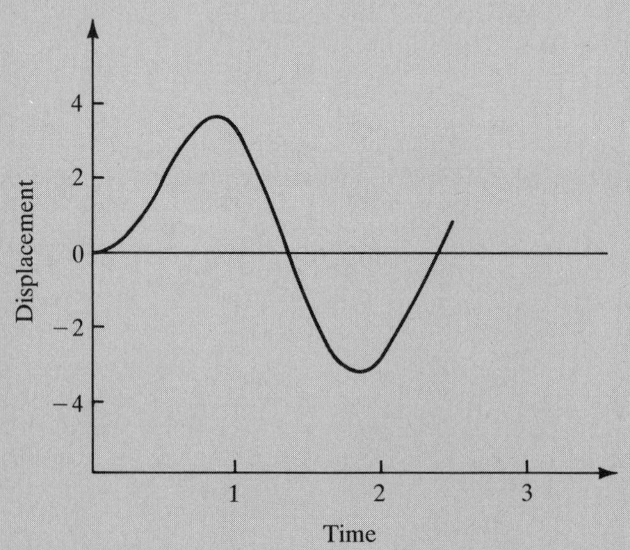


**Figure 12.17** Displacement versus time for example 12.14.

## 12.11 INITIAL-VALUE PROBLEMS INVOLVING FIRST-ORDER SYSTEMS

Many engineering problems involve the solution of a set of first-order differential equations. In this section, the direct procedure using the trapezoidal rule is developed for a system of first-order ordinary differential equations.

To illustrate this technique, consider the DC electric motor pictured below:

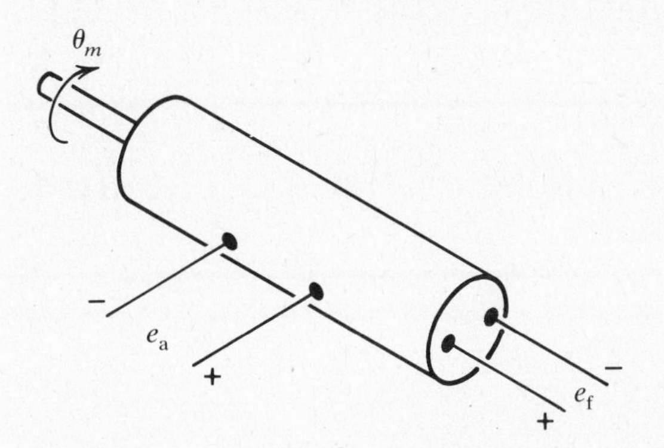


where $e_a(t)$ is the voltage applied to the armature winding, $e_f(t)$ is the voltage applied to the field coil, and $\theta_m(t)$ is the angular position of the armature shaft.

A set of four first-order differential equations governs the motor's behavior. These are

$$L_f \dot{I}_f + R_f I_f = e_f \tag{12.91a}$$

$$L_a \dot{I}_a + R_a I_a + k_b W_m = e_a \tag{12.91b}$$

$$J_m \dot{W}_m + B_m W_m - k_i I_a I_f = 0 \tag{12.91c}$$

$$\dot{\theta}_m - W_m = 0 \tag{12.91d}$$

Obviously, this set is nonlinear in that equation (12.91c) contains the terms $k_i I_a I_f$. Consequently the set is linearized by assuming either $I_a$ and $I_f$ to be constant. Therefore, assuming $I_a = c_1 = \text{constant}$ gives

$$\begin{bmatrix} L_f & 0 & 0 \\ 0 & J_m & 0 \\ 0 & 0 & 1 \end{bmatrix} \begin{Bmatrix} \dot{I}_f \\ \dot{W}_m \\ \dot{\theta}_m \end{Bmatrix} + \begin{bmatrix} R_f & 0 & 0 \\ -c_1 k & B_m & 0 \\ 0 & -1 & 0 \end{bmatrix} \begin{Bmatrix} I_f \\ W_m \\ \theta_m \end{Bmatrix} = \begin{Bmatrix} e_f \\ 0 \\ 0 \end{Bmatrix} \tag{12.92a}$$

This set can be expressed in compact matrix form as

$$[A]\{\dot{x}\}_i + [B]\{x\}_i = \{E\}_i \tag{12.92b}$$

Integrating using the trapezoidal rule, we obtain

$$\{x\}_1 = \{x\}_0 + \frac{h}{2}[\{\dot{x}\}_0 + \{\dot{x}\}_1] \tag{12.93}$$

Substituting equation (12.93) into (12.92b) and solving for the unknown vector $\{\dot{x}\}_1$ (note that $i = 1$) yields

$$[A]\{\dot{x}\}_1 + [B]\left[\{x\}_0 + \frac{h}{2}[\{\dot{x}\}_0 + \{\dot{x}\}_1]\right] = \{E\}_1$$

$$\{\dot{x}\}_1 = \left[[A] + \frac{h}{2}[B]\right]^{-1}\left[\{E\}_1 - [B]\{x\}_0 - \frac{h}{2}[B]\{\dot{x}\}_0\right]$$

In general,

$$\{\dot{x}\}_i = \left[[A] + \frac{h}{2}[B]\right]^{-1}\left[\{E\}_{i-1} - [B]\{x\}_{i-1} - \frac{h}{2}[B]\{\dot{x}\}_{i-1}\right] \tag{12.94}$$

where

$$\{\dot{x}\}_i = \begin{Bmatrix} \dot{I}_{fi} \\ \dot{W}_{mi} \\ \theta_{mi} \end{Bmatrix}$$

$$\{\dot{x}\}_{i-1} = \begin{Bmatrix} \dot{I}_{fi-1} \\ \dot{W}_{mi-1} \\ \dot{\theta}_{mi-1} \end{Bmatrix}$$

$$\{x\}_{i-1} = \begin{Bmatrix} I_{fi-1} \\ W_{mi-1} \\ \theta_{mi-1} \end{Bmatrix}$$

$$[A] = \begin{bmatrix} L_f & 0 & 0 \\ 0 & J_m & 0 \\ 0 & 0 & 1 \end{bmatrix}$$

$$[B] = \begin{bmatrix} R_f & 0 & 0 \\ -kc_1 & B_m & 0 \\ 0 & -1 & 0 \end{bmatrix}$$

$$\{E\}_i = \begin{Bmatrix} e_{fi} \\ 0 \\ 0 \end{Bmatrix}$$

The $\{x\}$ vector is determined by expressing equation (12.93) in matrix form as follows:

$$\{x\}_i = \{x\}_{i-1} + \frac{h}{2}[\{\dot{x}\}_{i-1} + \{\dot{x}\}_i] \tag{12.95}$$

where $h$ is an appropriate time interval.

Given the initial conditions, one can determine $\{\dot{x}\}_1$ and $\{x\}_1$ from equations (12.94) and (12.95), respectively. The procedure is continued until the total response is determined.

## 12.12 INITIAL-VALUE PROBLEMS INVOLVING SECOND-ORDER SYSTEMS

Nuclear power plants, spaceships, multistory buildings, automobiles, and electrical circuits are all systems whose design depends on solving a set of uncoupled second- and possibly higher-order differential equations. Some of these systems were first introduced in chapter 3. The resulting set of differential equations for a given system is often uncoupled. That is, any one of the differential equations involves more than one independent variable. Consequently, the object is to decouple the system. Generally, there are four different models that one may use in analyzing such systems. For an *m-c-k* system, these are summarized as follows.

Model 1: Undamped Free Response

$$[m]\{\ddot{x}\} + [k]\{x\} = \{0\} \tag{12.96a}$$

Model 2: Damped Free Response

$$[m]\{\ddot{x}\} + [c]\{\dot{x}\} + [k]\{x\} = \{0\} \tag{12.96b}$$

Model 3: Undamped Forced Response

$$[m]\{\ddot{x}\} + [k]\{x\} = \{P\} \tag{12.96c}$$

Model 4: Damped Forced Response

$$[m]\{\ddot{x}\} + [c]\{\dot{x}\} + [k]\{x\} = \{P\} \tag{12.96d}$$

In this section, two methods of solution are described. The first involves decoupling of the system of second-order equations into an equivalent set of independent second-order equations that can be solved individually by using the methods outlined earlier in this chapter. The second involves the application of the trapezoidal rule of integration.

### 12.12.1 Decoupling Procedure

Obviously, the model given by equation (12.96d) is the most general and therefore the most complex. The natural frequencies (eigenvalues) and the corresponding nodes (eigenvectors) are determined by using equation (12.96a) regardless of the model being used. For undamped free response, the displacement is simple harmonic, and may be expressed as

$$\{x\} = \{A\}\sin(wt + \alpha) \tag{12.97}$$

where

$\{A\}$ = amplitude vector

$w$ = natural frequency

$\alpha$ = phase angle

Therefore the acceleration is determined as

$$\{\ddot{x}\} = \{A\}(-w^2)\sin(wt + \alpha) \tag{12.98}$$

The natural frequencies and their corresponding modes are determined by substituting equations (12.98) and (12.97) into (12.96a):

$$-w^2[m]\{A\}\sin(wt + \alpha) + [k]\{A\}\sin(wt + \alpha) = \{0\}$$

Simplifying yields

$$[[k] - w^2[m]]\{A\} = \{0\} \tag{12.99}$$

It is clear that equation (12.99) is an eigenvalue problem. Therefore

$$|[k] - w^2[m]| = 0 \tag{12.100}$$

Equation (12.100) is called the frequency equation of the system. Corresponding to every frequency there is an eigenvector $\{\Phi\}$. These vectors can be assembled in matrix form as follows:

$$[\Phi] = [\{\Phi\}_1\{\Phi\}_2 \cdots \{\Phi\}_n] \tag{12.101}$$

where

$\{\Phi\}_1$ = first eigenvector

$\vdots$

$\{\Phi\}_n$ = $n$th eigenvector

Note that the eigenvectors must be normalized so that the largest number in any vector is 1.0. Using the orthogonality condition, one can show that

$$[\bar{m}] = [\Phi]^t[m][\Phi] \tag{12.102}$$

$$[\bar{k}] = [\Phi]^t[k][\Phi] \tag{12.103}$$

where

$[\bar{m}]$ = modal mass matrix and is diagonal

$[\bar{k}]$ = modal stiffness matrix and is diagonal

The implication is that the model given by (12.96a) can be decoupled. This can be accomplished by assuming the following transformation:

$$\{x\} = [\Phi]\{z\} \tag{12.104}$$

The acceleration vector is then given as

$$\{\ddot{x}\} = [\Phi]\{\ddot{z}\} \tag{12.105}$$

where

$\{z\}$ = modal displacement vector

$\{\ddot{z}\}$ = modal acceleration vector

Substituting equations (12.105) and (12.104) into (12.96a) gives

$$[m][\Phi]\{\ddot{z}\} + [k]\{\Phi\}\{z\} = \{0\} \tag{12.106}$$

Premultiplying (12.106) by $[\Phi]^t$ yields

$$[\Phi]^t[m][\Phi]\{\ddot{z}\} + [\Phi]^t[k][\Phi]\{z\} = \{0\} \tag{12.107}$$

Substituting equations (12.102) and (12.103) into (12.107) yields

$$[\bar{m}]\{\ddot{z}\} + [\bar{k}]\{z\} = \{0\} \tag{12.108}$$

Equation (12.108) is a decoupled differential equation. That is,

$$\begin{aligned} \bar{m}_1\ddot{z}_1 + \bar{k}_1 z_1 &= 0 \\ \bar{m}_2\ddot{z}_2 + \bar{k}_2 z_2 &= 0 \\ &\vdots \\ \bar{m}_n\ddot{z}_n + \bar{k}_n z_n &= 0 \end{aligned} \tag{12.109}$$

These can be solved individually for $z_1, \ldots, z_n$ in terms of modal initial conditions $z(0)$ and $\dot{z}(0)$:

$$\{z(0)\} = [\Phi]^{-1}\{x(0)\} \tag{12.110a}$$

$$\{\dot{z}(0)\} = [\Phi]^{-1}\{\dot{x}(0)\} \tag{12.110b}$$

The actual displacement vector is then determined by using equation (12.104).

**E** EXAMPLE 12.15

Determine the response of the following system:

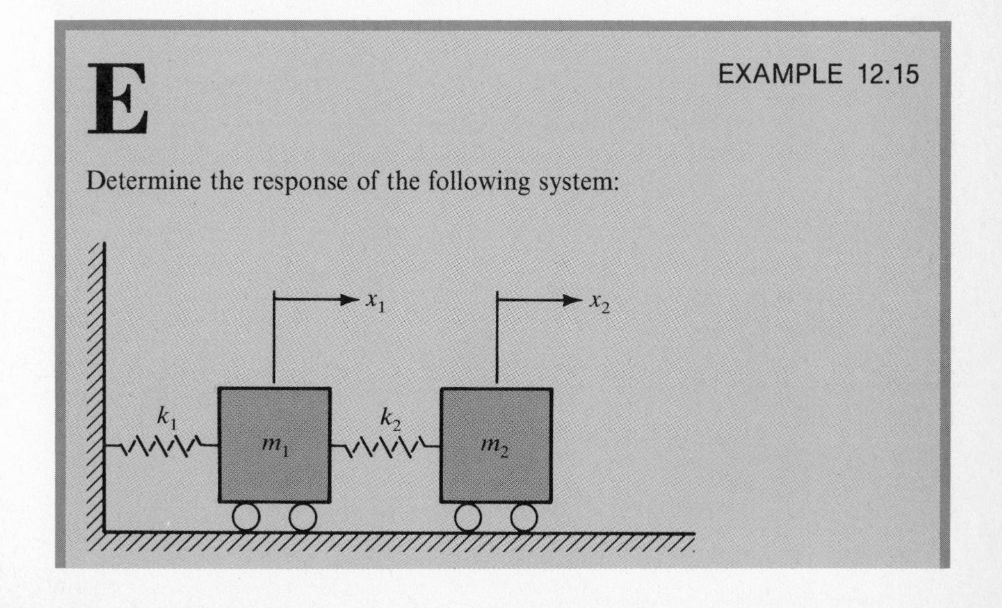

where

$$m_1 = 4 \frac{\text{kips-s}^2}{\text{in.}}$$

$$m_2 = 1 \frac{\text{kips-s}^2}{\text{in.}}$$

$$k_1 = 3 \frac{\text{kips}}{\text{in.}}$$

$$k_2 = 1$$

$$x_1(0) = 2 \text{ in.}$$

$$x_2(0) = -2 \text{ in.}$$

$$\dot{x}_1(0) = \dot{x}_2(0) = 0$$

## Solution

Expressing the model response in matrix form:

$$\begin{bmatrix} 4 & 0 \\ 0 & 1 \end{bmatrix} \begin{Bmatrix} \ddot{x}_1 \\ \ddot{x}_2 \end{Bmatrix} + \begin{bmatrix} 4 & -1 \\ -1 & 1 \end{bmatrix} \begin{Bmatrix} x_1 \\ x_2 \end{Bmatrix} = \begin{Bmatrix} 0 \\ 0 \end{Bmatrix}$$

Substituting the [k] and [m] matrices into equation (12.100) permits the determination of the frequencies:

$$\left| \begin{bmatrix} 4 & -1 \\ -1 & 1 \end{bmatrix} - w^2 \begin{bmatrix} 4 & 0 \\ 0 & 1 \end{bmatrix} \right| = 0$$

or

$$\begin{vmatrix} 4 - 4w^2 & -1 \\ -1 & 1 - w^2 \end{vmatrix} = 0$$

Expanding yields

$$(1 - w^2)^2 = \tfrac{1}{4}, \qquad w^2 = 1 \pm \tfrac{1}{2}$$

The first and second eigenvectors (frequencies) are given as

$$w_1 = \frac{1}{\sqrt{2}} \qquad \text{(first eigenvalue)}$$

$$w_2 = \sqrt{\tfrac{3}{2}} \qquad \text{(second eigenvalue)}$$

The corresponding vectors are determined by using equation (12.99). Thus

$$\left[\begin{bmatrix} 4 & -1 \\ -1 & 1 \end{bmatrix} - \frac{1}{2}\begin{bmatrix} 4 & 0 \\ 0 & 1 \end{bmatrix}\right]\begin{Bmatrix} A_1 \\ A_2 \end{Bmatrix} = \begin{Bmatrix} 0 \\ 0 \end{Bmatrix}$$

$$\begin{bmatrix} 2 & -1 \\ -1 & \frac{1}{2} \end{bmatrix}\begin{Bmatrix} A_1 \\ A_2 \end{Bmatrix} = \begin{Bmatrix} 0 \\ 0 \end{Bmatrix}$$

Note that the two equations are exactly the same, as they should be (why?). Hence

$$2A_1 - A_2 = 0$$

$$A_1 = \tfrac{1}{2}A_2$$

Since the eigenvectors are arbitrarily defined, assume $A_2 = 2$; then

$$\{\Phi\}_1 = \begin{Bmatrix} \frac{1}{2} \\ 1 \end{Bmatrix} \qquad \text{(first eigenvector)}$$

Similarly, the second eigenvector is determined thus

$$\{\Phi\}_2 = \begin{Bmatrix} \frac{1}{2} \\ -1 \end{Bmatrix} \qquad \text{(second eigenvector)}$$

These vectors are now assembled to give

$$[\Phi] = [\{\Phi\}_1 \quad \{\Phi\}_2]$$

$$= \begin{bmatrix} \frac{1}{2} & \frac{1}{2} \\ 1 & -1 \end{bmatrix}$$

The modal mass is determined by using equation (12.102):

$$[\bar{m}] = \begin{bmatrix} \frac{1}{2} & 1 \\ \frac{1}{2} & -1 \end{bmatrix}\begin{bmatrix} 4 & 0 \\ 0 & 1 \end{bmatrix}\begin{bmatrix} \frac{1}{2} & \frac{1}{2} \\ 1 & -1 \end{bmatrix}$$

$$= \begin{bmatrix} 2 & 0 \\ 0 & 2 \end{bmatrix}$$

The modal stiffness is determined by using equation (12.103):

$$[\bar{k}] = \begin{bmatrix} \frac{1}{2} & 1 \\ \frac{1}{2} & -1 \end{bmatrix}\begin{bmatrix} 4 & -1 \\ -1 & 1 \end{bmatrix}\begin{bmatrix} \frac{1}{2} & \frac{1}{2} \\ 1 & -1 \end{bmatrix}$$

$$= \begin{bmatrix} 1 & 0 \\ 0 & 3 \end{bmatrix}$$

Note that both $\bar{m}$ and $\bar{k}$ are diagonal. Consequently, substituting these matrices into equation (12.108) yields two decoupled second-order differential equations in $z$:

$$\begin{bmatrix} 2 & 0 \\ 0 & 2 \end{bmatrix} \begin{Bmatrix} \ddot{z}_1 \\ \ddot{z}_2 \end{Bmatrix} + \begin{bmatrix} 1 & 0 \\ 0 & 3 \end{bmatrix} \begin{Bmatrix} z_1 \\ z_2 \end{Bmatrix} = \begin{Bmatrix} 0 \\ 0 \end{Bmatrix} \tag{12.111}$$

Note that the modal response vector $\{z\}$ is related to the actual response vector $\{x\}$ by equation (12.104). That is,

$$\begin{Bmatrix} x_1 \\ x_2 \end{Bmatrix} = \begin{bmatrix} \frac{1}{2} & \frac{1}{2} \\ 1 & -1 \end{bmatrix} \begin{Bmatrix} z_1 \\ z_2 \end{Bmatrix}$$

Equations (12.111) can be expressed and solved individually as follows:

$$2\ddot{z}_1 + z_1 = 0 \tag{12.112a}$$

$$2\ddot{z}_2 + 3z_2 = 0 \tag{12.112b}$$

These equations can now be solved analytically as follows. Assume

$$z_1 = Ge^{st}$$

then

$$\ddot{z}_1 = Gs^2e^{st}$$

Substituting into (12.112a) yields

$$2Gs^2e^{st} + Ge^{st} = 0$$

or more simply

$$2s^2 + 1 = 0$$

$$s = \pm\frac{1}{\sqrt{2}}j$$

The solution to equation (12.112a) is now given as

$$z_1 = G_1e^{t/\sqrt{2}j} + G_2e^{-t/\sqrt{2}j}$$

Using the Euler identity, we have

$$z_1 = G_1\left(\cos\frac{t}{\sqrt{2}} + j\sin\frac{t}{\sqrt{2}}\right) + G_2\left(\cos\frac{t}{\sqrt{2}} - j\sin\frac{t}{\sqrt{2}}\right)$$

Combining terms, we have

$$z_1 = A\cos\frac{t}{\sqrt{2}} + B\sin\frac{t}{\sqrt{2}}$$

The constants $A$ and $B$ are determined from the initial conditions:

$$z_1(0) = A\cos 0 + B\sin 0$$

$$\dot{z}_1(0) = \frac{-A}{\sqrt{2}}\sin 0 + \frac{B}{\sqrt{2}}\cos 0$$

Solving for $A$ and $B$ gives

$$A = z_1(0), \qquad B = \sqrt{2}\dot{z}_1(0)$$

Therefore the solution to equation (12.112a) is given as

$$z_1 = z_1(0)\cos\frac{t}{\sqrt{2}} + \sqrt{2}\dot{z}_1(0)\sin\frac{t}{\sqrt{2}}$$

The solution to equation (12.112b) is obtained similarly:

$$z_2 = z_2(0)\cos\sqrt{\tfrac{3}{2}}t + \sqrt{\tfrac{2}{3}}\dot{z}_2(0)\sin\sqrt{\tfrac{3}{2}}t$$

Since initial conditions are given in terms of $\{x(0)\}$ and $\{\dot{x}(0)\}$, we need to transform them to get $\{z(0)\}$ and $\{\dot{z}(0)\}$. This is accomplished by using equations (12.109) and (12.110), respectively. Thus

$$\begin{Bmatrix} z_1(0) \\ z_2(0) \end{Bmatrix} = \begin{bmatrix} \frac{1}{2} & \frac{1}{2} \\ 1 & -1 \end{bmatrix}^{-1} \begin{Bmatrix} x_1(0) \\ x_2(0) \end{Bmatrix}$$

or more simply

$$\begin{Bmatrix} z_1(0) \\ z_2(0) \end{Bmatrix} = \begin{bmatrix} 1 & \frac{1}{2} \\ 1 & -\frac{1}{2} \end{bmatrix} \begin{Bmatrix} 2 \\ -2 \end{Bmatrix} = \begin{Bmatrix} 1 \\ 3 \end{Bmatrix}$$

The modal velocities are both zero. The modal displacement functions are now simplified to give

$$z_1 = \cos\frac{1}{\sqrt{2}}t$$

$$z_2 = 3\cos\sqrt{\frac{3}{2}}t$$

The actual displacement vector is given by equation (12.104) as

$$\begin{Bmatrix} x_1 \\ x_2 \end{Bmatrix} = \begin{bmatrix} \dfrac{1}{2} & \dfrac{1}{2} \\ 1 & -1 \end{bmatrix} \begin{Bmatrix} \cos\dfrac{1}{\sqrt{2}}t \\ 3\cos\sqrt{\dfrac{2}{3}}t \end{Bmatrix}$$

Multiplying the column vector by the coefficient matrix yields the desired closed-form solution:

$$x_1 = \frac{1}{2}\cos\frac{1}{\sqrt{2}}t + \frac{3}{2}\cos\sqrt{\frac{3}{2}}t$$

$$x_2 = \cos\frac{1}{\sqrt{2}}t - 3\cos\sqrt{\frac{3}{2}}t$$

Note that the solution satisfies the initial conditions. It is evident that when possible this solution technique is far superior to the traditional numerical solutions.

E

EXAMPLE 12.16

If in the previous example loads $P_1(t)$ and $P_2(t)$ were placed at masses 1 and 2, respectively, determine the set of uncoupled differential equations describing the response.

## Solution

Expressing the response in matrix form gives

$$\begin{bmatrix} 4 & 0 \\ 0 & 1 \end{bmatrix} \begin{Bmatrix} \ddot{x}_1 \\ \ddot{x}_2 \end{Bmatrix} + \begin{bmatrix} 4 & -1 \\ -1 & 1 \end{bmatrix} \begin{Bmatrix} x_1 \\ x_2 \end{Bmatrix} = \begin{Bmatrix} P_1 \\ P_2 \end{Bmatrix}$$

or in compact matrix form

$$[m]\{\ddot{x}\} + [k]\{x\} = \{P\}$$

Expressing the equation in terms of the modal response gives

$$[m][\Phi]\{\ddot{z}\} + [k][\Phi]\{z\} = \{P\}$$

Premultiplying by $[\Phi]^t$ gives

$$[\Phi]^t[m][\Phi]\{\ddot{z}\} + [\Phi]^t[k][\Phi]\{z\} = [\Phi]^t\{P\}$$

Simplifying yields

$$[\bar{m}]\{\ddot{z}\} + [\bar{k}]\{z\} = [\Phi]^t\{P\}$$

Substituting the various matrices, we obtain

$$\begin{bmatrix} 2 & 0 \\ 0 & 2 \end{bmatrix} \begin{Bmatrix} \ddot{z}_1 \\ \ddot{z}_2 \end{Bmatrix} + \begin{bmatrix} 1 & 0 \\ 0 & 3 \end{bmatrix} \begin{Bmatrix} z_1 \\ z_2 \end{Bmatrix} = \begin{bmatrix} \frac{1}{2} & 1 \\ \frac{1}{2} & -1 \end{bmatrix} \begin{Bmatrix} P_1 \\ P_2 \end{Bmatrix}$$

or more simply

$$2\ddot{z}_1 + z_1 = \tfrac{1}{2}P_1 + P_2$$

$$2\ddot{z}_2 + z_2 = \tfrac{1}{2}P_1 - P_2$$

Once again these are decoupled second-order differential equations. Consequently, the modal response can be determined if $P_1$ and $P_2$ are specified. The displacements are then determined as

$$\{x\} = [\Phi]\{z\}$$

E

EXAMPLE 12.17

Rework example 12.16 if damping is included as follows:

$$C_1 = 0.3\,\frac{\text{kips s}}{\text{in.}}$$

$$C_2 = 0.1\,\frac{\text{kips s}}{\text{in.}}$$

### Solution

Clearly this problem is the most general case that can be expected in practice relative to *m-c-k* systems. Hence

$$\begin{bmatrix} 4 & 0 \\ 0 & 1 \end{bmatrix}\begin{Bmatrix} \ddot{x}_1 \\ \ddot{x}_2 \end{Bmatrix} + \begin{bmatrix} 0.4 & -0.1 \\ -0.1 & 0.1 \end{bmatrix}\begin{Bmatrix} \dot{x}_1 \\ \dot{x}_2 \end{Bmatrix} + \begin{bmatrix} 4 & -1 \\ -1 & 1 \end{bmatrix}\begin{Bmatrix} x_1 \\ x_2 \end{Bmatrix} = \begin{Bmatrix} P_1 \\ P_2 \end{Bmatrix}$$

or

$$[m]\{\ddot{x}\} + [c]\{\dot{x}\} + [k]\{x\} = \{P\}$$

In terms of $z$,

$$[m][\Phi]\{\ddot{z}\} + [c][\Phi]\{\dot{z}\} + [k][\Phi]\{z\} = \{P\}$$

Premultiplying by $[\Phi]^t$ gives

$$[\Phi]^t[m][\Phi]\{\ddot{z}\} + [\Phi]^t[c][\Phi]\{\dot{z}\} + [\Phi]^t[k][\Phi]\{z\} = [\Phi]^t\{P\}$$

or more simply

$$[\bar{m}]\{\ddot{z}\} + [\bar{c}]\{\dot{z}\} + [\bar{k}]\{z\} = [\Phi]^t\{P\} \qquad \textbf{(12.113)}$$

where

$$[\bar{c}] = [\Phi]^t[c][\Phi] = \text{a diagonal modal damping matrix}$$

The $[\bar{c}]$ matrix is a diagonal matrix and can be calculated as follows:

$$[\bar{c}] = \begin{bmatrix} \frac{1}{2} & 1 \\ \frac{1}{2} & -1 \end{bmatrix}\begin{bmatrix} c_1 + c_2 & -c_2 \\ -c_2 & c_2 \end{bmatrix}\begin{bmatrix} \frac{1}{2} & \frac{1}{2} \\ 1 & -1 \end{bmatrix}$$

$$= \begin{bmatrix} 0.1 & 0 \\ 0 & 0.3 \end{bmatrix}$$

Substituting the values of $[\bar{m}]$, $[\bar{k}]$, and $[\bar{c}]$ into equation (12.140) gives

$$\begin{bmatrix} 2 & 0 \\ 0 & 2 \end{bmatrix}\begin{Bmatrix} \ddot{z}_1 \\ \ddot{z}_2 \end{Bmatrix} + \begin{bmatrix} 0.1 & 0 \\ 0 & 0.3 \end{bmatrix}\begin{Bmatrix} \dot{z}_1 \\ \dot{z}_2 \end{Bmatrix} + \begin{bmatrix} 1 & 0 \\ 0 & 3 \end{bmatrix}\begin{Bmatrix} z_1 \\ z_2 \end{Bmatrix} = \begin{bmatrix} \frac{1}{2} & 1 \\ \frac{1}{2} & -1 \end{bmatrix}\begin{Bmatrix} P_1 \\ P_2 \end{Bmatrix}$$

or in expanded form

$$2\ddot{z}_1 + 0.1\dot{z}_1 + z_1 = \tfrac{1}{2}P_1 + P_2$$

$$2\ddot{z}_2 + 0.3\dot{z}_2 + 3z_2 = \tfrac{1}{2}P_1 - P_2$$

These are two decoupled differential equations which can be solved easily once $P_1$ and $P_2$ are specified.

The solution procedure just outlined can be easily applied to the electrical circuit model discussed earlier and a variety of other systems.

### 12.12.2 Direct Integration Procedure

As in the case of a single second-order differential equation, a set can be solved in exactly the same manner with one exception. That is, instead of dealing with scalar properties, we now have matrices for the properties. That is, given

$$[m]\{\ddot{x}\} + [c]\{\dot{x}\} + [k]\{x\} = \{P\}$$

with initial conditions

$$\text{Displacements} = \{x(0)\}$$

$$\text{Velocities} = \{\dot{x}(0)\}$$

determine the initial acceleration vector:

$$\{\ddot{x}(0)\} = [m]^{-1}[\{P(0)\} - [c]\{\dot{x}(0)\} - [k]\{x(0)\}] \tag{12.114}$$

Then proceed by computing the acceleration vector, the velocity vector, and the displacement vector respectively using the following equations:

$$\{\ddot{x}\}_{i+1} = [A_0]^{-1}[[A_1]\{\ddot{x}\}_i + [A_2]\{\dot{x}\}_i + [A_3]\{x\}_i + \{P\}_{i+1}] \tag{12.115a}$$

$$\{\dot{x}\}_{i+1} = \{\dot{x}\}_i + \tfrac{1}{2}h[\{\ddot{x}\}_i + \{\ddot{x}\}_{i+1}] \tag{12.115b}$$

$$\{x\}_{i+1} = \{x\}_i + \tfrac{1}{2}h[\{\dot{x}\}_i + \{\dot{x}\}_{i+1}] \tag{12.115c}$$

where

$$[A_0] = [m] + \frac{h}{2}[c] + \frac{h^2}{4}[k] \tag{12.116a}$$

$$[A_1] = -\left[\frac{h}{2}[c] + \frac{h^2}{4}[k]\right] \tag{12.116b}$$

$$[A_2] = -[[c] + h[k]] \tag{12.116c}$$

$$[A_3] = -[k] \tag{12.116d}$$

These equations are similar to those developed for single second-order differential equations, except that the scalar quantities are replaced with matrices. This method of solution is extremely important, especially when the set of equations for a given system cannot be decoupled.

The $h$ value is now related to the largest frequency (eigenvalue) as given by

$$h \leq \frac{\pi}{10}\sqrt{\frac{1}{w_{\max}}}$$

where $w_{\max}$ is the largest circular frequency.

Similar expressions can be obtained for the electrical network model presented in section 1.7.7 by simply replacing the *m-c-k* matrices with their corresponding *L-R-C* matrices and $\{x\}, \{\dot{x}\}, \{\ddot{x}\}$ with $\{I\}, \{\dot{I}\}, \{\ddot{I}\}$. Other expressions for higher-order sets of ordinary differential equations can also be developed.

EXAMPLE 12.18

Determine the response of the *m-c-k* system given in example 12.17 for $t = 0.2$ s using the direct integration procedure.

Solution

The exact displacements are given by

$$x_1 = \frac{1}{2}\cos\frac{1}{\sqrt{2}}t + \frac{3}{2}\cos\sqrt{\frac{3}{2}}t$$

$$x_2 = \cos\frac{1}{\sqrt{2}}t - 3\cos\sqrt{\frac{3}{2}}t$$

These should serve as a check on the accuracy of the numerical method to be used in this case. Thus

$$h = \frac{\pi}{10\sqrt{W_{\max}}} = \frac{\pi}{10}\frac{1}{\sqrt{3/2}} = 0.2565$$

Taking $h = 0.20$ permits determination of the various matrices needed for the solution:

$$[A_0] = \begin{bmatrix} 4 & 0 \\ 0 & 1 \end{bmatrix} + \frac{0.20}{2}\begin{bmatrix} 0 & 0 \\ 0 & 0 \end{bmatrix} + \frac{0.04}{4}\begin{bmatrix} 4 & -1 \\ -1 & 1 \end{bmatrix}$$

$$= \begin{bmatrix} 4.04 & -0.01 \\ -0.01 & 1.01 \end{bmatrix}$$

$$[A_1] = -\left[\frac{0.2}{2}\begin{bmatrix}0 & 0\\0 & 0\end{bmatrix} + \frac{0.04}{4}\begin{bmatrix}4 & -1\\-1 & 1\end{bmatrix}\right]$$

$$= -\begin{bmatrix}0.04 & -0.01\\-0.01 & 0.01\end{bmatrix}$$

$$[A_2] = -\left[\begin{bmatrix}0 & 0\\0 & 0\end{bmatrix} + 0.2\begin{bmatrix}4 & -1\\-1 & 1\end{bmatrix}\right]$$

$$= -\begin{bmatrix}0.8 & -0.2\\-0.2 & 0.2\end{bmatrix}$$

$$[A_3] = -\begin{bmatrix}4 & -1\\-1 & 1\end{bmatrix}$$

and

$$[A_0]^{-1} = \begin{bmatrix}1.01 & 0.01\\0.01 & 4.04\end{bmatrix}\frac{1}{4.0803}$$

Substituting the various matrices into equation (12.115a) gives

$$\begin{Bmatrix}\ddot{x}_1\\\ddot{x}_2\end{Bmatrix}_{i+1} = \frac{1}{4.0803}\begin{bmatrix}1.01 & 0.01\\0.01 & 4.04\end{bmatrix}\left[\begin{bmatrix}-0.04 & 0.01\\0.01 & -0.01\end{bmatrix}\begin{Bmatrix}\ddot{x}_1\\\ddot{x}_2\end{Bmatrix}_i\right.$$
$$\left. + \begin{bmatrix}-0.8 & 0.2\\0.2 & -0.2\end{bmatrix}\begin{Bmatrix}\dot{x}_1\\\dot{x}_2\end{Bmatrix}_i + \begin{bmatrix}-4 & 1\\1 & -1\end{bmatrix}\begin{Bmatrix}x_1\\x_2\end{Bmatrix}_i + \begin{Bmatrix}0\\0\end{Bmatrix}\right]$$

Simplifying, we obtain

$$\begin{Bmatrix}\ddot{x}_1\\\ddot{x}_2\end{Bmatrix}_{i+1} = 10^{-3}\begin{bmatrix}-9.8767 & 2.4508\\9.8032 & -9.8767\end{bmatrix}\begin{Bmatrix}\ddot{x}_1\\\ddot{x}_2\end{Bmatrix}_i$$
$$+ \begin{bmatrix}-0.1975 & 0.0490\\0.1961 & -0.1975\end{bmatrix}\begin{Bmatrix}\dot{x}_1\\\dot{x}_2\end{Bmatrix}_i$$
$$+ \begin{bmatrix}-0.98767 & 0.24508\\0.98032 & -0.98767\end{bmatrix}\begin{Bmatrix}x_1\\x_2\end{Bmatrix}_i$$

The initial acceleration vector must now be determined from equation (12.114) as

$$\begin{Bmatrix}\ddot{x}_1\\\ddot{x}_2\end{Bmatrix}_0 = \begin{bmatrix}\frac{1}{4} & 0\\0 & 1\end{bmatrix}\left[\begin{Bmatrix}0\\0\end{Bmatrix} - \begin{bmatrix}0 & 0\\0 & 0\end{bmatrix}\begin{Bmatrix}0\\0\end{Bmatrix} - \begin{bmatrix}4 & -1\\-1 & 1\end{bmatrix}\begin{Bmatrix}2\\-2\end{Bmatrix}\right]$$
$$= \begin{Bmatrix}-2.5\\4.0\end{Bmatrix}$$

Therefore for $i = 0$, $t = 0.2$:

$$\begin{Bmatrix} \ddot{x}_1 \\ \ddot{x}_2 \end{Bmatrix}_1 = 10^{-3}\begin{bmatrix} -9.8767 & 2.4508 \\ 9.8032 & -9.8767 \end{bmatrix}\begin{Bmatrix} -2.5 \\ 4 \end{Bmatrix}_0 + \begin{bmatrix} 0 & 0 \\ 0 & 0 \end{bmatrix}$$

$$+ \begin{bmatrix} -0.98767 & 0.24508 \\ 0.98032 & -0.98767 \end{bmatrix}\begin{Bmatrix} 2 \\ -2 \end{Bmatrix}_0$$

$$= \begin{Bmatrix} -2.4310 \\ 3.8719 \end{Bmatrix}$$

The velocity vector is given by equation (12.115b):

$$\begin{Bmatrix} \dot{x}_1 \\ \dot{x}_2 \end{Bmatrix}_1 = \begin{Bmatrix} 0 \\ 0 \end{Bmatrix}_0 + \frac{0.2}{2}\left[\begin{Bmatrix} -2.5 \\ 4.0 \end{Bmatrix} + \begin{Bmatrix} -2.4310 \\ 3.8719 \end{Bmatrix}\right]$$

$$= \begin{Bmatrix} -0.4931 \\ 0.7872 \end{Bmatrix}$$

Finally, the displacement vector is calculated:

$$\begin{Bmatrix} x_1 \\ x_2 \end{Bmatrix}_1 = \begin{Bmatrix} 2 \\ -2 \end{Bmatrix}_0 + \frac{0.2}{2}\left[\begin{Bmatrix} 0 \\ 0 \end{Bmatrix} + \begin{Bmatrix} -0.4931 \\ 0.7872 \end{Bmatrix}\right]$$

$$= \begin{Bmatrix} 1.9507 \\ -1.9213 \end{Bmatrix}$$

This compares with the exact value of

$$\begin{Bmatrix} x_1 \\ x_2 \end{Bmatrix}_1 = \begin{Bmatrix} 1.9502329 \\ -1.9204325 \end{Bmatrix}$$

Note that greater accuracy can be attained by retaining more significant figures and reducing the step size $h$ further. This can be easily accomplished with a computer. It is evident that the direct numerical solution approach does not require the evaluation of the eigenvalues or their corresponding vectors. Furthermore, damped as well as undamped systems can be analyzed.

## 12.13 BOUNDARY-VALUE PROBLEMS

The analytical solution of differential equations using the methods of calculus is feasible, provided the values of the independent variable and/or its derivatives are specified at some point. That is, for $n$th-order differential equations a set of $n$ values must be specified. Furthermore, even when these conditions are met, there are situations in which the analytical solution might not be feasible. This situation arises when the function involved is a complicated one. In the preceding sections, we have introduced numerous methods for solving ordinary differential equations

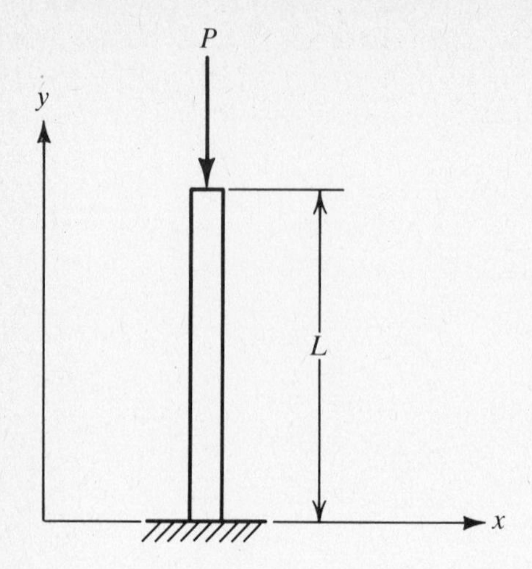


(a) Column buckling: $\dfrac{d^4x}{dy^4} + k^2\dfrac{d^2x}{dy^2} = 0$

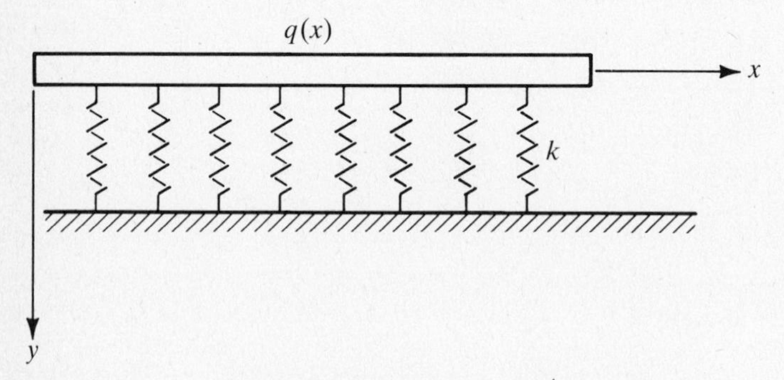


(b) Beam on elastic foundation: $EI\dfrac{d^4y}{dx^4} + ky = q$

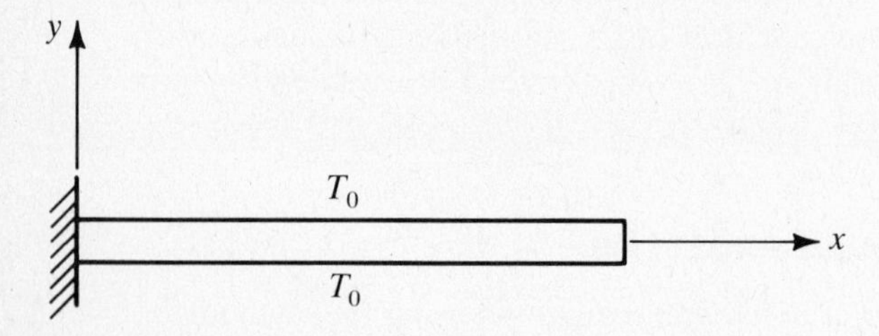


(c) Steady-state temperature problem: $\dfrac{d^2T}{dx^2} = 0$

**Figure 12.18** Examples of boundary-value problems.

numerically. In all of the cases discussed thus far, the functional values and/or its derivatives were specified at one point corresponding to the independent variable(s), generally at the beginning of the solution. These values are termed "initial values."

In this section, rather important classes of problems are introduced in which the functional values are specified at two different points. The location is normally specified at the physical boundaries of the engineering model being analyzed. For this reason, such problems are termed "boundary-value problems." Figure 12.18 illustrates different boundary-value problems along with their mathematical models (differential equations).

It is evident that these are only a few examples of the many practical boundary problems that we may encounter in practice. Furthermore, for the model presented in figure 12.18a, the boundary conditions are simply that the displacements and the slopes $dx/dy$ are all zero at $y = 0$ and $y = L$. These four conditions permit the solution of this particular problem.

While there are many methods that can be developed for solving boundary-value problems involving a single independent variable, the finite difference method is by far the easiest and most versatile. Consequently, it is the only technique described here.

Recall that in chapter 10, several numerical expressions for various derivatives were developed. In addition, these expressions involved different error terms. *As a matter of practice, for given ordinary differential equations, the error magnitude must be kept the same for all of the derivative terms appearing in that equation.* This is extremely important in ensuring that errors do not propagate and that accuracy is maintained at a fixed level.

Consider the steady-state beam deflection problem (figure 12.19). The deflection $y$ at any point $x$ is given by the following fourth-order ordinary differential equation:

$$EI(x)\frac{d^4y}{dx^4} = q(x) \tag{12.117}$$

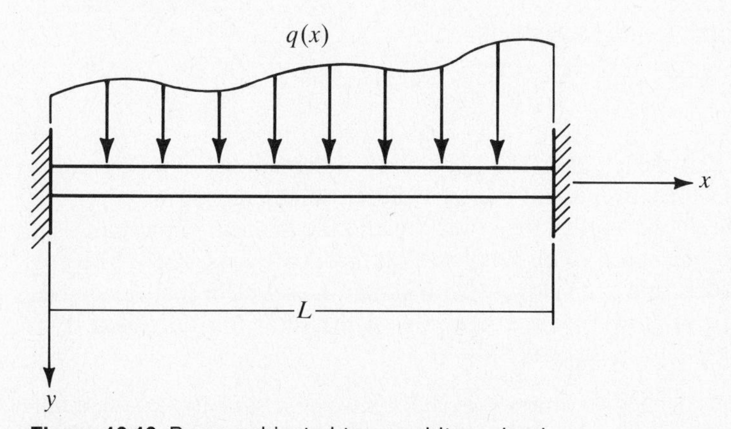


**Figure 12.19** Beam subjected to an arbitrary load.

where

$E$ = Young's modulus of elasticity of beam material

$I(x)$ = Moment of inertia of beam cross section at any point $x$

Equation (12.117) does not completely define deflection. This is because we have said nothing about the boundary conditions. Therefore one must specify whether the boundaries are fixed, hinged, free, or a combination of two. For illustrative purposes, suppose that the boundary at $x = 0$ is fixed and that at $x = L$ is hinged. This implies that the problem is statically indeterminate.

Equation (12.117) can be solved by first specifying the order of error then using the equivalent finite expression for the fourth derivative. Assume that the error order is $(O)h^2$. Thus, using table 10.4, we have

$$\frac{d^4y}{dx^4} = \frac{1}{h^4}\left( \textcircled{1} - \textcircled{-4} - \circledcirc\!6 - \textcircled{-4} - \textcircled{1} \right) \tag{12.118}$$

Substituting equation (12.118) into equation (12.117) gives

$$\frac{EI(x)}{h^4}\left( \textcircled{1} - \textcircled{-4} - \circledcirc\!6 - \textcircled{-4} - \textcircled{1} \right) = q(x) \tag{12.119}$$

Simplifying yields

$$\textcircled{1} - \textcircled{-4} - \circledcirc\!6 - \textcircled{-4} - \textcircled{1} = \frac{q(x)h^4}{EI(x)} \tag{12.120}$$

Note that the distance between any two adjacent circles is $h = \Delta x$. In addition, the load and the moment of inertia are taken as functions of the variable $x$. As a matter of fact, even $E$ can be taken as a variable if needed. Equation (12.120) is the finite difference equivalence of equation (12.117). The solution of the beam problem at hand may proceed as follows.

*STEP 1:* Select the length increment:

$$h = \frac{L}{n}$$

where $L$ is the length of the beam and $n$ is the number of nodes minus one. The length interval must be chosen small enough to reflect the changes in the applied load function and the moment of inertia. Also, $h$ must be chosen to be small so that errors are kept small. There is no golden rule as to how small $h$ should be selected. Just remember that the accuracy of the solution is increased as $h$ is decreased. For illustrative purposes, take $h = \frac{1}{3}L$.

*STEP 2:* Since the stencil of equation (12.120) has five circles, two of them fall outside the beam when the double-circle point is placed at $x = 0$ or $x = L$. Therefore we must create two imaginary points on each side of the beam and then number the nodes as shown below:

```
1     2     3     4     5     6     7     8
.    [.     .     .     .     .]    .     .
| L/3 | L/3 | L/3 | L/3 | L/3 | L/3 | L/3 |
```

*STEP 3:* Place the double circle at every node appearing on the actual beam (nodes 3, 4, 5, and 6).

```
1      2      3      4      5      6      7      8
.      .     [.      .      .      .]     .      .

1     −4      6     −4      1
       1     −4      6     −4      1
              1     −4      6     −4      1
                     1     −4      6     −4      1
```

*STEP 4:* Write down the linear algebraic equations relating the deflections at the various beam nodes to beam properties and $h$. So, if we call the deflection at node 1 $y_1$, and at node 2 $y_2$, etc., then

$$y_1 - 4y_2 + 6y_3 - 4y_4 + y_5 = \frac{q(3)h^4}{EI(3)} \quad \textbf{(12.121a)}$$

$$y_2 - 4y_3 + 6y_4 - 4y_5 + y_6 = \frac{q(4)h^4}{EI(4)} \quad \textbf{(12.121b)}$$

$$y_3 - 4y_4 + 6y_5 - 4y_6 + y_7 = \frac{q(5)h^4}{EI(5)} \quad \textbf{(12.121c)}$$

$$y_4 - 4y_5 + 6y_6 - 4y_7 + y_8 = \frac{q(6)h^4}{EI(6)} \quad \textbf{(12.121d)}$$

Note that $q(3)$ is the load magnitude per unit length at node 3 and $I(3)$ is the moment of inertia at node 3. Equations (12.121) involve four equations in eight unknowns. Therefore, four of the unknowns must be eliminated.

*STEP 5:* Introduce boundary conditions at both ends.

At $x = 0$:

(1) $y = 0 =$ deflection

(2) $\dfrac{dy}{dx} = 0 =$ slope

At $x = L$:

(3) $y = 0$

(4) $EI\dfrac{d^2y}{dx^2} = 0 =$ moment

The first boundary condition stipulates that the deflection at node 3 ($y_3$) is zero. Hence equation (12.121a) must be dropped. The second condition involves the first derivative. Consequently, a difference expression is needed. Note that the chosen expression must have error order of $(O)h^2$. Therefore, using table 10.1, we have

$$\frac{dy}{dx} = \frac{1}{2h}\left(\textcircled{-1}\!-\!\textcircled{\textcircled{0}}\!-\!\textcircled{1}\right) \tag{12.122}$$

At $x = 0$, equation (12.122) is written in terms of the deflections as

$$0 = \frac{1}{2h}[-y_2 + (O)y_3 + y_4]$$

Solving for $y_2$, we obtain

$$y_2 = y_4 \tag{12.123}$$

The third condition specifies that the deflection $y_6 = 0$. This implies that equation (12.121d) must be dropped. Finally, the fourth boundary condition involves the second derivative. Hence the finite difference equivalence of the second derivative is given as follows:

$$\frac{d^2y}{dx^2} = \frac{1}{h^2}\left(\textcircled{1}\!-\!\textcircled{\textcircled{-2}}\!-\!\textcircled{1}\right) \tag{12.124}$$

Applying equation (12.124) at node 6 gives

$$0 = \frac{1}{h^2}(y_5 - 2y_6 + y_7)$$

Solving for $y_7$, we obtain

$$y_7 = -y_5 \tag{12.125}$$

Substituting equations (12.123) and (12.125) into equations (12.121b) and (12.121c) and noting that $y_3 = y_6 = 0$, we get

$$7y_4 - 4y_5 = \frac{q(4)h^4}{EI(4)} \tag{12.126a}$$

$$-4y_4 + 5y_5 = \frac{q(5)h^4}{EI(4)} \tag{12.126b}$$

Equations (12.126) can now be solved for the unknown deflections $y_4$ and $y_5$. Keep in mind that the number of linear algebraic equations resulting from the finite difference procedure is dependent on our choice of the number of nodes.

The procedure just outlined can be employed to solve problems involving beams with different boundary conditions by simply introducing the appropriate boundary conditions. In addition, the interval $h$ can be reduced further if more accurate results are desired.

E

EXAMPLE 12.19

Derive modified stencils for beams having fixed boundaries, hinged boundaries, and free boundaries.

### Solution

These examples show how the stencils for an interior node is modified to handle boundary nodes. This is very helpful when dealing with beam problems in that no imaginary nodes need be established.

#### Fixed Ends

Consider the following beam with fixed boundaries and unknown load and cross section:

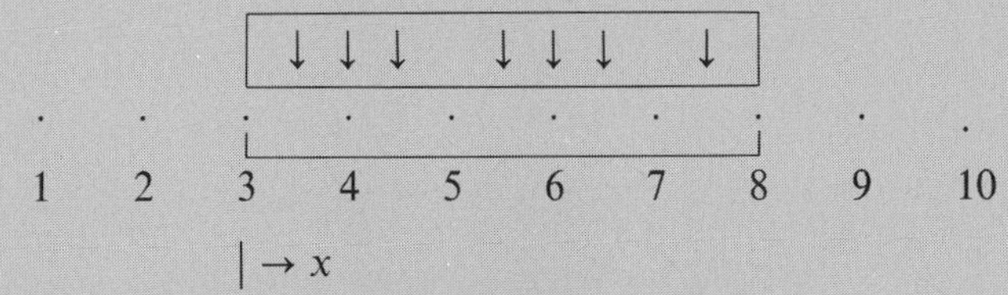

1 2 3 4 5 6 7 8 9 10

$|\rightarrow x$

Obviously, we have ten nodes, four of which (1, 2, 9, 10) are imaginary and two of which (3, 8) have zero deflection. Consequently, we would like to have four equations in terms of the unknown deflections at nodes 4, 5, 6, and 7. While this could be done by introducing the boundary condition, we wish to solve for the imaginary deflection in terms of interior nodes (4, 5, 6, 7) once and for all.

Applying equation (12.120) at nodes 5 and 6 yields two of the equations needed for the solution. However, applying the equation at node 4 yields an equation in terms of the deflection at node 2. That is,

$$y_2 - 4y_3 + 6y_4 - 4y_5 + y_6 = \frac{qh^4}{EI} \qquad \textbf{(12.127)}$$

Applying the boundary condition $dy/dx = 0$ at $x = 0$ shows that

$$y_2 = y_4 \qquad \textbf{(12.128)}$$

Substituting equation (12.128) into equation (12.127) gives

$$-4y_3 + 7y_4 - 4y_5 + y_6 = \frac{qh^4}{EI} \qquad \textbf{(12.129)}$$

But $y_3 = 0$. Hence, expressing equation (12.129) in stencil form, we get at node 4

$$⑦—(-4)—(1) = \frac{qh^4}{EI} \tag{12.130}$$

Similarly, at node 7 we have

$$(1)—(-4)—⑦ = \frac{qh^4}{EI} \tag{12.131}$$

Equations (12.130) and (12.131) can now be applied at nodes to the left and to the right of fixed boundaries respectively without the need for introducing the boundary conditions. This permits determination of the four equations in the four unknowns directly.

### Hinged Boundaries

Consider the following beam with two hinged ends, a uniform load, and uniform cross section.

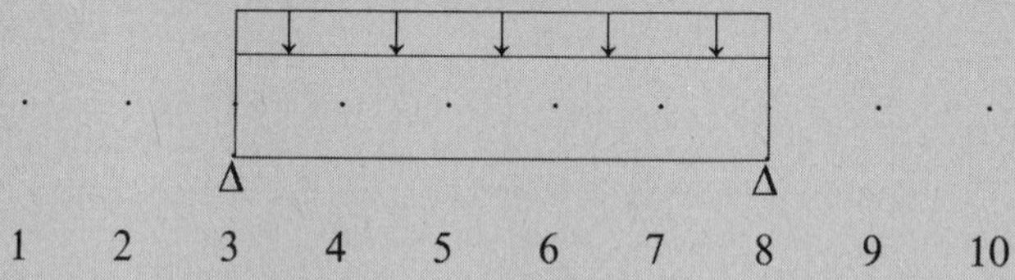

1 2 3 4 5 6 7 8 9 10

$$y_2 - 4y_3 + 6y_4 - 4y_5 + y_6 = \frac{qh^4}{EI} \tag{12.132}$$

Applying the boundary condition at $x = 0$, $d^2y/dx^2 = 0$ yields

$$y_2 = -y_4 \tag{12.133}$$

Substituting equation (12.133) into equation (12.132), we get

$$-4y_3 + 5y_4 - 4y_5 + y_6 = \frac{qh^4}{EI} \tag{12.134}$$

Note that $y_3$ is zero. Thus expressing equation (12.134) in stencil form at node 4 gives

$$⑤—(-4)—(1) = \frac{qh^4}{EI} \tag{12.135}$$

Similarly, at node 7 we have

$$(1)—(-4)—⑤ = \frac{qh^4}{EI} \tag{12.136}$$

Free Ends

When dealing with problems involving beams on elastic foundations or cantilever beams free ends must be dealt with. Consider the following cantilever beam.

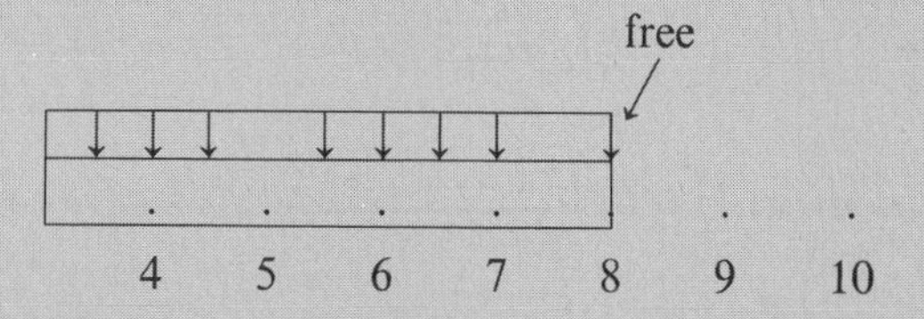


It is evident that the deflection at node 8 is no longer zero. Therefore applying equation (12.120) at node 7 gives

$$y_5 - 4y_6 + 6y_7 - 4y_8 + y_9 = \frac{qh^4}{EI} \quad \textbf{(12.137)}$$

Applying equation (12.120) at node 8 gives

$$y_6 - 4y_7 + 6y_8 - 4y_9 + y_{10} = \frac{qh^4}{EI} \quad \textbf{(12.138)}$$

Equation (12.137) involves one imaginary deflection ($y_9$) and equation (12.138) involves two imaginary deflections ($y_9$ and $y_{10}$). These can be eliminated by introducing the boundary conditions

$$x = L: \quad \frac{d^2y}{dx^2} = 0 \quad \textbf{(12.139)}$$

$$x = L: \quad \frac{d^3y}{dx^3} = 0 \quad \textbf{(12.140)}$$

From equation (12.139), we have at node 8

$$y_7 - 2y_8 + y_9 = 0 \quad \textbf{(12.141)}$$

Equation (12.140) states that at node 8

$$-y_6 + 2y_7 - 2y_9 + y_{10} = 0 \quad \textbf{(12.142)}$$

Note that equations (12.141) and (12.142) represent the finite difference definitions of the second and third derivatives with error order of $(O)h^2$. Hence, solving equation (12.141) for $y_9$ gives

$$y_9 = 2y_8 - y_7 \quad \textbf{(12.143)}$$

Substituting equation (12.143) into equation (12.142) and solving for $y_{10}$ yields

$$y_{10} = y_6 - 4y_7 + 4y_8 \quad \textbf{(12.144)}$$

Table 12.10 STENCILS FOR BEAMS HAVING DIFFERENT BOUNDARY CONDITIONS

| BOUNDARY | STENCIL | BEAM NODE LOCATION | EQUATION |
|---|---|---|---|
| Fixed | (−4)—((7))—(−4)—(1) | 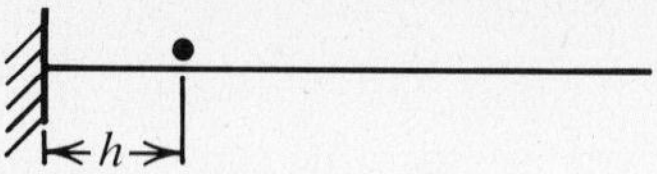 $h$ | (A1) |
| | (1)—(4)—((6))—(−4)—(1) | 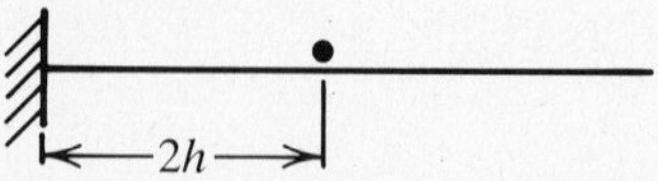 $2h$ | (A2) |
| | (1)—(−4)—((7))—(−4) |  $h$ | (A3) |
| | (1)—(4)—((6))—(−4)—(1) | 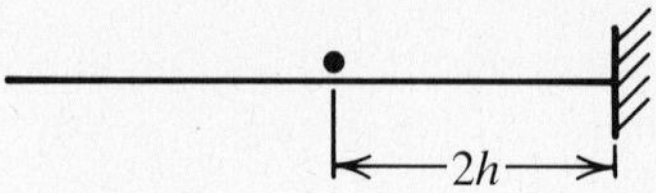 $2h$ | (A4) |
| Hinged | (−2)—((5))—(4)—(1) | 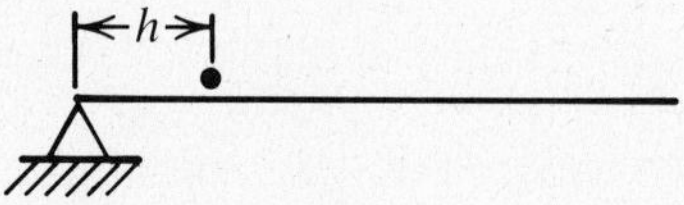 $h$ | (B1) |
| | (1)—(−4)—((6))—(−4)—(1) | 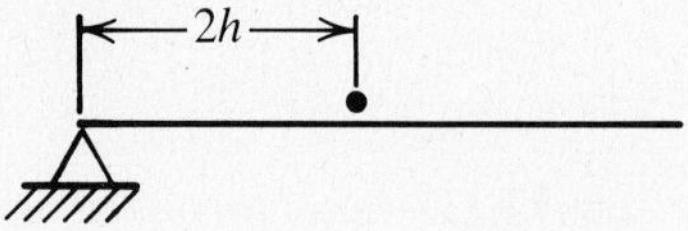 $2h$ | (B2) |
| | (1)—(−4)—((5))—(−2) | 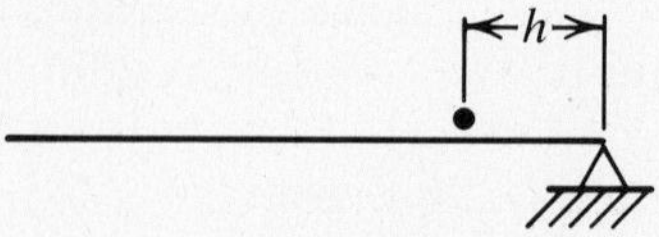 $h$ | (B3) |
| | (1)—(4)—((6))—(−4)—(1) | 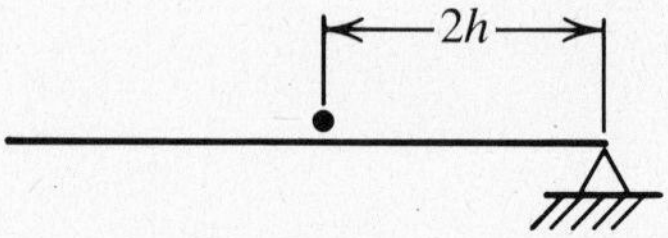 $2h$ | (B4) |
| Free | ((2))—(−4)—(2) | | (C1) |
| | (−2)—((5))—(−4)—(1) | 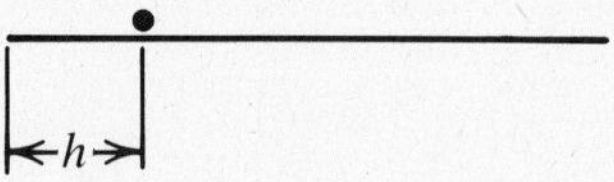 $h$ | (C2) |
| | (2)—(−4)—((2)) | | (C3) |
| | (1)—(−4)—((5))—(−2) |  $h$ | (C4) |

Interior nodes

(1)—(−4)—((6))—(−4)—(1) (D)

Equation (12.137) can now be modified by substituting (12.143) for $y_9$. Thus

$$y_5 - 4y_6 + 6y_7 - 4y_8 + 2y_8 - y_7 = \frac{qh^4}{EI}$$

Simplifying, we obtain

$$y_5 - 4y_6 + 5y_7 - 2y_8 = \frac{qh^4}{EI} \tag{12.145}$$

or in stencil form

$$\textcircled{1}-\textcircled{-4}-\textcircled{5}-\textcircled{-2} = \frac{qh^4}{EI} \tag{12.146}$$

Equation (12.146) is applied at a node located a distance $h$ from a free end. Consequently, substituting equations (12.143) and (12.144) into (12.138) gives the stencil for a node located at the free end. That is,

$$\textcircled{2}-\textcircled{-4}-\textcircled{2} = \frac{qh^4}{EI} \tag{12.147}$$

A summary of the various stencils relevant to beams is given in table 12.10.

# E

EXAMPLE 12.20

Determine the deflections using $h = \frac{1}{4}L$ for the following beam:

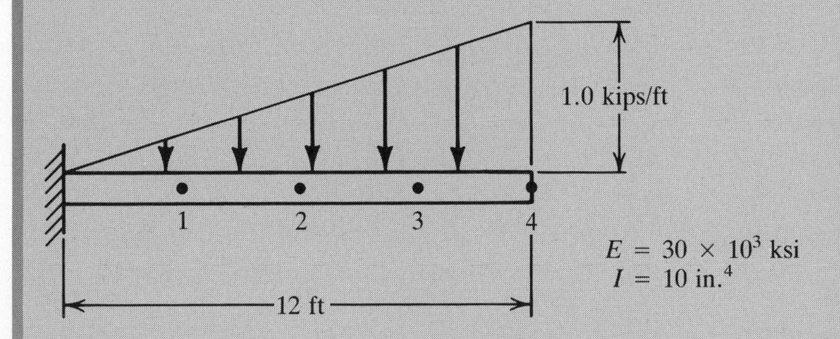


## Solution

As a first step we need to formulate four independent linear algebraic equations in the unknown deflections $y_1$, $y_2$, $y_3$, and $y_4$ at the four nodes. Therefore, applying equation (A1) from table 12.10 at node 1 and noting that at

$x = 0$ $y = 0$ gives

$$7y_1 - 4y_2 + y_3 = \left(\frac{qh^4}{EI}\right)_1$$

applying equation (A2) at node 2,

$$-4y_1 + 6y_2 - 4y_3 + y_4 = \left(\frac{qh^4}{EI}\right)_2$$

applying equation (C4) at node 3,

$$y_1 - 4y_2 + 5y_3 - 2y_4 = \left(\frac{qh^4}{EI}\right)_3$$

and, finally, applying (C3) at node 4, we obtain the fourth equation needed for determining the four deflections. That is,

$$2y_2 - 4y_3 + 2y_4 = \left(\frac{qh^4}{EI}\right)_4$$

Note that $d^2y/dx^2 = 0$ at $x = L = 12$ ft.

Expressing these equations in matrix form, we have

$$\begin{bmatrix} 7 & -4 & 1 & 0 \\ -4 & 6 & -4 & 1 \\ 1 & -4 & 5 & -2 \\ 0 & 2 & -4 & 2 \end{bmatrix} \begin{Bmatrix} y_1 \\ y_2 \\ y_3 \\ y_4 \end{Bmatrix} = \begin{Bmatrix} \left(\frac{qh^4}{EI}\right)_1 \\ \left(\frac{qh^4}{EI}\right)_2 \\ \left(\frac{qh^4}{EI}\right)_3 \\ \left(\frac{qh^4}{EI}\right)_4 \end{Bmatrix}$$

where

$$h = \tfrac{12}{4} = 3 \text{ ft}$$

$$q_1 = 0.25 \text{ kips/ft}$$

$$q_2 = 0.50 \text{ kips/ft}$$

$$q_3 = 0.75 \text{ kips/ft}$$

$$q_4 = 1.00 \text{ kips/ft}$$

and

$$EI = \frac{10 \times 30 \times 10^3}{144} = 2083.33 \text{ kips ft}^2$$

Substituting these values into the right-hand side of the above equations yields

$$\begin{bmatrix} 7 & -4 & 1 & 0 \\ -4 & 6 & -4 & 1 \\ 1 & -4 & 5 & -2 \\ 0 & 2 & -4 & 2 \end{bmatrix} \begin{Bmatrix} y_1 \\ y_2 \\ y_3 \\ y_4 \end{Bmatrix} = \begin{Bmatrix} 0.00972 \\ 0.01944 \\ 0.02916 \\ 0.03888 \end{Bmatrix}$$

Solving for the unknown vector, we get

$$y_1 = 0.10692 \text{ ft}$$

$$y_2 = 0.3499198 \text{ ft}$$

$$y_3 = 0.6609596 \text{ ft}$$

$$y_4 = 0.9914392 \text{ ft}$$

These values are consistent with our intuition in asserting that maximum deflection should occur at the free end.

**E** EXAMPLE 12.21

Evaluate the deflections at the various nodes using $h = L/5$ for the following beam on an elastic foundation.

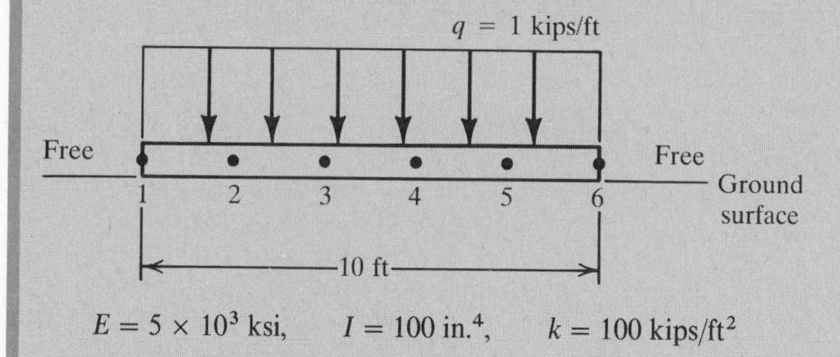


$$E = 5 \times 10^3 \text{ ksi}, \qquad I = 100 \text{ in.}^4, \qquad k = 100 \text{ kips/ft}^2$$

## Solution

The mathematical model describing the behavior of the beam in question is given in figure 12.18b. Thus

$$EI\frac{d^4y}{dx^4} + ky = q \tag{12.148}$$

Substituting the finite difference expression of order $(O)h^2$ into equation (12.86), we obtain

$$\frac{EI}{h^4}\left(\textcircled{1}-\textcircled{-4}-\textcircled{\textcircled{6}}-\textcircled{-4}-\textcircled{1}\right)+k\left(\textcircled{\textcircled{1}}\right)=q$$

Simplifying and combining terms gives

$$\textcircled{1}-\textcircled{-4}-\textcircled{\textcircled{6+\frac{kh^4}{EL}}}-\textcircled{-4}-\textcircled{1}=\frac{qh^4}{EI} \tag{12.149}$$

where

$$\frac{kh^4}{EI}=\frac{(100)(2)^4}{(5\times 10^3)\frac{100}{144}}=0.46$$

$$\frac{qh^4}{EI}=\frac{(1)(2)^4}{(5\times 10^3)\frac{100}{144}}=0.0046\text{ ft}$$

Substituting into (12.149) yields

$$\textcircled{1}-\textcircled{-4}-\textcircled{\textcircled{6.46}}-\textcircled{-4}-\textcircled{1}=0.0046 \tag{12.150}$$

Note that table 12.10 can be modified by adding $kh^4/EI$ to the double circles (pivotal points) of the various stencils. Hence the system of linear algebraic equations in $y$ can be readily formulated:

$$\begin{aligned}
2.46y_1 - 4y_2 + 2y_3 &= 0.0046\\
-2y_1 + 5.46y_2 - 4y_3 + y_4 &= 0.0046\\
y_1 - 4y_2 + 6.46y_3 - 4y_4 + y_5 &= 0.0046\\
y_2 - 4y_3 + 6.46y_4 - 4y_5 + y_6 &= 0.0046\\
y_3 - 4y_4 + 5.46y_5 - 2y_6 &= 0.0046\\
2y_4 - 4y_5 + 2.46y_6 &= 0.0046
\end{aligned}$$

Solving for the unknown deflections gives

$$\begin{aligned}
y_1 &= 0.01\text{ ft} \qquad & y_4 &= 0.01\text{ ft}\\
y_2 &= 0.01\text{ ft} \qquad & y_5 &= 0.01\text{ ft}\\
y_3 &= 0.01\text{ ft} \qquad & y_6 &= 0.01\text{ ft}
\end{aligned}$$

Once again, since the beam is assumed to be resting on elastic springs ("Winkler model"), the results are meaningful.

E

EXAMPLE 12.22

Redo example 12.21 if the beam is subjected to a concentrated load of 10 kips at node 3 in addition to the distributed load.

Solution

This problem can be easily solved without reformulating the set of algebraic equations. We need only adjust the right-hand side of the equation corresponding to node 3. That is,

$$\frac{qh^4}{EI} = 0.0046 + \frac{P}{h^2}\frac{h^4}{EI}$$

$$= 0.0046 + \frac{10(2)^2}{(5 \times 10^3)\frac{100}{144}}$$

$$= 0.01612$$

Therefore the third of the algebraic equations given in the previous example is now replaced by

$$y_1 - 4y_2 + 6.46y_3 - 4y_4 + y_5 = 0.01612$$

Solving for the unknown displacements yields

$$y_1 = 0.01359007 \qquad y_4 = 0.01599039$$

$$y_2 = 0.01648559 \qquad y_5 = 0.01256602$$

$$y_3 = 0.01855539 \qquad y_6 = 0.009302161$$

## Recommended Reading


*Modern Methods of Engineering Computation*, Robert L. Ketter and S. P. Prawel, McGraw-Hill Book Company, New York, 1969.

*Numerical Algorithms—Origins and Applications*, Bruce W. Arden and K. N. Astill, Addison-Wesley, Reading, Mass., 1970.

*Elementary Differential Equations*, Earl D. Rainville and P. E. Bedient, MacMillan Company, London, 1970.

*Numerical Methods for Scientists and Engineers*, R. W. Hamming, McGraw-Hill Book Company, New York, 1973.

*Dynamics of Structures*, Ray W. Clough, and J. Penzien, McGraw-Hill Book Company, New York, 1975.

*A First Course in Numerical Analysis*, Anthony Ralston and P. Rabinowitz, McGraw-Hill Book Company, New York, 1978.

*A Practical Guide to Computer Methods for Engineers*, Terry E. Shoup, Prentice-Hall, Inc., Englewood Cliffs, N.J., 1979.

*Finite Element Procedures in Engineering Analysis*, Klauss-Jurgen Bathe, Prentice-Hall, Inc., Englewood Cliffs, N.J., 1982.

*Applied Numerical Analysis*, Curtis F. Gerald and P. O. Wheatley, Addison-Wesley, Reading, Mass., 1984.

P

## PROBLEMS

**12.1** Given the following ordinary differential equation

$$\frac{dy}{dx} - x^2y = 0, \qquad y(0) = 1$$

Evaluate the solution at $x = 0.1$ using the method of

**(a)** Taylor series—include the sixth derivative term
**(b)** Least squares—use a second-order polynomial
**(c)** Galerkin—use a second-order polynomial
**(d)** Euler—use $h = 0.05$
**(e)** Modified Euler—use $h = 0.05$
**(f)** Runge–Kutta (fourth order)—use $h = 0.05$

**12.2** The number of bacterial cells ($P$) in a given reactor is related to time in days ($t$) as described by the following mathematical model

$$\frac{dP}{dt} = 0.3P - 0.0000007P^2$$

If at time $t = 0$, $P = 10^6$ determine the number of cells when $t = 30$ days. Use the Runge–Kutta second-order method and a time increment of 1 day.

**12.3** Rework problem 2 using the Runge–Kutta fourth-order method.

**12.4** The following ordinary differential equation

$$\frac{dy}{dx} = \frac{e^x}{y}, \qquad y(0) = \sqrt{2}$$

has initial values as follows

| $x$ | $y$ |
|---|---|
| 0 | 1.41421 |
| 0.2 | 1.56295 |
| 0.4 | 1.72732 |
| 0.6 | 1.90899 |
| 0.8 | 2.10976 |
| 1.0 | 2.33164 |
| 1.2 | 2.57687 |

**(a)** Use the Adams method to solve for $x = 2$, using $\Delta x = 0.2$.
**(b)** Use the Milne method to solve for $x = 3$, using $\Delta x = 0.40$.
**(c)** Use the Adams–Moulton method for $x = 3.2$ and $\Delta x = 1.0$.

**12.5** A one-story building is idealized as a rigid girder supported by weightless columns as shown below.

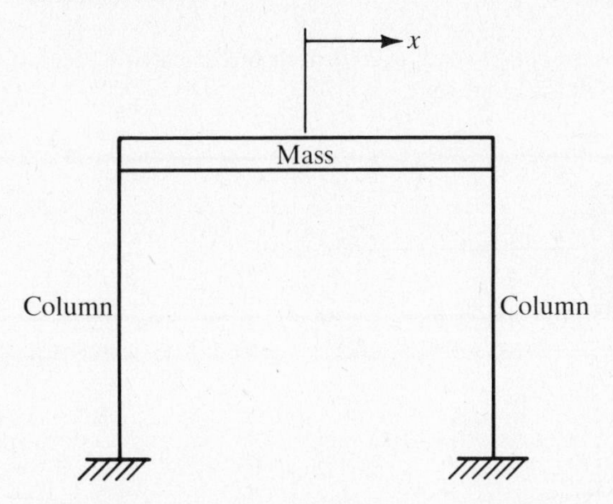


The response $x$ of the mass due to an external force $P(t)$ is given as follows.

$$m\frac{d^2x}{dt^2} + c\frac{dx}{dt} + kx = P(t)$$

where $m$ is mass, $c$ is damping, and $k$ is stiffness. For $t = 0$ to $t = 1.0$ s with

$$\Delta t = 0.1 \text{ s}$$

$$m = 4\,\frac{\text{kips-s}^2}{\text{in.}}$$

$$k = 100 \text{ kips/in.}$$

$$c = 1.6\,\frac{\text{kips-s}}{\text{in.}}$$

$$P(t) = 0$$

$$x(0) = 1 \text{ in.}$$

$$\dot{x}(0) = 10 \text{ in./s}$$

Determine the dynamic response using
**(a)** the finite difference method
**(b)** the trapezoidal rule method.

**12.6** Rework problem 5 with $P(t) = P_0 \sin 2t$ where $P_0 = 10$ kips and $t$ is in seconds.

**12.7** Rework problem 5 with the load given as follows

| $t$ (s) | 0 | 0.1 | 0.2 | 0.3 | 0.4 | 0.5 | 0.6 | 0.7 | 0.8 | 0.9 | 1.0 |
|---|---|---|---|---|---|---|---|---|---|---|---|
| $P$ (kips) | 1 | 3 | 2.5 | 1.2 | 0 | −1 | −1.5 | −0.7 | 0.1 | 0 | 0 |

**12.8** Rework problem 5b with the damping coefficient $c$ equal to 50 kips-s/in.

**12.9** Develop a direct integration procedure using the trapezoidal rule to solve the following differential equation:

$$A_0 \frac{d^3y}{dx^3} + A_1 \frac{d^2y}{dx^2} + y = 0$$

**12.10** When a vehicle travels over a rough road, a vertical displacement will result. Consider the following highly idealized model:

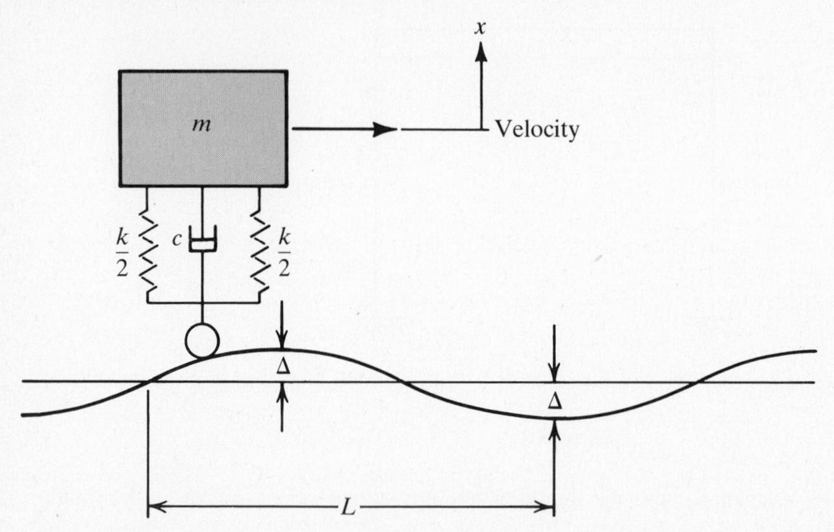


Suppose that a manufacturer is interested in determining the largest vertical displacement $x$ for a vehicle traveling at a velocity of 55 miles per hour over a sinusoidal surface, given

$m = 10$ lb-s²/in.

$k = 1210$ lb/in.

$c = 0$

$L = 50$ ft

$\Delta = 1$ in.

The mathematical model describing this behavior is given by

$$m \frac{d^2x}{dt} + c \frac{dx}{dt} + kx = m\bar{\omega}^2 \Delta \sin \bar{\omega} t$$

where $\bar{\omega} = 2\pi v/L$ where $v$ is the velocity of the vehicle. Determine the maximum displacement $x$ using the direct integration procedure and $\Delta t = 0.02$ s. Assume that the vehicle was initially at rest ($x(0) = \dot{x}(0) = 0$).

**12.11** Rework problem 10 using $\Delta t = 0.01$ s.

**12.12** Rework problem 10 with $c = 88$ lb-s/in. Note that the implication of introducing damping is that shock absorbers are provided.

**12.13** Rework problem 10 with $L = 25$ ft.

**12.14** Rework problem 10 with the velocity of the vehicle equal to
**(a)** 20 miles/hr  **(b)** 30 miles/hr
**(c)** 40 miles/hr  **(d)** 80 miles/hr

Use $\Delta t = 0.01$ s, and plot the displacement versus time and maximum displacement versus velocity.

**12.15** The voltage $V_{AB}$ is suddenly brought to 10 from an initial value of 0 by connecting a battery across the terminals $A$ and $B$. Determine the current versus time between $t = 0$ to $t = 0.1$ using a time increment of 0.02.

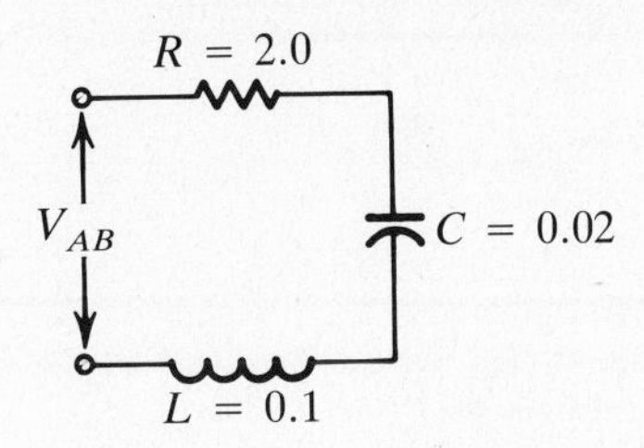


The mathematical model describing this circuit is given by

$$L\frac{d^2I}{dt} + R\frac{dI}{dt} + \frac{I}{C} = \frac{dV_{AB}}{dt}$$

Use the direct integration procedure and assume consistent units.

**12.16** Rework problem 15 with $V_{AB} = 10\sin 60\pi t$. Assume that the units are consistent.

**12.17** Solve the following differential equation using $\Delta t = 0.05$ for $t = 0$ to $t = 0.20$

$$2\frac{d^2y}{dt} + 4\frac{dy}{dt} + 20y = 0$$

**(a)** Use the Galerkin approach for $y(0) = 1$, $\dot{y}(0) = 0$, and assume

$$y = a_0 + a_1t + a_2t^2 + a_3t^3$$

**(b)** Use the Galerkin approach for $y(0) = 0$, $\dot{y}(0) = 1$, and assume

$$y = a_0 + a_1t + a_2t^2 + a_3t^3 + a_4t^4$$

**12.18** Rework problem 17 using the least squares method.

**12.19** Develop the finite difference stencils for solving the following beam problem:

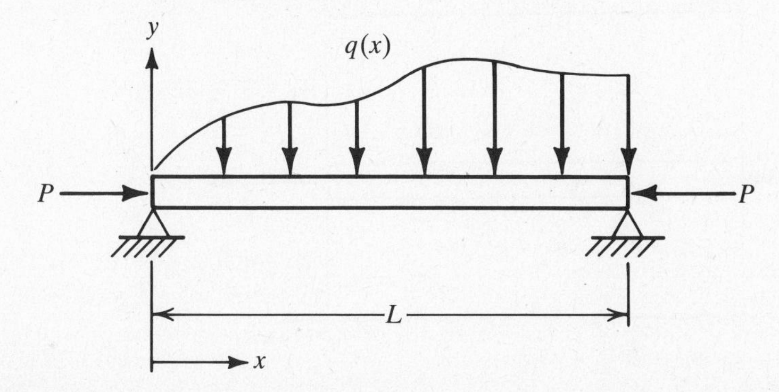

$$EI\frac{d^4y}{dx^4} + P\frac{d^2y}{dx^2} = q(x)$$

Assume that both boundaries are hinged.

**12.20** Determine the deflections of the beam given in problem 19 if $L = 15$ ft, $P = 0$, and $q(x) = 1.0$ kips/ft. Use $\Delta x = 3$ ft and assume a constant $EI$. Employ the finite difference method.

**12.21** Rework problem 20 with $q(x) = P_0x^2$ where $P_0 = 0.1$ kips/ft$^3$.

**12.22** Formulate the dynamic response model for the following *m-c-k* system.

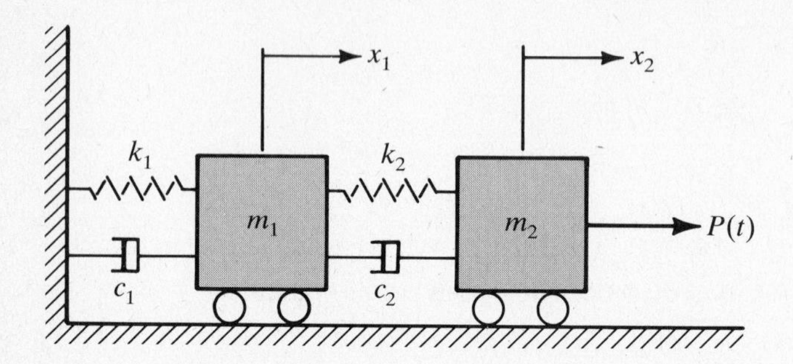


**12.23** Solve the dynamic response equations formulated in problem 22 in a closed form and by assuming

$$x_1(0) = 1, \quad x_2(0) = -1, \quad \dot{x}_1(0) = 0, \quad \dot{x}_2(0) = 0,$$

$$P(t) = 0, \quad m_1 = m_2 = 0.4 \text{ kips-s}^2/\text{in.} \quad c_1 = c_2 = 0, \quad k_1 = k_2 = 4 \text{ kips/in.}$$

**12.24** Decouple the system of differential equations obtained in problem 22 with $m_1 = m_2 = 0.4$, $c_1 = c_2 = 0.5$, $k_1 = k_2 = 4$, and $P(t) = 0$.

**12.25** Using the direct integration approach and $\Delta t = 0.1$, determine the displacements $x_1$ and $x_2$ in problem 22 with $m_1 = m_2 = 1$, $c_1 = c_2 = 1$, $k_1 = k_2 = 5$, and $P(t) = 0$ for $t = 0$ to $t = 0.2$, due to the initial conditions $\{\dot{x}(0)\} = \{0\}$, $x_1(0) = 1$, and $x_2(0) = 0$.

**12.26** The dynamic response of a two-story structure to an earthquake loading is determined by analzying the following model.

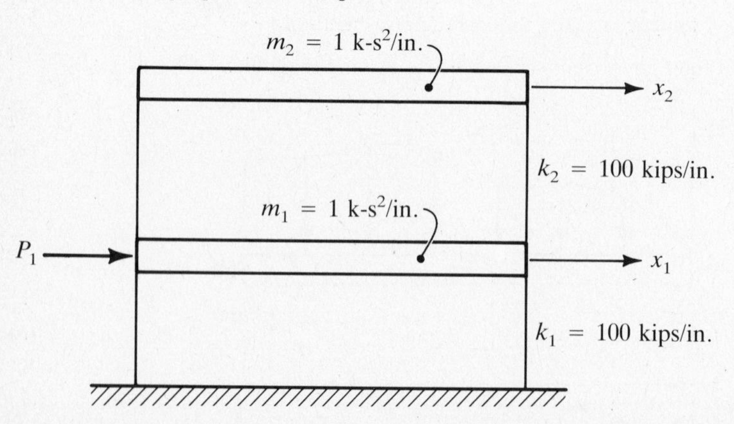

$$\begin{bmatrix} 200 & -100 \\ -100 & 100 \end{bmatrix} \begin{Bmatrix} x_1 \\ x_2 \end{Bmatrix} + \begin{bmatrix} 1 & 0 \\ 0 & 1 \end{bmatrix} \begin{Bmatrix} \ddot{x}_1 \\ \ddot{x}_2 \end{Bmatrix} = \begin{Bmatrix} P_1 \\ 0 \end{Bmatrix}$$

where the damping coefficients are assumed to be zero and the load $P_1$ is given as follows:

| $t$ (s) | 0 | 0.02 | 0.04 | 0.06 | 0.08 | 0.10 |
|---|---|---|---|---|---|---|
| $P_1$ (kips) | 0 | 1 | 2 | 0.8 | 0.2 | 0.1 |

Use the direct integration procedure to determine the displacements and take $\Delta t = 0.04$ s.

**12.27** Rework problem 26 with $P_1 = P_0 \sin 2t$ and $P_0 = 1$ kips.

**12.28** Rework problem 26 with $\Delta t = 0.02$ s.

**12.29** Formulate the finite difference solution for the following beam with free ends resting on an elastic foundation.

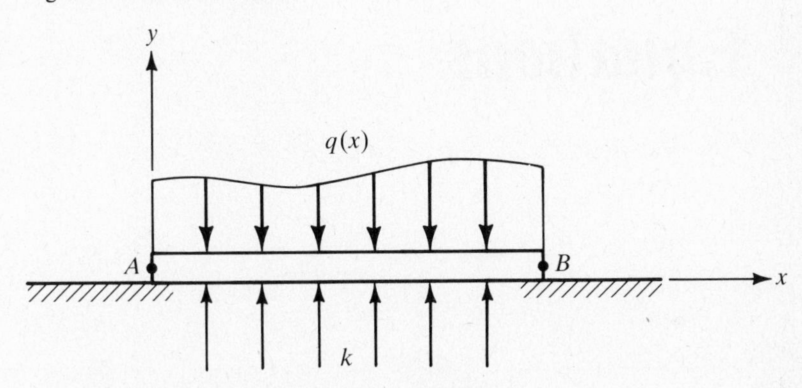


The model describing its behavior is given by

$$EI \frac{d^4y}{dx^4} = q(x) - ky$$

where $k$ and $EI$ are constants.

**12.30** Formulate the mathematical model pertaining to the following electrical circuit.

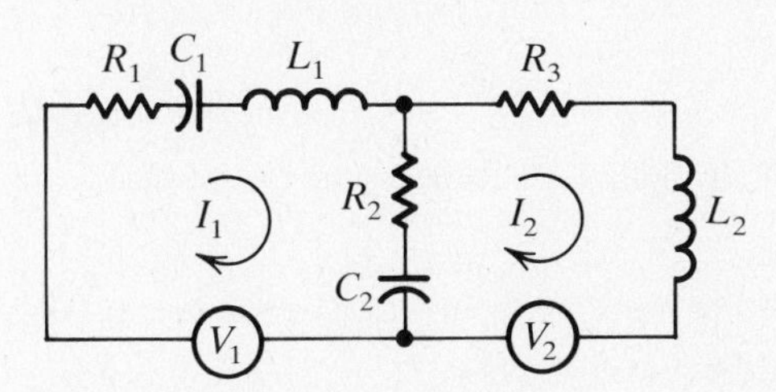


Then set up the direct integration matrices pertaining to the circuit.

# chapter 13

## Numerical Solution of Partial Differential Equations

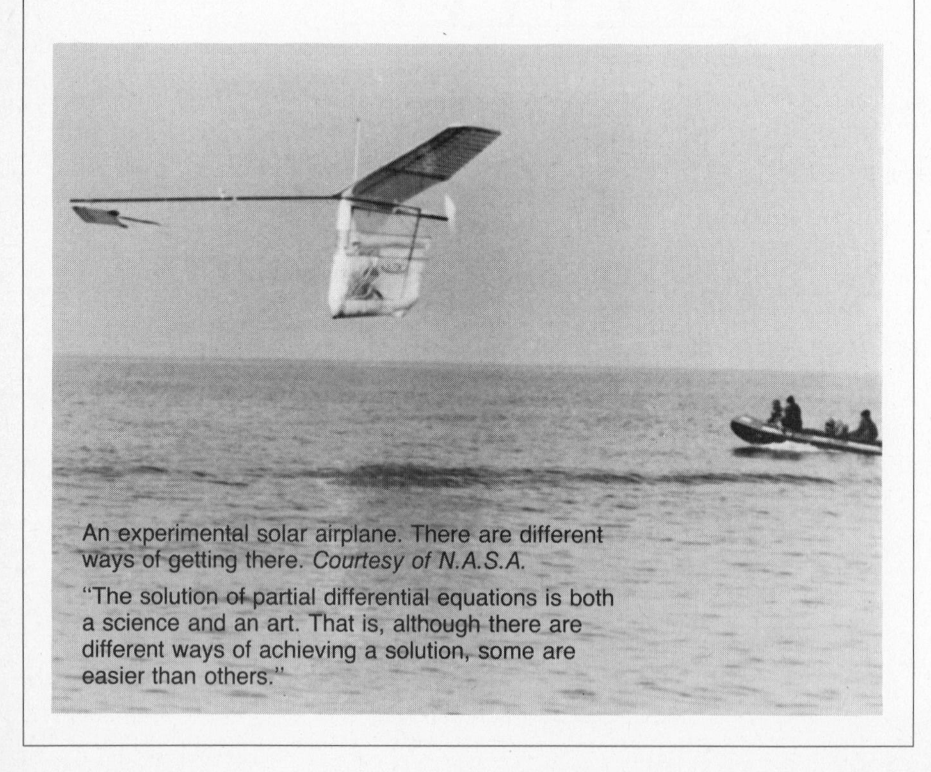

An experimental solar airplane. There are different ways of getting there. *Courtesy of N.A.S.A.*

"The solution of partial differential equations is both a science and an art. That is, although there are different ways of achieving a solution, some are easier than others."

# 13

## 13.1 INTRODUCTION

Differential equations involving two or more independent variables are referred to as partial differential equations. These equations may have only boundary conditions, in which case they are referred to as boundary value or steady-state equations. In some applications both boundary values and initial conditions are specified, in which case they are called transient problems. Note that ordinary equations are termed either initial or boundary value. A variety of steady-state and transient problems encountered in engineering practice are illustrated graphically in figure 13.1. These problems are only a few examples of the many situations in which partial differential equations may occur.

Partial differential equations are normally classified according to their mathematical form. However, in some cases they might be classified according to the particular physical problem being modeled. For example, consider the following second-order partial differential equation involving the two independent variables $x$ and $y$:

$$A\frac{\partial^2 w}{\partial x^2} + B\frac{\partial^2 w}{\partial x\,\partial y} + C\frac{\partial^2 w}{\partial y^2} + D = 0$$

Obviously, the parameters $A$, $B$, and $C$ can be functions of $x$, $y$, and $w$. Furthermore, the parameter $D$ may in some cases be a function of $x$, $y$, $\partial w/\partial x$, $\partial w/\partial y$, and $w$. However, in engineering, partial differential equations occur in a few common forms where $A$, $B$, and $C$ are constants. In this case, they are given the following mathematical classification:

$B^2 - 4AC < 0$: elliptic

$B^2 - 4AC = 0$: parabolic

$B^2 - 4AC > 0$: hyperbolic

In general, a partial differential equation can have both boundary values and initial values. Consequently, partial differential equations whose boundary conditions are specified are termed steady-state equations. On the other hand, if only initial values are specified, they are termed transient equations. Unfortunately, regardless of the differential equation type, very few analytical solutions are available. Consequently, numerical methods are used to overcome this difficulty. These are limited to the finite difference and finite element methods. In this chapter, the finite difference procedure is outlined.

## 13.2 FINITE DIFFERENCE METHODS AND GRID PATTERNS

The basis for the finite difference method of solving differential equations is that derivatives are replaced by difference expressions. The solution is obtained by

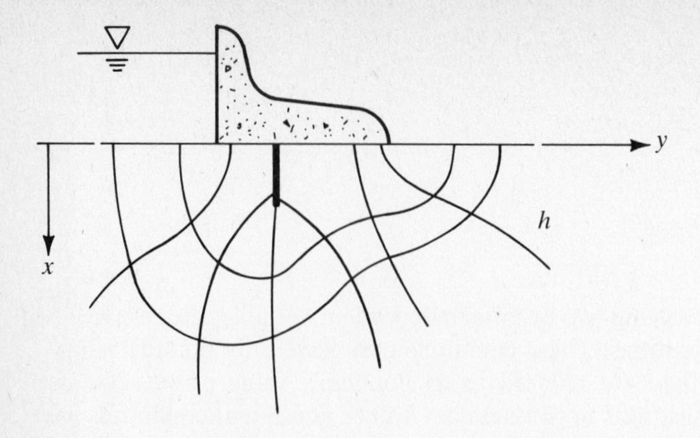


(a) Steady-state fluid flow under dams:

$$k_x \frac{\partial^2 h}{\partial x^2} + k_y \frac{\partial^2 h}{\partial y^2} = 0$$

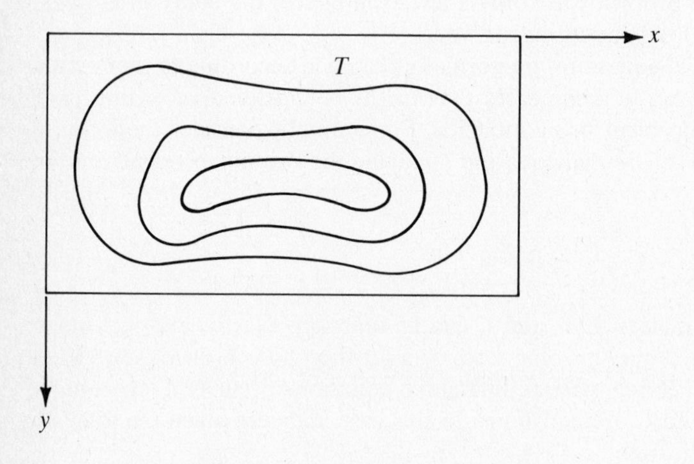


(b) Transient temperature distribution:

$$\frac{\partial^2 T}{\partial x^2} + \frac{\partial^2 T}{\partial y^2} = C \frac{\partial T}{\partial t}$$

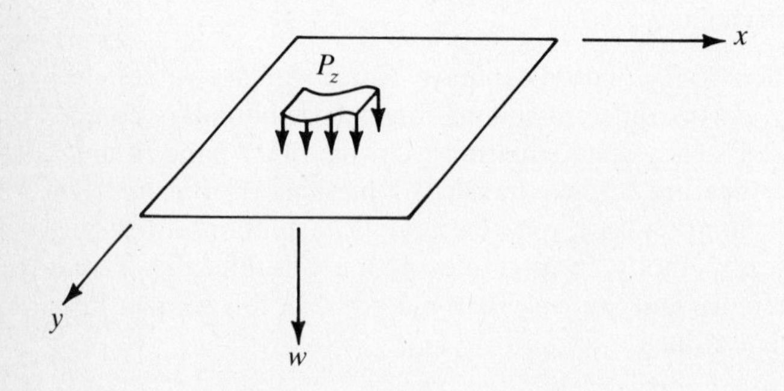


(c) Plate on elastic foundation:

$$\frac{\partial^4 w}{\partial x^4} + 2 \frac{\partial^4 w}{\partial x^2 \, \partial y^2} + \frac{\partial^4 w}{\partial y^4} = \frac{P_z}{D} - \frac{k}{D} w$$

**Figure 13.1** Practical problems encountered in engineering practice.

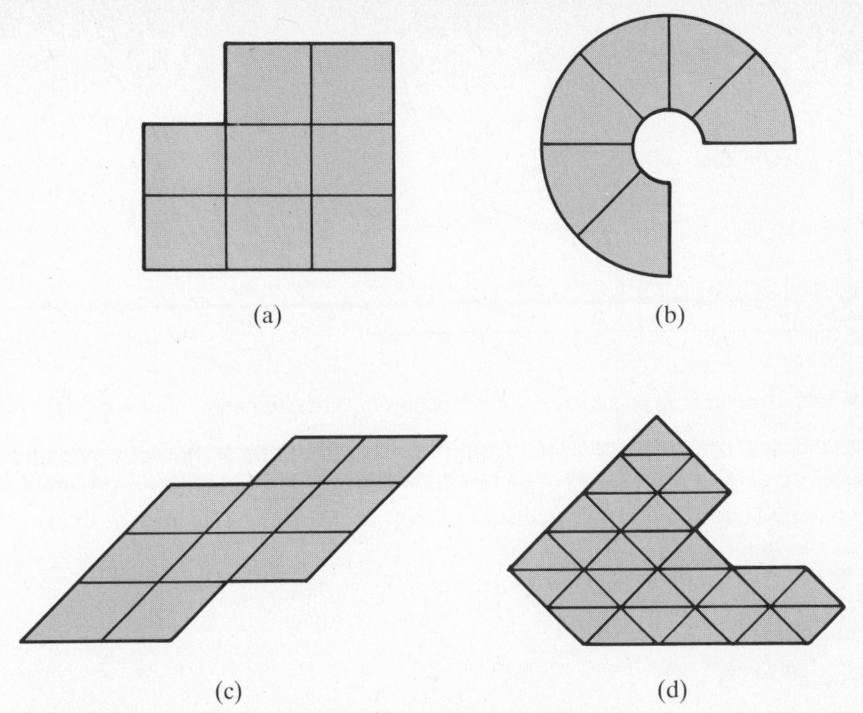


**Figure 13.2** Grid pattern for different physical regions.

first dividing the physical region into a grid of nodes. The grid shape depends to a large extent on the special nature of the physical problem being solved. A variety of these grid patterns are shown in figure 13.2 for steady-state temperatures in plates of different shapes.

Furthermore, the governing partial differential equation is transformed into the corresponding coordinate system that can best fit the chosen physical grid pattern and then expressed in finite difference form. The difference equation is then applied at each of the nodes in a given region and the functional values at these nodes are related to those nearby. Consequently, a set of linear algebraic equations is developed which can be solved for the unknown functional values after the proper boundary values have been applied.

## 13.3 TRANSFORMATION FROM CARTESIAN TO POLAR COORDINATES

While the Cartesian system of coordinates is by far the most widely used system, the polar system has proved very helpful in analyzing a variety of engineering problems. These problems include heat transfer, fluid flow, plate deflection, and many others. Consider, for example, a situation in which the variable $w$ describes a particular physical phenomenon such as deflection, pressure, etc. Assume that the partial derivatives of $w$ in the Cartesian system of coordinates are required, that these partials appear in the differential equation that describes the response of the system in question. Now let us see how that differential equation can be transformed to the polar form. For the two-dimensional case, we need the partial

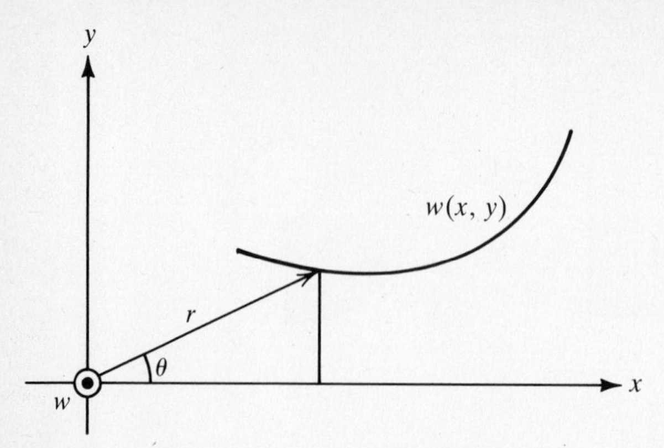


**Figure 13.3** Graphical representation of a function in Cartesian and polar systems.

derivatives of $w$ to any degree in the polar form. The situation is shown graphically in figure 13.3.

The function $w(x, y)$ can be defined at any point $(x, y)$ in polar form as follows:

$$\cos\theta = \frac{x}{r}, \qquad \sin\theta = \frac{y}{r}$$

In addition,

$$r^2 = x^2 + y^2 \tag{13.1}$$

and

$$\theta = \tan^{-1}\frac{y}{x} \tag{13.2}$$

The first partial derivatives of $w$ with respect to $x$ and $y$ are given by the chain rule as follows:

$$\frac{\partial w}{\partial x} = \frac{\partial w}{\partial r}\frac{\partial r}{\partial x} + \frac{\partial w}{\partial \theta}\frac{\partial \theta}{\partial x} \tag{13.3}$$

$$\frac{\partial w}{\partial y} = \frac{\partial w}{\partial r}\frac{\partial r}{\partial y} + \frac{\partial w}{\partial \theta}\frac{\partial \theta}{\partial y} \tag{13.4}$$

From equation (13.1) we have

$$2r\frac{\partial r}{\partial x} = 2x$$

$$\frac{\partial r}{\partial x} = \frac{x}{r} = \cos\theta$$

and

$$2r\frac{\partial r}{\partial y} = 2y$$

$$\frac{\partial r}{\partial y} = \frac{y}{r} = \sin\theta$$

From equation (13.2) we have

$$\frac{\partial \theta}{\partial x} = \frac{-y/x^2}{1 + (y/x)^2} = \frac{-y}{x^2 + y^2}$$

$$= -\frac{1}{r} \sin \theta$$

The partial of $\theta$ relative to $y$ is given by

$$\frac{\partial \theta}{\partial y} = \frac{1/x}{1 + (y/x)^2} = \frac{x}{x^2 + y^2}$$

$$= \frac{1}{r} \cos \theta$$

Substituting these relationships into equation (13.3) permits the determination of $\partial w/\partial x$ in polar form:

$$\frac{\partial w}{\partial x} = \frac{\partial w}{\partial r} (\cos \theta) + \frac{\partial w}{\partial \theta} \left( -\frac{1}{r} \sin \theta \right) \tag{13.5}$$

Similarly,

$$\frac{\partial w}{\partial y} = \frac{\partial w}{\partial r} (\sin \theta) + \frac{\partial w}{\partial \theta} \left( \frac{1}{r} \cos \theta \right) \tag{13.6}$$

The second partial derivatives are given as follows:

$$\frac{\partial^2 w}{\partial x^2} = \frac{\partial}{\partial x} \left( \frac{\partial w}{\partial x} \right) = \frac{\partial}{\partial x} (Q)$$

$$= \frac{\partial Q}{\partial r} \frac{\partial r}{\partial x} + \frac{\partial Q}{\partial \theta} \frac{\partial \theta}{\partial x}$$

$$= \cos \theta \frac{\partial Q}{\partial r} - \frac{1}{r} \sin \theta \frac{\partial Q}{\partial \theta}$$

$$= \cos \theta \frac{\partial}{\partial r} \left( \frac{\partial w}{\partial r} \cos \theta - \frac{1}{r} \sin \theta \frac{\partial w}{\partial \theta} \right)$$

$$- \frac{1}{r} \sin \theta \frac{\partial}{\partial \theta} \left( \frac{\partial w}{\partial r} \cos \theta - \frac{1}{r} \sin \theta \frac{\partial w}{\partial \theta} \right)$$

$$= \frac{\partial^2 w}{\partial r^2} \cos^2 \theta + \frac{\partial^2 w}{\partial \theta^2} \frac{\sin^2 \theta}{r^2} - 2 \frac{\partial^2 w}{\partial \theta \, \partial r} \frac{\sin \theta \cos \theta}{r}$$

$$+ \frac{\partial w}{\partial r} \frac{\sin^2 \theta}{r} + 2 \frac{\partial w}{\partial \theta} \frac{\sin \theta \cos \theta}{r^2} \tag{13.7}$$

and

$$\frac{\partial^2 w}{\partial y^2} = \frac{\partial^2 w}{\partial r^2}\sin^2\theta + \frac{\partial^2 w}{\partial \theta^2}\frac{\cos^2\theta}{r^2} + 2\frac{\partial^2 w}{\partial \theta\,\partial r}\frac{\sin\theta\cos\theta}{r}$$

$$+ \frac{\partial w}{\partial r}\frac{\cos^2\theta}{r} - 2\frac{\partial w}{\partial \theta}\frac{\sin\theta\cos\theta}{r^2} \quad \textbf{(13.8)}$$

The mixed partial derivative is determined by using the following approach:

$$\frac{\partial^2 w}{\partial x\,\partial y} = \frac{\partial}{\partial x}\left(\frac{\partial w}{\partial y}\right) = \frac{\partial R}{\partial x}$$

Therefore

$$\frac{\partial^2 w}{\partial x\,\partial y} = \frac{\partial R}{\partial x} = \frac{\partial R}{\partial r}\frac{\partial r}{\partial x} + \frac{\partial R}{\partial \theta}\frac{\partial \theta}{\partial x}$$

$$= \frac{\partial R}{\partial r}\cos\theta + \frac{\partial R}{\partial \theta}\left(-\frac{1}{r}\sin\theta\right)$$

$$= \cos\theta\,\frac{\partial}{\partial r}\left[\frac{\partial w}{\partial r}(\sin\theta) + \frac{\partial w}{\partial \theta}\left(\frac{1}{r}\cos\theta\right)\right]$$

$$- \frac{1}{r}\sin\theta\,\frac{\partial}{\partial \theta}\left[\frac{\partial w}{\partial r}(\sin\theta) + \frac{\partial w}{\partial \theta}\left(\frac{1}{r}\cos\theta\right)\right]$$

$$= \cos\theta\left(\sin\theta\,\frac{\partial^2 w}{\partial r^2} + \frac{1}{r}\cos\theta\,\frac{\partial^2 w}{\partial r\,\partial \theta} - \frac{\cos\theta}{r^2}\frac{\partial w}{\partial \theta}\right)$$

$$- \frac{1}{r}\sin\theta\left(\sin\theta\,\frac{\partial^2 w}{\partial r\,\partial \theta} + \cos\theta\,\frac{\partial w}{\partial r} + \frac{\partial^2 w}{\partial \theta^2}\frac{1}{r}\cos\theta - \frac{\partial w}{\partial \theta}\frac{1}{r}\sin\theta\right)$$

$$= \sin\theta\cos\theta\,\frac{\partial^2 w}{\partial r^2} + \frac{1}{r}(\cos^2\theta - \sin^2\theta)\frac{\partial^2 w}{\partial r\,\partial \theta} + \frac{1}{r^2}\sin\theta\cos\theta\,\frac{\partial^2 w}{\partial \theta^2}$$

$$+ \frac{1}{r^2}(\sin^2\theta - \cos^2\theta)\frac{\partial w}{\partial \theta} - \frac{1}{r}\sin\theta\cos\theta\,\frac{\partial w}{\partial r} \quad \textbf{(13.9)}$$

**E**

EXAMPLE 13.1

Transform the well-known Laplace equation of steady-state fluid flow from the Cartesian coordinate system to its equivalent polar form.

**Solution**

For two-dimensional steady-state fluid flow we have

$$\frac{\partial^2 h}{\partial x^2} + \frac{\partial^2 h}{\partial y^2} = 0$$

Hence letting $h = w$ and adding equations (13.7) and (13.8) gives

$$\begin{aligned}\frac{\partial^2 h}{\partial x^2} + \frac{\partial^2 h}{\partial y^2} &= (\cos^2\theta + \sin^2\theta)\frac{\partial^2 h}{\partial r^2} + \frac{(\sin^2\theta + \cos^2\theta)}{r^2}\frac{\partial^2 h}{\partial \theta^2} \\ &\quad + \frac{1}{r}(\sin^2\theta + \cos^2\theta)\frac{\partial h}{\partial r} \\ &= \frac{\partial^2 h}{\partial r^2} + \frac{1}{r^2}\frac{\partial^2 h}{\partial \theta^2} + \frac{1}{r}\frac{\partial h}{\partial r}\end{aligned} \tag{13.10}$$

## 13.4 TRANSFORMATION FROM CARTESIAN TO SKEWED COORDINATES

Many engineering problems involve parallelogrammatic boundaries. Fluid flow, temperature, and plate deflection are examples of such occurrences. The skewed coordinates are therefore used to replace the Cartesian form. Consider the function $w(x, y)$ shown in figure 13.4.

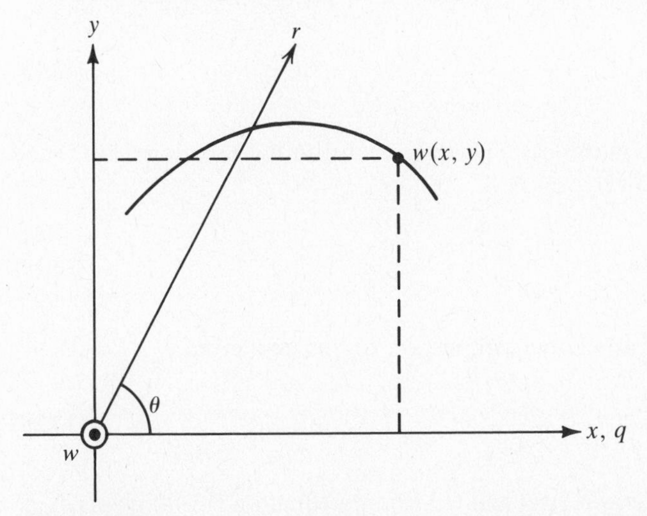


**Figure 13.4** Schematic representation of a function in skewed and Cartesian coordinates.

The position of any point $w(x, y)$ can be readily determined by using the skewed coordinate system as follows:

$$x = q + r\cos\theta \tag{13.11}$$

$$y = r\sin\theta \tag{13.12}$$

Let us approach this problem a little differently. Instead of finding the partials in the Cartesian system, we will determine the partials in the skewed coordinate system first. That is,

$$\frac{\partial w}{\partial q} = \frac{\partial w}{\partial x}\frac{\partial x}{\partial q} + \frac{\partial w}{\partial y}\frac{\partial y}{\partial q} \tag{13.13}$$

$$\frac{\partial w}{\partial r} = \frac{\partial w}{\partial x}\frac{\partial x}{\partial r} + \frac{\partial w}{\partial y}\frac{\partial y}{\partial r} \tag{13.14}$$

From equations (13.11) and (13.12) we have

$$\frac{\partial x}{\partial q} = 1 \tag{13.15}$$

$$\frac{\partial x}{\partial r} = \cos\theta \tag{13.16}$$

$$\frac{\partial y}{\partial q} = 0 \tag{13.17}$$

$$\frac{\partial y}{\partial r} = \sin\theta \tag{13.18}$$

Note that $\theta$ is a constant in this case. Therefore, substituting equations (13.15) and (13.16) into equation (13.13) gives

$$\frac{\partial w}{\partial q} = \frac{\partial w}{\partial x}(1) + \frac{\partial w}{\partial y}(0) \tag{13.19}$$

The second partial derivative of $w$ with respect to $q$ is now given as

$$\frac{\partial^2 w}{\partial q^2} = \frac{\partial}{\partial q}\left(\frac{\partial w}{\partial q}\right) = \frac{\partial}{\partial q}\left(\frac{\partial w}{\partial x}\right) = \frac{\partial^2 w}{\partial x^2} \tag{13.20}$$

Now substituting equations (13.16) and (13.18) into equation (13.14) gives

$$\frac{\partial w}{\partial r} = \frac{\partial w}{\partial x}(\cos\theta) + \frac{\partial w}{\partial y}(\sin\theta) \tag{13.21}$$

The second partial derivative of $w$ with respect to $r$ is given as follows:

$$\begin{aligned}
\frac{\partial^2 w}{\partial r^2} &= \frac{\partial}{\partial r}\left(\frac{\partial w}{\partial r}\right) = \frac{\partial Q}{\partial r} \\
&= \frac{\partial Q}{\partial x}\frac{\partial x}{\partial r} + \frac{\partial Q}{\partial y}\frac{\partial y}{\partial r} \\
&= \frac{\partial Q}{\partial x}(\cos\theta) + \frac{\partial Q}{\partial y}(\sin\theta) \\
&= \cos\theta\,\frac{\partial}{\partial x}\left(\frac{\partial w}{\partial x}\cos\theta + \frac{\partial w}{\partial y}\sin\theta\right) + \sin\theta\,\frac{\partial}{\partial y}\left(\frac{\partial w}{\partial x}\cos\theta + \frac{\partial w}{\partial y}\sin\theta\right) \\
&= \cos^2\theta\,\frac{\partial^2 w}{\partial x^2} + \sin\theta\cos\theta\,\frac{\partial^2 w}{\partial x\,\partial y} + \sin\theta\cos\theta\,\frac{\partial^2 w}{\partial x\,\partial y} + \sin^2\theta\,\frac{\partial^2 w}{\partial y^2} \\
&= \cos^2\theta\,\frac{\partial^2 w}{\partial x^2} + 2\sin\theta\cos\theta\,\frac{\partial^2 w}{\partial x\,\partial y} + \sin^2\theta\,\frac{\partial^2 w}{\partial y^2}
\end{aligned} \tag{13.22}$$

Obviously, we need to define the mixed partial derivative $\partial^2 w/\partial x\,\partial y$ if we are to determine the second partials of $w$ with respect to the skewed coordinates. Thus

$$\begin{aligned}
\frac{\partial^2 w}{\partial q\,\partial r} &= \frac{\partial}{\partial q}\left(\frac{\partial w}{\partial r}\right) = \frac{\partial R}{\partial q} \\
&= \frac{\partial R}{\partial x}\frac{\partial x}{\partial q} + \frac{\partial R}{\partial y}\frac{\partial y}{\partial q}
\end{aligned}$$

or

$$\begin{aligned}
\frac{\partial^2 w}{\partial q\,\partial r} &= \frac{\partial R}{\partial x} + \frac{\partial R}{\partial y}(0) \\
&= \frac{\partial}{\partial x}\left(\frac{\partial w}{\partial x}\cos\theta + \frac{\partial w}{\partial y}\sin\theta\right) \\
&= \cos\theta\,\frac{\partial^2 w}{\partial x^2} + \sin\theta\,\frac{\partial^2 w}{\partial x\,\partial y}
\end{aligned} \tag{13.23}$$

Equations (13.19) and (13.21) can be expressed in matrix form as

$$\begin{Bmatrix} \dfrac{\partial w}{\partial q} \\[2ex] \dfrac{\partial w}{\partial r} \end{Bmatrix} = \begin{bmatrix} 1 & 0 \\ \cos\theta & \sin\theta \end{bmatrix} \begin{Bmatrix} \dfrac{\partial w}{\partial x} \\[2ex] \dfrac{\partial w}{\partial y} \end{Bmatrix} \tag{13.24}$$

One can easily solve for the first partial derivatives in the Cartesian system using equation (13.24). The second partial derivatives in the Cartesian system can be solved for by using equations (13.20), (13.22), and (13.23) as follows:

$$\begin{Bmatrix} \dfrac{\partial^2 w}{\partial q^2} \\ \dfrac{\partial^2 w}{\partial q\, \partial r} \\ \dfrac{\partial^2 w}{\partial r^2} \end{Bmatrix} = \begin{bmatrix} 1 & 0 & 0 \\ \cos\theta & \sin\theta & 0 \\ \cos^2\theta & 2\sin\theta\cos\theta & \sin^2\theta \end{bmatrix} \begin{Bmatrix} \dfrac{\partial^2 w}{\partial x^2} \\ \dfrac{\partial^2 w}{\partial x\, \partial y} \\ \dfrac{\partial^2 w}{\partial y^2} \end{Bmatrix} \tag{13.25}$$

Equation (13.25) is a set of three equations in three unknowns that can be easily solved by using matrix inversion. Hence, rewriting equation (13.25) in compact matrix form, we have

$$\{W_s\} = [\theta]\{W_c\}$$

Solving for the unknown vector involving the partials in Cartesian coordinates gives

$$\{W_c\} = [\theta]^{-1}\{W_s\} \tag{13.26}$$

It is evident that for a given $\theta$ the relationship between the partial derivatives in the Cartesian system and the partial derivatives in the skewed system can be easily established by using equation (13.26).

E

EXAMPLE 13.2

Transform the Laplace equation of steady-state fluid flow into the skewed coordinate system using $\theta = 45°$.

Solution

Using the two-dimensional form we have

$$\frac{\partial^2 T}{\partial x^2} + \frac{\partial^2 T}{\partial y^2} = 0$$

Substituting $\theta = 45°$ into the matrix of equation (13.25) gives

$$\{W_s\} = \begin{bmatrix} 1 & 0 & 0 \\ 1/\sqrt{2} & 1/\sqrt{2} & 0 \\ 1/2 & 1 & 1/2 \end{bmatrix} \{W_c\}$$

Solving by the inverse procedure gives

$$\{W_c\} = \begin{bmatrix} 1 & 0 & 0 \\ -1 & \sqrt{2} & 0 \\ 1 & -2\sqrt{2} & 2 \end{bmatrix} \{W_s\}$$

Thus

$$\frac{\partial^2 w}{\partial x^2} = \frac{\partial^2 w}{\partial q^2} \tag{13.27}$$

$$\frac{\partial^2 w}{\partial x\, \partial y} = -\frac{\partial^2 w}{\partial q^2} + \sqrt{2}\,\frac{\partial^2 w}{\partial q\, \partial r} \tag{13.28}$$

$$\frac{\partial^2 w}{\partial y^2} = \frac{\partial^2 w}{\partial q^2} - 2\sqrt{2}\,\frac{\partial^2 w}{\partial q\, \partial r} + \frac{\partial^2 w}{\partial r^2} \tag{13.29}$$

Note that equation (13.28) is not needed in this case. Hence adding equations (13.27) and (13.29) gives

$$\frac{\partial^2 w}{\partial q^2} + \frac{\partial^2 w}{\partial q^2} - 2\sqrt{2}\,\frac{\partial^2 w}{\partial q\, \partial r} + 2\,\frac{\partial^2 w}{\partial r^2} = 0$$

Simplifying, we obtain

$$\frac{\partial^2 w}{\partial q^2} - \sqrt{2}\,\frac{\partial^2 w}{\partial q\, \partial r} + \frac{\partial^2 w}{\partial r^2} = 0 \tag{13.30}$$

Equation (13.30) is the Laplace equation in skewed coordinates with $\theta = 45°$.

## 13.5 TRANSFORMATION FROM CARTESIAN TO TRIANGULAR COORDINATES

This coordinate system is a very useful one in that it permits us to handle problems involving irregular regions. This is true because any area can be approximated by a set of triangles. Consider once again the function $w(x, y)$ shown in figure 13.5.

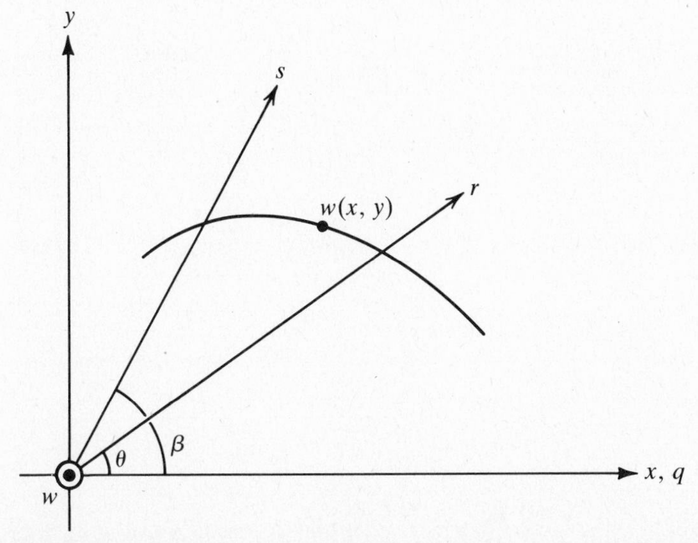


**Figure 13.5** Schematic representation of the triangular coordinate system.

In fact, triangular elements represent the building block for many finite element techniques.

$$x = q + r\cos\theta + s\cos\beta \tag{13.31}$$

$$y = r\sin\theta + s\sin\beta \tag{13.32}$$

Note that $x$ and $y$ are expressed in terms of three variables. Therefore

$$\frac{\partial w}{\partial q} = \frac{\partial w}{\partial x}\frac{\partial x}{\partial q} + \frac{\partial w}{\partial y}\frac{\partial y}{\partial q} \tag{13.33}$$

$$\frac{\partial w}{\partial r} = \frac{\partial w}{\partial x}\frac{\partial x}{\partial r} + \frac{\partial w}{\partial y}\frac{\partial y}{\partial r} \tag{13.34}$$

$$\frac{\partial w}{\partial s} = \frac{\partial w}{\partial x}\frac{\partial x}{\partial s} + \frac{\partial w}{\partial y}\frac{\partial y}{\partial s} \tag{13.35}$$

The first partial derivatives of $x$ and $y$ with respect to $q$, $r$, and $s$ are readily determined from equations (13.31) and (13.32):

$$\frac{\partial x}{\partial q} = 1 \qquad \frac{\partial y}{\partial q} = 0$$

$$\frac{\partial x}{\partial r} = \cos\theta \qquad \frac{\partial y}{\partial r} = \sin\theta$$

$$\frac{\partial x}{\partial s} = \cos\beta \qquad \frac{\partial y}{\partial s} = \sin\beta$$

Therefore equations (13.33), (13.34), and (13.35) are now given as

$$\frac{\partial w}{\partial q} = \frac{\partial w}{\partial x}$$

$$\frac{\partial w}{\partial r} = \cos\theta\frac{\partial w}{\partial x} + \sin\theta\frac{\partial w}{\partial y}$$

$$\frac{\partial w}{\partial s} = \cos\beta\frac{\partial w}{\partial x} + \sin\beta\frac{\partial w}{\partial y}$$

The second partial derivatives, computed as outlined in sections 13.3 and 13.4, are given directly as

$$\frac{\partial^2 w}{\partial q^2} = \frac{\partial^2 w}{\partial x^2}$$

$$\frac{\partial^2 w}{\partial r^2} = \cos^2\theta\frac{\partial^2 w}{\partial x^2} + 2\sin\theta\cos\theta\frac{\partial^2 w}{\partial x\,\partial y} + \sin^2\theta\frac{\partial^2 w}{\partial y^2}$$

$$\frac{\partial^2 w}{\partial s^2} = \cos^2\beta\frac{\partial^2 w}{\partial x^2} + 2\sin\beta\cos\beta\frac{\partial^2 w}{\partial x\,\partial y} + \sin^2\beta\frac{\partial^2 w}{\partial y^2}$$

These equations can be readily expressed in a matrix form as follows:

$$\begin{Bmatrix} \dfrac{\partial^2 w}{\partial q^2} \\ \dfrac{\partial^2 w}{\partial r^2} \\ \dfrac{\partial^2 w}{\partial s^2} \end{Bmatrix} = \begin{bmatrix} 1 & 0 & 0 \\ \cos^2\theta & 2\sin\theta\cos\theta & \sin^2\theta \\ \cos^2\beta & 2\sin\beta\cos\beta & \sin^2\beta \end{bmatrix} \begin{Bmatrix} \dfrac{\partial^2 w}{\partial x^2} \\ \dfrac{\partial^2 w}{\partial x\,\partial y} \\ \dfrac{\partial^2 w}{\partial y^2} \end{Bmatrix} \quad \textbf{(13.36a)}$$

or in compact matrix form

$$\{w_t\} = [A]\{w_c\} \quad \textbf{(13.36b)}$$

Equations (13.36) can be solved by using matrix inversion once $\theta$ and $\beta$ are specified. Note that we could easily invert the coefficient matrix for any $\theta$ and $\beta$; however, the resulting matrix is mathematically unattractive.

## 13.6 FINITE DIFFERENCE SOLUTION OF ELLIPTIC EQUATIONS

One of the simplest elliptic partial differential equations is that of the steady-state temperature distribution in a two-dimensional body. The partial differential equation describing such behavior is known as the Laplace equation and is given by

$$\frac{\partial^2 T}{\partial x^2} + \frac{\partial^2 T}{\partial^2 y} = 0 \quad \textbf{(13.37)}$$

This equation also arises in steady-state fluid flow in civil engineering and steady-state voltage distribution in a conductive medium in electrical engineering. Furthermore, if the right-hand side of equation (13.37) is equal to the scalar $-2.0$ then it is referred to as the Poisson equation, which is used to describe heat transfer problems for a body with internal heating source. In fact, Poisson's equation is frequently used in determining the stress function for a given cross section subjected to twisting. That is,

$$\frac{\partial^2 \Phi}{\partial x^2} + \frac{\partial^2 \Phi}{\partial y^2} = -2\Phi \quad \textbf{(13.38)}$$

Equation (13.38) is solved for the boundary conditions $\Phi_b = 0$, where $\Phi_b$ is the boundary stress function value.

It is evident that while it is possible to address each one of these elliptic partial differential equations separately, such an approach is repetitious. Therefore we shall limit our presentation to Laplace's equation for steady-state temperature distribution in a two-dimensional body.

### 13.6.1 Regions Involving Rectangular Elements

Recall that the solution of ordinary differential equations using the finite difference procedure involves replacing the derivatives with difference expression of known

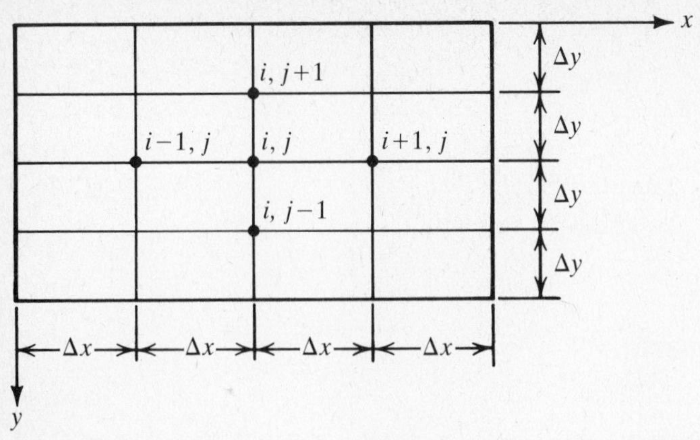


**Figure 13.6** Two-dimensional steady-state temperature distribution in a rectangular plate.

order or error. This concept can be easily extended to the solution of partial differential equations. To illustrate the procedure, let us consider the steady-state temperature problems modeled by equation (13.37). Furthermore, suppose that the physical region under consideration is a rectangle (figure 13.6). The object is to approximate the steady-state temperatures at any interior node $(x_i, y_i)$ in terms of boundary temperatures and/or their derivatives. The region is first divided into an appropriate grid of smaller rectangular elements and the resulting nodes are numbered. Consequently, for node $(i, j)$ the partial differential equation given by (13.37) is expressed in difference form by using table 10.2 to obtain the following difference expressions for local error order of $(O)h^2$:

$$\frac{\partial^2 T}{\partial x^2} \simeq \frac{1}{(\Delta x)^2}\left(\textcircled{1}\!-\!\textcircled{-2}\!-\!\textcircled{1}\right) \tag{13.39a}$$

$$\frac{\partial^2 T}{\partial y^2} \simeq \frac{1}{(\Delta y)^2}\left(\textcircled{1}\!-\!\textcircled{-2}\!-\!\textcircled{1}\right) \tag{13.39b}$$

These expressions are defined at node $(i, j)$ in terms of the temperatures as

$$\left(\frac{\partial^2 T}{\partial x^2}\right)_{i,j} = \frac{1}{(\Delta x)^2}(T_{i-1,j} - 2T_{i,j} + T_{i+1,j}) \tag{13.40a}$$

Note that the partial derivative in the $x$ direction is obtained by holding $y$ constant. This is precisely what equation (13.40a) states in that $(j)$ is held constant. Similarly,

$$\left(\frac{\partial^2 T}{\partial y^2}\right)_{i,j} = \frac{1}{(\Delta y)^2}(T_{i,j-1} - 2T_{i,j} + T_{i,j+1}) \tag{13.40b}$$

Substituting equation (13.40) into equation (13.37) gives the desired difference equation. That is,

$$\frac{1}{(\Delta x)^2}(T_{i-1,j} - 2T_{i,j} + T_{i+1,j}) + \frac{1}{(\Delta y)^2}(T_{i,j-1} - 2T_{i,j} + T_{i,j+1}) = 0 \tag{13.41}$$

Letting $\alpha = (\Delta y/\Delta x)^2$ and substituting into equation (13.41) gives

$$\alpha T_{i-1,j} - 2(\alpha)T_{i,j} + T_{i+1,j} + T_{i,j-1} + T_{i,j+1} - 2T_{i,j} = 0 \tag{13.42}$$

Equation (13.42) can be expressed more conveniently in stencil form, as described in chapter 10, to give

$$\left( \begin{array}{ccc} & 1 & \\ \alpha & -2(\alpha+1) & \alpha \\ & 1 & \end{array} \right) = 0 \tag{13.43}$$

$\Delta y$

$\Delta x$

For the special case in which $\Delta x = \Delta y$ the value of $\alpha$ is 1.0; thus the stencil is simplified to

$$\left( \begin{array}{ccc} & 1 & \\ 1 & -4 & 1 \\ & 1 & \end{array} \right) = 0 \tag{13.44}$$

It is interesting to note that the stencil (13.43) is determined by combining the difference expressions given by equations (13.39) as follows:

$$\frac{\partial^2 T}{\partial x^2} + \frac{\partial^2 T}{\partial y^2} = \frac{1}{(\Delta x)^2}\left( \begin{array}{ccc} 1 & -2 & 1 \end{array} \right) + \frac{1}{(\Delta y)^2}\left( \begin{array}{c} 1 \\ -2 \\ 1 \end{array} \right) = 0$$

$$= \begin{array}{ccc} & 1 & \\ \alpha & -2(\alpha+1) & \alpha \\ & 1 & \end{array} = 0$$

The difference equation can now be used to solve problems involving constant boundary temperatures. Keep in mind that the local errors are of order $(O)h^2$. Hence better results are obtained for smaller intervals.

E

EXAMPLE 13.3a

Determine the steady-state temperature of the following plate using $\alpha = 1.0$ and $\Delta x = 1$ ft:

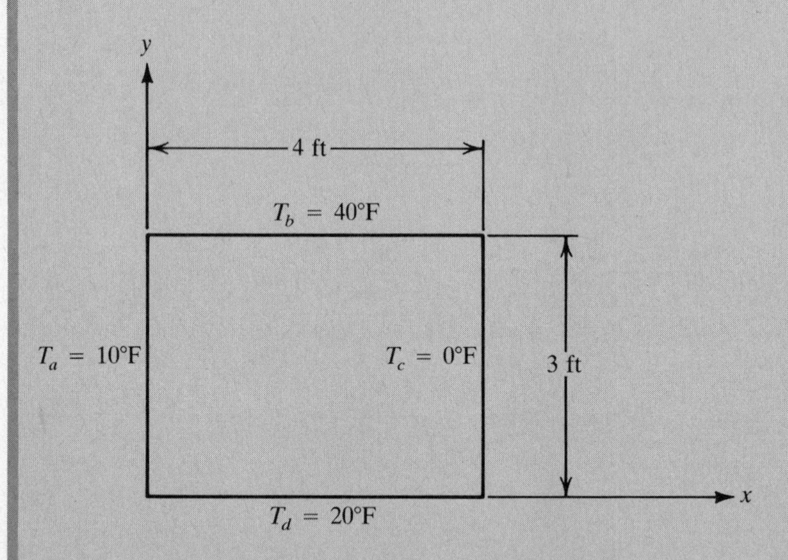


## Solution

As a first step, divide the region into the specified increments then number the resulting interior nodes. That is,

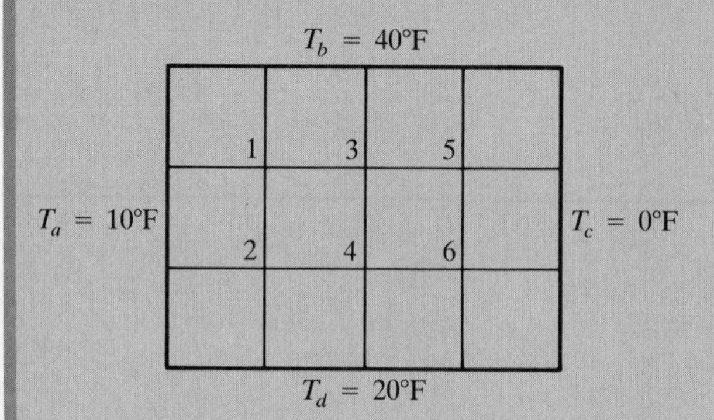


Note that one should proceed by numbering the short side first in order to reduce the bandwidth. This is especially important when one is dealing with a large number of nodes.

Placing the double circle on each numbered node results in a set of six linear algebraic equations as follows:

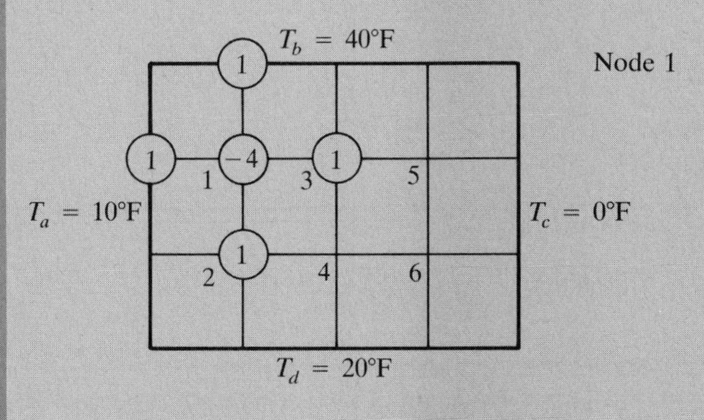


$$10 + 40 - 4T_1 + T_2 + T_3 = 0$$

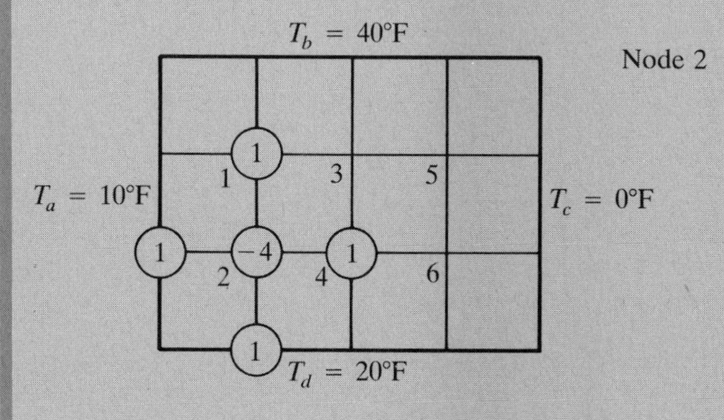


$$10 + 20 + T_1 - 4T_2 + T_4 = 0$$

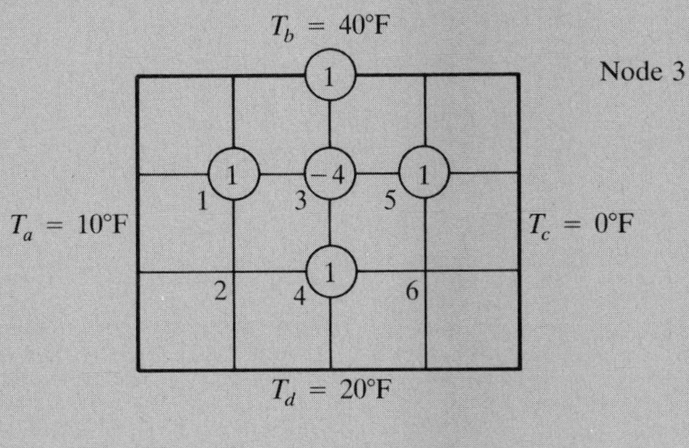


$$40 + T_1 - 4T_3 + T_4 + T_5 = 0$$

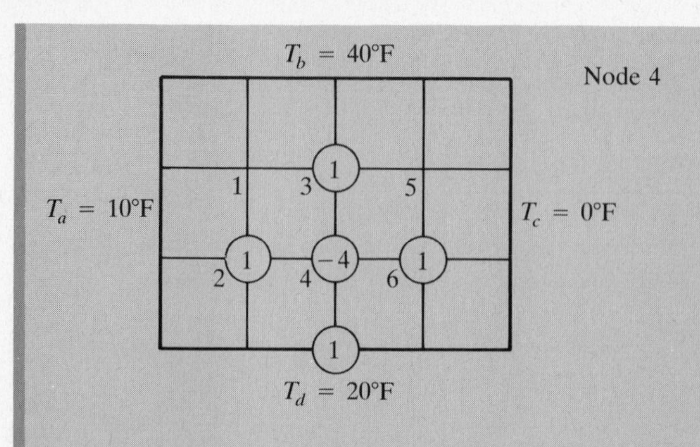


$$20 + T_2 + T_3 - 4T_4 + T_6 = 0$$

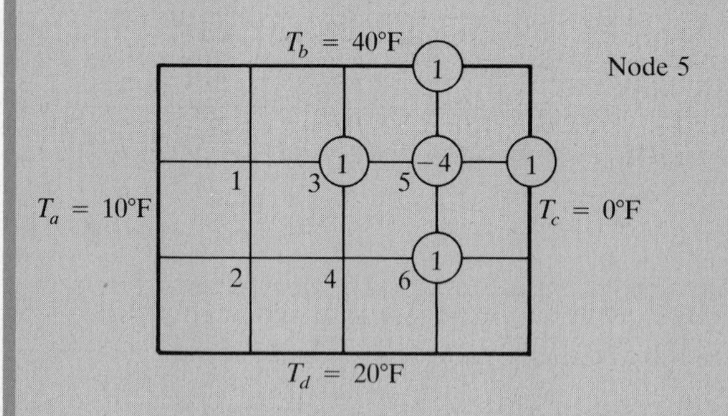


$$40 + T_3 - 4T_5 + T_6 = 0$$

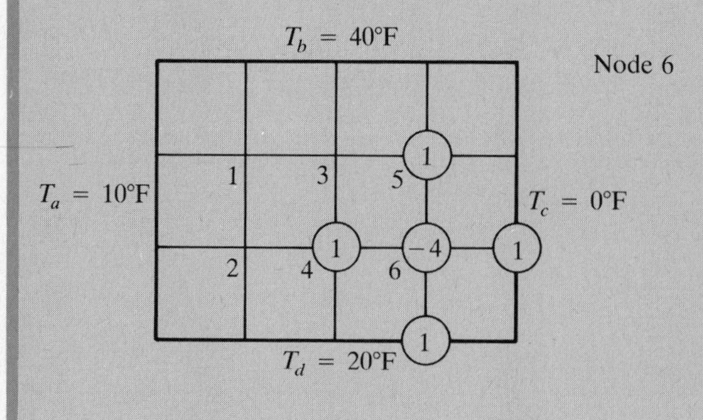


$$20 + T_4 + T_5 - 4T_6 = 0$$

The linear algebraic equations corresponding to each node can now be expressed in the following matrix form:

$$\begin{bmatrix} -4 & 1 & 1 & 0 & 0 & 0 \\ 1 & -4 & 0 & 1 & 0 & 0 \\ 1 & 0 & -4 & 1 & 1 & 0 \\ 0 & 1 & 1 & -4 & 0 & 1 \\ 0 & 0 & 1 & 0 & -4 & 1 \\ 0 & 0 & 0 & 1 & 1 & -4 \end{bmatrix} \begin{Bmatrix} T_1 \\ T_2 \\ T_3 \\ T_4 \\ T_5 \\ T_6 \end{Bmatrix} = \begin{Bmatrix} -50 \\ -30 \\ -40 \\ -20 \\ -40 \\ -20 \end{Bmatrix}$$

Solving for the steady-state temperatures yields

$$T_1 = 23.5611°\text{F} \qquad T_4 = 19.8136°\text{F}$$

$$T_2 = 18.3437°\text{F} \qquad T_5 = 20.2277°\text{F}$$

$$T_3 = 25.9006°\text{F} \qquad T_6 = 15.0104°\text{F}$$

These are the temperature values at the six nodes at time infinity. Note that better accuracy can be attained by using a smaller increment than the 1.0 ft used in arriving at this solution. However, this would require solving a larger set of equations.

# E

EXAMPLE 13.3b

Rework example 13.3a assuming arbitrary boundary temperatures $T_a$, $T_b$, $T_c$, and $T_d$.

## Solution

We begin by formulating the nodal equations as follows:

node 1: $T_a + T_b - 4T_1 + T_2 + T_3 = 0$

node 2: $T_a + T_d + T_1 - 4T_2 + T_4 = 0$

node 3: $T_b + T_1 - 4T_3 + T_4 + T_5 = 0$

node 4: $T_d + T_2 + T_3 - 4T_4 + T_6 = 0$

node 5: $T_b + T_c + T_3 - 4T_5 + T_6 = 0$

node 6: $T_c + T_d + T_4 + T_5 - 4T_6 = 0$

These equations can be easily expressed in matrix form to give:

$$\begin{bmatrix} -4 & 1 & 1 & 0 & 0 & 0 \\ 1 & -4 & 0 & 1 & 0 & 0 \\ 1 & 0 & -4 & 1 & 1 & 0 \\ 0 & 1 & 1 & -4 & 0 & 1 \\ 0 & 0 & 1 & 0 & -4 & 1 \\ 0 & 0 & 0 & 1 & 1 & -4 \end{bmatrix} \begin{Bmatrix} T_1 \\ T_2 \\ T_3 \\ T_4 \\ T_5 \\ T_6 \end{Bmatrix} = - \begin{bmatrix} 1 & 1 & 0 & 0 \\ 1 & 0 & 0 & 1 \\ 0 & 1 & 0 & 0 \\ 0 & 0 & 0 & 1 \\ 0 & 1 & 1 & 0 \\ 0 & 0 & 1 & 1 \end{bmatrix} \begin{Bmatrix} T_a \\ T_b \\ T_c \\ T_d \end{Bmatrix}$$

or in a compact form, we have

$$[A]\{T\} = [B]\{\bar{T}\}$$

Thus

$$\{T\} = [A]^{-1}[B]\{\bar{T}\}$$

Multiplying the inverse of $[A]$ times the $[B]$ matrix and then multiplying the resulting matrix by the boundary temperatures vectors $\{\bar{T}\}$ gives

$$T_1 = \tfrac{1}{2415}[920T_a + 1005T_b + 115T_c + 375T_d]$$

$$T_2 = \tfrac{1}{2415}[920T_a + 375T_b + 115T_c + 1005T_d]$$

$$T_3 = \tfrac{1}{2415}[345T_a + 1230T_b + 345T_c + 495T_d]$$

$$T_4 = \tfrac{1}{2415}[345T_a + 495T_b + 345T_c + 1230T_d]$$

$$T_5 = \tfrac{1}{2415}[115T_a + 1005T_b + 920T_c + 375T_d]$$

$$T_6 = \tfrac{1}{2415}[115T_a + 375T_b + 920T_c + 1005T_d]$$

The implication of these equations is rather significant in that one need not solve this particular problem again. This is because the solution is given for any boundary temperature. Also note that the sum of the coefficients of the boundary temperatures is equal to 2415. That is, these coefficients are weight functions whose sums for $T_1$ through $T_6$ are equal to 1.0.

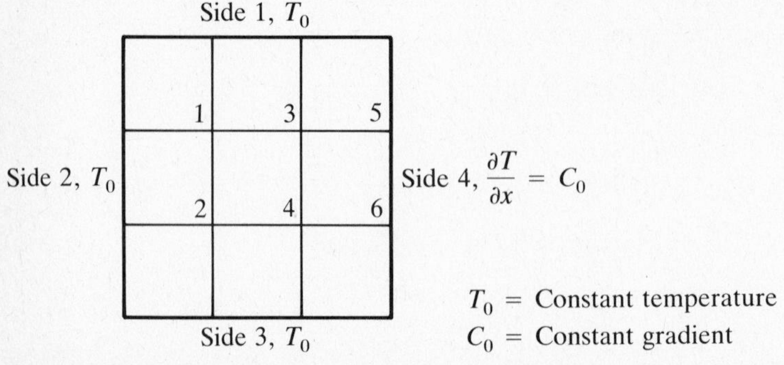


**Figure 13.7** Two-dimensional steady-state temperature problem with constant and derivative boundaries.

The boundary temperatures for certain problems may not be known; instead the derivative of the temperature may be specified. For such problems the stencil given by equation (13.43) must be modified for nodes adjacent to the derivative boundary. Consider figure 13.7.

The temperature is specified at three of the four boundaries, while the fourth boundary temperatures are not known. Therefore they must be calculated as if they were interior nodes. The difficulty with this is that when the double circle of the stencil given by equation (13.6) is placed at nodes 4 and 5 the circle on the left side will be outside the region of interest. Consequently, it must be eliminated by relating it to interior nodal temperatures. Consider side 4:

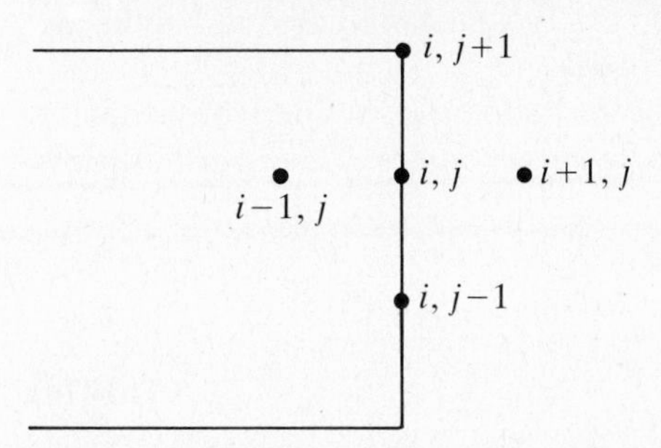


The first derivative at node $(i, j)$ is defined in finite difference form by using table 10.1:

$$\frac{\partial T}{\partial x} = \frac{1}{2\,\Delta x}\left( \textcircled{-1} - \textcircled{\textcircled{0}} - \textcircled{1} \right)$$

or simply

$$\frac{\partial T}{\partial x} = \frac{1}{2\,\Delta x}(-T_{i-1,j} + 0 + T_{i+1,j}) = C_0 \qquad \textbf{(13.45)}$$

Solving for $T_{i+1,j}$ yields

$$T_{i+1,j} = 2\,\Delta x\, C_0 + T_{i-1,j}$$

Consequently, equation (13.42) is modified for node $(i, j)$ by substituting equation (13.45) for $T_{i+1,j}$:

$$\alpha T_{i-1,j} - 2\alpha T_{i,j} + \alpha(2\,\Delta x\, C_0 + T_{i-1,j}) + T_{i,j-1} + T_{i,j-1} - 2T_{i,j} = 0$$

$$2\alpha T_{i-1,j} - 2(\alpha + 1)T_{i,j} + T_{i,j-1} + T_{i,j+1} - 2\alpha\,\Delta x\, C_0 = 0$$

or in stencil form at node $(i, j)$,

$$\frac{\partial^2 T}{\partial x^2} + \frac{\partial^2 T}{\partial y^2} = 0 = \left( \begin{array}{c} \textcircled{1} \\ | \\ \textcircled{2\alpha} - \textcircled{\textcircled{-2(\alpha + 1)}} + 2\alpha\,\Delta x\, C_0 \\ | \\ \textcircled{1} \end{array} \right) \qquad \textbf{(13.46a)}$$

For the special case in which $\alpha = 1$, the difference equation becomes

$$\begin{array}{cc} & 1 \\ 2 & -4 \\ & 1 \end{array} = -2\,\Delta x\, C_0 \qquad \textbf{(13.46b)}$$

The stencils given by equations (13.46) and (13.12) can now be used to handle any derivative boundary nodes, except corner nodes.

For the special case in which two adjacent boundaries are derivative boundaries, the stencil must be modified to handle corner nodes. This is given below.

$$\left( \begin{array}{cc} & 2 \\ 2\alpha & -2(\alpha + 1) \end{array} + 2\alpha\,\Delta x\, C_0 \right) = 0 \qquad \textbf{(13.47a)}$$

For problems in which $\Delta x = \Delta y$, the stencil is reduced to

$$\left( \begin{array}{cc} & 1 \\ 1 & -2 \end{array} + \Delta x\, C_0 \right) = 0 \qquad \textbf{(13.47b)}$$

EXAMPLE 13.4

Determine the nodal temperatures for the following plate assuming that $\Delta x = \Delta y = 1$ ft:

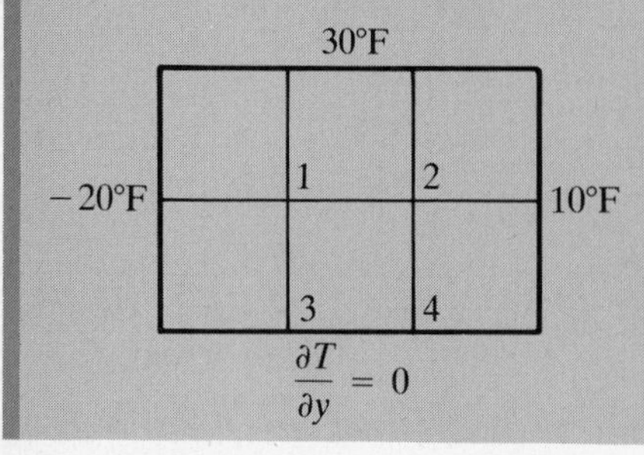

## Solution

Using the stencil given by equation (13.44) for nodes 1 and 2 gives

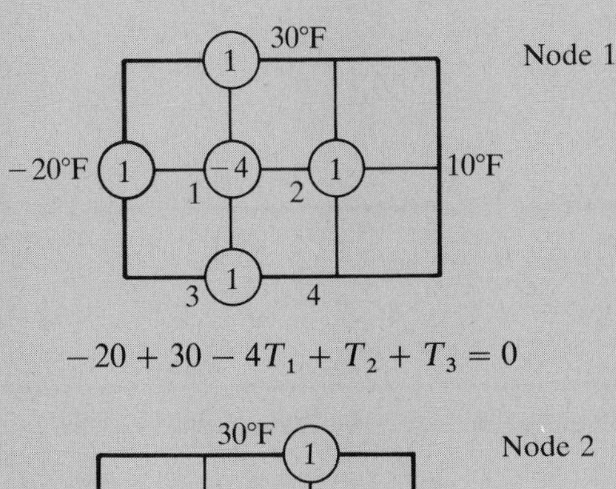


$$-20 + 30 - 4T_1 + T_2 + T_3 = 0$$

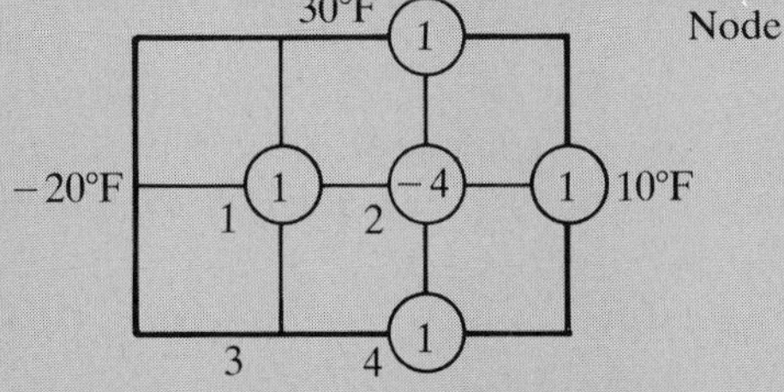


$$10 + 30 + T_1 - 4T_2 + T_4 = 0$$

The stencil given by equation (13.46b) is now applied at nodes 3 and 4 to give

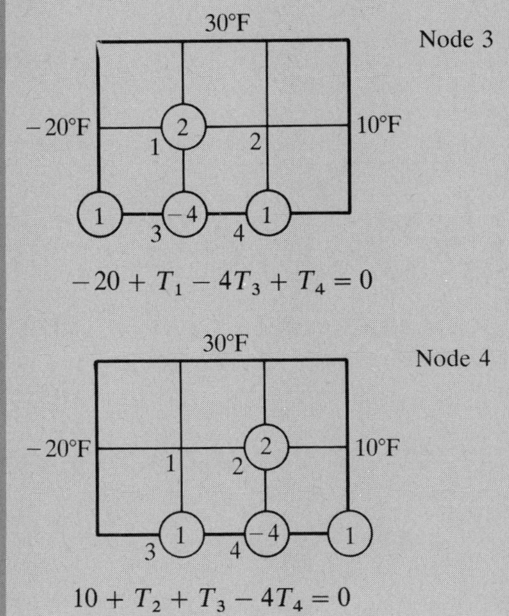


$$-20 + T_1 - 4T_3 + T_4 = 0$$

$$10 + T_2 + T_3 - 4T_4 = 0$$

It is evident that the corner nodes present a problem in that one may argue that these are common to both the derivative and the constant boundaries, hence making it inappropriate to asssume that they are constant. This problem can be resolved by increasing the number of nodes, thus reducing the errors involved. Thus, solving for the nodal temperatures yields

$T_1 = 5.00°F \qquad T_3 = -2.50°F$

$T_2 = 12.50°F \qquad T_4 = 5.00°F$

Better results can be obtained by using a finer mesh.

### 13.6.2 Circular Regions

Our scheme for solving the Laplace equation for steady-state temperature in a rectangular region can be extended to analyze circular bodies. For this special case, the region being analyzed is subdivided into sectors and the mathematical model [equation (13.37)] is transformed into an equivalent polar form. This is accomplished by using equation (13.10) and by simply replacing $\omega$ with $T$. Hence

$$\frac{\partial^2 T}{\partial x^2} + \frac{\partial^2 T}{\partial y^2} = \frac{\partial^2 T}{\partial r^2} + \frac{1}{r^2}\frac{\partial^2 T}{\partial \theta^2} + \frac{1}{r}\frac{\partial T}{\partial r} = 0 \qquad \textbf{(13.48)}$$

The polar form presented by equation (13.48) is to be transformed into a difference equation. Consider figure 13.8.

The partial derivative terms appearing in equation (13.48) can be defined at node 1 by using the difference expressions given in table 10.2:

$$\frac{\partial^2 T}{\partial r^2} = \frac{1}{A^2}(T_5 - 2T_1 + T_3) \qquad \textbf{(13.49a)}$$

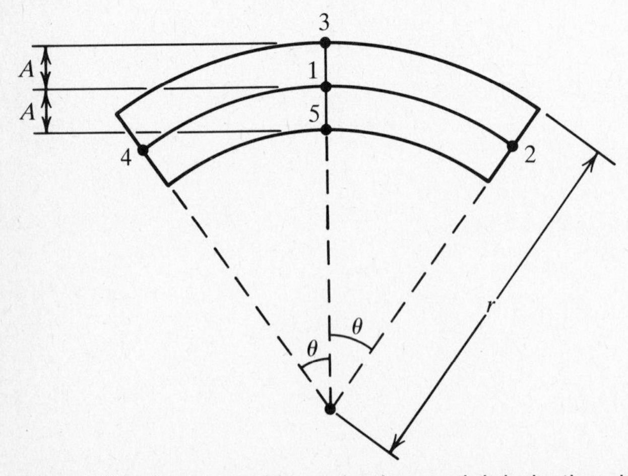


**Figure 13.8** Sector used in evaluating partial derivatives in polar form.

$$\frac{\partial^2 T}{\partial \theta^2} = \frac{1}{\theta^2}(T_4 - 2T_1 + T_2) \tag{13.49b}$$

$$\frac{\partial T}{\partial r} = \frac{1}{2A}(-T_5 + T_3) \tag{13.49c}$$

Substituting equation (13.49) into (13.48) yields the desired finite difference expression. Thus

$$\frac{\partial^2 T}{\partial x^2} + \frac{\partial^2 T}{\partial y^2} = \frac{1}{A^2}(T_5 - 2T_1 + T_3) + \frac{1}{r^2\theta^2}(T_4 - 2T_1 + T_2) + \frac{1}{2rA}(-T_5 + T_3)$$

Simplifying, we obtain

$$\frac{\partial^2 T}{\partial x^2} + \frac{\partial^2 T}{\partial y^2} = -\left(\frac{2}{A^2} + \frac{2}{r^2\theta^2}\right)T_1 + \left(\frac{1}{r^2\theta^2}\right)T_2 + \left(\frac{1}{A^2} + \frac{1}{2rA}\right)T_3 + \frac{1}{r^2\theta^2}T_4 + \left(\frac{1}{A^2} - \frac{1}{2rA}\right)T_5 \tag{13.50}$$

Equation (13.50) can be expressed in stencil form for the general case as follows:

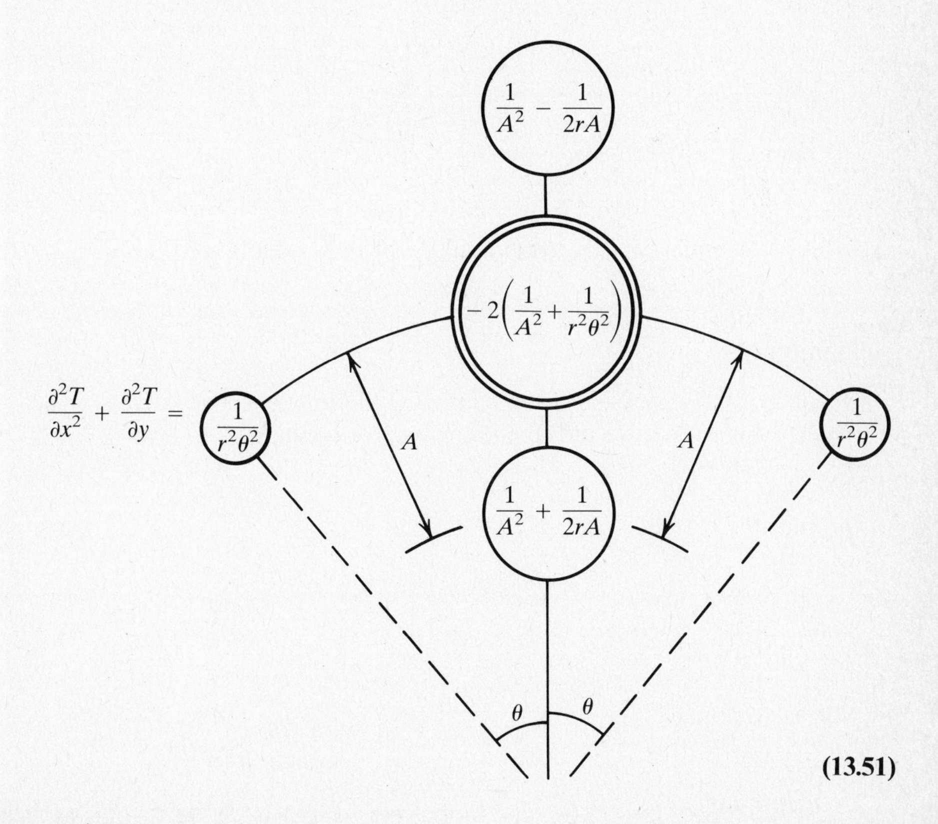


(13.51)

It is evident that for a given problem different stencils may have to be computed.

EXAMPLE 13.5

Determine the steady-state temperature for the following, at six nodes:

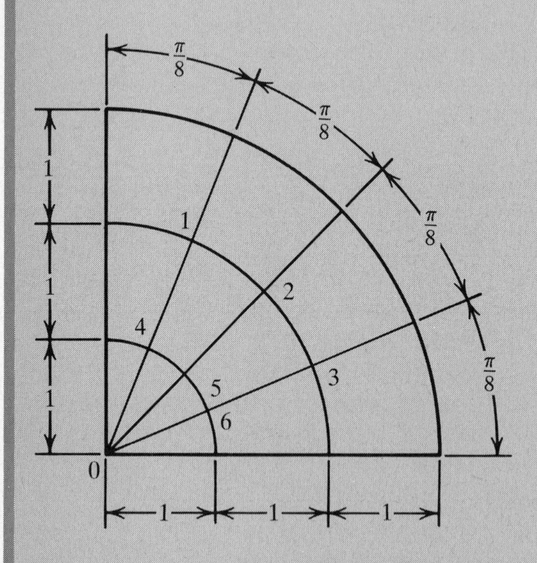


where the boundary temperature is 100°C at node 0 and zero elsewhere.

## Solution

Obviously, two stencils must be computed for this problem because even though $\theta$ and $A$ are constants the radius $r$ corresponding to nodes 1, 2, 3 is 2 and for nodes 4, 5, 6 is 1.0. Therefore

For nodes 1, 2, and 3:

$$r = 2$$

$$\theta = \frac{\pi}{8}$$

$$A = 1$$

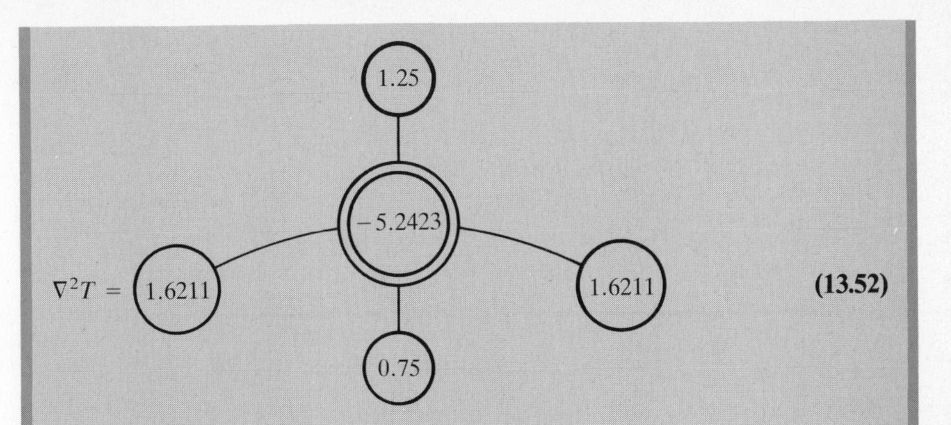


**(13.52)**

For nodes 4, 5, and 6:

$$r = 1$$

$$\theta = \frac{\pi}{8}$$

$$A = 1$$

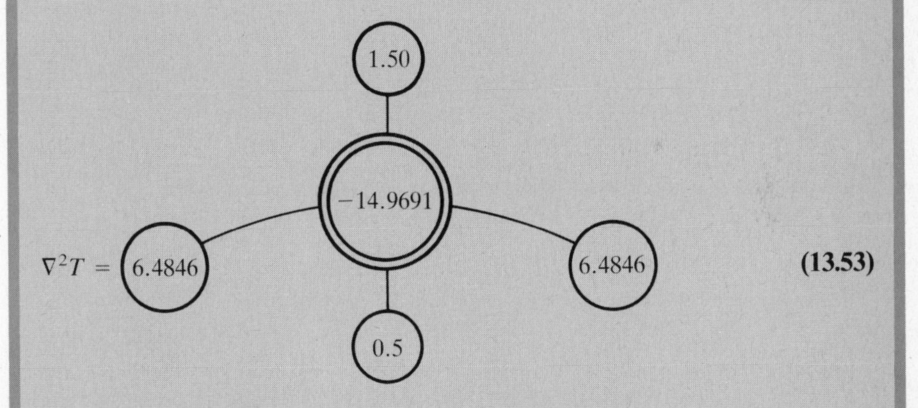


**(13.53)**

Applying the stencil given by equation (13.52) at nodes 1, 2, and 3 gives

Node 1:

$$-5.2423T_1 + 0.75T_4 + 1.6211T_2 = 0$$

Node 2:

$$-5.2423T_2 + 1.6211T_1 + 1.6211T_3 + 0.75T_5 = 0$$

Node 3:

$$-5.2423T_3 + 1.6211T_2 + 0.75T_6 = 0$$

Applying the stencil given by equation (13.53) at nodes 4, 5, and 6 yields the remaining three equations:

Node 4:

$$-14.9691T_4 + 1.5T_1 + 6.4846T_5 + 0.5(100) = 0$$

Node 5:

$$-14.9691T_5 + 6.4846T_4 + 6.4846T_6 + 1.5T_2 + 0.5(100) = 0$$

Node 6:

$$-14.9691T_6 + 6.4846T_5 + 1.5T_3 + 0.5(100) = 0$$

These equations are now arranged in matrix form to give

$$\begin{bmatrix} -5.2423 & 1.6211 & 0 & 0.75 & 0 & 0 \\ 1.6211 & -5.2423 & 1.6211 & 0 & 0.75 & 0 \\ 0 & 1.6211 & -5.2423 & 0 & 0 & 0.75 \\ 1.5 & 0 & 0 & -14.9691 & 6.4846 & 0 \\ 0 & 1.5 & 0 & 6.4846 & -14.9691 & 6.4846 \\ 0 & 0 & 1.5 & 0 & 6.4846 & -14.9691 \end{bmatrix}$$

$$\begin{Bmatrix} T_1 \\ T_2 \\ T_3 \\ T_4 \\ T_5 \\ T_6 \end{Bmatrix} = \begin{Bmatrix} 0 \\ 0 \\ 0 \\ -50 \\ -50 \\ -50 \end{Bmatrix}$$

Solving for the unknown temperatures gives

$$T_1 = 2.03°\text{C} \qquad T_4 = 8.18°\text{C}$$

$$T_2 = 2.79°\text{C} \qquad T_5 = 10.71°\text{C}$$

$$T_3 = 2.03°\text{C} \qquad T_6 = 8.18°\text{C}$$

It is interesting to note that the resulting temperatures are symmetrical about nodes 1 and 5, as they should be (why?).

### 13.6.3 Regions Involving Skewed Elements

In this section the steady-state temperature distribution problem for a parallelogram physical element is addressed. Suppose that a given region is of the form shown in figure 13.9. Then, subdividing it into elements having parallel boundaries

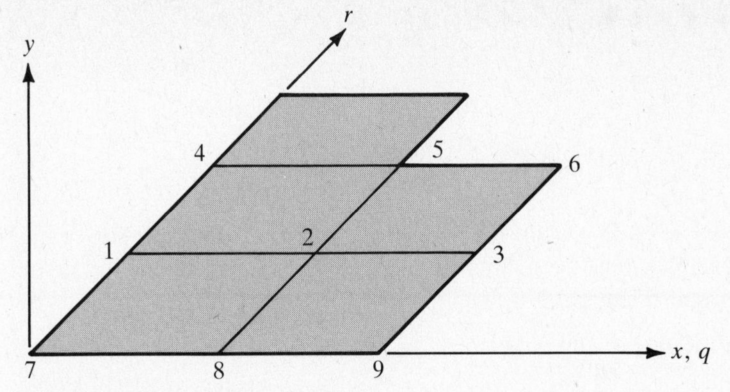

**Figure 13.9** Typical skewed region.

permits us to analyze the overall region using the skewed coordinate system. Our objective is to express the governing differential equation for steady-state temperature distribution [equation (13.37)] in skewed coordinates $(q, r)$. Although the transformation is possible for any $\theta$ value, the procedure is greatly simplified for a fixed $\theta$. Therefore, for the special case $\theta = 60°$ the transformation is readily given by equation (13.25) as follows:

$$\begin{Bmatrix} \dfrac{\partial^2 T}{\partial q^2} \\ \dfrac{\partial^2 w}{\partial q\,\partial r} \\ \dfrac{\partial^2 w}{\partial r^2} \end{Bmatrix} = \begin{bmatrix} 1 & 0 & 0 \\ \dfrac{1}{2} & \dfrac{\sqrt{3}}{2} & 0 \\ \dfrac{1}{4} & \dfrac{\sqrt{3}}{2} & \dfrac{3}{4} \end{bmatrix} \begin{Bmatrix} \dfrac{\partial^2 T}{\partial x^2} \\ \dfrac{\partial^2 w}{\partial x\,\partial y} \\ \dfrac{\partial^2 w}{\partial y^2} \end{Bmatrix}$$

Solving for the partial derivatives relative to $x$ and $y$ using matrix inversion yields

$$\frac{\partial^2 T}{\partial x^2} = \frac{\partial^2 T}{\partial q^2}$$

$$\frac{\partial^2 T}{\partial y^2} = \frac{1}{3}\frac{\partial^2 T}{\partial q^2} + \frac{4}{3}\frac{\partial^2 T}{\partial r^2} - \frac{4}{3}\frac{\partial^2 T}{\partial q\,\partial r}$$

Substituting these expressions into equation (13.37) yields the transformed Laplace equation for steady-state temperature. Hence

$$\frac{\partial^2 T}{\partial x^2} + \frac{\partial^2 T}{\partial y^2} = \frac{4}{3}\left(\frac{\partial^2 T}{\partial q^2} + \frac{\partial^2 T}{\partial r^2} - \frac{4}{3}\frac{\partial^2 T}{\partial q\,\partial r}\right) \tag{13.54}$$

Note that for a different $\theta$ value the transformed equation is different. In any case, we now approximate the partial derivatives appearing on the right-hand side of equation (13.54) using the finite difference procedure. Therefore, at node $z$ in figure

13.9, the partial derivatives are given as follows:

$$\frac{\partial^2 T}{\partial q^2} = \frac{1}{(\Delta q)^2}(T_1 - 2T_2 + T_3)$$

$$\frac{\partial^2 T}{\partial r^2} = \frac{1}{(\Delta r)^2}(T_5 - 2T_2 + T_8)$$

Obviously, the order of error is $(O)h^2$ for both of these approximations. Hence the mixed partial derivative must be evaluated using finite differences having error orders of $(O)h^2$ as well. That is,

$$\frac{\partial^2 T}{\partial q\,\partial r} = \frac{\partial}{\partial q}\left(\frac{\partial T}{\partial r}\right) = \frac{\partial}{\partial q}\left(\frac{1}{2\,\Delta r}(-T_8 + T_5)\right)$$

$$= \frac{1}{2\,\Delta r}\left(\frac{\partial T_8}{\partial q} + \frac{\partial T_5}{\partial q}\right)$$

$$= \frac{1}{2\,\Delta r}\left(-\frac{1}{2\,\Delta q}(-T_7 + T_9) + \frac{1}{2\,\Delta q}(-T_4 + T_6)\right)$$

$$= \frac{1}{4\,\Delta r\,\Delta q}(-T_4 + T_6 + T_7 - T_9)$$

The partial derivative approximations are now substituted into equation (13.54) to give the desired difference equation in skewed coordinates:

$$\frac{\partial^2 T}{\partial x^2} + \frac{\partial^2 T}{\partial y^2} = \frac{4}{3(\Delta q)^2}(T_1 - 2T_2 + T_3) + \frac{4}{3(\Delta r)^2}(T_5 - 2T_2 + T_8)$$
$$- \frac{4}{12\,\Delta q\,\Delta r}(-T_4 + T_6 + T_7 - T_9) \qquad \textbf{(13.55)}$$

Now, if we let $\alpha = \Delta q/\Delta r$, then equation (13.55) is given more conveniently in stencil form as

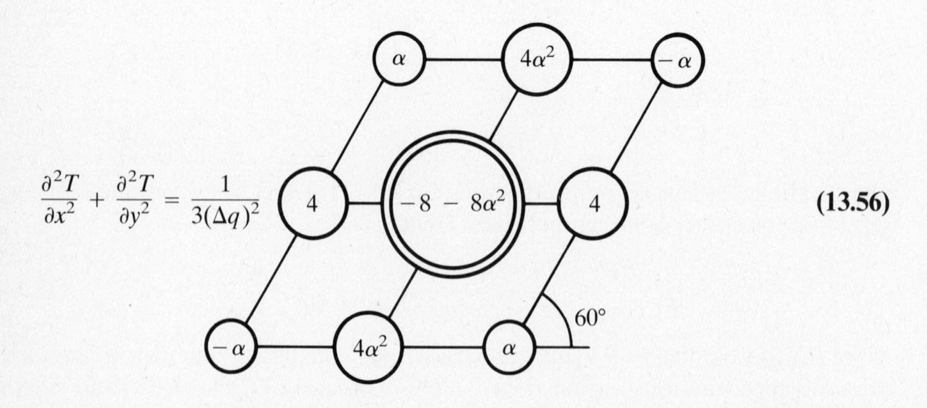


It is evident that equation (13.56) should only be applied for regions having $\theta = 60°$.

E

EXAMPLE 13.6

Calculate the nodal temperatures for the following region:

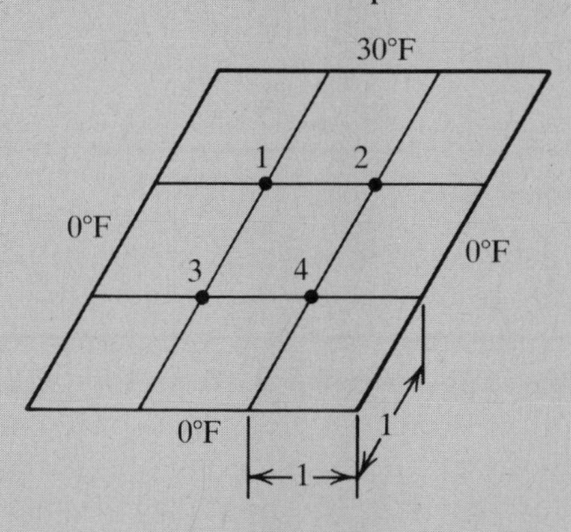


## Solution

It is clear that $\Delta q = \Delta r = 1$; therefore $\alpha = 1$. The corresponding stencil is readily determined from equation (13.56) as follows:

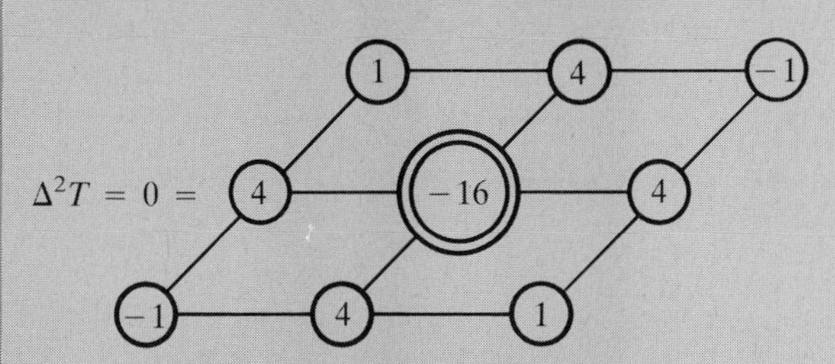


Applying the stencil at the four interior nodes yields four linear algebraic equations in the unknown temperatures.

Node 1:

$$30 + 30(4) - 30 - 16T_1 + 4T_2 + 4T_3 + T_4 = 0$$

Node 2:

$$30 + 30(4) - 30 - 4T_1 - 16T_2 - T_3 + 4T_4 = 0$$

Node 3:

$$4T_1 - T_2 - 16T_3 + 4T_4 = 0$$

Node 4:

$$T_1 + 4T_2 + 4T_3 - 16T_4 = 0$$

Solving for the temperatures at the interior nodes gives

$$T_1 = 11.38°\text{F} \qquad T_3 = 3.23°\text{F}$$

$$T_2 = 11.23°\text{F} \qquad T_4 = 4.32°\text{F}$$

Note that the degree of accuracy attained is directly related to the number of nodes used. That is, from the practical standpoint, we are only limited by computer memory.

### 13.6.4 Regions Involving Triangular Elements

It is evident that dividing a region into different grids of elements requires that the mathematical model be transformed by means of a coordinate system which best fits these elements. The triangular coordinate system plays a significant role in modern numerical methods in that any region can be approximated by a grid made up of triangular elements. In fact, such elements represent the foundation for many finite element schemes. For illustration purposes, let us consider the steady-state temperature mathematical model given by equation (13.37). Suppose that the region to be analyzed is of the form given in figure 13.10.

It is apparent that the triangular coordinate system is best suited for analyzing the region in question. Therefore equation (13.37) is readily transformed into its equivalent form by using the triangular coordinates $q$, $r$, and $s$ by specifying the angles $\beta$ and $\theta$. In this case, suppose that the individual elements are equilateral. The implication is that $\beta = 120°$ and $\theta = 60°$. Using equation (13.36a) we have

$$\begin{Bmatrix} \dfrac{\partial^2 T}{\partial q^2} \\ \dfrac{\partial^2 T}{\partial r^2} \\ \dfrac{\partial^2 T}{\partial s^2} \end{Bmatrix} = \begin{bmatrix} 1 & 0 & 0 \\ \dfrac{1}{4} & \dfrac{\sqrt{3}}{2} & \dfrac{3}{4} \\ \dfrac{1}{4} & -\dfrac{\sqrt{3}}{2} & \dfrac{3}{4} \end{bmatrix} \begin{Bmatrix} \dfrac{\partial^2 T}{\partial x^2} \\ \dfrac{\partial^2 T}{\partial x\, \partial y} \\ \dfrac{\partial^2 T}{\partial y^2} \end{Bmatrix}$$

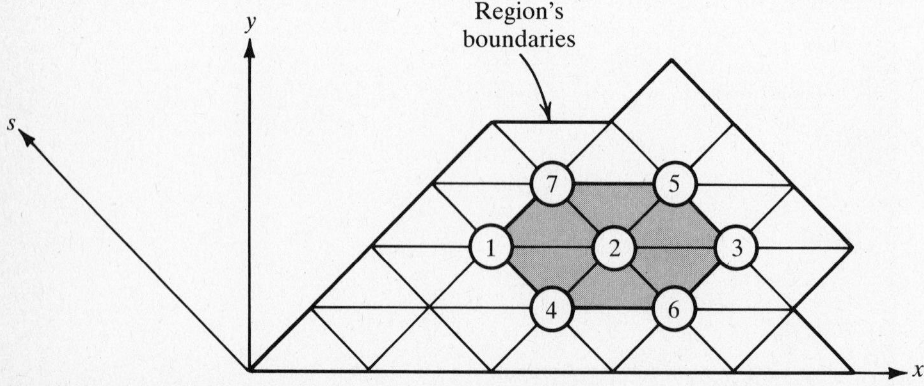


**Figure 13.10** Typical triangular grid constituting a region.

Solving for the partial derivatives in $x$ and $y$ using matrix inversion yields

$$\frac{\partial^2 T}{\partial x^2} = \frac{\partial^2 T}{\partial q^2}$$

$$\frac{\partial^2 T}{\partial y^2} = -\frac{1}{3}\frac{\partial^2 T}{\partial q^2} + \frac{2}{3}\frac{\partial^2 T}{\partial r^2} + \frac{2}{3}\frac{\partial^2 T}{\partial s^2}$$

Substituting into equation (13.57) gives the steady-state temperature equation in triangular coordinates. That is,

$$\frac{\partial^2 T}{\partial x^2} + \frac{\partial^2 T}{\partial y^2} = \frac{2}{3}\left(\frac{\partial^2 T}{\partial q^2} + \frac{\partial^2 T}{\partial r^2} + \frac{\partial^2 T}{\partial s^2}\right) \qquad \textbf{(13.57)}$$

As a result the partial derivatives relative to the coordinates $q$, $r$, and $s$ can now be approximated by using table 10.2. Therefore at node 2 (shown in figure 13.10) we have

$$\frac{\partial^2 T}{\partial q^2} = \frac{1}{(\Delta q)^2}(T_1 - 2T_2 + T_3)$$

$$\frac{\partial^2 T}{\partial r^2} = \frac{1}{(\Delta r)^2}(T_4 - 2T_2 + T_5)$$

$$\frac{\partial^2 T}{\partial s^2} = \frac{1}{(\Delta s)^2}(T_6 - 2T_2 + T_7)$$

Clearly these partial derivative approximations involve errors of order $(O)h^2$. Substituting into equation (13.57) and noting that for equilateral triangular elements $\Delta q = \Delta r = \Delta s = h$ yields the desired difference equation:

$$\frac{\partial^2 T}{\partial x^2} + \frac{\partial^2 T}{\partial y^2} = \frac{2}{3h^2}(T_1 - 6T_2 + T_3 + T_4 + T_5 + T_6 + T_7) \qquad \textbf{(13.58)}$$

Equation (13.58) is normally given more conveniently in the following stencil form:

$$\frac{\partial^2 T}{\partial x^2} + \frac{\partial^2 T}{\partial y^2} = \frac{2}{3h^2}\left\{\begin{array}{ccccc} & 1 & & 1 & \\ 1 & & -6 & & 1 \\ & 1 & & 1 & \end{array}\right\} = 0 \qquad \textbf{(13.59)}$$

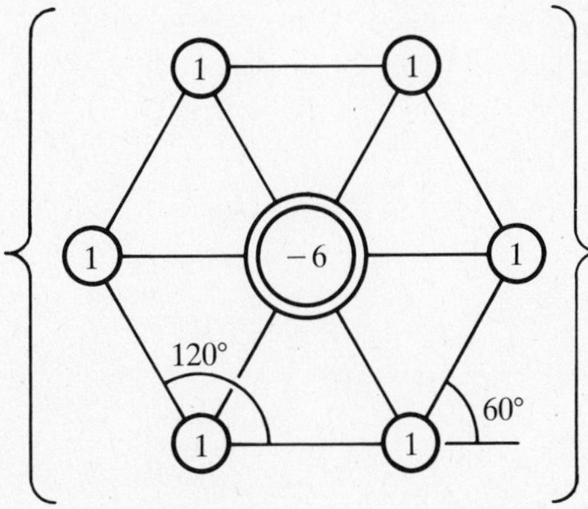

120°

60°

The reader is reminded that choosing different triangular elements is possible by simply specifying the angles $\beta$ and $\theta$.

E

EXAMPLE 13.7

Calculate the approximate steady-state temperatures at the interior nodes of the following triangular region:

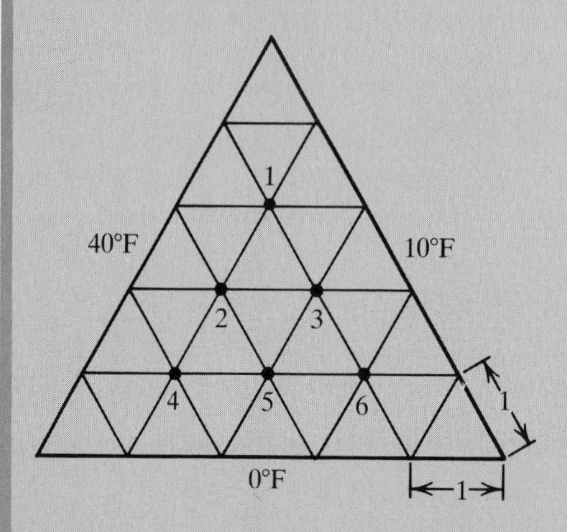


### Solution

Placing the stencil given by equation (13.59) at the various nodes yields six equations in the unknown temperatures at the six nodes.

Node 1:

$$40 + 10 + 40 + 10 - 6T_1 + T_2 + T_3 = 0$$

Node 2:

$$40 + 40 + T_1 - 6T_2 + T_3 + T_4 + T_5 = 0$$

Node 3:

$$10 + 10 + T_1 + T_2 - 6T_3 + T_5 + T_6 = 0$$

Node 4:

$$40 + 40 + T_2 - 6T_4 + T_5 + 0 + 0 = 0$$

Node 5:

$$T_2 + T_3 + T_4 - 6T_5 + T_6 + 0 + 0 = 0$$

Node 6:

$$10 + 10 + T_3 + T_5 - 6T_6 + 0 + 0 = 0$$

Solving for the unknown temperatures gives

$$T_1 = 23.17°\text{F} \qquad T_4 = 19.27°\text{F}$$

$$T_2 = 24.63°\text{F} \qquad T_5 = 10.98°\text{F}$$

$$T_3 = 14.39°\text{F} \qquad T_6 = 7.56°\text{F}$$

More accurate values can be obtained by using a finer mesh. Furthermore, the physical region need not be a triangular one for the method to work. Instead, as long as it is possible to establish a mesh whose individual elements are triangular in form, then this method of solution is applicable.

### 13.6.5 Special Considerations

The numerical solution techniques of elliptic partial differential equations are by no means limited to Laplace's equation ($\nabla^2 T = 0$). Nor are they limited to the specific regions illustrated throughout this section. Furthermore, these techniques are equally valid for analyzing three-dimensional bodies. For example, the finite difference solution of the steady-state temperature problem in two dimensions [equation (13.37)] is applicable to a variety of regions made up of grids of rectangles. The implication is that we may look at these rectangular elements as if they are building blocks from which a region can be built. Hence, the regions shown in figure 13.11 can easily be analyzed using the difference equation given by (13.43).

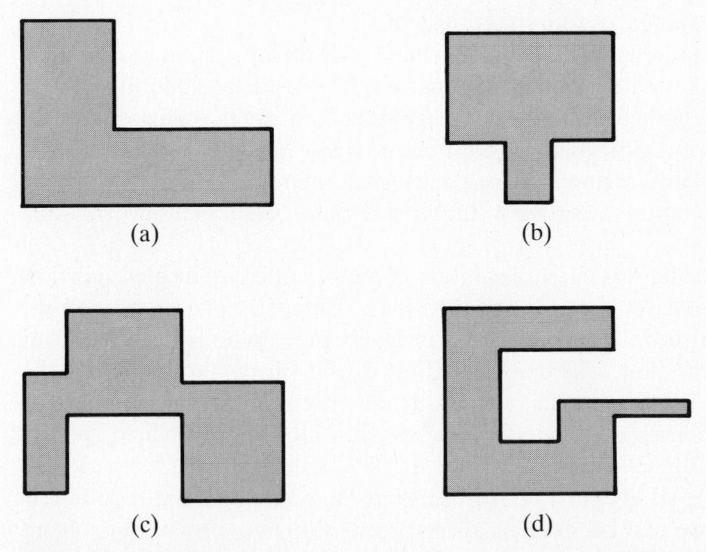


**Figure 13.11** Examples of physical regions that can be analyzed using the Cartesian coordinate system.

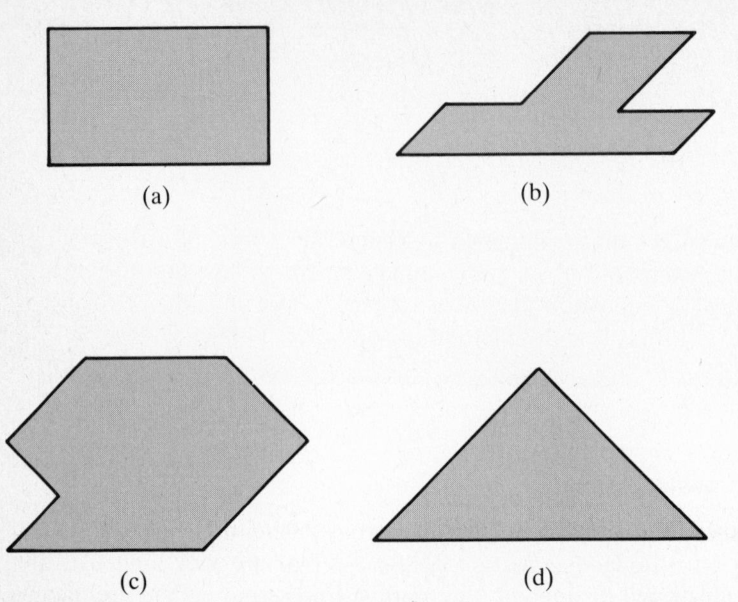


**Figure 13.12** Examples of physical regions that can be analyzed using the triangular coordinate system.

This is true since any one of these regions can be easily subdivided as shown into a set of smaller rectangular elements for which the Cartesian coordinate system is best suited. Furthermore, any region whose boundaries are made up of straight lines that intersect at $90^\circ$ angles can also be analyzed using the Cartesian system. This concept is also applicable in the case of polar, skewed, and triangular coordinate system. Figure 13.12 illustrates different physical regions that can be analyzed using the triangular coordinate system.

It is interesting to note that the region given by figure 13.12a can also be analyzed by using the skewed coordinate system (why?). The various coordinate systems covered are versatile tools that can be applied effectively in solving problems involving a variety of regions with smooth boundaries. While it is impossible for us to list every possible region that one may encounter, it suffices to say that imagination and common sense are valuable attributes when dealing with this type of problem.

One aspect of the finite difference solution of partial differential equations that has not been addressed yet is that of symmetry. The symmetry of a given problem can be a valuable asset in reducing the set of algebraic equations resulting from the application of the finite difference equation to the various nodes within the region in question. While it might not be obvious, the term symmetry refers to both physical boundaries and boundary values. Consider the rectangular region shown in figure 13.13.

Suppose that the steady-state temperatures at the nine nodes are required. It is evident that any one of these three cases can be easily analyzed by simply solving nine equations in the unknown temperatures. However, this may not be the most efficient method of tackling these problems. For example, figure 13.13a clearly

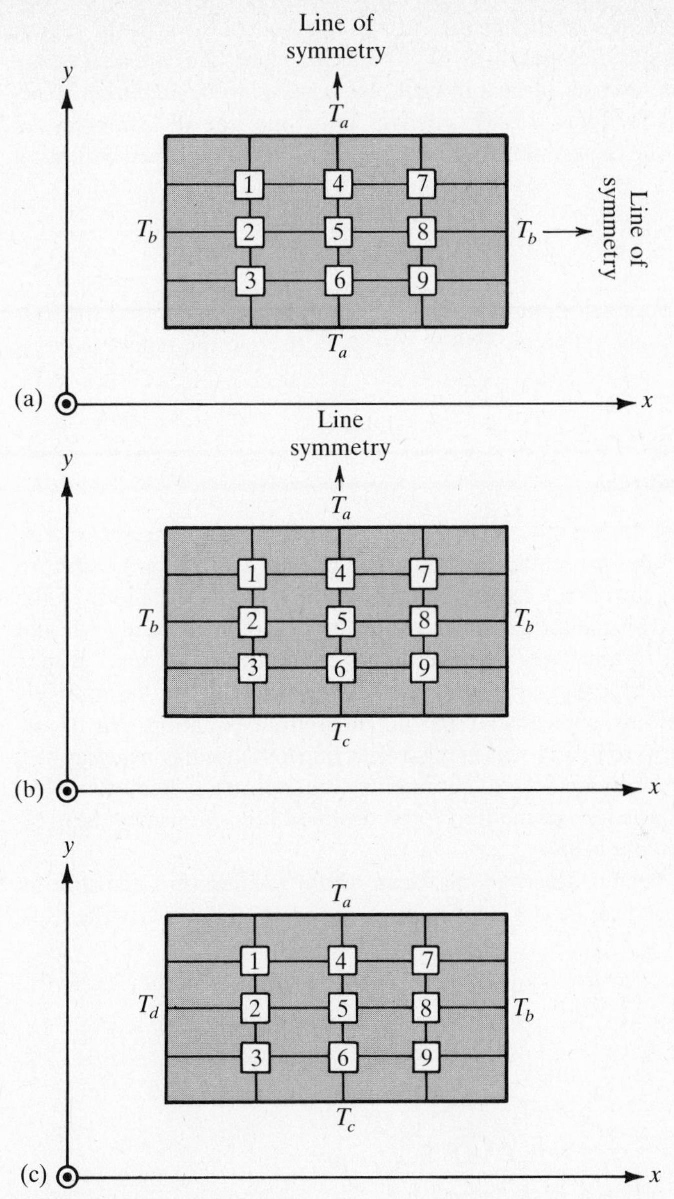


**Figure 13.13** Graphical illustration of symmetry determination.

indicates that, owing to the symmetrical nature of the physical boundaries and their boundary values, the number of unknowns may be reduced to only four nodes as follows:

$$T_1 = T_3 = T_7 = T_9$$

$$T_2 = T_4 = T_6 = T_8$$

This is because the boundary temperatures at the top and bottom of the region are equal and the boundary temperatures at the left and right sides are also equal. Consequently, only one-quarter of the physical region needs to be analyzed. However, the region given in figure 13.13b involves only one line of symmetry, in which case one-half of the region must be analyzed. This is accomplished by noting that

$$T_1 = T_7, \qquad T_2 = T_8, \qquad T_3 = T_9$$

Furthermore, the region given in figure 13.13c is not symmetrical in that the four boundary temperatures are different. This is true even though the region is physically symmetrical.

### 13.6.6 Irregular Boundaries

Occasionally, the region to be analyzed by means of the finite difference procedure may involve one or more irregular boundaries. That is, while it is feasible to subdivide the entire region into elements corresponding to one or more of the shapes considered thus far, the elements nearest to the irregular boundary will not necessarily correspond to any form we have discussed. With this in mind, it may be sensible to agree on a particular coordinate system whereby, the finite difference equation representing a particular partial differential equation can be developed in a general form to fit any region regardless of geometrical considerations. Since the Cartesian coordinate system represents the simplest system available, we shall use it as the basis for the method to be outlined here. Consider the arbitrary region given in figure 13.14.

Suppose that this region describes the steady-state temperature distribution problem whose mathematical model is given by equation (13.37). Clearly the solu-

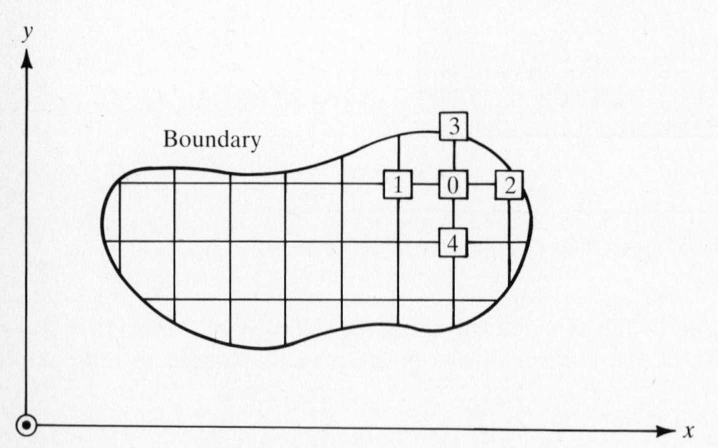


**Figure 13.14** Graphical illustration of an arbitrary region.

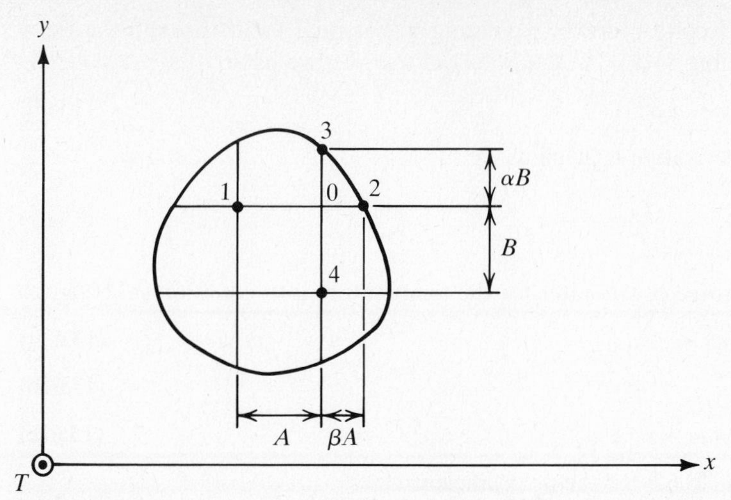


**Figure 13.15** Graphical illustration of a node adjacent to an arbitrary boundary.

tion developed earlier for this particular problem is still applicable for nodes not adjacent to the boundaries. However, for nodes that are adjacent to these arbitrary boundaries, the partial derivatives need to be defined in terms of uneven increments between nodes in both the $x$ and $y$ directions. For example, let us isolate node 0, which is adjacent to the boundary as shown in figure 13.15.

Note that the second partial derivative of the temperature relative to $x$ at node 0 must now be defined in terms of the increments $A$ and $\beta A$. Similarly, the second partial relative to $y$ can be expressed in terms of $B$ and $\beta B$. For example, consider figure 13.16.

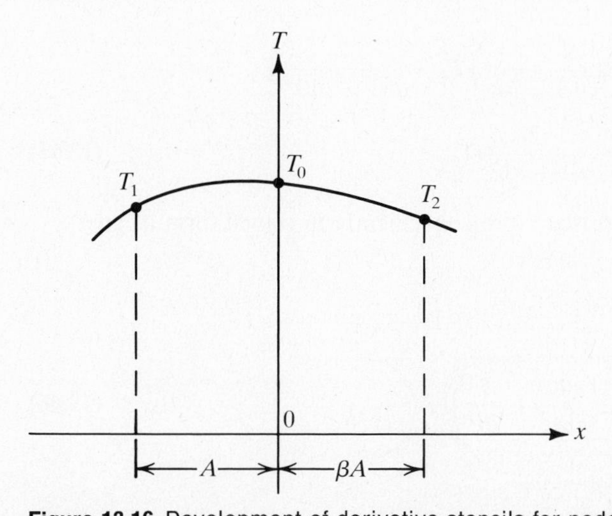


**Figure 13.16** Development of derivative stencils for nodes of unequal legs.

Assuming a second-order interpolating polynomial to fit through the three nodes corresponding to $x = -A$, $x = 0$, and $x = \beta A$, we have

$$T = a_0 + a_1 x + a_2 x^2 \tag{13.60}$$

and the second derivative is given as

$$\frac{d^2 T}{dx} = 2a_2 \tag{13.61}$$

Substituting the three coordinates for the temperature into equation (13.60) gives

$$T_1 = a_0 - Aa_1 + A^2 a_2 \tag{13.62a}$$

$$T_0 = a_0 \tag{13.62b}$$

$$T_2 = a_0 + \beta A a_1 + \beta^2 A^2 a_2 \tag{13.62c}$$

Solving equation (13.62) for the $a_2$ coefficient gives

$$a_2 = \frac{1}{(1+\beta)A^2}\left(T_1 - \frac{1+\beta}{\beta}T_0 + \frac{1}{\beta}T_2\right) \tag{13.63}$$

and

$$\frac{\partial^2 T}{\partial x^2} = \frac{2}{(1+\beta)A^2}\left(T_1 - \frac{1+\beta}{\beta}T_0 + \frac{1}{\beta}T_2\right) \tag{13.64a}$$

Similarly,

$$\frac{\partial^2 T}{\partial y^2} = \frac{2}{(1+\beta)B^2}\left(T_4 - \frac{1+\alpha}{\alpha}T_0 + \frac{1}{\alpha}T_3\right) \tag{13.64b}$$

Adding equations (13.63) and (13.64) yields the well-known Laplace equation for steady-state temperature:

$$\nabla^2 T = \frac{2}{(1+\beta)A^2}T_1 - 2\left(\frac{1}{\beta A} + \frac{1}{\alpha B}\right)T_0 + \frac{2}{\beta(1+\beta)A^2}T_2 + \frac{2}{(1+\alpha)B^2}T_4 + \frac{2}{\alpha(1+\alpha)B^2}T_3 \tag{13.65}$$

Equation (13.65) can be expressed more conveniently in stencil form to give

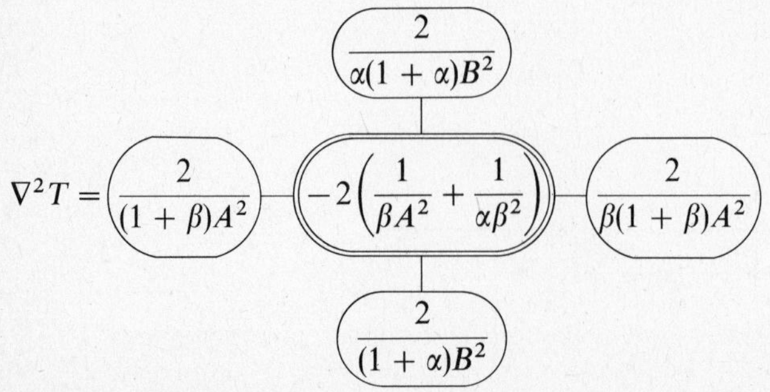

$$\nabla^2 T = \begin{array}{ccccc} & & \boxed{\dfrac{2}{\alpha(1+\alpha)B^2}} & & \\ & & | & & \\ \boxed{\dfrac{2}{(1+\beta)A^2}} & - & \boxed{\boxed{-2\left(\dfrac{1}{\beta A^2} + \dfrac{1}{\alpha\beta^2}\right)}} & - & \boxed{\dfrac{2}{\beta(1+\beta)A^2}} \\ & & | & & \\ & & \boxed{\dfrac{2}{(1+\alpha)B^2}} & & \end{array} \tag{13.66}$$

Note that this stencil reduces to that given by equation (13.44) for $\beta = \alpha$ and $A = B$ as it should. Furthermore, the order of error associated with this expression is $(O)h^2$.

## EXAMPLE 13.8

Determine the temperature at the nodes for the following region:

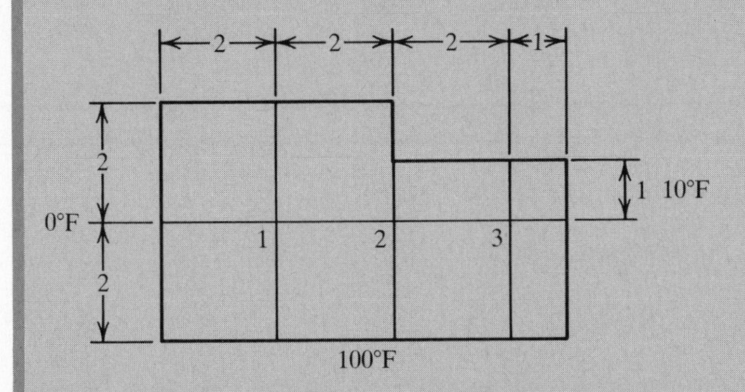


### Solution

Clearly nodes 2 and 3 require special stencils since they both involve unequal legs. Thus

Node 1:

$$10 + 100 - 4T_1 + T_2 = 0$$

Node 2:

Here $A = B = 2$ and $\alpha = \frac{1}{2}$; $\beta = 1$. Substituting into equation (13.66) gives

$$\nabla^2 T = \begin{matrix} & \left(\frac{2}{3}\right) & \\ \left(\frac{1}{4}\right) & \left(\!\left(\frac{3}{2}\right)\!\right) & \left(\frac{1}{4}\right) \\ & \left(\frac{1}{3}\right) & \end{matrix}$$

Applying the above stencil at node 2 gives

$$\tfrac{1}{4}T_1 - \tfrac{3}{2}T_2 + \tfrac{2}{3}(10) + \tfrac{1}{4}T_3 + \tfrac{1}{3}(100) = 0$$

Node 3:

In this case $A = B = 2$ and $\alpha = \beta = \frac{1}{2}$. Substituting into equation (13.66) yields the desired stencil. Applying the stencil at node 3 gives the third equation needed for the solution:

$$\tfrac{1}{3}T_2 + \tfrac{2}{3}(10) - 2T_3 + \tfrac{2}{3}(10) + \tfrac{1}{3}(100) = 0$$

These algebraic equations are now expressed in matrix form as follows:

$$\begin{bmatrix} -4 & 1 & 0 \\ \frac{1}{4} & -\frac{3}{2} & \frac{1}{4} \\ 0 & \frac{1}{3} & -2 \end{bmatrix} \begin{Bmatrix} T_1 \\ T_2 \\ T_3 \end{Bmatrix} = \begin{Bmatrix} -110 \\ -40 \\ -46.67 \end{Bmatrix}$$

Solving for the unknown nodal temperatures gives

$$T_1 = 36.94°\text{F}$$

$$T_2 = 37.76°\text{F}$$

$$T_3 = 29.63°\text{F}$$

Note that this problem could have been solved by subdividing the region differently using $1 \times 1$ elements without having to use the special stencil developed in this section. This stencil can now be used to handle any region, skewed, triangular, or polar. However, the advantage of the stencils developed in earlier sections is that they require much less work. Clearly, for an arbitrary region many different stencils may have to be computed by using equation (13.66).

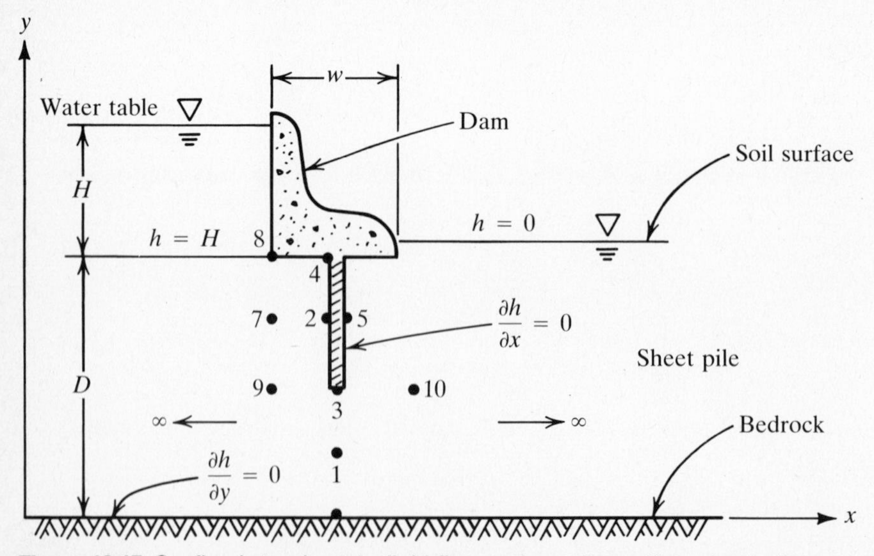


**Figure 13.17** Confined steady-state fluid flow under a dam with a sheet pile.

So far, we have purposely confined our discussions to the steady-state temperature so that the reader will be better able to relate the various methods correctly. However, the reader should be well aware of the fact that these methods are applicable to any partial differential equation of the form given by Laplace's equation. We mentioned earlier that this same partial differential equation is also used in describing steady-state fluid flow through soil. Steady-state fluid flow problems generally involve difficult boundary conditions and complex geometry. Consider, for example, the problem presented in figure 13.17.

It is evident that mixed boundary conditions are involved in this problem; that is, both constant and derivative boundary conditions are specified. Furthermore, the geometry of the region is rather involved. The problem to be solved generally deals with the calculation of the total head $h$ within the soil in terms of a total head difference of $H$. Generally there are four different nodes that can occur when one is dealing with a uniform mesh ($\Delta x = \Delta y = C$).

Node 1:

For any interior node within the soil mass, the stencil is given as follows:

$$\nabla^2 h = \frac{1}{C^2}\left(\begin{array}{ccc} & 1 & \\ 1 & -4 & 1 \\ & 1 & \end{array}\right) \tag{13.67}$$

Node 2:

For nodes 2, 5, and 6 the derivative boundary condition is used to modify equation (13.67) as follows:

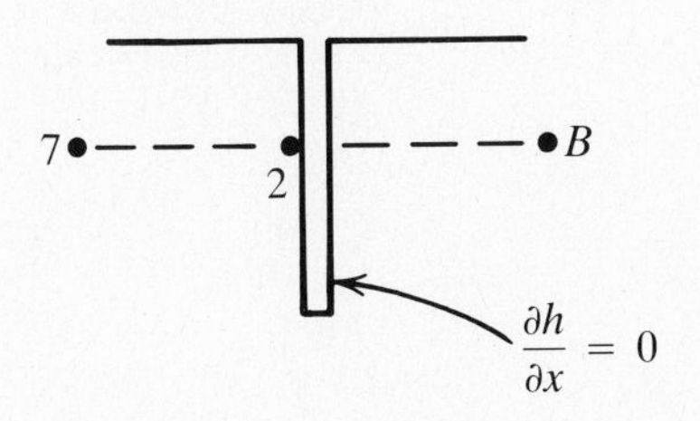


The boundary derivative is defined as

$$\frac{\partial h}{\partial x} = \frac{1}{2C}(-h_7 + h_B)$$

Obviously $h_B$ is imaginary in the sense that it does not physically exist. Consequently we must express it in terms of $h_7$. That is,

$$h_B = h_7$$

Therefore the stencil given by equation (13.67) is modified for nodes 2 and 3 as follows:

$$\nabla^2 h = \frac{1}{C^2}\left(\begin{array}{ccc} & & \textcircled{1} \\ & & | \\ \textcircled{2} & — & \textcircled{-4} \\ & & | \\ & & \textcircled{1} \end{array}\right) \tag{13.68}$$

Node 3:

For this node the stencil is developed rather easily by noting that $\partial^2 h/\partial y^2$ is defined in terms of four nodes rather than three. Consider node 3 isolated.

2| . . |5

9 . . 3 . 10

. 1

While the derivative in the $x$ direction is easily defined as

$$\frac{\partial^2 h}{\partial x^2} = \frac{1}{C^2}(h_9 - 2h_3 + h_{10})$$

the derivative in the $y$ direction must be defined in terms of nodes 1, 3, 2, and 5. That is,

$$\frac{\partial^2 h}{\partial y^2} = \frac{1}{C^2}\left(h_1 - 2h_3 + \frac{h_2 + h_5}{2}\right)$$

and

$$\nabla^2 h = \frac{1}{C^2}\left(\begin{array}{ccccc} & \textcircled{\tfrac{1}{2}} & & \textcircled{\tfrac{1}{2}} & \\ & \diagdown & & \diagup & \\ \textcircled{1} & — & \textcircled{-4} & — & \textcircled{1} \\ & & | & & \\ & & \textcircled{1} & & \end{array}\right) \tag{13.69}$$

Node 4:

This node is somewhat similar to node 2 except that the stencil contains two imaginary nodes that must be replaced in terms of interior nodes. Consider the following:

. $B$

8 . —— 4 —— . $A$

2 .

Obviously, both nodes $A$ and $B$ can be replaced by nodes 8 and 2 using boundary conditions. Thus

$$\frac{\partial h}{\partial x} = 0 = \frac{1}{2C}(-h_8 + h_A) \Rightarrow h_A = h_8$$

$$\frac{\partial h}{\partial y} = 0 = \frac{1}{2C}(-h_2 + h_B) \Rightarrow h_B = h_2$$

Therefore equation (13.67) is modified to give

$$\nabla^2 h = \frac{1}{C^2}\left(\textcircled{2}\!-\!\textcircled{\textcircled{-4}} \atop \textcircled{-2}\right) \qquad \textbf{(13.70)}$$

Note that the stencils given by equations (13.68)–(13.70) can be rotated to fit nodes positioned differently from those discussed.

Finally, the fact that the region to be analyzed is of infinite extent in the $x$ direction presents us with another problem; that is, how many nodes should we use? The authors suggest using a head of zero at a distance $3D$ in front of the dam and a head of $H$ at a distance $3D$ behind the dam.

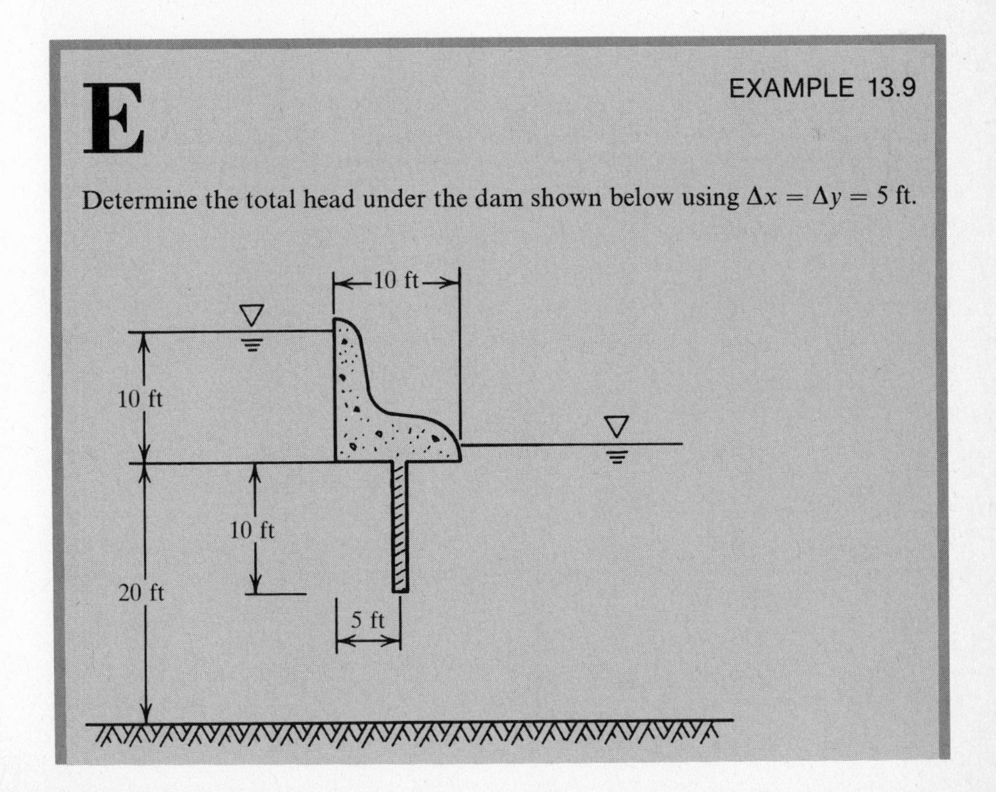

E
EXAMPLE 13.9
Determine the total head under the dam shown below using $\Delta x = \Delta y = 5$ ft.
10 ft
10 ft
10 ft
20 ft
5 ft

### Solution

We begin our solution by establishing a mesh and numbering the nodes:

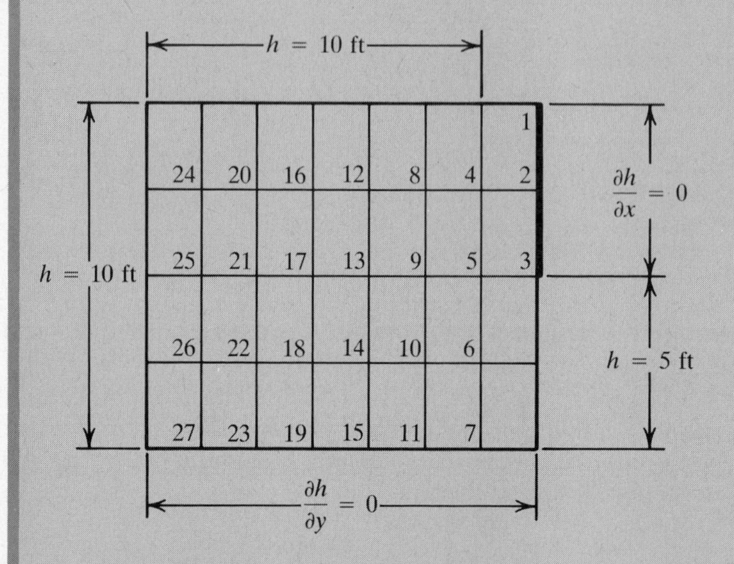


Note that owing to symmetry only one-half of the region need be considered. That reduces our nodes to almost one-half the original number. For practical consideration only nodes falling along the sheet pile are considered.

Node 1:

Apply the stencil given by equation (13.49):

$$2(10) - 4h_1 + 2h_2 = 0$$

Node 2:

Apply the stencil given by equation (13.47):

$$h_1 + 4h_2 + 2h_4 + h_3 = 0$$

Node 3:

Apply the stencil given by equation (13.47):

$$h_2 - 4h_3 + 5 + 2h_5 = 0$$

For the remaining nodes, equations (13.67) and (13.68) can be applied, resulting in the matrix shown on page 547. Solving this set of 27 equations for

(Total head $h_i$)

| Node | 1 | 2 | 3 | 4 | 5 | 6 | 7 | 8 | 9 | 10 | 11 | 12 | 13 | 14 | 15 | 16 | 17 | 18 | 19 | 20 | 21 | 22 | 23 | 24 | 25 | 26 | 27 | RHS |
|---|---|---|---|---|---|---|---|---|---|---|---|---|---|---|---|---|---|---|---|---|---|---|---|---|---|---|---|---|
| 1 | −4 | 2 | | | | | | | | | | | | | | | | | | | | | | | | | | −20 |
| 2 | 1 | −4 | 1 | 2 | | | | | | | | | | | | | | | | | | | | | | | | 0 |
| 3 | | 1 | −4 | | 2 | | | | | | | | | | | | | | | | | | | | | | | −5 |
| 4 | | 1 | | −4 | 1 | | | 1 | | | | | | | | | | | | | | | | | | | | −5 |
| 5 | | | 1 | 1 | −4 | 1 | | | 1 | | | | | | | | | | | | | | | | | | | 0 |
| 6 | | | | | 1 | −4 | 1 | | | 1 | | | | | | | | | | | | | | | | | | −5 |
| 7 | | | | | | 2 | −4 | | | | 1 | | | | | | | | | | | | | | | | | −5 |
| 8 | | | | 1 | | | | −4 | 1 | | | 1 | | | | | | | | | | | | | | | | −10 |
| 9 | | | | | 1 | | | 1 | −4 | 1 | | | 1 | | | | | | | | | | | | | | | 0 |
| 10 | | | | | | 1 | | | 1 | −4 | 1 | | | 1 | | | | | | | | | | | | | | 0 |
| 11 | | | | | | | 1 | | | 2 | −4 | | | | 1 | | | | | | | | | | | | | 0 |
| 12 | | | | | | | | 1 | | | | −4 | 1 | | | 1 | | | | | | | | | | | | −10 |
| 13 | | | | | | | | | 1 | | | 1 | −4 | | | | 1 | | | | | | | | | | | 0 |
| 14 | | | | | | | | | | 1 | | | 1 | −4 | 1 | | | 1 | | | | | | | | | | 0 |
| 15 | | | | | | | | | | | 1 | | | 2 | −4 | | | | 1 | | | | | | | | | 0 |
| 16 | | | | | | | | | | | | 1 | | | | −4 | 1 | | | 1 | | | | | | | | −10 |
| 17 | | | | | | | | | | | | | 1 | | | 1 | −4 | 1 | | | 1 | | | | | | | 0 |
| 18 | | | | | | | | | | | | | | 1 | | | 1 | −4 | 1 | | | 1 | | | | | | 0 |
| 19 | | | | | | | | | | | | | | | 1 | | | 2 | −4 | | | | 1 | | | | | 0 |
| 20 | | | | | | | | | | | | | | | | 1 | | | | −4 | 1 | | | 1 | | | | −10 |
| 21 | | | | | | | | | | | | | | | | | 1 | | | 1 | −4 | 1 | | | 1 | | | 0 |
| 22 | | | | | | | | | | | | | | | | | | 1 | | | 1 | −4 | 1 | | | 1 | | 0 |
| 23 | | | | | | | | | | | | | | | | | | | 1 | | | 2 | −4 | | | | 1 | 0 |
| 24 | | | | | | | | | | | | | | | | | | | | 1 | | | | −4 | 1 | | | −20 |
| 25 | | | | | | | | | | | | | | | | | | | | | 1 | | | 1 | −4 | 1 | | −10 |
| 26 | | | | | | | | | | | | | | | | | | | | | | 1 | | | 1 | −4 | 1 | −10 |
| 27 | | | | | | | | | | | | | | | | | | | | | | | 1 | | | 2 | −4 | −10 |

⟵ $R$ ⟶

Note: All blanks are zero.

the unknown total heads gives

| | | |
|---|---|---|
| $h_1 = 8.36$ | $h_{10} = 6.12$ | $h_{19} = 6.78$ |
| $h_2 = 6.72$ | $h_{11} = 6.05$ | $h_{20} = 7.11$ |
| $h_3 = 5.98$ | $h_{12} = 6.42$ | $h_{21} = 7.24$ |
| $h_4 = 6.27$ | $h_{13} = 6.17$ | $h_{22} = 7.56$ |
| $h_5 = 6.09$ | $h_{14} = 6.30$ | $h_{23} = 7.68$ |
| $h_6 = 5.71$ | $h_{15} = 6.36$ | $h_{24} = 8.96$ |
| $h_7 = 5.62$ | $h_{16} = 2.23$ | $h_{25} = 8.74$ |
| $h_8 = 7.28$ | $h_{17} = 5.55$ | $h_{26} = 8.78$ |
| $h_9 = 6.41$ | $h_{18} = 6.55$ | $h_{27} = 8.81$ |

Note that better results can be obtained by using a grid with 60 ft in front and 60 ft behind the dam. In addition, it is interesting to note that by numbering the nodes starting along the short side reduced our bandwidth $R$ to five. The $R$ value is calculated by finding the difference between the largest and the smallest node numbers for any of the squares constituting the grid. This is extremely helpful in solving large problems.

## 13.7 PARABOLIC PARTIAL DIFFERENTIAL EQUATIONS

This section deals with partial differential equations that are time dependent. Parabolic partial differential equations may involve one, two, or three spatial coordinates in addition to the time coordinate. Furthermore, both linear and nonlinear models may occur in practice. These models include the temperature distribution throughout a body with time and the pore water pressure distribution throughout a porous medium with time. Such problems involve both boundary and initial conditions. While the finite difference procedure is applied in solving these types of problems, the formulation is generally made using an explicit scheme or an implicit one, the difference being that in the explicit scheme no set linear algebraic equation would need to be solved; this is not the case with the implicit scheme.

### 13.7.1 Explicit Finite Difference Scheme

In order to illustrate the explicit finite difference scheme for solving parabolic partial differential equations we shall consider the temperature distribution problem along a rod at any given time. This problem is illustrated graphically in figure 13.18.

Here we wish to determine the temperature distribution for $t > t_0$ throughout the length of the rod $L$ subject to the boundary temperatures $T(x_0, t)$ and $T(x_1, t)$. It is evident that if the rod experiences no heat loss to the surrounding environment then the initial temperature distribution is expected to reach the steady state at

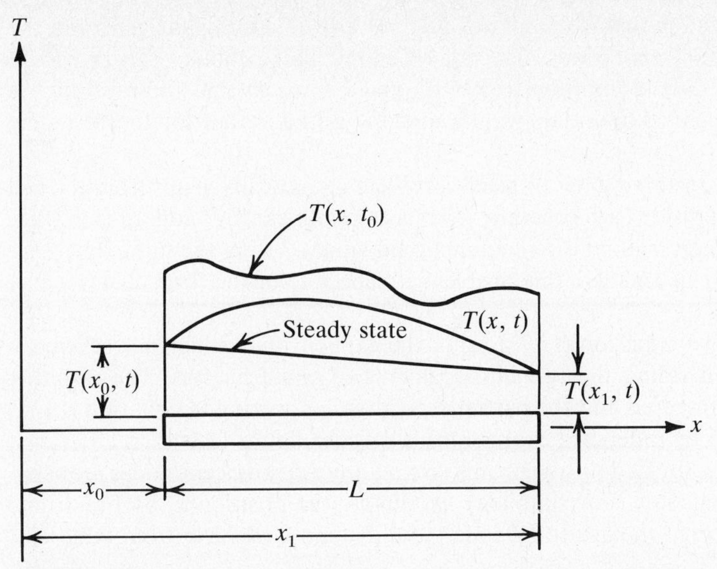


**Figure 13.18** Temperature distribution throughout a rod as a function of time.

time $t = \infty$. The steady-state distribution for a rod is given simply by the straight line connecting the boundary temperatures at the two ends. The question is how can we determine the distribution of the temperature at any given time $t$ such that $0 < t < \infty$? The partial differential equation whose solution provides the answer was addressed earlier in chapter 3 and is repeated here for convenience:

$$\frac{\partial T}{\partial t} = C \frac{\partial^2 T}{\partial x^2} \tag{13.71}$$

where $T$ is the temperature: this dependent variable is a function of the length $x$ (spatial dimension) and time $t$. For linear models the parameter $C$ is assumed to be constant throughout the length of the rod. This assumption is realistic for homogeneous isotropic mediums, but not so realistic when one is dealing with a composite medium.

The solution to equation (13.71) is obtained by first considering the initial and boundary conditions:

### Boundary Conditions

1. At $x = x_0$ and $t_0 \leq t \leq \infty$,

$$T(x_0, t) = T_0 \tag{13.72a}$$

2. At $x = x_1$ and $t_0 \leq t \leq \infty$,

$$T(x_1, t) = T_1 \tag{13.72b}$$

3. At $t = t_0$ and $x_0 \leq x \leq x_1$,

$$T(x, t_0) = T_x \tag{13.72c}$$

It is evident that at the two boundaries there is a problem in that generally the boundary and initial conditions may not be equal. This problem can be solved by adjusting the initial conditions at $x = x_0$ and $x = x_1$ so that these values are given as the average of boundary values and the initial values for the first time increment.

The difficulty in solving the transient problem analytically stems from the fact that the initial conditions are generally given as a nonlinear function. In fact, these may even be given as a set of discrete functional values. Using the finite difference procedure permits us to solve this problem irrespective of the form that $T_x$ may take.

The solution of equation (13.71) using the explicit finite difference scheme is attained by first dividing the rod into segments of equal lengths. The resulting nodes are then numbered and the partial derivatives appearing in equation (13.71) are approximated by difference expressions. Consider figure 13.19.

Here the rod is divided into segments so that any two adjacent nodes are separated by an increment $x$. For simplicity we denote the position $x_i$ by $i$ and time $t_j$ by $j$ only. Hence, at the arbitrary node $i$ and at time $j$, we have from table 10.2

$$\frac{\partial T}{\partial x^2} = \frac{1}{(\Delta x)^2}\left[T_{i-1,j} - 2T_{i,j} + T_{i+1,j}\right] \tag{13.73}$$

Obviously, the error associated with this approximation is of order $(O)h^2$. Therefore it would seem reasonable to approximate the partial derivative of $T$ relative to time using the same order of error as was recommended earlier. Unfortunately, this implies that the temperatures at $j + 1$ are skipped. Therefore we are forced to approximate $\partial T/\partial t$ using errors of $(O)h$. Hence, at node $i$ and for time $j$ to $j + 1$, we have

$$\frac{\partial T}{\partial t} = \frac{1}{\Delta t}(-T_{i,j} + T_{i,j+1}) \tag{13.74}$$

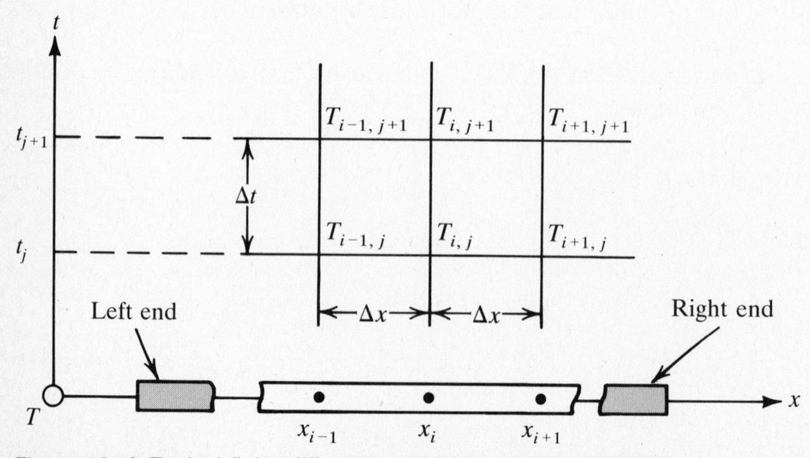


**Figure 13.19** Typical finite difference grid for equation (13.71).

Substituting equations (13.73) and (13.74) into (13.71) yields the desired difference formula. That is,

$$\frac{1}{\Delta t}(-T_{i,j} + T_{i,j+1}) = \frac{C}{(\Delta x)^2}(T_{i-1,j} - 2T_{i,j} + T_{i+1,j})$$

Simplifying then solving for the unknown temperature at position $i$ and time $j + 1$ gives

$$T_{i,j+1} = \alpha T_{i-1,j} + (1 - 2\alpha)T_{i,j} + \alpha T_{i+1,j} \tag{13.75a}$$

where

$$\alpha = \frac{C\,\Delta t}{(\Delta x)^2} \tag{13.75b}$$

Equation (13.75) permits the determination of time increment $T_{i,j+1}$ in terms of the preceding temperature values. This equation is stable as long as the following condition is met:

$$0 \le \alpha \le \tfrac{1}{2} \tag{13.75c}$$

Furthermore, setting $\alpha = \frac{1}{6}$ yields the best possible accuracy. Unfortunately, this tends to limit our ability to choose appropriate increments in the $x$ and $t$ directions. For this reason, we can only choose $\Delta x$ or $\Delta t$ then determine the remaining increment from equation (13.75b) by using an $\alpha$ value which meets the condition given by equation (13.75c). It is evident that the solution may involve a substantial amount of number crunching where unwanted intermediate temperature values must be computed. Equation (13.75a) is now conveniently expressed in stencil form as

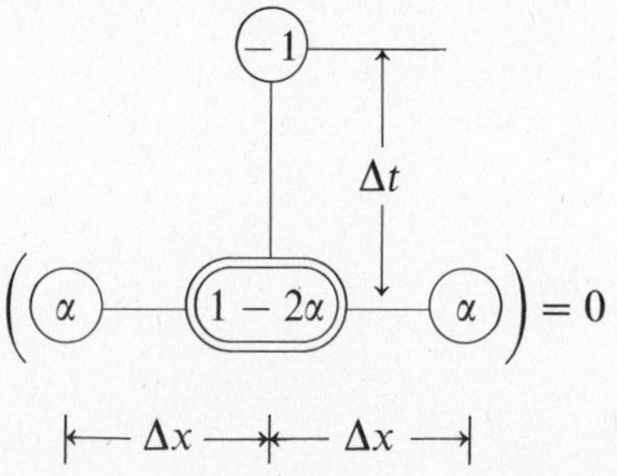

$$\left( \begin{array}{ccc} & -1 & \\ & \Big| \;\; \Delta t & \\ \alpha & (1-2\alpha) & \alpha \end{array} \right) = 0 \tag{13.76}$$

$\vert\leftarrow \Delta x \rightarrow\vert\leftarrow \Delta x \rightarrow\vert$

Obviously equation (13.76) is useful for hand calculations.

**E** EXAMPLE 13.10

Determine the temperature throughout a rod 10 ft long with boundary temperatures of 20 and 40°F. The initial temperature of the rod is 100°F with $C = 10$ ft$^2$/min. Use $\Delta x = 2$ ft and $t = 0.1$ min for $0 \le t = 0.3$.

## Solution

We begin our solution by investigating α to ensure a stable solution.

$$\alpha = \frac{10(0.1)}{(2)^2} = \frac{1}{4} < \frac{1}{2} \qquad \text{OK}$$

We now evaluate the stencil given by equation (13.65) as follows:

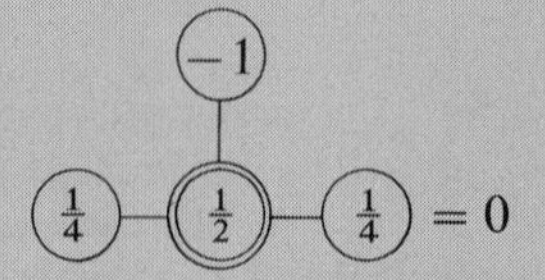


The rod is now subdivided appropriately using $\Delta x = 2$ ft and the resulting nodes are numbered arbitrarily:

$T_0 = 40°\text{F}$ 1 2 3 4 $T_L = 20°\text{F}$

100°F

We now proceed by adjusting the boundary values for the first increment. Hence

For $t = \Delta t$ use

$$\text{At} \quad x = x_0: \qquad T_0 = \frac{100 + 40}{2} = 70°\text{F}$$

$$\text{At} \quad x = x_L: \qquad T_L = \frac{100 + 20}{2} = 60°\text{F}$$

For $t > \Delta t$ use

$$\text{At} \quad x = x_0: \qquad T_0 = 40°\text{F}$$

$$\text{At} \quad x = x_L: \qquad T_L = 20°\text{F}$$

These conditions are illustrated graphically as follows:

| | | 1 | 2 | 3 | 4 | |
|---|---|---|---|---|---|---|
| $t = 0$ | 70 | 100 | 100 | 100 | 100 | 60 |
| | | $T_{1j}$ | $T_{2j}$ | $T_{3j}$ | $T_{4j}$ | |
| $t \geq \Delta t$ | 40 | * | * | * | * | 20 |

The computed values are shown appropriately in the following form:

| | | 1 | 2 | 3 | 4 | | |
|---|---|---|---|---|---|---|---|
| $t = 0$ | 70 | 100 | 100 | 100 | 100 | 60 | $j = 0$ |
| $t = \Delta t$ | 40 | 92.5 | 100 | 100 | 90 | 20 | $j = 1$ |
| $t = 2\,\Delta t$ | 40 | 81.25 | 98.13 | 97.5 | 75 | 20 | $j = 2$ |
| $t = 3\,\Delta t$ | 40 | 75.16 | 93.75 | 92.03 | 66.88 | 20 | $j = 3$ |

$\tfrac{1}{4}$ — $\tfrac{1}{2}$ — $\tfrac{1}{4}$; $-1$

For illustration purposes, the values for the first increment are calculated below:

$$T_{1,1} = \tfrac{1}{4}(70) + \tfrac{1}{2}(100) + \tfrac{1}{4}(100) = 92.5$$

$$T_{2,1} = \tfrac{1}{4}(100) + \tfrac{1}{2}(100) + \tfrac{1}{4}(100) = 100$$

$$T_{3,1} = \tfrac{1}{4}(100) + \tfrac{1}{2}(100) + \tfrac{1}{4}(100) = 100$$

$$T_{4,1} = \tfrac{1}{4}(100) + \tfrac{1}{2}(100) + \tfrac{1}{4}(60) = 90$$

Note that these values are then used to compute the nodal temperatures for $t = 2\,\Delta t$.

### 13.7.2 Implicit Method

One of the disadvantages of the explicit method is that the solution is unstable for $\alpha > \frac{1}{2}$. Thus a very small time increment is required to ensure accuracy. Furthermore, the derivative with respect to time involved errors of order $(O)h$ while that with respect to length is $(O)h^2$, thus resulting in further errors.

The implicit method does not have any of the disadvantages of the explicit method, but does involve solving a set of linear algebraic equations. The development of the difference equation is realized by noting that the nodal value at time $t$ is directly influenced by that at time $t + \Delta t$. This is explained by considering the following rod segment:

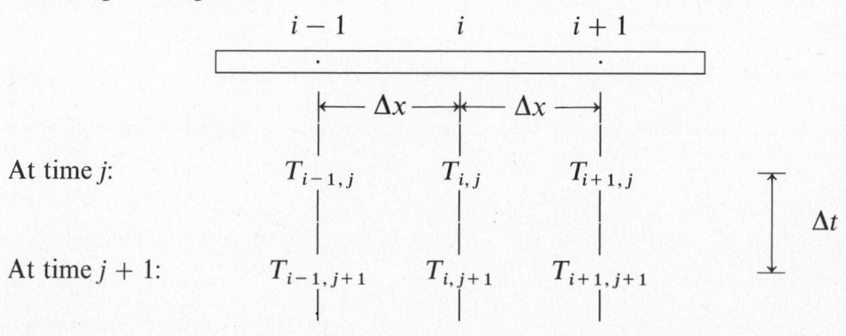

The partial derivative $\partial^2 T/\partial x^2$ is now taken as the average value of the approximations at time $j$ and $j+1$ to give

$$\frac{\partial^2 T}{\partial x^2} = \frac{1}{(\Delta x)^2}(T_{i-1,j} - 2T_{i,j} + T_{i+1,j})(1-\theta) + \frac{1}{(\Delta x)^2}(T_{i-1,j+1} - 2T_{i,j+1} + T_{i+1,j+1})\theta \tag{13.77}$$

where

$\theta$ = a weighting factor between 0 and 1.0

Furthermore, the partial derivative $\partial T/\partial t$ is approximated as

$$\frac{\partial T}{\partial t} = \frac{-T_{i,j} + T_{i,j+1}}{\Delta t} \tag{13.78}$$

$$C\left(\frac{1}{(\Delta x)^2}(T_{i-1,j} - 2T_{i,j} + T_{i+1,j})(1-\theta) + \frac{1}{(\Delta x)^2}(T_{i-1,j+1} - 2T_{i,j+1} + T_{i+1,j+1})\theta\right) = \frac{-T_{i,j} + T_{i,j+1}}{\Delta t}$$

Simplifying and collecting terms yields

$$\alpha(1-\theta)T_{i-1,j} + (1 - 2\alpha + 2\alpha\theta)T_{i,j} + \alpha(1-\theta)T_{i+1,j} + \alpha\theta T_{i-1,j+1} - (2\alpha\theta + 1)T_{i,j+1} + \alpha\theta T_{i+1,j+1} = 0 \tag{13.79}$$

Equation (13.79) is stable for any $\alpha$ value. However, greater accuracy is attained when using small $\alpha$ values of say $\alpha < 1$. Furthermore, the so-called Crank–Nicholson formula is obtained by setting $\theta = \frac{1}{2}$ into equation (13.79). The Crank–Nicholson formula is by far the most widely used implicit scheme. Hence

$$\alpha T_{i-1,j} + (2 - 2\alpha)T_{i,j} + \alpha T_{i+1,j} + \alpha T_{i-1,j+1} - (2 + 2\alpha)T_{i,j+1} + \alpha T_{i+1,j+1} = 0 \tag{13.80}$$

Equation (13.80) is expressed more conveniently in stencil form as follows:

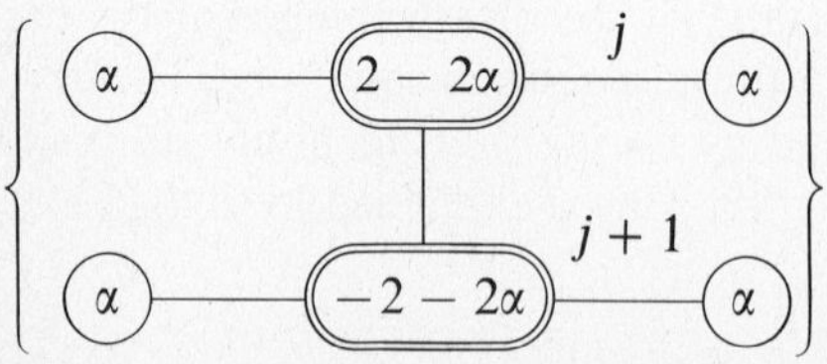

$$\left\{ \begin{array}{ccc} \alpha & (2-2\alpha) & \alpha \quad j \\ & | & \\ \alpha & (-2-2\alpha) & \alpha \quad j+1 \end{array} \right\} = 0 \tag{13.81}$$

**E**

**EXAMPLE 13.11**

Rework example 13.10 using the Crank–Nicholson implicit scheme for $t = 0.3$ min.

### Solution

We begin the solution by first computing the new $\alpha$ value

$$\alpha = \frac{10(0.3)}{2^2} = 0.75$$

The initial and boundary values are shown below for $t = 0$ to $t = 0.3$.

| | | 1 | 2 | 3 | 4 | |
|---|---|---|---|---|---|---|
| $t = 0$ | 100 | 100 | 100 | 100 | 100 | 100 |
| $t = 0.3$ | 40 | $T_{11}$ | $T_{21}$ | $T_{31}$ | $T_{41}$ | 20 |

The stencil corresponding to $\alpha = 0.75$ is given by equation (13.81):

$$\left\{ \begin{matrix} 0.75 & 0.50 & 0.75 \\ 0.75 & -3.5 & 0.75 \end{matrix} \right\} = 0$$

Applying this at the four nodes gives four equations in the unknown temperatures.

Node 1:

$$(0.75 + 0.50 + 0.75)(100) + 0.75(40) - 3.5T_{11} + 0.75T_{21} = 0$$

Node 2:

$$(0.75 + 0.50 + 0.75)(100) + 0.75T_{11} - 3.5T_{21} + 0.75T_{31} = 0$$

Node 3:

$$(0.75 + 0.50 + 0.75)(100) + 0.75T_{21} - 3.5T_{31} + 0.75T_{41} = 0$$

Node 4:

$$(0.75 + 0.50 + 0.75)(100) + 0.75T_{31} - 3.5T_{41} + 0.75(20) = 0$$

Expressing these equations in matrix form yields

$$\begin{bmatrix} -3.5 & 0.75 & 0 & 0 \\ 0.75 & -3.5 & 0.75 & 0 \\ 0 & 0.75 & -3.5 & 0.75 \\ 0 & 0 & 0.75 & -3.5 \end{bmatrix} \begin{Bmatrix} T_{11} \\ T_{21} \\ T_{31} \\ T_{41} \end{Bmatrix} = - \begin{Bmatrix} 230 \\ 200 \\ 200 \\ 215 \end{Bmatrix}$$

Solving for the unknown temperatures at time $t = 0.3$ gives

$$T_{11} = 86.296°\text{F} \qquad T_{31} = 95.262°\text{F}$$

$$T_{21} = 96.048°\text{F} \qquad T_{41} = 81.842°\text{F}$$

Note that these values do not compare very well with the values obtained for $t = 0.3$ in example 13.10. This is caused primarily by the discrepancy in the boundary values, at time $t = 0$ and $t = 0.30$. This problem can be lessened by using smaller $\alpha$ values.

**E**

EXAMPLE 13.12

Determine the excess pore water pressure distribution for the following soil strata after two days, using the implicit scheme.

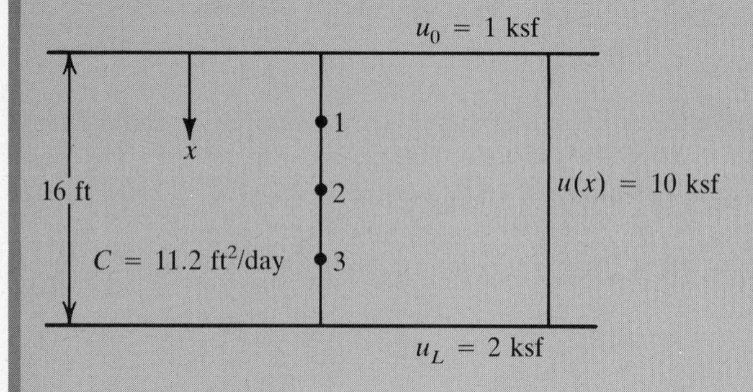


Use $\Delta x = 4$ ft and $\Delta t = 1$ day.

## Solution

This problem is solved by using equation (13.71) and replacing the temperature $T$ with the excess pore water pressure $u$.

We begin our solution by first calculating $\alpha$:

$$\alpha = \frac{C\,\Delta t}{(\Delta x)^2} = \frac{11.2(1)}{4^2} = 0.7$$

Note that unlike the explicit method, there is no need for adjusting boundary values (why?). Therefore applying the stencil given by equation (13.81) at three nodes yields three equations in the unknown pressures for the first time increment as follows:

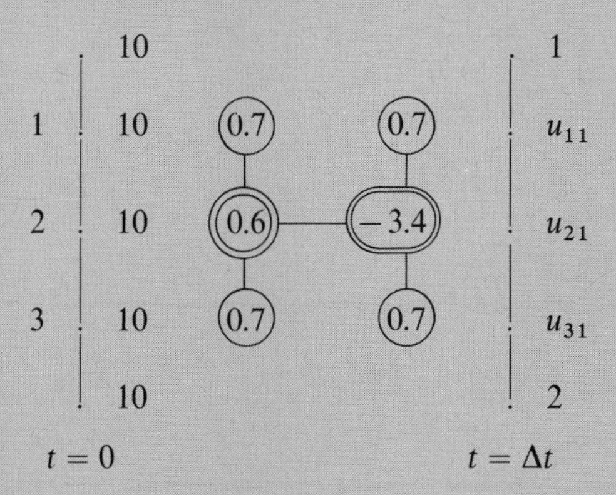


Node 1:

$$10(0.7) + 10(0.6) + 10(0.7) + 1(0.7) - 3.4u_{11} + 0.7u_{21} = 0$$

Node 2:

$$10(0.7) + 10(0.6) + 10(0.7) - 0.7u_{11} - 3.4u_{21} + 0.7u_{31} = 0$$

Node 3:

$$10(0.7) + 10(0.6) + 10(0.7) - 0.7u_{21} - 3.4u_{31} + 2(0.7) = 0$$

Expressing these equations in matrix form gives

$$\begin{bmatrix} 3.4 & 0.7 & 0 \\ 0.7 & -3.4 & 0.7 \\ 0 & 0.7 & 3.4 \end{bmatrix} \begin{Bmatrix} u_{11} \\ u_{21} \\ u_{31} \end{Bmatrix} = - \begin{Bmatrix} 20.7 \\ 20 \\ 21.4 \end{Bmatrix}$$

Solving for the unknown pressures using the square root procedure for symmetrical matrices yields

$$u_{11} = 7.985$$

$$u_{21} = 9.213$$

$$u_{31} = 8.191$$

in matrix form

$$\begin{bmatrix} 3.4 & 0.7 & 0 \\ 0.7 & 3.4 & 0.7 \\ 0 & 0.7 & 3.4 \end{bmatrix} \begin{Bmatrix} u_{12} \\ u_{22} \\ u_{32} \end{Bmatrix} = - \begin{Bmatrix} 12.640 \\ 16.851 \\ 14.164 \end{Bmatrix}$$

Solving for the pressures at the second increment gives

$$u_{12} = 5.198$$

$$u_{22} = 7.189$$

$$u_{32} = 5.646$$

Better results can be attained by increasing the number of nodes and reducing $\alpha$. The pore pressures for the two increments are shown graphically in figure 13.20.

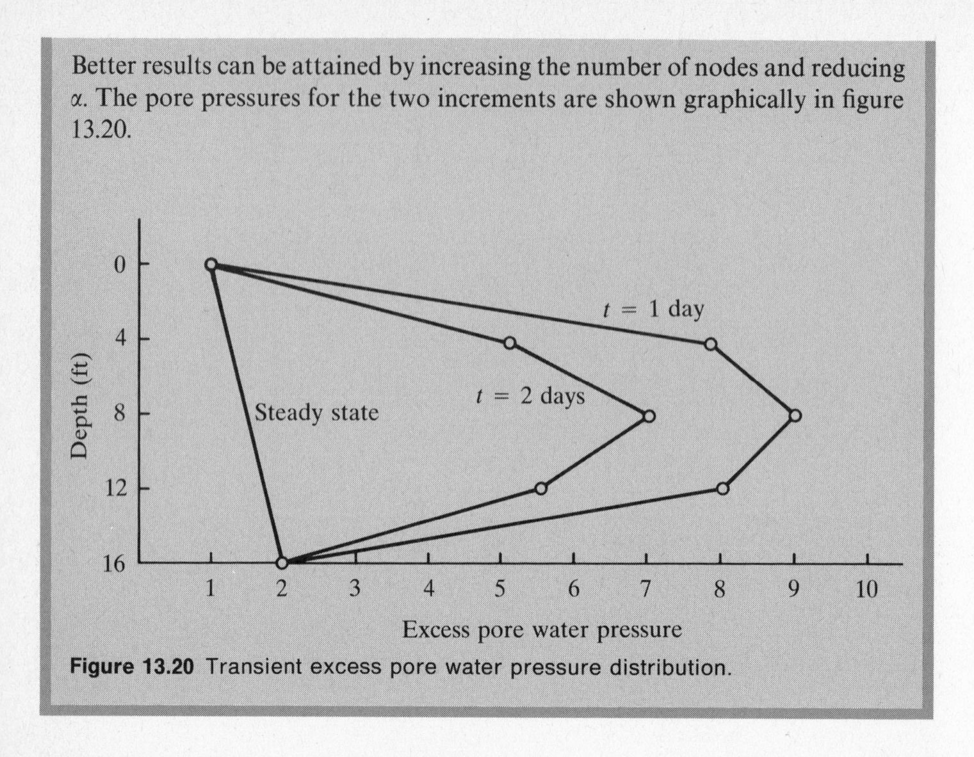


**Figure 13.20** Transient excess pore water pressure distribution.

### 13.7.3 Eigenproblem of the Explicit Scheme

The computations involved in using the finite difference procedure for solving parabolic differential equations may become significantly time consuming. This is especially true when the solution is required over a long period of time. In this section, an alternative approach is outlined that is more direct and efficient in evaluating the explicit formula than the algebraic technique. The basis for the method to be outlined is demonstrated by considering the temperature distribution problem addressed earlier in section 13.7.1. It is evident that at any given time $t \geq 0$ the temperature at the interior nodes is related to the past temperature values at the same nodes by equation (13.75a). Furthermore, the boundary temperatures remain constant throughout subsequent computations, the exception being in the first time increment, where the boundary values are adjusted so that the discrepancy between the boundary values and the initial values at the two ends is removed. With this in mind, suppose the temperature values for $n$ nodes at $t = \Delta t$ have been computed by using equation (13.75a) and are given as follows:

$$\{T\}_1 = \begin{Bmatrix} T_0 \\ T_{11} \\ T_{21} \\ \vdots \\ T_{n1} \\ T_L \end{Bmatrix}$$

where $T_0$ and $T_L$ are the boundary temperatures at the two ends. Therefore, for the subsequent time increment $t = 2\,\Delta t$, the solution is given by equation (13.75a) as

$$\begin{aligned}
T_{1,2} &= \alpha T_0 &&+ (1 - 2\alpha)\, T_{1,1} + \alpha T_{2,1} \\
T_{2,2} &= \alpha T_{1,1} &&+ (1 - 2\alpha)\, T_{2,1} + \alpha T_{3,1} \\
&\vdots && \\
T_{n,2} &= \alpha T_{n-1,2} &&+ (1 - 2\alpha)\, T_{n,1} + \alpha T_L
\end{aligned}$$

These equations are now expressed more conveniently in matrix form as

$$\begin{Bmatrix} T_{1,2} \\ T_{2,2} \\ \vdots \\ T_{n,2} \end{Bmatrix} = \begin{bmatrix} \alpha & 1 - 2\alpha & \alpha & & & \\ & \alpha & 1 - 2\alpha & \alpha & & \\ & & \ddots & \ddots & \ddots & \\ & & \alpha & 1 - 2\alpha & \alpha \end{bmatrix} \begin{Bmatrix} T_0 \\ T_{1,1} \\ T_{2,1} \\ \vdots \\ T_L \end{Bmatrix} \quad \textbf{(13.82)}$$

It is evident that for the special case in which the boundary values for $t > 0$ are both equal to zero equation (13.82) simplifies to

$$\begin{Bmatrix} T_{1,2} \\ T_{2,2} \\ \vdots \\ T_{n,2} \end{Bmatrix} = \begin{bmatrix} 1 - 2\alpha & \alpha & & 0 \\ \alpha & 1 - 2\alpha & \alpha & \\ \ddots & \ddots & \ddots & \\ & \alpha & 1 - 2\alpha & \alpha \\ & & \ddots & \ddots \\ 0 & & \alpha & 1 - 2\alpha \end{bmatrix} \begin{Bmatrix} T_{1,1} \\ T_{2,1} \\ \vdots \\ T_{n,1} \end{Bmatrix} \quad \textbf{(13.83)}$$

Equation (13.83) is given more generally for the $(j + 1)$th time increment as

$$\{T\}_{j+1} = [A]\{T\}_j, \qquad j = 1, \ldots, k \quad \textbf{(13.84)}$$

where

$\{T\}_{j+1}$ = nodal temperature vector at the next time increment $j + 1$

$\{T\}_j$ = nodal temperature vector at the previous time increment $j$

$[A]$ = matrix containing the $\alpha$ terms (symmetric)

Note that the coefficient matrix $[A]$ remains the same for any time increment. Hence

$$j = 1: \qquad \{T\}_2 = [A]\{T\}_1$$

$$j = 2: \qquad \{T\}_3 = [A]\{T\}_2 = [A][A]\{T\}_1 = [A]^2\{T\}_1$$

and in general

$$j = k: \qquad \{T\}_{k+1} = [A]^k\{T\}_1 \quad \textbf{(13.85)}$$

The implication is that for the $(k + 1)$st time increment, the solution is given directly in terms of the $\{T\}_1$ vector. The only remaining problem is raising the matrix $[A]$

to power $k$. This is a problem we addressed in chapter 7. In fact, we have shown that a square symmetrical matrix can be expressed in terms of its eigenvalues and eigenvectors as follows:

$$[A]^k = [\hat{V}][\lambda]^k[\hat{V}]^t \tag{13.86}$$

where $[\hat{V}]$ is the normalized eigenvectors matrix and $[\lambda]$ is the eigenvalue matrix. Hence, substituting equation (13.86) into (13.85) yields the following simple formula:

$$\{T\}_{k+1} = [\hat{V}][\lambda]^k[\hat{V}]^t\{T\} \tag{13.87}$$

Fortunately, owing to the special nature of matrix $[A]$ the eigenvalues are given directly by the formula

$$\lambda_i = 1 - 4\alpha \sin^2\left(\frac{i\pi}{2n+2}\right), \qquad i = 1, \ldots, n \tag{13.88}$$

The corresponding vectors are evaluated by solving the following equation:

$$[[A] - \lambda_i[I]]\{V\}_i = \{0\} \tag{13.89}$$

The significance of equation (13.87) is quite apparent in that the solution for any time increment $k$ is given by multiplying three matrices and the initial vector. This is so since raising the $[\lambda]$ matrix to the power $k$ is accomplished by raising its diagonal elements (eigenvalues) to the power $k$. The implication is that we no longer have to compute intermediate solutions at every single time increment as was the case with algebraic procedures for evaluating the explicit scheme. Obviously, equation (13.87) is valid for zero boundary values only.

One question we have purposely avoided is what if the boundary values are not zero? For this more general case, the initial vector $\{T\}_1$ must be transformed so that the boundary values are forced to be zero. Fortunately, this is possible if we realize that at $t \to \infty$ the temperature distribution reduces to the linear steady-state distribution. Recall that in chapter 3 we stated that the general solution of time-dependent response is given as the sum of the steady-state response $\{T_s\}$ and the transient response $\{\bar{T}\}$. Hence at any time $j$ we have

$$\{T\}_j = \{\bar{T}\}_j + \{T_s\}_j \tag{13.90}$$

Furthermore, we have indicated that the transient portion of the solution tends to zero as $t \to \infty$, in which case the total solution is equal to the steady-state solution. It is evident that this is precisely what this problem involves. That is, equation (13.87) is used for determining the transient portion of the total solution in terms of the initial transient vector $\{T\}_1$. Clearly $\{T\} = \{\bar{T}\}$ if and only if the steady-state solution $\{T_s\}$ is zero. Consequently, in order to apply equation (13.87) when the steady-state distribution is not zero we need to calculate the transient portion of the total solution at time $t = \Delta t$ ($j = 1$) using equation (13.90) as follows:

$$\{\bar{T}\}_1 = \{T\}_1 - \{T_s\}_1 \tag{13.91}$$

Note that $\{T\}_1$ corresponds to the interior node temperatures and the steady-state temperature vector $\{T_s\}_1$ at these nodes is readily evaluated by passing a first-order polynomial through the boundary values. That is,

$$T_s(x) = a_0 + a_1 x \tag{13.92}$$

where $T_s(x)$ is the steady-state temperature at a node located a distance $x$ from the left end of the rod. The unknowns $a_0$ and $a_1$ are evaluated by noting that

At $x = x_0$: $T = T_0$

At $x = x_L$: $T = T_L$

Solving for the unknown coefficients in equation (13.92) gives

$$T_s(x) = \frac{x_L T_0 - x_0 T_L}{x_L - x_0} + \frac{T_L - T_0}{x_L - x_0} x \tag{13.93}$$

Since $x_0$ can be taken at the left end of the rod to be analyzed, equation (13.93) is simplified by substituting $x_0 = 0$ to give

$$T_s(x) = T_0 + \frac{T_L}{L} x \tag{13.94}$$

Note that $x_L = L$, where $L$ is the length of the rod. Clearly the steady-state vector can now be easily formed and equation (13.87) can be used in general, regardless of the boundary values. That is,

$$\{\bar{T}\}_{k+1} = [\hat{V}][\lambda]^k[\hat{V}]^t\{\bar{T}\}_1 \tag{13.95}$$

Equation (13.95) is more general than equation (13.87) in that it presupposes that the steady-state solution is not zero. Consequently, once the transient solution at time $k + 1$ is determined, then the total solution is readily determined by using equation (13.90) for $j = k + 1$.

EXAMPLE 13.13

Rework example 13.10 using the procedure just outlined. Solve for $t = 0.30$ min.

### Solution

Let us begin by using the set of calculated temperature values at the interior nodes. Hence, using equation (13.94) and taking $x_0 = 0$ gives

$$T_2(x) = 40 - 2x$$

The adjusted temperatures are readily calculated by subtracting the steady-state values at the different nodes from $\{T\}_1$ to give

At $x = 0'$ (left boundary):

$$\bar{T}_0 = 40 - [40 - 2(0)] = 0$$

At $x = 2'$ (node 1):

$$\bar{T}_{1,1} = 92.5 - [40 - 2(2)] = 92.5 - 36 = 56.5$$

At $x = 4'$ (node 2):

$$\bar{T}_{2,1} = 100 - [40 - 2(4)] = 100 - 32 = 68$$

At $x = 6'$ (node 3):

$$\bar{T}_{3,1} = 100 - [40 - 2(6)] = 100 - 28 = 72$$

At $x = 8'$ (node 4):

$$\bar{T}_{4,1} = 90 - [40 - 2(8)] = 90 - 24 = 66$$

At $x = 10'$ (right boundary):

$$\bar{T}_L = 20 - [40 - 2(10)] = 0$$

Clearly the adjusted boundary values are now equal to zero. Therefore the transformed vector at $t = 0.1$ is given as follows:

$$\{\bar{T}\}_1^t = \{56.5 \quad 68 \quad 72 \quad 66\}$$

Note that the boundary values are not included in the transformed vector because they are both zero and will remain so for any increment. Consequently, the solution is given directly by equation (13.95). That is,

$$\{\bar{T}\}_{k+1} = [\hat{V}][\lambda]^k[\hat{V}]^t\{\bar{T}\}_1 \tag{13.96}$$

The eigenvalues are determined from equation (13.88) by noting that $\alpha = \frac{1}{4}$, $n = 4$, and $i = 1, 2, 3, 4$ to give

$$\lambda_1 = 1 - 4\frac{1}{4}\sin^2\left(\frac{\pi}{2(4) + 2}\right) = 0.9456$$

$$\lambda_2 = 1 - 4\frac{1}{4}\sin^2\left(\frac{2\pi}{2(4) + 2}\right) = 0.6545$$

$$\lambda_3 = 1 - 4\frac{1}{4}\sin^2\left(\frac{3\pi}{2(4) + 2}\right) = 0.3455$$

$$\lambda_4 = 1 - 4\frac{1}{4}\sin^2\left(\frac{4\pi}{2(4) + 2}\right) = 0.0955$$

The corresponding vectors are determined using equation (13.89) for $i = 1, 2, 3, 4$ to give

$$\{V\}_1 = \begin{Bmatrix} 1 \\ 1.618 \\ 1.618 \\ 1 \end{Bmatrix} \Rightarrow \{\hat{V}\}_1 = \begin{Bmatrix} 0.3718 \\ 0.6015 \\ 0.6015 \\ 0.3718 \end{Bmatrix}$$

$$\{V\}_2 = \begin{Bmatrix} -1 \\ -0.618 \\ 0.618 \\ 1.0 \end{Bmatrix} \Rightarrow \{\hat{V}\}_2 = \begin{Bmatrix} 0.6015 \\ -0.3718 \\ 0.3718 \\ 0.6015 \end{Bmatrix}$$

$$\{V\}_3 = \begin{Bmatrix} 1 \\ -1.618 \\ -1.618 \\ 1 \end{Bmatrix} \Rightarrow \{\hat{V}\}_3 = \begin{Bmatrix} 0.6015 \\ -0.3718 \\ -0.3718 \\ 0.6015 \end{Bmatrix}$$

$$\{V\}_4 = \begin{Bmatrix} -1 \\ 1.618 \\ -1.618 \\ 1 \end{Bmatrix} \Rightarrow \{\hat{V}\}_4 = \begin{Bmatrix} -0.3718 \\ 0.6015 \\ -0.6015 \\ 0.3718 \end{Bmatrix}$$

Therefore the normalized vectors matrix is readily formed to give

$$[\hat{V}] = \begin{bmatrix} 0.3718 & 0.6015 & 0.6015 & -0.3718 \\ 0.6015 & -0.3718 & -0.3718 & 0.6015 \\ 0.6015 & 0.3718 & -0.3718 & -0.6015 \\ 0.3718 & 0.6015 & 0.6015 & 0.3718 \end{bmatrix}$$

Substituting into equation (13.96) gives

$$\{\bar{T}\}_{k+1} = [\hat{V}] \begin{bmatrix} 0.9456^k & & & \\ & 0.6545^k & & \\ & & 0.3455^k & \\ & & & 0.0955^k \end{bmatrix} [\hat{V}]^t \{\bar{T}\}_1$$

For $t = 0.3$ min and $\Delta t = 0.1$ this implies that $k = 2$. Hence,

$$\{\bar{T}\}_3 = \begin{Bmatrix} 38.78 \\ 62.42 \\ 63.39 \\ 42.88 \end{Bmatrix}$$

The total solution is given by equation (13.90) as the sum of the transient and the steady-state solutions:

$$\{T\}_3 = \begin{Bmatrix} 38.78 \\ 62.42 \\ 63.39 \\ 42.88 \end{Bmatrix}_3 + \begin{Bmatrix} 36 \\ 32 \\ 28 \\ 24 \end{Bmatrix}_3 = \begin{Bmatrix} 74.78 \\ 94.42 \\ 91.39 \\ 66.88 \end{Bmatrix}$$

These values compare reasonably well with the values computed earlier. Better accuracy is achieved by retaining more significant figures.

It is interesting to note that the computed temperature values were obtained directly without having to calculate $\{T\}_2$. The implication is that if one wishes to compute the temperature after 1000 increments, then the amount of effort needed is precisely the same as that for two increments. Therefore this method is recommended when dealing with a large number of time increments. In practice, the solution at $t = 1000\,\Delta t$ or more may be needed, in which case the eigenproblem approach is by far the most efficient.

### 13.7.4 Eigenproblem of the Implicit Scheme

The procedure outlined in the preceding section is equally applicable to the implicit scheme for solving parabolic partial differential equations. For clarity of presentation, let us again consider the temperature distribution problem along a rod of length $L$. Using the implicit scheme we express the solution as given by equation (13.80) for $n$ nodal points and for the $(j+1)$th time increment as follows:

$$\begin{bmatrix} 2+2\alpha & -\alpha & & 0 \\ -\alpha & 2+2\alpha & -\alpha & \\ & \ddots & \ddots & -\alpha \\ 0 & & -\alpha & 2+2\alpha \end{bmatrix} \begin{Bmatrix} \bar{T}_1 \\ \bar{T}_2 \\ \vdots \\ \bar{T}_n \end{Bmatrix}_{j+1}$$

$$= \begin{bmatrix} 2-2\alpha & \alpha & & 0 \\ \alpha & 2-2\alpha & \alpha & \\ & \ddots & \ddots & \alpha \\ 0 & & \alpha & 2-2\alpha \end{bmatrix} \begin{Bmatrix} \bar{T}_1 \\ \bar{T}_2 \\ \vdots \\ \bar{T}_n \end{Bmatrix}_{j}$$

Clearly this equation is applicable regardless of the boundary conditions and gives the transient portion of the total solution. In compact matrix form, we have

$$[A]\{\bar{T}\}_{j+1} = [B]\{\bar{T}\}_j, \qquad j = 1, \ldots, k \tag{13.97}$$

Equation (13.97) corresponds to the Crank–Nicholson scheme. Furthermore, the matrices $[A]$ and $[B]$ can be easily evaluated for given $\alpha$ values, in which case we can solve for the $(j+1)$th time increment as follows:

$$\{\bar{T}\}_{j+1} = [A]^{-1}[B]\{\bar{T}\}_j$$

or more simply

$$\{\bar{T}\}_{j+1} = [C]\{\bar{T}\}_j, \qquad j = 1, \ldots, k \qquad \textbf{(13.98)}$$

Comparison of equation (13.98) with that of the explicit scheme [equation (13.84)] clearly indicates that they are of the same general form, the difference being that here we have a different coefficient matrix (matrix $[C]$). Note that since the $[A]$ and $[B]$ matrices are time independent, then $[C]$ is time independent as well. The implication is that the implicit form can be reduced to an equivalent explicit form by using equation (13.98); for any time increment the solution is thus given directly without the need for solving a set of linear algebraic equations. In any case, using equation (13.98) we have

$$j = 1: \qquad \{T\}_2 = [C]\{\bar{T}\}_1$$

$$j = 2: \qquad \{T\}_3 = [C]\{\bar{T}\}_2 = [C][C]\{\bar{T}\}_1 = [C]^2\{\bar{T}\}_1$$

and in general

$$j = k: \qquad \{\bar{T}\}_{k+1} = [C]^k\{\bar{T}\}_1 \qquad \textbf{(13.99)}$$

Equation (13.99) is analogous to equation (13.85); therefore we may write directly that

$$\{\bar{T}\}_k = [\hat{V}_c][\lambda_c]^k[\hat{V}_c]^t\{\bar{T}_1\} \qquad \textbf{(13.100)}$$

Note that the $[C]$ matrix is symmetrical and that equation (13.100) corresponds to equation (13.87). Furthermore, $[\hat{V}_c]$ is the matrix containing the normalized eigenvectors of matrix $[C]$ and $[\lambda_c]$ is the matrix containing the eigenvalues of matrix $[C]$. Fortunately, the eigenvalues can be evaluated directly in terms of $\alpha$ and the number of interior nodes $n$ as follows:

$$\lambda_{ci} = \frac{2 - 4\alpha \sin^2(i\pi/2n - 2)}{2 + 4\alpha \sin^2(i\pi/2n - 2)}, \qquad i = 1, \ldots, n \qquad \textbf{(13.101)}$$

The corresponding vectors are determined as

$$[[C] - \lambda_i[I]]\{V_{ci}\} = \{0\}, \qquad i = 1, \ldots, n \qquad \textbf{(13.102)}$$

Consequently, the solution is now complete in that once again the transient response for any time increment is given directly by equation (13.100).

### 13.7.5 Stability and Convergence

Earlier we stated that while the Crank–Nicholson implicit scheme is stable regardless of the $\alpha$ value used, the stability of the explicit scheme can only be ensured if $0 \le \alpha \le \frac{1}{2}$. In this section, the stability and convergence of the two schemes are discussed.

The term convergence is used in the sense that as $\Delta t$ and $\Delta x$ approach zero, the dervative approximations used in developing the finite difference schemes approach their analytical values. For both the explicit and implicit schemes this can be attained by simply increasing the number of nodes, in which case the time increment is reduced. Recall that

$$\alpha = \frac{c\,\Delta t}{(\Delta x)^2}$$

or more simply

$$\Delta t = \frac{\alpha(\Delta x)^2}{c}$$

Therefore increasing the number of nodes forces $\Delta x$ to be small and $\Delta t$ to be even smaller. Consequently the difference approximations tend to the analytical values. The question is what values of $\alpha$ should be used so that the solution will be stable?

The term stability refers to the study of errors made at one stage of the computations and how they may affect subsequent computations. This problem can be easily addressed for the explicit scheme by noting that the approximate solution at any stage can be expressed as a linear combination of the exact values and associated errors. That is,

$$\{T\}_j = \{T_e\}_j + \{E\}_j \tag{13.103a}$$

$$\{T\}_{j+1} = \{T_e\}_{j+1} + \{E\}_{j+1} \tag{13.103b}$$

where $\{T\}_j$ and $\{T\}_{j+1}$ are the approximate solutions at time $t = j\,\Delta t$ and $t = (j+1)\,\Delta t$, respectively. In addition, $\{T_e\}_j$ and $\{T_e\}_{j+1}$ are the exact solutions at the times $t = j\,\Delta t$ and $t = (j+1)\,\Delta t$. Furthermore, the errors at these times are given by the vectors $\{E\}_j$ and $\{E\}_{j+1}$. Substituting equation (13.103) into the explicit formula given by (13.84) gives

$$\{T_e\}_{j+1} + \{E\}_{j+1} = [A]\{T_e\}_j + [A]\{E\}_j$$

The exact solutions at $j+1$ and $j$ are related by

$$\{T_e\}_{j+1} = [A]\{T_e\}_j$$

Therefore

$$\{E\}_{j+1} = [A]\{E\}_j \tag{13.104}$$

The implication of equation (13.104) is that the errors distribute in exactly the same fashion as the solution itself. That is,

$$\begin{aligned}
j = 1: &\quad \{E\}_2 = [A]\{E\}_1 \\
j = 2: &\quad \{E\}_3 = [A]^2\{E\}_1 \\
&\quad \vdots \qquad\quad \vdots \\
j = 3: &\quad \{E\}_{k+1} = [A]^k\{E\}_1
\end{aligned}$$

It is evident that since the error vector $\{E\}_1$ is not zero when using the difference approximation the matrix $[A]^k$ must vanish as $k \to \infty$. In section 13.7.3 we showed that the $[A]^k$ matrix can be expressed in terms of its eigenvalues and eigenvectors as given by equation (13.86). Therefore

$$\{E\}_{k+1} = [\hat{V}][\lambda]^k[\hat{V}]^t\{E\}_1$$

It is evident that the only way the $[A]$ matrix would approach the null matrix is if $[\lambda]^k \to [0]$ for a large $k$ value. However, since the eigenvalue matrix is a diagonal matrix containing the $n$ eigenvalues, it will converge to the null matrix only if the eigenvalues have magnitudes between $-1.0$ and $+1.0$. Our problem is now reduced to investigating the eigenvalues for the explicit scheme as given by equation (13.88). Hence

$$-1 \leq 1 - 4\alpha \sin^2 \frac{i\pi}{2n+2} \leq 1$$

The limiting values are established by noting that the $\sin^2$ function is always positive and can have a maximum value of 1.0. This is indeed the critical value that needs to be looked at. Hence we may write

$$-1 \leq 1 - 4\alpha \leq 1$$

Since $\alpha$ is always positive, then

$$1 - 4\alpha \leq 1 \Rightarrow \alpha \geq 0$$

$$1 - 4\alpha \geq -1 \Rightarrow \alpha \leq \tfrac{1}{2}$$

Similarly for the implicit scheme we have from equation (13.101)

$$-1 \leq \frac{2 - 4\alpha \sin^2[i\pi/(2n-2)]}{2 + 4\alpha \sin^2[i\pi/(2n-2)]} \leq 1$$

which clearly indicates that any values of $\alpha$ will satisfy these two conditions. Hence the implicit scheme is unconditionally stable when using the Crank–Nicholson scheme.

### 13.7.6 Derivative Boundary Conditions

In the preceding sections, we have developed two different schemes for solving parabolic partial differential equations numerically where the boundary conditions are assumed to be constants. In this section, the explicit scheme is illustrated for problems involving derivative boundary conditions. Consider the temperature distribution along a rod of a finite length whose left boundary condition involves $\partial T/\partial x = m$ as shown in figure 13.21.

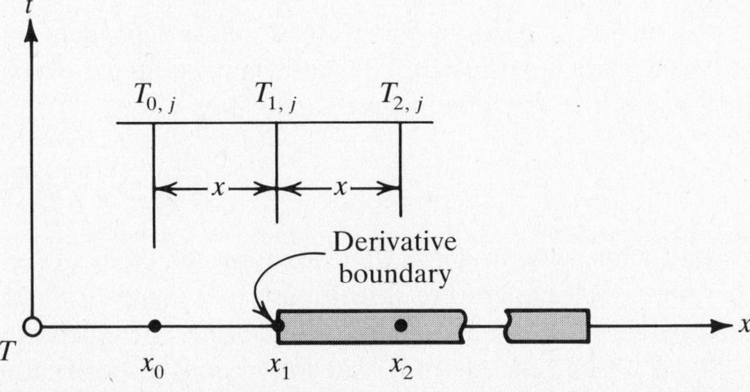


**Figure 13.21** Derivative boundaries for a parabolic partial differential equation with one spatial coordinate.

It is evident that since the temperature at node 1 is an unknown, equation (13.75a) must be applied at this node to give

$$T_{1,j+1} = \alpha T_{a,j} + (1 - 2\alpha)T_{1,j} + \alpha T_{2,j} \tag{13.105}$$

It is clear that equation (13.105) introduces an additional unknown, namely $T_{0,j}$. Consequently, while equation (13.75a) can be easily applied at the remaining nodes $i = 2, \ldots, n$ the resulting set of equations will involve $n + 1$ unknowns. Hence we must introduce an additional equation in order to solve for the unknown temperatures. This additional equation is given by the derivative boundary condition at node 1. That is,

$$\frac{\partial T}{\partial x} = m$$

which can be approximated by the finite difference method to give

$$\left.\frac{\partial T}{\partial x}\right|_{\text{at node 1}} = m = \frac{-T_{0,j} + T_{2,j}}{2\,\Delta x}$$

Note that the order of error is $(O)h^2$ since this is equivalent to the error order used in approximating $\partial^2 T/\partial x^2$. Therefore, solving for the $T_{0,j}$ value, we get

$$T_{0,j} = T_{2,j} - 2\,\Delta x\, m$$

Substituting into equation (13.105) and simplifying yields

$$T_{1,j+1} = (1 - 2\alpha)T_{1,j} + 2\alpha(1 - m\,\Delta x)T_{2,j} \tag{13.106a}$$

Equation (13.106a) can now be combined with the remaining $n - 1$ equations for nodes $2, \ldots, n$ to give an $n \times n$ set which can be solved as before. This equation is given in a stencil form as follows:

(−1)
|
((1 − 2α))—(2α(1 − m Δx))

**(13.106b)**

### 13.7.7 Nonlinear Problems

The time rate settlement of structures placed on fine-grained soils is dependent on changes in the excess pore water pressure (Epwp). The mathematical model which relates Epwp distribution to depth at any time is given as follows:

$$\frac{\partial u}{\partial t} = C_v(x)\frac{\partial^2 u}{\partial x^2} \tag{13.107}$$

where $C_v(x)$ is a variable that changes with depth depending on the types of soil in a given profile. Furthermore, the temperature distribution in a composite rod would involve solving exactly the same partial differential equation. The difference is that $u$ is now replaced by $T$. While the analysis that follows deals specifically with the Epwp problem, the method is equally applicable to the transient temperature problem involving variable coefficients. Consider figure 13.22.

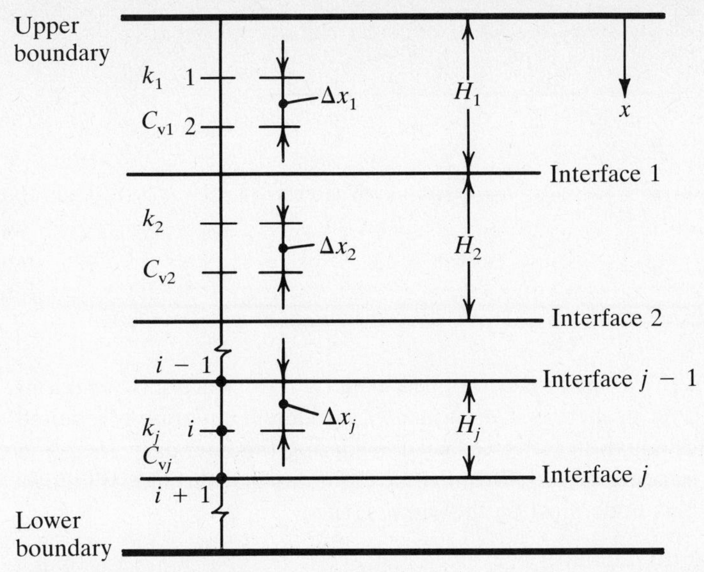


**Figure 13.22** Schematic representation of one-dimensional consolidation for multilayered soil system.

It is assumed that there are $n$ layers of soil having $n$ permeabilities $k$ and $n$ coefficients of consolidation $C_v$. For a given node $i$ within a given layer $j$ the finite difference expression for the equation can be derived rather easily on the assumption that within that layer both $C_{vj}$ and $k_j$ are constants. Thus

$$\frac{\partial u}{\partial t} = \frac{-u_{i,t} + u_{i,t+\Delta t}}{\Delta t} + (O)h$$

and

$$\frac{\partial^2 u}{\partial x^2} = \frac{u_{i+1,t} - 2u_{i,t} + u_{i-1,t}}{(\Delta x)^2} + (O)h^2$$

Substituting into equation (13.107) then solving for $u_{i,t+\Delta t}$ gives

$$u_{i,t+\Delta t} = \alpha_j u_{i-1,t} + (1 - 2\alpha_j)u_{i,t} + \alpha_j u_{i+1,t} \tag{13.108}$$

where

$$\alpha_j = \frac{C_{vj}\Delta t}{(\Delta x)^2} \leq \frac{1}{2} \tag{13.109}$$

and

$$\Delta t_1 = \Delta t_2 = \cdots = \Delta t_n = \Delta t \tag{13.110}$$

Note that there are $n$ $\alpha$-values that must satisfy the condition that $\alpha$ is less than one-half. Equation (13.108) can be applied at any node falling inside any of the $n$ layers. The question that must be addressed is, what about nodes falling on the interfaces? To answer this question, let us consider the $j$th interface:

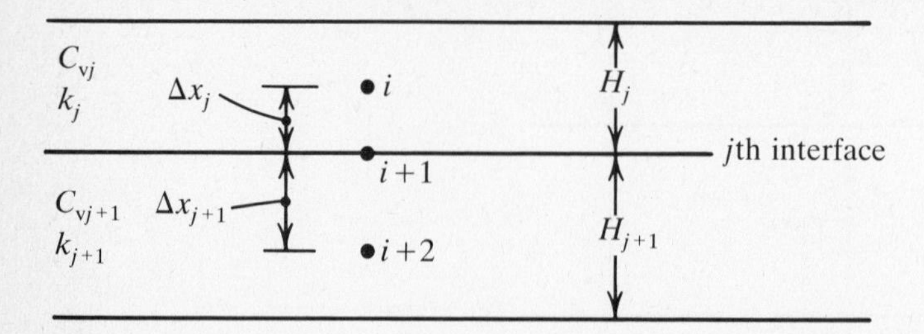


Obviously, equation (13.98) cannot be applied at node $i + 1$ since this would involve two different layer properties. Consequently, a different equation is required for the interface. This equation is derived by considering the boundary condition at node $i + 1$. That is, continuity of flow must be maintained at this line. It follows that the velocity at this node must be the same. Hence

$$v_j = v_{j+1} \qquad \text{(at } j\text{th interface)}$$

or

$$k_j \frac{\partial u}{\partial x} = k_{j+1} \frac{\partial u}{\partial x} \tag{13.111}$$

Setting equation (13.111) in a difference form gives

$$\frac{k_j}{\Delta x_j}(-u_{i,t+\Delta t} + u_{i+1,t+\Delta t}) = \frac{k_{j+1}}{\Delta x_{j+1}}(-u_{i+1,t+\Delta t} + u_{i+2,t+\Delta t})$$

Solving for $u_{i+1,t+\Delta t}$ gives

$$u_{i+1,t+\Delta t} = \frac{\beta_j u_{i+2,t+\Delta t} + u_{i,t+\Delta t}}{1 + \beta_j} \tag{13.112}$$

Equation (13.112) simply states that for a given time increment the interior nodal pressures are computed by using equation (13.108) and skipping the interfaces nodal pressures, then substituting these newly computed values into equation (13.112). The interface pressures are calculated. Equations (13.108) and (13.112) may be expressed in stencil form as follows:

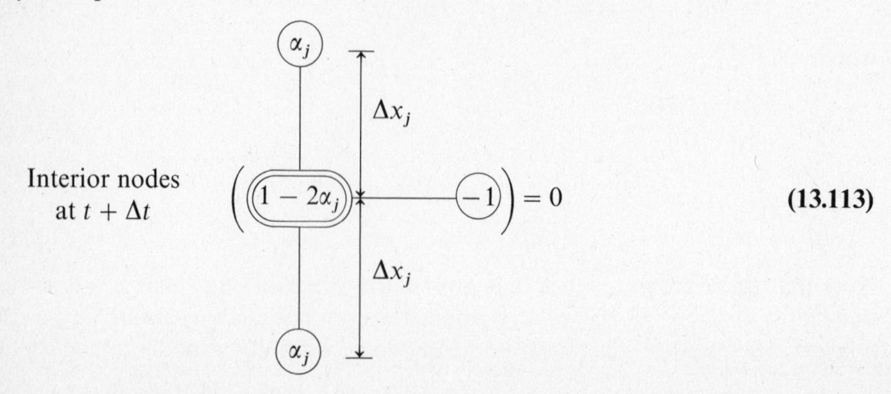

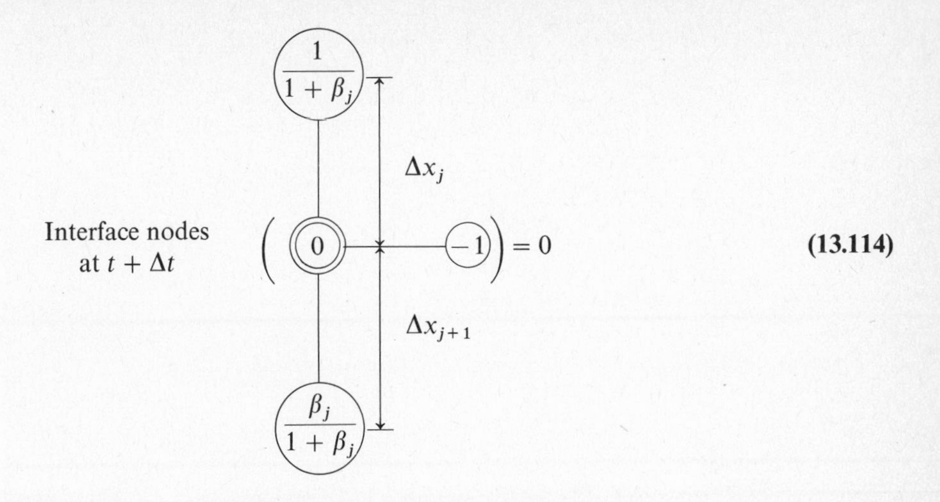


(13.114)

These equations provide the answer to the nonlinear transient problem. For cases in which $C_v$ and $k$ are not constant even within a given layer, that layer is divided into sublayers and the same equations are applied. However, these equations are applicable only when both boundaries are previous (not derivative).

**E**

EXAMPLE 13.14

Determine the excess pore water pressure distribution for the following profile assuming both boundaries are previous and initial conditions of 1.0 ksf throughout the depth:

| Thickness | Properties | Layer | |
|---|---|---|---|
| 5 ft | $k_1 = 5$ ft/year, $C_{v1} = 5$ ft²/year | Layer 1 | |
| | | | Interface 1 |
| 10 ft | $k_2 = 2$ ft/year, $C_{v2} = 4$ ft²/year | Layer 2 | |
| | | | Interface 2 |
| 5 ft | $k_1 = 5$ ft/year, $C_{v3} = 5$ ft²/year | Layer 3 | |

Use 5 nodes in layer 1, 10 nodes in layer 2, 5 nodes in layer 3, and $\Delta t = 0.05$ yr. Calculate all values up to $t = 10$ yr.

## Solution

We first must check $\alpha_j$ for the three layers. However, because of symmetry only $\alpha_1$ and $\alpha_2$ are needed:

$$\alpha_1 = \frac{5(0.05)}{(1)^2} = 0.25 < \frac{1}{2} \qquad \text{OK}$$

$$\alpha_2 = \frac{4(0.05)}{(1)^2} = 0.20 < \frac{1}{2} \qquad \text{OK}$$

We now compute the stencils needed for interior nodes. That is,

Layer 1:

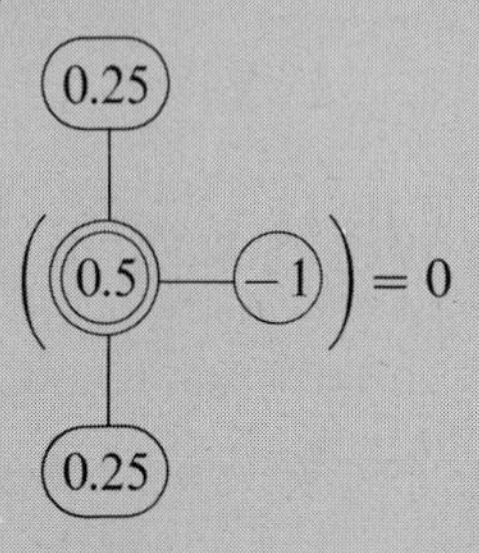


Layer 2:

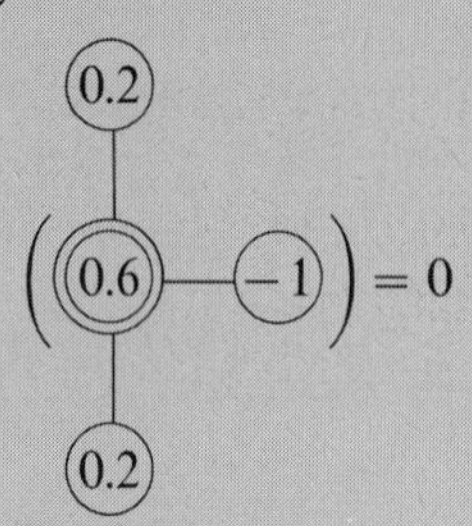


Now compute $\beta_1$ and its corresponding stencil:

Interface 1:

$$\beta_1 = \frac{k_2}{k_1}\frac{\Delta x_1}{\Delta x_2} = \frac{(2)(1)}{(5)(1)} = \frac{2}{5} = 0.4$$

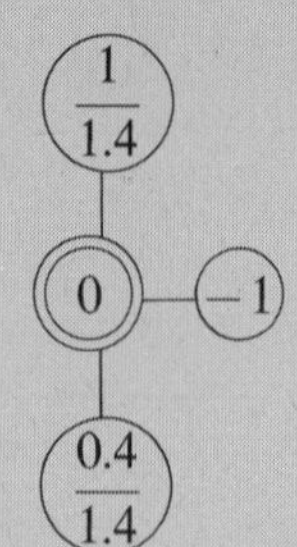

Now we proceed by computing the average boundary values for the first increment. Note that only the first 10 increments are listed in table 13.1; the rest of the increments are shown graphically in figure 13.23.

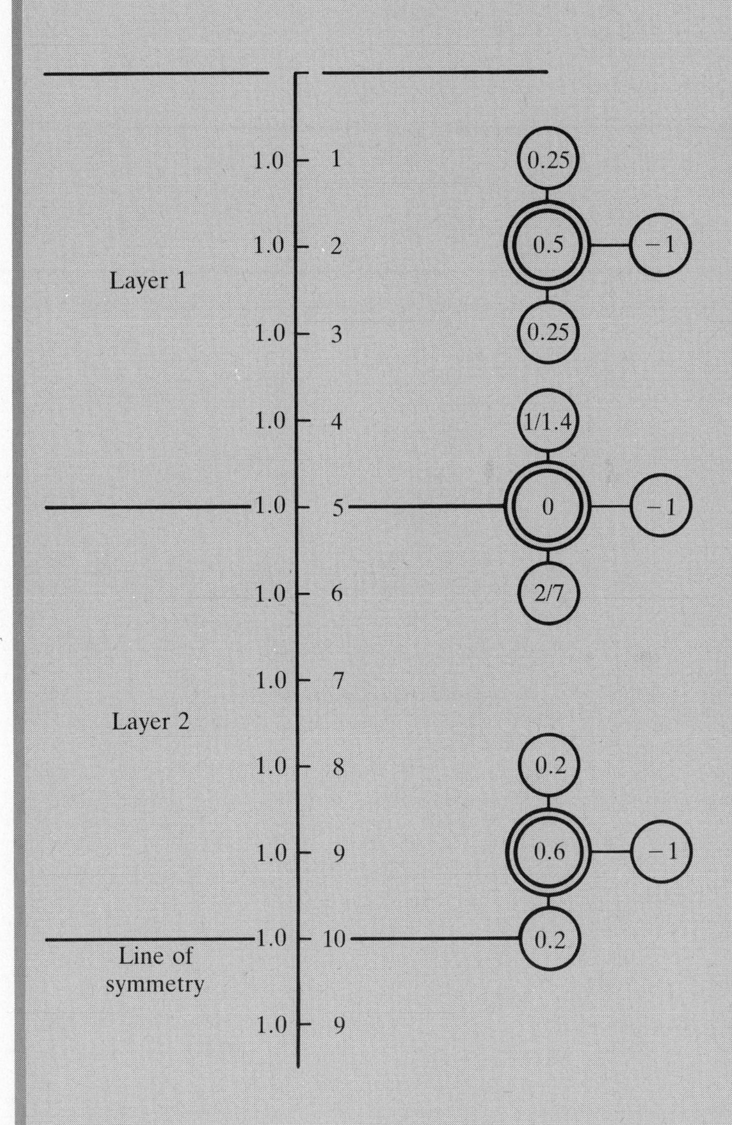


Note that nodes appearing on either side of the line of symmetry must have the same Epwp.

Table 13.1 COMPUTED EPWP FOR EXAMPLE 13.14 FOR THE FIRST TEN TIME INCREMENTS

| DEPTH | INITIAL | EPWP (KSF) | | | | | | | | | |
|---|---|---|---|---|---|---|---|---|---|---|---|
| (FT) | EPWP | 0.05 | 0.1 | 0.15 | 0.2 | 0.25 | 0.30 | 0.35 | 0.4 | 0.45 | 0.50 |
| 0 | 0.5 | 0 | 0 | 0 | 0 | 0 | 0 | 0 | 0 | 0 | 0 |
| 1 | 1 | 0.875 | 0.6875 | 0.5859 | 0.5195 | 0.4717 | 0.4350 | 0.4058 | 0.3818 | 0.3616 | 0.3443 |
| 2 | 1 | 1 | 0.9688 | 0.9063 | 0.8477 | 0.7969 | 0.7533 | 0.7157 | 0.6829 | 0.6539 | 0.6280 |
| 3 | 1 | 1 | 1 | 0.9922 | 0.9726 | 0.9477 | 0.9211 | 0.8942 | 0.8679 | 0.8425 | 0.8182 |
| 4 | 1 | 1 | 1 | 1 | 0.9980 | 0.9918 | 0.9814 | 0.9676 | 0.9514 | 0.9337 | 0.9152 |
| 5 | 1 | 1 | 1 | 1 | 0.9986 | 0.9941 | 0.9865 | 0.9763 | 0.9642 | 0.9507 | 0.9365 |
| 6 | 1 | 1 | 1 | 1 | 1 | 0.9999 | 0.9993 | 0.9981 | 0.9961 | 0.9933 | 0.9896 |
| 7 | 1 | 1 | 1 | 1 | 1 | 1 | 1.0 | 0.9999 | 0.9997 | 0.9994 | 0.9988 |
| 8 | 1 | 1 | 1 | 1 | 1 | 1 | 1 | 1.0 | 1.0 | 1.0 | 0.9999 |
| 9 | 1 | 1 | 1 | 1 | 1 | 1 | 1 | 1 | 1.0 | 1.0 | 1.0 |
| 10 | 1 | 1 | | | | | | | | | |
| 9 | 1 | 1 | 1 | 1 | 1 | 1 | 1 | 1.0 | 1.0 | 1.0 | 0.9999 |

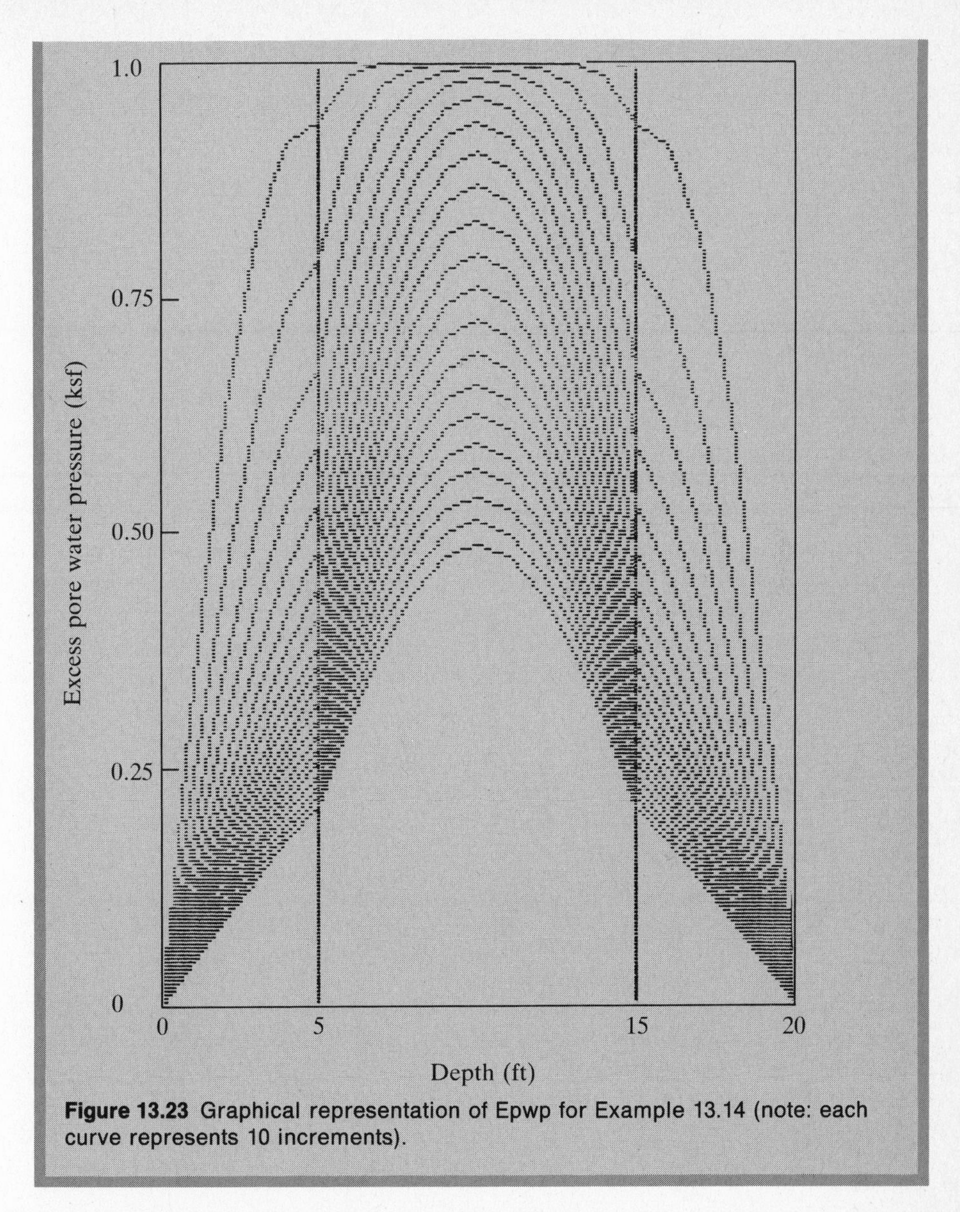


**Figure 13.23** Graphical representation of Epwp for Example 13.14 (note: each curve represents 10 increments).

In conclusion of this section, it should be obvious to the student that derivative boundary conditions can be handled in exactly the same way as the linear problems covered in the preceding section.

The solution of linear and nonlinear parabolic partial differential equations may involve two or three spatial coordinates. That is, the physical region to be analyzed is that of an area or volume. Although only one spatial dimension is considered in this section, the procedures outlined are equally applicable to higher dimensions.

## 13.8 HYPERBOLIC PARTIAL DIFFERENTIAL EQUATIONS

Hyperbolic equations represent the final category of second-order partial differential equations occurring in engineering practice. These equations are encountered mostly in the following form:

$$c^2 \frac{\partial^2 w}{\partial x^2} = \frac{\partial^2 w}{\partial t^2} \tag{13.115}$$

This equation describes various types of behavior such as the vibration of strings, membranes, and stress waves in driven piles. In all cases, the governing equation given by (13.115) is called the wave equation. More recently, the wave equation has been used successfully in the design of piles and pile-caps as well as hammers used in driving piles for building foundations. While this is not the proper place for illustrating its various uses, it suffices to say that equation (13.115) is important for solving a variety of problems encountered in the different engineering disciplines.

The solution to these problems may be needed in one, two, or three spatial coordinates. For clarity, we shall limit our presentation to one spatial dimension only. Furthermore, let us consider the problem of pile driving. Piles are normally driven by a hammer impacting the head of a pile, in which case a compressive wave is sent down the pile at a velocity $c$ in the pile material. That is,

$$c = (E/\gamma)^{1/2}$$

where

$c$ = velocity

$E$ = modulus of elasticity of pile material

$\gamma$ = mass density of pile material

The amount of compression that the pile undergoes at any instance is termed $\phi$. Consequently, at time $j$ and for the $i$th node in a given pile the solution to equation (13.115) can be easily approximated by means of the finite difference procedure. Consider figure 13.24.

The partial derivatives appearing in equation (13.115) can be approximated at node $i$ and time $j$ by using the finite difference approximation to give

$$\frac{\partial^2 w}{\partial x^2} = \frac{1}{(\Delta x)^2}(w_{i-1,j} - 2w_{i,j} + w_{i+1,j})$$

$$\frac{\partial^2 w}{\partial t^2} = \frac{1}{(\Delta t)^2}(w_{i,j-1} - 2w_{i,j} + w_{i,j+1})$$

These approximations have an error order of $(O)h^2$. Substituting into equation (13.115) gives after simplification

$$w_{i,j+1} = \beta^2 w_{i-1,j} + 2(1 - \beta^2)w_{i,j} + \beta^2 w_{i+1} - w_{i,j-1} \tag{13.116}$$

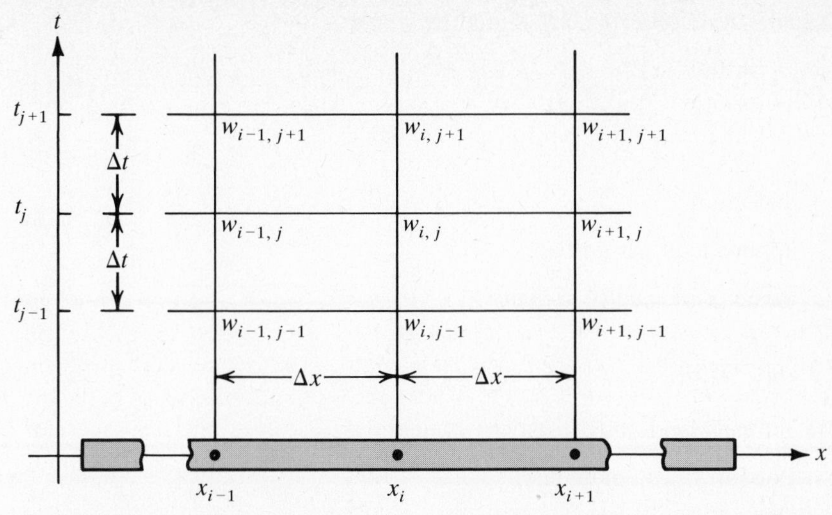


**Figure 13.24** Typical Cartesian grid for approximating the derivatives of the wave equation at node *i* and time *j*.

where

$$\beta = \frac{c\,\Delta t}{\Delta z}$$

The solution given by equation (13.116) includes the forward movement in time without the need for solving systems of linear algebraic equations. That is, this scheme is equivalent to the explicit scheme used earlier in solving parabolic equations. The solution achieved by equation (13.116) can be shown to be stable for $\beta \leq 1$. However, the solution is most accurate when $\beta = 1$. Note that for $\beta < 1$ the solution is stable but the accuracy is reduced. Consequently, equation (13.116) is simplified for $\beta = 1$ to

$$w_{i,j+1} = w_{i-1,j} + w_{i+1,j} - w_{i,j-1} \tag{13.117}$$

Introducing the initial and boundary conditions permits the evaluation of the wave equation numerically.

## 13.9 BIHARMONIC EQUATION

The analysis of certain engineering systems may require solving partial differential equations of the third, fourth, and higher order. While a fourth-order partial differential equation can adequately describe plate deflection, an eighth-order partial differential equation is required for describing shell behavior. This section deals with the plate equation, but is by no means an exhaustive treatment of this broad subject.

Plates are straight, flat surface structures whose thickness is very small compared to their other dimensions. They are normally supported at their edges or even in the interior. For plates of linearly elastic isotropic material, the deflection is described by the following differential equation:

$$\frac{\partial^4 w}{\partial x^4} + 2\frac{\partial^4 w}{\partial x^2\,\partial y^2} + \frac{\partial^4 w}{\partial y^4} = \frac{P_z}{D} - \frac{k}{D}w \tag{13.118}$$

where

$w$ = deflection

$P_z$ = applied load per unit area

$D$ = modulus of rigidity of plate

$= \dfrac{Et^3}{12(1 - \nu^2)}$

$E$ = Young's modulus of elasticity of material

$\nu$ = Poisson's ratio of material

$t$ = plate thickness

$k$ = modulus of subgrade reaction

A schematic representation of a typical plate is shown in figure 13.25.

While closed-form solutions can be obtained for a few combinations of loading and boundary conditions, these solutions involve infinite series. Consequently, the so-called exact solutions are only approximations to the exact ones.

The finite difference method is a powerful tool that can be used for solving any plate problem regardless of the loading and boundary conditions. For clarity of presentation, let us assume that $\Delta x = \Delta y = h$ and use the central difference expressions developed in chapter 10. Furthermore, assume an error order of $(O)h^2$. Thus, for a given interior node,

$$\frac{\partial^4 w}{\partial x^4} = \frac{1}{h^4}\left( \textcircled{1} - \textcircled{-4} - \textcircled{\textcircled{6}} - \textcircled{-4} - \textcircled{1} \right) \tag{13.119}$$

$$\frac{\partial^2 w}{\partial y^4} = \frac{1}{h^4}\left( \textcircled{1} - \textcircled{-4} - \textcircled{\textcircled{6}} - \textcircled{-4} - \textcircled{1} \right) \tag{13.120}$$

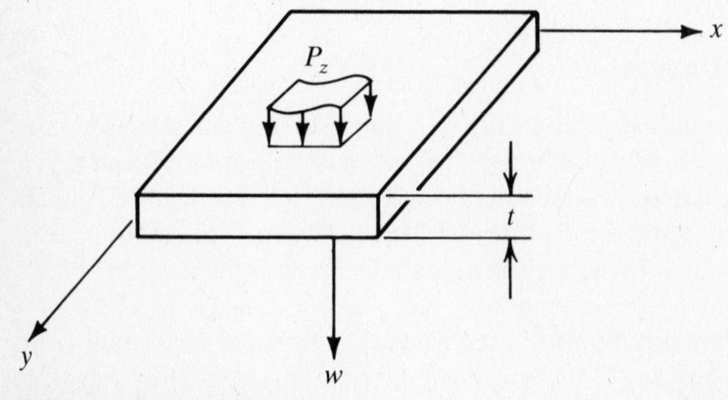


**Figure 13.25** A typical plate.

$$\frac{\partial^2 w}{\partial x^2} = \frac{1}{h^2}\left(\textcircled{1} - \textcircled{\textcircled{-2}} - \textcircled{1}\right) \tag{13.121}$$

$$\frac{\partial^2 w}{\partial y^2} = \frac{1}{h^2}\left(\textcircled{1} - \textcircled{\textcircled{-2}} - \textcircled{1}\right) \tag{13.122}$$

The mixed partial derivative term $\partial^4 w/\partial x^2\,\partial y^2$ can now be determined by applying the stencil given by equation (13.121) to that given by (13.122) as follows:

$$\begin{aligned}
2\frac{\partial^4 w}{\partial x^2\,\partial y^2} &= \frac{2}{h^4}\left(\begin{Bmatrix} 1 \\ -2 \\ 1 \end{Bmatrix}\{1 \quad -2 \quad 1\}\right) \\
&= \frac{2}{h^4}\begin{bmatrix} 1 & -2 & 1 \\ -2 & 4 & -2 \\ 1 & -2 & 1 \end{bmatrix} \\
&= \frac{1}{h^4}\left(\begin{array}{ccccc} \textcircled{2} & - & \textcircled{-4} & - & \textcircled{2} \\ | & & | & & | \\ \textcircled{-4} & - & \textcircled{\textcircled{8}} & - & \textcircled{-4} \\ | & & | & & | \\ \textcircled{2} & - & \textcircled{-4} & - & \textcircled{2} \end{array}\right)
\end{aligned} \tag{13.123}$$

Substituting equations (13.119), (13.120), and (13.123) into (13.118) yields the finite difference equation needed for solving the plate problem:

$$\nabla^4 w = \frac{1}{h^4}\left(\textcircled{1} - \textcircled{-4} - \textcircled{\textcircled{6}} - \textcircled{-4} - \textcircled{1}\right) + \frac{1}{h^4}\left(\begin{array}{ccccc} \textcircled{2} & - & \textcircled{-4} & - & \textcircled{2} \\ | & & | & & | \\ \textcircled{-4} & - & \textcircled{\textcircled{8}} & - & \textcircled{-4} \\ | & & | & & | \\ \textcircled{2} & - & \textcircled{-4} & - & \textcircled{2} \end{array}\right)$$

$$+ \frac{1}{h^4}\left(\begin{array}{c} \textcircled{1} \\ | \\ \textcircled{-4} \\ | \\ \textcircled{\textcircled{6}} \\ | \\ \textcircled{-4} \\ | \\ \textcircled{1} \end{array}\right) = \frac{P_z}{D} - \frac{k}{D}\textcircled{\textcircled{1}}$$

Combining the corresponding pivotal points yields the following expressions:

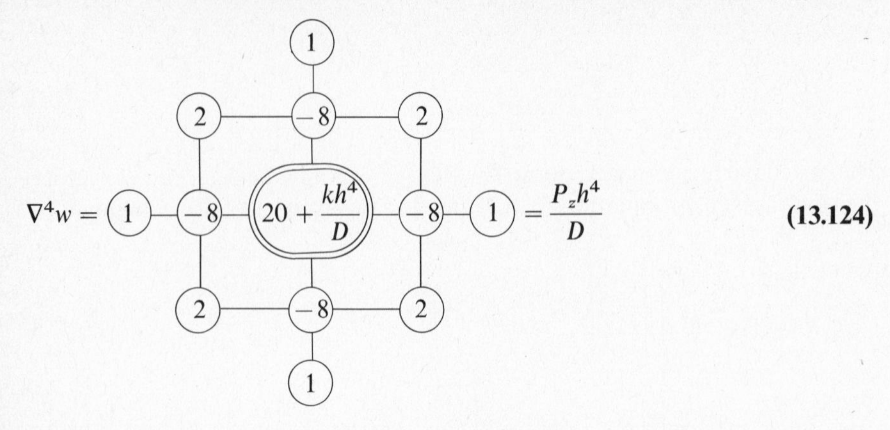


(13.124)

The stencil given by equation (13.125) can be applied at any interior nodal point. However, it must be altered for nodes adjacent to the boundaries. Furthermore, for the special case in which a plate is supported by its edges, the $k$ value is zero. It is interesting to note that derivative boundary conditions must be applied in oıder to eliminate unknown displacements resulting from applying the stencil at nodes adjacent to the edges. Fortunately, these conditions are precisely the same as those developed for a beam in chapter 12. This is true for hinged and fixed boundaries since these result in a single circle falling outside the region to be analyzed.

EXAMPLE 13.15

Determine the deflections at the various nodes of the following plate using $h = 5$ ft, $k = 100$ kips/ft$^3$, $t = 1$, $\nu = 0.2$, and $E = 20 \times 10^5$ psi.

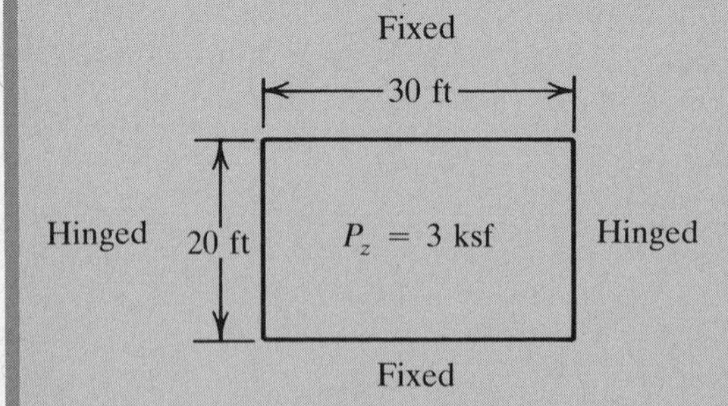


### Solution

This problem is greatly simplified by observing that the loads and boundary conditions are symmetrical about the middle planes. Therefore, noting that for a beam with fixed end the value of the imaginary node is equal to the value of the first node on the other side of the boundary and that for a hinged beam end it is equal to the negative value of the first interior node makes it possible to establish the following grid:

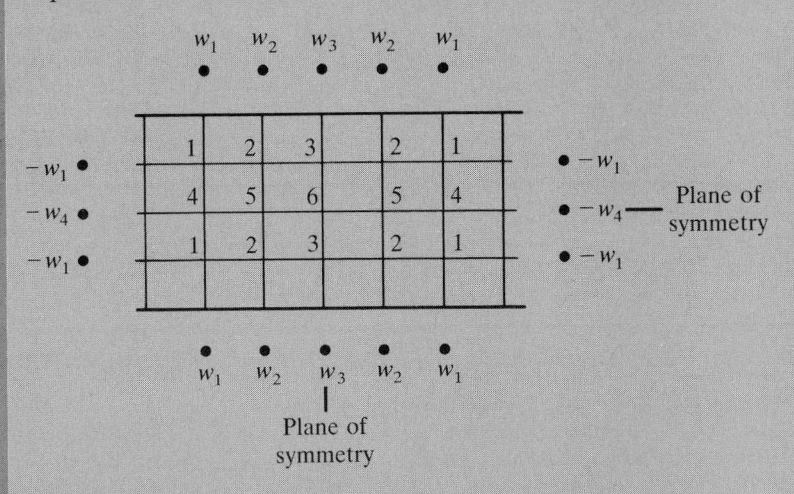


Now calculate the following:

$$D = \frac{Et^3}{12(1 - \nu^2)} = \frac{(20 \times 10^5)(12)^3}{12(1 - 0.2^2)} = 3 \times 10^5 \text{ kips-in.}$$

$$\frac{kh^4}{D} = \frac{(100)(\frac{1}{12})^3(5 \times 12)^4}{3 \times 10^5} = 2.50$$

$$\frac{P_z h^4}{D} = \frac{(3)(\frac{1}{12})^2(5 \times 12)^4}{3 \times 10^5} = 0.075 \text{ in.}$$

Substituting into the stencil given by equation (13.124) gives

$$\nabla^4 w = \begin{array}{ccccc} & & 1 & & \\ & 2 & -8 & 2 & \\ 1 & -8 & 22.5 & -8 & 1 \\ & 2 & -8 & 2 & \\ & & 1 & & \end{array} = 0.075$$

Node 1:

$$22.5w_1 - 8(w_2 + w_4) + w_1 + 2w_5 + w_3 = 0.075$$

Node 2:

$$22.5w_2 - 8(w_1 + w_3 + w_5) + 2(w_4 + w_6) + w_2 + w_2 + w_2 = 0.075$$

Node 3:

$$22.5w_3 - 8(w_2 + w_2 + w_6) + 2(w_5 + w_5) + w_1 + w_1 + w_1 + w_3 + w_3 = 0.075$$

Node 4:

$$22.5w_4 - 8(w_1 + w_5) + 2(w_2 + w_2) + w_6 - w_4 = 0.075$$

Node 5:

$$22.5w_5 - 8(w_2 + w_4 + w_2 + w_6) + 2(w_1 + w_3 + w_1 + w_3) + w_5 = 0.075$$

Node 6:

$$22.5w_6 - 8(w_5 + w_3 + w_3 + w_5) + 2(w_2 + w_2 + w_2 + w_2) + w_4 + w_4 = 0.075$$

Note that deflections of the boundary nodes are equal to zero. Consequently, solving for the unknown deflections gives

$w_1 = 0.009386$ in. $\quad w_4 = 0.010731$ in.

$w_2 = 0.014411$ in. $\quad w_5 = 0.020095$ in.

$w_3 = 0.015385$ in. $\quad w_6 = 0.022486$ in.

This problem illustrates clearly the advantages gained by using symmetry. However, it should be emphasized that for plates involving general loading, the nodes must be numbered in such a way that unique deflections are obtained.

So far we have not dealt with the problem of free boundary conditions. While a complete treatment of the subject is beyond the scope of this text, the various stencils resulting for such an edge are summarized in figure 13.26. These were obtained by introducing appropriate boundary conditions at the boundaries. Furthermore, the stencils are functions of the Poisson ratio ($\nu$), and were developed by assuming equally spaced nodes in the $x$ and $y$ directions.

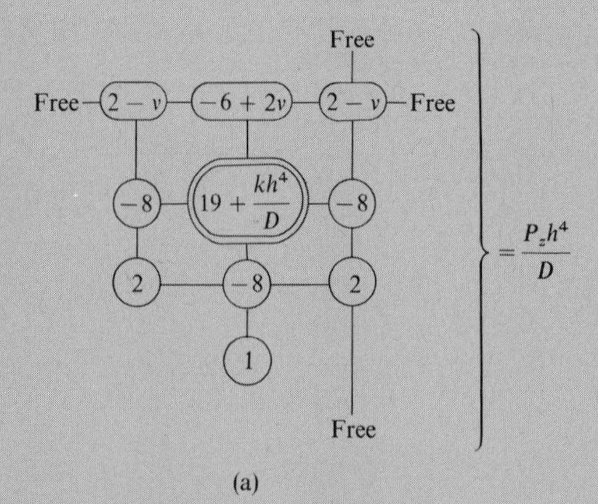


**Figure 13.26** Stencils for a plate resting on elastic springs with free edges.

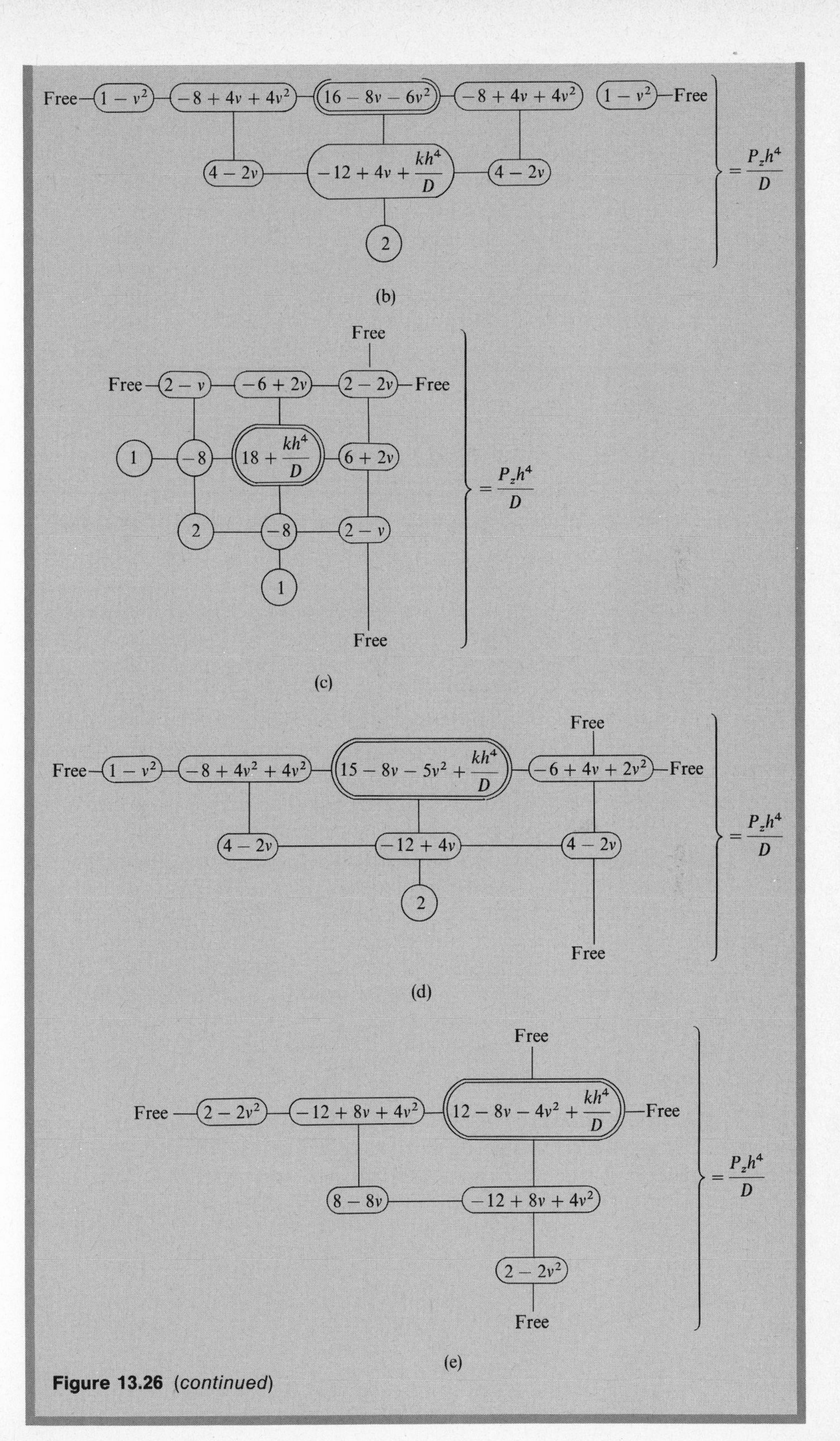


**Figure 13.26** *(continued)*

Note that the stencil given by equation (13.124) is used for interior nodes whose molecules fall inside the region. It is evident that these stencils correspond to different node locations within a plate. Furthermore, since for a free boundary more than one circle falls outside the region, it is necessary to introduce additional boundary conditions to obtain the solution presented in figure 13.26. The problem of plates on elastic foundations with free edges is important in analyzing nuclear power plants, high-rise buildings, and radar station foundation systems. Furthermore, the transformation of the biharmonic equation from the Cartesian to polar coordinate systems is necessary in solving circular plate regions. In fact, skewed and triangular coordinates may be used to analyze problems involving skewed and triangular grids.

## 13.10 INTRODUCTION TO THE FINITE ELEMENT METHOD

### 13.10.1 Introduction

The finite element method is a numerical procedure for solving problems occurring in engineering and related sciences. While this method in one form or another has existed for a number of years, it did not gain enough attention until powerful computers became available. This is because the method basically relies on solving a large set of algebraic equations and entails considerable manipulation, as is the case with the finite difference procedure. The method was formulated by civil engineers. In its present form the finite element method is simply a variation of the well-known Raleigh–Ritz procedure used for solving structural problems.

The fundamental concept of the finite element method is that any region is made up of elements; therefore, the general behavior of a system can be determined by considering the behavior of its components (subsystem).

The finite element method has all of the advantages of the finite difference method and more. That is, material properties of adjacent elements do not have to be the same. Additonally, irregular boundaries can be approximated with connected elements in the same way as regular boundaries are treated! Recall that the finite difference procedure requires establishing different stencils for different boundary conditions and for different properties. This is not the case with the finite element method.

### 13.10.2 Discretization of a Region

The first step in the finite element method is to subdivide the region under consideration into smaller regions (elements). Hence, a rod is subdivided into smaller rods connected at the nodes, a plate is subdivided into a set of smaller plates connected at the nodes, etc. Consider the following one-dimensional element.

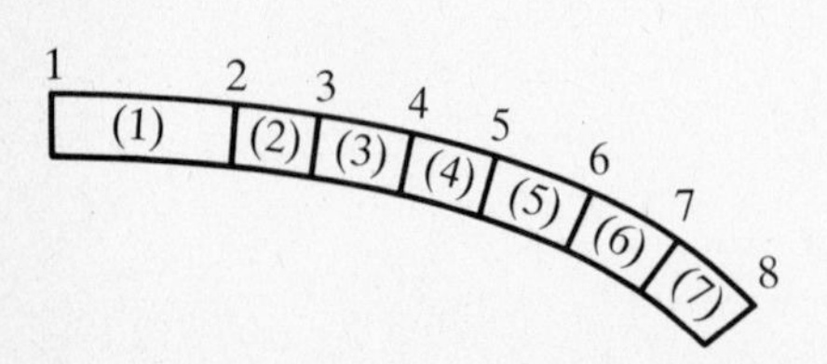

Here the rod is cut into seven smaller rods. The first element is connected to the second element at node 2, the second element connected to the third at node 3, etc. The implicaton is that the original rod can be put back together by connecting the seven smaller rods. These smaller rods are called finite elements. It is evident that the elements do not have to be of equal lengths nor does the rod have to be straight.

Two-dimensional regions can be subdivided into smaller subregions in a variety of ways. Consider the following region.

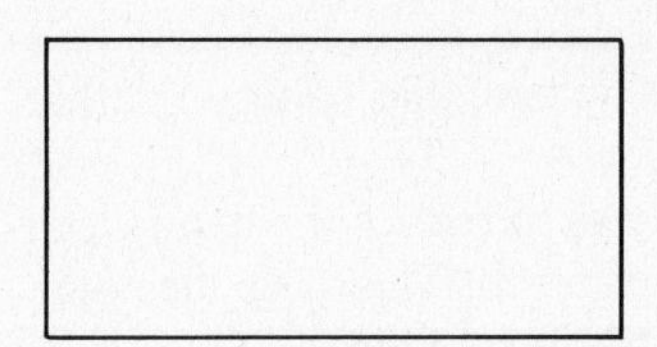

Obviously, one way of subdividing the region is to form rectangular or square or even triangular elements as indicated below:

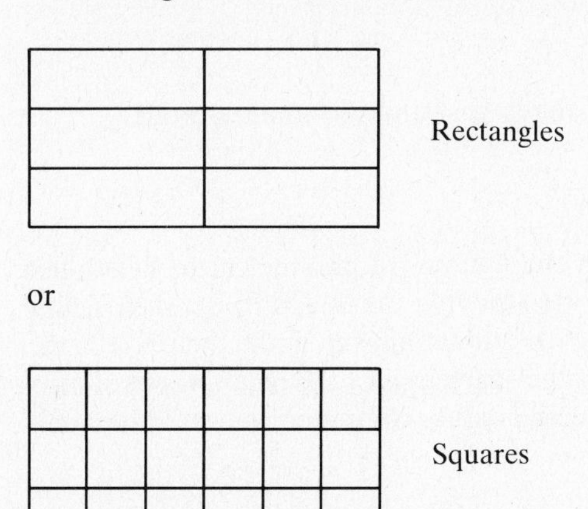


or

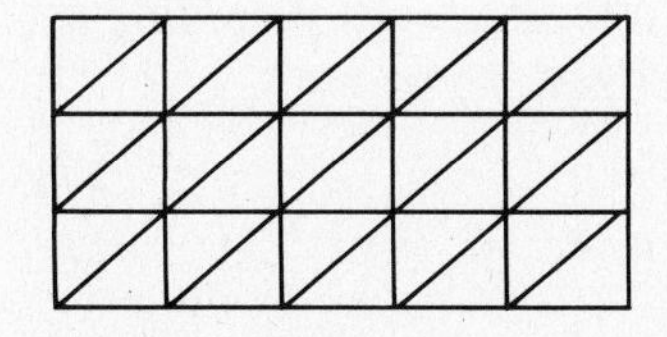

Triangles

It is evident that even the triangular elements can be varied in size and shape. This concept can be extended to three-dimensional systems. Fundamentally, the basic idea here is that a region is subdivided in such a way as to increase the number of elements in the region where accurate results are desired and to reduce the number of elements where results are not so important. Furthermore, the resulting element should be fat rather than thin. That is, ideally we want to get elements that have equal sides.

Good element

Bad element

These concepts are extremely important in getting a reasonable solution rather than a bad one.

Finally, the nodes must be numbered in such a way as to reduce the bandwidth of the resulting set of algebraic equations that must be solved. The basic idea here is to number the nodes by going through the short side first. Thus

$$R = (1 + M)F \tag{13.125}$$

where

$R$ = bandwidth

$M$ = maximum difference between the largest and the smallest node numbers for any element

$F$ = number of degrees of freedom

The $F$ parameter is nothing more than the number of parameters to be studied at any given node. Consequently, if we are studying the temperature, then $F = 1$; if we are concerned with the slope and deflection at a given node, then $F = 2$, etc. The bandwidth is important in reducing the sparseness of the resulting set of algebraic equations and for a large problem could reduce computer time and ultimately the cost of analyzing a given problem.

### 13.10.3 Interpolation and Shape Functions

The finite element method of solution is dependent on the choice of the approximating function for the unknown parameter(s) to be determined in the vicinity of an element. Finite elements are therefore classified into three groups according to the order of the interpolating function used in the analysis. These are

1. simplex elements
2. complex elements
3. multiplex elements

The simplex element has an approximating polynomial with the number coefficient equal to dimension of the coordinate plus one. The complex element has an approximating polynomial that includes a constant, linear, and higher-order terms. Finally, the multiplex element is a complex element where the boundaries are parallel, such as a rectangle. These elements are shown in figure 13.27.

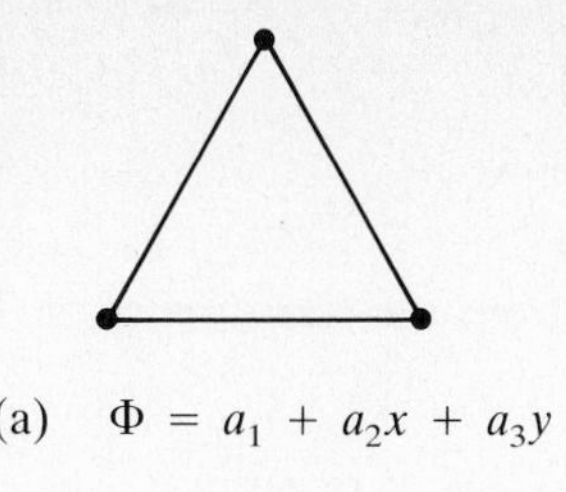

(a) $\Phi = a_1 + a_2 x + a_3 y$

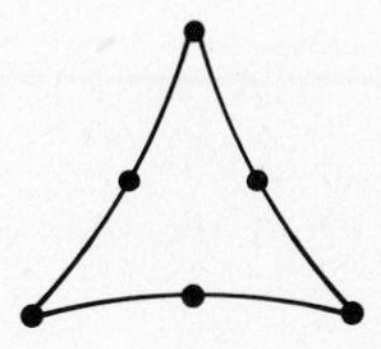

(b) $\Phi = a_1 + a_2 x + a_3 y + a_4 x^2 + a_5 y^2 + a_6 xy$

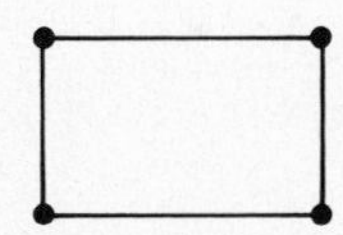

(c) $\Phi = a_1 + a_2 x + a_3 y + a_4 xy$

**Figure 13.27** Different elements used in the finite element methods.

Consider a one-dimensional simplex element for analyzing steady state temperature, steady state seepage, etc.

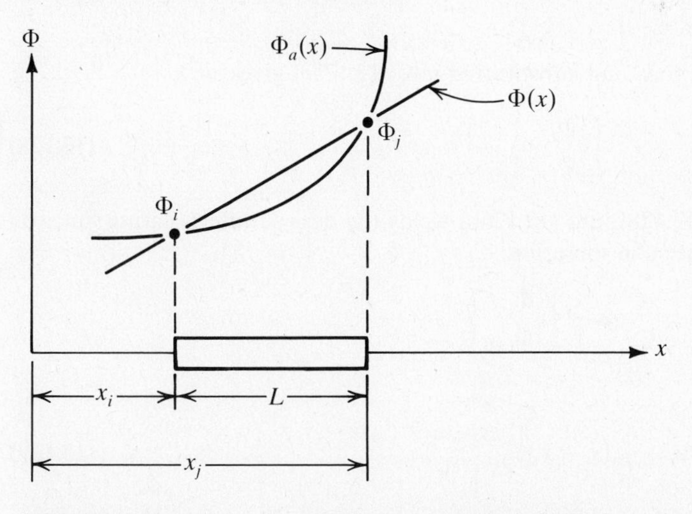


Here $\Phi_i$ is the value of a parameter at node $i$, $\Phi_j$ is the value of the parameter at node $j$, $\Phi_a(x)$ is the actual function describing the behavior of $\Phi$ between the two nodes, and $\Phi$ is the approximation to $\Phi_a$. That is,

$$\Phi = a_1 + a_1 x \quad \textbf{(13.126a)}$$

or in matrix form

$$\Phi = \{1 \quad x\}\begin{Bmatrix} a_1 \\ a_2 \end{Bmatrix} \tag{13.126b}$$

The coefficients $a_1$ and $a_2$ are determined from the coordinates of the two nodes. Thus

$$\Phi = \Phi_i \qquad \text{at} \quad x = x_i$$

and

$$\Phi = \Phi_j \qquad \text{at} \quad x = x_j$$

Substituting these conditions into equation (13.126a) and then expressing the results in a matrix form gives

$$\begin{Bmatrix} \Phi_i \\ \Phi_j \end{Bmatrix} = \begin{bmatrix} 1 & x_i \\ 1 & x_j \end{bmatrix} \begin{Bmatrix} a_1 \\ a_2 \end{Bmatrix} \tag{13.127a}$$

or

$$\{\Phi\} = [A]\{a\} \tag{13.127b}$$

and

$$\{a\} = [A]^{-1}\{\Phi\} \tag{13.127c}$$

These two equations in the two unknowns can be solved for using matrix inversion. Hence

$$[A]^{-1} = \frac{1}{x_j - x_i}\begin{bmatrix} x_j & -x_i \\ -1 & 1 \end{bmatrix}$$

Noting that $L = x_j - x_i$ and substituting into (13.127c) gives

$$\begin{Bmatrix} a_1 \\ a_2 \end{Bmatrix} = \frac{1}{L}\begin{bmatrix} x_j & -x_i \\ -1 & 1 \end{bmatrix}\begin{Bmatrix} \Phi_i \\ \Phi_{i+1} \end{Bmatrix} \tag{13.128}$$

Now substituting (13.128) into (13.126b) yields the desired interpolating function for the element under consideration.

$$\Phi = \frac{1}{L}\{1 \quad x\}\begin{bmatrix} x_j & -x_i \\ -1 & 1 \end{bmatrix}\begin{Bmatrix} \Phi_i \\ \Phi_{i+1} \end{Bmatrix}$$

or

$$\Phi = \frac{1}{L}(x_j - x)\Phi_i + \frac{1}{L}(x - x_i)\Phi_j \tag{13.129}$$

The value of the parameter $\Phi$ is expressed in terms of the nodal values and two linear functions. Equation (13.129) is generally expressed in the following form:

$$\Phi = N_i\Phi_i + N_j\Phi_j \tag{13.130a}$$

where $N_i$ and $N_j$ are referred to as the shape functions for the element; hence

$$N_i = \frac{1}{L}(x_j - x) \tag{13.130b}$$

$$N_j = \frac{1}{L}(x - x_i) \tag{13.130c}$$

The shape functions have unique characteristics in that $N_i$ has a value of 1.0 at node $i$ and zero at node $j$ and $N_j$ has a value of zero at node $i$ and 1.0 at node $i + 1$. One can look at these functions as weight factors having values between 0 and 1.0. Thus

$$0 \le N_i \le 1, \qquad 0 \le N_j \le 1$$

These properties are common to all shape functions regardless of the type of interpolating function used. The gradient of $\Phi$ can be easily defined as follows:

$$\frac{\partial \Phi}{\partial x} = \frac{\partial N_i}{\partial x}\Phi_i + \frac{\partial N_j}{\partial x}\Phi_j$$

and

$$\frac{\partial \Phi}{\partial x} = -\frac{1}{L}\Phi_i + \frac{1}{L}\Phi_j$$

Consider the following example.

**E**

EXAMPLE 13.16

Determine the steady-state temperature at $x = 12$ for the following rod and the gradient.

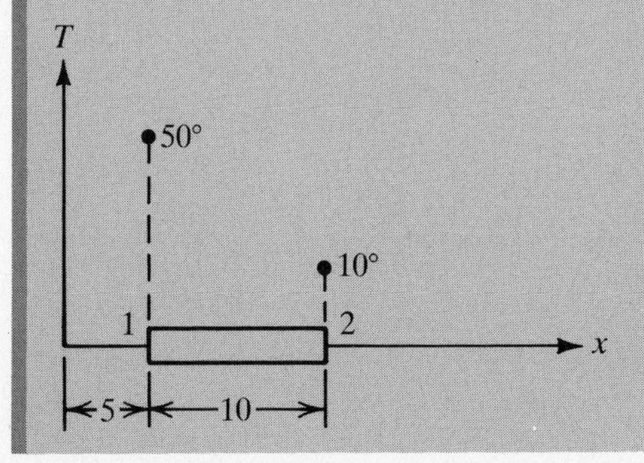

## Solution

Obviously, equation (13.130) can be used to solve this simple problem. Thus

$$N_1 = \tfrac{1}{10}(15 - 2) = \tfrac{3}{10}, \qquad N_2 = \tfrac{1}{10}(12 - 5) = \tfrac{7}{10}$$

and

$$\begin{aligned} T &= N_1 T_1 + N_2 T_2 \\ &= \tfrac{3}{10}(50) + \tfrac{7}{10}(10) \\ &= 22^\circ \end{aligned}$$

$$\frac{\partial T}{\partial x} = \frac{1}{10}(50) + \frac{1}{10}(10) = -4^\circ/\text{ft}$$

The two-dimensional simplex element is a triangular element with three nodes. Consider the following:

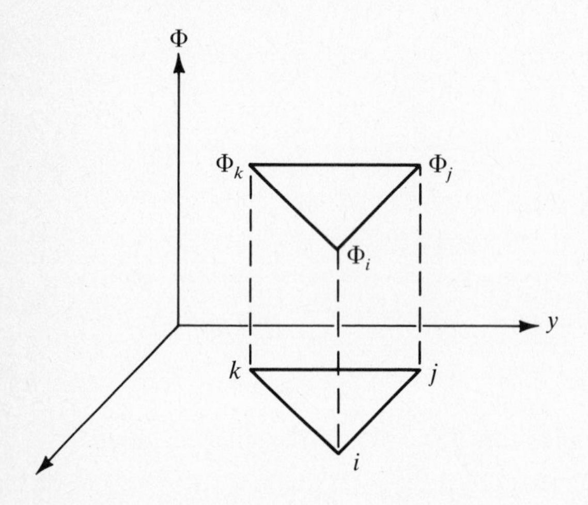


The interpolating function is of the form

$$\Phi = a_1 + a_2 x + a_3 y \tag{13.131a}$$

$$= \{1 \quad x \quad y\} \begin{Bmatrix} a_1 \\ a_2 \\ a_3 \end{Bmatrix} \tag{13.131b}$$

Using the following nodal conditions,

$$\Phi = \Phi_i \quad \text{at} \quad x = x_i \quad \text{and} \quad y = y_i$$

$$\Phi = \Phi_j \quad \text{at} \quad x = x_j \quad \text{and} \quad y = y_j$$

$$\Phi = \Phi_k \quad \text{at} \quad x = x_k \quad \text{and} \quad y = y_k$$

the $\{a\}$ vector can be determined by substituting these values into equation (13.131b). That is,

$$\begin{Bmatrix} \Phi_i \\ \Phi_j \\ \Phi_k \end{Bmatrix} = \begin{bmatrix} 1 & x_i & y_i \\ 1 & x_j & y_j \\ 1 & x_k & y_k \end{bmatrix} \begin{Bmatrix} a_1 \\ a_2 \\ a_3 \end{Bmatrix} \tag{13.132a}$$

or

$$\{\Phi\} = [A]\{a\} \tag{13.132b}$$

Solving for the $\{a\}$ vector gives

$$\{a\} = [A]^{-1}\{\Phi\} \tag{13.132c}$$

where

$$[A]^{-1} = \frac{1}{2A_0} \begin{bmatrix} x_j y_k - x_k y_j & x_k y_i - x_i y_k & x_i y_j - x_j y_i \\ y_j - y_k & y_k - y_i & y_i - y_j \\ x_k - x_j & x_i - x_k & x_j - x_i \end{bmatrix} \tag{13.133}$$

and

$$2A_0 = \det \begin{vmatrix} 1 & x_i & y_i \\ 1 & x_j & y_j \\ 1 & x_k & y_k \end{vmatrix} \tag{13.134}$$

Note that $A_0$ is the area of the triangle in question. Therefore, substituting equation (13.133) into (13.134) gives

$$\begin{aligned} \Phi = {} & \frac{1}{2A_0}[(x_j y_k - x_k y_j) + (y_j - y_k)x + (x_k - x_j)y]\Phi_j \\ & + \frac{1}{2A_0}[(x_k y_i - x_i y_k) + (y_k - y_i)x + (x_i - x_k)y]\Phi_j \\ & + \frac{1}{2A_0}[(x_i y_j - x_j y_i) + (y_i - y_j)x + (x_j - x_i)y]\Phi_k \end{aligned} \tag{13.135}$$

or

$$\Phi = N_i \Phi_i + N_j \Phi_j + N_k \Phi_k \tag{13.136}$$

The gradient of $\Phi$ is once again defined with respect to the $x$ and $y$ directions.

$$\frac{\partial \Phi}{\partial x} = \frac{\partial N_i}{\partial x}\Phi_i + \frac{\partial N_j}{\partial x}\Phi_j + \frac{\partial N_k}{\partial x}\Phi_k$$

or

$$\frac{\partial \Phi}{\partial x} = (y_i - y_k)\Phi_i + (y_k - y_i)\Phi_j + (y_i - y_j)\Phi_k \tag{13.137a}$$

Similarly

$$\frac{\partial \Phi}{\partial y} = (x_k - x_j)\Phi_i + (x_i - x_k)\Phi_j + (x_j - x_i)\Phi_k \quad \textbf{(13.137b)}$$

Shape functions for other types of elements can be derived in precisely the manner just outlined.

EXAMPLE 13.17

Determine the temperature at $x = 5$ and $y = 3$ for the following triangular element and the gradients in $x$ and $y$ directions.

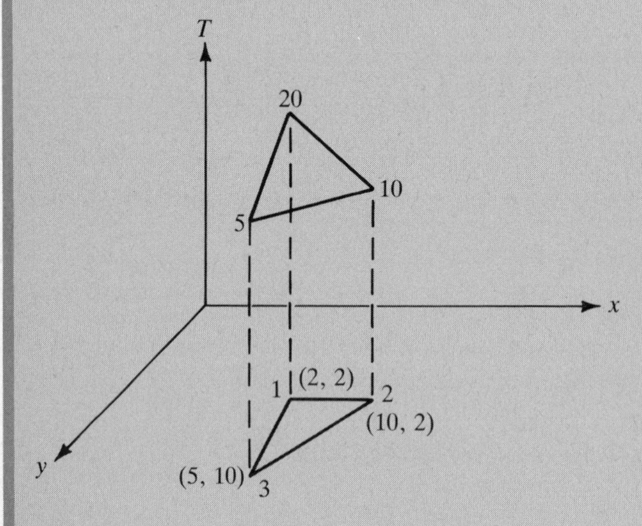


Solution

We begin by numbering the nodes and determining the shape functions.

$x_1 = 2, \qquad y_1 = 2$

$x_2 = 10, \qquad y_2 = 2$

$x_3 = 5, \qquad y_3 = 10$

Therefore

$$N_1 = \frac{1}{2A_0}[(10 \times 10 - 5 \times 2) + (2 - 10)x + (5 - 10)y]$$

$$= \frac{1}{2A_0}[90 - 8x - 5y]$$

$$N_2 = \frac{1}{2A_0}[(5 \times 2 - 2 \times 10) + (10 - 2)x + (2 - 5)y]$$

$$= \frac{1}{2A_0}[-10 + 8x - 3y]$$

$$N_3 = \frac{1}{2A_0}[(2 \times 2 - 10 \times 2) + (2 - 2)x + (10 - 2)y]$$

$$= \frac{1}{2A_0}[-16 + 8y]$$

and

$$2A_0 = \begin{vmatrix} 1 & 2 & 2 \\ 1 & 10 & 2 \\ 1 & 5 & 10 \end{vmatrix} = 64$$

At $x = 5$ and $y = 3$ the shape functions are constants that can be determined readily. Hence

$$N_1 = \tfrac{1}{64}[90 - 8 \times 5 - 5 \times 3] = \tfrac{35}{64}$$

$$N_2 = \tfrac{1}{64}[-10 + 8 \times 5 - 3 \times 3] = \tfrac{21}{64}$$

$$N_3 = \tfrac{1}{64}[-16 + 8 \times 3] = \tfrac{8}{64}$$

The temperature is then given in terms of the nodal values and the calculated shape functions as

$$\begin{aligned} T &= N_1 T_1 + N_2 T_2 + N_3 T_3 \\ &= \tfrac{1}{64}[35T_1 + 21T_2 + 8T_3] \\ &= \tfrac{1}{64}[35(20) + 21(10) + 8(5)] \\ &= 14.84^\circ \end{aligned}$$

The gradients are determined by taking the partial derivatives in $x$ and $y$ direction of the three shape functions:

$$\frac{\partial T}{\partial x} = \tfrac{1}{64}[(-8)(20) + (8)(10) + 0] = -1.25$$

$$\frac{\partial T}{\partial y} = \tfrac{1}{64}[(-5)(20) - (3)(10) + (8)(5)] = -1.41$$

Note that a complex triangular element could have been used if the temperature between adjacent node is known to vary in a nonlinear fashion.

### 13.10.4 One-Dimensional Formulation

Fluid flow and heat flow through a solid is generally a three-dimensional problem. For some problems a one-dimensional model can sufficiently describe the general behavior. Therefore, let us begin by considering the transient one-dimensional

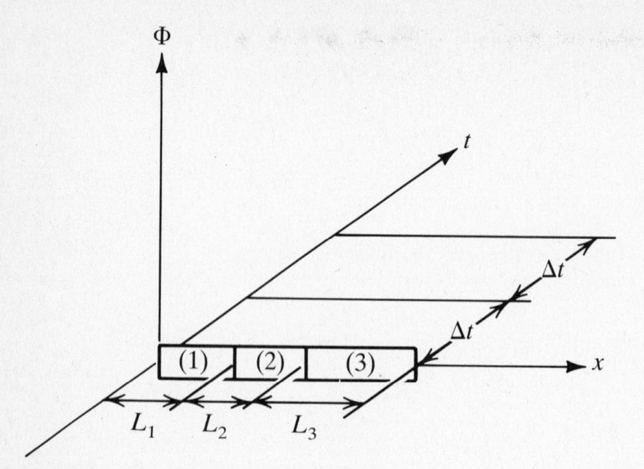


**Figure 13.28** Graphical interpretation of a one-dimensional transient problem.

model

$$\frac{\partial \Phi}{\partial t} = C \frac{\partial^2 \Phi}{\partial x^2} \tag{13.138}$$

Note that we have used $\Phi$ instead of $T$ or $u$ so that our solution is more general in that $\Phi$ can be replaced by whatever parameter we choose. Consider figure 13.28. It is evident that the region to be analyzed is divided into three elements of different lengths. Consequently, the total region can simply be put back together by connecting the three elements as shown in figure 13.28.

In order to show how the finite elements method works, let us start by analyzing element 1. Hence

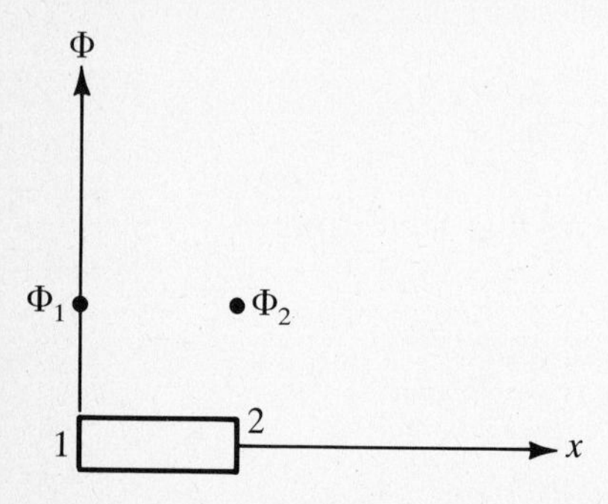


For simplicity, we choose $x_1$ to be zero. Consequently, $x_2 = L_1$. The function $\Phi$ can be expressed conveniently in terms of the two shape functions $N_1$ and $N_2$ and the nodal values $\Phi_1$ and $\Phi_2$ as follows:

$$\Phi = N_1 \Phi_1 + N_2 \Phi_2 \tag{13.139}$$

Recall from chapter 12 that by using the Galerkin procedure a function can be minimized using weight functions. In the finite elements method the shape functions are used as the weight functions for minimizing a residual $R$, where

$$R = C_1 \frac{\partial^2 \Phi}{\partial x^2} - \frac{\partial \Phi}{\partial t}$$

where $C_1$ is a $C$ value for element 1. Consequently, a solution is obtained if $R$ is minimized over the domain $x_1$ to $x_2$. Hence

$$\int_{x_1}^{x_2}\left(C_1\frac{\partial^2\Phi}{\partial x^2}-\frac{\partial\Phi}{\partial t}\right)N_1\,dx=0 \tag{13.140a}$$

and

$$\int_{x_1}^{x_2}\left(C_1\frac{\partial^2\Phi}{\partial x^2}-\frac{\partial\Phi}{\partial t}\right)N_2\,dx=0 \tag{13.140b}$$

Since $x_1 = 0$ and $x_2 = L_1$ these equations can be rewritten in terms of the new limits. Thus

$$\int_0^{L_1} C_1\frac{\partial^2\Phi}{\partial x^2}N_1\,dx-\int_0^{L_1}\frac{\partial\Phi}{\partial t}N_1\,dx=0 \tag{13.141a}$$

and

$$\int_0^{L_1} C_1\frac{\partial^2\Phi}{\partial x^2}N_2\,dx-\int_0^{L_1}\frac{\partial\Phi}{\partial t}N_2\,dx=0 \tag{13.141b}$$

The first term appearing in equation (13.141a) can be readily integrated by parts as follows:

$$uv-\int v\,du \tag{13.142}$$

where

$$u=N_1\Rightarrow\frac{\partial u}{\partial x}=\frac{\partial N_1}{\partial x}$$

$$dv=C_1\frac{\partial^2\Phi}{\partial x^2}\,dx\Rightarrow v=C_1\frac{\partial\Phi}{\partial x}$$

Therefore

$$\int_0^{L_1} C_1\frac{\partial^2\Phi}{\partial x^2}N_1\,dx=C_1\frac{\partial\Phi}{\partial x}N_1\Big|_0^{L_1}-\int_0^{L_1}C_1\frac{\partial\Phi}{\partial x}\frac{\partial N_1}{\partial x}\,dx \tag{13.143}$$

Substituting equation (13.143) into equation (13.141a) gives

$$C_1\frac{\partial\Phi}{\partial x}N_1\Big|_0^{L_1}-\int_0^{L_1}C_1\frac{\partial\Phi}{\partial x}\frac{\partial N_1}{\partial x}\,dx-\int_0^{L_1}\frac{\partial\Phi}{\partial t}N_1\,dx=0 \tag{13.144a}$$

Similarly equation (13.141b) is expressed as

$$C_1\frac{\partial\Phi}{\partial x}N_2\Big|_0^{L_1}-\int_0^{L_1}C_1\frac{\partial\Phi}{\partial x}\frac{\partial N_2}{\partial x}\,dx-\int_0^{L_1}\frac{\partial\Phi}{\partial t}N_2\,dx=0 \tag{13.144b}$$

These two equations are now rearranged for reasons to be explained later.

$$\int_0^{L_1}C_1\frac{\partial\Phi}{\partial x}\frac{\partial N_1}{\partial x}\,dx+\int_0^{L_1}\frac{\partial\Phi}{\partial t}N_1\,dx=C_1\frac{\partial\Phi}{\partial x}N_1\Big|_0^{L_1} \tag{13.145a}$$

and

$$\int_0^{L_1} C_1 \frac{\partial \Phi}{\partial x}\frac{\partial N_2}{\partial x}\,dx + \int_0^{L_1} \frac{\partial \Phi}{\partial t} N_2\,dx = C_1 \frac{\partial \Phi}{\partial x} N_2 \Big|_0^{L_1} \tag{13.145b}$$

Substituting equation (13.139) into the right-hand sides of equations (13.145) gives

$$\int_0^{L_1} C_1 \frac{\partial N_1}{\partial x}\left[\frac{\partial N_1}{\partial x}\Phi_1 + \frac{\partial N_2}{\partial x}\Phi_2\right]dx + \int_0^{L_1} N_1\left[N_1 \frac{\partial \Phi_1}{\partial t} + N_2 \frac{\partial \Phi_2}{\partial t}\right]dx$$
$$= C_1 \frac{\partial \Phi}{\partial x} N_1 \Big|_0^{L_1} \tag{13.146a}$$

and

$$\int_0^{L_1} C_1 \frac{\partial N_2}{\partial x}\left[\frac{\partial N_1}{\partial x}\Phi_1 + \frac{\partial N_2}{\partial x}\Phi_2\right]dx + \int_0^{L_1} N_2\left[N_1 \frac{\partial \Phi_1}{\partial t} + N_2 \frac{\partial \Phi_2}{\partial t}\right]dx$$
$$= C_1 \frac{\partial \Phi}{\partial x} N_2 \Big|_0^{L_1} \tag{13.146b}$$

These equations are now expressed in matrix form to give

$$\int_0^{L_1} C_1 \begin{bmatrix} \frac{\partial N_1}{\partial x}\cdot\frac{\partial N_1}{\partial x} & \frac{\partial N_1}{\partial x}\cdot\frac{\partial N_2}{\partial x} \\ \frac{\partial N_2}{\partial x}\cdot\frac{\partial N_1}{\partial x} & \frac{\partial N_2}{\partial x}\cdot\frac{\partial N_2}{\partial x} \end{bmatrix} \begin{Bmatrix} \Phi_1 \\ \Phi_2 \end{Bmatrix} dx$$
$$+ \int_0^{L_1} \begin{bmatrix} N_1^2 & N_1N_2 \\ N_2N_1 & N_2^2 \end{bmatrix} \begin{Bmatrix} \frac{\partial \Phi_1}{\partial t} \\ \frac{\partial \Phi_2}{\partial t} \end{Bmatrix} dx = C_1 \begin{Bmatrix} \frac{\partial \Phi}{\partial x} N_1 \\ \frac{\partial \Phi}{\partial x} N_2 \end{Bmatrix} \Bigg|_0^{L_1} \tag{13.147}$$

The various integrals are now evaluated separately as follows

$$\int_0^{L_1} \frac{\partial N_1}{\partial x}\cdot\frac{\partial N_1}{\partial x}\,dx = \int_0^{L_1}\left(-\frac{1}{L_1}\right)\left(-\frac{1}{L_1}\right)dx = \frac{1}{L_1}$$

$$\int_0^{L_1} \frac{\partial N_2}{\partial x}\cdot\frac{\partial N_2}{\partial x}\,dx = \int_0^{L_1}\left(\frac{1}{L_1}\right)\left(\frac{1}{L}\right)dx = \frac{1}{L_1}$$

$$\int_0^{L_1} \frac{\partial N_1}{\partial x}\frac{\partial N_2}{\partial x} = \int_0^{L_1}\left(-\frac{1}{L_1}\right)\left(\frac{1}{L_1}\right)dx = -\frac{1}{L_1}$$

$$\int_0^{L_1} N_1N_2\,dx = \int_0^{L_1}\left(-\frac{x}{L_1}\right)\left(\frac{x}{L_1}-1\right)dx = \frac{L_1}{6}$$

$$\int_0^{L_1} N_2N_1\,dx = \frac{L_1}{6}$$

$$\int_0^{L_1} N_1^2\,dx = \int_0^{L_1}\left(-\frac{x}{L_1}\right)^2 dx = \frac{L_1}{3}$$

and

$$\int_0^{L_1} N_2^2\,dx = \int_0^{L_1}\left(\frac{x}{L}-1\right)^2 dx = \frac{L_1}{3}$$

Substituting these quantities into equation (13.147) yields the element equations. That is,

$$C_1\begin{bmatrix} \frac{1}{L_1} & -\frac{1}{L_1} \\ -\frac{1}{L_1} & \frac{1}{L_1}\end{bmatrix}\begin{Bmatrix}\Phi_1\\ \Phi_2\end{Bmatrix} + \frac{1}{6}\begin{bmatrix}2L_1 & L_1\\ L_1 & 2L_1\end{bmatrix}\begin{Bmatrix}\dot{\Phi}_1\\ \dot{\Phi}_2\end{Bmatrix}$$

$$= C_1\begin{Bmatrix}\frac{\partial\Phi}{\partial x} N_1\\ \frac{\partial\Phi}{\partial x} N_2\end{Bmatrix}\Bigg|_0^{L_1} \tag{13.148}$$

Note that the right-hand side of equation (13.148) represents the flux at the nodes. That is,

$$C_1\frac{\partial\Phi}{\partial x}N_1\Big|_0^{L_1} = C_1\left(\frac{\partial\Phi}{\partial x}\right)_1, \qquad N_1 = 1, \text{ at } x = 0$$

$$C_1\frac{\partial\Phi}{\partial x}N_2\Big|_0^{L_1} = C_1\left(\frac{\partial\Phi}{\partial x}\right)_2, \qquad N_2 = 1 \text{ at } x = L_1$$

The derivative of the $\Phi$ vector with respect to time is approximated as follows:

$$\frac{\partial\Phi_1}{\partial t} = \dot{\Phi}_1 = \frac{\Phi_1(t) - \Phi_1(t-\Delta t)}{\Delta t}$$

$$\frac{\partial\Phi_2}{\partial t} = \dot{\Phi}_2 = \frac{\Phi_2(t) - \Phi_2(t-\Delta t)}{\Delta t}$$

Substituting into equation (13.148) gives

$$\frac{C_1}{L_1}\begin{bmatrix}1 & -1\\ -1 & 1\end{bmatrix}\begin{Bmatrix}\Phi_1\\ \Phi_2\end{Bmatrix}_{t+\Delta t} + \frac{L_1}{6\,\Delta t}\begin{bmatrix}2 & 1\\ 1 & 2\end{bmatrix}\begin{Bmatrix}\Phi_1\\ \Phi_2\end{Bmatrix}_{t+\Delta t} - \frac{L_1}{6\,\Delta t}\begin{bmatrix}2 & 1\\ 1 & 2\end{bmatrix}\begin{Bmatrix}\Phi_1\\ \Phi_2\end{Bmatrix}_t$$

$$= C_1\begin{Bmatrix}\left(\frac{\partial\Phi}{\partial x}\right)_1\\ \left(\frac{\partial\Phi}{\partial x}\right)_2\end{Bmatrix}_t \tag{13.149}$$

Combining terms and letting $\alpha_1 = C_1\,\Delta t/L_1^2$ gives the following simple expression:

$$\begin{bmatrix}\alpha_1 + \frac{1}{3} & -\alpha_1 + \frac{1}{6}\\ -\alpha_1 + \frac{1}{6} & \alpha_1 + \frac{1}{3}\end{bmatrix}\begin{Bmatrix}\Phi_1\\ \Phi_2\end{Bmatrix}_{t+\Delta t} = \frac{1}{6}\begin{bmatrix}2 & 1\\ 1 & 2\end{bmatrix}\begin{Bmatrix}\Phi_1\\ \Phi_2\end{Bmatrix}_t + \alpha_1 L_1\begin{Bmatrix}\left(\frac{\partial\Phi}{\partial x}\right)_1\\ \left(\frac{\partial\Phi}{\partial x}\right)\end{Bmatrix} \tag{13.150}$$

Consequently, for any element $i$ one can write the generalized elemental equations as follows:

$$\begin{bmatrix} 6\alpha_i + 2 & -6\alpha_i + 1 \\ -6\alpha_i + 1 & 6\alpha_i + 2 \end{bmatrix} \begin{Bmatrix} \Phi_i \\ \Phi_{i+1} \end{Bmatrix}_{t+\Delta t} = \begin{bmatrix} 2 & 1 \\ 1 & 2 \end{bmatrix} \begin{Bmatrix} \Phi_i \\ \Phi_{i+1} \end{Bmatrix}_t + 6\alpha_i L_i \begin{Bmatrix} \left(\dfrac{\partial \Phi}{\partial x}\right) \\ \left(\dfrac{\partial \Phi}{\partial x}\right) \end{Bmatrix}_t \tag{13.151}$$

where

$$\alpha_i = \frac{C_i \Delta t}{L_i^2}$$

Note that $\Delta t$ must be the same for all elements. The implication is that one is interested in studying the total response of the entire system at a given time and not the individual element response. Additionally, one may use either of the forms given in equation (13.150) or (13.151). Consequently, for element 2

$$\begin{bmatrix} \alpha_2 + \frac{1}{3} & \frac{1}{6} - \alpha_2 \\ \frac{1}{6} - \alpha_2 & \alpha_2 + \frac{1}{3} \end{bmatrix} \begin{Bmatrix} \Phi_2 \\ \Phi_3 \end{Bmatrix}_{t+\Delta t} = \frac{1}{6} \begin{bmatrix} 2 & 1 \\ 1 & 2 \end{bmatrix} \begin{Bmatrix} \Phi_2 \\ \Phi_3 \end{Bmatrix}_t + \alpha_2 L_2 \begin{Bmatrix} \left(\dfrac{\partial \Phi}{\partial x}\right)_2 \\ \left(\dfrac{\partial \Phi}{\partial x}\right)_3 \end{Bmatrix}_t$$

and for the third element, the element equations are given as

$$\begin{bmatrix} \alpha_3 + \frac{1}{3} & \frac{1}{6} - \alpha_3 \\ \frac{1}{6} - \alpha_3 & \alpha_3 + \frac{1}{3} \end{bmatrix} \begin{Bmatrix} \Phi_3 \\ \Phi_4 \end{Bmatrix}_{t+\Delta t} = \frac{1}{6} \begin{bmatrix} 2 & 1 \\ 1 & 2 \end{bmatrix} \begin{Bmatrix} \Phi_3 \\ \Phi_4 \end{Bmatrix}_t + \alpha_3 L_3 \begin{Bmatrix} \left(\dfrac{\partial \Phi}{\partial x}\right)_3 \\ \left(\dfrac{\partial \Phi}{\partial x}\right)_4 \end{Bmatrix}_t$$

Connecting element 1 to element 2 at node 2 and element 2 to element 3 at node 3 results in the total region in question. For example, assuming constant boundary condition would result in the following set of equations for the region given in figure 13.28:

$$\begin{bmatrix} \alpha_1 + \frac{1}{3} & \frac{1}{6} - \alpha_1 & & \\ \frac{1}{6} - \alpha_1 & \alpha_1 + \alpha_2 + \frac{2}{3} & \frac{1}{6} - \alpha_2 & \\ & \frac{1}{6} - \alpha_2 & \alpha_2 + \alpha_3 + \frac{2}{3} & \frac{1}{6} - \alpha_3 \\ & & \frac{1}{6} - \alpha_3 & \alpha_3 + \frac{1}{3} \end{bmatrix} \begin{Bmatrix} \Phi_1 \\ \Phi_2 \\ \Phi_3 \\ \Phi_4 \end{Bmatrix}_{t+\Delta t}$$

$$= \frac{1}{6} \begin{bmatrix} 2 & 1 & & \\ 1 & 4 & 1 & \\ & 1 & 4 & 1 \\ & & 1 & 2 \end{bmatrix} \begin{Bmatrix} \Phi_1 \\ \Phi_2 \\ \Phi_3 \\ \Phi_4 \end{Bmatrix}_t$$

Obviously, there is no limit to the number of elements that one may have in a given problem except that with more elements a larger set of linear algebraic equations would have to be solved.

E

## EXAMPLE 13.18

Determine the temperature distribution for the following rod at $t = 0.1$ hr for a rod having $C = 30\ \text{ft}^2/\text{hr}$ and length of 15 ft. Use $\Delta x = 3$ ft.

1 2 3 4 5 6

$\Phi_1 = 18°$ (1) • (2) • (3) • (4) • (5) $\Phi_6 = 0°$

$\leftarrow \Delta x \rightarrow$

$\Phi(x) = 100°$

## Solution

We begin by formulating the element equations. Note that the rod is uniform and the properties of each of the five elements are constant. We need to determine the element equations for a single element. Thus

Element $i$

$$\alpha_i = \frac{\Delta t C_i}{L_i^2} = \frac{(0.1)(30)}{3^2} = \frac{1}{3} \qquad i = 1, 2, 3$$

and

$$\begin{bmatrix} \frac{2}{3} & -\frac{1}{6} \\ -\frac{1}{6} & \frac{2}{3} \end{bmatrix} \begin{Bmatrix} \Phi_i \\ \Phi_{i+1} \end{Bmatrix}_{t+\Delta t} = \frac{1}{6} \begin{bmatrix} 2 & 1 \\ 1 & 2 \end{bmatrix} \begin{Bmatrix} \Phi_i \\ \Phi_{i+1} \end{Bmatrix}_t$$

for $i = 1, 2, 3, 4, 5$. The global matrix is now formed to give

$$\begin{bmatrix} \frac{2}{3} & -\frac{1}{6} & & & & \\ -\frac{1}{6} & \frac{4}{3} & -\frac{1}{6} & & & \\ & -\frac{1}{6} & \frac{4}{3} & -\frac{1}{6} & & \\ & & -\frac{1}{6} & \frac{4}{3} & -\frac{1}{6} & \\ & & & -\frac{1}{6} & \frac{4}{3} & -\frac{1}{6} \\ & & & & -\frac{1}{6} & \frac{2}{3} \end{bmatrix} \begin{Bmatrix} \Phi_1 \\ \Phi_2 \\ \Phi_3 \\ \Phi_4 \\ \Phi_5 \\ \Phi_6 \end{Bmatrix}_{t=0.1}$$

$$= \frac{1}{6} \begin{bmatrix} 2 & 1 & & & & \\ 1 & 4 & 1 & & & \\ & 1 & 4 & 1 & & \\ & & 1 & 4 & 1 & \\ & & & 1 & 4 & 1 \\ & & & & 1 & 2 \end{bmatrix} \begin{Bmatrix} \Phi_1 \\ \Phi_2 \\ \Phi_3 \\ \Phi_4 \\ \Phi_5 \\ \Phi_6 \end{Bmatrix}_{t=0}$$

Note that at $t = 0$, $\Phi_1 = \Phi_2 = \cdots = \Phi_6 = 100°$. At $t = 0.1$, $\Phi_1 = 18$ and $\Phi_6 = 0$. Substituting into the above and removing the first and sixth rows gives

$$\begin{bmatrix} -\frac{1}{6} & \frac{4}{3} & -\frac{1}{6} & & & \\ & -\frac{1}{6} & \frac{4}{3} & -\frac{1}{6} & & \\ & & -\frac{1}{6} & \frac{4}{3} & -\frac{1}{6} & \\ & & & -\frac{1}{6} & \frac{4}{3} & -\frac{1}{6} \end{bmatrix} \begin{Bmatrix} 18 \\ \Phi_2 \\ \Phi_3 \\ \Phi_4 \\ \Phi_5 \\ 0 \end{Bmatrix}_{t=0.1}$$

$$= \frac{1}{6} \begin{bmatrix} 1 & 4 & 1 & & & \\ & 1 & 4 & 1 & & \\ & & 1 & 4 & 1 & \\ & & & 1 & 4 & 1 \end{bmatrix} \begin{Bmatrix} 100 \\ 100 \\ 100 \\ 100 \\ 100 \\ 100 \end{Bmatrix}_{t=0}$$

Simplifying gives

$$\begin{bmatrix} \frac{4}{3} & -\frac{1}{6} & & \\ -\frac{1}{6} & \frac{4}{3} & -\frac{1}{6} & \\ & -\frac{1}{6} & \frac{4}{3} & -\frac{1}{6} \\ & & -\frac{1}{6} & \frac{4}{3} \end{bmatrix} \begin{Bmatrix} \Phi_2 \\ \Phi_3 \\ \Phi_4 \\ \Phi_5 \end{Bmatrix}_{t=0.1} = \begin{Bmatrix} 103 \\ 100 \\ 100 \\ 100 \end{Bmatrix}_{t=0}$$

Solving for the nodal temperatures gives

$$\Phi_2 = 89.56, \qquad \Phi_3 = 98.47, \qquad \Phi_4 = 98.22, \qquad \Phi_5 = 87.28$$

It is important that the right-hand side is equal to the initial conditions. This is because the computed values can be substituted for the initial condition vector for the next time increment.

The procedure just outlined can be used to solve problems involving composite regions, constant and derivative boundary conditions, and elements of different lengths. This is the essence of the finite element method. In short, we no longer need to make simplifying assumptions in order to arrive at a solution that may or may not describe the actual system being analyzed. We can deal with the real system as it is.

This section is by no means a complete treatment of the finite element method nor is it intended to be. It merely introduces the basic concepts and illustrates its power.

## Recommended Reading

*Modern Methods of Engineering Computation*, Robert L. Ketter and S. P. Prawel, McGraw-Hill Book Company, New York, 1969.

*Numerical Algorithms—Origins and Application*, Bruce W. Arder and K. N. Astill, Addison-Wesley Publishing Co., Reading, Mass., 1970.

*Theory and Analysis of Plates*, R. Szilard, Prentice-Hall, Inc., Englewood Cliffs, N.J., 1974.

*Applied Finite Element Analysis*, L. J. Segerland, John Wiley and Sons, New York, 1976.

*A Practical Guide to Computer Methods for Engineers*, Terry E. Shoup, Prentice-Hall, Inc., Englewood Cliffs, N.J., 1979.

*Numerical Mathematics and Computing*, W. Cheney and K. David, Brooks/Cole Publishing Co., Monterey, Calif., 1980.

*Numerical Solution of Partial Differential Equations in Science and Engineering*, L. Lapidus and F. P. George, John Wiley and Sons, New York 1982.

*Introduction to Groundwater Modeling—Finite Difference and Finite Element Methods*, Herbert F. Wang and M. P. Anderson, W. H. Freeman and Company, San Francisco, 1982.

## P PROBLEMS

**13.1** Transform the following partial differential equation from the Cartesian to the polar coordinate system for $\theta = 30°$.

$$\frac{\partial^2 h}{\partial x^2} + 2\frac{\partial^2 h}{\partial x\,\partial y} + \frac{\partial^2 h}{\partial y^2} = 0$$

(*Hint:* use the results obtained in example 13.1)

**13.2** Transform the following partial differential equation from the Cartesian to the skewed coordinate system. Use equation (13.25) and $\theta = 30°$.

$$\frac{\partial^2 T}{\partial x^2} - 2\frac{\partial^2 T}{\partial x\,\partial y} + \frac{\partial^2 T}{\partial y^2} = 0$$

**13.3** Transform the following partial differential equation from the Cartesian to the triangular coordinate system. Use equation (13.36a), $\theta = 60°$, and $\beta = 120$.

$$\frac{\partial^2 w}{\partial x^2} + \frac{\partial^2 w}{\partial x\,\partial y} + \frac{\partial^2 w}{\partial y^2} = 0$$

What kind of a region would the newly derived differential equation be applicable to?

**13.4** Express the following differential equation in a finite difference form using the central difference expression with $(O)h^2$ error order.

$$\frac{\partial^2 \phi}{\partial x^2} + \frac{\partial^2 \phi}{\partial y^2} = -2\phi$$

**13.5** Transform the partial differential equation given in problem 4 to skewed coordinates; then express it in a difference form using central difference expressions of $(O)h^2$.

**13.6** Determine the steady-state temperature at the nodes of the following plate. The plate is divided into square elements measuring 5 ft by 5 ft.

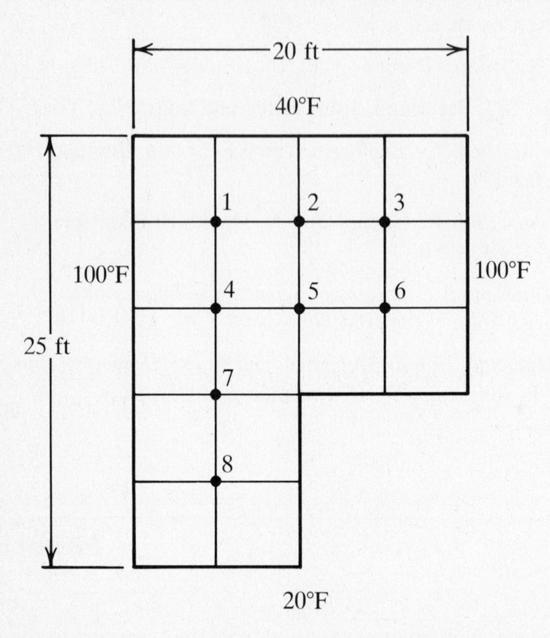


**13.7** Determine the steady-state temperature at the nodes of the following triangular plate. The plate is divided into triangular elements. The increments in the $x$ and $y$ directions are equal.

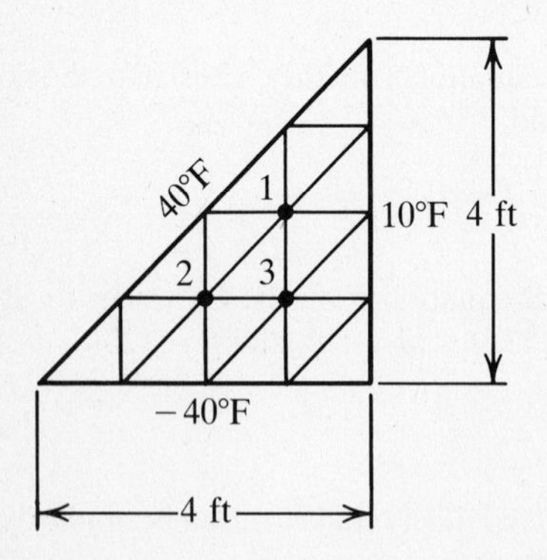


**13.8** Rework problem 7 assuming arbitrary boundary temperatures. Calculate the steady-state temperatures for shown boundary temperatures.

**13.9** Determine the steady-state temperatures for the following triangular plate. Note that triangular elements are equilateral.

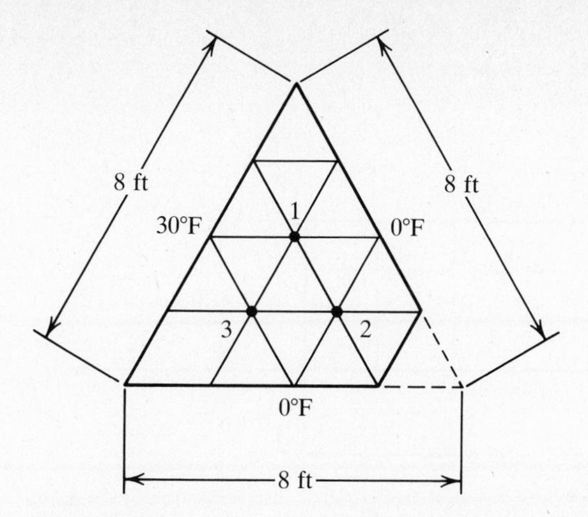


**13.10** Determine the steady-state temperatures for the skewed plate shown below. Assume that the temperatures at the corners are equal to the average temperature value of adjacent boundary temperatures.

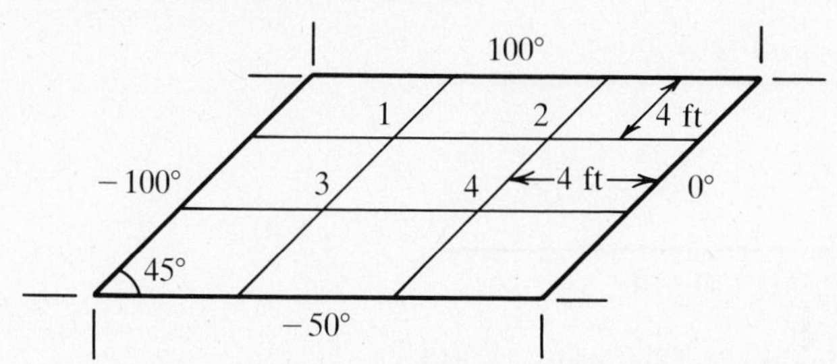


**13.11** Rework problem 10 with arbitrary boundary temperatures.

**13.12** Determine the steady-state head for the following dam. Use $\Delta x = \Delta y = 10$ ft and assume that the head is constant with depth at distance of 60 ft upstream. Note that symmetry can be used to reduce the size of linear algebraic equations needed for solving this problem.

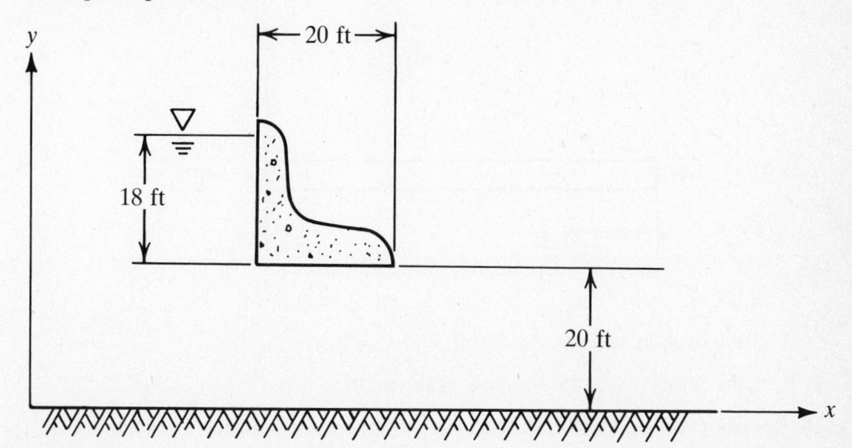

**13.13** Consider the following plate; then set up the algebraic equations required for determining the steady-state temperature at the nodes.

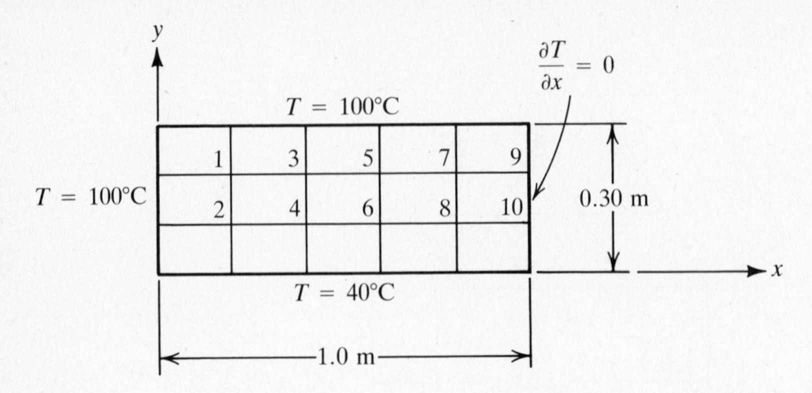


**13.14** Determine the excess pore water pressure for the following soil profile using a depth increment of 2 ft and a time increment of 1 day. Calculate all values for the first 5 days using the explicit scheme.

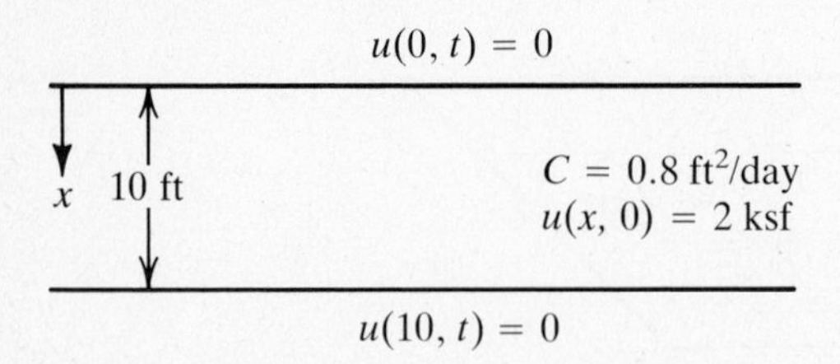


[*Hint:* use equation (13.75a) and replace $T$ with $u$.]

**13.15** Rework problem 14 using the implicit scheme. Note that equation (13.80) can be used to solve this problem if proper variable substitution is made.

**13.16** Determine the temperature distribution for the rod shown below for first five time increments using $\alpha = \frac{1}{6}$. Assume that the rod is insulated along its length.

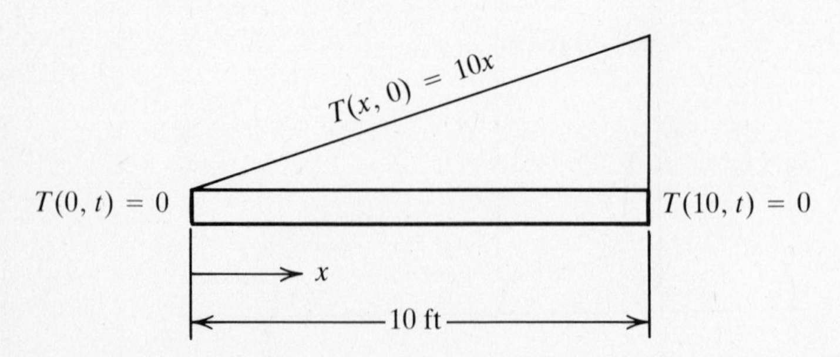


Use the explicit scheme, $\Delta x = 2$ ft, and $C = \frac{4}{3}$.

**13.17** Use the implicit scheme to solve problem 16.

**13.18** Rework problem 16 with the boundary temperatures of $T(0, t) = 40$ and $T(10, t) = 20$.

**13.19** Determine the excess pore water pressure distribution for the following soil profile after one year and using $\Delta x = 4$ ft, $\Delta t = 0.5$ year. Assume that the pore water pressure at the boundaries is equal to zero and that the profile is subjected to a sudden increase in the pressure of 1.0 ksf.

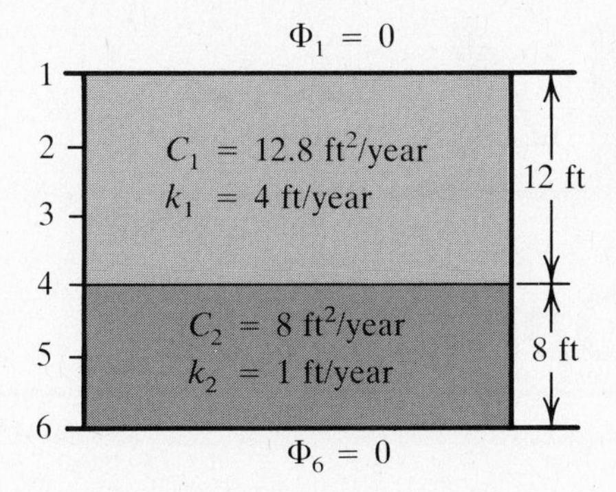


**13.20** Determine the deflection at the nodes for the following plate supported at its edges. Assume that $\Delta x = \Delta y = 5$ ft, $E = 20 \times 10^5$ psi, $P_z = 1.2$ ksf, a plate thickness of 1.0 ft, and a Possion's ratio of 0.20.

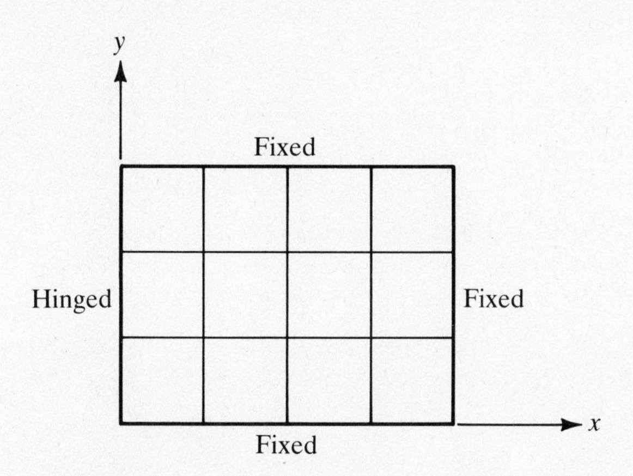


**13.21** Determine the temperature at $t = 0.2$ hour, for the following rod, using the finite element method. Assume that $C = 83.33$ ft$^2$/hour.

x
1 2 3 4
$\Phi = 10°$ (1) (2) (3) $\Phi = -10°$
5 ft 5 ft 5 ft
$\Phi(x) = -100°$

# chapter 14

# Analytic Optimization

Skylab mapping earth's surface. *Courtesy of N.A.S.A.*
"Engineers are skilled in getting the most efficient use of resources."

# 14

## 14.1 INTRODUCTION

The formulation, analysis, design, and operation of modern engineering systems often involves social, political, economical, and technical constraints. In general, an engineering system consists of many components, each of which performs a specific task, but all of which are intended for a common purpose. The effectiveness of a system's performance is measured by the degree of success attained in achieving the common purpose. Furthermore, in order to optimize the effectiveness of either the design or operation of an engineering system, the decison making process must be placed on a rational and objective basis. Unfortunately, since many engineering systems are too complex to be treated analytically, it is often necessary to introduce simplifying assumptions by including only the components deemed important by the decision maker. In general, we shall seek to minimize the effort (cost) and/or to maximize the benefit (profit) associated with a given engineering system. An optimization problem in which each decision is known to lead invariably to a specific outcome is termed deterministic, otherwise, nondeterministic. Our presentation in this chapter is limited to a few important analytical techniques of optimizing deterministic models. Both linear and nonlinear systems are discussed.

## 14.2 CHARACTERISTICS OF OPTIMIZATION PROBLEMS

The formulation of optimization problems involves the transformation of the physical engineering system into an equivalent mathematical model. This concept was treated in chapter 3. Consider, for example, the problem an auto manufacturer might face. That is, given the demand for compact, subcompact, and medium size cars and the profits associated with each unit, the question is how many units of each type should be built so that profit is maximized, given the limitations on equipment, labor, and other constraints. A feasible solution to such a model represents the selection of a set of variables which satisfy the intended functional relationships. While in practice no model is truly deterministic, the probabilistic nature of the decision variables can be ignored, at least in the exploratory stage of the investigation. For our example, the profit associated with each car type may be assumed to be a constant in a given year.

Deterministic models can be generally assumed to have the following five basic characteristics:

### Decision Variables

These are variables that can be freely controlled by the decisionmaker such as the number of subcompact, compact, and medium size cars that an automaker may have to build.

### Objective Function

This is a function relating the various decision variables and is used in ranking the desirability of the various possible solutions to the model. For example, the total profit associated with the number of different auto types is an objective function that we wish to maximize.

### Constraint Functions

These are conditions imposed on the decision variables, and may include economical, political, and physical constraints. For example, the automaker cannot manufacture a negative number of cars. This implies that the decision variables must be greater than zero.

### Feasible Solution

Any solution to a mathematical model in question which satisfies the constraints but may or may not satisfy the objective function.

### Optimal Solution

This is a feasible solution that satisfies the constraints and the objective function. Note that in a given problem it is possible to have more than one optimal solution.

EXAMPLE 14.1

Three electric generators have costs which are functions of power output as follows:

$$C_1 = 1 + 0.5x + x^2$$

$$C_2 = 1 - y + y^2$$

$$C_3 = 1 + 0.5z + 0.5z^2$$

where $x$, $y$, and $z$ are the power outputs provided by generators 1, 2, and 3, respectively. Formulate a model that will minimize the cost if five units of power is required.

### Solution

The objective function can be easily formed in terms of the decision variables $x$, $y$, and $z$ as follows:

$$\text{MINIMIZE} \quad Q = C_1 + C_2 + C_3$$

or more explicitly

$$\text{MINIMIZE} \quad Q = 3 + 0.5x + x^2 - y + y^2 + 0.5z + 0.5z^2$$

The object is to minimize the cost $Q$ subject to the constraint that the total power output must be equal to five units. That is,

$$x + y + z = 5$$

Obviously, this is an oversimplified nonlinear model that needs to be solved for the most appropriate $x$, $y$, and $z$ values so that the cost is minimum. One feasible solution to the problem is to choose $x = 5$ and $y = z = 0$. However, is this the best choice to minimize the cost? The answer is most probably not, unless we are lucky.

EXAMPLE 14.2

An automaker decided to discontinue production of a line of large size cars. This created considerable excess production capacity in three different plants. Management is considering devoting the excess capacity to one or more of three new cars, types 1, 2, and 3. The available capacities of the three plants A, B, and C in man–hours are summarized as follows:

| PLANT | MAN–HOURS PER DAY |
|---|---|
| A | 4000 |
| B | 1000 |
| C | 600 |

The number of man–hours required for producing each unit are given below:

| PLANT | TYPE 1 | TYPE 2 | TYPE 3 |
|---|---|---|---|
| A | 8 | 2 | 3 |
| B | 4 | 3 | 0 |
| C | 2 | 0 | 1 |

Furthermore, the sales department indicated that the profits per unit associated with car types 1, 2, and 3 are \$1400, \$600, and \$1200. Formulate a mathematical model for determining the optimum solution.

### Solution

Obviously the model to be formulated should provide answers to the problem of determining the most appropriate combination of car types to be manufactured that will maximize the profit. Let us begin by first assuming a set of decision variables as follows:

$x_1$ = Number of type 1 cars

$x_2$ = Number of type 2 cars

$x_3$ = Number of type 3 cars

The constraint functions can now be easily established as follows:

Plant A: $8x_1 + 2x_2 + 3x_3 \leq 4000$

Plant B: $4x_1 + 3x_2 \leq 1000$

Plant C: $2x_1 + x_3 \leq 600$

Note, that instead of an equals sign, we have used a less than or equals sign. This is because the given man–hours are the maximum that any of the three plants can provide; in reality, these values may be reduced by work stoppages.

The only remaining task is to formulate the objective function. That is, let $z$ be the total profit to be generated, in thousands of dollars, from the sales of cars; then we need to

$$\text{MAXIMIZE} \quad z = 1.4x_1 + 0.6x_2 + 1.2x_3$$

It is evident that one feasible solution to the problem is to produce $x_1 = x_2 = x_3 = 0$. This solution satisfies the constraints, but not the objective function. Therefore we need to develop methods that will enable us to determine the optimal solution, that is, maximize profit and satisfy the constraints. Note that since the mathematical model involves linear equations only, it is referred to as a "linear programming model"; such models are treated later in this chapter.

## 14.3 UNCONSTRAINED OPTIMIZATION—METHOD OF CALCULUS

The method of calculus for determining the maxima and minima of functions involving real variables can be readily applied to solve certain types of optimization problems. These include mathematical models that can be reduced to a single function to be minimized or maximized and no constraints. The implication is that if the constraints can be solved explicitly in terms of some variables appearing in the objective function, then these variables can be eliminated from the objective function, in which case one or more decision variables remain. The optimization of functions involving more than one decision variable is generally very complex.

However, functions involving a single decision variable can be easily solved, especially when no constraints are present in the model.

### 14.3.1 Functions of One Variable

In order to gain some insight into the more general problem of optimization involving more than one variable, let us begin by reviewing some basic concepts pertaining to extrema of functions in one real variable. The simplest case of such optimization problems is that of unconstrained objective functions. That is, given the objective function $y = f(x)$, we seek to determine an $x$ value such that $y$ is either a maximum or a minimum. The necessity condition for a local maximum or a local minimum at a given point $x = x_i$ for a continuous $f(x)$ in the neighborhood of $x_i$ is that

$$\left.\frac{dy}{dx}\right|_{x=x_0} = f'(x_i) = 0 \qquad \textbf{(14.1)}$$

This necessity condition provides no information on the global maximum or global minimum, which are sometimes referred to as the absolute optima. That is, a function may have several points at which it attains minimum or maximum local values in a given region. Consequently, the absolute maximum or minimum (global) of a function can only be ascertained if all of its extremities within the region are known. If this is so, the largest and smallest of all these values are the global maximum and minimum, respectively. These concepts are shown graphically in figure 14.1.

It is interesting to note that the local minimum value at $x = x_3$ is higher than the local maximum at $x = x_0$. This is indeed possible in a given region. Furthermore, local maxima and local minima may be separated by using the second

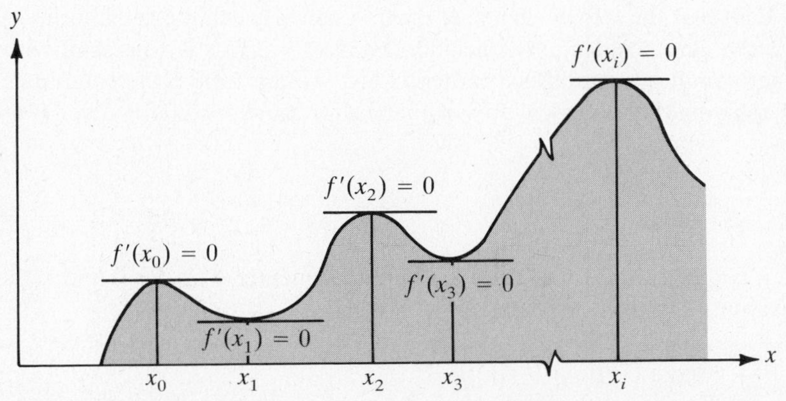


**Figure 14.1** Local minimum and maximum of a function involving one real independent variable.

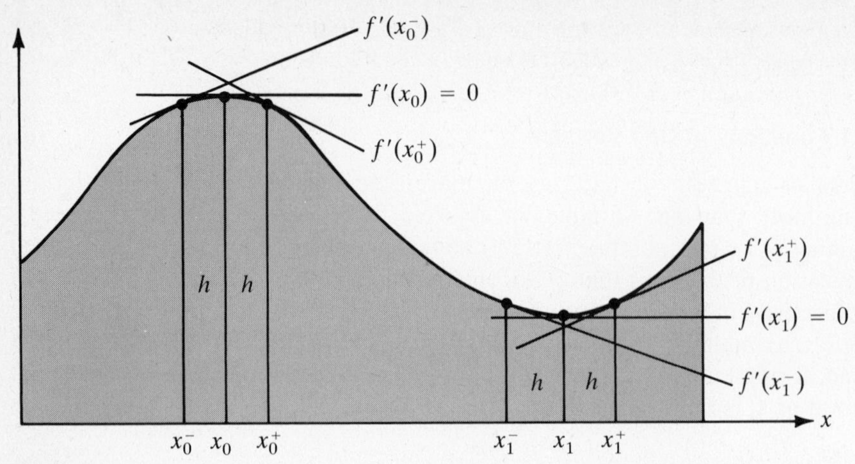


**Figure 14.2** Illustration of the sufficiency conditions for local maximum and local minimum.

derivative as follows:

Local maximum:

$$\left.\frac{d^2y}{dx^2}\right|_{x=x_i} = f''(x_i) \leq 0 \tag{14.2a}$$

Local minimum:

$$\left.\frac{d^2y}{dx^2}\right|_{x=x_i} = f''(x_i) \geq 0 \tag{14.2b}$$

Clearly these sufficiency conditions for extrema hold if the first derivative is equal to zero at $x = x_i$. These concepts are illustrated in figure 14.2.

It is evident that the rate of change of the first derivative in the neighborhood of $x_0$ is negative and positive in the neighborhood of $x_1$. This is consistent with the sufficiency conditions established earlier. In fact, we can verify these conditions by directly applying the numerical approximation for the second derivative. That is, at $x = x_0$:

$$f''(x_0) = \frac{1}{h^2}\,|f(x_0^-) - 2f(x_0) + f(x_0^+)|$$

Clearly $f(x_0)$ is greater than $f(x_0^-)$ and $f(x_0^+)$. Consequently, $f''(x_0) < 0$ and $x_0$ is a local maximum. On the other hand, at $x = x_1$:

$$f''(x_1) = \frac{1}{h^2}\,|f(x_1^-) - 2f(x_1) + f(x_1^+)|$$

We have $f(x_1^-) > f(x_1)$ and $f(x_1^+) > f(x_1)$, in which case $f''(x_1) > 0$. Therefore the point $x_1$ is a local minimum.

In some problems, both the first and the second derivatives are equal to zero, in which case higher derivatives must be considered. For example, suppose that the first through $n$th derivatives are all equal to zero at $x = x_i$ but the $(n + 1)$th derivative is not. Then the function has a local maximum at $x = x_i$, if $n$ is odd and $f^{(n+1)}(x_i) > 0$; $f(x)$ has neither a local maximum nor a local minimum at $x = x_i$ if $n$ is even.

E

EXAMPLE 14.3

A farmer has material sufficient to build a 200-ft-long fence of appropriate height. Determine the maximum rectangular area that he can enclose.

Solution

The decision variables are first specified by letting $x$ be the length and $y$ be the width of the enclosed area. Therefore the constant equation is given by the circumference as follows:

$$2x + 2y = 200$$

The objective function is given in terms of the decision variables as the enclosed area $A$ to be maximized. That is,

$$\text{MAXIMIZE} \quad A = xy$$

Obviously this is a constrained optimization problem. Consequently the methods of calculus cannot be used directly, unless the model is transformed into an unconstrained one. This can be easily accomplished by solving the constraint equation for $y$ then substituting into the objective function to give

$$\text{MAXIMIZE} \quad A = 100x - x^2$$

Therefore applying the necessity condition for local minimum or local maximum given by equation (14.1) yields

$$\frac{dA}{dx} = 100 - 2x = 0$$

Solving for $x$ gives

$$x = 50 \text{ ft}$$

The implication is that the area is either maximized or minimized at $x = 50$. Consequently using the sufficiency conditions for the extrema of a function as given by equation (14.2) yields

$$\frac{d^2A}{dx^2} = -2$$

Since the second derivative is less than zero, $x = 50$ is a point at which the area is maximized. Obviously the solution is $x = y = 50$ ft, and the maximum of $A$ is 2500 ft$^2$.

E

EXAMPLE 14.4

Determine the points of local maximum and local minimum in the closed interval $0 \leq x \leq 4$ for the following function:

$$y = 2x^3 - 12x^2 + 18x + 3$$

### Solution

Applying the necessity conditions for local maximum or local minimum, we have

$$\frac{dy}{dx} = 6x^2 - 24x + 18 = 0$$

Solving for the two roots gives

$$x_1 = 1, \qquad x_2 = 3$$

Obviously these points fall within the closed interval $0 \leq x \leq 4$. Therefore they must be tested by using the sufficiency condition for local maximum and local minimum by taking the second derivative:

$$\frac{d^2y}{dx^2} = 12x - 24$$

For $x_1 = 1$:

$$\frac{d^2y}{dx^2} = -12 \qquad \text{local maximum}, \quad y = 11$$

For $x_2 = 3$:

$$\frac{d^2y}{dx^2} = 12 \qquad \text{local minimum}, \quad y = 3$$

Note that the boundary values must be included in determining all of the local maximum and local minimum values within the interval. That is,

For $x = 0$, we have $y = 3$ and

For $x = 4$, we have $y = 11$

Clearly the first derivatives at the boundaries are not zero, in which case the second derivative test is not used. The results are now summarized as follows:

$x_0 = 0, \quad y = 3$: local minimum

$x_1 = 1, \quad y = 11$: local maximum

$x_2 = 3, \quad y = 3$: local minimum

$x_3 = 4, \quad y = 11$: local maximum

These values along with the function being optimized are shown in figure 14.3.

Note here that since point $A$ has a functional value smaller than 3 and point $B$ has a functional value greater than 11, the four points investigated for the closed interval (shaded area) are indeed local points.

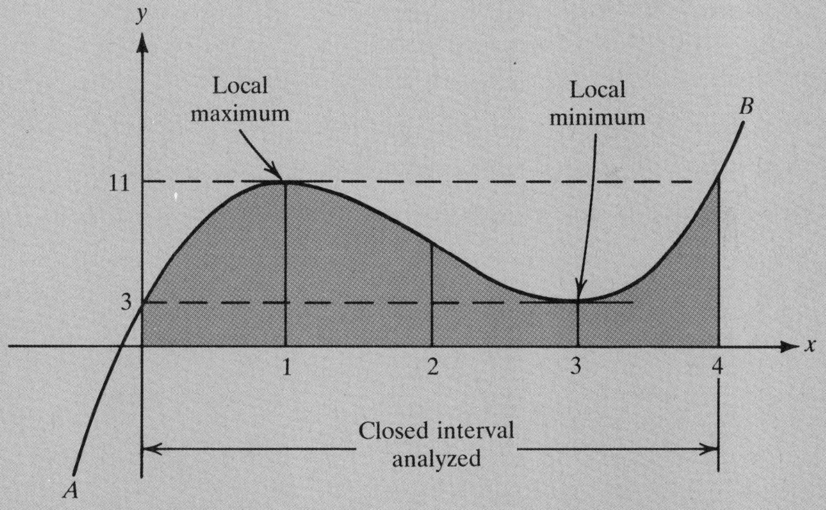


**Figure 14.3** Local maximum and local minimum within a closed interval for the function given by example 14.4.

### 14.3.2 Functions of Many Variables

In this section, we deal with the more general unconstrained optimization of a real-valued function involving two or more variables. That is, given

$$y = f(x_1, x_2, x_3, \ldots, x_n)$$

we seek to determine the points of local maximum and local minimum within a given domain. The necessity condition for the existence of extrema is

$$\frac{\partial f}{\partial x_i} = 0, \qquad i = 1, \ldots, n \tag{14.3}$$

Equation (14.3) is of precisely the same form as that used for functions involving one variable, except that we now may have to solve a system of nonlinear algebraic equations. Furthermore, the sufficiency condition for a local maximum or a local minimum requires that a matrix be tested for being negative definite or positive definite, respectively. The matrix in question is given below:

$$[A] = \begin{bmatrix} \frac{\partial^2 f}{\partial x_1^2} & \frac{\partial^2 f}{\partial x_2 \, \partial x_1} & \cdots & \frac{\partial^2 f}{\partial x_n x_1} \\ \frac{\partial^2 f}{\partial x_1 \, \partial x_2} & \frac{\partial^2 f}{\partial x_2^2} & \cdots & \frac{\partial^2 f}{\partial x_n \, \partial x_2} \\ \vdots & \vdots & & \vdots \\ \frac{\partial^2 f}{\partial x_1 \, \partial x_n} & \frac{\partial^2 f}{\partial x_2 \, \partial x_n} & \cdots & \frac{\partial^2 f}{\partial x_n^2} \end{bmatrix} \tag{14.4}$$

Obviously this matrix can be defined at a point of extremum provided the various partial derivatives are continuous and exist. Therefore, suppose $\{x_i^*\} = \{x_1^*, x_2^*, \ldots, x_n^*\}$ is a point of extremum determined from the necessary condition given by equation (14.3); the following conclusions can be drawn:

1. The point $\{x_i^*\}$ represents a local maximum if $[A]$ is negative definite.
2. The point $\{x_i^*\}$ represents a local minimum if $[A]$ is positive definite.

The test for whether a matrix is positive or negative definite can be easily made by resolving the $[A]$ matrix into a lower $[L]$ and an upper unit triangular matrix $[T]$ using the Crout's method to give

$$[A] = \begin{bmatrix} L_{11} & & & 0 \\ L_{12} & L_{22} & & \\ \vdots & \vdots & \ddots & \\ L_{1n} & L_{2n} & \cdots & L_{nn} \end{bmatrix} \begin{bmatrix} 1 & t_{12} & \cdots & t_{1n} \\ & 1 & \cdots & t_{2n} \\ & & \ddots & \vdots \\ & 0 & & 1 \end{bmatrix}$$

In which case $[A]$ is said to be positive definite if

$$L_{11} < 0, \qquad L_{22} > 0, \qquad \ldots, \qquad L_{nn} < 0 \qquad \text{for } n \text{ odd}$$

and

$$L_{11} < 0, \qquad L_{22} > 0, \qquad \ldots, \qquad L_{nn} > 0 \qquad \text{for } n \text{ even}$$

Furthermore, the matrix $[A]$ is said to be negative definite if

$$L_{ij} > 0 \qquad \text{for} \quad i = 1, 2, \ldots, n$$

If the diagonal elements in the lower-triangular matrix do not obey any one of these rules, then the point $\{x_i^*\}$ is neither a local maximum nor a local minimum. Clearly this procedure is to be applied repeatedly for each extremum.

**E**

EXAMPLE 14.5

Determine the displacements $x_1$ and $x_2$ for the following mechanical system caused by an external static force $P$.

$$k_1 = 2 \text{ kips/in.}$$
$$k_2 = 3 \text{ kips/in.}$$
$$P = 5 \text{ kips}$$

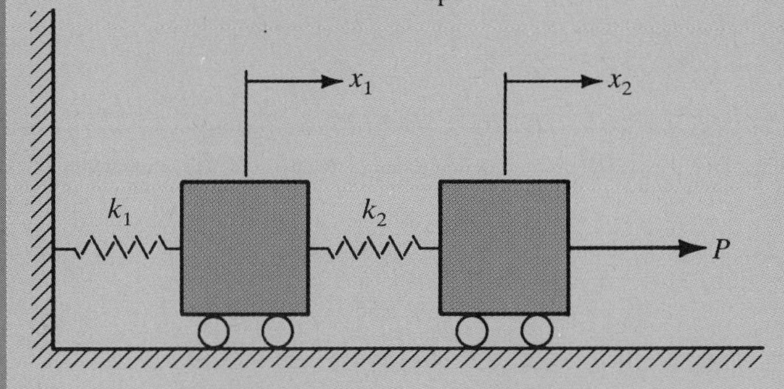


## Solution

Since the force is applied statically, the inertial forces corresponding to $m_1$ and $m_2$ are zero. While the solution of this problem can be easily accomplished by using the conditions of static equilibrium for each of the masses, let us approach the solution a little differently by minimizing the potential energy of the assembly. The potential energy $E$ is given as follows:

$$E = \tfrac{1}{2}K_1x_1^2 + \tfrac{1}{2}K_2(x_2 - x_1)^2 - Px_2$$

or more simply

$$E = x_1^2 + 1.5(x_2 - x_1)^2 - 5x_2$$

The necessary condition for a local maximum or local minimum is given by equation (14.3) as follows:

$$\frac{\partial E}{\partial x_1} = 2x_1 - 3(x_2 - x_1) = 0$$

$$\frac{\partial E}{\partial x_2} = 3(x_2 - x_1) - 5 = 0$$

Note that for this special case the resulting system involves linear equations and that in general a nonlinear system may result. In any case, solving for the unknowns $x_1$ and $x_2$ gives

$$x_1 = \tfrac{15}{6} \text{ in.,} \qquad x_2 = \tfrac{25}{6} \text{ in.}$$

In order to check whether this solution represents a maximum or minimum, we need to apply the sufficiency condition. That is, we form the matrix given by equation (14.4) to give

$$[A] = \begin{bmatrix} \dfrac{\partial^2 E}{\partial x_1^2} & \dfrac{\partial^2 E}{\partial x_2 \, \partial x_1} \\ \dfrac{\partial^2 E}{\partial x_1 \, \partial x_2} & \dfrac{\partial^2 E}{\partial x_2^2} \end{bmatrix} = \begin{bmatrix} 5 & -3 \\ -3 & 3 \end{bmatrix}$$

Note that $[A]$ is independent of $x_1$ and $x_2$. The implication is that the solution is a global minimum or maximum. Using the Crout method, we decompose $[A]$ into $[L]$ and $[T]$ to give

$$[A] = \begin{bmatrix} 5 & 0 \\ -3 & \frac{6}{5} \end{bmatrix} \begin{bmatrix} 1 & -\frac{3}{5} \\ 0 & 1 \end{bmatrix}$$

Since $L_{11} = 5 > 0$ and $L_{22} = \frac{6}{5} > 0$ the matrix $[A]$ is positive definite and the solution of $x_1 = \frac{15}{6}$ and $x_2 = \frac{25}{6}$ represents a minimum of the potential energy.

## 14.4 CONSTRAINED OPTIMIZATION—LAGRANGE MULTIPLIERS

In this section, we deal with optimization problems involving constraint equations. Obviously, all practical problems dealing with optimization are constrained in terms of the magnitude of the decision variables they can assume. However, here we deal with problems in which the constraint equations may or may not be readily expressed in terms of the decision variables (see example 14.3).

The method of Lagrange multipliers is of considerable importance in the area of constrained optimization of nonlinear models. The basis for the method can be easily explained by considering the following simple model:

Given the objective function

$$\text{MINIMIZE} \quad y = f(x_1, x_2, x_3) \qquad \textbf{(14.5)}$$

subject to the constraint

$$G(x_1, x_2, x_3) = 0 \qquad \textbf{(14.6)}$$

Suppose that it is possible to solve equation (14.6) for $x_3$; then substituting into equation (14.5) reduces the problem to an unconstrained one. That is,

$$\text{MINIMIZE} \quad y = F(x_1, x_2)$$

where $F(x_1, x_2)$ is the new function in which $x_3$ is eliminated. The necessary condition that $y$ be a minimum is that

$$dy = \frac{\partial F}{\partial x_1} dx_1 + \frac{\partial F}{\partial x_2} dx_2 = 0 \qquad \textbf{(14.7)}$$

Furthermore, consideration of the partial differentials of the original problem gives

$$dy = \frac{\partial f}{\partial x_1}\,dx_1 + \frac{\partial f}{\partial x_2}\,dx_2 + \frac{\partial f}{\partial x_3}\,dx_3 \tag{14.8}$$

and

$$\frac{\partial G}{\partial x_1}\,dx_1 + \frac{\partial G}{\partial x_2}\,dx_2 + \frac{\partial G}{\partial x_3}\,dx_3 = 0 \tag{14.9}$$

Now, multiplying equation (14.9) by a multiplier $\lambda$ of unknown value then adding to equation (14.8) yields

$$dy = \left(\frac{\partial f}{\partial x_1} + \frac{\partial G}{\partial x_1}\right)dx_1 + \left(\frac{\partial f}{\partial x_2} + \frac{\partial G}{\partial x_2}\right)dx_2 + \left(\frac{\partial f}{\partial x_3} + \frac{\partial G}{\partial x_3}\right)dx_3 \tag{14.10}$$

Equating (14.10) and (14.7) readily yields the following relationships:

$$\frac{\partial F}{\partial x_1} = \frac{\partial f}{\partial x_1} + \frac{\partial G}{\partial x_1} = 0$$

$$\frac{\partial F}{\partial x_2} = \frac{\partial f}{\partial x_2} + \frac{\partial G}{\partial x_2} = 0$$

$$\frac{\partial f}{\partial x_3} + \frac{\partial G}{\partial x_3} = 0$$

Consequently, if we introduce a function $Q$ such that

$$Q(x_1, x_2, x_3, \lambda) = f(x_1, x_2, x_3) + \lambda G(x_1, x_2, x_3)$$

then the optimization procedure can be performed directly using the following:

$$\frac{\partial Q}{\partial x_1} = 0 = \frac{\partial f}{\partial x_1} + \frac{\partial G}{\partial x_1}$$

$$\frac{\partial Q}{\partial x_2} = 0 = \frac{\partial f}{\partial x_2} + \frac{\partial G}{\partial x_2}$$

$$\frac{\partial Q}{\partial x_3} = 0 = \frac{\partial f}{\partial x_3} + \frac{\partial G}{\partial x_3}$$

$$\frac{\partial Q}{\partial \lambda} = 0 = G(x_1, x_2, x_3)$$

Clearly these four equations can be readily solved for the unknowns $x_1$, $x_2$, $x_3$, and $\lambda$. This concept can be extended to the general case of constrained optimization of functions involving many variables and many constraints. Consider

MINIMIZE $y = f(x_1, x_2, \ldots, x_n)$

subject to the following $m$ constraints $(m < n)$:

$$G_i(x_1, x_2, \ldots, x_n) = 0, \qquad i = 1, 2, \ldots, m$$

where all functions are assumed to be differentiable. Then the Lagrange function is determined as follows:

$$Q = f(x_1, \ldots, x_n) + \sum_{i=1}^{m} \lambda_i G_i(x_1, \ldots, x_n) \tag{14.11}$$

and the necessary conditions for minimizing $y$ are given by

$$\frac{\partial Q}{\partial x_j} = 0, \qquad j = 1, 2, \ldots, n \tag{14.12}$$

$$\frac{\partial Q}{\partial \lambda_i} = 0, \qquad i = 1, 2, \ldots, m \tag{14.13}$$

These $m + n$ equations can then be solved for the unknowns. Note that the resulting set may be nonlinear, in which case many solutions are possible. That is, many points of local minimum can be determined.

EXAMPLE 14.6

A country imports the oil it needs from three different sources, A, B, and C. The costs per unit amount imported are given as follows:

| SOURCE | COST $\phi_i$ |
|---|---|
| A | $\phi_1 = 2x + 2x^2$ |
| B | $\phi_2 = 3y + 1.5y^2$ |
| C | $\phi_3 = z + 3z^2$ |

Determine the number of units $x$, $y$, and $z$ to be imported from sources A, B, and C so that the cost is minimized. The total amount to be imported is 8.5 units.

Solution

The objective function is easily determined as the sum of all cost functions. That is,

$$\begin{aligned} \text{MINIMIZE} \quad \phi &= \phi_1 + \phi_2 + \phi_3 \\ &= 2x + 2x^2 + 3y + 1.5y^2 + z + 3z^2 \end{aligned}$$

The constraint equation is given by the stipulation that the total amount imported is equal to 8.5 units. Thus

$$x + y + z = 8.5$$

This problem can be easily solved by forming the Lagrange function to give

$$\text{MINIMIZE} \quad Q = 2x + 2x^2 + 3y + 1.5y^2 + z + 3z^2 + \lambda(x + y + z - 8.5)$$

Consequently the minimization of the Lagrange function is established by setting the partial derivatives of $Q$ relative to $x$, $y$, $z$, and $\lambda$ equal to zero and then solving the resulting equations for the unknowns.

$$x = 2.8889 \text{ units}$$

$$y = 3.5185 \text{ units}$$

$$z = 2.0926 \text{ units}$$

$$\lambda = -13.555$$

The minimum cost is then given by substituting these values into the objective function to yield $Q = 66.82$.

EXAMPLE 14.7

Suppose that the amount to be imported in the preceding example is not known exactly, but instead is given as a range of 7–14 units. Develop expressions relating the amount imported from each source as a function of the total amount.

Solution

This problem is easily solved by recognizing that the model is precisely the same with one exception; the total amount can be assumed to be a variable $T$. Hence, solving the following set using matrix inversion:

$$2 + 4x + \lambda = 0$$

$$3 + 3y + \lambda = 0$$

$$1 + 6z + \lambda = 0$$

$$x + y + z - T = 0$$

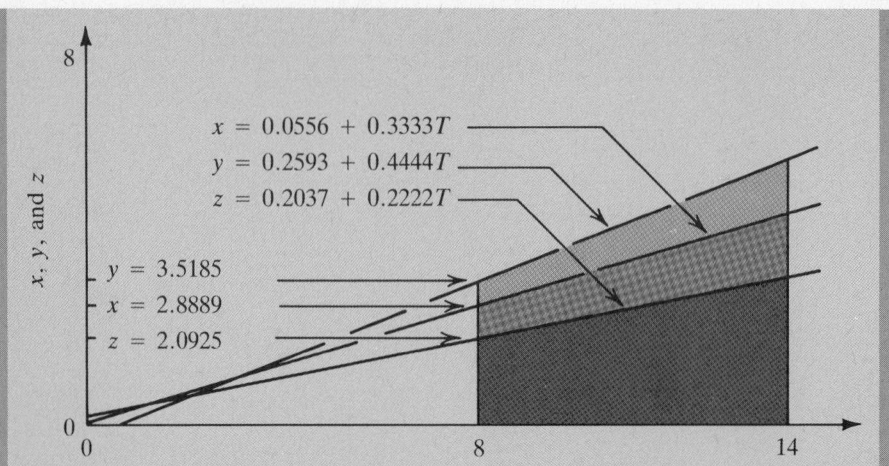


**Figure 14.4** Graphical illustration of quantities imported from the various sources versus total quantity.

yields the following relationships:

$$x = 0.0556 + 0.3333T$$

$$y = 0.2593 + 0.4444T$$

$$z = 0.2037 + 0.2222T$$

$$\lambda = -2.2222 - 1.3333T$$

Substituting these functional relationships into the objective function gives the minimum cost in terms of the total amount to be imported as follows:

$$\text{MINIMIZE} \quad Q = 2x + 2x^2 + 3y + 1.5y^2 + z + 3z^2$$

or more simply

$$\text{MINIMIZE} \quad Q = -0.23158 + 2.222T + 0.66666T^2$$

Figure 14.4 shows the relationship between $x$, $y$, $z$, and the total number of units imported where the cost is minimum.

Note that the solution indicates that as long as $T > 0$, the functional relationship holds regardless of the magnitude of the total amount of oil imported. This is true in this particular case because of the special nature of the cost functions. In general, no such simple solution is possible because the set of equations resulting from the partial derivatives is generally nonlinear.

## 14.5 OPTIMIZATION OF LINEAR MODELS

In this section, we deal with optimization problems involving a linear objective function subjected to a set of linear constraint functions. Furthermore, the constraints generally involve inequalities. These problems occur frequently when dealing with cost and profit optimization. Such problems are referred to as "linear programming problems."

### 14.5.1 Standard Equality Form

All linear programming problems are generally expressed in a so-called standard form of $m$ linear equations involving $n$ unknowns. Furthermore, the decision variables involved are assumed to be non-negative and the objective function is minimized. The problem can be generally stated as follows:

$$\text{MINIMIZE} \quad Q = C_1x_1 + C_2x_2 + \cdots + C_nx_n$$

subject to:

$$\left.\begin{array}{l} a_{11}x_1 + a_{12}x_2 + \cdots + a_{1n}x_n = b_1 \\ a_{21}x_1 + a_{22}x_2 + \cdots + a_{2n}x_n = b_2 \\ \quad\vdots \qquad\quad \vdots \qquad\qquad\quad \vdots \qquad\quad \vdots \\ a_{m1}x_1 + a_{m2}x_2 + \cdots + a_{mn}x_n = b_m \end{array}\right\} \qquad x_j \geq 0, \quad j = 1, \ldots, n$$

That is, given the costs $C_1, \ldots, C_n$ associated with a set of decision variables $x_1, \ldots, x_n$, we seek to determine the solution that minimizes the objective function and at the same time satisfies the set of linear constraints. Such a solution is termed "optimal." In many linear programming problems, the objective function is a maximization rather than a minimization. For such problems, the standard form can be easily satisfied by multiplying the objective function by a negative one. That is, given

$$\text{MAXIMIZE} \quad z = C_1x_1 + C_2x_2 + \cdots + C_nx_n$$

The standard form is attained as follows:

$$\text{MINIMIZE} \quad -z = \text{MINIMIZE} \quad Q = -C_1x_1 - C_2x_2 - \cdots - C_nx_n$$

Furthermore, some decision variables may not be constrained. That is, they may take on negative values. Since such a variable violates the standard form just outlined, it must be replaced by two non-negative decision variables as follows.

Suppose $x_1$ is a variable whose value might be a negative number, then we can easily replace it by

$$x_1 = x_1' - x_1''$$

where $x_1' \geq 0$ and $x_1'' \geq 0$. Finally, constraint equations are normally given in an inequality form such as the following:

$$a_{k1}x_1 + a_{k2}x_2 + \cdots + a_{kn}x_n \geq b_k$$

here we simply subtract a so-called slack variable from the left-hand side in order to achieve the standard form. That is,

$$a_{kn}x_1 + a_{k2}x_2 + \cdots + a_{kn}x_n - x_{n+1} = b_k$$

on the other hand, if the inequality is of the form

$$a_{kn}x_1 + a_{k2}x_2 + \cdots + a_{kn}x_n \leq b_k$$

then we can easily achieve the standard form by adding a slack variable to the right-hand side to give

$$a_{kn}x_1 + a_{k2}x_2 + \cdots + a_{kn}x_n + x_{n+1} = b_k$$

The purpose of the presentation thus far is to outline a unified scheme for presenting the linear mathematical model so that a solution procedure can be developed.

EXAMPLE 14.8

Express the following linear programming model in the standard equality form. Assume that $x_1$ and $x_2$ are known to take on positive values only.

$$\text{MAXIMIZE} \quad Z = 3x_1 + x_2$$

subject to

$$\left.\begin{aligned} -4x_1 + x_2 &\leq 6 \\ 2x_1 + 3x_2 &\geq 7 \\ 4x_1 + x_2 &\leq 10 \end{aligned}\right\} \qquad x_1 \geq 0, \quad x_2 \geq 0$$

Solution

It is clear that the objective function does not meet the standard form of minimization. Multiplying by $-1$ yields the following more suitable form:

$$\text{MINIMIZE} \quad -z = \text{MINIMIZE} \quad Q = -3x_1 - x_2$$

Note that $Q = -z$. In addition, the inequality constraints can be easily transformed into the equality form by introducing three slack variables to give

$$\left.\begin{aligned} -4x_1 + x_2 + x_3 &= 6 \\ 2x_1 + 3x_2 - x_4 &= 7 \\ 4x_1 + x_2 + x_5 &= 10 \end{aligned}\right\} \qquad x_j \geq 0, \quad j = 1, \ldots, 5$$

### 14.5.2 Mathematical Search for an Optimal Solution

It is evident that the standard form used in solving linear programming problems generally results in a set of linear algebraic equations involving more unknowns than equations. That is, given a set of $m$ equations in $n$ unknowns, where $n > m$, we can easily achieve a solution by arbitrarily setting $n - m$ variables equal to zero, then solving the resulting $m \times m$ set for the remaining unknowns. Such a solution may or may not be optimal. In fact, it may not even be feasible. The implication is that we need to find all possible solutions to the set, then test each one for optimality by simply substituting into the objective function. The general procedure of searching for the optimal solution may be simplified by ignoring all solutions involving negative values, since such solutions violate the constraint that all variables must be positive. Since any set of $m$ variables can be chosen as basic variables (variables to be solved for), there are $k$ possible solutions where

$$k = \frac{n!}{m!(n-m)!} \tag{14.14}$$

Any one of the $k$ solutions that does not satisfy all of the constraints is called "nonfeasible." Therefore a solution that satisfies all of the constraints is referred to as a "feasible solution." Furthermore, the first feasible solution obtained for a given problem is termed an "initial feasible solution."

EXAMPLE 14.9

Determine the optimal solution to the following linear programming model using the mathematical search method just outlined.

$$\text{MINIMIZE} \quad Q = 2x_1 + x_2$$

subject to

$$\left.\begin{aligned} x_1 + x_2 &\leq 5 \\ 2x_1 + 4x_2 &\geq 4 \end{aligned}\right\} \qquad x \geq 0, \quad x_2 \geq 0$$

Solution

As a first step, the model must be expressed in the standard form to give

$$\text{MINIMIZE} \quad Q = 2x_1 + x_2$$

subject to

$$\left.\begin{aligned} x_1 + x_2 + x_3 &= 5 \\ 2x_1 + 4x_2 - x_4 &= 4 \end{aligned}\right\} \qquad x_j \geq 0, \quad j = 1, \ldots, 4$$

Using equation (14.14), we can easily determine the number of solutions that must be calculated. Thus, for $m = 2$ and $n = 4$, we have

$$k = \frac{4!}{2!(4-2)!} = \frac{4(3)2!}{2(1)2!} = 6$$

Therefore there are six solutions that must be investigated for optimality. These are

First solution ($x_3 = x_4 = 0$):

$$\left.\begin{aligned} x_1 + x_2 &= 5 \\ x_1 + 4x_2 &= 4 \end{aligned}\right\} \Rightarrow \left\{\begin{aligned} x_1 &= \tfrac{16}{3} \\ x_2 &= -\tfrac{1}{3} \end{aligned}\right.$$

Second solution ($x_2 = x_4 = 0$):

$$\left.\begin{aligned} x_1 + x_3 &= 5 \\ 2x_1 \quad &= 4 \end{aligned}\right\} \Rightarrow \left\{\begin{aligned} x_1 &= 2 \\ x_3 &= 3 \end{aligned}\right.$$

Third solution ($x_2 = x_3 = 0$):

$$\left.\begin{aligned} x_1 \quad &= 5 \\ 2x_1 - x_4 &= 4 \end{aligned}\right\} \Rightarrow \left\{\begin{aligned} x_1 &= 5 \\ x_4 &= 6 \end{aligned}\right.$$

Fourth solution ($x_1 = x_4 = 0$):

$$\left.\begin{aligned} x_2 + x_3 &= 5 \\ 4x_2 \quad &= 4 \end{aligned}\right\} \Rightarrow \left\{\begin{aligned} x_2 &= 1 \\ x_3 &= 4 \end{aligned}\right.$$

Fifth solution ($x_1 = x_3 = 0$):

$$\left.\begin{aligned} x_2 \quad &= 5 \\ 4x_2 - x_4 &= 4 \end{aligned}\right\} \Rightarrow \left\{\begin{aligned} x_2 &= 5 \\ x_4 &= 16 \end{aligned}\right.$$

Sixth solution ($x_1 = x_2 = 0$):

$$\left.\begin{aligned} x_3 &= 5 \\ -x_4 &= 4 \end{aligned}\right\} \Rightarrow \left\{\begin{aligned} x_3 &= 5 \\ x_4 &= -4 \end{aligned}\right.$$

Obviously, the first and the sixth solutions are not feasible, since they violate the stipulation that all decision variables must be non-negative. Hence let us consider the remaining solutions and determine the one that minimizes the objective function.

| SOLUTION | $x_1$ | $x_2$ | OBJECTIVE FUNCTION $z = 2x_1 + x_2$ |
|---|---|---|---|
| 2 | 2 | 0 | 4 |
| 3 | 5 | 0 | 10 |
| 4 | 0 | 1 | 1 |
| 5 | 0 | 5 | 5 |

Obviously the fourth solution is the optimal solution to the linear programming problem. Unfortunately, such an approach to solving problems is not practical in that a significant amount of number crunching is required, especially when one is dealing with a large number of constraints.

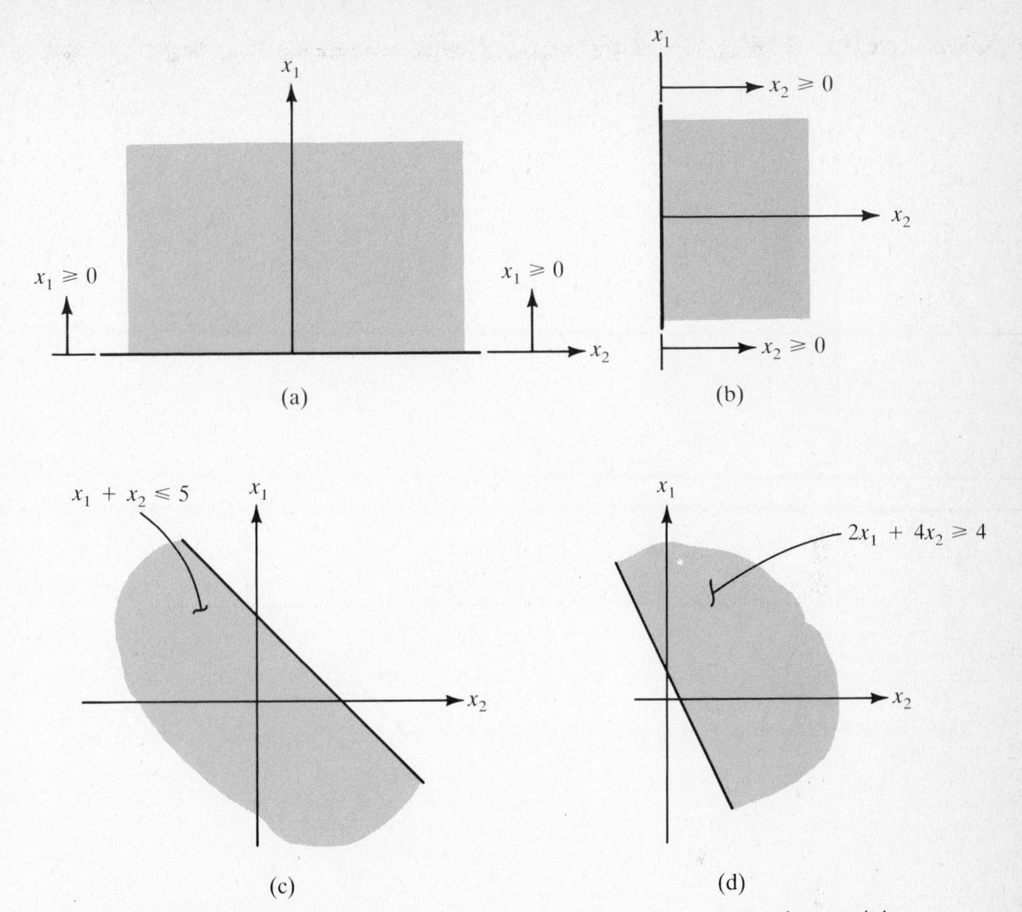


**Figure 14.5** Graphical illustration of constraints of a linear programming model.

### 14.5.3 Graphical Interpretation of an Optimal Solution

In order to gain insight into the process that permits the determination of optimal solutions, let us consider the problem presented in example 14.9 once again. We begin by plotting the constraint equations as shown in figure 14.5.

Note that any point inside the shaded area satisfies that particular constraint. That is, figure 14.5a satisfies the condition that $x_1$ is non-negative, etc. Consequently, the area common to all of the constraints is the area of feasible solutions. This is shown in figure 14.6.

The remaining question is how can we determine the optimal solution to the problem since there is generally an infinite number of solutions? Fortunately, this is accomplished rather easily by plotting the objective function for an arbitrary value

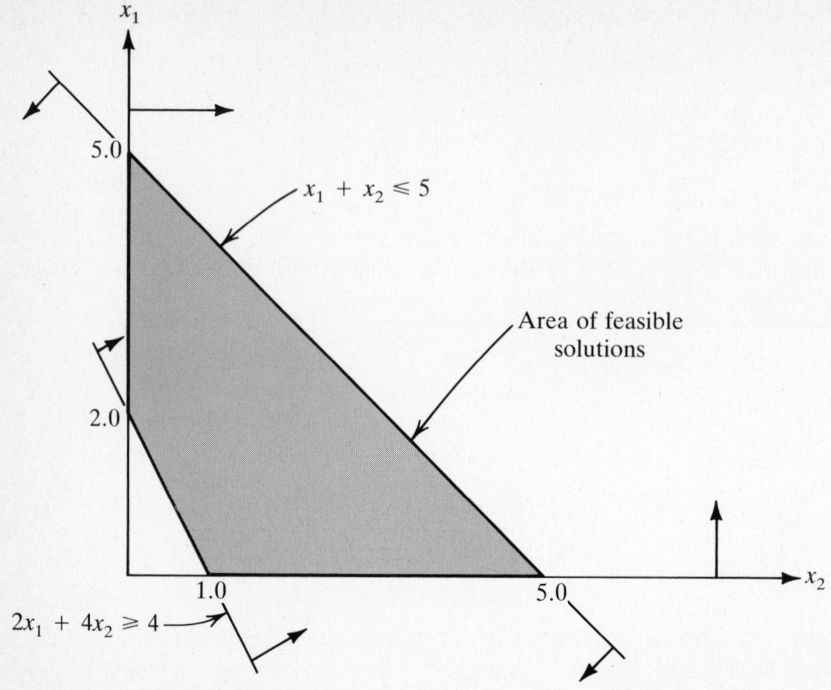


**Figure 14.6** Determination of area of feasible solutions for a linear programming model.

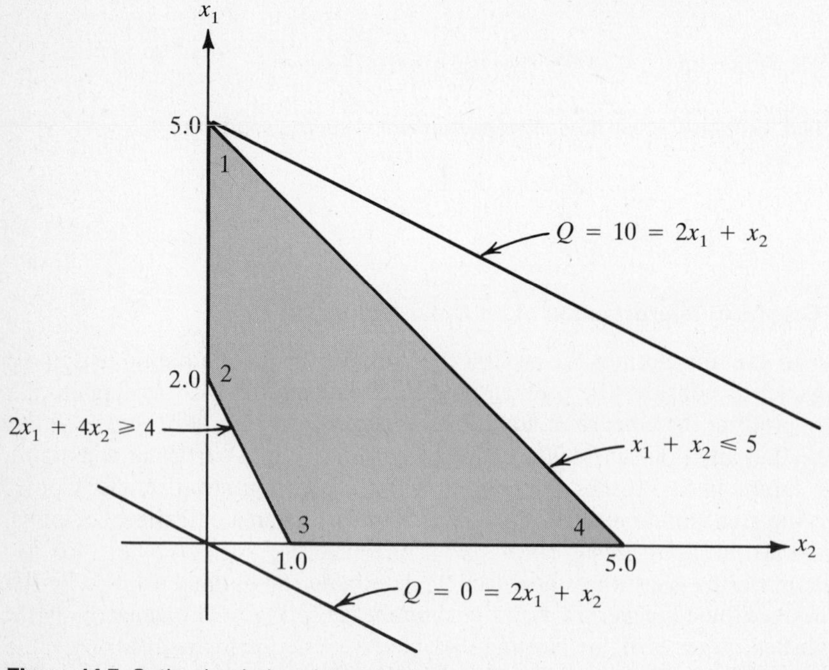


**Figure 14.7** Optimal solution determination.

of $Q$. For example, the common area along with assumed values of $Q = 10$ and 0 are shown in figure 14.7.

It is interesting to note that the objective function attains its maximum and minimum values when it intersects the vertices of the shaded area. That is, if we are to slide the objective function for $Q = 0$ until it intersects the area of feasible solution, maintaining its slope, then it will intersect node 3 first. Obviously this node has the coordinates $x_1 = 0$ and $x_2 = 1$, which gives $Q = 1$. This is precisely the optimal solution we obtained mathematically for example 14.9. Furthermore, sliding the objective function further in the direction of nodes 1, 2, and 4 gives higher values for $Q$. In fact, the objective function of $Q = 10$ gives the maximum value of $Q$ at node one. The implication is rather significant in that the optimal solution is found by searching the vertices of the area of feasible solutions. Recall that in example 14.9, we have determined only four feasible solutions and these correspond exactly to coordinates of nodes 1, 2, 3, and 4 shown in figure 14.6. Despite its limitations, the graphical method provides a great deal of information relative to the linear programming problem solving. The reader is therefore urged to understand its application.

EXAMPLE 14.10

Optimize the following linear programming model using the graphical procedure:

$$\text{MAXIMIZE} \quad z = x_1 + x_2$$

subject to

$$\left.\begin{aligned} x_1 + 5x_2 &\leq 15 \\ 2x_1 + 4x_2 &\geq 4 \\ x_1 + x_2 &\leq 5 \end{aligned}\right\} \qquad x_1 \geq 0, \quad x_2 \geq 0$$

Solution

The constraints along with the objective function are shown graphically in figure 14.8.

The objective function is clearly parallel to the constraint $x_1 + x_2 \leq 5$. Therefore it will intersect the area of feasible solutions at an infinite number of points. Therefore there are infinite optimal solutions to this example problem. In fact, the optimal solution is given directly by

$$x_1 + x_2 \leq 5$$

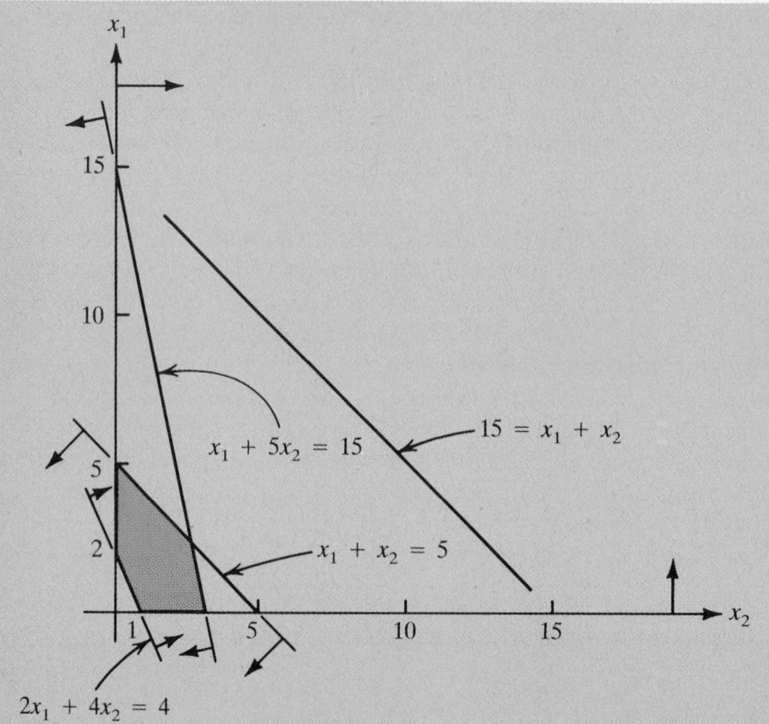


**Figure 14.8** Solution of example 14.10.

with the condition that

$$0 \le x_2 \le 2.5$$

The limits on $x_2$ are determined readily by noting that $x_1 + 5x_2 \le 15$ and $x_1 + x_2 \le 5$ intersect at $x_2 = 2.5$ and $x_1 = 2.5$. Therefore this constitutes the line segment of optimal solutions. For example, taking $x_1 = 5$ and $x_2 = 0$ gives the maximum of $z = 5$; similarly taking $x_1 = 4$ and $x_2 = 1$ gives the maximum of $z = 5$ again. Note that these assumed optimal solutions satisfy all of the constraints. However, letting $x_1 = 2$ and $x_2 = 3$ gives $Q = 5$, but violates the constraints $x_1 + 5x_2 \le 15$, in which case it is not feasible.

At this stage, it is important to note that if no common area can be found the problem has no solution. Furthermore, unbounded constraints result in an unbounded solution. Consider, for example, the constraints

$$\left.\begin{aligned} x_1 + x_2 &\le 5 \\ x_1 + 2x_2 &\ge 12 \end{aligned}\right\} \quad x_1 \ge 0, \quad x_2 \ge 0$$

These clearly do not have a common area. Therefore a solution to the problem is not feasible.

## 14.6 SIMPLEX METHOD

This method is one of the most practical and efficient mathematical techniques of solving linear programming problems. While a thorough treatment of the theory behind the method is beyond the scope of this text, we can simply state that the simplex method is a search method. That is, it simply searches for the optimal solution by going from one vertex to the next vertex of the area of feasible solutions until the optimal solution is found. Consequently, there is an inherent requirement for an initial feasible solution. Consider the following maximization problem:

$$\text{MAXIMIZE} \quad z = C_1x_1 + C_2x_2$$

subject to

$$\left.\begin{aligned} a_{11}x_1 + a_{12}x_2 \leq b_1 \\ a_{21}x_1 + a_{22}x_2 \leq b_2 \end{aligned}\right\} \quad x_1 \geq 0, \quad x_2 \geq 0, \quad b_1 \geq 0, \quad b_2 \geq 0$$

Using the simplex method, the following steps must be followed in order to achieve an optimal solution if one does exist.

*STEP 1:* Express the model in the standard form discussed earlier to give

$$\text{MINIMIZE} \quad Q = -C_1x_1 - C_2x_2$$

subject to

$$a_{11}x_1 + a_{12}x_2 + x_3 = b_1$$

$$a_{21}x_1 + a_{22}x_2 + x_4 = b_2$$

Note that $Q = -z$ and $x_3$ and $x_4$ are "slack variables."

*STEP 2:* Determine the initial feasible solution by setting the nonbasic variables $x_1 = x_2 = 0$, then letting the basic variables $x_3 = b_1$ and $x_4 = b_2$.

*STEP 3:* Examine the objective function and determine the most negative coefficient. That is, if $C_1 < C_2$, then select $C_1$ as the coefficient, indicating that $x_1$ is going to replace either $x_3$ or $x_4$ in evaluating a new feasible solution.

*STEP 4:* Suppose that $C_1 < C_2$. Then determine the following ratios:

$$R_1 = \frac{b_1}{a_{11}}, \qquad R_2 = \frac{b_2}{a_{21}}$$

Ignore negative and undefined $R$ values; then select the least positive $R$ value. Suppose it is $R_2$; then this indicates that $x_1$ is replacing $x_4$, since $x_4$ appears in the equation with the least positive $R$ value.

*STEP 5:* Eliminate $x_1$ from all equations including the objective function.

*STEP 6:* Repeat steps 2–5 until the coefficients of variables appearing in the objective function are all positive.

*STEP 7:* Set the variables appearing in the objective function to zero. Then solve the constraint equations for the optimal solution.

Generally, when dealing with maximization problems, the constraint equations normally involve less than signs, in which case, an initial feasible solution can be readily determined. However, when dealing with minimization problems greater than signs are normally encountered in the constraints. Consequently, slack variables must be subtracted rather than added to the left-hand side. The implication is that we no longer can assume the slack variables as our basic variables as was the case with the maximization problem. This is because negative values will be obtained. Therefore a different approach is needed in determining an initial feasible solution. Consider the following model:

$$\text{MINIMIZE} \quad Q = C_1 x_1 + C_2 x_2$$

subject to

$$\left.\begin{aligned} a_{11}x_1 + a_{12}x_2 \geq b_1 \\ a_{21}x_1 + a_{22}x_2 \geq b_2 \end{aligned}\right\} \quad x_1 \geq 0, \quad x_2 \geq 0, \quad b_1 \geq 0, \quad b_2 \geq 0$$

*STEP 1:* Express the model in the standard form. That is,

$$\text{MINIMIZE} \quad Q = C_1 x_1 + C_2 x_2$$

$$a_{11}x_1 + a_{12}x_2 - x_3 = b_1$$

$$a_{21}x_1 + a_{22}x_2 - x_4 = b_2$$

*STEP 4:* Since we cannot let $x_1 = x_2 = 0$, because this gives $x_3 = -b_1$ and $x_4 = -b_2$, we need to introduce so-called artificial variables $x_5$ and $x_6$ to the constraints to give

$$\text{MINIMIZE} \quad w = x_5 + x_6$$

subject to

$$a_{11}x_1 + a_{12}x_2 - x_3 + x_5 = b_1$$

$$a_{21}x_1 + a_{22}x_2 - x_4 + x_6 = b_2$$

Note that our objective is now to reduce $x_5 + x_6$ to zero. Clearly an initial feasible solution is obtained by letting $x_5 = b_1$, $x_6 = b_2$, and $x_1 = x_2 = x_3 = x_4 = 0$.

*STEP 5:* Eliminate $x_5$ and $x_6$ from the objective function to give

$$\begin{aligned} \text{MINIMIZE} \quad & w - b_1 - b_2 \\ & = -(a_{11} + a_{21})x_1 - (a_{12} + a_{22})x_2 + x_3 + x_4 \end{aligned}$$

*STEP 6:* Proceed with the optimization procedure as before until MINIMIZE $w = 0$. If MINIMIZE $w \neq 0$, then the problem has no feasible solution.

E

EXAMPLE 14.11

Solve the following linear programming model using the simplex method:

$$\text{MAXIMIZE} \quad z = 2x_1 + 2x_2$$

subject to

$$\left.\begin{aligned} 4x_1 + 2x_2 &\leq 12 \\ x_1 + 3x_2 &\leq 6 \end{aligned}\right\} \qquad x_1 \geq 0, \quad x_2 \geq 0$$

Solution

Expressing the model in the standard form gives

$$\text{MINIMIZE} \quad Q = -2x_1 - 2x_2 \qquad \textbf{(A1)}$$

subject to

$$\left.\begin{aligned} 4x_1 + 2x_2 + x_3 &= 12 \\ x_1 + 3x_2 + x_4 &= 6 \end{aligned}\right\} \qquad x_j \geq 0, \quad j = 1, \ldots, 4 \qquad \begin{aligned} &\textbf{(B1)} \\ &\textbf{(C1)} \end{aligned}$$

The initial basic feasible solution is given as $x_3 = 12$, $x_4 = 6$, and $x_1 = x_2 = 0$. Since $c_1 = -2$ and $c_2 = -2$, it makes no difference which one is chosen. Therefore, if we select $c_1$, then $x_1$ is the entering basic variable. The departing variable is determined by calculating the $R$ values. Thus

$$R_1 = \tfrac{12}{4} = 3, \qquad R_2 = \tfrac{6}{1} = 6$$

This implies that $x_1$ is the entering variable and $x_3$ is the departing variable, since $R_1$ is less than $R_2$. Therefore $x_1$ must be eliminated from all equations, except of course equation (B1). That is,

$$-x_2 + 0.5x_3 = \text{MIN}(Q + 6) \qquad \textbf{(A2)}$$

$$x_1 + 0.5x_2 + 0.5x_3 = 3 \qquad \textbf{(B2)}$$

$$2.5x_2 - 0.25x_3 + x_4 = 3 \qquad \textbf{(C2)}$$

Clearly $x_2$ is the entering variable. The departing variable is determined by once again examining the $R$ values:

$$R_1 = \frac{3}{0.5} = 6, \qquad R_2 = \frac{3}{2.5} = 1.2$$

Hence $x_4$ is the departing variable and $x_2$ is the entering variable. Therefore $x_2$ must be eliminated from all equations and should only appear in equation

(C2):

$$0.4x_3 + 0.4x_4 = \text{MIN}(Q + 7.2) \quad \textbf{(A3)}$$

$$x_1 + 0.3x_3 - 0.2x_4 = 2.4 \quad \textbf{(B3)}$$

$$x_2 - 0.1x_3 + 0.4x_4 = 1.2 \quad \textbf{(C3)}$$

Clearly the objective function given by equation (A3) involves only positive coefficients. Therefore an optimal solution has been achieved. That is,

$$x_3 = x_4 = 0 \Rightarrow \text{MIN}(Q + 7.2) = 0$$

$$x_1 = 2.4$$

$$x_2 = 1.2$$

Note, that $Q = 7.2 = 0$ gives $Q = -7.2$ and that since $Q = -z$ the maximum of $z = 7.2$.

**E**

EXAMPLE 14.12

Solve the following linear programming problem using the simplex method:

$$\text{MINIMIZE} \quad q = 2x_1 + 3x_2$$

subject to

$$\left.\begin{aligned} 4x_1 + 2x_2 &\geq 12 \\ x_1 + 4x_2 &\geq 6 \end{aligned}\right\} \quad x_1 \geq 0, \quad x_2 \geq 0$$

Solution

We begin by expressing the model in the standard equality form:

$$\text{MINIMIZE} \quad Q = 2x_1 + 3x_2$$

subject to

$$\left.\begin{aligned} 4x_1 + 2x_2 - x_3 &= 12 \\ x_1 + 4x_2 - x_4 &= 6 \end{aligned}\right\} \quad x_j \geq 0, \quad j = 1, \ldots, 4$$

Consequently the constraint equations as well as the objective function must be modified by introducing the two artificial variables $x_5$ and $x_6$. This is because a feasible solution is not readily available since the slack variables

$x_3$ and $x_4$ have negative coefficients. Thus

$$\text{MINIMIZE} \quad w = x_5 + x_6$$

subject to

$$4x_1 + 2x_2 - x_3 + x_5 = 12$$

$$x_1 + 4x_2 - x_4 + x_6 = 6$$

The objective function is readily modified by eliminating $x_5$ and $x_6$ to give

$$\text{MINIMIZE} \quad w - 18 = -5x_1 - 6x_2 + x_3 + x_4 \qquad \textbf{(A1)}$$

subject to

$$4x_1 + 2x_2 - x_3 + x_5 = 12 \qquad \textbf{(B1)}$$

$$x_1 + 4x_2 - x_4 + x_6 = 6 \qquad \textbf{(C1)}$$

Clearly equations (B1) and (C1) can be easily used to obtain an initial feasible solution. That is,

$$x_5 = 12$$

$$x_6 = 6$$

$$x_j = 0, \qquad j = 1, \ldots, 4$$

Examination of equation (A1) clearly shows that $x_2$ is the entering variable. Furthermore, the $R$ values for the constraints are determined to be

$$R_1 = \tfrac{12}{2} = 6, \qquad R_2 = \tfrac{6}{4} = 1.5$$

Since $R_x < R_1$, $x_6$ is the departing variable. That is, it can be dropped from all expressions. Hence, eliminating $x_2$ from equations (A1) and (B1), we have

$$\text{MINIMIZE} \quad W - 9 = -3.5x_1 + x_3 - 0.5x_4 \qquad \textbf{(A2)}$$

subject to

$$3.5x_1 - x_3 + 0.5x_4 + x_5 = 9 \qquad \textbf{(B2)}$$

$$0.25x_1 + x_2 - 0.25x_4 = 1.5 \qquad \textbf{(C2)}$$

Obviously $x_1$ has the smallest coefficient; therefore $x_1$ is the entering variable. The departing variable is determined by considering the $R$ value and dropping $x_6$ from all equations:

$$R_1 = \frac{9}{3.5} = 2.57, \qquad R_2 = \frac{1.5}{0.25} = 6$$

Thus $x_5$ is the departing variable since $R_1$ is smaller than $R_2$. Now, eliminating $x_1$ from (C2) and (A2) gives

$$\text{MINIMIZE} \quad W = 0$$

subject to

$$x_1 - 0.28x_3 + 0.14x_4 = 2.57$$

$$x_2 + 0.07x_3 - 0.21x_4 = 0.857$$

The solution is now given as

$$x_1 = 2.57 \qquad x_3 = 0$$

$$x_2 = 2.36 \qquad x_4 = 0$$

and

$$\text{MIN } Z = 2(2.57) + 3(2.36) = 12.22$$

Note that by minimizing $W$ we were able to minimize $Z$ indirectly.

The simplex method can be applied more efficiently if a tabular form is used. This is shown in the following example.

**E** EXAMPLE 14.13

Redo example 14.12 using the tabular form.

## Solution

| BASIC VARIABLE | $x_1$ | $x_2$ | $x_3$ | $x_4$ | $b_i$ | | $R_i$ |
|---|---|---|---|---|---|---|---|
| $x_3$ | 4 | 2 | 1 | | 12 | ← | $\frac{12}{4} = 3$ |
| $x_4$ | 1 | 3 | | 1 | 6 | | $\frac{6}{1} = 6$ |
| Cycle 1: | $\boxed{-2}$ | −2 | | | $Q$ | | |
| $x_1$ | 1 | 0.5 | 0.25 | | 3 | | 3/0.5 = 6 |
| $x_4$ | | 2.5 | −0.25 | 1 | 3 | ← | 3/2.5 = 1.2 |
| Cycle 2: | | $\boxed{-1}$ | 0.5 | | $Q + 6$ | | |
| $x_1$ | 1 | | 0.3 | −0.2 | 2.4 | | |
| $x_2$ | | 1 | −0.1 | 0.4 | 1.2 | | |
| | | | 0.4 | 0.4 | $Q + 7.2$ | | |

The first cycle begins by using the initial feasible solution in which the basic variables are simply the slack variables. Note that all basic variables have coefficients of 1.0 in one of the constraints (row) and zero elsewhere in that column. Furthermore, once the most negative (least value) coefficient is determined a square is used to indicate the column of the entering variable that will replace one of the basic variables $x_3$ or $x_4$. The $R$ values are then calculated and an arrow is used to indicate the departing basic variable. For example, $R_1 = 3$ indicates that $x_3$ is replaced by $x_1$ as shown in the second cycle. Clearly the solution is attained after the second cycle. That is, $x_3 = x_4 = 0$ and $Q = -7.2$ ($z = 7.2$). This is because the objective function involves positive coefficients only.

## Recommended Reading


*Introduction to Operation Research*, Hillier Lieberman, Holden-Day, Inc., San Francisco, 1967.

*Introduction to Systems Engineering Deterministic Models*, Tung Au and T. E. Stelson, Addition-Wesley Publishing Co., Reading, Mass., 1969.

*Modern Methods of Engineering Computation*, Robert L. Ketter and S. P. Prawel, McGraw-Hill Book Company, New York, 1969.

*Finite Mathematics with Business Applications*, John G. Kemeny et al., Prentice-Hall, Inc., Englewood Cliffs, N.J., 1972.

*Optimization in Systems Engineering*, Marlin H. Mickle and T. W. Sze, Intext Educational Publishers, Scranton, Pa., 1972.


P

## PROBLEMS

**14.1** Evaluate the point at which the following objective function attains its global minimum:

$$f(x) = x^2 - 3x + 7$$

**14.2** Evaluate all points of local minimum and local maximum for the following function in the interval $-1 \leq x \leq 8$:

$$f(x) = x^3 - 8x^2 + 7x - 16$$

**14.3** Determine whether or not the following matrix is positive definite:

$$[A] = \begin{bmatrix} 3 & -1 & 2 \\ -3 & 8 & -1 \\ 2 & -1 & 2 \end{bmatrix}$$

**14.4** Solve the model developed in example 14.1.

**14.5** Solve the following linear programming model graphically:

$$\text{MINIMIZE} \quad Z = x_1 + 2x_2$$

subject to

$$2x_1 + x_2 \leq 10$$

$$x_1 + 2x_2 \geq 10$$

$$x_1 \geq 0, \qquad x_2 \geq 0$$

**14.6** Solve the following linear programming model graphically:

$$\text{MAXIMIZE} \quad Z = 3x_1 + 5x_2$$

subject to

$$x_1 + x_2 \leq 4$$
$$5x_1 + 3x_2 \geq 8$$
$$x_1 \geq 0, \qquad x_2 \geq 0$$

**14.7** Express the following linear programming model in the standard form:

$$\text{MAXIMIZE} \quad Z = 2x_1 + 5x_2 + x_3$$

subject to

$$2x_1 - x_2 + 7x_3 \leq 6$$
$$x_1 + 3x_2 + 4x_3 \leq 9$$
$$3x_1 + 6x_2 + x_3 \leq 3$$
$$x_1 \geq 0, \qquad x_2 \geq 0, \qquad x_3 \geq 0$$

**14.8** Express the following linear programming model in the standard form (modified):

$$\text{MINIMIZE} \quad Z = 2x_1 + x_2$$

subject to

$$x_1 + 3x_2 \geq 6$$
$$2x_1 + x_2 \geq 4$$
$$x_1 + x_2 \leq 3$$
$$x_1 \geq 0, \qquad x_2 \geq 0$$

**14.9** Use the graphical method to solve problem 8.

**14.10** Use the simplex method to solve problem 7.

**14.11** Solve the model developed in example 14.8 using the simplex model.

**14.12** Solve the model developed in example 14.2, using the simplex method.

# Index


**A**

Adams methods, 429
Adams – Moulton method, 436
Addition and subtraction
  of matrices, 16
  of partitioned matrices, 27
Algebraic equations
  linear, 85 – 136
  nonlinear, 161 – 202
Analytic optimization, 606 – 638
Application of matrices to the rotation of a coordinate system, 31
Area calculation using determinants, 44
Associative law of matrix multiplication, 22
Augmented matrices, 13

**B**

Back-substitution, 91
Backward derivatives, 346, 351
Backward differences, 284
Backward interpolation, 274
Bairstow's method, 182 – 190
Biharmonic partial differential equations, 577
Boundary conditions
  derivative, 567
  irregular, 538
Boundary-value problems
  civil engineering systems, 69
  electrical engineering systems, 56
  mechanical engineering systems, 62
  ordinary differential equations, 479
  partial differential equations, 74
Buckling, column, 71, 215 – 218

**C**

Calculus, method of optimization, 610
Central differences, 286
Characteristic equation, 206
Circular regions, 524
Civil engineering systems, 68
  boundary-value problems, 69
  eigenproblems, 71
Coefficient of multiple determination, 327
Cofactors, 35
Column buckling, 215
Combined operations of matrices, 30
Comparison of methods (linear algebraic equations), 131
Complex coefficients
  linear algebraic equations, 129
  matrix inversion, 151
Constraint functions, 608
Convergence of partial differential equations, 565
Cramer's rule
  linear algebraic equations, 86
  matrix inversion, 139
Crank – Nicholson method, 554
Crout's method, 98
Cubic splines, 299
Curve fitting
  coefficients of multiple determination, 327
  functional approximations, goodness of, 326
  least squares, method of, 310
  linearization, 315
  linear regression, 312
  multiple regression, 320
  nonlinear regression, 318
  orthogonal polynomials, 322
  standard error, 329

**D**

D'Alembert's principle, 228
Data differentiation, 357
Decision variables, 607
Decoupling procedure, 467
Deflation of matrices, 244
Derivative boundary conditions, 567
Determinants
  area calculation using, 44
  Laplace method of cofactors, 35
  pivotal condensation, 40
  properties of, 34
  upper-triangle elimination method, 37
  volume calculation using, 44
Diagonal matrix, 10
Difference operators and tables
  backward differences, 284
  central differences, 286
  cubic splines, 299
  difference operators, relationships between, 287
  forward differences, 281
  interpolating polynomials, 290
    uneven intervals, 294
  interpolation errors, 297
  inverse interpolation, 299
Differential equations (ordinary), numerical solution
  Adams method, 429
  Adams – Moulton method, 436
  boundary-value problems, 479
  decoupling procedure, 467
  direct integration procedure, 476
  Euler and modified Euler methods, 419

Differential equations (*continued*)
finite difference methods, 444
first-order systems (initial-value problems), 465
Galerkin method, 417
high-order equations
initial-value problems, 444
and systems of first order, 440
least-squares method, 415
Milne method, 433
predictor – corrector methods, 428
Runge – Kutta methods, 424
second-order systems (initial-value problems), 467
step size and errors, 438
stiff equations, 440
Taylor series method, 413
trapezoidal rule method, 449
Differentiation, numerical
data, 357
errors, 354
interpolating polynomial methods, 343
mixed derivatives, 357
special derivative approximations, 360
stencil representation of derivatives, 361
mixed, 369
Taylor series method, 345
review of, 336
undetermined coefficients method, 349
Direct determinant expansion, 207
Direct integration procedure, 476
Double integration, 406
Dynamic response, 227

**E**

Eigenproblems, 59, 60, 71, 204
characteristic equation determination, 206
column buckling, 215
direct determinant expansion, 207
dynamic response, 227
eigenvalues, 212
eigenvectors, 212
electrical circuits, 223
functions of matrix, 261
Householder method, 257
indirect determinant expansion, 208
intermediate eigenvalue method, 244
Jacobi method, 251
mechanical vibrations, 218
polynomial iteration method, 247
smallest eigenvalue method, 241
transformation methods, 250
vector iteration techniques, 234
Eigenvalues, 212
Eigenvectors, 212
Electrical circuits, 223
Electrical enginering systems, 55
boundary-value problems, 56
eigenproblems, 59
initial-value problems, 60
Elimination method, 143
Elliptic equations, finite difference solution of, 513
Engineering systems
civil, 68
electrical, 55
mechanical, 61
response, 72
Errors
general, 4
nonlinear algebraic equations, 198
numerical differentiation, 354
Explicit scheme
eigenproblem, 558
finite difference, 548
Euler and modified Euler methods, 419

**F**

False-position method, 164
Feasible solutions, 608
Finite difference
boundary-value problems, 479
explicit scheme, 548
first derivative approximations, 363 – 364
fourth derivative approximations, 368
grid patterns, 501
initial-value problems, 444
second derivative approximations, 365 – 366
solution of elliptic equations, 513
third derivative approximations, 367
Finite element method
discretization of a region, 584
interpolation and shape functions, 586
introduction to, 584
one-dimensional formulation, 593
First-order systems (initial-value problems), 465
Forward differences, 281
Functional approximations, goodness of, 326
Functions of matrix
Caley – Hamilton method, 261
special matrix decomposition method, 265
static condensation method, 266

**G**

Galerkin method, 417
Gauss – Jordan elimination method, 94
Gauss quadrature formulas, 393
Gauss – Seidel method, 118
Gauss's elimination method, 90
Graphical method
nonlinear equations, 161 – 162
optimization, 627 – 629
Grid patterns, 501

**H**

Halving the intervals, 162
High-order equations
initial-value problems, 444
and systems of first order, 440
Homogeneous differential equations, 412
Homogeneous linear equations, 86
Householder method, 257
Hyperbolic partial differential equations, 576 – 577

**I**

Ill-conditioned linear algebraic equations, 121
Implicit scheme, 553
eigenproblem, 564
Indirect determinant expansion, 208
Initial-value problems, 59, 60, 64, 77, 444, 465, 467
Integration, numerical, 373
double integration, 406
Gauss quadrature formulas, 393
Romberg's integration, 390
Simpson's 1/3 rule, 380
special integration forumlas, 382
stencil representation of integration formulas, 388
trapezoidal rule, 374
unevenly spaced data points, 385

Intermediate eigenvalue method, 244
Interpolating polynomials
  difference operators and tables, 290, 294
  methods, 343
Interpolation, 272, 297, 299
  backward, 274
  central, 275
  errors, 297
  forward, 273
Interval-halving method, 162
Inverse interpolation, 229
Inverse of a matrix
  complex coefficients, 151
  Cramer's rule, 139
  elimination method, 143
  partitioning method, 148
  reducing matrix method, 145
  symmetrical matrices, 155
  triangular matrices, 153
Irregular boundaries, 538
Iterative methods
  eigenvector, 234 – 244
  false-position method, 164 – 167
  Gauss – Seidel method, 118 – 121
  Jacobi's method, 114 – 118
  nonlinear algebraic equations, 161 – 202
  polynomial, 247 – 250

J

Jacobian, 193
Jacobi's method, 114, 251

L

Lagrange multipliers, 618
Laplace method of cofactors, 35
Least-squares method
  curve fitting, 310
  differential equations, 415
Lin – Bairstow method for roots of polynomials, 182
Linear algebraic equations, 85
  comparison of method efficiencies, 131
  complex coefficients involving, 129
  Cramer's rule, 86
  Crout's method, 98
  Gauss – Jordan elimination method, 94
  Gauss – Seidel method, 118
  Gauss's elimination method, 90
  ill-conditioned sets, 121
  iterative methods, 113
  Jacobi method, 114
  more equations than unknowns, 126
  more unknowns than equations, 124
  reducing matrix method, 106
  scaling, 121
  square root method, 103
  tridiagonal systems, 110
Linearization, 315
Linear regression, 312

M

Mathematical models, 2, 53 – 83
Mathematical search, 624 – 627
Matrices
  applied to rotation of coordinate systems, 31
  partitioned, 25
  rules for combined operations, 30
  special, 8
Matrix addition, 16
Matrix associative law, 22
Matrix commutativity, 21
Matrix equality, 15
Matrix inversion, 138
  complex coefficients involving, 151
  Cramer's rule, 139
  elimination method, 143
  partitioning matrix method, 148
  reducing matrix method, 145
  symmetrical matrices, 155
  triangular matrices, 153
Matrix multiplication, 17 – 21
Matrix subtraction, 16
Maximum, local
  functions of many variables, 615
  functions of one variable, 611 – 615
Mechanical engineering systems, 61
  boundary-value problems, 62
  eigenproblems, 67
  initial-value problems, 64
Mechanical vibrations, 218
Milne method, 433
Minimum, local
  functions of many variables, 615
  functions of one variable, 611 – 615
Mixed derivatives, 357
Models, 54
Models involving partial differential equations, 74
  boundary-value problems, 74
  comparison of, 78
  initial-value problems, 77
Modified Newton – Raphson methods, 177
More equations than unknowns, 126
More unknowns than equations, 124
Multiple regression, 320
Multipliers, Lagrange, 618 – 622

N

Natural circular frequency
  beam, 231
  electrical system, 224 – 227
  mechanical system, 219 – 220, 222
Newton – Raphson first method, 167
Newton – Raphson second method, 174
Newton – Raphson systems of equations, 190
Nonlinear algebraic equations, 161
  errors, 198
  false-position method, 164
  interval-halving method, 162
  Lin – Bairstow method for roots of polynomials, 182
  modified Newton – Raphson methods, 177
  Newton – Raphson first method, 167
  Newton – Raphson second method, 174
  Newton – Raphson systems of equations, 190
  numerical solutions, 2 – 3
  root multiplicity, 199
Nonlinear problems, partial differential equations, 568
Nonlinear regression, 318
Null matrix, 8
Numerical differentiation
  data, 357
  errors, 354 – 356
  interpolating polynomial methods, 343 – 345
  introduction, 336
  mixed derivatives, 357 – 360
  undetermined coefficients, 349 – 354
  using Taylor's series, 345 – 349
Numerical integration
  double integration, 406
  Gauss quadrature formulas, 393

Numerical integration (*continued*)
  Romberg's integration, 390
  Simpson's 1/3 rule, 380
  special integration formulas, 382
  stencil representation of integration formulas, 388
  trapezoidal rule, 374
  unevenly spaced data points, 385

**O**

Objective function, 608
Orthogonal polynomials, 322
Optimal solution
  graphical interpretation, 627
  search for, 624
Optimization (constrained), Lagrange multipliers, 618
Optimization of linear models, 622
  optimal solution, graphical interpretation, 627
  optimal solution, search for, 624
Optimization problems, 607
  characteristics of, 607
  constraint functions, 608
  decision variables, 607
  feasible solution, 608
  objective function, 608
  optimal solution, 608
  simplex method, 631
Optimization (unconstrained), method of calculus, 610
  functions of many variables, 615
  functions of one variable, 611

**P**

Partial differential equations
  circular regions, 524
  elliptic equations, finite difference solution of, 513
  finite difference methods and grid patterns, 501
  finite element method
    discretization of a region, 584
    interpolation and shape functions, 586
    introduction to, 584
    one-dimensional formulation, 593
  irregular boundaries, 538
  rectangular elements, regions involving, 513
  skewed elements, regions involving, 528
  special elements, regions involving, 535
  triangular elements, regions involving, 532
  transformations
    from Cartesian to polar coordinates, 503
    from Cartesian to skewed coordinates, 507
    from Cartesian to triangular coordinates, 511
Partial differential equations, biharmonic, 577
Partial differential equations, hyperbolic, 576 – 577
Partial differential equations, parabolic
  derivative boundary conditions, 567
  explicit scheme
    eigenproblem, 558
    finite difference, 548
  implicit scheme, 553
    eigenproblem, 564
  nonlinear problems, 568
  stability and convergence, 565
Partitioned matrices, 25
  addition, 27
  inversion, 148
  multiplication, 28
  subtraction, 27
  transpose, 25
Pivotal condensation, 40
Polynomial iteration method, 247
Predictor – corrector methods, 428

**R**

Rectangular elements, 513
Reducing matrix method, 106
  matrix inversion, 145
Regression analysis, *see* Curve fitting
Romberg's integration formula, 390
Root multiplicity, 199
Roots of equations, *see* Nonlinear algebraic equations
Rotation of coordinate systems, 31
Runge – Kutta methods, 424

**S**

Scaling (linear algebraic equations), 121
Second-order systems (initial-value problems), 467
Simplex method, 631
Simpson's 1/3 rule, 380
Skewed elements (partial differential equations), 528
Smallest eigenvalue method, 241
Special derivative approximations, 360
Special elements (partial differential equations), 535
Special integration formulas, 382
Special matrices, 8
Splines, cubic, 299
Square root method (linear algebraic equations), 103
Stability of partial differential equations, 565
Standard error (curve fitting), 329
Stencil representation
  of derivatives, 361
    mixed, 369
  of integration, 388
Step size and errors (differential equations), 438
Stiff equations, 440
Symmetrical matrices (matrix inversion), 155

**T**

Taylor series, 5
  review of, 336
Taylor series method
  differential equations, 413
  numerical differentiation, 345
Transformation from Cartesian to polar coordinates, 503
Transformation from Cartesian to skewed coordinates, 507
Transformation from Cartesian to triangular coordinates, 511
Transformation methods (eigenproblems), 250
Trapezoidal rule, 374
  method, 449
Triangular elements, 532
Triangular matrices, 153
Tridiagonal systems, 110

**U**

Undetermined coefficients method, 349
Unevenly spaced data points, 385
Upper-triangle elimination method, 37

**V**

Vector iteration techniques, 234
Vectors and scalars, 8
Volume, calculation using determinants, 44